D1744349

Geoff Desa • Xiangping Jia

Editors

Social Innovation
and Sustainability Transition

Previously published in *Agriculture and Human Values* Volume 37,
Issue 4, December 2020

 Springer

Editors
Geoff Desa
Lam Family College of Business
San Francisco State University
San Francisco, CA, USA

Xiangping Jia
Chinese Academy of Agricultural Sciences
Beijing, China

ISBN 978-3-031-18562-5   ISBN 978-3-031-18560-1   (eBook)
https://doi.org/10.1007/978-3-031-18560-1

© The Editor(s) (if applicable) and The Author(s), under exclusive license to Springer Nature
Switzerland AG 2022
Chapters "Palatable disruption: the politics of plant milk", "Feeding the melting pot: inclusive strategies for the
multi-ethnic city", "A carrot isn't a carrot isn't a carrot: tracing value in alternative practices of food exchange",
"Virtualizing the 'good life': reworking narratives of agrarianism and the rural idyll in a computer game" and
"'Workable utopias' for social change through inclusion and empowerment? Community supported agriculture
(CSA) in Wales as social innovation" are licensed under the terms of the Creative Commons Attribution 4.0
International License (http://creativecommons.org/licenses/by/4.0/). For further details see license information in the
chapters.

This work is subject to copyright. All rights are solely and exclusively licensed by the Publisher, whether the whole or
part of the material is concerned, specifically the rights of translation, reprinting, reuse of illustrations, recitation,
broadcasting, reproduction on microfilms or in any other physical way, and transmission or information storage and
retrieval, electronic adaptation, computer software, or by similar or dissimilar methodology now known or hereafter
developed.
The use of general descriptive names, registered names, trademarks, service marks, etc. in this publication does not
imply, even in the absence of a specific statement, that such names are exempt from the relevant protective laws and
regulations and therefore free for general use.
The publisher, the authors, and the editors are safe to assume that the advice and information in this book are believed
to be true and accurate at the date of publication. Neither the publisher nor the authors or the editors give a warranty,
expressed or implied, with respect to the material contained herein or for any errors or omissions that may have been
made. The publisher remains neutral with regard to jurisdictional claims in published maps and institutional affiliations.

This Springer imprint is published by the registered company Springer Nature Switzerland AG
The registered company address is: Gewerbestrasse 11, 6330 Cham, Switzerland

Contents

Agriculture and Human Values (2020) 37:935–943
https://doi.org/10.1007/s10460-020-10163-0

Radical roots and twenty-first century realities: rediscovering the egalitarian aspirations of Land Grant University Extension

Marcia Ostrom[1] 🆔

Accepted: 26 September 2020 / Published online: 20 October 2020
© Springer Nature B.V. 2020

Abstract

Anniversaries and funding crises prompt periodic calls to reevaluate the mission and public perceptions of the U.S. Land-Grant University system. One such call was issued by the Kellogg Commission on the Future of State Colleges and Land Grant Universities in their 1999 report, "Returning to Our Roots: the Engaged Institution." Written by leaders of state universities and land-grant colleges, this report urges these institutions to engage more authentically and equitably in two-way relationships with their local constituents. Twenty years later, Land-Grant Universities continue to struggle with building widespread public support for their missions and equitable involvement in research, teaching, and extension functions across diverse constituencies. While largely discounted by the Kellogg Commission, a fresh look at the role originally envisioned for the extension arm of the trifold land-grant mandate suggests that we may be conceiving of this system too narrowly. The establishment of statewide extension systems was once seen as a way to ensure that Land-Grant Universities would be accessible and responsive to all of a state's residents. Extension systems continue to offer a front-door to a major public university in almost every county of the United States, but they tend to be viewed primarily as a way to translate science or distribute information from the university to the public. This discussion uses a historical and modern lens to reimagine the role that Extension could potentially play in catalyzing reciprocal, co-learning relationships between Land-Grant Universities and their diverse local constituencies.

Keywords Land-Grant University · Extension · Cooperative Extension · Community engagement · Community development · Educational organizing

Introduction

A major university in each U.S. county. This is the promise of Land-Grant University Extension: placing the best of science within local reach and holding science accountable to local needs. Today, this vision faces unprecedented challenges. Loss of funding, public indifference, environmental disruption, and social inequity all threaten the democratic aims of our national Extension system. In the face of contemporary challenges, I find it instructive to consider the origins of this system and the bold aspirations on which it was built. Two recent addresses at past AFHVS annual meetings asked hard questions about the consequences of the commercialization of our public agricultural research system for society and the environment: Leland Glenna's 2017 Presidential Address, "The Purpose-Driven University: the role of university research in the era of science commercialization" (Glenna 2017) and the 2018 Keynote by agronomist Ricardo Salvador, "Science is not Neutral."

While these speakers posed questions about the direction of our public research system, I ask questions about what is at stake if we lose our vision of a rigorous, publicly-driven extension system. First, I consider the historical purpose and ideals of early Land-Grant University Extension based on the words and work of its founding leaders. Second, I take a look inside today's Extension and its publicly stated purpose. Finally, I compare and contrast these sketches from the past and present to inform consideration of future possibilities. How should we conceive of this system and its purpose going forward? What are its roles and responsibilities? I

Presidential Address, Agriculture Food and Human Values Society, Anchorage Alaska, June 29, 2019.

✉ Marcia Ostrom
mrostrom@wsu.edu

[1] School of the Environment, Washington State University, Wenatchee, WA, USA

Reprinted from the journal

approach this topic from my personal experience as a long-time, Land-Grant University faculty member with an Extension appointment.

You may have seen periodic calls to reexamine the mission of our land-grant system, particularly at anniversaries or when budget cuts loom. One of the more notable calls in recent years was the 1999 Kellogg Commission report on the Future of State and Land-Grant Universities, "Returning to our Roots: the Engaged Institution." Written by 25 state and Land-Grant University presidents, the report warned of "growing public frustration with what is seen to be our unresponsiveness" and "a perception that we are out of touch and out of date." The proposed solution was to profoundly re-envision the concept of "university engagement" based on principles of sharing and reciprocity. The Commission set out seven-part guidelines for transforming "inherited concepts" of extension and outreach that "emphasize a one-way process in which the university transfers its expertise to key constituents" to "two-way streets defined by mutual respect among the partners for what each brings to the table" (1999, p. 9). These guidelines emphasized accessibility and working in partnership with diverse community groups to bring university knowledge resources to bear on the actual problems faced by communities.

More recently, my university brought in Steven Gavazzi to talk about his book, *The Land Grant Universities of the Future* written with Gordon Gee, a Land-Grant University president and former member of the Kellogg Commission (Gavazzi and Gee 2018). In their book, Gavazzi and Gee ask how Land-Grant Universities can reclaim their place as the "people's universities." Their recommendation for renewing public support is to ensure widespread access and prioritize community-based work as "mission-critical" across all university functions and disciplines, an organizational orientation they refer to as being a "servant university" (p. 160).

While some might think that this is exactly what the Extension system was set up to accomplish (Smith 1949), it is barely mentioned in this book. Making sense of why these publications do not turn towards this century-old organization for leading community engagement efforts requires revisiting its past and present and the ways that it has been portrayed.

Historical roots and aspirations

What is a Land-Grant University? The series of Land Grant Acts was designed to democratize access to higher education in three ways: Teaching, Research, and Extension. The Morrill Act of 1862 was about teaching. This Act apportioned tracts of land to each state to finance a public college that would "promote the liberal and practical education of the industrial classes." This was followed by a research measure,

the Hatch Act of 1887 that established public Agricultural Experiment stations to be managed by the Land-Grant Universities in each state. The Morrill Act of 1890 established historically black colleges and universities (HBCU). It was the Smith-Lever Act of 1914 that formalized a national Cooperative Extension System to be jointly financed by county, state, and federal governments. A system of tribal colleges and universities was established in 1994 (Land Grant Impacts 2019a; NIFA 2008a, 2008b, 2009, 2019a).

Despite stated goals of countering the elitism of private universities by creating widespread access to higher education, this successive legislation reflected deeply flawed politics of domination, discrimination, and segregation by class, gender, and race (INFAS 2018; Stein 2017). Much of the land comprising the original "grants" had been recently appropriated from tribes (Stein 2017). The 1890 land grant colleges had to be created because Black residents in the Confederate States were being shut out of the original 1862 colleges. Even then funds for research and extension were not allocated fairly across the 1862 and 1890 institutions (Comer et al. 2006; Lee and Keys 2013). The tribal colleges and universities were also added because of a lack of access. An account of the land-grant systems' founding ideals would be incomplete without acknowledging these foundational contradictions. Such recognition is imperative when considering how the past informs future opportunities and responsibilities in Extension. To explore some of these questions in more human terms, I look to the example of its first leaders.

Cornell University extension

Liberty Hyde Bailey of Cornell has the distinction of leading the first continuous, government-funded Extension program at a Land-Grant College. New York State allocated its first funds in 1894, well before the Smith-Lever Act. I was first introduced to Bailey's ecological approach to agriculture and his democratic views of education from a talk by Scott Peters at the (2001) annual AFHVS meetings. A well-known horticulturist and prolific publisher of seminal works such as the *Encyclopedia of Plants in Agriculture* and *How Plants get their Names*, Bailey's extension programs followed his philosophy that the "getting of information is but the beginning of education" (1903, p. 14). He believed that the study of nature would inspire people to investigate for themselves and wrote that "knowledge is not the peculiar property of the teacher, but is the right of anyone who seeks it" (1903, p. 29,145). For Bailey, the purpose of Extension was to improve the quality of individual and civic life through building connections with nature, community, and a scientific spirit of observation and experimentation (Peters 2006). His programs engaged tens of thousands of rural people in locally-led reading circles, farmer institutes and on-farm research and demonstration projects. Members of the public were invited to campus for "Farm and home Weeks." He eventually became Dean of the College of Agriculture and chaired President Roosevelt's 1908 Commission on Country Life that recommended a national extension system.

In his assessment of the Land Grant Acts, Bailey extolled their inclusive spirit. He considered the system of public agricultural colleges foundational to "the future welfare and peace of the people" (1915, p. 94). At the founding of Smith-Lever Extension, he observed:

> No such national plan on such a scale has ever been attempted and it almost staggers one when one even partially comprehends the tremendous consequences that in all likelihood will come of it. The significance of it is not yet grasped by the great body of people. Now the problem is to relate all of this public work to the development of a democracy (1915:94).

Bailey's ambitious vision illustrated the pivotal role he hoped that Extension and Land-Grant education would play in enhancing civic participation.

In his Extension program, Bailey hired the first women faculty at Cornell: Martha Van Rensselaer and Anna Botsford Comstock. Comstock pioneered the field of nature study in the 1890s that spread across the country and helped to popularize a conservation ethic (Smith 1949; Armitage 2009). The first report submitted by Cornell Extension to the legislature listed nature study as the most important agricultural extension method (Smith 1949, p. 66). Comstock taught farmers, academics, teachers, and youth about nature study as a method of everyday, place-based scientific inquiry and self-discovery (Fig. 1).

Comstock's popular nature study leaflets led to a junior naturalist magazine, teaching curricula, and a Nature Study Handbook. Dubbed the "Nature Bible," Comstock's Nature Study Handbook became Cornell's best-selling publication (Comstock 1911; Smith 1949, p. 39).

From her involvement with the nature study movement as a county education commissioner, Martha Van Rensselaer was recruited by Comstock and Bailey in 1900 to organize Extension programs for rural women. The daughter of a suffragette, Van Rensselaer was committed to empowering women through access to education. Within 5 years, she had enrolled 20,000 women in reading courses and trained local women to lead study clubs (CU 2001; Smith 1949). Initially hired to start a reading program, she went on to found the Department of Home Economics that later became Cornell's College of Human Ecology. Van Rensselaer co-directed the Home Economics program with her life and professional partner, Flora Rose, with whom she shared a professorship (Fig. 2).

Sometimes called the "Miss Van Rose," they led efforts both on and off-campus to mentor and empower women through organizing local civic organizations and creating access to education, professional positions, and leadership opportunities for even "that last forgotten woman hidden away on a back-roads farm" (Rose in Smith 1949, p. 89; CU 2001). Some of you may know that Cornell's Martha Van Rensselear Hall was built with funds raised by her friend Eleanor Roosevelt (Smith 1949). Roosevelt and Van Rensselaer's views were in keeping with other leaders of the emerging home economics movement who defined the concept of "home" broadly in order to include health and nutrition, the environment, the community, and the city as arenas for women's civic participation and leadership (Clancy 1999).

Fig. 1 Comstock (left) and Bailey (right) favored nature study as an extension method (Cornell University Library 2019a)

Fig. 2 The Cornell Department of Home Economics, 1914, with Rose second from left and Van Rensselaer third from left on bottom row (Cornell University Library 2019b)

Tuskegee University extension

While Cornell's is recognized as the first government-funded extension program at a Land-Grant University, extension was already well underway at what is now Tuskegee University in Alabama at the same time. With $2000 in state funds and a church outbuilding for a meeting room, Booker T. Washington established a normal school in Tuskegee for formerly enslaved families in 1881 (TU 2019; Mayberry 1989). Over time, he used his public relations skills to persuade donors and legislators to invest in his school creating a 2000 acre campus and an endowment. In response to the extreme poverty and landlessness of surrounding Black farmers, he began inviting them to monthly meetings and conferences on campus. He hoped to hear from farmers themselves about their conditions and "get their ideas for remedies" (Mayberry 1989, p. 39; Jones 1975). At Tuskegee's first annual "Negro Conference" held in 1892, participants identified a list of the problems they faced and generated an action plan for addressing them. As these ideas spread, the conference added a women's day and attracted thousands of farm families and other community leaders within a few years. This was considered the beginning of "Negro Agricultural Extension" and a self-empowerment movement that spread across the southern Black Belt (Jones 1975; Mayberry 1989, p. 39). Women's clubs were added to Tuskegee's Extension programs by Margaret Murray Washington, the head of women's education at Tuskegee and Booker T. Washington's wife (TU 2019).

In 1896, George Washington Carver was recruited to start a formal agricultural department. His first bulletins also emphasized nature study methods for teaching skills of observation and investigation and restoring farmland (Carver 1897). To reach the poorest farmers, Washington and Carver raised funds from donors to build a mule-drawn wagon, the Jesup Wagon, and hired the first cooperatively-funded agricultural extension agent in the country, Thomas Monroe Campbell, a recent Tuskegee graduate, to operate it (Fig. 3).

In his book, "The Moveable School goes to the Negro Farmer," Campbell described his methods for organizing interactive teaching and demonstration sites in the most marginalized communities and the attention that this attracted from international visitors (Campbell *in* Mayberry 1989, p. 104; Campbell 1936, p. 145). His topics ranged from food production and preservation to carpentry. The wagon eventually became a truck in 1918. When it broke down, 30,000 different Black farm families contributed the $5000 needed to buy a new truck named after Booker T. Washington (Mayberry 1989, p. 98). The success of this grassroots fundraiser shows the value that communities placed on Extension. For his part, Booker T. Washington's advocacy and fundraising for his school and his commitment to education as a means of social reform

Fig. 3 The Jesup Wagon Moveable School (Tuskegee University Archives 2019)

led to national prominence, including speaking engagements and an invitation to dine at the White House (TU 2019).

Views from across the national extension system

We can ask whether these early Extension programs were anomalies or typical of the programs being established throughout the country. A government report covering this period, written by the first director of the National Extension Service under Smith-Lever, Dr. Alfred True, offers some insights (True 1928). As in my examples, True observed that successful agricultural extension agents were functioning "largely as organizers" in bringing together small community groups "to study local problems" and participate in building solutions. He found that the classes in food production and preservation, home demonstrations, and women's clubs organized by women extension agents had led to community improvements and increased social intercourse (128–129).

Scott Peters' archival research supports these themes. A 1930 report on the national extension system states:

> There is a new leavening at work in rural America. It is stimulating to better endeavor farming and home-making, bringing rural people together in groups for social intercourse and study, solving community and neighborhood problems, fostering better relations and common endeavor (Smith and Wilson 1930:1 *in* Peters 2014).

Peters also quotes from a speech by the former Chief of the Office of Extension.

Probably the biggest thing that adult agricultural extension and 4-H club work are doing for individuals and the nation is not so much the growing of better crops, or the rearing of better livestock, or the making of better kitchens, but rather the giving of actual experience in the practice of democracy down to the smallest community and individual farm (Smith 1939:2 *in* Peters 2014).

These assessments suggest that the earliest practitioners of Extension conceived of its purpose in broad and aspirational ways. In contrast to the one-way delivery of expertise to key constituents described in the Kellogg Commission report (1999, p. 9), the history of Extension shows a focus on individual empowerment and community capacity-building through interactive learning methods and an emphasis on serving the most marginalized members of society.

LGU extension today

In light of this history, how many Extension personnel today would describe their work as building "experience in the practice of democracy"? Deciphering the purpose and function of our present-day land-grant Extension system is far from straightforward. This decentralized system spans all U.S. states and territories and involves over a hundred universities and 2900 county offices (Land Grant Impacts 2019b). The areas of work described by the national leadership, the Extension Committee on Operations and Policy (ECOP), have broadened. In addition to the original areas of agriculture, 4-H, nutrition, and family and consumer sciences; programs such as community development, environment, and climate have been added (ECOP 2019). Funding remains a combination of federal, state and county government, however, the relative proportions vary by state (ECOP 2019). Federal capacity funds for Extension have been cut significantly in recent years, as have state and county funds leading to profound staffing and budgeting challenges (Wang 2014; Coppess et al. 2018; CRS 2019). As regular government funding has eroded, most systems have increased their reliance on grants, donations, fees, and contracts to varying degrees (Wang 2014; Coppess et al. 2018; ECOP 2019).

In the face of these resource challenges, what are the dominant narratives used to explain and promote Extension to the public? On their Extension homepage, the federal Extension partner, the National Institute of Food and Agriculture (NIFA), states "the Cooperative Extension System (CES), in partnership with NIFA, is translating research into action: bringing cutting-edge discoveries from research laboratories to those who can put knowledge into practice" (NIFA 2019b). The NIFA page on Extension History celebrates its role in bringing about the American agricultural revolution through "extending new technologies" that dramatically increased agricultural production (NIFA 2019c). Similarly, the recent Gavazzi and Gee book describes Extension as a system designed "to take findings from the research laboratory and apply them directly to farms, fields, families and related subjects." Given these portrayals of a one-way system of technology and knowledge transfer, primarily for agriculture, perhaps it is not surprising that the Kellogg Commission and Gavazzi and Gee did not place Extension at the forefront of land-grant community engagement efforts.

But how does the public perceive Extension? To help find out, I want you to consider a few questions. For those of you who live in the United States, how many of you work at a Land-Grant University? How many of you received a degree from a Land-Grant University? Can you name a Land-Grant University in your state? From the most recent research I could find on this topic, a national poll by my university in the mid 1990s, only about 25% of U.S. residents could name their Land-Grant University (Christenson et al. 1995). Continuing this exploration, how many of you who live in the United States know where your county extension office is? Do you know the name of your county extension agent or educator? It doesn't seem like many you are in touch with your local extension office. Maybe there is a problem if most U.S. residents don't know that they have a land-grant university or where its Extension offices are located. A common story of why Extension is so little known goes something like this. The Extension Service was created at a time when 30 to 40% of Americans worked in farming and around half the population lived in rural areas. Today, less than 2% of the workforce is involved with farming and over 80% of the population is urban. The few farmers that remain are so technologically and digitally sophisticated that they no longer rely on Extension for their information. As a result, Extension has a relevancy problem (*see* Coppess et al. 2018).

Peters argues, however, that part of the problem lies in how we *perceive* Extension and its history (2014). Because today Extension is commonly associated with a tradition of one-way "delivery" of peer-reviewed research or science-based technologies from university experts to clientele seeking information, it is easy to assume that it has always been this way (*see* Raison 2014, p. 1). But, from his historical research, Peters sees multiple traditions at work. He contrasts a "technocratic tradition" designed largely to improve economic efficiencies, with another tradition that, while less recognized, also has deep roots in Extension's history. He calls this an "educational organizing" tradition (2002). In an educational organizing role, rather than holding "trainings" to induce the adoption of particular technologies or behavior changes with predetermined ends in mind, Extension personnel serve as facilitators and co-learners. They build relationships that bring people together to identify and investigate problems for themselves and build their own capacity to implement solutions by working in partnership with university experts (Peters 2002, p. 2). In this method, Extension seeks to catalyze the best of local and university knowledge to flexibly respond to community-identified needs and build civic participation. Consistent with the historic examples provided here, Peters concludes that an educational organizing approach was the "official" Extension policy in its early years, but has since been marginalized (5). Regardless of its original purpose, Extension has clearly been effective at transferring scientific research and technologies from universities for agricultural industrialization. But is this all it has done or could do well?

I would argue that while an educational organizing approach may be less visible on national and university Extension land-grant websites, it is still common. The ECOP website states the importance of having county extension agents living and working in the communities they serve "to respond to local needs" (2019). One study used the national reporting database to show that over half of Extension personnel reported working in areas such as family and consumer sciences, nutrition, and 4-H youth development (Wang 2014, p. 4); areas of work that typically employ interactive learning models and strong community relationships. For instance, the 4-H model has long relied upon local volunteer organizing, hands-on learning through demonstrations, and community partnerships to foster leadership development and civic participation among youth and adults. In nutrition education, the national EFNEP program utilizes dialogic learning techniques and peer-to-peer coaching for improving health in low-income communities (EFNEP 2019).

New national initiatives demonstrate Extension's capacity to flexibly respond to emerging community needs through organizing respectful learning partnerships and knowledge sharing environments. The eXtension system was set up to create online learning communities among Extension personnel (2020). A survey of the eXtension Community Practice, "Community, Local and Regional Food Systems," a learning community of Extension practitioners who work across the country to develop socially just, healthy, and ecologically sustainable food and farming systems (CLRFS 2019), found that over three-fourths worked in semi-urban or urban areas and they characterized their skillsets as facilitation, network building, community building, capacity building, and forming partnerships. They listed their focus as urban farmers, small and beginning farmers, community gardens, school gardens, non-profits, and community food systems (Diekmann et al. 2016). A "Racial Equity in the Food System" workgroup has emerged from this Extension Community of Practice to provide national leadership on developing racially just food systems in partnership with local practitioners (MSU 2020).

In another example, a national initiative emerged to respond to the racial injustices brought to attention by the Black Lives Matter movement by building Extension's capacity to facilitate internal and external civil discourse on racism and racial equity. This pilot program called "Coming Together for Racial Understanding (CTRU)" was first organized as an ECOP "Rapid Response Team" in 2016. Led by the Southern Rural Development Center in partnership with the non-profit, Everyday Democracy, each state was encouraged to send a core team of Extension personnel from their 1862, 1890, and 1994 Land-Grant Universities for a weeklong train-the-trainer at the national 4-H Headquarters. These teams learned self-awareness, antiracism strategies, facilitation, and civil dialogue-to-action techniques to share with their home states (CTRU 2019; Walcott et al. 2020). Our Washington State CTRU team recently offered a multiday workshop for Extension on facilitating dialogue-to-action methods using the Everyday Democracy curriculum, Facing Racism in a Diverse Nation (Abdullah and McCormack 2019). Participants generated action plans and recommendations for acknowledging and addressing racism within their Land-Grant University, their extension programs, and their communities (Fig. 4). They requested that recommendations be shared with university administrators.

Extension for the future

While educational organizing and community-driven Extension programming may not be the work that is most prominently featured on USDA or Land-Grant University websites, you do not have to look very far to find it in practice. In the face of substantial budgetary and related personnel challenges, county offices continue to welcome the public. The

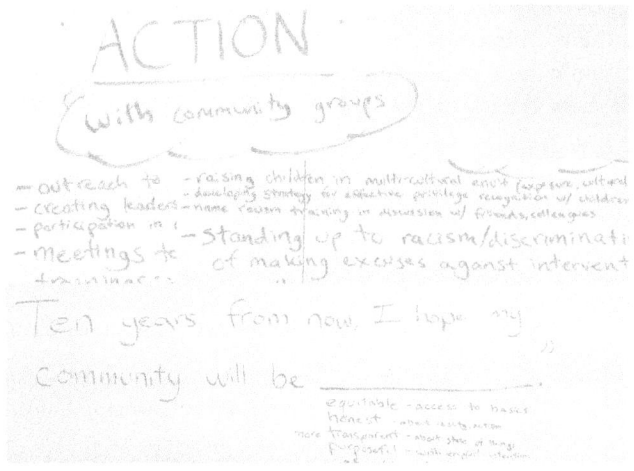

Fig. 4 Coming Together for Racial Understanding Civil Dialogue Workshop (WSU Extension 2019)

educational organizing approach can be found in emerging community food systems work, urban agriculture work, and racial equity work. It has likely been present, to at least some degree, all along in 4-H and family living programs, and food and nutrition programs. Bailey, Comstock, and Carver's focus on nature study methods, experiential learning, and peer-to-peer networks remain popular under other names in extension programs on sustainable agriculture, diversified small farming, gardening, and environmental conservation. Efforts to address inequities of access are gathering momentum. The persistence of this work and the reasons it is not more prominently featured or rewarded deserve further attention. Research is needed to analyze the conditions that foster equitable access and two-way community learning partnerships and the obstacles to such work.

Despite resource challenges, nearly all counties still have a staffed Extension office, a meeting room, and virtual technologies that offer a front-door to a Land-Grant University. Substantive deliberations about how to conceive of this system and its roles will be required to re-center Extension and its community relationships within the land-grant mission and the public funding portfolio. At the same time, cultivating and valuing relationships with diverse constituents to carry out participatory learning and problem-solving will be required in order for them to perceive that they have a stake in the well-being of this system. From revisiting the aspirations and methods of Extension's founders, it is clear that they saw building authentic, reciprocal relationships with communities, including the most marginalized communities, as the heart of the university, not an optional add-on to research and teaching. These early Extension leaders were not relegated to the sidelines of university leadership or scholarship, but rather rewarded with departments, colleges, buildings, and monuments. National politicians conferred with them. There is much to learn from retelling these stories. As Peters points out, there are many different ways to tell the story of Extension and the land-grant mission, some of them critical and some of them celebratory (2013). But it is only through getting people to see themselves as part of the story that it will be possible to imagine a different future.

References

Abdullah, Carolyne Miller and Susan McCormack. 2019. Facing racism in a diverse nation: a guide for public dialogue and problem solving. https://www.everyday-democracy.org/resources/facing-racism

Armitage, Kevin. 2009. *The nature study movement: the forgotten popularizer of America's conservation ethic*. Lawrence: University Press of Kansas.

Bailey, Liberty Hyde. 1903. *The nature-study idea*. New York: Doubleday, Page and Company.

Bailey, Liberty Hyde. 1915/1980. *The holy earth*. Ithaca, NY: New York State College of Agriculture and Life Sciences.

Campbell, Thomas M. 1936. *The movable school goes to the negro farmer*, 145, *in* Mayberry, Bennie D. 1989. *Role of Tuskegee University in the origin, growth and development of the Negro Cooperative Extension System, 1881-1990*, 99. Montgomery: Brown Printing Company

Carver, George Washington. 1897. *Progressive nature studies*, 4. Tuskegee, AL: Tuskegee Institute Printing.

Christenson, James A., Don A. Dillman, Paul D. Warner, and Priscilla Salant. 1995. The public view of Land Grant Universities: Results from a national survey. *Choices* 10: 37–39.

Clancy, Kate. 1999. Reclaiming the social and environmental roots of nutrition education. *Journal of Nutrition Education* 31 (4): 190–193.

Comer, Marcus M., Thasya Campbell, Kelvin Edwards, and John Hillison. 2006. Cooperative extension and the 1890 land-grant institution: The real story. *Journal of Extension* 44 (3): 1–6.

Cornell University (CU). 2001. What was Home Economic? Martha Van Rensselaer, Faculty Biographies, Division of Rare & Manuscript Collections, 2B Carl A. Kroch Library, Cornell University, Ithaca, NY. https://rmc.library.cornell.edu/homeEc/bios/marthavanrensselaer.html. Accessed 9 May 2019.

Comstock, Anna B. 1911. *Handbook of nature-study*. Ithaca: Comstock Publishing Company.

Congressional Research Service (CRS). 2019. The U.S. land-Grant university system: An overview,Report # R45897, August 29, 2019. https://crsreports.congress.gov/product/pdf/R/R45897. Accessed 20 Mar 2020.

Cornell University Library. 2019a. Digital collections, Liberty Hyde Bailey Photographs with Anna Comstock (5), #8330032, 1905-05-29, New York State College of Agriculture records, Division of Rare Book and Manuscript Collections, Cornell University Library, Ithaca, NY. https://digital.library.cornell.edu/catalog/ss:8330032. Accessed 9 May 2019.

Cornell University Library. 2019b. Human Ecology Historical Photographs, New York State College of Home Economics records, #23-2-749. Division of Rare Book and Manuscript Collections, Cornell University Library, Ithaca, NY. https://he-photos.library.cornell.edu/dbgfx/2000/AP-P-03.jpg. Accessed 9 May 2019.

Coppess, J., G. Schnitkey, N. Paulson, and C. Zulauf. 2018. Reviewing the CBO Baseline for 2018 Farm Bill Debate. *Farmdoc Daily* (8): 65. Department of Agricultural and Consumer Economics, University of Illinois at Urbana-Champaign. https://farmdocdaily.illinois.edu/2018/08/farm-bill-food-for-thought-research-and-extension.html. Accessed 20 May 2019.

CTRU. 2019. Coming together for racial understanding (CTRU) pilot initiative summary report, Southern Rural Development Center and Cooperative Extension. https://srdc.msstate.edu/civildialogue/2019-YearEnd-Coming-Together-training%2520report.pdf. Accessed 5 Feb 2020.

CLRFS. 2019. Community, Local, and Regional Food Systems (CLRFS) Community of Practice (CoP). eXtension, USDA National Institute of Food and Agriculture. https://foodsystems.extension.org/. Accessed 7 Dec 2019.

Diekmann, L., Dawson, J., Kowalski, Ostrom, M., Raison, B, Bennaton, B. and C. Fisk (2016). Survey of Extension's role in urban agriculture and local food systems, Community, Local & Regional Food Systems (CLRFS), eXtension Community of Practice. https://foodsystems.extension.org/wp-content/uploads/2019/03/UA-Survey-Final-6-10-16.pdf. Accessed 1 Sept 2020.

ECOP. 2019. Facts about Cooperative Extension, Extension Committee on Organization and Policy, Cooperative Extension Section, Association of Public and Land Grant Universities. https://www.aplu.org/members/commissions/food-environment-and-renewable-resources/board-on-agriculture-assembly/cooperative-extension-section. Accessed 1 Nov 2019.

EFNEP. 2019. The Expanded Food and Nutrition Program, National Institute of Food and Agriculture, United States Department of Agriculture https://nifa.usda.gov/program/expanded-food-and-nutrition-education-program-efnep. Accessed 1 Nov 2019.

eXtension. 2020. Cooperative extension system. https://impact.extension.org/. Accessed 7 June 2019.

Gavazzi, Stephen, and Gordon Gee. 2018. *Land-grant Universities for the Future: Higher Education for the Public Good.* Baltimore: John Hopkins University Press.

Glenna, Leland. 2017. AFHVS 2017 presidential address. *Agriculture and Human Values* 34 (4): 1021–1031.

INFAS. 2018. A deeper *Challenge of Change:* The role of land-grant universities in assessing and ending structural racism in the US food system, Inter-institutional Network for Food, Agriculture, and Sustainability (INFAS). February https://asi.ucdavis.edu/networks/infas/a-deeper-challenge-of-change-the-role-of-land-grant-universities-in-assessing-and-ending-structural-racism-in-the-us-food-system. Accessed 28 Nov 2019.

Jones, Allen. 1975. The role of Tuskegee Institute in the education of black farmers. *The Journal of Negro History* 60 (2): 252–267.

Kellogg Commission. 1999. *Returning to our roots: The engaged institution,* Kellogg Commission on the Future of State, Land-Grant universities, National Association of State universities, & Land-Grant Colleges (Vol. 3). National Association of State Universities and Land-Grant Colleges, Office of Public Affairs. https://www.aplu.org/library/returning-to-our-roots-the-engaged-institution/file. Accessed 5 May 2019.

Land Grant Impacts. 2019a. What are Land Grant universities? https://www.landgrantimpacts.org/about/. Accessed 11 Nov 2019.

Land Grant Impacts. 2019b. We are extension. https://landgrantimpacts.org/extension-2. Accessed 11 Nov 2019.

Lee, John Michael and Samaad, Wes Keys. 2013. Land-Grant but Unequal: State One-to-one Match Funding for 1890 Land-Grant Universities, APLU Office for Access and Success Policy Brief, Report No. 3000-PBI. https://www.aplu.org/library/land-grant-but-unequal-state-one-to-one-match-funding-for-1890-land-grant-universities/file. Accessed 1 June 2019.

Mayberry, Bennie D. 1989. *The Role of Tuskegee University in the origin, growth and development of the Negro Cooperative Extension System, 1881–1990.* Montgomery: Brown Printing Company.

MSU. 2020. Racial equity in the food system workgroup, community, local, and regional food systems community of practice, eXtension, Michigan State University (MSU). https://www.canr.msu.edu/racial-equity-workgroup/index. Accessed 1 Jan 2020.

NIFA. 2008a. Hatch Act of 1887, Section 3, 8-1, National Institute of Food and Agriculture, United States Department of Agriculture. https://nifa.usda.gov/sites/default/files/program/Compilation%2520The%2520Hatch%2520Act%2520of%25201887.pdf. Accessed 30 June 2019.

NIFA. 2008b. Smith-Lever Act of 1914, Section 3, 13-1, National Institute of Food and Agriculture, United States Department of Agriculture. https://nifa.usda.gov/sites/default/files/Smith-Lever%2520Act.pdf. Accessed 30 June 2019.

NIFA. 2009. Land-Grant Colleges, 9–1. Act of 1862, Section 4 and Act of 1890 Section 9, National Institute of Food and Agriculture, United States Department of Agriculture. https://nifa.usda.gov/sites/default/files/asset/document/First%2520and%2520Second%2520Morrill%2520Act.pdf. Accessed 30 June 2019.

NIFA. 2019a. Land Grant Colleges and Universities, the National Institute of Food and Agriculture, United States Department of Agriculture. https://nifa.usda.gov/land-grant-colleges-and-universities-partner-website-directory. Accessed 30 June 2019.

NIFA. 2019b. The cooperative extension system, National Institute of Food and Agriculture, United States Department of Agriculture. https://nifa.usda.gov/cooperative-extension-system. Accessed 1 Nov 2019.

NIFA. 2019c. Cooperative Extension history, National Institute of Food and Agriculture, United States Department of Agriculture. https://nifa.usda.gov/cooperative-extension-history. Accessed 1 Nov 2019

Raison, B. 2014. Doing the work of extension: Three approaches to identify, amplify, and implement outreach. *Journal of Extension* 52 (2): 2.

Peters, Scott. 2001. *"Liberty Hyde Bailey's Prophetic Outlook on Sustainability in Agriculture", presentation at annual meetings of the agriculture, food, and human values society.* Minneapolis, MN: University of Minnesota.

Peters, Scott J. 2002. Rousing the people on the land: The roots of the educational organizing tradition in extension work. *Journal of Extension* 40 (3): 3FEA1.

Peters, Scott J. 2006. Every farmer should be awakened: Liberty Hyde Bailey's vision of agricultural extension work. *Agricultural History* 80: 190–219.

Peters, Scott. 2013. Storying and restorying the land-grant system. In *The land-grant colleges and the reshaping of American higher education,* ed. Roger Geiger and Nathan Sorber, 335–353. New Brunswick, Penn State: Transaction Publishers.

Peters, Scott J. 2014. Extension reconsidered. *Choices Agriculture & Applied Economics Association* 29 (1): 1–6.

Smith, Clarence B. 1939. Twenty-five years of 4-H experience as a basis for our vision of 4-H twenty-five years ahead. Washington, D.C.: U.S. Department of Agriculture, miscellaneous publications *in*Peters, Scott J. 2014. Extension Reconsidered. *Choices.*Agriculture & Applied Economics Association29(1): 1–6

Smith, Clarence B. and Meredith Chester Wilson. 1930. *The Agricultural Extension System of the United States.* New York: John Wiley & Sons, 1, *in*Peters, Scott J. 2014. Extension Reconsidered. *Choices.*Agriculture & Applied Economics Association29(1): 1–6

Smith, Ruby Green. 1949. *The People's Colleges: A history of the New York State Extension Service in Cornell University and the State 1876–1948,* 2013rd ed. Ithaca, NY: Cornell University Press and Fall Creek Books.

Stein, Sharon. 2017. A colonial history of the higher education present: Rethinking land-grant institutions through processes of accumulation and relations of conquest, *Critical Studies in Education,* December, 1–17.

True, Alfred C. 1928. *A history of agricultural extension work in the United States, 1785–1923* with foreword by C.B. Smith, Chief Office of Cooperative Extension, USDA Publication #15, Government Printing Office, Washington D.C.

TU. 2019. Dr. Booker Taliaferro Washington, Founder and First President of Tuskegee Normal and Industrial Institute, Tuskegee University (TU) Website. https://www.tuskegee.edu/discover-tu/tu-presidents/booker-t-washington. Accessed 10 Nov 2019.

Tuskegee University Archives. 2019. The Jesup Wagon, Tuskegee Library Digital Collections, https://www.tuskegee.edu/libraries/archives/digital-collections. Accessed 9 June 2019.

Walcott, Eric, Raison, Brian, Welborn, Rachel, Pirog, Rich, Emery, Mary, Stout, Mike, Hendrix, Laura, and Marcia Ostrom. 2020. We (All) Need to Talk About Race: Building Extension's Capacity for Dialogue and Action. *Journal of Extension.* https://www.joe.org.

Wang, Sun Ling. 2014. The cooperative extension system: Trends and economic impacts on US agriculture. *Choices* 29 (316): 1–8.

Washington State University (WSU) Extension. 2019. Coming together for racial understanding (CTRU) Facing Racism workshop, Leavenworth, WA.

Publisher's Note Springer Nature remains neutral with regard to jurisdictional claims in published maps and institutional affiliations.

Marcia Ostrom is an Associate Professor of Sustainable Food and Farming Systems in the School of the Environment and the College of Agricultural, Human, and Natural Resource Sciences at Washington State University (WSU). Her assigned responsibilities include teaching and extension. She teaches courses in agroecology, food systems analysis, and the environment. She leads participatory research and extension programs to improve the sustainability and equity of regional food and farming systems. At WSU she established the Cultivating Success Sustainable Farming Education Program, the Small Farms Program, the Farm Walk Program, and the Immigrants in Agriculture Program. Long-time research interests include values-based food supply chains, agrifood movements, and access to land-grant resources for diverse farmers.

Agriculture and Human Values (2020) 37:945–962
https://doi.org/10.1007/s10460-020-10022-y

Palatable disruption: the politics of plant milk

Nathan Clay[1] · Alexandra E. Sexton[1] · Tara Garnett[2,3] · Jamie Lorimer[1]

Accepted: 24 January 2020 / Published online: 30 January 2020
© The Author(s) 2020

Abstract

Plant-based milk alternatives–or *mylks*–have surged in popularity over the past ten years. We consider the politics and consumer subjectivities fostered by mylks as part of the broader trend towards 'plant-based' food. We demonstrate how mylk companies inherit and strategically deploy positive framings of milk as wholesome and convenient, as well as negative framings of dairy as environmentally damaging and cruel, to position plant-based as the 'better' alternative. By navigating this affective landscape, brands attempt to (re)make mylk as simultaneously palatable and disruptive to the status quo. We examine the politics of mylks through the concept of *palatable disruption*, where people are encouraged to care about the environment, health, and animal welfare enough to adopt mylks but to ultimately remain consumers of a commodity food. By encouraging consumers to reach for "plant-based" as a way to cope with environmental catastrophe and a life out of balance, mylks promote a neoliberal ethic: they individualize systemic problems and further entrench market mechanisms as solutions, thereby reinforcing the political economy of industrial agriculture. In conclusion, we reflect on the limits of the current plant-based trend for transitioning to more just and sustainable food production and consumption.

Keywords Alternative food network · Dairy · Food industry · Neoliberal · Protein · Vegan

The rise of plant milks

"If you want to change the world change your milk" (Plenish Drinks 2019).
"The subtle sweet and creamy flavour of Alpro Soya will brighten any breakfast. It isn't plain, it's plain delicious!" (Alpro 2019).

Plant-based milk alternatives (or *mylks*[1]) are booming. In the US, sales rose by 61% between 2012 and 2017 (Mintel 2018), reaching $1.9 billion by 2019 (Good Food Institute 2019). Varieties have expanded beyond the traditional soymilk to include mylks made from almond, oat, coconut, pea, hemp, and other grains, seeds, nuts, and legumes. Mylks now account for 13% of total retail milk sales in the US (Good Food Institute 2018) and around 8% in the UK (Mintel 2019). Other plant-based dairy substitutes (ice cream, yogurt, creamer, and cheese) have seen similarly rapid growth, with US sales doubling over the past 2 years to $920 million in 2019 (Good Food Institute 2019).

Once sidelined in natural food stores and health food aisles, plant mylk has 'gone mainstream,' as a recent piece in *The Economist* affirms, proclaiming 2019 'the year of the vegan' (Parker 2018). Yet, the recent surge of plant-based milk and meat may owe less to people adopting vegan diets and more to the emerging *flexitarian* trend (Wohl 2019).

✉ Nathan Clay
   nathan.clay@zoo.ox.ac.uk

Alexandra E. Sexton
alexandra.sexton@zoo.ox.ac.uk

Tara Garnett
taragarnett@fcrn.org.uk

Jamie Lorimer
jamie.lorimer@ouce.ox.ac.uk

1   School of Geography and the Environment, University of Oxford, South Parks Road, Oxford OX1 3QY, UK

2   Food Climate Research Network, Oxford, UK

3   Environmental Change Institute, University of Oxford, South Parks Road, Oxford OX1 3QY, UK

---

[1] Since 2013, as a result of pressure from the dairy sector, EU regulations have stated that designations like milk, butter, cheese, cream and yogurt can only be used to market products derived from animal milk. Plant milk companies have responded with a set of neologisms including 'mylk' (Rebel Kitchen), 'm*lk' (Minor Figures) and 'malk' (Malk Organics). We use mylk in this paper as shorthand for the range of plant-based milk alternatives. The term also captures the industry's vision of the 'disruptive possibilities' of these beverages (Gambert and Linné 2019, p. 65). We recognize that in using this term we may be subtly reinforcing its visibility.

Flexitarians are people actively reducing meat and dairy consumption for environmental, ethical, and health concerns (Wood 2019). Fittingly, these are the cares promoted by mylk marketing.

This article considers the politics surrounding this mainstreaming of plant-based products to question how mylks are positioned as alternative to dairy milk. By exploring the narrative framings employed to position mylks as the better milk, we consider the consumer subjectivities fostered and the political economy that this reinforces. We examine the politics of plant milk by developing the concept of *palatable disruption*, which posits that people are encouraged to care about the environment, health, and animal welfare enough to adopt mylks but to ultimately remain consumers of a commodity food. The rise of flexitarianism marks a change in how many people see their relationships to the environment through food. While this could have important implications for sustainability, we argue that it has created an opportunity for the food industry to reposition milk as a fix to environmental and health crises caused by overproduction. We follow work on alternative food networks (AFNs) such as local, organic, and fair trade (DuPuis and Goodman 2005; Goodman 2004; Guthman 2008) to critically assess the politics enabled by the rise of mylks. Our motivation is to explore how plant-based foods have been de-politicized and naturalized as solutions to climate change, animal welfare, and human health challenges. Our analysis is not meant to be dismissive but to urge caution against any implicit assumption that plant-based offers food futures that are better for the environment, health, and animal welfare.

Dairy crisis

The rise of mylks comes at a particularly fraught moment for the dairy industry. Dairy is experiencing a pronounced economic crisis as a result of overproduction and decreasing consumer demand (Clay et al. 2020). After 50 years of policies pushing dairy intensification and retailer-controlled milk pricing,[2] profit margins for milk are extraordinarily thin. Production costs (including feed, land, and water) have ramped up in recent years (Hadrich et al. 2017) and dairy farm concentration has accelerated over the past decade, with thousands of smaller farms in the US and Europe going out of business every year and herd sizes on larger farms growing exponentially (Clay et al. 2020). Fluid milk consumption has been declining since the 1970s in the US and UK. Fluid milk consumption in the UK is about half of

1970s levels (Defra 2017). Consumption is particularly low among younger generations. In the UK, only 73% of people aged 16 to 24 now drink milk, compared to 92% of those over 45 years (Mintel 2019). From 2017 to 2018, fluid dairy milk sales in the US declined by 8%, a loss of $1.1 billion (Dairy Farmers of America 2019). Mylk sales increased by 9% that year (Plant-Based Foods Association 2018).

One driver of decreased dairy consumption is that people—particularly younger generations (ages 16 to 24)—increasingly associate dairy farming with environmental degradation (Mintel 2019). Recent studies reveal a large water, land, and greenhouse gas footprint of dairy relative to other foods (Poore and Nemececk 2018; Springmann et al. 2018). Others suggest that reducing consumption of animal protein may both decrease human mortality and reduce environmental impacts (Westhoek et al. 2014; Springmann et al. 2016; Clark et al., 2019). This story of dairy's environmental impacts has circulated widely in the UK and US. It was covered by eight articles in *The Guardian* in 2018, including an article headlined "avoiding meat and dairy is the single biggest way to reduce your impact on earth" (Carrington 2018), which ran on the front page and at one time amassed more than 900,000 shares via social media. In short, scientific research and public messaging about the multiple benefits of reducing meat and dairy consumption has never been stronger.

The ascent of plant mylks has been propelled and shaped by sizeable marketing investment, much of which speaks to these environmental and health concerns. The excerpts opening this paper are taken from cartons of oat and soymilk. These quotes capture the spectrum of current narratives that are used by companies to position mylks as the better alternative to milk. One significant story presents plant mylks as a *disruption*. The UK company Plenish Drinks tells us that 'if you want to change the world change your milk' (Fig. 1). This slogan appeared alongside images of milky explosions and an almond taking the form of a hand grenade: a 'weapon of mass disruption'. Similarly, the Swedish company Oatly ran a marketing campaign that foretells of the rise of a 'Post Milk Generation': 'a non-profit mindset that works to inform the public about the health and sustainability advantages of eating a plant-based diet'.

In contrast, the formulations of alternativeness by longstanding mylk brands such as Alpro and Silk (both owned since 2017 by the dairy multinational Danone) are comparatively docile. The second quote, on a carton of Alpro soymilk, captures this more measured approach. The language mobilizes inherited framings of milk as wholesome, promising a creamy texture, sweet taste, and familiar role in a convenient breakfast. Alpro's and Silk's imagery of flowing white liquid (Fig. 2) and their wide availability in supermarket dairy aisles celebrates mylk's continued milkiness.

---

[2] Vertical integration in the milk sector has given retailers power to set milk prices at levels that are substantially lower than the cost of production for most smaller farms (Jay and Morad 2007; MacDonald et al. 2016).

Fig. 1 PlenishDrinks advertisement

Fig. 2 Alpro and silk packaging

The politics of plant milk

AFNs such as organic, local, and community supported agriculture were established to counteract problems linked to globalized industrial agriculture. Such alternatives often seek to reformulate social and ecological relations underlying food production, distribution, and consumption to rebuild trust that had been eroded under corporate food regimes (Goodman et al. 2012). In contrast, mylks have emerged as an alternative that is already conspicuously within the food industry. The rapid scaling up and corporate control of plant-based testifies to this. Mylk companies are attracting investments from the likes of Goldman Sachs and from venture capital firms on the order of tens of millions of USD (Fields 2019). The plant-based trend is celebrated by one prominent investor network as indication of 'an appetite for disruption' (Ramachandran et al. 2019). Yet, even though mylk brands have proliferated, the majority of market share

is concentrated with a few large beverage-focused multinationals—most prominently dairy giant Danone through its Whitewave/Alpro holdings. Danone recorded over $1.9 billion in plant-based beverage sales in 2018 and has promised to triple sales within 5 years (Camacho 2018), a goal that attracted a flurry of investor interest in 2019. The Coca-Cola Company is expanding mylk offerings through their Innocent and AdeS brands. PepsiCo launched an 'Oat Beverage' in 2019 through their Quaker brand.

In this paper, we examine how the plant mylk sector employs narrative frames and affective sensibilities to shape food palatability, re-make milk as plant-based, and responsibilize 'ethical' consumers. We critically explore how mylks are positioned as the 'better milk' through simultaneously securing past framings of milk as 'good' (wholesome, healthy, tasty, and convenient) while mobilizing narratives of dairy as 'bad' (environmentally damaging and cruel). In different ways, mylks navigate these contrasting narratives with their marketing campaigns. As with AFNs such as organic, fair-trade, or local, mylk's claims of alterity are founded on a range of cares, including for the environment, bodily health and 'wellness,' animal welfare, taste, and convenience. As we will argue in this paper, mylk companies deploy these cares in ways that uphold the political economy of industrial agriculture and grant food industry further power to shape food futures.

We explore the politics of mylks through the concept of palatable disruption. This builds on work by Jesse Goldstein (2018) on the 'non-disruptive disruptions' that he argues are at the heart of the 'new green spirit of capitalism'. Non-disruptive disruptions are 'technologies that can deliver "solutions" without actually changing much of what causes the underlying problems' (10). The palatable disruption concept offers a way to critically assess the "ethic of care" that is promoted in a post-milk utopia. Our assessment of the politics of plant milk speaks to efforts to transition to more environmentally sustainable and socially just agri-food systems by responsibilizing consumer-citizens (Lockie 2009; Johnston and Szabo 2011; Roe and Buser 2016). Mylks, we argue, encourage people to rebel just enough to switch from dairy milk to mylk while entreating them to remain devoted consumers of commodity mylk (*and* dairy milk).

The post-milk imaginary procures an "unreflexive politics" (c.f. DuPuis and Goodman 2005, p. 361) and neoliberal consumer-subjects by individualizing systemic problems of environment, health, and animal welfare. This serves to bolster corporate knowledge claims about sustainability, entrench market mechanisms, and reify commodity food as a solution.

To critically assess plant milk, we engage with food studies literature on AFNs. This work has demonstrated the challenges of delivering more ethical, environmentally sustainable, healthy, and just food systems through niche production-consumption networks that can exclude producers and consumers (DuPuis and Goodman 2005; Guthman 2008; Alkon and Mares 2012). AFNs are critiqued for the ease with which they are subsumed into productivist logics and coercive politics that undermine the ideals behind food movements (Guthman 2004; Goodman et al. 2012). In particular, we speak to studies that conceptualize consumption as empowering the retail preferences of rational and/or emotional economic actors (Clarke et al. 2007; Swaffield 2016; Evans et al. 2017). We acknowledge work demonstrating that consumption is a complex affective, social and political act, and one that is entangled in networks of concern that stretch well beyond the individual in the here and now (Miller and Rose 1997; Hayes-Conroy 2013, 2010; Carolan 2016). As DuPuis (2000) demonstrates in her work on organic milk, consumers are neither entirely sovereign in making decisions nor entirely victims of marketing. At the same time, we recognize that this green consumerism 'responsibilizes' consumers to make environmentally sustainable choices in ways that can entrench the political economic status quo by positioning market exchange as the solution to problems caused by excessive consumption and corollary overproduction (Goodman 2004; Shove 2010; Jones et al. 2011).

Our case study expands on this work by examining how plant-based food futures are shaped, by whom, and to what ends. We describe how palatability is choreographed to secure affective continuity in user experience while conferring an aspirational sense of novelty and disruption. In this way, mylks resemble white middle-class social improvement efforts that constructed dairy milk as a "perfect food" (DuPuis 2002). Promises of perfection, purity, and social change appear in mylk marketing, such as Oatly's promotion of a post-milk future, which presumes that avoiding milk will rectify issues stemming from agro-industrial dairy production. We suggest that within this post-milk utopia lies a dichotomous ethic of care: an assumption that avoiding dairy can address these issues in the dairy system. This procures an "unreflexive politics" (DuPuis and Goodman 2005, p. 361), privileging food companies to enact a post-milk world. We point to the contradictions and restrictive contingencies that arise in establishing an ethic of care based on such a consumer-company relationship.

Researching palatability

This paper's argument derives from an analysis of the existing literatures on the framings of dairy milk, market research information on the mylk sector, the packaging and marketing campaigns of a range of brands, and interviews with 12 representatives of mylk companies, their suppliers, and associations promoting plant-based. We concentrate on the four mylk varieties with the largest market share: almond, soy, coconut, and oat (The Good Food Institute 2019). We selected brands to capture a range of company sizes and histories, including: Alpro and Silk (large companies more than 30 years old); Oatly and Liquats (mid-size companies more than 20 years old); Califia Farms; Plenish Drinks, Rude Health, Rebel Kitchen, and Ripple (smaller companies less than 10 years old).

Our methodology develops critical food studies work on consumer choice, food system transitions, and AFNs, particularly that which focuses on food companies' 'mobilization of affective and emotional registers' (Doyle et al. 2019, p. 3). DuPuis (2000) has analyzed the various cultural framings found on cartons of organic milk and what they say to consumers. Carolan (2015) demonstrates the value of interviewing what he calls 'the tastemakers' in food companies and how the food industry aims to activate consumers' emotional registers in product development and marketing. Sexton et al. (2019) examine how cultured meat is framed with narratives of alternativeness that entice consumers through stories of what is both present and what is absent in the products. This work, in turn, builds on a long history of research in cultural studies that takes marketing as the 'poetry of capitalism' (Barthes 1972; Williamson, 1978) and seeks to 'lay bare the prejudices that lie behind the smooth surface of the visible' (Rose 2007, p. 76). Such visual methodology requires close critical reading of marketing materials—attending as much to absences as to presences—and awareness of the vital role of intertextuality in creating meaning, cultural value, and emotional experience (Lorimer 2010).

Building on these studies, we take palatability to be a multisensory, affective experience that emerges from both the visual experience of the 'affective surfaces' (cf. Forsyth et al. 2013) of milk packaging and marketing, and the taste experience of the mylks themselves. We examine taste and its disruption as both material and semiotic processes (Roe 2006; Hayes-Conroy 2010; Evans and Miele 2012). In keeping with other work in this vein (Longhurst et al. 2008; Mann et al. 2011; Sexton 2016), we tasted the products we describe and explored how the affective experience of drinking mylk is conceived and modified by those in the trade. Informed by thinking in the industry science of behavior change (Marteau 2018), we trace how the claimed disruption of mylk involves both pre-discursive and discursive interventions that work on consumers' 'slow' and 'fast' thinking (Kahneman 2011) in ways that far exceed narrow understandings of rational economic action. This element of our methodology is important as mylks have excelled in their ability to create emotional connections through their social media allure and appeal to consumers' palates and habits (Levitt 2018).

The paper starts with a discussion of the historical narratives surrounding dairy milk to establish how plant mylks inherit framings of milk as: (i) pure, wholesome, and healthy; (ii) tasty and convenient; (iii) risky and environmentally damaging; and (iv) cruel and inhumane. We then explore how mylks navigate these framings to remake the milk experience as palatable and disruptive. Our analysis illustrates three techniques of palatable disruption, documenting how plant mylks: (i) 'taste good' by securing affective continuity in taste experience; (ii) 'feel good' by affirming and facilitating broadcast (i.e. virtue signaling) of one's environmental and health values while avoiding unpalatable political registers of disgust and agonism; and thereby (iii) maintain the political and cultural economic status quo through the consumption of agro-industrial food. In conclusion, we identify the political economic characteristics and implications of this model of change in the food system, which we flag as priorities for future research.

The discursive landscape of dairy milk

Contemporary forms of commodity milk production-consumption are the result of discursive and material processes; the work of innumerable actors and institutions (Smith-Howard 2014). We focus on the discursive constructions of milk, which are entangled with political economies and ecologies of production (DuPuis 2002). While milk's cultural, political, and economic importance spans the world (Valenze 2012; Wiley 2008), this article concentrates on the US and UK. Over the past 200 years, milk has been continuously re-framed in response to changing societal values about food, animals, and the environment (DuPuis 2002; Freidberg 2009; Atkins 2010). Claims about milk's healthiness, ethics, wholesomeness, and worth have been repeatedly contested and various 'better' alternative dairy production-consumption practices have emerged to counter skepticism. Here we briefly outline the framings of milk that mylk producers in North America and the UK inherit, navigate and repurpose to justify their products' alterity. Drawing in part on existing literatures on milk's cultural and political history (e.g. DuPuis 2002; Freidberg 2009; Atkins 2010; Valenze 2012; Wiley 2014), we suggest that mylks curate affirmative cultural signifiers of milk's palatability as: (i) wholesome and healthy; and (ii) tasty and convenient. But

mylks must also depart from the framings that make dairy disgusting, in which: (iii) milk is risky and environmentally destructive; and (iv) milk is murder.

Milk as wholesome and healthy

Humans have relied on animal milk as a source of calories and nutrients in many regions of the world for 3000 to 7000 years (Salque et al. 2013). Milk is the only food that human bodies are also capable of producing to feed offspring, and animal milk has frequently been associated with motherhood, vitality, and the sacred (DuPuis 2002; Valenze 2012; Wiley 2014). Its dietary importance among some societies generated reverence for milk, the animals producing it, and the people tending to them; a mythical status that is captured in art and literature (Kurlansky 2018). In places where a majority of people could digest lactose, dairy milk was perhaps seen as the original 'superfood'. In medieval England, dairy was a crucial food source for the rural poor, for whom it was regarded as 'white meat' (Freidberg 2009). The socially-constructed image of milk as wholesome and pastoral amongst urban, industrial consumers is a more recent abstraction, yet one that mobilizes similar narratives of health and motherhood towards milk's commodification (DuPuis 2002; Wiley 2014).

Historians suggest that this imaginary took hold once urban consumers were alienated from relations of production. The framing of milk as a nourishing, healthy food for urban citizens dovetailed with the separation of cities from rural areas, the rise of refrigerated train car technology, and laws that favored larger dairies that had the ability to pasteurize at scale (Freidberg 2009; Atkins 2010). Milk marketing campaigns have repeatedly developed milk's wholesome palatability through use of the nostalgic conception of the pastoral (Marx 2000; Wiley 2008; Paxson 2013). These positive traits persist to this day in the advertising of dairy milk, particularly in its organic variant (DuPuis 2000) as well as almond milk (Bladow 2015). This affirmative connotation of milk has overlapped with a parallel discourse of milk as nutritious, which has repeatedly targeted 'weak' women and infants—promoting the substitution of breast milk for cow's milk (Atkins 2010; Dupuis 2002; Valenze 2012). Latterly—for example in the high-profile US 'got milk' campaign—this gave way to a more general mobilization of the archetypal healthy bodies of celebrity sportsmen and women.

Milk as tasty, affordable, and convenient

Milk's commodification in twentieth century North America and Europe was entangled with a story of progress that centered on the modernization of production processes to increase the availability of pasteurized milk and to ensure healthy, strong bodies. Milk's idealized position as a 'perfect food' (DuPuis 2002) merged with a narrative of modern nation building through increasing production and reliable year-round supply. These were enabled by transportation networks that made milk good value in terms of price to nutrition (DuPuis 2002; Freidberg 2009). Cheap commodity milk was upheld through government subsidy structures that furthered dairy farm specialization. These political economic and cultural processes continued to fuse with the imaginary of milk as wholesome and pure and were central to milk becoming a household staple in the US and UK following the second World War (Valenze 2012). Surplus milk and corn were conjoined in the marketing of the cereal breakfast that became a US institution (Kellogg's Corn Flakes being a prime example). Standardizing the lipid, protein, and sugar content was a key aspect of milk's imaginary as modern (Atkins 2010). In the twenty-first century, milk processing and distribution were even more heavily standardized to ensure consistent taste and reliable food safety (Smith-Howard 2014). This standardization, together with an emphasis on reducing cost through larger farms and the homogenization of cattle genetic diversity and feed sources, led to a toning down of the intensity and diversity of milk flavors (Freidberg 2009; Levitt 2018). Tasty milk is promoted more through the absence of the flavors associated with fermentation, which might indicate spoiling and the possibility of food poisoning, in marked contrast to the promotion of cheese (Paxson 2013). The convenient qualities of milk are characterized more by a consistent 'mouth-feel' and dependable material performance in bowl, pot and cup. The taste of this bulk milk can then be enhanced, and economic value added, through a proliferating range of added flavors and forms of 'fortification'.

Milk as risky and environmentally damaging

As milk consumption in urban areas grew at the start of the twentieth century, so did the distance that milk travelled, leading to increased risk of milk spoiling and of potentially fatal disease.[3] Some urban consumers were skeptical of milk coming from outside the city, which could go off or be skimmed or adulterated by unscrupulous merchants. These practices were widespread in the UK and US (Freidberg 2009; Atkins 2010). This established a framing of milk as risky and unhealthy, which has been reinvigorated as a result of more recent public health research establishing links between cardiovascular disease and the consumption of saturated fats.[4] By the end of the twentieth century, a perception had emerged in some circles that milk and dairy products were unhealthy (Valenze 2012). Meanwhile, concerns in the US about the use of antibiotics and growth hormones in milk production led to renewed disquiet about the purity of milk and drove demand for organic milk (DuPuis 2000). Over the past decade, there has been increasing anxiety about the negative environmental impacts of dairy production. These concerns initially centered on the impacts on water, land use, and biodiversity caused by intensive dairy systems and have since expanded to focus on the greenhouse gas emissions of industrial dairy production (FAO 2006; Foote et al. 2015; Springmann et al. 2016; Poore and Nemecek 2018). Prominent examples of this negative framing include campaigning films like *Cowspiracy* (Andersen and Kuhn 2014) and *What the Health* (Kuhn et al. 2017) that make visible the health impacts and ecological relations of meat and dairy to create doubt and unease amongst consumers.

Milk as inhumane

Concerns about animal welfare in dairy production systems can be traced back to at least the nineteenth century (Fisher 2018). In the USA, images and descriptions of urban 'swill dairies' (which fed cows mainly with by-products of beer brewing) circulated in newspapers in Boston and New York. These exposés led to public outcry, government regulations, and eventually the closure of urban dairies (Freidberg 2009). In the twentieth and twenty-first centuries, animal welfare and animal rights campaigns have persistently criticized the dairy industry for animal abuse. Led by non-governmental organizations such as People for the Ethical Treatment of Animals (PETA) and the Vegan Society, such campaigns have engendered a strongly negative view of dairy's impacts on animal welfare (Mylan et al. 2018). This negative perception further established itself in the public consciousness with the use of high-profile advertising, demonstrations, social media, and an accelerated film campaign with titles such as *Food, Inc.* (Kenner et al. 2008) and *Eating Animals* (Foer et al. 2017). These framings tend to accentuate the corporeal affinities between human and bovine bodies and the physical and emotional violence associated with the dairy industry (Tulloch and Judge 2018). They reference the severed maternal-infant bond ('not your mom, not your milk') and use shocking images to present 'milk as murder'. In so doing they engender disgust, subverting the more traditional perceptions of the wholesome palatability of milk.

---

[3] Infant mortality was particularly high in urban areas in the US during this time, with diarrhoea often blamed on bacteria in milk, as well as other diseases such as strep throat and tuberculosis (Freidberg 2009).

[4] These links continue to be the subject of vigorous dispute.

Framing plant mylks

Over the past few decades, campaigns for or against milk have strategically mobilized these conceptualizations in ways that shape and connect with consumers' emotions. It is from this contested melee of arguments and feelings that the current framings of plant mylks have emerged. In taking their products into the mainstream, mylk companies must navigate a contentious material and semiotic terrain to curate the palatability of milk, while also promising to disrupt the status quo to address consumers' concerns. We focus our analysis on three prominent framings.

Looks, acts, and tastes like dairy milk

'When should you use it? Whenever you would use old-school *milk* from cows—chilled in a glass, for cooking or baking—in exactly the same amounts.' (Oatly 2019, emphasis in original).

Perhaps the most obvious feature of the mylk companies' efforts to maintain palatability is the ways in which they strive to mimic how dairy milk looks, acts and tastes. Mylks exist invariably as white 'milky' liquids, and plants are selected and processed with this end in mind. Good Hemp Barista Seed Milk, for example, claims on the carton that it is 'naturally white.' Three Ones Almond Milk notes on the carton that it is 'pure white.' Imagery of milky liquid feature prominently on packaging. Sometimes, as in the Plenish advertisement in Fig. 1, milk is exploding or otherwise emanating from almonds, soybeans or other plant components. In the case of the less self-consciously 'radical' mylk brands, such as Alpro and Silk, advertisements depict a pitcher of creamy plant milk pouring into a bowl filled with breakfast cereal. The quintessential modern Western breakfast.

When it comes to taste, some mylks (such as Rude Health and Plenish) claim to replicate the refreshing, 'pure' taste of (chilled) dairy milk, and to derive this purity from using only few ingredients. Many brands include flavorings, stabilizers, emulsifiers, and refined sugars in an effort to mimic the texture or mouth feel of milk. A carton of Rebel Kitchen Mylk exemplifies this, claiming 'what we all really want from a dairy free alternative is that it tastes & looks just like real milk. Right?' In trends comparable with dairy milk, many mylks are also flavored, most often with vanilla, in ways that explicitly depart from the claimed blandness of pure milk. As we examine in detail below, the form and style of the packaging often purposively resembles that of commodity dairy milk. Mylks are also sold in ways that resemble dairy milk, located in refrigerated aisles adjacent to dairy milks (Fig. 3) or next to the UHT milks in the ambient aisles. Likewise, in coffee shops, which have proven to be crucial spaces in which consumers first try mylk, they are sold with

Fig. 3 Refrigerated mylk in supermarket. Photo by Nathan Clay, 2020

a promise of comparable or even enhanced frothability. Several industry respondents emphasized that it is these 'flavor cues' that drive mylk consumers far more than environment or health claims.[5]

Nutritious, powerful, pure

"This is THE almond milk YOU DESERVE" (Califia Farms 2019).

"Plant powered" is a ubiquitous phrase among mylks. It harnesses preexisting understandings of the health benefits of protein while overcoming the potential harms of dairy milk. Mylks' wellness claims span myriad definitions of health and multiple epistemologies through which it might be known and achieved. Many brands reference nutrition, claiming to maintain or even enhance milk's calcium and protein, although this is inevitably through the addition of supplements and patented processes of protein extraction. Califia Farms Ubermilk, for example, claims 45% more

---

[5] See also a recent interview with Califia Farms (almond milk company) Greg Stentenpohl, in which he discusses the importance of what he calls 'flavor cues': https://www.foodnavigator-usa.com/Article/2019/11/15/PepsiCo-drops-Quaker-Oat-Beverage-less-than-a-year-after-launch.

Fig. 4 Califia Farms Ubermilk

calcium than dairy milk and a range of other essential nutrients (Fig. 4). Some brands pick up on concerns over heart health and dairy fats. Hearts (a regulated symbol of the American Heart Association) adorn PepsiCo's Quaker Oat Beverage. A Plenish Drinks mylk carton says 'When you replace saturated fats with heart-healthy monounsaturated fats found in this hazelnut m*lk, you can reduce blood cholesterol levels. High cholesterol can lead to heart disease, so make the switch for good'.

The Plenish package goes on to claim that 'by proactively filling up on natural, healthy ingredients and harnessing the mighty power of plants, you can press on and crush it!' This appeal channels the current infatuation with protein and a prevalent mantra of overcoming challenges of ordinary life through food.[6] Indeed, many mylks leverage a generic image of strength and power. Slogans such Alpro's 'enjoy

---

[6] The trope of overcoming privileged 'adversity' is represented, for instance, in the You and Almonds Vs. marketing campaign by the California Almond Board, in which almonds provide the strength to fix a printer out of toner or locate a misplaced television remote control.

plant power' convey an image that bodily strength is possible through plants, a discourse which resonates with the highly physical, masculinist vegan identity that is coming to prominence amongst some high-profile vegan advocates (Sexton et al. 2019). Here mylks eschew past gendering of milk as female while amplifying earlier framings that associate dairy milk with sporting prowess and the bodies of celebrities known for their physique.

Many mylks are marketed with more diffuse notions of health, likely because mylks lack the quantity and variety of nutrients found in dairy milk. These abstract health claims are captured in words like 'wellness' and 'cleanliness', as well as by Instagram-friendly iconography. Such promises of holistic health are epitomized by slogans like 'eat right, stay brilliant' (Rude Health) and 'feel good food' (Happy Planet Oatmilk). As a signifier of wellness, Mylks frequently gesture to what is absent. Often this is dairy. For example, Rebel Kitchen says of its semi-skimmed mylk: 'why the y? It's made from plants, not cows.' Notions of purity similarly abound. As Rude Health Ultimate Almond states, 'we only use the kinds of ingredients you'd have in your own kitchen—nothing artificial, nothing refined. We source our ingredients from fields, orchards and vines—not laboratories.' This dovetails with the health as purity discourse that underlies the recent 'clean eating' trend. More overtly, mylk packaging frequently states what is absent in terms of calories and added sugar.

Green and compassionate

"We make compassionate food for passionate people" (So Delicious CoconutMilk).

Certifications abound on mylk cartons. 'Vegan' is ubiquitous. 'GMO-free' adorns most mylk cartons in the US. A vital storyline across brands is the lower environmental footprint of plant milks relative to dairy milk. Yet this is displayed to different degrees. Dairy-owned brands like Alpro appear more reserved, relegating discussion of environmental footprint to a small text box on the side of the carton. Others are louder. Plenish challenges: 'if you want to change the world change your milk.' Industry respondents informed us that environmental claims are regulated far less than are health claims. Many brands use relatively vague imagery to convey sustainability. For example, Milkadamia states on the side of the carton that it uses 'free range trees, trees supporting life, not trees on life support... in total harmony with the earth, nurtured by natural rainfall and sunshine.' Others display environmental footprints based on product life cycle assessments. Oatly displays its oatmilk's carbon footprint on each container. Plenish Drinks has an environmental footprint calculator on its website.

Some European almond and soymilk brands address environmental concerns through the identification of geographic origin. For example, Provamel Almond [beverage] has a map of the world with an arrow pointing to Europe alongside the text: 'We care about where we source our ingredients. That's our promise to you.' Such coding tacitly acknowledges, while also distancing from, the controversial intensive almond supply chains that are booming in drought-prone California.[7] For reasons that we go into below, mylk brands devote little explicit attention to animal welfare in their marketing. A few, such as Good Karma (owned by US dairy corporation Dean Foods), make implicit reference to ethics, but the majority choose to emphasize their 'plant-based' rather than 'animal-free' constitution. They leave animals—with their powerful affective associations with cruelty and disgust—absent and unsaid.

To summarize, through this work plant milk companies successfully inherit framings of milk as wholesome and convenient, while circumventing framings of milk as cruel and environmentally damaging. Some frame their products as the refined continuation of milk tradition, while others present a disruptive break with an anachronistic dairy past and a step toward a post-milk future.

Palatable disruption

"Good for you products that are also good for the planet" (Califia Farms).

In this section, we explore the politics currently afforded by mylk. We present the mainstreaming of plant-based dairy as an example of Goldstein's non-disruptive disruptions—in which grandiose claims of challenging an environmentally damaging status quo provide 'moral legitimacy and affective force for proposals to irrevocably transform capitalism into a more environmentally virtuous economy; still capitalism just a better, greener version' (2018, p. 30). We advance this assertion and its relevance to food by developing the concept of palatable disruption. A palatable disruption is a widely promulgated claim for a change in the food system that: (i) maintains continuity in taste experience; (ii) performs a politics that feels good to citizen-consumers; and (iii) works to sustain or even amplify elements of the political economic status quo that are palatable to corporate interests. We explore these three themes below with attention to how flexitarian consumer-subjects are produced through neoliberal mechanisms that underwrite the palatable disruption of

[7] This is also true for soya, where European (over US or South America) provenance is coded as an environmental good due to perceived GM-free status.

mylk. The case of mylk provides a window onto the broader trend of plant-based foods.

Consumption continuity

Considering all the talk of disruption, it is perhaps striking how much mylks look and taste like milk. Yet, it is this interchangeability that has made mylks such effective commodities. As we demonstrated above, mylk companies have worked to secure continuity in their users' experience, even as they shift mylk's material composition and herald its disruptive potential. There is no necessary reason why mylk should be white or served cold. Liquid plant products don't need to be used to dilute coffee, bulk out smoothies, or moisten cereals. But this is how and where they invariably end up. As such, they testify to an inertia in the North America and European food system; a cultural economy of western breakfast that resists transformation. Like Goldstein's (2018) Cleantech entrepreneurs, mylk companies ultimately seek marginal gains in established markets for commodity accompaniments to cereal and coffee. By the time mylks get to the coffee shop or supermarket shelves, these products are not intended to shift ingrained habits to create new markets. Companies are aware of how hard it is to get consumers to try new products, and instead seek to replicate the palatable experience of milk consumption.

This continuity can be understood by attending to how the practical and affective dimensions of the mylk consumption experience are choreographed by the applied sciences of food product formulation, packaging, and retail. As critics have observed, these are established knowledge practices that have mastered how to create, shape, and ultimately govern consumer desires, often through techniques that work more on bodily feelings and habits than through appeals to rational choice (Moss 2013; Carolan 2015; Schatzker 2015). Making mylks palatable involves drawing on gastronomic science and the technical skills of food processing to reformulate their taste away from the 'mealy' and 'beany' flavor of earlier plant mylks that catered to vegans and the lactose intolerant. Mylks have been smoothened, sweetened and refined to match the taste and mouth feel of dairy. The carton of Rebel Kitchen Mylk that we encountered above, which claims that 'what we all really want from a dairy free alternative is that it tastes & looks just like real milk', goes on to note the various tastes of milk which its team of taste profilers sought to emulate with plant-based ingredients. In response to the rhetorical question 'how do we make it so mylky?' the package lists 'coconut cream for creaminess... brown rice for sweetness... cashew for earthiness... nutritional yeast for grassiness'. This strategy also speaks to a marketing trend of incorporating 'tasting notes' on various foods, a practice imported from wine and designed to flatter consumers for their appreciation of flavor subtleties. Some

Fig. 5 Rebel kitchen cartons

brands, such as Oatly and Rebel Kitchen, offer 'skimmed' and 'whole fat' versions, in the latter of which fats are added to mylks (Fig. 5). This process is reminiscent of industrial dairy milk production, which skims all milk and adds fat back as needed.

These techniques often center on producing sweetness in mylks to resemble the lactose flavor of dairy milk. Often, sweetness is produced through adding refined sugars. Many mylk brands offer unsweetened versions, which claim to include no added sugar. Yet, these claims may be misleading. Oatly's oat milk, for example, relies on enzymes to break down plant starch into simple sugars. Oatly recently stopped advertising no added sugars following a complaint by rival Campbell Foods, whose oatmilk (sold through their Pacific Foods subsidiary) contains 17 g of sugar per serving (Watson 2019). Mylks also seek to replicate the frothability of dairy milk that is so valued in contemporary urban coffee culture. Oatly pioneered this with their 'barista' edition. Many other companies have since followed, blending mylks with plant oils as well as acidity regulators that suppress separation once coffee is added. For example, a mylk company representative explained how brand loyalty was built through this consistent re-creation of coffeeshop rituals: foaming to the right consistency, mixing with coffee at the right ratio without separating, and offering the capacity for latte art. He explained how the consumers' affective experience of their product is preeminent, and it is only later that a story of health or environmental sustainability comes into play. Haptic and olfactory consistency may trump ethical exhortation as a driver of mylk sales.

This attention to embodied practice and affective experience also informs the science of product location and the choreography of the supermarket shopping experience. Ethnographic studies on supermarket design and use have revealed the subtle and sophisticated ways in which consumers are trained and habituated to navigate grocery stores and fill their baskets (Colls and Evans 2008;

Johnston and Szabo 2011; Carolan 2018). A growth-oriented corporate mindset often underlies these supermarket strategies, even among 'alternative' outlets such as Whole Foods Market, where retail spaces are imbued with feminized notions of care through food (Johnston 2008; Cairns and Johnston 2015). This work suggests that much consumer 'choice' is habituated and subconscious, and that consumption acts are choreographed to limit conscious decision making (Sexton 2018). Food companies pay a premium to have their products placed at the end of aisles (or 'endcaps') and at desired heights on shelves. These premium locations are understood to matter as much as price and special offers in driving sales. Dairy milk is commonly a loss-leading staple in supermarkets and its location is carefully planned: far enough from the entrance that consumers must pass other tempting aisles to reach it, but not so far or so hidden as to be inconvenient.

In contrast, plant mylks have historically been found in peripheral 'alternative' or health food aisles where there is limited chance of serendipitous encounter. To normalize their brands, some mylk companies have paid a premium to have their products located in the refrigerated dairy aisle. To enhance this affective continuity in shopping experience, in 2016 Tesco and Alpro teamed up to sell their milk chilled, although this is not required for food safety (White 2016). The US companies Silk and Almond Breeze similarly relied on connections with prominent dairy milk companies to gain access to privileged refrigerated shelf-space (Franklin-Wallis 2019). Companies like Rebel Kitchen have even sought to emulate the packaging of dairy milk, using rectangular cartons with caricatured bovine white and black text, and a familiar range of single color tones (red, green, blue) to denote to UK consumers skimmed, semi-skimmed, or whole mylk options (Fig. 5).

The commercial success of this making palatable is evidenced in both the quantity of sales and the crossover between consumers of both dairy and plant milks. Market research suggests that consumers' adoption of mylks has not involved the like for like replacement of dairy products. Around 80% of households that purchase mylks also buy dairy milk (Mintel 2019). This fact is not lost on mylk companies, who target flexitarian consumers rather than vegans. Interview participants at one mylk company noted market research that plant-based mylk consumers actually consume *more* dairy milk. The reason given was that mylk drinkers tend to be 'foodies,' that is, people who are more interested in food in general. Despite claims of a post-milk generation, for the time being at least it appears that the rise of plant mylks represents a net increase in the consumption of packaged white liquid drinks. This would cast doubts on mylk companies' claims for environmental sustainability through the consumption of 'less' milk.

Cozy politics for flexitarians

This choreography of the mylk experience is given meaning through the range of storytelling practices we encountered above. Together these interventions work to create, shape and subjectify a 'good consumer'. Advertising has long been invested in reflecting and morphing modes of social distinction, channeling cultural identities to create affirmative product associations. The framings of milk and its alternatives are animated through a range of affective styles.[8] Advertisements reflect, refract, and sometimes forge social norms and identities. For example, we have seen in the images above how the framing of milk as wholesome is enabled by an affective style that conjoins rural iconography of white bodies, traditional technology, and sunlit landscapes with retro visual filters, pastoral music, and linear editing. The result is a nostalgic sense of continued social order. In contrast, framings of milk as murder feature animal head shots and industrial technology conjoined with guttural animal sounds. This harsh soundscape, frequent jump cuts, and shaky low-fi image quality suggests covert provenance and shocks viewers, who are disgusted with the palpable sense of social disorder.

Dairy-owned mylk brands, like Alpro, have chosen to persevere with the nostalgic pastoral style to promote plant milks as the healthy continuation of wholesome animal milk. We are to believe that there is nothing radical about their products; that they offer a logical technological innovation that replaces cows with plants. The tenor of this *cozy marketing* is exemplified by the absence of reference to animals and animal welfare. These mylk brands calculate that long-standing vegan consumers do not need reminding of this, while new flexitarian consumers prefer affirmative connections with ideals such as cleanliness, power, and wellness. Mylk companies do not want to invoke powerful gut feelings of disgust at animal suffering, even if no animals are harmed in plant mylk production. Our interviewees at the more ostensibly disruptive brands expressed reservations about referencing animals due to the risk of alienating the 98% of their consumers that are not vegan. Mylks thus inherit and benefit from the unpalatable framings of milk offered by campaigning vegan and animal rights organizations, without needing to give them explicit publicity: they are compassionate by default. Refraining from revolting and shaming consumers is especially important given that the vast majority of mylk drinkers also consume dairy. Even the most overt mention of animals (such as Oatly's slogan 'wow, no cow!') are not

explicitly related to animal rights but rather to the absence of animals, or animals as 'non-stuff' (Sexton 2016).

The vegan studies scholar Richard White argues that this rise of the 'vegan-consumer' and the flexitarian food subject represents a radical departure from 'vegan-activism'. The latter is commonly associated with abstinence, a withdrawal from mainstream food cultures, and an antagonistic politics of protest. Veganism was commonly sidelined by mainstream media as extremist (Cole and Morgan 2011). In contrast, most mylks are promoted as 'plant-based' or 'plant-forward' rather than animal-free. This offers a seemingly cozy, harmless and aspirational coding for mylk consumption, untainted by associations with 'reactionary' animal rights movements (White 2018; Davis Undated).

Those promoting mylk as a radical break from animal milk develop a different, affirmative style of disruption: their mylks are neither cozy, nor revolting, but revolutionary. One tactic is to channel the longer history in advertising of building product associations with the celebrities, music, fashion and iconography of youthful rebellion. Twentieth century shifts in social values like the hippy, rock or punk movements have long been deployed by advertisers seeking to differentiate their products away from mass marketing and towards lifestyle marketing to rebuild trust through co-opting elements of counterculture (Binkley 2003). Today, mylks deploy self-aware advertising to reach a millennial generation that is not only skeptical of corporate power but also adept at decoding and dismissing traditional advertising. Campaigns have shifted from celebrity endorsement to relying on social media and the established advertising tactic of irony, playing with intertextuality in their images and discourse to acknowledge their viewers' cultural sophistication (Jackson and Taylor 1996) and speak to their multiple identities as both citizens and consumers.

Arguably the most effective tactic used by disruptive brands to tap into the millennial zeitgeist to drive sales has been to combine irony with transparency in effort to build a more authentic, trusting relationship with consumers. Oatly's Creative Director John Schoolcraft captures this sentiment in describing Oatly as a 'challenger brand':

> Being a challenger is having a mindset of realizing you're trying to change something, rather than be a challenger to be cool and help sell more product. Because consumers will be able to feel it. Of course we want to sell our product, but we want to challenge the norms at the same time, and that's bigger. If you can get that right, you're going to sell a lot of product, and we need to sell product so that can continue to do what we're doing. (The Challenger Project 2016).

Michael Lee, the strategic director for international markets at Oatly, notes that this type of branding requires both an ironic 'Oatly tone of voice' that 'flexes on the nonsensical'

---

[8] By affective style we are referring to the emotional tenor of the advertisement that is produced through the planned juxtaposition of sound and imagery, alongside spoken and written content.

Fig. 6 Oatly billboards

as 'it just becomes lame when you start preaching it in your communications', but it also requires a veneer of transparency to acknowledge consumers' distrust of advertising (Rogers 2019). Advertising commentator Jamie Williams (2019) explains how Oatly has pioneered a tactic known as 'unadvertising', which mocks the traditional advertising formula. This style takes the intertextuality of ironic advertising to another level, repurposing the subversive tactics of the 1990s anti-capitalist Adbuster movement in an effort to overcome widespread cynicism about the social role of advertising (Lasn 1999).

Unadvertising is compelling because it makes manifest the ubiquity of advertising, while celebrating the individual ability of millennials to deconstruct and reflect on their own subjectification (Fig. 6). But the disruptive power of this style is ultimately limited by the absences it is willing to make present and the cozy types of affect it finds palatable. As a result, the palatable coziness or bounded self-reflexive edginess of this mode of consumer-led disruption lacks the affective agonism that political theorists hold to be central to the successful functioning of democracy.[9]

The palatable, 'feel good' food politics of disruptive mylk advertising—that eschews disgust at animal death and the 'extremism' of vegan activism—also evades the disagreeable opinions of those who stand to lose out from this

reorganization of the dairy system. It avoids unpalatable ruminations over whose economic interests it serves, and the social relations involved in producing almonds, soy and oats, for example. It certainly can't stomach questioning the claim that buying more will save the world. In short, there is little space for debate here, in spite of the proliferation of rhetorical ethical questions. While this criticism no doubt sets the bar too high for what we might realistically expect of fast-moving consumer goods, it does allow us to dispel the more outlandish claims that these products will necessarily catalyze political economic disruption.

The spectacle of care

The careful choreography and sophisticated marketing of the mylk experience is geared towards the creation, subjectification and governance of a set of 'good' or 'ethical' consumers. In AFNs such as local food (DuPuis and Goodman 2005), organic (Guthman 2004), and fair trade (Goodman 2004), consumers are responsibilized through concepts such as food miles or through third-party certification, underlying which are often relatively rigid norms and imaginaries that can exclude as much as they include. In this final analytical section, we explore the "normative pre-set 'standards'" (DuPuis and Goodman 2005) that arise through the utopian model of plant mylks. What consumer subjects are required to make plant mylks palatable to the 'new green spirit of capitalism' (c.f. Goldstein 2018)? What models of production do these standards promote?

Ethical food consumerism has a checkered history that is well reported in the academic literature. As discussed in the introduction, AFNs emerged by drawing attention to the social, animal and environmental harms caused by agro-industrial systems. Through organic or fair-trade certification, or through community supported agriculture, they enable consumers to 'vote with their wallets' and support alternative social-environmental ideals that better align with their values. Studies have traced these food movements' complicated relationship with 'conventional' food systems. AFNs have often been subsumed within a model of neoliberal agro-capitalism which places the consumer as the sovereign political agent in determining how, what, and where food is grown, distributed, and consumed (Goodman et al. 2010, 2012; Guthman 2011; Alkon and Mares 2012). The manifold imperfections of AFNs are seen to boil down to inequalities in 'who gets to the table' to eat 'good' food and to make political decisions and whether food movements absolve the state of responsibilities to ensure healthy food and environments (Guthman 2008; Hinrichs and Allen 2008; McMahon 2011; Johnston 2017).

The story of mylk as a palatable disruption allows us to develop two strands of this literature. For one, the centrality afforded to the consumer-citizen as the locus of change

---

[9] Chantel Mouffe (2013) and Jacques Rancière (2010), for example, argue that politics requires disagreement, claiming that dissensus is necessary for robust and accountable decision making.

in the politics of palatable disruption relegates other models of food system transformation that might address more systemic issues. But even more importantly, mylks foster a reliance upon food companies as ethical food system actors. Mylk companies' promises of disruption hinge on establishing their legitimacy as conduits of food system change and as custodians of a diverse range of consumers' cares. In short, it is up to the companies whether they adopt practices that stimulate changes in supply chains and yield benefits to social, environmental, and animal welfare dimensions of production.

Some companies go as far as to present themselves as social movements, endeavoring to replicate the sociology of their non-profit precursors. This trend is shown in Oatly's efforts to forge a 'post milk generation'. Mylk consumers are invited to imagine themselves as part of a radical social movement, united by their demographic (aspirationally coded as young and enlightened) and counter-posed to an older section of the population (coded as conservative, ignorant and/or reckless). Identification with this 'neo-tribe' (Maffesoli 1995), attached to a generational divide, serves to solidify the bond between consumer and company. This is enhanced by the provision of branded goods (t-shirts, loyalty cards, stickers), the creation of social media communities (blog posts, giveaways, and Instagram friendly imagery), and visibility at music festivals and other archetypal generational rights-of-passage events (Rogers 2019). Oatly implores consumers to invest in this relationship—and therefore a collective future—by drinking their milk:

> You are one of us now. You are now part of a growing group of people that understand the benefits of eating and drinking plants so your body feels good and so the planet can better cope with the impact we humans place on it (Oatly 2019).

In so doing these brands leverage the radical history of 'new social movements' and their politics—including those that sought change through alternative consumption—while sterilizing their potential for democratic transformation. Indeed, disparaging (often older) critics take issue with the ways in which such brands co-opt activist-inspired discourse to stimulate a feeling of urgency and to cultivate the sense of a collective agenda (White 2018). They variously dismiss this social movement simulation as feel-good, techno-optimistic 'slacktivism' that helps further entrench the fundamentally neoliberal project of ethical consumption by attaching a revolutionary air to it (Morozov 2013; Dennis 2018). Such consumers stand accused of narcissistic, 'virtue signaling,' that is, posting images of their consumption choices to social media in order to depict themselves as ethical. Indeed, social media has been crucial to inscribing the alternativeness of mylks in the collective consciousness. Unlike AFNs, mylk's alternativeness centers not on networks with distant producers and landscapes but on interactions with the brand. As an Oatly representative discussed of its advertising campaign in the London Underground, this consumer interaction with the brand is unprecedented.

We might view the palatable disruption presented in claims for a post-milk generation as premised on a mode of what Goodman et al (2016) term 'spectacular environmentalism'. This concept develops Guy Debord's analysis of the rise of the 'society of the spectacle', in which 'visual commodity fetishism' supplants 'real forms of human connection and sociality' (Goodman et al 2016, p. 678). Goodman et al. apply this work to present modes of green-mediated consumption to help understand how consumers reflexively engage with advertising, especially on interactive social media. Mylks offer one such spectacular environmentalism. Here mylk becomes a green commodity fetish: an object alienated from the social and ecological relations of its production.

This fetishism is displayed in how companies engage with questions of sustainability and wellness. The environmental promises made of mylks often center on *outcomes* ascertained via life cycle assessments (LCAs). The rise of mylk is thus linked to the incursion of scientific expertise—both health claims from nutritional sciences and environmental claims from LCAs—into domestic spaces. Through these mechanisms, food companies posture as scientific experts through food choice. These calculations furthermore dissuade consideration of sustainability as a dynamic social-environmental *process* that involves multiple actors and locations. Seeing sustainability as an outcome rather than a process encourages technological fixes and standards (such as organic) to govern at a distance. These have been demonstrated to undermine attempts to improve environmental outcomes through food production-consumption (Guthman 2004; Mansfield 2004). Corporations, which excel at incremental technological changes (Goldstein 2018), have seized in mylks an opportunity to write themselves as heroes in food system change. In coordinating the green fetishizing of mylk, brands perform a spectacular form of environmental care.

This fetishism is similarly articulated in the health and wellness claims of mylk. Underlying these claims are at times specific statements about the nutritional qualities of mylk that are bolstered discursively by nutritional sciences and materially through supplementary injections of calcium, fats, and nutrients. At other times, health claims are tied to a relatively vague notion of wellness that is upheld more by what is absent in mylks; often dairy, soy, or additives. In selling wellness attached to convenience, consumers are puzzlingly encouraged to cure the negative psychological effects of a societal drive towards hyper-productivity by consuming an on-the-go product that enables them to continue to be ultra-productive. A carton of

Califia Farms' Ubermilk captures this with a thank-you note to itself: 'thank you Ubermilk for being so extra. You go ABOVE and BEYOND so we can too.' As with environmental claims, these health framings of mylks evoke a spectacle of care. This form of care through industry relies upon a neoliberal consumer-subject that desires to use food to cope with environmental catastrophe and a life out of balance.

These framings have served to remove mylks from social-ecological contexts. Mylks do not engender consumer connections with specific places, landscapes, farmers, environments, or animals. This represents a significant departure from the raft of AFNs that arose with an explicit mission of contesting placeless agri-industrial food by rebuilding trust through embedding food systems in places, as with local food movements (DuPuis and Goodman 2005) or connecting consumers to distant producers, as in fair trade (Goodman 2004). With mylks, the consumer relationship ends at the brand. This keeps politics firmly within the realm of consumption and power with corporations. Moreover, while AFNs such as organic entail standards and verification to regulate production practices (Guthman 2004), mylks are verified simply by the absence of animal products. As a result, prospects for governance of food production rely upon existing agricultural laws or the discretion of food companies.

In these ways, mylks sustain undemocratic production-consumption dynamics. Consumers are encouraged to disrupt their patterns by choosing foods marketed as better for their health and the environment. Yet, despite the premiums paid for mylks, these products often rely on commodity production systems that uphold the market logics embedded in late agrarian capitalism. While some mylk companies devote time to verifying the source of their plant ingredients, these products have added premiums. The bulk of mylk sales accrues to large companies that purchase ingredients on commodity markets. And, while commodity markets may be under increasing pressure to become more sustainable, environmental regulation through markets has inherent limits (Freidberg 2018). Almond milk, the continued leader among mylks, is a key example of these limits. More than 80% of global almond production occurs in drought-prone areas of California on mega-farms in monoculture systems. These systems have drained aquifers during droughts to irrigate almonds (Reisman 2019) and use copious herbicides (most notably glyphosate), which have contributed to the decimation of honeybee populations (McGivney 2020). These industrial almond production systems supply world-leading brand Almond Breeze and Silk Almondmilk, which together grossed just under $1 billion in US sales in 2019 (Shahbandeh 2019). Such agro-industrial production systems are effectively hidden with the claims of alternativeness and disruption discussed above.

Nature's perfect neoliberal food

This paper assessed how plant mylks have been de-politicized and naturalized as solutions to problems of climate change, animal welfare, and human health. Mylks write new chapters in what DuPuis (2002) has called the 'perfect stories of milk,' or the narratives of degradation and salvation that have been foundational in middle-class efforts of social reform since the nineteenth century. The 'downfall' story highlights the deleterious effects of industrial animal agricultural systems on the environment and human health. The 'salvation' narrative stars plant-based as a promise to cope with both environmental catastrophe and psychological distress of a hectic work life where there is little time to pause for meals. Mylks address this confluence of environment and health concerns by doubling down on the individual as the locus of change; effectively neoliberalizing governance of global environmental and public health issues.

With a historical study of milk in the US, DuPuis (2002) has demonstrated how middle-class social reformers, dairy farmers, politicians, and health experts worked together to frame milk as 'nature's perfect food' in the twentieth century. This article depicts a similar politics of perfection and purification that serves to make mylks palatable in the twenty-first century. In contrast to the Fordist political economic structures undergirding dairy milk's becoming a perfect food (DuPuis 2002), plant mylks are enmeshed in a neoliberal political economy. In offering a way to purchase imaginaries of wellness and climate change mitigation, mylks promote a neoliberal ethic of care. Like other forms of green consumerism, mylks identify the individual as the key actor and global markets as the platform for solving environmental and health problems. A politics of perfection depoliticized milk by attaching it to powerful social narratives of purity (DuPuis 2002). Similarly, by curating palatability and a food ethic based on absence, mylks depoliticize what might be a contested terrain of food system change. This serves to reinforce the political economy of agro-industry. Mylks thus appear to be a neoliberal articulation of food perfection. By strengthening corporate control over 'alternatives', mylks risk foreclosing on other potential pathways of food system change. As Guthman (2008, p. 442) suggests of AFNs in California, the mainstreaming of veganism through mylks reflects a 'limited politics of the possible'.

Our aim in this article has been to open discussion to the limits of the current plant-based trend in hoped-for transitions to more just, ethical, and sustainable food futures. Our analysis traced how palatable disruption was achieved, identifying the importance of affective continuity in users experience of milk, the role played by cozy marketing to flexitarians, and the importance of spectacular modes of green commodity fetishism. As others put forth, mylks do have disruptive possibilities (Gambert and Linné 2019). Yet, a 'post-milk' future will not automatically address problems caused by the overproduction of industrial foods. Mylks excel in their ability to make food placeless. With further legitimacy gained from nutritionism and LCAs—and without animals getting in the way—mylk may be even more effective as a commodity than dairy milk. By merely grafting plant milks into existing production-consumption practices, agro-industrial problems are not so much fixed as they are diverted, obscured, or even forgotten. Mylks may afford at best an interruption to the challenges they claim to resolve. At worst, they could distract from the need for systemic changes by virtue of fitting so well within the contours of globalized industrial agri-food.

Increased consumption of plant mylk could in theory drive change in dairy systems through decreased demand for dairy milk. Yet such a trajectory is far from given. Dairy systems are highly heterogeneous (Clay et al. 2020). Water use, land use, and greenhouse gas emissions various enormously across farms and regions (Poore and Nemececk 2018). A post-milk imaginary does not necessarily exert influence over the type of dairy system. If past trajectories of intensification in the dairy sector are an indication, a likely response to decreased milk demand could be for the industrial dairy industry to further intensify production. Even though fluid milk consumption is decreasing in the US and Europe, it is increasing worldwide. One possible outcome is that mylk consumption will encourage industrial dairy systems that are environmentally harmful and of limited benefit to rural livelihoods. Continued consolidation into mega-farms has been driven in the past by price competition that privileges economies of scale. At the same time, dairy operations with a lower environmental footprint, higher animal welfare, and value to rural livelihoods and cultural landscapes will likely continue disappearing.

This interpretation of the politics of plant-based milk is meant as cautionary rather than dismissive. Plant-based milk and meat are flourishing. As these products to grow and diversify, it is crucial to consider how they might enable more democratic food futures. Flexitarianism presents a potentially open, inclusive, and democratic form of consumption that could drive food system change in just and sustainable ways. Its crux may be its mutability, which makes it readily co-optable. The corporate mylk regime that was the focus of this article does not exhaust the cultural, political, and economic forms that configure how milk alternatives can and do arise. Much less does it encapsulate the pathways by which we might transition to plant-rich diets. This is the crucial point. Despite this industrial incarnation, plant mylk can in fact be made at home with relative ease (at least compared to dairy milk, which requires a lactating mammal). There is no necessary reason why liquids derived from plants cannot give rise to environmentally beneficial, socially just, ethical, and nutritious ways of feeding people. Yet, assuring that they do requires attention to processes of production, distribution, and consumption.

Acknowledgements The research informing this article was supported by the Wellcome Trust, Our Planet Our Health (Livestock, Environment and People–LEAP) project, Award Number 205212/Z/16/Z, at the University of Oxford. Many thanks to Dan Clay and Elissa Dickson, as well as three anonymous reviewers, for comments on earlier versions of this manuscript. The authors alone bear responsibility for any errors in fact or interpretation.

Compliance with ethical standards

Ethical approval All procedures performed in studies involving human participants were in accordance with the ethical standards of the institutional and/or national research committee (University of Oxford Central University Research Ethics Committee, SOGE 1A-19-10) and with the 1964 Helsinki declaration and its later amendments or comparable ethical standards.

Open Access This article is licensed under a Creative Commons Attribution 4.0 International License, which permits use, sharing, adaptation, distribution and reproduction in any medium or format, as long as you give appropriate credit to the original author(s) and the source, provide a link to the Creative Commons licence, and indicate if changes were made. The images or other third party material in this article are included in the article's Creative Commons licence, unless indicated otherwise in a credit line to the material. If material is not included in the article's Creative Commons licence and your intended use is not permitted by statutory regulation or exceeds the permitted use, you will need to obtain permission directly from the copyright holder. To view a copy of this licence, visit http://creativecommons.org/licenses/by/4.0/.

References

Alkon, A.H., and T.M. Mares. 2012. Food sovereignty in US food movements: Radical visions and neoliberal constraints. *Agriculture and Human Values* 29: 347–359.
Andersen, K., and K. Kuhn. 2014. *Cowspiracy: The sustainability secret*. Santa Rosa: A.U.M. Films.
Atkins, P. 2010. *Liquid materialities: A history of milk, science and the law*. Burlington: Ashgate.
Barthes, R. 1972. *Mythologies* (trans: Lavers, A.). London: Jonathan Cape.
Binkley, S. 2003. Cosmic profit: Countercultural commerce and the problem of trust in American marketing. *Consumption Markets and Culture* 6 (4): 231–249.
Bladow, K. 2015. Milking it: The pastoral imaginary of california's (non)dairy farming. *Gastronomica* 15 (3): 9–17.
Cairns, K., and J. Johnston. 2015. *Food and femininity*. New York: Bloomsbury.
Camacho, F. 2018. *Taking plant-based to the max*. Presentation at Danone Investor Seminar, 22 October 2018. https://www.danone.com/content/dam/danone-corp/investors/en-investor-seminars/2018/EDP.pdf. Accessed 6 Aug 2019.

Carolan, M. 2015. Affective sustainable landscapes and care ecologies: Getting a real feel for alternative food communities. *Sustainability Science* 10 (2): 317–329.

Carolan, M. 2016. Adventurous food futures: Knowing about alternatives is not enough, we need to feel them. *Agriculture and Human Values* 33 (1): 141–152.

Carolan, M. 2018. Big data and food retail: Nudging out citizens by creating dependent consumers. *Geoforum* 90: 142–150.

Carrington, D. 2018. Avoiding meat and dairy is 'single biggest way' to reduce your impact on Earth. The Guardian, 31 May. https://www.theguardian.com/environment/2018/may/31/avoiding-meat-and-dairy-is-single-biggest-way-to-reduce-your-impact-on-earth. Accessed 6 Aug 2019.

Clark, M.A., M. Springmann, J. Hill, and D. Tilman. 2019. Multiple health and environmental impacts of foods. *Proceedings of the National Academy of Sciences* 116 (46): 23357–23362.

Clarke, N., C. Barnett, P. Cloke, and A. Malpass. 2007. Globalising the consumer: Doing politics in an ethical register. *Political Geography* 26: 231–249.

Clay, N., T. Garnett, and J. Lorimer. 2020. Dairy intensification: Drivers, impacts and alternatives. *Ambio* 49: 35–48. https://doi.org/10.1007/s13280-019-01177-y.

Cole, M., and K. Morgan. 2011. Vegaphobia: Derogatory discourses of veganism and the reproduction of speciesism in UK national newspapers 1. *The British Journal of Sociology* 62 (1): 134–153.

Colls, R., and B. Evans. 2008. Embodying responsibility: Children's health and supermarket initiatives. *Environment and Planning A: Economy and Space* 40: 615–631.

Dairy Farmers of America. 2019. DFA reports 2018 financial results. 20 March 2019. https://www.dairyherd.com/sites/default/files/inline-files/DFA-2018%20Financial%20Release-FINAL.pdf. Accessed 6 Aug 2019.

Davis, J. (undated) Are you a positive or a negative veg*n? International Vegetarian Union. https://ivu.org/index.php/blogs/john-davis/67-are-you-a-positive-or-a-negative-veg-n. Accessed 31 July 2019.

Department for Environment, Food & Rural affairs (Defra). 2017. Family food Datasets: Household Purchases 2016/2017. https://www.gov.uk/government/statistical-data-sets/family-food-datasets. Accessed 6 Aug 2019.

Dennis, J. 2018. *Beyond Slacktivism: Political participation on social media*. London: Palgrave Macmillan.

Doyle, J., N. Farrell, and M.K. Goodman. 2019. The cultural politics of climate branding: Poject Sunlight, the biopolitics of climate care and the socialisation of the everyday sustainable consumption practices of citizens-consumers. *Climatic Change*. https://doi.org/10.1007/s10584-019-02487-6.

DuPuis, E.M. 2000. Not in my body: BGH and the rise of organic milk. *Agriculture and Human Values* 17 (3): 285–295.

DuPuis, E.M. 2002. *Nature's perfect food: How milk became America's drink*. New York: NYU Press.

DuPuis, E.M., and D. Goodman. 2005. Should we go "home" to eat? Toward a reflexive politics of localism. *Journal of Rural Studies* 21 (3): 359–371.

Evans, A., and M. Miele. 2012. Between food and flesh: How animals are made to matter (and not matter) within food consumption practices. *Environment and Planning D: Society and Space* 30 (2): 298–314.

Evans, D., D. Welch, and J. Swaffield. 2017. Constructing and mobilizing 'the consumer': Responsibility, consumption and the politics of sustainability. *Environment and Planning A: Economy and Space* 49: 1396–1412.

FAO. 2006. Livestock's long shadow. Rome: Food and agriculture organization of the united nations. https://www.fao.org/3/a0701e/a0701e.pdf. Accessed 4 Mar 2019.

Fields, D. 2019. Investors thirst for plant-based milks. Forbes, January 31. https://www.forbes.com/sites/mergermarket/2019/01/31/investors-thirst-for-plant-based-milks/#73981d574184. Accessed 6 Aug 2019.

Fisher M (2018) *Animal Welfare Science, Husbandry and Ethics: The Evolving Story of Our Relationship with Farm Animals*. 5M Publishing Ltd: Sheffield

Foer JS, Quinn CD, Portman N (2017) Eating Animals. Ro*co Films Educational: Sausalito

Foote, K.J., M.K. Joy, and R.G. Death. 2015. New Zealand dairy farming: Milking our environment for all its worth. *Environmental Management* 56 (3): 709–720.

Forsyth, I., H. Lorimer, P. Merriman, and J. Robinson. 2013. What are surfaces? *Environment and Planning A: Economy and Space* 45 (5): 1013–1020.

Franklin-Wallis, O. 2019. White gold: the unstoppable rise of alternative milks. The Guardian, 29 January. https://www.theguardian.com/news/2019/jan/29/white-gold-the-unstoppable-rise-of-alternative-milks-oat-soy-rice-coconut-plant. Accessed 31 July 2019.

Freidberg, S. 2009. *Fresh: A perishable history*. Cambridge: Harvard University Press.

Freidberg, S. 2018. Assembled but unrehearsed: corporate food power and the 'dance' of supply chain sustainability. *The Journal of Peasant Studies*. https://doi.org/10.1080/03066150.2018.1534835.

Gambert, I., and T. Linné. 2019. Got mylk? The disruptive possibilities of plant milk. *Brooklyn Law Review* 84 (3): 801–871.

Goldstein, J. 2018. *Planetary Improvement: Cleantech Entrepreneurship and the Contradictions of Green Capitalism*. Cambridge: MIT Press.

The Good Food Institute. 2018. Plant-based Alternatives Data Sheet. https://www.gfi.org/images/uploads/2018/09/Good-Food-InstitutePlant-Based-Nielsen-Data-Sheet-2018-0911-v3.pdf. Accessed 28 July 2019.

The Good Food Institute. 2019. Plant-based food retail sales are growing 5x total food sales. 16 July. https://www.gfi.org/spins-data-release-2019. Accessed 9 Nov 2019.

Goodman, M.K. 2004. Reading fair trade: Political ecological imaginary and the moral economy of fairy trade foods. *Political Geography* 23 (7): 891–915.

Goodman, M.K., D. Maye, and L. Holloway. 2010. Ethical foodscapes? Premises, promises, and possibilities. *Environment and Planning A: Economy and Space* 42: 1782–1796.

Goodman, D., E.M. DuPuis, and M.K. Goodman. 2012. *Alternative food networks: Knowledge, practice, and politics*. London: Routledge.

Goodman, M.K., J. Littler, D. Brockington, and M. Boykoff. 2016. Spectacular environmentalisms: Media, knowledge and the framing of ecological politics. *Environmental Communication* 10 (6): 677–688.

Guthman, J. 2004. *Agrarian dreams: The paradox of organic farming in California*. Berkeley: University of California Press.

Guthman, J. 2008. "If they only knew": Color blindness and universalism in California alternative food institutions. *The Professional Geographer* 60 (3): 387–397.

Guthman J (2011) *Weighing in: Obesity, food justice, and the limits of capitalism*. University of California Press: Berkeley

Hadrich, J.C., C.A. Wolf, and K.K. Johnson. 2017. Characterizing US dairy farm income and wealth distributions. *Agricultural Finance Review* 77 (1): 64–77. https://doi.org/10.1108/AFR-04-2016-0040.

Hayes-Conroy, A. 2010. Feeling slow food: Visceral fieldwork and empathetic research relations in the alternative food movement. *Geoforum* 41 (5): 734–742.

26

Hayes-Conroy, J., and A. Hayes-Conroy. 2013. Veggies and visceralities: A political ecology of food and feeling. *Emotion, Space and Society* 6: 81–90.

Hinrichs, C., and P. Allen. 2008. Selective patronage and social justice: Local food consumer campaigns in historical context. *Journal of Agricultural and Environmental Ethics* 21 (4): 329–352.

Jackson, P., and J. Taylor. 1996. Geography and the cultural politics of advertising. *Progress in Human Geography* 20: 356–371.

Jay, M., and M. Morad. 2007. Crying over spilt milk: A critical assessment of the ecological modernization of New Zealand's dairy industry. *Society & Natural Resources* 20 (5): 469–478.

Johnston, J. 2008. The citizen-consumer hybrid: ideological tensions and the case of Whole Foods Market. *Theory and Society* 37 (3): 229–270.

Johnston, J. 2017. Can consumers buy alternative foods at a big box supermarket. *Journal of Marketing Management* 33 (7–8): 662–671.

Johnston, J., and M. Szabo. 2011. Reflexivity and the whole foods market consumer: The lived experience of shopping for change. *Agriculture and Human Values* 28 (3): 303–319.

Jones, R., J. Pykett, and M. Whitehead. 2011. Governing temptation: Changing behaviour in an age of libertarian paternalism. *Progress in Human Geography* 35: 483–501.

Kahneman, D. 2011. *Thinking*. Fast and Slow: Penguin.

Kenner, R., E. Pearlstein, and K. Roberts. 2008. *Food Inc*. London: Dogwoof Ltd.

Kuhn, K., K. Andersen, and K. Kuhn. 2017. *What the health*. United States: A.U.M. Films & Media.

Kurlansky, M. 2018. *Milk! A 10,000-year food fracas*. New York, NY: Bloomsbury.

Lasn, K. 1999. *Culture Jam: The Uncooling of America*. New York, NY: William Morrow and Company.

Levitt, T. 2018. Put a label on it: why the future of milk is a branded one. Nuffield UK. https://static1.squarespace.com/static/564cbb1ee4b0ff765b5ae062/t/5be451094d7a9c2fe6767194/1541689613190/Tom+Levitt+Report.pdf. Accessed 6 Aug 2019.

Lockie, S. 2009. Responsibility and agency within alternative food networks: Assembling the "citizen consumer". *Agriculture and Human Values* 26 (3): 193–201.

Longhurst, R., E. Ho, and L. Johnston. 2008. Using 'the body' as an 'instrument of research': Kimch'i and pavlova. *Area* 40 (2): 208–217.

Lorimer, J. 2010. Moving image methodologies for more-than-human geographies. *Cultural Geographies* 17: 237–258.

MacDonald, J. M., C. Jerry, and M. Roberto. 2016. Changing structure, financial risks, and government policy for the U.S. dairy industry. Research in agricultural and applied economics, https://ageconsearch.umn.edu/record/262200. Accessed 6 Aug 2019.

Maffesoli, M. 1995. *The time of the tribes: The decline of individualism in mass society*. London: SAGE.

Mann A, Mol A, Satalkar P, Savirania A, Selim A, Sur M, Yates-Doerr E (2011) Mixing methods, tasting fingers: Notes on an ethnographic experiment. *HAU: Journal of Ethnographic Theory* 1(1):221–243.

Mansfield, B. 2004. Rules of privatization: Contradictions in neoliberal regulation of North Pacific Fisheries. *Annals of the Association of American Geographers* 94 (3): 565–584.

Marteau, T.M. 2018. Changing minds about changing behaviour. *The Lancet* 391: 116–117.

Marx, L. 2000. *The machine in the garden: Technology and the pastoral ideal in America*. Oxford: Oxford University Press.

McGivney, A. 2020. "'Like sending bees to war': the deadly truth behind your almond-milk obsession." The Guardian, January 8. https://www.theguardian.com/environment/2020/jan/07/honey-bees-deaths-almonds-hives-aoe?CMP=fb_gu. Accessed 10 Jan 2020.

McMahon, M. 2011. Standard fare or fairer standards: Feminist reflections on agri-food governance. *Agriculture and Human Values* 28 (3): 401–412.

Miller, P., and N. Rose. 1997. Mobilizing the consumer: Assembling the subject of consumption. *Theory, Culture & Society* 14: 1–36.

Mintel. 2018. US non-dairy milk sales grow 61% over the last five years. Mintel, 4 January. https://www.mintel.com/press-centre/food-and-drink/us-non-dairy-milk-sales-grow-61-over-the-last-five-years. Accessed 4 Mar 2019.

Mintel. 2019. Milking the vegan trend: a quarter (23%) of Brits use plant-based milk. Mintel, 19 July. https://www.mintel.com/press-centre/food-and-drink/milking-the-vegan-trend-a-quarter-23-of-brits-use-plant-based-milk. Accessed 6 Aug 2019.

Morozov, E. 2013. *To save everything, click here: Technology, solutionism, and the urge to fix problems that don't exist*. London: Penguin.

Moss, M. 2013. *Salt, sugar, and fat: How the food giants hooked us*. New York, NY: Random House.

Mouffe, C. 2013. *Agnostics: Thinking the world politically*. London: Verso.

Mylan, J., C. Morris, E. Beech, and F.W. Geels. 2018. Rage against the regime: Nich-regime interactions in the societal embedding of plant-based milk. *Environmental Innovation and Societal Transitions* 31: 233–247.

Oatly. 2019. Oat drink. https://www.oatly.com/uk/products/oat-drink. Accessed 8 July 2019.

Parker, J. 2018. The year of the vegan. The economist, December. https://worldin2019.economist.com/theyearofthevegan?utm_source=412&utm_medium=COM. Accessed 10 Jan 2020.

Paxson, H. 2013. *The life of cheese: Crafting food and value in America*. Berkeley: University of California Press.

Plant-Based Foods Association. 2018. Plant-based food sales grow 20%. https://plantbasedfoods.org/wp-content/uploads/2018/07/PBFA-Release-on-Nielsen-Data-7.30.18.pdf. Accessed 6 Aug 2019.

Poore, J., and T. Nemecek. 2018. Reducing food's environmental impacts through producers and consumers. *Science* 360: 987–992.

Ramachandran, A., J. Raven, and R. Wardle. 2019. Appetite for disruption. How leading food companies are responding to the alternative protein boom. Fairr: A Coller Initiative. https://cdn.fairr.org/2019/07/24112310/FAIRR_Appetite_For_Disruption_Public_24_July_20191.pdf. Accessed 8 Aug 2019.

Rancière, J. 2010. *Dissensus: On politics and aesthetics*. London: Bloomsbury.

Reisman, E. 2019. The great almond debate: A subtle double movement in California water. *Geoforum* 104: 137–146.

Roe, E.J. 2006. Things becoming food and the embodied, material practices of an organic food consumer. *Sociologia Ruralis* 46 (2): 104–121.

Roe, E., and M. Buser. 2016. Becoming ecological citizens: Connecting people through performance art, food matter and practices. *Cultural Geographies* 23 (4): 581–598.

Rogers, C. 2019. How marketing is fueling the 'post-milk generation'. Marketing Week, 2 January 2. https://www.marketingweek.com/marketing-fuelling-post-milk-generation/. Accessed 31 July 2019.

Rose, G. 2007. *Visual methodologies: An introduction to the interpretation of visual materials*. London: SAGE.

Salque, M., P.I. Bogucki, J. Pyzel, I. Sobkowiak-Tabaka, R. Grygiel, M. Szmyt, and R.P. Evershed. 2013. Earliest evidence for cheese making in the sixth millennium BC in northern Europe. *Nature* 493: 522–525.

Schatzker, Mark. 2015. *The dorrito effect: The surprising new truth about food and flavor*. New York: Simon and Schuster.

Sexton, A. 2016. Alternative proteins and the (non)stuff of "meat". *Gastronomica* 16 (3): 66–78.

Sexton, A.E. 2018. Eating for the post-Anthropocene: Alternative proteins and the biopolitics of edibility. *Transactions of the Institute of British Geographers* 43: 586–600.

Sexton, A.E., T. Garnett, and J. Lorimer. 2019. Framing the future of food: The contested promises of alternative proteins. *Environment and Planning E: Nature and Space* 2 (1): 47–72.

Shahbandeh, M. 2019. "Leading U.S. refrigerated almond milk brands based on dollar sales 2019". https://www.statista.com/statistics/372461/leading-us-rtd-almond-milk-brands-based-on-dollar-sales/. Accessed 10 Jan 2020.

Shove, E. 2010. Beyond the ABC: climate change policy and theories of social change. *Environment and Planning A: Economy and Space* 42 (6): 1273–1285.

Smith-Howard, Kendra. 2014. *Pure and modern milk: An environmental history since 1900*. Oxford: Oxford University Press.

Springmann, M., H.C.J. Godfray, M. Rayner, and P. Scarborough. 2016. Analysis and valuation of the health and climate change cobenefits of dietary change. *Proceedings of the National Academy of Sciences* 113: 4146–4151.

Springmann, M.M., D. Clark, K. Mason-D'Croz, B.L. Wiebe, L. Bodirsky, W. de Lassaletta, S.J. Vries, M.Herrero Vermeulen, et al. 2018. Options for keeping the food system within environmental limits. *Nature* 562 (7728): 519.

Swaffield, J. 2016. After a decade of critique: Neoliberal environmentalism, discourse analysis and the promotion of climate-protecting behaviour in the workplace. *Geoforum* 70: 119–239.

The Challenger Project. 2016. An interview with John Schoolcraft, the Creative Director of Oatly. https://thechallengerproject.com/blog/2016/oatly. Accessed 31 July 2019.

Tulloch, L., and P. Judge. 2018. Bringing the calf back from the dead: Video activism, the politics of sight and the New Zealand dairy industry Video. *Journal of Education and Pedagogy* 3: 1–20.

Valenze, D. 2012. *Milk: a local and global history*. New Haven: Yale University Press.

Watson, E. 2019. Food Navigator USA, 28 June 2019. https://www.foodnavigator-usa.com/Article/2019/06/29/Oatly-challenged-over-no-added-sugars-claims-on-unsweetened-oatmilk. Accessed 10 Jan 2020.

Westhoek, H.J.P., T. Lesschen, S. Rood, A. De Wagner, D. Marco, A. Murphy-Bokern, H.G. Leip, et al. 2014. Food choices, health and environment: Effects of cutting Europe's meat and dairy intake. *Global Environmental Change* 26: 196–205.

White, K. 2016. Tesco and Alpro develop chilled free-from dairy alternatives fixture. The Grocer, 24 October 2016. https://www.thegrocer.co.uk/tesco-and-alpro-develop-chilled-free-from-fixture/543928.article. Accessed 31 July 2019.

White, R. 2018. Looking backward/ moving forward. Articulating a "Yes, BUT…!" response to lifestyle veganism, and outlining post-capitalist futures in critical veganic agriculture. *EuropeNow* (20). https://www.europenowjournal.org/2018/09/04/looking-backward-moving-forward-articulating-a-yes-but-response-to-lifestyle-veganism/. Accessed 29 Jan 2020.

Wiley, A.S. 2008. Transforming milk in a global economy. *American Anthropologist* 109 (4): 666–677.

Wiley, A.S. 2014. *Cultures of Milk. The biology and meaning of dairy products in the United States and India*. Cambridge: Harvard University Press.

Williams, J. 2019. How Oasis, Oatly and BrewDog are benefiting from 'unadvertising'. The Grocer. 19 June 2019. https://www.thegrocer.co.uk/marketing/how-oasis-oatly-and-brewdog-are-benefiting-from-unadvertising/594509.article. Accessed 31 July 2019.

Williamson, J. 1978. *Decoding advertisements: Ideology and meaning in advertising*. Ann Arbor, MI: University of Michigan Press.

Wohl, Jessica. 2019. How the rise of 'flexitarians' is powering plant-based foods. Adage, 1 April. https://adage.com/article/cmo-strategy/power-plant-based-food/317167. Accessed 9 Nov 2019.

Wood, Z. 2019. Plant-based milk the choice for almost 25% of Britons now. The Guardian, 19 July. https://www.theguardian.com/food/2019/jul/19/plant-based-milk-the-choice-for-almost-25-of-britons-now. Accessed 25 July 2019.

Publisher's Note Springer Nature remains neutral with regard to jurisdictional claims in published maps and institutional affiliations.

Nathan Clay is a postdoctoral researcher at the University of Oxford in the School of Geography and the Environment. His work concentrates on climate change, food systems, and environmental governance, exploring power and justice in relationships among humans, animals, and the environment.

Alexandra E. Sexton is a postdoctoral researcher at the University of Oxford in the School of Geography and the Environment. Her research examines the politics of food technology and food security with emphasis on alternative approaches to livestock production.

Tara Garnett runs the Food Climate Research Network at the University of Oxford. She has a particular interest in the debates around livestock and their role in sustainable food systems.

Jamie Lorimer is an Associate Professor in the School of Geography and the Environment at the University of Oxford. His research explores the social dimensions of environmental management.

Agriculture and Human Values (2020) 37:963–981
https://doi.org/10.1007/s10460-020-10023-x

To the market and back? A study of the interplay between public policy and market-driven initiatives to improve farm animal welfare in the Danish pork sector

Lars Esbjerg[1]

Accepted: 28 January 2020 / Published online: 4 February 2020
© Springer Nature B.V. 2020

Abstract

This article discusses the interplay of public policy and market-driven initiatives to improve farm animal welfare (FAW). Over the last couple of decades, the notion of 'market-driven animal welfare' has become popular, but can the market deliver the FAW that consumers and politicians expect? Using the Danish pork sector as the empirical setting, this article studies efforts to improve private FAW standards following changes to general regulations. The analysis shows that ethical misgivings regarding the adequacy of current and prospective FAW standards are tempered by the economic considerations that guide the practices of some actors. The study also shows that efforts to improve FAW standards are contingent on collaboration and coordination across globalised markets among actors with divergent interests. The findings have important implications for market practices and public policy in relation to FAW.

Keywords Farm animal welfare · Market practices · Public policy · Private standards

Abbreviations

EEC	European Economic Community
EU	European Union
FAW	Farm animal welfare

Introduction

Farm animal welfare (FAW) is an important and highly contentious issue in modern industrialised food production. Similar to issues such as food safety and quality, setting and monitoring FAW requirements are increasingly the domain of private-sector actors like retailers and animal welfare organisations. Through the development, implementation and monitoring of private standards, they impose requirements on farmers that go beyond the requirements prescribed by public policy and regulations (Christensen et al. 2012; Richards et al. 2013). The growing importance of private standards represents a significant change in governance practices with profound implications for who has the responsibility for setting and enacting standards for the welfare of farm animals (Carey et al. 2017; Parker et al. 2017).

The aim of this article is to study the interplay of public policy and market-driven initiatives when negotiating and enacting FAW standards. The notion of 'market practices' (Kjellberg and Helgesson 2007) is used to explore how FAW standards are negotiated and enacted through materially interwoven activities of business and non-business actors in the global food marketing system. This is important due to the growing importance of private standards and because standards are not just neutral technologies for classifying and ordering reality—standards play an important role in constituting reality as they mandate particular practises to be followed in the production and selling of animal products (Busch 2000, 2011; Mol 2002).

Through a study of the market practices of business and non-business actors in the Danish pork sector, the article investigates three related research questions: (1) How do actors in and around the Danish pork sector understand the market for pork, and in particular the market for pork with better animal welfare than conventional pork? (2) How do various actors try to change the market for (welfare) pork

✉ Lars Esbjerg
   lae@mgmt.au.dk

[1] Department of Management, Aarhus University, Fuglesangs
    Allé 4, 8210 Aarhus V, Denmark

through the development and use of particular standards?[1] (3) What are the exchange practices among actors on the domestic market, particularly with regard to welfare pork? The three research questions are inspired by Kjellberg and Helgeson's (2007) distinction between three types of closely related market practices—representational practices, normalising practices and exchange practices (see below for details). The study focuses on how actors reacted to new regulations requiring sows to be kept in group or loose housing during gestation (popularly referred to as 'loose sows'), which came into effect in 2013.

Attention is thus on the role of standards for market practices, in particular efforts to change the market and how it functions, such as when ethical considerations regarding the adequacy of current and prospective FAW standards are tempered by other considerations, including economic considerations, affecting the practices of various actors. The article demonstrates how ethical issues—in this case the welfare of Danish piglets and sows—are often global in nature and therefore cannot easily be isolated to specific markets. Making ethical improvements to business practices is systemic and requires participation of both business and non-business actors along the entire value chain. The article concludes that left to its own devices, the market cannot be expected to deliver the general improvements in animal welfare that proponents of market-based animal welfare hope for, and that therefore there is still an important role to play for public policy.

The remainder of this article is organised into five main sections. First, the shift from public policy to private standards for FAW is discussed in more detail to provide some background. Second, it is discussed what a market is and how it is constituted through the practices of business and non-business actors. Next, the setting and methodology used in the empirical study are presented. Fourth, the findings are presented. Here it is discussed how different actors' understandings of the market for welfare pork influence their efforts to change the market through standards, thereby showing that the development, implementation and monitoring of FAW standards mobilise a large number of actors and imply collective work involving both public policy and marketers. The article ends with a discussion of the implications for market practices and public policy in relation to FAW.

Background

In Danish law, farm animals have historically been seen as property, serving the interests of their owners, not as having any rights on their own.[2] During the 1800s, the need for rules to prevent cruelty to animals became obvious. Cruelty to animals became a criminal offence in 1857 and was incorporated in the Danish penal code in 1866. Over time, it became clear that this was not sufficient, and in 1916 parliament passed the first Danish law for the prevention of cruelty to animals. The law made it a criminal offence to abuse animals or to treat them irresponsibly due to neglect or over-exertion.

With minor changes and additions, the law remained effective until 1950. Since the first law was enacted, a consensus had been established that, as living creatures, animals had a right to proper treatment and protection against suffering. §1 of the law of 1950 thus stated that animals must be treated properly and not be subject to neglect, over-exertion or any other uncessary suffering. Suffering was thus accepted if deemed necessary. Although the legal standing of animals was strengthened compared to the law of 1916, the new law did not concern the behaviour of animals. With minor changes, this law was in effect until 1991. In the meantime, concern about modern farming practices was spreading.

The development and diffusion of intensive factory farming practices after World War II, first in poultry and then in pig and cattle production (Singer 1976), made animal production much more efficient in terms of increasing output while lowering production costs (Sandøe et al. 2003). However, this increased efficiency came at the cost of the welfare of farm animals, as initially highlighted by Ruth Harrison (1964) in her seminal book *Animal Machines*. The exposition of the disturbing conditions in which animals were kept in factory farms caused an uproar in the United Kingdom (Singer 1976), forcing the U.K. government to form *The Technical Committee to Enquire into the Welfare of Animals Kept under Intensive Livestock Husbandry Systems*. The findings and recommendations of this committee were published in 1965 in the Brambell Report.

Without questioning the right to raise animals for food, the Brambell Report recommended that, at a minimum, an animal should have the freedom of movement to stand up, lie down, turn around, groom itself and stretch its limbs (Brambell 1965, p. 13). These 'five freedoms' have since been expanded and revised to include freedom from hunger and thirst; from discomfort; from pain, injury and disease; to express normal behaviour; and from fear and distress

---

[1] Welfare pork is defined as pork that meets higher FAW requirements than conventional production.

[2] The following brief historic overview of animal welfare regulation in Denmark is based on Broberg (2016).

(FAWC 2009). The notion of the five freedoms has also had substantial impact on animal welfare legislation and the development of welfare codes and schemes throughout Europe (McCulloch 2013; Webster 1994), although its focus on reducing what is now called 'negative welfare'—harmful experiences and outcomes such as pain and frustration—has been criticised (Sandøe and Christensen 2018). For instance, the focus on the absence of suffering means that there is no mention of positive welfare states such as pleasure (Palmer and Sandøe 2018). Furthermore, the focus in most animal welfare legislation is on 'unnecessary' suffering, understood as suffering that can be avoided without compromising the level of production, which means that in practice some welfare problems are accepted as necessary (Palmer and Sandøe 2018; Sandøe and Jensen 2011).

The Brambell Report saw national legislation as the main tool for securing animal welfare, and made recommendations for changes to British animal welfare legislation. Recognising that scientific knowledge about FAW was limited and that flexibility was required to be able to take new developments in agriculture into account, the committee suggested a flexible legal structure of animal welfare legislation based on enabling acts[3]—a principle that has since become central to both national and international regulation (Brambell 1965; Sandøe and Christensen 2018).

Enabling acts have also become central to regulation of FAW in Denmark, where a new enabling act on animal welfare was passed in 1991 (it replaced the old 1950 law), creating the legal basis for detailed regulation to be imposed administratively through ministerial orders. In §1, the law recognises that animals are living creatures with the right to being treated properly and protected in the best possible way against pain, suffering, anxiety, permanent impairment and significant nuisance. The phrase 'in the best possible way' implies that not all pain is avoidable, whereas the focus on suffering and anxiety goes beyond the physical wellbeing of animals. §2 obligates all who keep animals to ensure that their animals are treated appropriately and that their physiological, behavioural and health-related needs are respected in line with approved practical and scientific experience. Since practical and scientific experience evolves, this means that what is regarded as appropriate treatment is dynamic.

Development of common European rules and standards

The Brambell Committee realised that its recommendations might increase costs in certain sectors of the food industry and that this might hamper the competitiveness of domestic farmers, leading to an increase in imports from countries with lower animal welfare requirements (Brambell 1965). To avoid such unintended consequences of national animal welfare legislation, efforts have since the 1970s been made to set up common rules and standards for FAW on an European level (Sandøe and Christensen 2018). Since Denmark became a member of the European Economic Community (EEC) in 1973, European rules and regulations have become important for the development of Danish FAW regulations.

The Council of Europe[4] put the prevention of cruelty against animals on the agenda in the late 1960s, and in 1976 issued the 'European Convention for the Protection of Animals kept for Farming Purposes' outlining basic principles for the proper treatment of farm animals (Council of Europe 1976). The Convention states that animals should be sheltered and provided for in a manner appropriate to their physiological and ethological needs, and called for the establishment of a Standing Committee responsible for making recommendations on more specific subjects based on available scientific knowledge. The Standing Committee made specific recommendations concerning pigs in 1986 and 2004.

The EEC approved the Convention in 1978 and ratified it by 1988 in order to ensure free trade within its borders recognizing that "there are disparities between existing national laws on the protection of animals kept for farming purposes which may give rise to unequal conditions of competition and which may consequently have an indirect effect on the proper functioning of the common market" (Council of the European Communities, 1978).

Subsequently, minimum standards for the main animal species farmed in Europe were developed, including pigs. In 1991, a Council Directive outlined minimum requirements for unobstructed floorspace available to pigs dependent on weight and banned the construction or conversion of facilities in which sows and gilts were tethered after 31 December 1995 (to be phased out in existing installations by the end of 2005) (Council of the European Communities 1991). Later regulation has focused on the ability of pigs and sows to move around. Especially

---

[3] An enabling act is a law that empowers an administrative body (typically a government minister or public authority) to regulate activities within a certain area, e.g., to set more specific regulations of animal welfare through ministerial orders.

---

[4] The Council of Europe is a supranational collaboration of 47 European countries set up in 1949 to promote democracy and protect human rights and the rule of law in Europe. It should not be confused with the European Union, which currently has 28 members. The European Union was preceded by the European Economic Community (until 1993) and the European Community (until 2009).

pertinent for the present study, as of 1 January 2013, the European Union requires that all pregnant sows and gilts be kept in 'loose systems' during gestation (Council of the European Union 2008).[5]

With these and other directives, European regulation has set minimum standards but allowing member states to have higher FAW requirements. An option used by Sweden, for instance, whereas Denmark has been reluctant to impose stricter regulations.

From public policy to private standards

European standards have primarily aimed at ensuring that the internal market is efficient and works satisfactorily and to prevent a race to the bottom. However, there has been a marked slow-down in animal welfare legislation initiatives in the EU since the beginning of the 2000s (Sandøe and Christensen 2018), as EU policy makers have become reluctant to impose stricter regulatory demands. The reluctance to impose strict regulation reflects that, as the market for food products has become more global, supply chains often span several national boundaries and regulatory regimes. It is therefore difficult for politicians and public authorities to effectively regulate food safety and quality practices (Hatanaka et al. 2005; Tennent and Lockie 2012). Rapid changes in production practices and the concurrent expansion of quality attributes also hinder effective regulatory solutions, as these often take long time to negotiate and implement.

To overcome these challenges, policy makers in the EU and elsewhere have looked to the market as a means to improve FAW (Heerwagen et al. 2013; Ingenbleek et al. 2012; Parker et al. 2017). Focus has been on making markets work by providing consumers with information that will enable them to express latent needs and demand products with higher animal welfare, thus providing market opportunities to primary producers. The rationale for doing so is outlined as follows:

> Improved information to consumers offers the prospect of a virtuous cycle where consumers create a demand for food products sourced in a more animal welfare

friendly manner which is transmitted through the supply chain back to the primary producer, who may be able to receive a premium price for their product and thus recoup a portion of any associated higher production costs (European Commission 2009, p. 2).

Based on this line of reasoning, business and non-business actors have increasingly been given (and taken) the responsibility to set and monitor standards for FAW. To ensure that the these standards are credible and that consumers believe they make a difference, the use of private standards and third-party certification has increased substantially in the past 20 years (Guthman 2007; Henson 2008).

In the meat sector, current private standards are partly a response to consumers' ever-increasing concern about the safety and quality of fresh meat due to a string of food scandals since the 1990s (Blokhuis et al. 2003; Verbeke 2009). To reassure consumers and meet their demands, actors in the meat sector have developed, implemented and communicated several quality management systems (Blokhuis et al. 2010; McEachern and Warnaby 2004; Main et al. 2007). These are distinguished by various labels in order to capture and/or retain value for certain actors (Guthman 2007). Some private standards are broad in scope, covering many different aspects of the production process, including animal welfare, while others focus mainly on FAW. Examples of the latter include Animal Welfare Approved (United States) and RSPCA Assured (United Kingdom).

The development of private standards can be seen as market driven, as private standards are responses to consumer perceptions and behaviour. Thus, private welfare standards can help retailers and producers meet already existing consumer demands (for examples, see Elzen et al. 2011; Uzea et al. 2011). However, since private standards typically do not attempt to fundamentally reshape market structures and/or the behaviour of market actors in relation to FAW, they are not 'driving markets' (Jaworski et al. 2000). That is, in themselves private standards are not likely to increase supply or demand for welfare products.

Retailers and other actors, including public authorities, use third-party certification to ensure that firms live up to a particular private standard for food safety and production processes. Third-party certification is regarded as independent and effective because the firm or organisation responsible for the certification (at least on paper) is independent of the buyer or seller. This verification makes labelling claims believable, which is crucial if consumers are to pay a premium for products meeting particular standards (Guthman 2007).

As a governance mechanism, third-party certification offers retailers (1) an opportunity to differentiate foods on attributes of interest to consumers (e.g., production methods, product quality, animal welfare or origin), (2) consistent

---

[5] The directive lays down the minimum standards for the protection of pigs confined for rearing and fattening. For instance, the use of tethers is prohibited and members states are required to ensure that sows and gilts are kept in groups during a period starting from 4 weeks after the service to 1 week before the expected time of farrowing. Furthermore, the directive sets minimum requirements for the total unobstructed floor area available to each gilt after service and to each sow when gilts and/or sows are kept in groups, as well as to the flooring surface. Sows and gilts must have permanent access to manipulable material. Certain exemptions are possible for small producers and pigs that have to be kept in groups that are particularly aggressive, that have been attacked by other pigs or that are sick or injured may temporarily be kept in individual pens.

implementation of a standard, regardless of the country of origin, and (3) a means to minimise their transaction costs and economic responsibility, by transferring them to producers and auditors (Hatanaka et al. 2005). Similarly, private standards enable retailers to coordinate and control other actors in their supply chain and hedge against network risks (Hatanaka et al. 2005; Rindt and Mouzas 2015; Tennent and Lockie 2012). Private standards help retailers standardise their direct and indirect relationships with suppliers, transfer liability to suppliers and third-party auditors, and balance control over suppliers with the flexibility to replace suppliers to get better deals (Rindt and Mouzas 2015).

Despite these advantages for retailers, other actors view private standards in a less positive light. For example, standards might undermine farmers' right and flexibility to make decisions about their own land, company and production processes (Tennent and Lockie 2012). In addition, meeting certain standards rarely translates directly into higher prices paid to farmers; instead, standards often impose large administrative burdens to document that their practices meet set requirements (Hubbard et al. 2007). Although meeting private standards is, in principle, voluntary, for many producers it is a prerequisite for continued production (Hubbard et al. 2007; Richards et al. 2013). Furthermore, the effectiveness of such initiatives to improve the welfare of farm animals in general has been questioned because they only appeal to the minority of consumers willing to pay a premium for animal welfare products (Lusk 2011). Thus, resarch suggest that private standards may lead to only modest improvement over legal minimum requirements for some animals (Parker et al. 2018). Finally, it has been noted that private standards and third-party certification establish barriers to entry as they inevitably rest on the presumption that only some producers and products can meet the requirements of any given standard (Guthman 2007).

Through the growth of private standards and related market-based initiatives, recent decades have thus seen a 'marketisation' (Çalışkan and Callon 2009) of animal welfare, where animal welfare has become an economic and not just ethical concern, subject to processes of qualification by a range of economic actors in and around the value chain for animal products (Miele and Lever 2014). The marketisation of animal welfare has often been welcomed by both policy makers and NGOs as a way to get actors to voluntarily make socially desirable changes in value chain and consumption practices rather than through new regulations (Miele and Lever 2013).

The introduction of animal welfare labelled products can be seen as a consequence of the resurgence of neoliberal policies (Busch 2014; Guthman 2007). However, because governance of private standards usually goes on behind closed doors and where popular partcipation is restricted (Busch 2014), their democratic legitimacy in terms of criteria such as participation, transparency and accountability has been questioned (Fuchs et al. 2011), as has the success of markets in delivering public goods such as animal welfare (Guthman 2008; Renard 2005).

Markets and market practices

As noted above, the idea of 'market-driven animal welfare' has become something of a mantra. But what is this 'market' in which politicians and others place their hopes for improving FAW? It is rarely defined explicitly but rather taken for granted, reflecting an underlying belief in an abstract, perfect market in which anonymous market forces, such as the famous 'invisible hand' (Smith 1970 [1776]), cause the demand and supply for a good, like FAW, to meet and set a price that can serve as a signal for the behaviours of buyers and sellers (Marshall 1920).

Economists have put forward powerful arguments for the idea that markets serve as an efficient economic and social mechanism for setting prices, coordinating behaviour and promoting individual choice and freedom (Satz 2010). Nevertheless, it is also widely recognised that markets sometimes fail to provide adequate levels of certain goods. Previous research thus discusses the lack of animal welfare as a 'market failure' (Harvey and Hubbard 2013), as consumers have limited opportunities for expressing preferences for improved FAW in the current market environment (Lusk 2011). Central to discussions of market failure is the concept of externalities, which refers to costs and benefits imposed on a third party not involved in the original transaction (Satz 2010). Animal welfare (or suffering) is an example of an externality generated by the production of animal products (Lusk 2011).

This paper dispenses with this neoclassical view of markets as quasi-natural realities that exist independently of the actors involved. Instead, markets are seen as constituted and reconfigured through the concrete activities of numerous actors engaging in their different everyday activities and using different forms of expertise and market devices (Berndt and Boeckler 2009; Callon 2007; Callon et al. 2002; Kjellberg and Helgesson 2007). Markets thus "only exist in the doing of them" (MacKenzie 2006, p. 34).

Kjellberg and Helgesson (2007, p. 141) define market practices as "all activities that contribute to constitute markets" and suggest that we distinguish three interlinked types of market practices: exchange, representational and normalising practices (see Table 1). *Exchange practices* are concrete activities associated with performing economic transactions, whether specific to a particular transaction or more general activities that also shape individual transactions. *Representational practices* contribute to the formation of understandings of the market and how it functions, which

Table 1 Market practices. *Source* Adapted from Kjellberg and Helgesson (2007)

	Representational practices	Normalising practices	Exchange practices
Definition	Activities that contribute to the construction of understandings and models about how the market "looks" and how it "works"	Activities that contribute to establishing guidelines for how a market should be (re)shaped or work according to some (group of) actor(s)	The concrete activities related to performing discrete economic transactions
Examples	Gathering and analysis of sales statistics in order to evaluate the effect of advertising and promotion activities Customer segmentation based on importance of animal welfare	Efforts to change markets (e.g., liberalisation) Efforts to specify and enforce general guidelines (e.g., legal requirements for animal welfare) Establishment of voluntary standards (both public and private) for animal welfare	Activities linked to specific transactions: Product specifications Price negotiations General activities: Organisation of marketing channels Generic promotion of animal welfare Annual negotiations

can help make it more visible to other actors. Research shows that different actors construct quite different understandings of their environment and therefore have different approaches to doing business (Esbjerg 2004; Finch and Acha 2008). Finally, *normalising practices* contribute to establishing normative guidelines for how a market should function or how certain market actors should act. Through these activities, different actors try to change market practices, such that they contribute to the formation and implementation of certain norms and guidelines for action, including the private standards increasingly important for regulating FAW. Because actors such as farmers, slaughterhouses, retailers and animal welfare NGOs often have conflicting ideals and interests, negotiating and implementing FAW standards can be a highly charged political process.

In this study, this typology of market practices is used to analyse the negotiation and enactment of FAW standards through the practices of business and non-business actors in the Danish pork sector. The three types of market practices are entangled and linked through chains of translation. Translation in this context refers to basic social processes through which something spreads across time and space (Callon 1986; Latour 1987). This ongoing process involves intermediaries, such as rules, tools, measures and measurements (Callon and Muniesa 2005). The market is thus shaped and continually evolves in day-to-day interactions among actors (Kjellberg and Helgesson 2006).

Market actors frame and perform markets by defining standards for FAW, monitoring exchange processes, benchmarking goods, calculating costs and prices, etc. (cf. Callon 2005). For a transaction to take place, all elements that are not to be taken into account have to be excluded from the market frame, at least for the time being. Framing involves selecting certain elements, severing links and finally making some trajectories irreversible, at least temporarily (Callon 2007). Any framing can be challenged and is thus temporary.

Framing is a delicate process that is neither complete nor perfect and that easily gets out of control. To capture this, Michel Callon argues that framings are sources of overflowing (Callon 1998; see also Parker et al. 2018). Overflowings of goods and the activities related to them "occur when goods act unpredictably, transgressing the frameworks set for them and the passivity imposed on them" (Callon 2007, p. 144). Economic agents can also overflow, resulting in "the creation of new identities, concerns and forms of action" (Callon 2007, p. 145). Overflowings can have both positive and negative consequences or connotations. They trigger the formulation of new questions or problems (also referred to as matters of concern), which can stimulate the creation of collectives or groups (Callon 2007). In the case of pork, examples of issues that are sometimes excluded from consideration, but trigger the formulation of new matters of concerns, include the effects of meat consumption on climate change and pollution resulting from animal husbandry or animal welfare.

In the attempted qualification and singularisation of products, the actors involved—here, pig producers, slaughterhouses, retailers, auditors, animal welfare organisation, etc.—weave a web of entanglements between them (Callon et al. 2002). As entanglements proliferate, the disentanglement of goods becomes increasingly problematic and difficult to obtain (Callon 2005). However, disentanglement from producers, former users or prior contexts is necessary for market transactions to take place (Callon 2005; Thomas 1991).

It is with these insights about markets and market practices in mind that the Danish pork sector was studied in order to explore the three research questions inspired by the distinction between representational, normalising and exchange practices: how actors understand the market for (welfare) pork, how they try to change how the market for (welfare) pork works through the development and use of particular standards and how these are enacted in exchange practices trough the ongoing interactions and relationships between actors.

Methods

Research setting

FAW is a recurring theme in relation to the increasingly concentrated and highly industrialised production of pigs in Denmark, making the Danish pork sector an interesting setting in which to study how FAW standards are negotiated and enacted.

Around 19 million pigs and sows are slaughtered in Denmark annually, and roughly 13 million piglets are exported for rearing and slaughtering in other countries (primarily Germany and Poland). The sector is very export oriented, as approximately 90% of the pork produced in Denmark is sent elsewhere. Conventional pigs account for around 94% of Danish pig production. In conventional pig production, sows are almost always kept in farrowing crates when they farrow and suckle (or the last 20% of the production cycle), although most stakeholders agree that this practice is not optimal from an animal welfare perspective. Alternative production systems that leave sows loose while they farrow and suckle exist, but they often involve substantial extra costs for farmers.

As of 1 January 2013, the European Union requires that all pregnant sows and gilts be kept in 'loose systems' during gestation (Council of the European Union. 2008), a period that accounts for about 60% of the production cycle. Central to this study is how market actors reacted to this, in particular how actors discussed how to handle the marketing of intermediate products with the lifting of the floor.

In 2013, the Danish Parliament passed a bill stipulating that sows should be kept in loose systems during the service phase (mating or insemination), where they spend around 20% of the production cycle. All new units built after 2014 must meet this requirement; full implementation is expected in 2035.

In an agreement reached at a summit in 2014 with the Danish Minister for Food, representatives from Danish agriculture, slaughterhouses, animal welfare organisations, consumer organisations, veterinarians and retail chains promised to work toward several objectives: reduce piglet mortality, keep 10% of sows in loose systems during farrowing as of 2020, halt the castration of male piglets, reduce the number of tail dockings, reduce the prevalence of ulcers among sows and pigs, provide consumers with more information about animal welfare, and give consumers a means to choose products produced with better animal welfare standards (Minister for Food 2014).

The institutional setting in Denmark combines elements of the *(super)market* and *welfare state* models discussed by Kjærnes et al. (2009). Retailers try to differentiate pork sold under their brands through labelling initiatives, but this is combined with a strong regulatory regime and widespread trust in government regulation and monitoring of agriculture, processing and retailing, which means that animal welfare is not a matter of concern for the majority of Danish consumers.

Data

To develop an understanding of how FAW standards are negotiated and enacted in particular market practices, semi-structured interviews with stakeholders on the Danish market and five export markets were conducted. This article primarily draws on twenty interviews that were conducted with actors in and around the domestic market during fall 2012 and winter and spring 2013. Informants came from along the entire Danish value chain from primary producers to retailers, as well as from representatives of three stakeholder organisations working with animal welfare in various ways (see Appendix 1 for details).[6]

The interview guide was tailored to each informant, but in general, informants were asked about their own role and job functions, the company/organisation they worked for, their understanding of the pork sector in general and how they developed this representation of the pork sector, how they anticipated the market to change and if it would have implications for FAW; also there were questions about which other actors they collaborated with and how. All interviews ended with questions about who else might provide insights in further interviews. Thus, snowballing was used to identify potential informants. The informants are not necessarily representative, but they represent different perspectives on FAW. Interviews in Denmark were conducted in person. To supplement interviews, relevant documentary material was collected about FAW practices by the companies studied and store visits conducted to see how pork was being sold.

Analysis and interpretation

Interviews typically lasted about an hour and were recorded and transcribed verbatim. Analysis began by careful reading of transcripts and documents, looking for overall themes and patterns in the empirical material (Miles and Huberman 1994). Then the formal analysis started with coding the transcripts, using categories developed from the market practice perspective. Additional categories emerged inductively from the empirical material. Because coding involves "separating

---

[6] In addition to the domestic market, twenty interviews were conducted in relation to market practices in Australia, China/Hong Kong, Great Britain, Sweden and the United States. These are all important export markets for Danish pork, and discussions with Danish practitioners suggested that FAW differed in importance across these countries.

data extracts from their context" (Coffey and Atkinson 1995, p. 30), there is a risk of data being decontextualised (Alvesson and Sköldberg 2000). The empirical material therefore was also analysed using 'meaning condensation' (Kvale and Brinkmann 2009), which involves reducing empirical material into more concise statements.

Market practices related to farm animal welfare

For the sake of clarity, the presentation of findings is organised around the three types of market practices introduced above. First, representational practices are discussed in order to show how actors in the Danish pork sector understand the market for welfare pork in domestic and global market contexts. Next, normalising practices are described, including the ways that FAW is framed by different standards and how private standards are renegotiated following the increase in general welfare standards. Finally, it is outlined how actors enact animal welfare in everyday exchange practices. As discussed earlier, it is important to keep in mind that the different types of practices are related and influence other, e.g. that representations are translated into normalising and exchange practices.

Representational practices: understanding markets for farm animal welfare

Although less than 10% of the pork produced in Denmark is consumed domestically, it is important to look at how actors view the local Danish market context because how the production of pigs is perceived by the wider public is important for the legitimacy of continued production. However, due to the international nature of the pork industry, viewing the domestic market in isolation is fraught with difficulties as FAW issues 'overflow' geographic boundaries. Danish actors, not least farmers and slaughterhouses, take this overflow into account in various ways when they talk about the challenges related to improving animal welfare for Danish sows and pigs.

Animal welfare in a domestic market context

According to a retail category manager, Danish consumers can be divided into two distinct segments in relation to FAW: those who don't care that much, for whom it is all about price, and then the rest. According to the category manager, this second group accounts for only a small percentage of all consumers. Although this segmentation is a bit crude, it reflects a widespread sentiment among informants, namely that FAW is not a major concern for most Danish consumers, at least if they have to pay more for their pork chops. Danish

consumers are represented as mainly interested in tasty and lean pork at as low a price as possible. For example, one informant describes consumers as follows[7]:

> If they can choose between welfare pork and conventional pork, and the latter is a few kroner cheaper, consumers will choose the cheaper product [conventional pork]. [Store Manager, Soft Discount Retailer]

Only a very small consumer segment is understood to demand pork produced with better FAW standards than conventional pork and willing to pay a (small) price premium. However, at least some informants think this segment is growing:

> I would say that, over the last few years, there has been a development so that some people are willing to pay a premium for better quality or higher animal welfare. Quality can be measured in many ways [...]. We clearly sense that there are more consumers willing to pay [a small premium] if they have the opportunity and feel certain that it is better, whether it is in terms of animal welfare or something else. But you feel you get something for your money. [Sales Manager, Slaughterhouse]

Because the represenation that most consumers are focused on price is widely shared, FAW is subordinated to other concerns in the day-to-day practices of most actors on the Danish market.

Some actors argue that the niche status of welfare meat is rooted in the limited supply of organic, free-range and other types of welfare pork. Others hold that supply simply mirrors demand: *if only* consumers would demand welfare meat, *then* the pork sector would be happy to oblige and produce it. For instance, a primary producer says "when the customer wants to buy a welfare product, then we make it," while a representative from an industry body argues that "if [the slaughterhouses] can see a market potential [in higher FAW], then this is a driver for things happening in production." However, consumers have to be willing to pay a premium that would enable farmers to recoup the extra costs that improving FAW is argued to entail.

Rather than seek to drive the market for FAW, actors in the pork sector can be seen as market driven (cf. Jaworski et al. 2000). They wait for consumers to act and thereby assign considerable responsibility to consumers, who have to demand higher levels of animal welfare and be willing to pay extra, not just state that FAW is important when surveyed. This type of circular reasoning explains why farmers

---

[7] All quotes have been translated from Danish to English. An effort has been made to stay as close as possible to the words used during the interviews.

and slaughterhouses are reluctant to make dramatic animal welfare improvements and ensures that change is slow.

A key Danish animal welfare organisation asserts growing interest in FAW among Danish consumers and argues that for Danish consumers to demand welfare pork, they may need to be better informed about animal welfare (thus echoing the European Commission). However, other informants suggest consumers already have so much information at their disposal that they have difficulty making sense of it all. These informants do not consider more information an appropriate option, which suggests a link between representations of the market and normalising practices. That positive attitudes towards FAW do not always translate into consumer behaviour is known from previous studies (e.g., Verbeke 2009).

Overflows between domestic and export markets

When discussing opportunities for improving the welfare of Danish pigs and sows, it is remarkable that many Danish informants have constructed views that explicitly link the local Danish market to the global nature of the Danish pork industry. Because it accounts for 5% of total Danish exports, informants widely recognize the pork sector as being of significant economic importance for the country. Most actors therefore acknowledge that improvements in welfare for Danish sows and pigs must not undermine the international competitiveness of the sector as a whole, although the animal welfare NGO interviewed maintains that their main concern is the welfare of animals, not the pork sector. Thus, FAW cannot be isolated to the domestic market; it transgresses geographic boundaries and attempts to frame it unequivocally. It 'overflows' (Callon 1998, 2007).

The competitiveness of Danish production is something different actors discuss, with labour costs being seen as a particularly important issue:

> It is far too expensive to produce [pigs] in Denmark [...]. Our production costs mean that it is difficult to be competitive. [...] We have hit a wall in terms of costs in Denmark. [...] Every day we try to rationalise our production of pigs so that we become more competitive, i.e., lower the labour costs per kilogram pork produced. [Primary producer, Free Range Pigs]

> Wages are one of the biggest challenges we have as an industry in Denmark. Not that I'm saying that we don't deserve the wages we're paid, generally speaking, but the countries we are competing with simply have a different structure when it comes to taxes and wages. A couple of years ago, a benchmark was made [...] and the result that came out was a number that said index 44 on wages. That is, every time you do the exact same task in relation to slaughter, slicing, etc., you have to

pay 44 per cent [in Germany] of what you pay in Denmark. [Sales Manager, Slaughterhouse]

Informants that are mainly active in the Danish market tended to refer to 'the global market' in an undifferentiated and general manner, glossing over the differences that other informants identify between markets. Interviews with export managers and actors in key export markets for Danish pork thus suggest that 'the global market' is not homogenous but rather quite heterogeneous and that they have to take differences between countries into account when operating on export markets.[8] Some export markets are seen as more important than others in terms of how much Danish pork is bought and at what prices. Different cuts of pork are also sold in different markets, each with different demands. This heterogeneity poses a serious challenge to Danish pig producers and slaughterhouses, which must take the unique demands of customers in different countries into account, including their demands for FAW. To complicate matters further, only some cuts command a price premium for higher FAW, and only on some markets. Other cuts sell at world-market prices on markets where FAW is not important, particularly pork sold for industrial use or further processing.

> [The big cooperatives] divide the carcass and sell the different cuts in 130 countries. This means that if we in Denmark want something [in terms of FAW], then the cuts sold in Denmark have to pay for all the extra costs, also for the parts that are sold in other countries not willing to pay a premium [for FAW]. [Area Manager, Trade Organisation]

That only a small segment of Danish consumers are seen as willing to pay a premium for FAW and that FAW is seen as playing a small role on export markets is reflected in efforts to change market practices, to which attention now turns.

Normalising practices: efforts to change how the market for farm animal welfare works

In this section focus is on the normalising practices different actors engage in as they try to change (or maintain) how the market for FAW works. Efforts to either change or maintain how the market works are informed by the representations of the market and understandings of whether current FAW standards are adequate or insufficient. Because FAW standards are constituted as central, we begin by considering how FAW is framed in current standards before proceeding to discuss efforts to negotiate changes to FAW standards.

---

[8] It is beyond the scope of this paper to discuss the differences seen to exist between export markets in detail.

Framing farm animal welfare through standards

FAW is a vague, contested concept that means different things to different actors, who disagree about what constitutes good FAW and how it should be regulated and monitored (Verhoog et al. 2004). FAW is framed, objectified and singularised in different ways in different markets and by different actors. Market framing is a powerful mechanism of exclusion (Callon et al. 2002), allowing some products to be defined as welfare meat while others are not. What constitutes welfare meat thus depends on the framing, including the dimensions and thresholds used in the qualification of pig husbandry systems.

In modern agriculture, a central element in the framing process is the development of standards. A standard refers to any set of agreed-upon rules for the production of certain objects (Bowker and Star 1999). Standards enable things and actors to work together by spanning more than one community of practice. To ensure compliance with the standard, some mechanism of enforcement is required. Inevitably, any standard also valorises some points of view and silences others, but the politics of this fade into the background and become invisible once an agreement about a standard has been formed (Bowker and Star 1999). In turn, standards generally exhibit significant inertia, so changing them can be very difficult and expensive. Finally, standards are not only symbolic but also material and embodied in infrastructure. Thus, FAW standards are embodied in material artefacts such as pits, fixtures, slaughterhouses and transportation equipment. This creates a time-bind for farmers (Esposito 2011; Laursen and Noe 2018), who are constrained in relation to what they can do to improve FAW by prior investments in particular production facilities that are compliant with a certain standard. Thus, they can only produce the particular type of animal welfare materially embodied in these facilities.

In the Danish pork sector, different actors represent FAW in different ways that are never neutral but that generally favour the actor's own interests or point of view. For instance, the Danish Agriculture & Food Council's 'benchmark' of how pigs are produced in four European countries offers a broad frame in which production systems are compared on a wide range of properties (Agriculture & Food 2014), with animal welfare as just one aspect among several (e.g., quality and control, health and use of medicines, feed, the environment, transport, slaughterhouses, food safety). As Table 2 shows, in this benchmark FAW is operationalized in concrete terms such as the width of farrowing pens, the nature of rooting and enrichment materials, whether floors are fully slatted and whether tail docking is permitted on a routine basis. In this comparison, the product—conventional Danish pork—is singularised, made calculable and positioned relative to its main European competitors and so becomes both comparable and different. Considering the organisation performing the comparison, it is perhaps not surprising that, on most accounts, conventional Danish production standards come out as just a little bit better than competitors' production standards (cf. Bowker and Star 1999).

A narrower framing, strictly focused on FAW, is provided by Dyrenes Beskyttelse (2012), a key Danish animal welfare organisation, which compares Danish pig production systems in terms of outdoor access, use of farrowing pens, tail docking, weaning of piglets, access to rooting materials, medication, space, transportation to slaughter, use of electric prods to drive pigs, control of farmers and where meat is sold. Dyrenes Beskyttelse also distinguishes different types of pork: conventional, organic, free range and special productions. Each 'production' is associated with certain standards, including FAW. Dyrenes Beskyttelse is unequivocal in its assessment of the FAW levels of the various productions, deeming only organic and free-range productions satisfactory. FAW in conventional pork production systems, which as we just saw was assessed favourably by the Danish Agriculture & Food Council, is summarily dismissed as "simply not good enough" (Dyrenes Beskyttelse 2012).

These two examples illustrate how the Danish Agriculture & Food Council and Dyrenes Beskyttelse have very different assumptions about the nature of the world, including about the prominence that should be assigned to FAW. That is, they have different 'organisational paradigms' (Brown 1989) that provide them with particular roles that are enacted in particular ways, in particular settings and in particular relations with other actors. The Danish Agriculture and Food Council see themselves as representing the interests of the entire (mainstream) pork sector across numerous issues, while Dyrenes Beskyttelse is an organisation focused specifically on the wellbeing of animals.

Organisations can have an interest in promoting their own standards at the expense of other standards or labelling schemes. Dyrenes Beskyttelse thus endorses products that live up to their requirements for FAW. This endorsement is viewed as very important by pork suppliers and retailers, as Dyrenes Beskyttelse is regarded as having high credibility among consumers.

These framings also illustrate how the abstract concept of FAW becomes concrete and comparable, able to be passed from hand to hand and across time and space (cf. Callon 2005). Through the various standards, certain qualities become associated with particular goods, and vice versa, and the goods are converted into commodities that can be exchanged.

Pork qualifies as welfare meat by living up to certain standards. What these standards are or should be is contested by farmers that seek to protect their investments in sties and fixtures, animal welfare organisations driven by FAW ideals,

Table 2 Comparison of housing and welfare between Denmark and European competitors. *Source* Agriculture & Food (2014)

	Denmark/DANISH	Denmark/UK contract	England	Holland	Germany
Pregnant sows	Housed in accordance with EU legislation. The pen must not be narrower than 3 m. There must be straw on the solid or drained flooring	No confinement from weaning until 7 days before predicted date of farrowing—otherwise requirements as per Danish standard	No confinement from weaning until 7 days before predicted date of farrowing. At least 2.8 m between sides of pen in indoor systems. Around 40% of UK breeding herd is kept outdoors	Housed in accordance with EU legislation. At least 2.8 m between the sides of the pen	Housed in accordance with EU legislation. At least 2.8 m between the sides of the pen
Farrowing pens	Housed in line with EU legislation. Appropriate nest building material in sufficient quantity is required, unless this is technically impossible because of the slurry system used at the farm. The piglets must have an area that is separate from the sow. If necessary, there must be a source of heat. The Danish pig industry's aim is for 10% of sows to be loose in the farrowing pens by 2020. After 2021 all newly built farrowing units must be designed as loose systems		Housed in line with EU legislation. Appropriate nest building material in sufficient quantity is required, unless technically impossible because of the slurry system used at the farm. The piglets must have a thermally comfortable and dry lying area	Housed in line with EU legislation. Appropriate nest building material in sufficient quantity is required, unless technically impossible because of the slurry system used at the farm. The piglets must have an area that is separate from the sow. There must be a source of heat	Housed in line with EU legislation. Appropriate nest building material in sufficient quantity is required, unless technically impossible because of the slurry system used at the farm. The piglets must have an area that is separate from the sow. There must be a source of heat
Weaning of piglets	After 28 days. The average in 2012 was 31 days		Not before 28 days or 21 days for batch production	After 28 days or 21 days for batch production	After 28 days or 21 days for batch production
Enrichment and rooting materials	All pigs must have permanent access to sufficient quantities of straw or other manipulable rooting and enrichment material. Enrichment and rooting material must be of natural materials and in contact with the floor. Chains alone are not acceptable		All pigs must have permanent access to sufficient quantities of enrichment or other rooting material. Chains alone are not acceptable	All pigs must have permanent access to manipulable materials. Chains with plastic hooks are permitted	All pigs must have permanent access to manipulable materials. The materials must be harmless and adequate. Chains with plastic hooks are permitted
Flooring for piglets and finishers	Since 2000, it has been forbidden to build sties with fully slatted floors. With regard to newly built sties, at least half of the floor for piglets and at least one-third of the floor for finishers must be solid or drained. Applies to all systems from 2015		Fully slatted floors are permitted provided minimum slat widths and opening widths are observed	40% solid floor for piglets and finishers required	Fully slatted floors are permitted
Sprinkling systems	All pigs over 20 kg (including sows) must have access to a sprinkling system or another system to keep cool		No regulation	No regulation	No regulation
Castration	Pain relief must be administered before castration takes place. Anaesthetic must be used if castration is carried out 7 days after farrowing		Castration is not permitted by RTA standards. According to UK legislation, castration is permitted up to the seventh day after farrowing	Pain relief must be administered before castration takes place. Anaesthetic must be used if castration is carried out 7 days after farrowing	Pain relief must be administered before castration takes place. Anaesthetic must be used if castration is carried out 7 days after farrowing
Tail docking	Not permitted on a routine basis, but permitted if it can be documented that measures must be taken to prevent tail biting. Only permitted between day 2 and 4 after birth and no more than half of the tail may be docked		Only within the first 72 h after birth and not on a routine basis	Docking of part of the tail no later than seven days after birth	Docking of part of the tail no later than four days after birth

Table 2 (continued)

	Denmark/DANISH	Denmark/UK contract	England	Holland	Germany
Tooth reduction	Tooth clipping is not permitted. Tooth grinding is allowed, but not on a routine basis. Tooth grinding must take place within the first four days after birth		Piglet teeth clipping is allowed up to 72 h after birth, but not on a routine basis	Tooth clipping is allowed within the first seven days after birth	Tooth clipping is allowed within the first seven days after birth

retailers that want to be able to source products globally but still seek the legitimacy associated with selling local or regional pork (see also Esbjerg 2004) and other actors interested in preserving production, jobs and export earnings by maintaining the competiveness of the Danish pork sector.

Negotiating new and revised farm animal welfare standards

Negotiating and implementing new or revised FAW standards takes time. The requirement that as from 1 January 2013, all pregnant sows and gilts must be kept in groups during gestation is an example of this (Council of the European Union 2008), as this issue has been discussed since the 1970s (Elzen et al. 2011).

Because the rules contained in Council Directive 2008/120/EC were known for many years, market practices did not just change over night at the beginning of 2013. Nevertheless, because the requirement that sows and gilts be housed in loose systems during gestation was one of the key differences between conventional pork and intermediate production systems, the new rules substantially eroded the difference between standard/conventional production and intermediate specialty productions with regard to FAW. Hence, several actors engaged in normalising practices to change the standards according to which intermediate products are produced in order to re-establish a 'difference' compared to conventional products and thereby make intermediate products 'marketable' again. If there are no differences in FAW levels, it becomes difficult to use FAW in marketing:

> I just think that if you have to have something to tell consumers, then you have to have something tangible. As I say about the free-range pig, there you are willing to pay a little bit more per pack as that pig after all has had a better life. [Sales Manager, Pork Supplier]

Identifying and implementing a difference that makes a difference to the animals and consumers can be difficult. For instance, one retail informant noted that he was in discussions with a supplier about which FAW requirements to incorporate in the production of the retailer's brand of pork to once again singularise the retailer's private brand and re-establish a difference relative to conventional pork. However, the retailer had found it difficult to reframe the welfare aspects of the pork sold under its own brand because the supplier balanced this particular retailer's demands and wishes against other concerns, including the preferences of other customers and the sector as a whole. This again demonstrates how individual relationships are entangled and can overflow, and hence that renegotiating standards is far from straightforward, as actors can have conflicting interests and might have made investments in brands and production facilities that they want to protect. Different actors engage in different, sometimes competing normalising practices to influence the market for FAW products in specific directions (cf. Kjellberg and Helgesson 2007).

Some informants described one of the largest Danish suppliers of pork as lethargic, arguing that as a large bureaucratic organisation it finds it difficult, or is unwilling, to change its practices and cope with the hassles involved in serving niches, such as the minority of Danish consumers willing to pay extra for higher FAW levels:

> There is no doubt that the stronger your position on the market, the more complacent you become. [Category Manager, Soft Discount Retailer]

> [The producer] is very big, and they are super focused on being big. [Category Manager, Supermarket Retailer]

The major actors in the pork sector appear satisfied that Danish pig producers live up to Danish legislation, which they see as more than sufficient to meet the demands from world markets. Because of high relative costs in Denmark, these actors are afraid that unilateral FAW improvements will undermine the competitiveness of the sector. In their opinion, on world markets Danish products are the standard against which other producers are measured when it comes to FAW, food safety, standardisation, legislation and monitoring. The representation of the international nature of the market is performative (Kjellberg and Helgesson 2007; MacKenzie 2006), as it offers both an explanation and an excuse for the current state of affairs and (lack of) initiatives for improving FAW.

The Danish Agriculture & Food Council affects FAW, according to the interviews, though it is also accused of foot dragging by one retail informant:

> The Danish Agriculture & Food Council works for animal welfare, but they don't lead. They are perfectly able to see where Denmark makes money. It would be nice to see them step up [in relation to farm animal welfare], but you don't cut off the branch you are sitting on. [Category Manager, Supermarket Retailer]

This informant wants the Council to do more, even while acknowledging that Danish pork competes on global markets and that the sector cannot be expected to undermine its own position. Using similar arguments, an industry body emphasises that improvements must be made step by step in order to maintain competitiveness (and protect jobs and export earnings). It prefers voluntary improvements initially, to gain experience with new production systems and routines before they are implemented on a larger scale but is against big unilateral changes in Danish legislation:

> With regard to legislation, we have a pretty clear policy, according to which we are willing to take small

steps ahead of what other countries do [in terms of FAW], but not further than we can maintain our competitiveness, production and jobs in Denmark. [Trade Organisation]

To get pig producers to change their farm-level practices voluntarily, some informants suggest that incentive schemes are necessary to make testing and converting to new production systems appealing. Thus, to test new welfare initiatives in real life practice, and not just on research stations, industry bodies are willing to offer financial support to farmers:

> At the moment we know that [improving FAW] requires extra investment, because it require more space, and it costs in terms of mortality and it costs in terms of labour hours. If any producers are to take a leap, and there is a substantial risk that things can go wrong on all three accounts, and that perhaps it costs even more than we think, then it is good to be able to say to some [producers], that if you are willing to run this risk, then we are willing to help you. [Manager, Trade Organisation]

Implementing new standards on a large scale takes time, as animal husbandry is very capital intensive. Thus, one informant mentioned that producers need at least a rough idea of the conditions they will have to meet and of the impending requirements because a new sty will have a life-time of 20–25 years. New production facilities require substantial financial investments, and therefore there is widespread acceptance that welfare requirements cannot change every 2 years: "A certain stability is necessary, if producers are to be able to live with [FAW] requirements" (NGO). In addition, it can be difficult for farmers that want to improve FAW to finance improvements:

> We have been, and are still, looking for more free-range producers. [However,] it is not so attractive to produce pigs at the moment. Therefore, it can be difficult to get farmers to switch to free-range [production]. Many have difficulty borrowing the money it costs, as the economics of producing pigs has been strained for a while. [Sales Manager, Pork Supplier]

Speaking from a different organisational paradigm, an animal welfare NGO is dissatisfied with the status quo and demands significant improvements in relation to FAW issues such as the castration of male piglets, tail docking and fixation of farrowing and suckling sows. The normalising practices that the NGO engages in include working on a political level to improve overall FAW standards. It does not favour leaving improvements to voluntary initiatives and instead prefers binding legislation. Because 'market-driven animal welfare' is a mantra for the Danish government, the NGO also takes responsibility for informing consumers about animal welfare issues in general and the the animal welfare levels associated with different production standards in particular.

The normalising efforts of some smaller retailers to improve FAW are positively acknowledged by other informants. In contrast, a category manager for a small supermarket retailer lamented that major retailers did not do more to improve FAW. He suggested that even though retailers wield substantial power over suppliers, generally they do not want to use it to improve FAW:

> It is possible to add something in the supply chain, such as animal welfare, and to get some consumers to pay a premium, but it requires that retail buyers such as me believe in it. If everyone thinks that it all revolves around price, there is only one thing that can be done, and that is to buy where prices are low. [Category Manager, Supermarket Retailer]

Other informants agree that major retailers do not emphasise FAW because they perceive consumers as more interested in low prices and taste than FAW and thus unwilling to pay a premium for better welfare. The focus on prices also gets linked to the growth of the discount sector, which has continuously gained market shares in Denmark. Instead of making FAW a priority, the three main retail chains are described as focusing on stable supplies and (low) prices, which reverberates back through the supply chain to slaughterhouses and pig producers, who must find ways to cut costs not improve FAW. Several informants assert that larger retailers should take greater responsibility for FAW because they are in a position to influence consumers. Instead, the informants indicate that these retailers only expect Danish pork to meet current FAW requirements. The large retailers are described as skilled negotiators, interested only in making money, even when it comes to FAW. This is illustrated in the following quotes:

> The Danish market is dominated by three big retail chains that are ruthless in their trading approach and that do not emphasise animal welfare in their general policies. It's all about price. But they also sell specialty products that focus on animal welfare in some way. [NGO]

> Those big corporations on the retail market dominate many issues. They are not always governed by the same concerns they would be if they asked their customers what was important to them. They are driven by what's good for them [i.e., the retailer], and they influence what the customers think more than the other way around. [NGO]

Retailers say that they are simply selling the products that consumers demand, thus assigning responsibility for improving FAW to consumers (thus shirking it themselves), the implication being that if consumers would demand products with higher FAW, retailers would supply it. Along similar lines, informants from the pork sector argue that if consumers placed greater emphasis on FAW (and were willing to pay a premium), the supply of welfare meat would increase.

Focusing on FAW in communication with consumers can also have negative consequences, however. One informant thus suggested that it could lead to question such as "Why have pigs not always been treated well?".

Informants expect FAW to be more important for consumers in future but also acknowledge that many consumers will continue not to care. Yet consumers played a role in establishing standards for the transport of live animals, even as they have remained unaware of the issue of confinement of sows. Several informants assert that retailers should take more responsibility for FAW and for making consumers aware that their consumption choices have consequences, that is, that they overflow.

Exchange practices: enacting animal welfare in ongoing relationships

On the domestic market, retailers and pork suppliers are engaged in long-term relationships. Retailers and suppliers typically negotiate annual contracts that serve as frameworks, regulating their day-to-day operations and interactions. Exchange practices between pork suppliers and retailers are characterised by frequent interaction and complex relationships, with numerous contacts at various organisational levels. For instance, the sales manager of a large pork supplier described how he was in ongoing dialogue with the retailer he was key account manager for. At HQ he talks with the CEO of the customer, two operations managers and four regional managers:

> So, you can say that, already there, I have seven entrances [or contacts to the retailer]. Every day I speak to one or more of them. Of course, each of them has their own focus areas, which have to come to together in the end. There's not a day where I don't talk to at least one of them. Often, I talk to two to four of them, or talk to them several times during day. It all depends on what we have going on. [Sales Manager, Slaughterhouse]

In addition, the sales manager interacts directly with store managers and store-level butchers.

Relationships are generally described in positive terms, even as the informants acknowledge differing interests. According to a large meat processor, it has developed special productions because retail chains want to offer their customers something unique. The processor continuously discusses the demands that these productions must meet with the respective retailers. For all special productions, FAW is framed as better than that for conventional pork, though the extent of the improvement differs.

Retail buyers emphasise frequent interaction and good, long-term relationships with their Danish pork suppliers. For example, the category manager of one retail chain described the chain's co-operation with its supplier as follows:

> I regularly sit down with people from [Pork Supplier] to exchange ideas. We talk about what we can do differently, e.g., by giving the pigs a certain feed. [...] Factory managers at the slaughterhouses think in terms of logistics, rational production and efficiency. The people that I do business with are thinking in terms of sales. They then have to go back and make things happen. Some things can be done, others cannot. [...] We have a very, very close relationship with [Pork Supplier]. We recognise each other's mission in this world. It all revolves around 'what we can do' in order for both us to run a better business. [Category Manager, Supermarket Retailer]

Because welfare pork constitutes a small part of the domestic market, FAW does not feature prominently in day-to-day interactions, which are focused on operational issues and ongoing marketing initatives. FAW is a more general issue that is taken up at irregular intervals when discussing more strategic initiatives although the sales manager of one of the slaughterhouses opined that it is difficult to talk long-term initiatives with big retailers, because retail buyers are evaluated on their ability to purchase cheaply. In contrast, some small retail chains view FAW as an issue that they can use to differentiate themselves from large competitors. Hence it it easier for suppliers to discuss more strategic initiatives with smaller retail chains.

Thus, the sales manager of large pork suppliers had recently been involved in developing different training programmes in collaboration with two smaller retail chains in order to prepare butchers for their interactions with consumers. Through these training programmes butchers get insight into how free range pigs are bred and the role that Dyrenes Beskyttelse, the main animal rights organisation, played in monitoring FAW. For instance, he had devised a training programme where the store-level butchers of a supermarket retailer visited and shadowed farmers for a day in order to get a glimpse of what daily life on a farm was like.

Both retailers and pork suppliers reported having good working relationships with Dyrenes Beskyttelse about the organic and free range pork products meeting their requirements for FAW and are endorsed by the organisation.

In summary, FAW is not an issue that features in exchange practices related to day-to-day interactions. These are focused on operational issues related to specific transactions. FAW is a more general issue that is dealt with in annual negotiations and various discussion about how the relationship between pork supplier and retailer is to develop.

Discussion

The increasing use of private standards and third-party certification to regulate and monitor FAW constitutes a significant change in 'market practices' (Kjellberg and Helgesson 2007), as authorities have become reluctant to impose stricter regulations to increase FAW standards due the difficulty of enforcing regulations in global supply chains.

This article has investigated the shift to market-driven animal welfare in the context of the Danish pork sector. Although there is a strong ideological belief in the power of the invisible hand of the market, this study has shown that it is far from certain that the market can deliver the levels of FAW that consumers and politicians expect. One reason is that when animal welfare is subject to marketisation, it becomes one among several isses for actors to take into account in their entangled interactions. Although some actors have misgivings about the adequacy of current standards, the importance afforded to economic considerations such as the global competitiveness of the Danish pork sector mean that progress has been very slow. A similar conclusion has been reached by Parker et al. (2018), who only found small and incremental improvements in FAW following to the introduction of ethical labelling of chicken-meat in Australia.

The slow progess also reflects that FAW is a contested concept in the context of Danish pork sector, as it has also been shown to be in other contexts (e.g., Carey et al. 2017; Fraser 2008; Parker et al., 2018). Different actors having different ideas about what constitutes good animal welfare and the importance that should be afforded animal welfare compared to other issues such as the competitiveness of the Danish pork sector. The way that actors approach FAW depends on what good or outcome is sought: FAW, jobs, export earnings or sensory experiences. Some actors prioritise FAW over other concerns in their normalising practices to influence how the market for pork works, whereas for other actors, it is the other way around—they believe that improvements in FAW cannot come at the expense of the competitiveness of the overall industry.

Any standard represents a particular view of FAW. Developing a FAW standard involves establishing a metrological network that measures and objectifies certain aspects of animal welfare. For pigs, FAW is thus measured in terms of hay, space, naturalness, and (the absence of) tail docking and castration. Certain ideas or values accordingly transform into social facts, and the abstract notion of FAW becomes visible, audible, tangible and knowable. Ideas about FAW also translate into standards and procedures, material devices such as fixtures, maximum travel times and so forth. Through this standardisation process, pigs become commodities that can be classified into certain categories (conventional, organic, free-range) and easily exchanged among economic actors (farmers, slaughterhouses, importers, retailers, consumers). As a consequence of the marketisation of animal welfare, FAW becomes one product attribute among many that consumers and other actors take into account in their exchange practices (Kjellberg and Helgesson 2007; Miele and Lever 2014).

As socio-technical devices and procedures, FAW standards organise encounters between goods and agencies (Callon et al. 2002). They are negotiated over time and enacted in the daily practices of various actors. Compliance can be monitored in different ways, whether by the actors themselves engaging in routine self-monitoring efforts, or by third-party auditors that make a living from checking private standards. Third-party certification is important for convincing consumers that there is a difference between labelled and un-labelled products that they should pay a difference for (Guthman 2007).

One of the issues highlighted by the present study is that enacting and complying with a FAW standard often involves investing in material devices. Implementing FAW standards often requires substantial remodelling of existing pits and fixtures or the construction of new production facilities. Improvements in FAW are thus contingent on changes made by farmers who invest in new pits and new furnishings and alter their management practices. It is more labour intensive to have loose sows, for instance, at least in the short run when new routines and skills must be developed and learned. As FAW is seen as important to only a small segment of Danish consumers and the majority of consumers are regarded as unwilling to pay extra for better animal welfare, most actors in the Danish pork sector are reluctant to make the investments in new pigsties, furnishings and marketing necessary to improve FAW. Many actors are playing a waiting game: they express interest in higher FAW but are waiting for other actors to take the first step, not least for more consumers to be willing to pay extra.

The analyses show that the development, implementation and monitoring of FAW standards mobilise many actors working together (and sometimes against each other) to reconcile divergent practices and make compromises among various justifications for standards (Busch 2011). Through their normalising practices, these actors engage in collective work (Callon 2005), attempting to accommodate or make compromises among different interests and ideas — not just about what constitutes good animal welfare and how it should be improved but also about the costs involved, supply, demand and the international competitiveness of the Danish pork industry. In the attempted qualification and singularisation of some pork products as welfare products,

the key actors involved—producers, slaughterhouses, retailers, auditors, animal welfare organisations—weave a 'web of entanglements' (Callon et al. 2002). It is impossible for actors to disentangle from these complex relationships, but it is relevant for the various participants to reflect on how they can be more perceptive and responsive to the relationships they are in (Gruen 2015). Except for noting that changes in FAW standards have to make a difference to pigs and sows, the data shows that there is very little room for the experiences of animals themselves in the market practices of actors in the Danish pork sectors. Furthermore, the data shows that ethical reflections on the appropriateness of current FAW standards and related practices are tempered by economic considerations.

Agreeing on a standard is only the start of a range of other decisions that must be made by the actors. A farmer has to choose whether to invest in new pigsties and furnishings that conform with a particular standard; retailers must decide whether the standard is right for them; animal welfare organisations have to determine whether they want to endorse a particular standard as animal friendly or not, and so on. Improvements in FAW is thus entangled in a complex of relationships between numerous actors working together or against each other.

Previous work on how FAW standards are negotiated has been narrower in scope than the present study. For instance, the illuminating work by Carey, Parker and Scrinis on the development of welfare labelling in Australia mainly focuses on what we have classified as normalising practices—how different actors in and around the Australian egg industry tried to influence the definition of "free range" eggs in the face of consumer-oriented challenges to prevailing labelling practices (Carey et al. 2017; Parker et al. 2017). Using the market practice perspecitve, the present article extends this work by studying how representations of the market are taken into consideration when developing FAW standards and how these are enacted in ongoing exchanges between different, deeply entangled actors in the Danish pork sector and beyond.

Because the Danish pork sector is geared towards world markets, where animal welfare is seen as playing a minor role, the demands of individual Danish retail customers wanting to develop a private standard are balanced against other concerns by slaughterhouses. Furthermore, actors often have divergent interests and views of the market. Therefore, they sometimes actively work against each other, such as when animal welfare NGOs rail against intermediate levels of FAW sold under retailer brands.

This study has focused on the Danish pork sector, and many of the findings are very context-specific. Nevertheless findings are relevant for other export-oriented sectors, as it highlights how improvements in national FAW levels is dependent on entanglements with global markets, especially when production is very export-oriented. By showing how some sectors are tied into global networks that influence the ability and willingness to make FAW improvements, this article differs from previous work. A different take on internationalisation of FAW is provided by the work by Parker and colleagues on the Australian chicken-meat sector, in which they show how stakeholders refer to developments in EU regulation in order to push the Australian government to follow and have challenged local supermarket chains to follow the example of UK retailers and be more proactive in fostering increased consumption of higher-welfare meat (Parker et al. 2018).

Left to its own devices, there is a risk that 'the market' will focus on economies of scale and lowering costs, not on serving the small, if growing, niche willing to pay extra for better FAW or improving FAW in general. This is especially true if different actors continue to wait on others to take the next step: for consumers to be willing to pay extra, producers to increase supply or retailers to take a leading role in marketing products with better FAW. There was thus no evidence to suggest that any of the major actors were striving to drive markets for FAW products. The conclusion must therefore be that public policy must continue to set acceptable minimum standards that leave room for actors to offer products that offer products meeting higher requirements at an acceptable extra cost. Although the ability of the market to successfully deliver sufficient levels of public goods like animal welfare can be questioned (e.g., Busch 2014; Guthman 2008), there is also evidence suggesting that a market for animal friendly products can be created and sustained (Miele and Lever 2013). In the case of the Danish pork sector, the gradual introduction of new, stricter requirements through public policy is likely to stimulate market-driven efforts to offer products meeting higher requirements because retailers view FAW as an important issue that they can use for differentiation purposes. However, such change is by necessity slow, as improvements in animal welfare involve significant investments in production facilities and marketing.

Acknowledgements This research was made possible by a grant from the Danish Pig Levy Foundation. The author thanks Maja Pedersen and Kathrine Nørgaard Hansen for their invaluable assistance with conducting the interviews and initial analyses. Furthermore,the author thanks Klaus Brønd Laursen and the two anonymous reviewers for their constructive comments and feedback.

Appendix 1: Informants

Twenty informants were interviewed about the Danish market during fall 2012 and winter and spring 2013. Interviews were conducted with three primary producers (a conventional pig producer, a free-range pig producer and an organic pig producer), sales managers from two large Danish slaughterhouses, a project manager from a company marketing organic and free-range meat and the corporate communications manager of a large meat processor. Furthermore, interviews were conducted with three retailers: a soft discount chain, a mid-market supermarket chain and an upmarket supermarket chain. For each retailer, the interviews involved the relevant category manager/retail buyer at the corporate level, a store manager and a store-level butcher/category manager. Finally, we interviewed representatives of three stakeholder organisations working with animal welfare in different ways: the communication manager and the project manager responsible for pigs at a large Danish animal rights organisation, the area manager of an organisation representing Danish farmers and the Danish food industry and the area manager for housing and environment of an organisation in charge of research and development tasks related to live pigs communicating knowledge obtained through these activities to practitioners.

References

Agriculture & Food. 2014. Danish pig production in a European context. A benchmarking exercise: Denmark, UK, Holland and Germany. Copenhagen: Danish Agriculture & Food Council.

Alvesson, M., and K. Sköldberg. 2000. *Reflexive methodology*. London: Sage.

Berndt, C., and M. Boeckler. 2009. Geographies of circulation and exchange: Constructions of markets. *Progress in Human Geography* 33 (4): 535–551.

Blokhuis, H.J., R.B. Jones, R. Geers, M. Miele, and I. Veissier. 2003. Measuring and monitoring animal welfare: Transparency in the food product quality chain. *Animal Welfare* 12 (4): 445–455.

Blokhuis, H.J., I. Veissier, M. Miele, and B. Jones. 2010. The Welfare Quality® project and beyond: Safeguarding farm animal well-being. *Acta Agriculturae Scandinavica, Section A—Animal Science* 60(3): 129–140.

Bowker, G., and S.L. Star. 1999. *Sorting things out: Classification and its consequences*. Cambridge, MA: MIT Press.

Brambell, F.W.R. 1965. *Report of the technical committee to enquire into the welfare of animals kept under intensive livestock husbandry systems*. London: Her Majesty's Stationary Office.

Broberg, B. 2016. Dyreværnslovens 100 års fødselsdag [100 years of the act for the prevention of cruelty against animals]. In *Dyrevelfærd i Danmark 2016* [Animal welfare in Denmark 2016], 7–13. Copenhagen: Ministry of Environment and Food, Danish Veterinary and Food Administration.

Brown, R.H. 1989. *Social science as civic discourse*. Chicago: University of Chicago Press.

Busch, L. 2000. The moral economy of grades and standards. *Journal of Rural Studies* 16 (3): 273–283.

Busch, L. 2011. How animal welfare standards create and justify realities. *Animal Welfare* 20 (1): 21–27.

Busch, L. 2014. Governance in the age of global markets: Challenges, limits, and consequences. *Agriculture and Human Values* 31 (3): 513–523.

Çalışkan, K., and M. Callon. 2009. Economization, part 1: shifting attention from the economy towards processes of economization. *Economy and Society* 38 (3): 369–398.

Callon, M. 1986. Some elements of a wociology of translation: Domestication of the scallops and the fishermen of St Brieuc Bay. In *Power, action and belief: A new sociology of knowledge?*, ed. J. Law, 196–223. London: Routledge.

Callon, M. 1998. *The laws of the markets*. London: Blackwell.

Callon, M. 2005. Why virtualism paves the way to political impotence: A reply to Daniel Miller's critique of The laws of the markets. *Economic Sociology: European Electronic Newsletter* 6 (2): 3–20.

Callon, M. 2007. An essay on the growing contribution of economic markets to the proliferation of the social. *Theory, Culture & Society* 24 (7–8): 139–163.

Callon, M., C. Méadel, and V. Rabeharisoa. 2002. The economy of qualities. *Economy and Society* 31 (2): 194–217.

Callon, M., and F. Muniesa. 2005. Economic markets as calculative collective devices. *Organization Studies* 26 (8): 1229–1250.

Carey, R., C. Parker, and G. Scrinis. 2017. Capturing the meaning of "free range": The contest between producers, supermarkets and consumers for the higher welfare egg label in Australia. *Journal of Rural Studies* 54: 266–275.

Christensen, T., L. Esbjerg, and P. Sandøe. 2012. Kan markedet give plads til flere glade grise? [Can the market make room for more happy pigs?]. In *Dyrevelfærd i Danmark* [Animal welfare in Denmark], 40–49. Glostrup: Fødevarestyrelsen.

Coffey, A., and P. Atkinson. 1995. *Making sense of qualitative cata*. Newbury Park, CA: Sage.

Council of Europe 1976. European Convention for the Protection of Animals kept for Farming Purposes. European Treaty Series - No. 87.

Council of the European Communities. 1978. COUNCIL DECISION of 19 June 1978 concerning the conclusion of the European Convention for the protection of animals kept for farming purposes (78/923/EEC). Official Journal of the European Communities. L 323/12.

Council of the European Communities. 1991. COUNCIL DIRECTIVE of 19 November 1991 laying down minimum standards for the protection of pigs (91 / 630 / EEC). *Official Journal of the European Communities* 11 (12): 1991.

Council of the European Union. 2008. COUNCIL DIRECTIVE 2008/120/EC of 18 December 2008 laying down minimum standards for the protection of pigs. Official Journal of the European Union 18.2.2009, L 47/5.

Dyrenes Beskyttelse. 2012. *Den store svinekødsguide* [The extensive guide to buying pork]. Copenhagen: Dyrenes Beskyttelse.

Elzen, B., F.W. Geels, C. Leeuwis, and B. Mierlo. 2011. Normative contestation in transitions 'in the making': Animal welfare concerns and system innovation in pig husbandry. *Research Policy* 40 (2): 263–275.

Esbjerg, L. 2004. Retailer buying as meaningful action. PhD dissertation. Aarhus: Aarhus School of Business.

Esposito, E. 2011. *The future of futures: The time of money in financing and society*. London: Edward Elgar.

European Commission. 2009. Options for animal welfare labelling and the establishment of a European Network of Reference Centres for the protection and welfare of animals. Report from the Commission to the European Parliament, the Council, the European Economic and Social Committee and the Committee of the Regions COM. 584 final. Brussels: The Council of the European Union.

FAWC. 2009. *Farm animal welfare in Great Britain*. London: Farm Animal Welfare Council.

Finch, J.H., and V.L. Acha. 2008. Making and exchanging a secondhand oil field, considered in an industrial marketing setting. *Marketing Theory* 8 (1): 45–66.

Fraser, D. 2008. Understanding animal welfare. *Acta Veterinaria Scandinavica* 50 (1): S1.

Fuchs, D., A. Kalfagianni, and T. Havinga. 2011. Actors in private food governance: The legitimacy of retail standards and multistakeholder initiatives with civil society participation. *Agriculture and Human Values* 28: 353–367.

Gruen, L. 2015. *Entangled empathy: An alternative ethic for our relationships with animals*. New York: Lantern Books.

Guthman, J. 2007. The Polanyian way? Voluntary food labels as neoliberal governance. *Antipode* 39 (3): 456–478.

Guthman, Julie. 2008. Thinking inside the neoliberal box: The micropolitics of agro-food philanthropy. *Geoforum* 39 (3): 1241–1253.

Harrison, R. 1964. *Animal machines*. London: Vincent Stuart.

Harvey, D., and C. Hubbard. 2013. The supply chain's role in improving animal welfare. *Animals* 3: 767–785.

Hatanaka, M., C. Bain, and L. Busch. 2005. Third-party certification in the global agrifood system. *Food Policy* 30 (3): 354–369.

Heerwagen, L.R., T. Christensen, and P. Sandøe. 2013. The prospect of market-driven improvements in animal welfare: Lessons from the case of grass milk in Denmark. *Animals* 3: 499–512.

Henson, S. 2008. The role of public and private standards in regulating international food markets. *Journal of International Agricultural Trade and Development* 4 (1): 63–81.

Hubbard, C., M. Bourlakis, and G. Garrod. 2007. Pig in the middle: Farmers and the delivery of farm animal welfare standards. *British Food Journal* 109 (11): 919–930.

Ingenbleek, P.T.M., V.M. Immink, H.A.M. Spoolder, M.H. Bokma, and L.J. Keeling. 2012. EU animal welfare policy: Developing a comprehensive policy framework. *Food Policy* 37 (6): 690–699.

Jaworski, B., A.K. Kohli, and A. Sahay. 2000. Market-driven versus driving markets. *Journal of the Academy of Marketing Science* 28 (1): 45–54.

Kjellberg, H., and C.-F. Helgesson. 2006. Multiple versions of markets: Multiplicity and performativity in market practice. *Industrial Marketing Management* 35 (7): 839–855.

Kjellberg, H., and C.-F. Helgesson. 2007. On the nature of markets and their practices. *Marketing Theory* 7 (2): 137–162.

Kjærnes, U., B.B. Bock, and M. Miele. 2009. Improving farm animal welfare across Europe: Current initiatives and venues for future strategies. In *Farm animal welfare within the supply chain*, ed. U. Kjærnes, B.B. Bock, M. Higgins, and J. Roex, 1–69. Cardiff: Cardiff University Press.

Kvale, S., and S. Brinkmann. 2009. *InterViews*. Thousand Oaks, CA: Sage.

Latour, B. 1987. *Science in action*. Cambridge, MA: Harvard University Press.

Laursen, K.B., and E. Noe. 2018. The paradox of stability and dynamics in organizational couplings. *Cybernetics and Human Knowing* 25 (1): 71–90.

Lusk, J.L. 2011. The market for animal welfare. *Agriculture and Human Values* 28: 561–575.

MacKenzie, D. 2006. Is economics performative? Option theory and the construction of derivatives markets. *Journal of the History of Economic Thought* 28 (1): 29–55.

McCulloch, S.P. 2013. A critique of FAWC's five freedoms as a framework for the analysis of animal welfare. *Journal of Agricultural and Environmental Ethics* 26: 959–975.

McEachern, M.G., and G. Warnaby. 2004. Retail 'quality assurance' labels as a strategic marketing communication mechanism for fresh meat. *International Review of Retail, Distribution and Consumer Research* 14 (2): 255–271.

Main, D.C.J., H.R. Whay, C. Leeb, and A.J.F. Webster. 2007. Formal animal-based welfare assessment in UK certification schemes. *Animal Welfare* 16 (2): 233–236.

Marshall, A. 1920. *Principles of economics*. London: MacMillan.

Miele, M., and J. Lever. 2013. Civilizing the market for welfare friendly products in Europe? The techno-ethics of the Welfare Quality® assessment. *Geoforum* 48: 63–72.

Miele, M., and J. Lever. 2014. Improving animal welfare in Europe: Cases of comparative bio-sustainabilities. In *Sustainable food systems: Building a new paradigm*, ed. T. Marsden and A. Morley, 143–165. Abingdon: Routledge.

Miles, M.B., and A.M. Huberman. 1994. *Qualitative data analysis: An expanded sourcebook*. Newbury Park, CA: Sage.

Minister for Food. 2014. *Topmødeerklæring: Bedre velfærd for svin* [Summit statement: Better welfare for pigs]. Copenhagen: Ministry for Food, Agriculture and Fisheries.

Mol, A. 2002. *The body multiple*. Durham, NC: Duke University Press.

Palmer, C., and P. Sandøe. 2018. Welfare. In *Critical terms for animal studies*, ed. L. Gruen. Chicago: Chicago of University Press.

Parker, C., R. Carey, J. De Costa, and G. Scrinis. 2017. Can the hidden hand of the market be an effective and legitimate regulator? The case of animal welfare under a labeling for consumer choice policy approach. *Regulation & Governance* 11 (4): 368–387.

Parker, C., R. Carey, and G. Scrinis. 2018. The meat in the sandwich: Welfare labelling and the governance of meat-chicken production in Australia. *Journal of Law and Society* 45 (3): 341–369.

Renard, M.-C. 2005. Quality certification, regulation and power in fair trade. *Journal of Rural Studies* 21 (4): 419–431.

Richards, C., H. Bjørkhaug, G. Lawrence, and E. Hickman. 2013. Retailer-driven agricultural restructuring—Australia, the UK and Norway in comparison. *Agriculture and Human Values* 30 (2): 235–245.

Rindt, J., and S. Mouzas. 2015. Exercising power in asymmetric relationships: The use of private rules. *Industrial Marketing Management* 48: 202–213.

Sandøe, P., and T. Christensen. 2018. Farm animal welfare in Europe: From legislation to labelling. Working Paper. Copenhagen: Department of Food and Resource Economics, University of Copenhagen.

Sandøe, P., S.B. Christiansen, and M.C. Appleby. 2003. Farm animal welfare: The interaction of ethical questions and animal welfare science. *Animal Welfare* 12: 469–478.

Sandøe, P., and K.K. Jensen. 2011. The idea of animal welfare—developments and tensions. In *ICVAE—First International Conference on Veterinary and Animal Ethics 2011*, ed. C. Wathes, S. May, S. McCulloch, M. Whiting, and S. Corr, 11–17. London: The Royal Veterinary College, University of London.

Satz, D. 2010. *Why some things should not be for sale: The moral limits of markets*. Oxford: Oxford University Press.

Singer, P. 1976. *Animal liberation*. London: Jonathan Cape.

Smith, A. 1970 [1776]. *The wealth of nations*. London: J.M. Dent & Sons.

Tennent, R., and S. Lockie. 2012. Production relations under GLOBALG.A.P: The relative importance of standards and retail structure. *Sociologia Ruralis* 52 (1): 31–47.

Thomas, N. 1991. *Entangled objects*. Cambridge, MA: Harvard University Press.

Uzea, A.D., J.E. Hobbs, and J. Zhang. 2011. Activists and animal welfare: Quality verifications in the Canadian pork sector. *Journal of Agricultural Economics* 62 (2): 281–304.

Verbeke, W. 2009. Stakeholder, citizen and consumer interests in farm animal welfare. *Animal Welfare* 18 (4): 325–333.

Verhoog, H., V. Lund, and H.F. Alrøe. 2004. Animal welfare, ethics and organic farming. In *Animal health and welfare in organic agriculture*, ed. M. Vaarst, S. Roderick, V. Lund, and W. Lockeretz, 73–94. Wallingford: CABI Publishing.

Webster, J. 1994. *Animal welfare: A cool eye towards Eden*. Oxford: Blackwell.

Publisher's Note Springer Nature remains neutral with regard to jurisdictional claims in published maps and institutional affiliations.

Lars Esbjerg is associate professor of marketing at Aarhus University in Denmark. He has studied collaborative innovation, inter-organisational relations, retailer decision-making and transnational value chains in the food sector. His current research interests are related to how market practices constitute markets and influence consumer practices in relation to animal welfare; third-party certification; how inter-organisational relationships develop and evolve; markets for local food; and retailer buying practices.

Agriculture and Human Values (2020) 37:983–997
https://doi.org/10.1007/s10460-020-10030-y

How farmers "repair" the industrial agricultural system

Matthew Houser[1] · Ryan Gunderson[2] · Diana Stuart[3] · Riva C. H. Denny[4]

Accepted: 24 March 2020 / Published online: 31 March 2020
© Springer Nature B.V. 2020

Abstract

Scholars are increasingly calling for the environmental issues of the industrial agricultural system to be addressed via eventual agroecological system-level transformation. It is critical to identify the barriers to this transition. Drawing from Henke's (Cultivating science, harvesting power: science and industrial agriculture in California, MIT Press, Cambridge, MA, 2008) theory of "repair," we explore how farmers participate in the reproduction of the industrial system through "discursive repair," or arguing for the continuation of the industrial agriculture system. Our empirical case relates to water pollution from nitrogen fertilizer and draws data from a sample of over 150 interviews with row-crop farmers in the midwestern United States. We find that farmers defend this system by denying agriculture's causal role and proposing the potential for within-system solutions. They perform these defenses by drawing on ideological positions (agrarianism, market-fundamentalism and techno-optimism) and may be ultimately led to seek system maintenance because they are unable to envision an alternative to the industrial agriculture system.

Keywords Agroecology · Agriculture · Nitrogen · Non-point source pollution · Ideology

Abbreviations

N   Nitrogen
IA Iowa
IN Indiana
MI Michigan

✉ Matthew Houser
  mkhouser@iu.edu

  Diana Stuart
  Diana.stuart@nau.edu

  Riva C. H. Denny
  rchdenny@umich.edu

[1] Department of Sociology, Environmental Resilience Institute, Indiana University, 717 E 8th Street, Bloomington, IN 47404, USA

[2] Department of Sociology and Gerontology, Miami University, 375 Upham Hall, 100 Bishop Circle, Oxford, OH 45056, USA

[3] Program in Sustainable Communities, School of Earth Sciences and Sustainability, Northern Arizona University, Room 280 Building 70, SBS West, 19 W. McConnell, PO Box: 6039, Flagstaff, AZ 86011-6039, USA

[4] School for Environmental and Sustainability, University of Michigan, Dana Building, 440 Church Street, Ann Arbor, MI 48109, USA

Introduction

The industrial agricultural system—typified by its high capital intensity and low ecological diversity—causes significant levels of nutrient-related water-pollution across the world and is prone to soil erosion and degradation (Montgomery 2007; Veenstra and Burras 2015). It also contributes and is highly vulnerable to climate change (Schlenker and Lobell 2010). Within-system, technological fixes, provide the limited potential to significantly mitigate industrial agriculture's contributions to environmental degradation, neither do they achieve climate change resilience (Frison 2016). Consequently, scholars have increasingly called for addressing industrial agriculture's issues through system-transformation, toward a bio-diverse, low-input agroecological system that relies on ecologically-based management approaches (Altieri and Nicholls 2012; DeLonge and Basche 2017; Ponisio et al. 2015).

If we assume the industrial system has the above noted inherent flaws, and that agroecological transition is a viable alternative, the critical question becomes: What are the forces that enable the persistence of this system? Most past work on the maintenance of the industrial agriculture system has focused on barriers to an agroecological transition in the United States at the macro-level, including: policy incentives and funding opportunities that reward and maintain the industrial system and its institutions (e.g. seed companies; Kloppenburg 2005; Mendez et al. 2013); federal crop insurance that supports the more vulnerable simplified, industrial system (USDA 2006); and a general technology fetish in agricultural production and society, that may privilege the high-tech solutions proposed to fix industrial agricultural issues (Altieri 1989; Montenegro de Wit and Iles 2016).

This macro-level attention misses an important area of consideration, the individual level. Past research on the individual-level has generally focused on why farmers are unable to adopt agroecological practices (Roesch-McNally et al. 2018a, b). Few studies have explored if farmers in developed countries perceive agroecological approaches or a transition as necessary to address the issues of the industrial agriculture system, nor how they defend or critique the industrial system in response to its apparent faults (c.f. Dentzman and Jussaume 2017). We feel this research area demands further attention. In many cases, individuals can either (re)enact or deviate from structurally set patterns of behaviors. Individuals can thus be a source of 'bottom-up' change, as individual deviation at a significant enough scale is the precedent for system-level change (Archer et al. 2013; Bhaskar 1998). While the structural factors constrain or create the perception of constraint for some farmers, farmers' willingness to deviate from their industrial methods is a necessary (yet insufficient) precondition for the widespread transformation of the agricultural sector. Therefore, we must be attuned to the role farmers play in participating in or barring a transition to an agroecological system.

In this article, we focus on farmers' participation in the process of social reproduction. Our central intention is to develop a preliminary understanding of how farmers participate in perpetuating the industrial system, particularly as they become aware of its faults and their contributions to these faults and how they perform this support/critique. We give particular attention to how they maintain the systems' perceived legitimacy in the face of external threats.

Toward this end, we interviewed farmers who are deeply entrenched in the industrial agricultural regime: over 150 large-scale United-States midwestern corn-soy farmers, focusing on the use of and environmental effects of nitrogen (N) fertilizer. Drawing from Henke's (2008) theory of "repair," we explore support for system-maintenance at the individual level—how farmers' "discursively repair," or argue for the continuation of the industrial agriculture system as they acknowledge its flaws, its environmental impacts, and their role as contributors.

Conceptual background: a theory of repair

In Henke's (2008) *Cultivating Science, Harvesting Power*, he articulates his "repair" focused theory of social maintenance/change. Henke's first focus is to describe the nature

of social-ecological systems of capital production. He sees systems, like industrial, capitalist agriculture as constantly threatened by disruptions: "anything from a dry year or a failed experiment to a budget crisis or a war" challenges the continued functioning of the industrial agricultural system (Henke 2008, p. 7). Disruptions, like the recognition of the system's environmental issues, provide the opportunity for system change. But whether change occurs, or the status quo is maintained depends on "repairs" undertaken by system actors (including farmers) in response to disruption. Repair "is the work of maintaining this system in the face of constant change" (Henke 2008, p. 10). Critically, repair is not a singular process; it involves distinct means of enacting repair and opposing strategies of repair can be pursued. Related to the means for enacting repair, Henke offers two: discursive and ecological.

Discursive repair deals with the social construction of issues and solutions: Do actors perceive a problem? If so, what do they see as its cause and the best solution? This is the realm of maintaining or eroding the legitimacy of the agricultural system. *Ecological repair* is the realm of action, where material changes to the system of relations are made. Related to these two repair practices are two larger strategies of repair to be pursued using them: maintenance or transformation. *Maintenance repair* refers to changes that are intended to keep the structure functioning as is. Modest adjustments to elements within an established structure or system are pursued when this strategy is followed. *Transformation* refers to complete alterations to the system, in "which the relationships between culture, practice, and environment are substantially reordered" (p. 11). This strategy is pursued when identified problems are seen to be endemic to an established system, thus indicating that system-level change is the required solution.

In his empirical application, Henke focuses on ecological repair and on the role that agricultural advisors play in ecological repair. We extend his model both conceptually and empirically by focusing on the realm of *discursive repair* and how farmers perform this activity to actively maintain the industrial agricultural system in response to disruptions. Below we provide additional conceptual background toward this end.

The ideological basis of discursive maintenance

Henke identifies three questions discursive repair addresses: (1) Does a problem exist? (2) What is the nature of the problem? (3) And what is the best way of solving it? As he states, "The answers to these questions represent practical attempts to shape the discursive frame for meaning and action, which, in turn, leads to ideas about what form a structure can or should take" (Henke 2008, p. 13). For our study, we borrow from the social movement framing literature and call responses to these questions the "tasks" of discursive repair (Benford and Snow 2000). Henke provides little detail on *how* actors perform these tasks. To accomplish the tasks of "discursive maintenance," we expect farmers will draw on more widely available ideological positions. In consequence, we pull from and describe addition theory on ideology and social reproduction (i.e. maintenance).

Although ideology takes "material" form in taken-for-granted practices structured by institutions (Althusser 1971), here we interpret *ideology* as the beliefs and worldviews that shape everyday thoughts and decisions. In many cases, these ideologies also match the "negative" Marxist definition (Larrain 1982), as they often conceal underlying contradictions about the dominant capitalist system (or in this case, the industrial [capitalist] agricultural system). Our broader definition here matches what is articulated by Therborn (1980), who also describes an ideological logic of change: to change the state of something one must (1) know that it exists, (2) decide if it is good or bad, and (3) believe there is an actual chance of changing it. The first two align well with Henke's (2008) questions of discursive repair. Here we also stress the importance of believing that another way is possible. This also relates to Marxist scholarship on social reproduction, or the purposeful and strategic maintenance of the current system by those in power (Wright 2010). Convincing people, as Margaret Thatcher famously stated, "there is no alternative" has been effective means of maintaining the status quo. Therefore, pursuing social transformation relies on overcoming these ideologies and confronting the forces of social reproduction/maintenance (Wright 2010).

Farmers in the industrial agriculture system face accelerating pressures to maximize profit and increase production (Ashwood et al. 2014; Hendrickson et al. 2019; Levins and Cochrane 1996). At the same time, public recognition and response to the growing environmental issues associated with industrial agriculture are expanding in our region of study (see below). Achieving production imperatives can be at odds with mitigating agriculture's environmental impacts (Magdoff et al. 2000; Roesch-McNally et al. 2018b). To deal with their limited capacity to meet environmental demands given production pressures, farmers develop specific ideologies that justify their continued practices, while also legitimizing and naturalizing the contradictions of the industrial agriculture system (Ellis 2013; Emery 2015; Cilia 2020; Dentzman and Jussaume 2017; Hendrickson and James 2005). For instance, Dentzman and Jussaume (2017) find that farmers use a "techno-optimist" ideological position to explain their faith in future chemical treatments to herbicide resistance, but they argue this framing is a means to cope with their limited capacity to enact less productivist, but more effective practices.

To return to Henke's framework, we use the term "discursive maintenance repair tasks" to refer

to *ideological* responses to the questions: (1) Does a problem exist? (2) What is the nature of the problem (including the problem's *cause*)? (3) What is the best way of solving it? The tasks of discursive maintenance repair are to answer these questions in ways that reproduce the system through its legitimation of and/or naturalization. Reflecting Therborn (1980) and other scholars (Wright 2010), we also give attention to farmers' beliefs about (4) if alternatives are possible, or conversely not possible. In short, these discursive maintenance repair tasks frame[1] the problem and solutions in ways that naturalize and/or legitimate the system, thereby contributing to its reproduction.

Using this approach (see Table 1), we examine how farmers perform discursive maintenance in response to acknowledging water-pollution issues associated with the industrial corn-soy agricultural system's reliance on nitrogen fertilizer, focusing on the reasons they provide for system maintenance. Farmers face constrained choices in what practices they ultimately can pursue (Stuart 2009), and moving away from nitrogen fertilizer is limited by numerous forces in the industrial agricultural system (Stuart et al. 2012; Stuart and Houser 2018; Stuart and Schewe 2017). In this context, we expect farmers will have drawn on preexisting ideological positions that justify their continued support for industrial agriculture (and nitrogen use). More specifically, we expect these positions will be used to perform discursive maintenance repair tasks. Our application context is outlined below.

Application context

Synthetic nitrogen use and loss in the Midwest

Today's industrial agriculture system does not exist without synthetic nitrogen (N) fertilizer. The industrial agricultural system relies very little on organic sources of N (e.g. legumes, manure, compost) and instead meets crop requirement for N through the application of synthetic N fertilizer (Smil 2002). Synthetic N use is particularly prevalent in the United States midwestern "corn-belt" agricultural system. Corn receives the majority of N applied in the US. In total, about 96% of all US corn receives synthetic N application and approximately 50% of all applied N in the US is applied to corn (Cassman et al. 2002; ERS 2018).

Due to this high level of synthetic N use, the midwestern industrial system is inherently "leaky," as synthetic N is highly prone to loss (Drinkwater and Snapp 2007). On average, around 38% of all applied N being lost as environmental

---

[1] Though similar to Goffman's (1974) notion of "frame," Henke's discursive repair concept points more directly to the significant role that individuals' or groups' construction of problems and solutions can have in system maintenance or transformation.

Table 1 Conceptual framework for discursive repair

Definition: The social construction of a system's issues and their solution, toward either maintaining or eroding the legitimacy of the system		Types: Maintenance or Transformation
Questions discursive repair is tasked with addressing*	Answers if performing *discursive maintenance repair*	Answers if performing *discursive transformation repair*
1. Does a problem exist (Henke 2008)?	No; or if yes, then the problem is insignificant/easily solvable	Yes, and it is significant
2. What is the nature of the problem (including the problem's *cause*) (Henke 2008)?	The problem is not systematic/structural	The problem is inherent given the design of the system
3. What is the best way of solving it (Henke 2008)?	Minor adjustments within the current structure of the system	The system must be replaced if the problem is to be addressed
4. Is another system/solution actually possible (Therborn 1980; Wright 2010)?	No; or even if another way is desired, it cannot be envisioned	Yes; can potentially articulate a vision for this system

*As we note, ideological positions will be drawn on and (re)created in actors' answers to these questions

pollution (Gardner and Drinkwater 2009). Given this amount of loss, synthetic N from midwestern agriculture is associated with significant environmental issues (Robertson and Vitousek 2009; Riabudo et al. 2011). We focus on water pollution. N lost as nitrate pollutes fresh water, leading to eutrophication, or hypoxia (Conley et al. 2009). Most notably, synthetic agricultural N from the Midwest is *the* major contributor to hypoxia, or the "dead zone" in the Gulf of Mexico, contributing 41% of all N found in the dead zone (USGS 2017). Waterways and groundwater in Midwestern states like Iowa and Illinois have been found to contain some of the highest concentrations of N, particularly in agricultural areas (USGS 2010).

N pollution: a systematic issue

The possibility of mitigating the industrial corn-soy system's loss of synthetic N fertilizer via within-system technological solutions is frequently promoted (e.g. Flis 2017; Robertson et al. 2013; Ribaudo et al. 2011; 2012). However, reflecting the inherent, system-level flaws with an agricultural system based on synthetic N use, the potential of technologies to mitigate contributions to water pollution and N loss to the environmental overall are limited, with studies finding little to no reduction in agricultural N loss to waterways from their use (Blesh and Drinkwater 2013; Sprague et al. 2011). After their analysis of the impact of technological approaches in reducing N loss to waterway from midwestern corn-soy farms, Blesh and Drinkwater (2013, p. 1031) summarize: "Our results suggest that the dominant management emphasis on adjusting the timing and placement of [synthetic] N inputs (i.e. within-system approaches) has biogeochemical limitations in terms of the degree to which N retention can be increased." In other words, it is increasingly clear that significantly mitigating the issue of N water pollution requires agroecological approaches and eventual system

transformation toward a much more ecological diverse approach to agriculture (Blesh and Drinkwater 2013).

Movement toward this approach is limited in the Midwest. For instance, cover crops were planted on only 2.5% of farmland acres in Iowa 2017 (Juchems 2018) and only 6% of Indiana's acres in 2018 (ISDA 2018). The vast majority of agricultural land in the Midwest produces only two crops—corn and soybeans—and this state of simplified agriculture has intensified in recent years, despite evidence of the benefits of a more diversified, agroecological approach (Plourde et al. 2013; UCS 2014).

We build on the past work that has examined barriers to agroecological transitions or practice adoption (see introduction) by focusing on how farmers might be active participants in constraining an agroecological response to the issues of the industrial agriculture system, like endemic N loss. We focus on how farmers justify their continued faith in the system (i.e. perform discursive maintenance) by drawing on multiple ideological positions. Our study provides preliminary evidence of farmers' active role in maintaining the industrial agriculture system in the realm of discourse, something few studies have previously addressed (c.f. Dentzman 2018; Dentzman and Jussaume 2017). We see this process to be (re)constructing an ideology that ultimately limits a more widespread emergence of agroecological practices/transitions to address N pollution, even in the preliminary state of farmers' interest/desire.

Methods

We examined qualitative data gathered from 154 interviews with corn farmers in three US states: 53 interviews in Iowa (IA), 51 in Indiana (IN) and 48 in Michigan (MI). Qualitative methods are ideal at providing insights into little-studied topics (Kreuger and Casey 2009) and these methodological benefits have been noted in past work in the agricultural context (Prokopy et al. 2017). Given that we have limited understanding of how US farmers defend the industrial agriculture system, we use a qualitative approach to develop a preliminary understanding of this process.

Interviews to assess farmers' N use decisions and the factors that shape them were conducted on a one-on-one basis between a researcher and the farmer between May 2014 and December 2014. The majority of interviews were done in person on-farm, with a small number conducted over the phone. Aside from one interview (in which the participant requested that we take hand notes only), all interviews were audio-recorded with the permission of participants.

Initial interview participants were primarily recruited through University Extension and other state resource professionals, with a reliance on snowball sampling after initial contacts. Snowball sampling is considered a good method to contact subjects who are difficult to access (Faugier and Sargeant 1997). Across all three states, Extension was our main source of farmer contacts (48%). All farmers interviewed were white, English-speaking males. Though not specifically asked to identify their age, very few farmers in our sample just started farming or were nearing retirement. These features generally match that of the broader US farming population (NASS 2012). Farm sizes of interviewed farmers ranged from 170 to 14,000 acres. Over 70% grew only corn or just corn and soybeans. Around 18% of interviewees were currently using cover crops to at least some extent, a much higher percentage than the general farming population in much of the Midwest (see above). This bias toward an agroecological approach likely reflects our contact strategy, i.e. Extension based. We feel this bias makes for a particularly telling study of farmers' maintenance of the industrial system. Even among our more environmentally minded sample, support for maintaining the industrial system was high. We give attention to the dissenters after our results section.

Interviews lasted between 22 min and 2.5 h. Upon completion, interviews were transcribed and analyzed using NVivo software. Our analysis for this paper focused on farmers' responses to questions in the section of the semistructured interview regarding their awareness and perception of N's contribution to water pollution, specifically (1) whether they had heard of or personally seen nitrogen related water pollution; (2) what they believe caused these environmental effects and (3) what they believed needed to be done to address these issues.

After identifying if farmers believed agriculture contributed to water pollution issues to any extent, coding focused on how they discussed the severity, nature and solutions to this problem to identify their discursive repair strategy (e.g. maintenance or transformation). Driven by the data, we focused on identifying the major points farmers made about N water pollution issues (*Eclectic Coding*; Saldana

2015), then performed a more specific process of thematic grouping from this initial list of points (code "mapping"; Saldana 2015). This thematic grouping was performed in two stages. First, farmer comments were categorized within Henke's (2008) discursive repair categories (e.g. maintenance or transformation). Following this initial grouping, we categorized codes considered as maintenance repair by "ideological position." While we were aware of past literature on farmer ideology, we aimed to let our positions emerge organically from the data. Initial electric and thematic coding was performed by a single coder, with the other study authors evaluating the thematic codes subsequently. Disagreements led to a discussion, and if necessary, a re-coding of a farmer's quote. We now turn to discussing the thematic ideological positions farmers evoked in their discursive repair that emerged during this analysis.

Results

Acknowledging a disruption to the industrial agricultural system

The year that our interviews took place was, in many ways, particularly well-suited to discussing water pollution from industrial agriculture in the Midwest. The Iowa Nutrient Reduction Strategy, a suite of policies aimed at reducing agricultural contributions to Gulf of Mexico hypoxia from the state, was in full swing, being implemented only 2 years prior (IASTATE n.d.). The city of Toledo's "Water Crisis" also occurred over our interview period. As Lake Erie was flooded with toxic algae blooms, local and national media sites were widely covering the issue of non-point source pollution from agricultural nutrient loss in the region (e.g. Wines 2014). Farmers were highly attuned to the issue and every farmer we interviewed was at least aware that agriculture was considered to contribute greatly to N pollution issues in local and national waterways—though as we discuss below, there was considerable variation in the level of responsibility farmers were willing to assign to agriculture.

Reflecting the "tasks" of discursive maintenance repair, after assessing if farmers were aware of N-related pollution in local or national waterways, we asked them to discuss what they felt was the cause of these issues and the solutions to them. This was where farmers began to preform discursive maintenance tasks. As one Indiana farmer put it, referring to nitrogen water pollution: "Yeah, nobody can deny that the problem is there, the argument is what's causing it" (IN06). In terms of our theoretical framework, though farmers (1) acknowledged there was a problem with N pollution, most responded to questions about N water pollution in ways that effectively performed two tasks of discursive maintenance repair—(2) denying agriculture's responsibility for causing these issues and/or (3) emphasizing that minor, within system solutions would or were solving what portion agriculture did contribute to N pollution. Farmers drew on ideological positions to achieve these maintenance tasks. Below we illustrate the three most common thematic positions farmers used in their performance of discursive maintenance.

Performing discursive maintenance: what ideological positions maintain the system?

Agrarianism: responsibility denial via framing urban areas as the problem

Agrarianism refers to farmers' expressions of the belief that agriculture was relatively blameless as a source of environmental pollution, and instead urban activities were the culprits. Agrarianism captures a long-standing ideological position in American history that is prevalent enough that it has been called by many names including the "agrarian myth," (Hofstadter 1955), "the myth of the garden" (Smith 1950) or the "pastoral ideal," (Marx 1964) to name a few. To put it concisely, agrarianism is an ideology that frames agriculture, and rural life in general as "good," especially in comparison to urban, non-agricultural areas/citizens.

This position has long been held by and influential amongst American farmers, and when used by this group it particularly tends to frame farmers as distinct from and the victims of urban life/actors in cities. As Hofstadter (1955, p. 35) wrote about early to mid-twentieth century farmers, agrarianism led them, "to believe that they were not themselves an organic part of the whole order of business enterprise and speculation that flourished in the city [...] but rather the innocent pastoral victims of a conspiracy hatched in the distance." Similarly, Mooney and Hunt (1996, p. 183), in their historical analysis of US farmer social movements, identify that agrarianism is one of the most long-standing ideological framing of protests and reflects a "conflicting or exploitive relationship between certain townspeople and farmers," while also tending to portray agriculture and farmers in a positive light—that farming is the "natural life" and necessary for urban places to exist. There was and is then a tendency among farmers to depict agriculture in a highly positive light, especially compared to urban areas.

This appears to hold true for at least some farmers today. Among the 154 farmers in our sample, 53 expressed an "agrarianism" position as they considered industrial agriculture's contributions to environmental degradation. This position formed a basis for which to discursively maintain the industrial system by denying responsibility for the system's role in water pollution. Farmers expressing agrarianism acknowledged that agriculture contributed to N-related water pollution, but they argued this contribution was relatively minor compared to sources from urban areas. For

the most part, farmers asserted that residential lawn care and golf courses were the primary cause. They frequently argued that N fertilizer was greatly over-used in these areas, compared to it being well managed by the vast majority of farmers. As one Indiana farmer put it:

> I think the real polluters, of especially city water, is not us. It is because [city residents] sprinkle their lawn, it runs in the sewer, then it goes directly back to the [city water treatment plant]... Because that's how they collect the rainwater to fill the reservoir, from the sewer water. Well where's the nitrogen coming from? It's coming off that guy's yard, it ain't coming out of [agriculture]...The ground is a great filter of nitrates; it will filter out nitrates before it gets to a tile. Tiles goes in the river anyhow, and they don't drink the river (IN14).

The numerous inaccuracies in this farmers' comment suggest how agrarianism masked several realities for farmers that enabled them to feel and position agriculture as relatively blameless compared to urban sources for N water pollution.

Farmers expressing agrarianism did not only tend to argue that urban areas were the major source of nitrogen that was polluting waterway, they expressed frustration for being blamed for this issue, arguing that the rural/agricultural areas were unjustly being negatively framed. This injustice component is especially well captured by one farmer in saying, "I get a little bit miffed when, you know, the subdivisions [i.e. suburban residents] that over-apply products from Greenlawn and all those places, and the golf courses, they seem to be immune to criticism in the press. And most farmers, honestly, I think, are trying to do a pretty good job" (IN40). Others commented similarly reflecting the presence of agrarianism (see Table 2).

Other farmers evoked a relative form of agrarianism in arguing that agriculture was a major contributor of N to water-pollution, but that these contributions were already as minor as possible or were justified. One farmer's comment illustrates the relative form of agrarianism well. He rather accurately described the proportion of agricultural contributions (NRDC 2008) but seems to justify agriculture's role to some extent in saying: "I would say that agriculture is probably 70% [or more] of the total [cause], but if you look at the landmass [...] the other 20 some percent that's not ag that does contribute to nitrogen [lawn care and golf courses, it] is not near as many acres" (IN07). Like this farmer, others acknowledged agriculture's significant role, but still evoked agrarianism in suggesting that the industrial agriculture system was actually very efficient, performing well especially compared to the inefficient urban sources (see Table 2).

Farmers may develop or draw on ideological positions that help them avoid cognitive dissonance given the conflicting pressures of maximizing production and the need

Table 2 Agrarianism as discursive maintenance (illustrative quotes)

"I have a tough time believing that the corn farmers in the Midwest is the cause of it [...]

I look at what a farmer applies on a per acre basis compared to what a lotta of people in town apply to their lawns or gardens in town. And oh, those rates are a lot more drastic and it doesn't get used, it doesn't get consumed necessarily [...] to pin point it on the Midwest corn farmer or wheat farmer, take your pick, I have a tough time with that" (IA18)

"I think we need to realize that if they really tested things and went out here how much of this nitrate is really coming from the guy out here that spraying his lawn? There's a lot comes from that; it's not us" (IN13)

"We need to educate the public. And because the first thing, you know like the Lake Erie situation, that's all coming back to the farmers, the farmers the farmers, and you probably understand it, 90% of the people, they only know what they're told, and whatever they're told they believe" (IN10)

"They're finding out [farmers] calibrate and regulate what we put on soils, much more than somebody who goes to Earl May and buys a bag of lawn fertilizer and throws it on their lawns [...] So they're finding, they have backed off a little bit that it was strictly agriculture. A lot of it was home owner and industrial use also" (IA20)

"I think that ag is not the only source. You have golf courses, home owners, the amount of fertilizer that gets spread on lawns in towns, that's a huge industry, not only the people applying it, but the people selling it" (IA26)

"I didn't feel as a farmer that I could be the solution to that problem. Because home owners and golf courses use a much higher level than I would ever dream of using in a farm situation" (IA30)

"I think our municipalities are a lot to blame there. You know, electrics going out, sewer being dumped out over the river... You know, I think that's where we need to probably make more improvements then to think that coming back to the farmer's the big problem" (IN50)

to reduce N related pollution (Ellis 2013). One Iowa farmer passionately defended farmers and agriculture against evidence of its contributions to water pollution, emphasizing farmers' love for the land as a reason why agriculture could not be a significant source of N loss:

> [N]o one values the soil more than a farmer. [...] We wouldn't poison it, destroy it, or abuse it for anything. It is life. [...] I wouldn't poison the Gulf [of Mexico] for nothing. I want that land, I want the soil [i.e. nutrients] on my land. I don't want it down there. And if anyone thinks that we're throwing this away on a whim, they believe in some agenda that does not exist. [...] I know that I speak for the American farmer. Are there corporations out there that are abusing? Yes. Guess what? We hate them worse than you do (IA38).

This heartfelt message is hard to ignore—farmers care deeply about the environment. But this does not change the fact that they are caught up in an agricultural system that has inherent flaws in its design. The industrial system will always be "leaky" when it comes to synthetic N loss (Drinkwater and Snapp 2007). As the above quote illustrates, agrarianism is one position that farmers draw on that enables them to continue to meet the challenge of accelerating production (often via using more N) without having to experience the cognitive dissonance from how this task contradicts their desire to also be environmental minded.

Toward discursive maintenance, the agrarianism position achieves the maintenance task of "responsibility denial" (Stuart and Worosz 2013). Responsibility denial is a means to de-attribute industrial agriculture's or farmers' role in causing pollution issues, and thereby to suggest that transformation change is not necessary, as the "problem" does not

merit this type of solution. In denying the corn-soy agriculture's significant and inherently high contributions to water pollution through agrarianism, farmers are practicing discursive maintenance, as they are constructing a narrative that legitimizes the system even in contradiction to their personal beliefs about how that system should be.

Market fundamentalism: within-system solutions and responsibility denial via the invisible hand of the market

Another discursive strategy for maintenance repair was what we refer to as "market fundamentalism." Market fundamentalism describes the ideological position that the free-market, the "invisible hand" of capitalism (Smith 1776) will solve most social, economic and environmental problems (Malin et al. 2017). This self-correcting nature of the market has been argued to be a fundamental ideological element of the capitalist socio-economic system (Gladwin et al. 1997). When applied specifically to environmental issues, market-fundamentalism, or sometimes the "free-market ideology" (Heath and Gifford 2006), discourages acknowledgment of environmental degradation caused by capitalist production and eschews any environmental regulation. If a problem is acknowledged, this position is used to frame environmental degradation as solvable and being solved through price mechanisms that increase the costs of undertaking environmentally harmful activities (Shrivastava 1995). In our case, farmers were considered to be expressing market-fundamentalism when discussing how the market, economic system or price mechanisms would or had prevented agriculture from significantly contributing to N loss to waterways.

Table 3 Market fundamentalism as discursive repair (illustrative quotes)

"You know, the free market determines" (IN34)

"The fertilizer and seed is so expensive that, you know, we don't want to… We're not out here… This isn't a charity, we want to make money, you know, we don't have unlimited resources to throw out there, we just want to, you know, get the bang for our buck so to speak" (MI17)

"No. I don't think anybody can afford to put gobs amounts on, too much" (IN08)

"We're doing it because we want to save the money, and you know take good care of what we bought, not that were against the environment, but they both work pretty hand-in-hand" (IN07)

"Yeah, I mean, how can you afford to do that? You can't afford to fertilize Lake Huron, or any other lake, the creek or whatever" (MI13)

Farmers' widely drew upon a market fundamentalist ideological position (35 out of the 154 interviewees) when discussing if there was a need to address agriculture's role in N pollution issues. Their comments centeredaround the belief that the mechanism of the market—particularly the high cost of N—would ensure or already had ensured farmers were not inefficiently using N fertilizer.[2] In other words, market-fundamentalism was drawn upon to perform the latter two tasks of discursive maintenance: (2) *responsibility denial*, suggesting that N loss from agriculture could not be a major, systematic issue because the market had assuredly prevented an inefficient system (though again, these farmers did not deny the existence of N-related water pollution) or was a foundation for belief in (3) *within-system solutions*, positioning the market as the means to correct over-use of N if it was occurring. Though most farmers used market fundamentalism to deny responsibility for agriculture's role (see Table 3), occasionally market-fundamentalism was expressed to accomplish both the tasks of the denial of responsibility and the potential for within-system solutions simultaneously. One farmer responded to the question about what should be done to address N loss from agriculture in a way that illustrates how market fundamentalism was paradoxically used to achieve both tasks:

[Farmers are] not going to spend more money than what's going to make them an economical return, so… Maybe this thing has been blown out of proportion, this runoff thing, maybe there is some runoff but maybe there isn't, I don't know. Why would a guy spend more money than his crop can use? That just doesn't make any sense to me.
Interviewer: "So you sort of see this economics of the system taking care of it a little bit there?
Sure. Yeah, I think so…The economics will definitely take care of the over-use of the stuff (IN38).

As this suggests, this farmer espoused a market-fundamentalism position—that price mechanisms will prevent/solve environmental issues—in arguing that the profit-based nature of agricultural activities was ensuring (responsibility denial) or would ensure the best possible outcomes for all (within-system solutions), even if environmentalism is not the goal of individual farmers or the agricultural system itself. In describing how he defends the agricultural system and farmers to the public when discussing N-related water pollution, another interviewee expressed this particularly well:

I try to tell people this: it's not financially good for us to use any more [nitrogen fertilizer] than we have to […] I'm environmentally concerned but I'm more… I am profit driven, you know, which puts me in that same boat, you know, so I can't pat myself too hard on the back, because it's more about profit really. You know, but if you're truly profit driven, you're going to be environmentally driven because you're not going to [be an inefficient user of N] (MI31).

In other words, farmers argued that the industrial agriculture system was blameless because N pollution from a profit-oriented activity was a market-impossibility and/or that this system was fixable because the market would ensure efficient use of N, thus eliminating what contributions agriculture makes to N-related water pollution. These sentiments were not only occasionally paradoxical but ignore the severity of agriculture's contributions and inherent leakiness of synthetic N. By masking these realities, farmers perform the aforementioned tasks of discursive repair and they do so through evoking positions that reflect the ideology of market-fundamentalism, a widely prevalent position among interviewed farmers (see Table 3 for more evidence).

Techno-optimism: the basis of belief in within-system solutions

Many farmers were willing to offer solutions to agriculture's (perceived minor) contributions to water pollution. The potential of technological solutions to reform systems, solving environmental issues without changing the systems or requiring a change in human behavior has been previously

---

[2] However, there are several reasons why the price of nitrogen does not limit farmers to applying as little as possible. Indeed, it is sometimes considered economically rational for farmers to over-apply nitrogen (Sheriff 2005) or, if not national, at least the inefficient use of N is not a significant economic cost to farmers (Pannell 2017).

referred to as "techno-optimism" (Weinberg 1966/1981) a position that has been previously identified to influence the views of row-crop farmers' related to their herbicide use (Dentzman et al. 2016) and climate change adaptation decisions (Gardezi and Arbuckle 2018), as well as large-scale beekeepers perspective on the viability of the industry (Cilia 2020).

In total, 28 of the 154 interviewed farmers expressed a techno-optimism ideology in positioning technological solutions as the most feasible and effective means to addressing N's contributions to water pollution issues (despite evidence to the contrary; Blesh and Drinkwater 2013). Farmers often pointed to the potential of current management technologies to limit N pollution, such as bio-reactors (i.e. catchment of N runoff) or equipment that allows for more efficient timing and placement of N applications (e.g. sidedress). They generally expressed optimism that this technology would be increasingly adopted/improved to curb whatever portion of N loss to water-ways agriculture did contribute. The following response to a question about what needs to be done to address N pollution issues was typical of this vein:

> Timing. The nitrogen stabilizers. The timing. Give the farmer and the equipment people some credit, our ability to put the stuff on when it needs to be put on is light years ahead of where it used to be. I think we've come light years as far as where we place it, our timing for placing it and our ability to measure and control what we place. Much much. In the last 10 years that has changed a lot. And it will change that much again (IA62).

Others focused more on the potential of future technologies to solve agricultural N run-off issues. In particular, farmers expressed the belief that corn plants would be developed by seed-companies that did not require N use:

> I don't know a lot about it, I've just heard a little bit about it, about low, I don't even know what the word is for it, but corn that needs a lot less nitrogen, and I'd be all about that. I know that… I don't know that there's any on the market yet necessarily, but the way I understand it is the seed companies will eventually come out with some corn… And I'm sure you still have to put some nitrogen on just not… maybe you could cut it back 100 pounds or 50 pounds or something like that, and yeah, I'd be all the way in on that… (IN15).

A techno-optimist position was present even as farmers questioned the practical, political limitations to this solution in the current, agro-business dominated system. As the above farmer continued on later: "…I don't know who's going to fund the university research on that. I suppose the seed companies will, but the fertilizer… Mosaic fertilizer sure as hell isn't going to, or Koch because that would piss

them off, all of a sudden we don't need to use as much nitrogen." Other farmers expressed a similar faith in the development of this type of corn, or other similar technological solutions to N loss (see Table 4).

The degree to which farmers truly expected this "magic bullet" type of technology is well illustrated by one farmer. He notes how a technology that could read corn-tissue plants was previously proposed and this would eliminate over-use of N (though even this will not effectively curb N loss without agroecological approaches; Blesh and Drinkwater 2013). As he states, this technology, "never materialized" but he still felt like in response to industrial agriculture's contributions to N pollution, "They'll do something. They'll come up with something!" (IA57).

Techno-optimism was a widely drawn upon ideological position among farmers in our sample. Farmers expressed a great amount of faith in the potentials of current or future technologies to reduce or eliminate industrial agriculture's contributions to N pollution. This position, therefore, performed the discursive maintenance task of proposing within-system solutions and framed system maintenance, rather than transformation, as the appropriate response.

Cracks in the system: frustration and confusion in the face of recognizing system limits

Reflecting our conceptual model, disruptions such as recognizing the environmental consequences of industrial agriculture can also promote transformative discursive repair, where farmers point to the systematic nature of problems and call for new-system solutions, such as agroecological approaches. The performance of discursive transformation repair was limited. While discursive transformation repair is not our central focus, we provide some evidence on its occurrence below to suggest how environmental consequences can disrupt the ideological positions that are (re) structuring the industrial agriculture system.

Farmers in Iowa, in particular, were speaking out in opposition to the agrarianism position, saying instead that agriculture was the primary cause of N loss. Indiana and Michigan farmers rarely accepted the extent to which agriculture contributed to these issues. This may point to the impact of Iowa's Nutrient Reduction Strategy (ISU, n.d.) in catalyzing awareness among some farmers of their role in N-pollution. In particular, as farmers were enrolled in nitrate monitoring groups, it became more difficult for them to maintain their faith in an agrarianist position, that N loss was primarily from urban areas:

> So I have nitrate awareness. If there is such a thing. Where most farmers don't. And I didn't until I started doing this [nitrate monitoring group]. Cause I thought where's all this nitrate coming from? But in reality, it's

Table 4 Techno-optimism as discursive repair

Current technologies as solutions (illustrative quotes)

"Protecting environmental quality…They're state-of-the-art sprayers now and the technology and the GPS has really attributed to this […] our planters are doing the same thing, you know, it shuts off so that we're not over seeding and over fertilizing, because if the seed's going down the fertilizer's going down, in its setting that stuff off as it goes across the field, you know" (MI14)

"Well, I definitely think we need to do more sidedressing" (IA12)

"I think by using the nitrogen stabilizer, I think by several split application of nitrogen, different times throughout the season, would probably help" (IA16)

"Of course, with today's technology of guidance systems and those kinds of things, some of those things can be overcome" (IA38)

Future technologies as solutions (illustrative quotes)

"It's a big step in the right direction/ And you need technology. I'm hoping the technology will get better or easier to somehow predict what you need while you're going through the field. There are tools out there that will do that, but they're a little bit unwieldy. And weather depending too." (IA12)

"Well, yeah if we know where they might need nitrogen and if they need more nitrogen or if they… Hopefully we get some hybrids developed pretty soon that don't need as much nitrogen. I think that's the road they're trying to be going down, but whether we get there or not…" (IN17)

"If we could do automatic variable rate […], that would be great if they could come up with that" (IA57)

"They are working on better hybrids that will take less nitrogen to raise a big crop, and I think that will probably be the next thing in the works I'm thinking is the work on hybrids that won't need as much N, and I think that will be a biggie" (IN24)

"They're developing corn that's going to be a lot more efficient at utilizing nitrogen, so I would say their policies are going to drive that. Fertilizer companies, again, looking at denitrification inhibitors, that sort of thing, I think that's probably going to be a higher interest level, and there may be better products out. I think Instinct's probably better than NServe for example, so all that goes into the equation" (MI28)

coming from my own farm. I always talked about, it's the golf courses, it's city people, it just washes off the lawns and goes down the gutter. But in this watershed, as I mentioned earlier, 88% of the acres in the watershed are agriculture. So the amount of nitrogen put on in the towns in minimal (IA01).

Others performed discursive transformative repair by casting doubt on technological solutions (being "techno-pessimistic"). For instance, in discussing some technological approaches to reducing N loss, one respondent said, "You know, I think maybe on a short-term, but on a long-term basis, on a large-scale basis, I'm not sure that they are the answer" (IA17). Another farmer recognized the systematic nature of nutrient loss in the industrial system, focusing on its linear (inputs → outputs) nature:

You will always have to take some resource from some other place to replace what you've removed. You will always have to do that, and that's not sustainable because you have depleted that other source. And so many people don't get that; you know, they say 'well yeah, okay, I know, but…I'll tell you what, you know, we'll do this and then we'll spread seaweed!' Where the hell do you get the seaweed?! Where did that come from? (MI09).

This farmer is correct in many ways—linear systems are fundamentally problematic, as Karl Marx pointed out in his now-famous "metabolic rift" critique of industrial agriculture (Foster 1999). Given this reasoning, the respondent felt that the industrial agriculture system was as "sustainable" as

any other form of agriculture. These quotes are only illustrative. Particularly, more farmers expressed a recognition that agriculture was the primary culprit of N loss to waterways. But this acknowledgment rarely led to an outright performance of discursive transformation repair, with only five farmers expressing this view.

Another way is (not) possible

As we noted earlier, proposing outside the system solutions is a critical dimension of seeking transformation change. Among the already small group of farmers who saw agriculture as the primary culprit of N loss to waterways, only a very small group proposed agroecological approaches to address these issues:

From my way of thinking, I'm starting to shift from the idea of being, basically all we're doing is importing nutrients and exporting them. And I'm very interested in how to grow those nutrients at home with a diverse crop rotation, and I think there has to be a shift sooner or later away from commercial fertilizer and more towards what nature can produce for you (IA13).

However, even among those who felt that the industrial system was not "maintainable," farmers commenting on the need to pursue agroecological changes were the exception. This group of farmers discussed the limited potential to solve the loss of N within the industrial agricultural system, i.e. they recognized that as long as synthetic N was being used, there would be problems, even if technologies were widely adopted. However, their performance of this discursive transformational repair rarely made it to proposing actual alternatives to the industrial agricultural system.

The limited number of farmers who felt like system-change was necessary to address industrial agriculture's contributions to N loss often concluded their statements with comments indicating their uncertainty of what the actual change could be (13 of the 154 farmers). As one respondent stated, "Well I think, the [nutrient] erosion situation needs to be addressed.[…] the corn-soybean rotation is not the best way […]. But I'm…not quite sure what the answer [is]" (IA39). Similarly, the inability to see another type of system, a more wholistic society, and agrarian system, is illustrated by another comment from the above linear-system farmer, who felt that this is the only form an agricultural system can take: "As long as we bury people in cemeteries, as long as we have septic tanks and public sewer systems and all of those things we will never have sustainable agriculture. It's impossible!" (MI09). Lastly, one respondent explained that to address environmental problems the only solution is to "quit farming. Which isn't going to happen, people need to eat" (MI45). Even when problems are acknowledged, there was no recognition of alternative ways to produce food with less pollution. Farmers overlook that another system is possible, one where society-environment relations are remodeled in a way to enable a more fully incorporated and regenerative agriculture system (Frison 2016).

These comments reveal one of the most significant ideological positions promoting discursive maintenance: the inability to envision an alternative system (Wright 2010). Farmers have developed and can draw on multiple ideological positions to perform the tasks of discursive maintenance repair. At this point, however, they appear to be unable to draw upon an ideological position that enables them to envision a new-system solution to industrial agriculture's loss of N. In effect, they cannot offer discursive transformation repair at its fullest when they cannot see the solutions. Even as farmers begin to see the significant flaws in the industrial agriculture system, like the loss of N to waterways, their capacity to critique and pursue systematic change is *in part* constrained by their inability to think outside of the industrial system's positions and structure.

Discussion/conclusion

We explored how farmers actively participated in maintaining the industrial agricultural system, especially as they became aware of its environmental faults. In response to widely acknowledging the issue of N-related water pollution, we identified how farmers performed the tasks of discursive maintenance through drawing on a variety of ideological positions. Farmers defended the current system through evoking agrarianism and market-fundamentalism to achieve blame-shifting that denied agriculture's responsibility, and through expressing techno-optimism, a reliance on technological fixes that represent small tweaks to the current system. Even when serious issues with the current system were identified, farmers were unable to envision or articulate what an alternative might be, with some not believing a viable alternative existed.

Returning to Henke's (2008) questions related to discursive repair, we find that many farmers believe there is a problem with water pollution, yet the cause or nature of the problem is not perceived as related to farmers' actions or their responsibility. Ideologies that focus on the role of urban polluters (agrarianism), conceal the significant role of agricultural N loss. In terms of the best way of solving N pollution, farmers largely believed that technological changes could reduce pollution (techno-optimism). While some suggested that markets had already or were the best way to solve problems (market-fundamentalism), none proposed an increase in fertilizer price or a fertilizer tax as a solution—although this solution has been effective where adopted (Hamblin 2009)—nor did they consider other policies that would at least encourage a reduction in total N use (Kanter et al. 2015). Solutions proposed do not represent a significant transition towards ecological practices but justify maintaining the current system.

We enhanced Henke's (2008) framework by focusing on how ideology is drawn upon to achieve discursive maintenance. The ideological positions interviewed farmers drew upon to defend the legitimacy of the system reflect broader ideological frames (see above). Our results suggest these ideological contexts are both drawn upon and likely shape farmers' views, promoting their pursuit of system maintenance and the practices that align with this vision. This finding engages with long-standing and more recent literature that also reveals the significant role of ideology in farmers' decision-making (Ellis 2013; Emery 2015) and continues to build a case for the presence of two specific ideological positions in the agricultural context: techno-optimism (Dentzman 2018; Dentzman et al. 2016; Dentzman and Jussaume 2017; Gardezi and Arbuckle 2018) and agrarianism (Mooney and Hunt 1996).

Drawing from Therborn (1980) and Wright (2010), we can see how these ideological positions serve to prevent social transformation and maintain the current system, a finding that reflects prior work examining the role of agrarian ideologies as forces of social reproduction in the face of environmental changes (Dentzman 2018). In our case, a widespread belief that another way to produce food, with far less pollution, is not possible prevents farmers from engaging in transformative projects, be it agricultural or political. Whether large corporations selling seed, fertilizer, and other key components to the industrial system are propagating narratives that "there is no alternative" is beyond the scope of this paper, but others have identified that seed and fertilizer companies use information, marketing strategies and personal contact with farmers to encourage the use of their products (Bell et al. 2015), as well as the belief that "high yield" is the ultimate production goal (Stuart and Houser 2018; Stuart et al. 2018), despite evidence of persistently low prices due to over-production (Blank 2018). Given this evidence, it can be reasonably suggested that input giants strategically promote industrial and technological pathways that preserve the current system and at the same time ignore or refute the possibility of transformation toward an agroecological system. Regardless of the role of input companies, we find that farmers adopt ideological positions that rationalize their role in the current system and serve to maintain this system and fail to envision or consider agroecological alternatives that may more effectively reduce environmental degradation.

A lack of belief that another system is possible represents a considerable barrier to transforming the agricultural system to address environmental impacts. As Therborn (1980) explains, if one cannot see that there is a real possibility for change it will not occur. While technological fixes can reduce nitrogen loss (Robertson et al. 2013), nitrogen loss is a system issue and agro-ecological transition is required to adequately address it (Blesh and Drinkwater 2013). Given our findings, increasing awareness about agroecological practices and visions and policies for an agroecological system is a paramount first step in supporting efforts toward pursuing this transition.

For this study, we considered the widespread adoption of agroecological approaches "transformative" in the sense that it would radically alter how crops are grown in the industrial agriculture system. Yet, we recognize agroecological practices do not shift every dimension of the current agri-food system. The productivist values that are foundational to the industrial system (Hendrickson and James 2005) can accord with and even justify the adoption of agroecological approaches, as has been shown with other "sustainable" approaches to agriculture (Guthman 2004; Jaffee and Howard 2010). Relatedly, agro-ecological transformation of the industrial agriculture system will not address other key flaws in the global agro-food system, such as the unequal distribution of food, power, and profits. Addressing these issues likely requires more widespread efforts to reform (or replace) the broader political economy of capitalist production in which the food system is embedded (Magdoff et al. 2000; McMichael 2009). Our findings suggest that it is unlikely many US row-crop farmers are interested in calling for these more radical transformations yet. But, given the accelerating economic and environmental contradictions of industrial agriculture (Blank 2018; Houser and Stuart 2020), farmers' critique of not only the industrial method of production but the

structural economic context it is embedded within, may emerge more forcefully. Future studies should build on our analysis by examining if these dynamics are giving rise to farmers' interest in more transformative critiques of and visions for the agro-food system.

Acknowledgements We would like to especially thank the farmers who participated in these interviews. Our work is in no way intended to reflect poorly upon them. Only to point out how current modes of thinking considerably shape their views and intended actions. Consciousness of the possibility for a new and better system is needed in our efforts to address societal environmental issues—something everyone (including the authors) must continue to develop. We would also like to thank Dr. Adam Reimer for conducting a number of these interviews in Michigan. Finally, Matthew Houser would like to acknowledge Dr. Elizabeth Grennan-Browning for her willingness to provide source material on agrarianism and to thank his co-authors for lending their expertise to this project.

Funding This work was supported by the National Science Foundation's (NSF) Dynamics of Coupled Natural and Human Systems program under Grant [1313677], the NSF's Kellogg Biological Station Long Term Ecological Research Site. Grant Number [DEB 1027253] and the Environmental Resilience Institute, funded by Indiana University's Prepared for Environmental Change Grand Challenge initiative.

Compliance with ethical standards

Conflict of interest The authors declare that they have no conflict of interest.

Ethical approval All procedures performed in studies involving human participants were in accordance with the ethical standards of the institutional and/or national research committee (include name of committee + reference number) and with the 1964 Helsinki declaration and its later amendments or comparable ethical standards.

Informed consent Informed consent was obtained from all individual participants included in the study.

References

Althusser, L. 1971. Ideology and ideological state apparatuses. In *Lenin and philosophy and other essays*, ed. L. Althusser, 127–186. New York: Monthly Review Press.

Altieri, M. 1989. Agroecology: A new research and development paradigm for world agriculture. *Agriculture Ecosystems and the Environment* 27 (1–4): 37–46. https://doi.org/10.1016/0167-8809(89)90070-4.

Altieri, M.A., and C.I. Nicholls. 2012. Agroecology scaling up for food sovereignty and resiliency. In *Sustainable agriculture reviews*, ed. E. Lichtfouse, 1–29. Dordrecht: Springer.

Archer, M., R. Bhaskar, A. Collier, T. Lawson, and A. Norrie. 2013. *Critical realism: Essential readings*. New York: Routledge.

Ashwood, L., D. Diamond, and K. Thu. 2014. Where's the farmer? Limiting liability in Midwestern industrial hog production. *Rural Sociology* 79 (1): 2–27.

Bell, S.E., A. Hullinger, and L. Brislen. 2015. Manipulated masculinities: Agribusiness, deskilling, and the rise of the businessman-farmer in the United States. *Rural Sociology*. 80 (3): 285–313.

Benford, R.D., and D.A. Snow. 2000. Framing processes and social movements: An overview and assessment. *Annual Review of Sociology* 26 (1): 611–639.

Bhaskar, R. 1998. Philosophy and scientific realism. In *Critical realism: Essential readings*, ed. R. Bhaskar, M. Archer, A. Collier, T. Lawson, and A. Norrie, 16–47. New York: Routledge.

Blank, S.C. 2018. The profit problem of American agriculture: What we have learned with the perspective of time. *Choices* 33 (3): 1–7.

Blesh, J., and L.E. Drinkwater. 2013. The impact of nitrogen source and crop rotation on nitrogen mass balances in the Mississippi River Basin. *Ecological Applications* 23 (5): 1017–1035.

Cassman, K.G., A. Dobermann, and D.T. Walters. 2002. Agroecosystems, nitrogen-use efficiency, and nitrogen management. *AMBIO: A Journal of the Human Environment* 31 (2): 132–141.

Cilia, L. 2020. 'We don't know much about Bees!' Techno-Optimism, Techno-Scepticism, and Denial in the American large-scale Beekeeping Industry. *Sociologia Ruralis* 60 (1): 83–103.

Conley, D.J., H.W. Paerl, R.W. Howarth, D.F. Boesch, S.P. Seitzinger, K.E. Havens, C. Lancelot, and G.E. Likens. 2009. Controlling eutrophication: Nitrogen and phosphorus. *Science* 323 (5917): 1014–1015.

DeLonge, M., and A. Basche. 2017. Leveraging agroecology for solutions in food, water and energy. *Elementa: Science of the Anthropocene* 5: 6.

Dentzman, K. 2018. "I would say that might be all it is, is hope": The framing of herbicide resistance and how farmers explain their faith in herbicides. *Journal of Rural Studies* 57: 118–127.

Dentzman, K., R. Gunderson, and R. Jussaume. 2016. Techno-optimism as a barrier to overcoming herbicide resistance: Comparing farmer perceptions of the future potential of herbicides. *Journal of Rural Studies* 48: 22–32.

Dentzman, K., and R. Jussaume. 2017. The ideology of US agriculture: How are integrated management approaches envisioned? *Society & Natural Resources* 30 (11): 1311–1327.

Devine, J. 2008. Beating a dead zone. National Resource Defense Council. https://www.nrdc.org/experts/jon-devine/beating-dead-zone

Drinkwater, L.E., and S.S. Snapp. 2007. Nutrients in agroecosystems: rethinking the management paradigm. *Advances in Agronomy* 92: 163–186.

Ellis, C. 2013. The symbiotic ideology: Stewardship, husbandry, and dominion in beef production. *Rural Sociology* 78 (4): 429–449.

Emery, S.B. 2015. Independence and individualism: Conflated values in farmer cooperation. *Agriculture and Human Values* 32 (1): 47–61.

[ERS] Economic Research Service. 2018. Fertilizer use and price. US Department of Agriculture, Oct 18. Retrieved from ers.usda.gov/data-products/ fertilizer-use-and-price.aspx

Faugier, J., and M. Sargeant. 1997. Sampling hard to reach populations. *Journal of Advanced Nursing* 26 (4): 790–797.

Flis, S. 2017. 4R nutrient stewardship and nitrous oxide losses. *Crops and Soils* 51 (1): 10–12.

Frison, E.A. 2016. From uniformity to diversity: A paradigm shift from industrial agriculture to diversified agroecological systems. Louvain-la-Neuve (Belgium): IPES, 96. https://hdl.handle.net/10568/75659

Gardezi, M., and J.G. Arbuckle. 2018. Techno-optimism and farmers' attitudes toward climate change adaptation. *Environment and Behavior* 52 (1): 81–105.

Gardner, J.B., and L.E. Drinkwater. 2009. The fate of nitrogen in grain cropping systems: A meta-analysis of ^{15}N field experiments. *Ecological Applications* 19: 2167–2184.

Gladwin, T.N., W.E. Newburry, and E.D. Reiskin. 1997. Why is the northern elite mind biased against community, the environment, and a sustainable future. In *Environment, ethics and behaviour: The psychology of environmental valuation and degradation*, ed.

M.H. Bazerman, D.M. Messick, A.E. Tenbrunsel, and K.A. Wade-Benzoni, 234–274. San Francisco: New Lexington Press.

Guthman, J. 2004. The trouble with 'organic lite' in California: A rejoinder to the 'conventionalisation' debate. *Sociologia ruralis* 44 (3): 301–316.

Hamblin, A. 2009. Policy directions for agricultural land use in Australia and other post-industrial economies. *Land Use Policy* 26: 1195–1204.

Hendrickson, M.K., P.H. Howard, and D.H. Constance. 2019. Power, food and agriculture: Implications for farmers, consumers and communities. In *Defense of farmers: The future of agriculture in the shadow of corporate power*, ed. J.W. Gibson and S.E. Alexander, 13–61. Lincoln, NE: University of Nebraska Press.

Hendrickson, M.K., and H.S. James. 2005. The ethics of constrained choice: How the industrialization of agriculture impacts farming and farmer behavior. *Journal of Agricultural and Environmental Ethics* 18 (3): 269–291.

Henke, C.R. 2008. *Cultivating science, harvesting power: Science and industrial agriculture in California*. Cambridge, MA: MIT Press.

Hofstadter, R. 1955. *The age of reform: From Bryan to FDR*, vol. 95. New York: Vintage.

Houser, M., and D. Stuart. 2020. An accelerating treadmill and an overlooked contradiction in industrial agriculture: Climate change and nitrogen fertilizer. *Journal of Agrarian Change*. 20: 215–237.

[ISDA] Indiana State Department of Agriculture 2018. Cover Crop and Tillage Transect Data. Indiana State Department of Agriculture. https://www.in.gov/isda/2383.htm

Jaffee, D., and P.H. Howard. 2010. Corporate cooptation of organic and fair-trade standards. *Agriculture and Human Values* 27 (4): 387–399.

Juchems, L. 2018. Iowa cover crop acres grow, but rate declines in 2017. Iowa State University. https://www.extension.iastate.edu/news/iowa-cover-crop-acres-grow-rate-declines-2017

Kanter, D.R., X. Zhang, and D.L. Mauzerall. 2015. Reducing nitrogen pollution while decreasing farmers' costs and increasing fertilizer industry profits. *Journal of Environmental Quality* 44 (2): 325–335.

Kloppenburg, J.R. 2005. *First the seed: The political economy of plant biotechnology*. Madison, WI: University of Wisconsin Press.

Kreuger, R.A., and M.A. Casey. 2009. *Focus groups: A practical guide for applied research*. New York: SAGE Publications.

Larrain, J. 1982. On the character of ideology Marx and the present debate in Britain. *Theory, Culture & Society* 1 (1): 5–22.

Levins, R.A., and W.W. Cochrane. 1996. The treadmill revisited. *Land Economics* 72 (4): 550–553.

Magdoff, F., J.B. Foster, and F. Buttel. 2000. *Hungry for profit: The agribusiness threat to farmers, food, and the environment*. New York: Monthly Review Press.

Malin, S.A., A. Mayer, K. Shreeve, S.K. Olson-Hazboun, and J. Adgate. 2017. Free market ideology and deregulation in Colorado's oil fields: Evidence for triple movement activism? *Environmental politics* 26 (3): 521–545.

Marx, L. 1964. *The machine in the garden: Technology and the pastoral ideal in America*. Oxford: Oxford University Press.

McMichael, P. 2009. A food regime genealogy. *The Journal of Peasant Studies* 36 (1): 139–169.

Méndez, V.E., C.M. Bacon, and R. Cohen. 2013. Agroecology as a transdisciplinary, participatory, and action-oriented approach. *Agroecology and Sustainable Food Systems* 37 (1): 3–18.

Montenegro de Wit, M., and A. Iles. 2016. Toward thick legitimacy: Creating a web of legitimacy for agroecology. *Elementa: Science of the Anthopocene*. 4: 000115. https://doi.org/10.12952/journal.elementa.000115.

Montgomery, D.R. 2007. Soil erosion and agricultural sustainability. *Proceedings of the National Academy of Sciences* 104 (33): 13268–13272.

Mooney, P.H., and S.A. Hunt. 1996. A repertoire of interpretations: Master frames and ideological continuity in US agrarian mobilization. *Sociological Quarterly* 37 (1): 177–197.

[NASS] U.S. Department of Agriculture-National Agricultural Statistics Service 2012. 2012 U.S. Census of Agriculture: United States Summary and State Data, volume 1.

Pannell, D.J. 2017. Economic perspectives on nitrogen in farming systems: Managing trade-offs between production, risk and the environment. *Soil Research* 55 (6): 473–478.

Plourde, J.D., B.C. Pijanowski, and B.K. Pekin. 2013. Evidence for increased monoculture cropping in the Central United States. *Agriculture, ecosystems & environment* 165: 50–59.

Ponisio, L.C., L.K. M'Gonigle, K.C. Mace, J. Palomino, P. de Valpine, and C. Kremen. 2015. Diversification practices reduce organic to conventional yield gap. *Proceedings of the Royal Society B: Biological Sciences* 282 (1799): 20141396.

Prokopy, L.S., J.S. Carlton, T. Haigh, M.C. Lemos, A.S. Mase, and M. Widhalm. 2017. Useful to usable: Developing usable climate science for agriculture. *Climate Risk Management* 15: 1–7.

Ribaudo, M., J.A. Delgado, L. Hansen, M. Livingston, R. Mosheim, and J. Williamson. 2011. *Nitrogen in agricultural systems: Implications for conservation policy*. Washington, D.C., USA: U.S. Department of Agriculture, Economic Research Service.

Ribaudo, M., M. Livingston, and J. Williamson. 2012. *Nitrogen management on us corn acres, 2001–10*. Washington, D.C., USA: United States Department of Agriculture, Economic Research Service.

Robertson, P.G., T.W. Bruulsema, R.J. Gehl, D. Kanter, D.L. Mauzerall, C. Rotz, and C.O. Williams. 2013. Nitrogen-climate interactions in US agriculture. *Biogeochemistry* 114: 41–70.

Robertson, G.P., and P.M. Vitousek. 2009. Nitrogen in agriculture: Balancing the cost of an essential resource. *Annual Review of Environment and Resources* 34: 97–125.

Roesch-McNally, G.E., J.G. Arbuckle, and J.C. Tyndall. 2018b. Barriers to implementing climate resilient agricultural strategies: The case of crop diversification in the US Corn Belt. *Global Environmental Change* 48: 206–215.

Roesch-McNally, G.E., A.D. Basche, J.G. Arbuckle, J.C. Tyndall, F.E. Miguez, T. Bowman, and R. Clay. 2018a. The trouble with cover crops: Farmers' experiences with overcoming barriers to adoption. *Renewable Agriculture and Food Systems* 33 (4): 322–333.

Saldaña, J. 2015. *The coding manual for qualitative researchers*. London: Sage.

Schlenker, W., and D.B. Lobell. 2010. Robust negative impacts of climate change on African agriculture. *Environmental Research Letters* 5 (1): 014010.

Sheriff, G. 2005. Efficient waste? Why farmers over-apply nutrients and the implications for policy design. *Review of Agricultural Economics* 27: 542–557.

Shrivastava, P. 1995. The role of corporations in achieving ecological sustainability. *Academy of management review* 20 (4): 936–960.

Smil, V. 2002. Nitrogen and food production: Proteins for human diets. *Ambio* 31: 126–131.

Smith, A. 1776/1976. An inquiry into the nature and causes of the wealth of nations. Chicago: University of Chicago Press

Smith, H.N. 1950/1978. Virgin land: The American West as symbol and myth. Cambridge. MA: Harvard University Press

Sprague, L.A., R.M. Hirsch, and B.T. Aulenbach. 2011. Nitrate in the Mississippi River and its tributaries, 1980 to 2008: Are we making progress? *Environmental Science & Technology* 45 (17): 7209–7216.

Stuart, D. 2009. Constrained choice and ethical dilemmas in land management: Environmental quality and food safety in California agriculture. *Journal of agricultural and environmental ethics* 22 (1): 53.

Stuart, D., R.C.H. Denny, M. Houser, A.P. Reimer, and S. Marquart-Pyatt. 2018. Farmer selection of sources of information for nitrogen management in the US Midwest: Implications for environmental programs. *Land Use Policy* 70: 289–297.

Stuart, D., and M. Houser. 2018. Producing compliant polluters: Seed companies and nitrogen fertilizer application in US corn agriculture. *Rural Sociology* 83 (4): 857–881.

Stuart, D., and R.L. Schewe. 2016. Constrained choice and climate change mitigation in us agriculture: Structural barriers to a climate change ethic. *Journal of Agricultural and Environmental Ethics* 29 (3): 369–385.

Stuart, D., R.L. Schewe, and M. McDermott. 2012. Responding to climate change: Barriers to reflexive modernization in US agriculture. *Organization & Environment* 25: 308–327.

Stuart, D., and M.R. Woroosz. 2013. The myth of efficiency: Technology and ethics in industrial food production. *Journal of agricultural and environmental ethics* 26 (1): 231–256.

Therborn, G. 1980. *The ideology of power and the power of ideology*. London: Verso.

[UCS] Union of Concerned Scientist. 2014. Healthy farm practices: Crop rotation and diversity. Union of Concerned Scientists, Food & Agriculture. https://www.ucsusa.org/food_and_agriculture/solutions/advance-sustainable-agriculture/crop-diversity-and-rotation.html

[USDA] United States Department of Agriculture Economic Research Service. 2006. Environmental effects of agricultural land use: The role of economics and policy. https://www.ers.usda.gov/publications/err-economic-research-report/err25.aspx.

[USGS] United States Geological Survey. 2010. The quality of our nation's water—Nutrients in the nation's streams and groundwater, 1992–2004. National Water-Quality Assessment Program Circular 1350. Washington, DC: U.S. Geological Survey.

[USGS] United States Geological Survey. 2017. The quality of our nation's waters. U.S. Geological Survey Circular 1225. https://pubs.usgs.gov/circ/circ1225/

Veenstra, J.J., and C.L. Burras. 2015. Soil profile transformation after 50 years of agricultural land use. *Soil Science Society of America Journal* 79 (4): 1154–1162.

Weinberg, A.M. 1966/1981. Can technology replace social engineering? Bulletin of the Atomic Scientists 22(10): 4–8.

Wines, M. 2014. Behind Toledo's water crisis, a long-troubled Lake Erie. *The New York Times*. https://www.nytimes.com/2014/08/05/us/lifting-ban-toledo-says-its-water-is-safe-to-drink-again.html

Wright, E.O. 2010. *Envisioning real utopias*. London: Verso.

Publisher's Note Springer Nature remains neutral with regard to jurisdictional claims in published maps and institutional affiliations.

Matthew Houser is an Assistant Research Scientist and Faculty Fellow at Indiana University's Environmental Resilience Institute. He is primarily interested in understanding the processes constraining or motivating societal response to environmental changes on how social contexts (e.g. policy, markets, socials norms) drive environmental inaction at the individual level (or the persistence of humans' environmentally/self-destructive behavior) through constraining behavioral change or limited the desire to act.

Ryan Gunderson is an Assistant Professor of Sociology and Social Justice Studies in the Department of Sociology and Gerontology and Affiliate of the Institute for the Environment and Sustainability at Miami University. His research interests include environmental sociology, the sociology of technology, social theory, political economy, and animal studies.

Diana Stuart is an Associate Professor in the Sustainable Communities Program and in the School of Earth and Sustainability at Northern Arizona University. Her work focuses on climate change mitigation and adaptation, agriculture, conservation, animal studies, political economy, and social theory.

Riva C. H. Denny is a Post-Doctoral Research Fellow in the School for Environment and Sustainability at the University of Michigan. Her research focuses on human decision-making in a range of environmental contexts and scales and includes agricultural conservation practice adoption, climate change perceptions, and food security.

Agriculture and Human Values (2020) 37:999–1012
https://doi.org/10.1007/s10460-020-10026-8

Agencing an innovative territorial trade scheme between crop and livestock farming: the contributions of the sociology of market agencements to alternative agri-food network analysis

Ronan Le Velly[1] · Marc Moraine[1]

Accepted: 18 February 2020 / Published online: 22 February 2020
© Springer Nature B.V. 2020

Abstract

The aim of this article is to show the relevance of the sociology of market agencements (an offshoot of actor–network theory) for studying the creation of alternative agri-food networks. The authors start with their finding that most research into alternative agri-food networks takes a strictly informative, cursory look at the conditions under which these networks are gradually created. They then explain how the sociology of market agencements analyzes the construction of innovative markets and how it can be used in agri-food studies. The relevance of this theoretical frame is shown based on an experiment aimed at creating a local trade scheme between manure from livestock farms and alfalfa grown by grain farmers. By using the concepts of the sociology of market agencements, the authors reveal the operations that are required to create an alternative agri-food network and underscore the difficulties that attend each one of these operations. This enables them to see the phenomena of lock-ins and sociotechnical transition in a new light.

Keywords Actor–network theory · Agroecology · Alternative agri-food networks · Crop-livestock integration · Marketization · Innovation

Abbreviations

ANT	Actor-Network Theory
FNAB	Fédération Nationale de l'Agriculture Biologique (French National Federation of Organic Agriculture)
GAB	Groupement d'Agriculture Biologique (Organic Agriculture Group)
INRA	Institut National de la Recherche Agronomique (French National Institute of Agricultural Research

✉ Ronan Le Velly
  levelly@supagro.fr

Marc Moraine
marc.moraine@inrae.fr

[1] UMR Innovation, Univ. Montpellier, Cirad, INRAE, Montpellier SupAgro, 2 place Viala, 34060 Montpellier, France

Introduction

Research into alternative agri-food networks, which began in the late 1990s, now makes up a substantial body of work in the field of rural studies (Goodman et al. 2012). This research has focused since its beginnings on the "alternativeness" of initiatives such as organic farming, fair trade, local produce, and short food supply chains. The initial aims of such research were to underscore the fact that these approaches could be answers to the many injustices in the dominant agri-food system (Kloppenburg et al. 1996; Renard 1999) and lay the foundations for a new rural development model (van der Ploeg et al. 2000; Renting et al. 2003). The researchers then delved deeper and went farther afield by looking at the phenomena of hybridization (Ilbery and Maye 2005) and conventionalization (Guthman 2004). So, alternative agri-food networks were no longer seen as separate worlds functioning totally independently and differently from conventional networks. Instead, their practices were acknowledged to be partly alternative and partly conventional (Hinrichs 2003; Kneafsey et al. 2008) and to face various difficulties maintaining their alternativeness, due in particular to the pressure exerted by players from conventional networks (Raynolds et al. 2007; Jaffee and Howard 2010).

To account for these phenomena of hybridization and conventionalization, the researchers worked on the actors' characteristics, for example, to determine whether organic farmers' motivations were different from those of conventional farmers and whether these motivations differed between pioneers and newcomers (Hall and Mogyorody 2001; Lockie and Halpin 2005; Best 2008). They also worked on the ways that alternative agri-food networks were organized, putting emphasis in particular on the roles of standards and certification (DuPuis and Gillon 2009; Raynolds 2009; Fouilleux and Loconto 2017) or showing the importance of processing and distribution infrastructures (Klein 2015; Clark and Inwood 2016). Finally, some of these studies led to analysis of the connections between alternative agri-food networks and the agri-food system's dominant operating patterns. They then described the lock-in and transition phenomena that could be observed through the theoretical framework of Geels's multi-level perspective (Geels 2004; see in particular Wiskerke 2003; Bui et al. 2016) and discussed the close ties between alternative initiatives and the dominant neo-liberal regime (Allen and Guthman 2006; Laforge et al. 2017).

This necessarily quick and incomplete overview attests to the diversity and richness of the work done to date. Yet, we noticed one thing on which relatively little research has been done, namely, how alternative agri-food networks are created. The aforementioned investigations often include sections relating the histories of the initiatives being studied, but these histories are only contextual elements required for the reader's comprehension. However, special stakes ride on working on alternative initiative creation processes. Michel Callon and Bruno Latour's sociology of innovation is instructive in this regard (Callon 1986, 1987; Latour 1987; Akrich et al. 2002; see also Kristensen and Kjeldsen 2016). It shows that creating a new product or organizational scheme requires a considerable mass of operations going from the definition of the problems to solve to the recruitment of allies in different social worlds, with the construction of prototypes in between. These creation phases are marked by uncertainty, for example, regarding how the technology will perform or consumers will behave. Only little by little does the action stabilize around rules and identities that become "black boxes" and are no longer challenged. What is more, the innovation process very often does not end in this stabilization, but in failure or abandonment. Whereas a retrospective report centered on successful innovations tends to lose sight of the uncertainties and difficulties encountered, the monitoring of innovations "in the making" that Callon and Latour promote makes it possible to grasp all the complexity of these processes.

In this article we shall place ourselves in this vein of work. We shall show that economic sociology research, which we refer to under the term "sociology of market agencements" and of which Callon is a key protagonist

(Callon 1998; Callon et al. 2002; Çalişkan and Callon 2010) allows in-depth analysis of the creation of alternative agri-food networks. This research has grown considerably over the past ten years and been the subject of numerous publications, especially in the form of special issues and collective works (Cochoy and Dubuisson-Quellier 2000; MacKenzie et al. 2007; Muniesa et al. 2007; Geiger et al. 2014; Cochoy et al. 2015, 2016; Kjellberg et al. 2015). Since just recently it has also been used to study agri-food networks, especially to reconstitute situations of innovation (Miele and Lever 2013; Ouma et al. 2013; Buller and Roe 2014; Hébert 2014; Ouma 2015; Le Velly and Dufeu 2016; Phillips 2016; Henry 2017; Henry and Prince 2018; Wang 2018, 2019). We shall thus devote a large part of this article to presenting this work.

This article will also make use of a particularly rich and original case study, that of an experiment that we conducted in Tarn-et-Garonne (France) aimed at creating a system of exchanges of manure and crops from livestock and grain farmers. The complementarity between cropping and livestock farming that has existed on farms for centuries is currently declining in France (Chatellier and Gaigné 2012). The modernization of agriculture has given rise to the spread of highly specialized farms that rely on chemical inputs. That has generated countless problems for the farmers (especially when it comes to controlling pests and maintaining soil quality) and environmental problems that are just as alarming (for example, the pollution of water by nitrates in areas heavily specialized in livestock) (Peyraud et al. 2014). One way to handle these crises is to reinstate a more agroecological model combining multicropping and livestock farming on these farms. However, many farmers deem transitioning to this model impossible because of the investments that must be made and skills that must be acquired (Meynard et al. 2013). An original way to overcome this hurdle is thus to create trade schemes between the livestock and crop farmers from the same area, whereby the former provide the latter with manure and slurry and the latter provide the former with feed for their livestock (forage, straw, and meslin). In this article we propose to consider such trade schemes alternative agri-food networks. That may be surprising, since these initiatives do not involve food consumption by humans. Yet they share the core characteristic of the alternative agri-food networks that are usually studied by researchers, namely, the ambition of establishing renewed production, exchange, and/or consumption relations that correct conventional agri-food networks' perceived malfunctions.

The rest of the article will be organized as follows: In the first part, we shall present the theoretical framework of the sociology of market agencements, stressing the way it can be used to study innovation in agri-food networks. In the second part we shall present the methodology of the experiment in which we engaged to build the territorial network of livestock and crop farmers in Tarn-et-Garonne. We shall see that this experiment was not a total success and did not culminate in a scheme of true exchanges between livestock and crop farmers. Far from weakening our arguments, this finding bolsters our idea that the creation of alternative agri-food networks is a difficult operation, the process of which deserves thorough analysis. Indeed, we shall perform this analysis in the third part of the article. In referring to the main concepts of the sociology of market agencements we shall describe the various operations that are required to create a new agri-food network, namely, identifying "overflowing" that justifies new "framings," defining the quality of the goods involved, producing references for "qualculation," organizing the "market encounter," setting prices, and finally "attaching" the actors to the new network. This will enable us to come back, in the conclusions, to the contributions that the theoretical framework makes to analyzing alternative agri-food network creation. We shall stress how it renews the ways sociotechnical lock-ins are analyzed.

The sociology of market agencements: a theoretical framework for analyzing agri-food network creation

A concern for innovation processes

The sociology of market agencements, inspired by the science and technology studies, pays particular attention to the innovation processes in markets. It is meaningful that Callon's research focuses on the "marketization" processes that allow markets' creation rather than on markets as already constituted objects (Çalişkan and Callon 2010). The same can be said for the articles that underscore the importance of analyzing the "agencing of markets" (Onyas and Ryan 2015; Cochoy et al. 2016), "market work" (Cochoy and Dubuisson-Quellier 2000; Mason et al. 2017), "market design" (MacKenzie 2009), or "market innovation" (Kjellberg et al. 2015). The vocabulary changes from one text to the next, but the desire to understand how new markets take shape remains.

The study of marketization processes has two goals. The first one is to underscore the mass of operations necessary for markets to form. In applying the sociology of market agencements to analyze the development of mango export markets in Ghana, Stefan Ouma and his colleagues write that this work makes it possible to "de-essentialize markets, unveil their often messy and compromised construction from below, and reconstruct the diversity of legitimate arrangements that finally evolve from the encounters of a diverse range of market makers" (Ouma et al. 2013, p. 234). This finding is borne out in other studies mobilizing this sociology. The fragility or failures of innovation processes are thus underscored convergently for the construction of lettuce

export markets between Taiwan and Japan (Wang 2018), development of tools to identify products that respect animal welfare (Miele and Lever 2013), and shifts in the operations of a commercial salmon fishery in southwest Alaska (Hébert 2014). As the end of the preceding quotation already attests, the second aim of studying marketization processes is to put emphasis on the very great diversity of actual markets that can result from market formation processes. Marketization can thus be motivated by very diverse ends. For example, it is possible to test new market agencements to reduce greenhouse gas emissions (MacKenzie 2009) or improve the accessibility of health for all (Geiger and Gross 2018).

When it is applied to agri-food studies, the sociology of market agencements thus enables one to challenge the stable ontologies of the market, economy, or capitalism that can be observed in economics but also in critical political economy (Ouma 2015; Henry and Prince 2018). This is not new, either: It already happened with Actor–Network Theory (ANT), which inspired the sociology of market agencements (Busch and Juska 1997). Seen in this way, the market cannot be defined as the impersonal, self-seeking, competitive space that has come in for as much praise from neoclassical economists as denunciation by the critics of commodification. "Markets organize the conception, production and circulation of goods, as well as the voluntary transfer of some sorts of property rights attached to them," Çalişkan and Callon write more openly (Çalişkan and Callon 2010, p. 3). This definition does not rule out the possibility that actual markets may take the form of neoclassical economic models (Callon 1998), but it does not exclude their taking any other form, either. Note as well that the scope of the definition includes the conception and production of goods. To understand why, it is useful to clarify the link with ANT.

The connection with Actor–Network theory

ANT (for an introduction, see Latour 2005) has inspired the sociology of market agencements strongly. This is particularly true for three of its major teachings.

The first one concerns taking non-human entities into account in analyzing marketization processes. The sociology of market agencements has thus put great emphasis on the action of what it calls "market devices" (Muniesa et al. 2007). The creation of new markets and their ability to achieve the ends that their promoters give them depend on material devices such as standards and labels, logistic infrastructure and points of sale, mathematical algorithms and quality assessment guides, and so on (Karpik 1996; Kjellberg and Helgesson 2007; Cochoy 2008). The problem for rural studies is then to grasp the "materiality" of agri-food markets. The sociology of market agencements is close in this respect to the studies that have shown the impacts of standards and labels (Raynolds 2009; Fouilleux and Loconto 2017) and the importance of distribution infrastructure (Ilbery and Maye 2005; Klein 2015) in alternative agri-food networks. However, in taking up the matter of materiality systematically, it also makes it possible to look at other market devices, the importance of which is less well known, such as packaging (Phillips 2016), charts of price lists (Henry 2017), displays urging farmers to save (Onyas et al. 2018), and so on.

A second set of non-human entities must be spotlighted, namely, natural entities. To tell the truth, these entities are less present in marketization research than material devices are. Yet, when it comes to agri-food markets, attention must be paid to them. In taking the actions of these natural entities into account, the sociology of market agencements, like ANT more generally, sheds an original light on things compared with other theoretical frameworks used in rural studies (Goodman 1999; Dwiartama and Rosin 2014; Henry 2017; Wang 2019). Agricultural crops, be they animal or vegetable, are living things that cannot be marketed willy-nilly: they are usually seasonal; they are perishable, if not hazardous when improperly stored; and they are characterized by greater variability and unpredictability than industrial products (Bernard de Raymond et al. 2013). In ANT terms, they act, they impact marketization. The actions of other natural entities, such as insects and bacteria, are also vital, to the point that their behavior must often be kept in check to succeed in creating a market. It is not by chance that rural studies articles referring to the sociology of market agencement so frequently underscore the importance of packing that keep produce cool or refrigerated transport infrastructure (Ouma 2015; Le Velly and Dufeu 2016; Phillips 2016; Henry 2017). These objects, which could seem insignificant, occur at the intersection of the actions of the human beings (who produce the devices), material devices (which keep the food's properties intact), and natural entities (which must be channeled).

The second teaching, which has already been affirmed in science and technology studies (Callon 1987; Akrich et al. 2002), is that market innovation is a process that creates supply, demand, and the market in the same movement. The human and non-human entities that slip in between the producers and consumers act as "mediators" (Latour 2005) that modify these two opposite ends of the network. The emphasis is then put on all the operations, such as taste tests for food products, by means of which the supply adjusts to what it perceives to be the demand (Callon et al. 2002). Even more obvious examples are those of guidebooks such as the *Michelin Guide* and purchasing guides drafted by environmental groups, which act on the demand by educating and equipping consumers but also act on the supply by establishing reference frames of practices that consumers are assumed to expect (Karpik 2000; Dubuisson-Quellier 2013). Talking about "market mediation" thus amounts to

saying that market agencements are not simple meeting places between pre-existing blocs of supply and demand but sociotechnical arrangements, the different components of which define each other (Cochoy et al. 2016). When it is applied to the creation of alternative agri-food markets, this concept then enables one to underscore that producers' characteristics and/or consumers' expectations must not be considered to be prior givens for innovation, but rather elements that will be defined as the marketization process advances (Le Velly and Dufeu 2016; see also Lockie and Kitto 2000; Wang 2019).

The third teaching derives from the first two. Callon explains that he chose the French term "agencements" rather than "arrangement" because it is closer to "agency." A market agencement is a sociotechnical arrangement that is capable of productive and market action (Çalişkan and Callon 2010). This presentation has the advantage of underscoring the distributed nature of agency: The ability to act is never restricted to humans alone. One implication of this is that understanding the motivations or attitudes of alternative agri-food networks' actors, as is usually done in research (for example, Best 2008) does not suffice to understand their actions. The sociology of market agencements forces one to take the actions of the material devices and natural entities in the networks into account as well. What an alternative agri-food network is capable of doing to guarantee remunerative prices for farmers or improved welfare for livestock, for example, is the fruit of a network composed of human, natural, and material entities (Buller and Roe 2014; Le Velly and Dufeu 2016). Speaking of the agencing, as we do in the title of this article, thus means envisioning an operation that creates a new entity with the ability to act.

Six marketization processes

The sociology of market agencements has listed the operations that are required for agencing new markets with great care. Several proposals have been made in this regard (see, for example, Kjellberg and Helgesson 2007). We, for our part, shall rely on the one made by Callon (Çalişkan and Callon 2010; Callon 2017), who breaks agencing down into six processes, which we shall present in four stages.

Çalişkan and Callon (2010) call the first process "pacifying goods." Before being tradable, a good must acquire well-identified qualities that enable the market players to have expectations about its characteristics. This is particularly important for agricultural goods, which are biological entities characterized by great variety (Henry 2017). Without standards for the generic characteristics of agricultural commodities or more specific qualities linked to one or the other option of sustainable agriculture, the large-scale trade in these commodities as we know it could not exist (Ouma 2015; Onyas et al. 2018). Symmetrically, a failure to design new quality standards hobbles the development of a new market (Miele and Lever 2013). The second process, which is closely linked to the first one, concerns the construction of "qualculating agencies." The term "qualculation" was coined by Cochoy (2008) to underline the continuity between operations of numerical calculus and qualitative judgment. For the sociology of market agencements, these two types of operation belong to the same logic of ranking things. Both are also the fruit of distributed cognition: Market players never do their computations alone; they do them with the cognitive equipment of their sociotechnical networks.

The third and fourth processes, namely, ensuring "market encounters" and "price setting," are often linked. Research on financial markets has been particularly useful in revealing the agencing work that makes encounters between supply and demand and market price setting possible. It has shown in particular that economics has not just described these markets, but has done a great deal in recent years to configure them (MacKenzie et al. 2007; Muniesa et al. 2007). Ouma has likewise shown how the supply chain management approach—an operational application of critical research on global commodity chains—has supported and oriented the structuring of agricultural commodity export markets in the countries of the Southern Hemisphere (Ouma et al. 2013). The sociology of market agencements also leads to comprehension of the vital role of sales intermediaries. These actors, with their knowledge and infrastructure, do not just move products from production to consumption. They play a vital mediating role by enabling products to be available to consumers under good conditions, solving seasonality problems, guaranteeing compliance with quality standards, and so on (Le Velly and Dufeu 2016; Wang 2018). A device such as the price schedule setting a list of prices for various qualities of lamb in New Zealand is also a good example of this (Henry 2017). Matthew Henry shows that this device is central to the creation of a national and export market for this type of meat, but he also shows the huge amount of work needed to design it as well as the controversies associated with its content.

The fifth process, which is absent from the text written with Çalişkan but was developed in Callon's 2017 book (and in Callon et al. 2002), concerns the way "attachments" between the sellers, their products, and their customers are created. This has been the subject of many articles, notably to understand how trade loyalties form and dissolve (Cochoy et al. 2015; Le Velly and Goulet 2015). Beyond this, other attachments are often necessary for new market creation. Depending on the case, the market's promoters must also convince banks to finance them, government authorities to change regulations, scientists to back the proposed changes, and so on (Kjellberg and Helgesson 2007; Doganova and Karnøe 2015; Mason et al. 2017). The attachment processes then strongly recall the processes of enrollment and

mobilization such as they have previously been described in the sociology of innovation (Callon 1986; Akrich et al. 2002).

Finally, through the term "market maintenance," Çalişkan and Callon (2010) stress that marketization is a process that combines stability and change (see also Kjellberg et al. 2015). The first five processes stabilize market relations, but they may be constantly challenged. In this connection, Callon stresses the centrality of the dynamics known as "framing" and "overflowing." A market agencement is stable when the marketization processes under way are not controversial. Yet this situation is fragile nonetheless, for the qualities chosen to pacify goods, elements taken into account to "qualculate," conditions of the market encounter, price-setting mechanisms, and established attachments are always partial and biased. They can then be the subjects of controversy and generate changes in the market agencements (Callon 2009; Miele and Lever 2013; Henry 2017). The desire to create alternative agri-food networks can moreover be analyzed as a will to contest the existing marketization processes. Highlighting instances of overflowing, as fair trade activists do when they call for fair payments for small farmers, boils down to challenging the framings that have been done. In other words, the introduction of new "concerns" leads their proponents to contest the stability of the existing market agencements (Geiger et al. 2014; Geiger and Gross 2018).

Methodology

A sociological analysis of an "in vivo experiment"

The case that we shall mobilize in the next part of this article is an in vivo experiment (Callon, 2009) to create a territorial system of exchanges between organic livestock and crop farmers in Tarn-et-Garonne (France). This experiment was conducted by the second author between 2013 and 2016 under a PhD thesis on which he was working at the French National Institute of Agricultural Research (INRA). His aim was by no means to test one or the other sociological analytical frame, but to work on designing ways to increase the territorial integration of crop farming and livestock operations. This is an issue that arose in agricultural research in the late 2000s (see Martin et al. 2016). The sociological analysis proposed in this article was thus done afterwards.

To do this analysis, the first author relied primarily on the second author's research reports. He then supplemented this data by conversing with the second author. The main limitation of this article is then that it is strongly influenced by the second author's experience and the first author's analytical grid. Regarding the first point, our approach can be perfected. It would have benefited from being bolstered by additional interviews to capture the players' feelings at the moment. It is nevertheless compatible with the methodological orientation of ANT, which gives priority to monitoring innovation process "in the making" ethnographically (Onyas et al. 2018). Let us also point out that experiments, whether in a laboratory or "in vivo," are classic subjects of ANT (Callon 1986; Kristensen and Kjeldsen 2016) and the sociology of market agencements (MacKenzie et al. 2007; Callon 2009). The methodology of this article belongs to this tradition for this reason as well. Regarding the second point, we take full responsibility for our choice of a specific analytical grid. Our aim is not to reveal various possible interpretations of this experiment, but to show the relevance of the sociology of market agencement to studying the creation of alternative agri-food networks. Describing in detail the steps and vicissitudes of this experiment will enable us to show how new ways of organizing agri-food networks develop and underscore the magnitude of the practical operations that this requires.

A four-step participatory design

Before proceeding with this analysis, we feel it is useful to give a quick description of the history of this experiment. It was initiated in 2013 by the second author, who was a PhD student in agronomy at the time, as part of a European research project involving several INRA researchers. It involved ten crop farmers and fourteen livestock farmers who were engaged in organic farming and belonged to Groupement d'Agriculture Biologique (Organic Agriculture Group) Bio 82 (GAB Bio 82), an association working to promote organic agriculture in Tarn-et-Garonne Department. These farmers volunteered to work with the researchers in order to think about developing close cooperation between their farms. The researchers' main aim was to contribute to the emerging field of research on territorial integration between cropping and animal husbandry, with the secondary aim of developing tools that would be used by other groups of farmers. Finally, the investigators also felt it was important to implement participatory methods that would involve the farmers in order to integrate their objectives, preferences and constraints into the design process.

The first step in this participatory design exercise was to hold a workshop with all twenty-four farmers in order to determine the issues and goals to target. The researchers wanted to work on in-depth forms of collaboration between the livestock and grain farmers, but the actual forms that such collaboration would take was not set at the outset. This workshop first revealed that the livestock farmers and grain farmers were united in their desire to diversify and make their farms more self-sufficient. Yet, all of these farmers were also of the opinion that they could not envision developing mixed multicropping and livestock models on their farms due to a lack of skills and land (many were working

small acreages that did not belong to their families). Next, the grain farmers explained that they found it interesting to add legume forage crops (alfalfa, clover, etc.) in order to improve soil fertility and further organic pest control, but did not do so because they had problems selling those crops. They also explained that the organic regulation-compliant fertilizers that they bought to fertilize their fields were very expensive and often of unknown origin. They thus thought that replacing them with manure coming from the neighborhood would make sense. The livestock farmers, for their part, likewise stressed that buying feed for their stock was very expensive in organic farming. What is more, feeding their stock with soybeans imported from other countries seemed to be highly inconsistent with the short supply-chain models that most of them were developing for their products. At the end of these exchanges the farmers in both groups agreed that it would be relevant to develop trade with each other, with stable prices and quantities that would enable them to plan their production levels.

In the second step, the doctoral student then took control to draw up a scenario that would make it possible to achieve these goals. To do that, he conducted individual interviews to determine each livestock and grain farmer's needs as well as what each would be willing to exchange within the group. Then, based on these findings, he did a computer simulation of a scenario that included (i) growing alfalfa and legumes on the grain farms, (ii) selling alfalfa to the livestock farmers, and (iii) producing composted manure on the livestock farms and selling it to the grain farmers. This scenario met the farmers' expectations and tied in with the researchers' knowledge: Adding alfalfa and manure to the grain farming systems adds nitrogen and organic matter to the soil, reduces the incidence of disease, and makes controlling weeds easier. Alfalfa, for its part, is a high-protein crop that can reduce reliance on soybean cattle cake. Finally, composting manure facilitates its transport and reduces the risk of introducing the seeds of adventitious plants.

The aim of the third step was to discuss ways of organizing the exchanges. The PhD student proposed three main ways of organizing these exchanges (which we shall describe in the next section) at a half-day workshop with the farmers. The farmers weighed the pros and cons of each scheme, and then came to an agreement on the "polycentric model" and split up into several territorial subgroups tasked with finding practical solutions to ordering and delivery issues.

The fourth step in this participatory design phase was to do a multifactor evaluation of the sustainability that would result from the new organization of farming and trade. To do this, the PhD student and farmers agreed on an analytical grid containing five fields: the metabolic efficiency of the system, the production of ecosystem services for agriculture, socioeconomic performance, increases in the farmers' knowledge, and the social and territorial embeddedness of the farming operations. A series of indicators was developed for each of these fields for a total of thirty-one in all (see Table 1). Then these indicators gave rise to a scoring system to compare the baseline situation with the situation that the new system might generate. For the first two fields this score was based on quantitative references from agricultural research. For the other three fields the assessments took a more qualitative turn and were based on discussions with the farmers. This work resulted in figures that revealed the benefits to be gleaned from the new organization in each of the five fields (see Moraine et al. 2017).

Two hapless attempts to implement the experiment

At the end of this work, the farmers thus agreed on a roadmap for setting up the new agri-food network. This roadmap spelled out the changes to make in both production planning and the cropping and animal husbandry systems on the farms. It also established a preferential way to coordinate the exchanges that was suited to small territorial groups. Finally, it assigned the role of running the exchanges for the coming year to GAB Bio 82's technical adviser.

However, not everything transpired as expected. The changes planned for 2015 were not made. So, the farmers and researchers found that many organizational issues had not been well-enough prepared for. In some groups of farmers, responsibility for haying the alfalfa and composting the manure had not been clearly established. Transport was another, particularly ticklish, point: Some of the farmers felt that the transportation costs linked to the experiment were too high; others simply were unable to find transportation, since certain hauliers refused to take manure, even after it was composted.

Table 1 Concerns, criteria, and indicators used for the ex-ante assessment of the territorial trade scheme between livestock and crop farmers

Concerns	Criterion	Indicator
Metabolic efficiency of the system	Crop systems	Amount of exogenous N-source fertilizers (t N/year)
	Livestock systems	Amount of exogenous fodder (t/year)
		Amount of exogenous concentrates (t/year)
		Amount of exogenous straw (t/year)
Production of ecosystem services	Organic manure application	Area of arable land receiving organic manure (ha)
	Symbiotic fixation of N	Percentage of legume crops in the crop rotation (%)
	Diversity of crops at field level	Duration of crop rotations (years)
		Number of botanical families
	Diversity of land use at landscape level	Abundance of grasslands in landscape*
Socioeconomic performances	Workload	Amount of work*
	Work quality	Difficulty of work*
	Stability of costs	Stability of supply and prices*
	Added value of products	Development of quality labels*
		Direct sales and collective shops*
		Use of by-products*
	Profitability	Gross margins in crop rotations (€/ha)
		Costs of animal feeding systems (€)
Social learning and capacity building	Autonomy of farmers	Independence from commercial organizations*
		Institutionalization of groups*
		Structure of exchanges*
	Knowledge capitalization	Exchange of practices and results of trials*
	Adaptive capacity	Strategic planning and tactical adaptation to annual conditions*
Social and territorial embeddedness	Social acceptability of agriculture	Landscape quality*
		Direct producer–consumer relationships*
		Animal welfare*
		Quality of products*
	Contribution to local economic dynamism	Tourism activities*
		Development of local supply chains and new activities*
	Contribution to local and global sustainability issues	Establishment of new organic farmers*
		Conversion to organic farming*
		Impact of farming on water quality*

Asterisks denote indicators that were estimated by producers qualitatively. Other indicators are quantitative indicators estimated at the individual level and averaged for the group. Adapted from Moraine et al. (2017)

In the face of this failure, the decision was made in 2016 to concentrate on a single group of two crop and four cattle farmers who were no more than 20 km from each other and volunteered to go on the experiment (Ryschawy et al. 2017). Two preliminary meetings were held to adjust the scenario. The farmers reached an agreement on the crop rotations needed to grow alfalfa and feed corn, which were both intended for sale to the cattle farmers. Price adjustments were also made to allow for the farmers' unequal involvement in the storage and delivery tasks. One last meeting was held before sowing the winter crops to allow for the year's weather conditions. These compromises seemed acceptable during the discussions, but, ultimately, not all the farmers stuck to their commitments. One grain farmer failed to deliver grain to a livestock farmer in the project, which created tension between the farmers. In addition, GAB Bio 82 lacked financing during this period and its technical adviser ultimately was not able to participate fully in the project. She was unable to guarantee that each participant met his commitments and to get after those who were remiss.

Analyzing the conditions and difficulties of "agencing" alternative agri-food networks

How can we explain these difficulties, if not these failures? A first response, which can be found in the agronomy literature (Meynard et al. 2013; Martin et al. 2016; Asai et al. 2018), is inspired by research on sociotechnical transition (Geels 2004). It consists in underscoring the constraints that come from the institutional rapports in the dominant sociotechnical regime. One then talks about a "lock-in" to underscore the fact that institutional rapports built in the past hobble the development of innovative types of organization, even if the latter seem to be more satisfactory. Agricultural research on crop-livestock integration has thus shown repeatedly that the knowledge, networks, and equipment that farmers have due to their inclusion in conventional agri-food networks make it difficult to establish complementarity between cropping and livestock farming. Next, depending on the case, other types of lock-in have been identified. These include criteria of professional excellence that do not value this complementarity, agricultural policies conducive to specialization in specific crops, and very demanding health regulations (Meynard et al. 2013; Martin et al. 2016; Asai et al. 2018).

Explaining these difficulties by the lock-ins that come from the dominant sociotechnical regime obviously is not unfounded. Yet, the sociology of market agencements enables one to analyze things somewhat differently. Rather than envisioning the weight of external constraints on the innovation process, it enables one to develop an analysis that focuses on the innovation process itself. Rather than explaining the difficulties of innovation by the lock-ins produced by the existing system, it shows that creating something new is difficult in itself. We shall take this perspective in reconstructing the marketization processes that were carried out one by one.

Introducing new concerns

The sociology of market agencements beckons the researcher to analyze the experiment conducted as being motivated by the criticism of the framings that stabilize conventional agri-food networks and the inclusion of new concerns (Geiger et al. 2014). In the case under study, this work was largely accomplished in the first participatory workshop involving the twenty-four volunteer farmers and the researchers. In this first step the farmers came to an agreement on what, in their view, was problematic and the aims of the experiment. Stressing this is important. Launching a process to change market mediations does not consist in finding solutions to pre-existing problems. The creation of alternative agri-food networks benefits from being understood to be an operation in which problems as well as solutions are identified. This teaching, which was already present in Callon's first writings on the sociology of innovation (Callon 1986), can be reiterated in experiments aimed at constructing new market agencements (Callon 2009).

The participatory workshops and, more generally, the participatory design method that we chose bring to mind what Callon calls a "hybrid forum" (Callon et al. 2002). Callon uses this term to describe situations in which the actors discuss market agencements explicitly. Their hybridness refers first of all to the fact that the concerns included in the discussions can be economic, environmental, moral, ethical, etc., with all these dimensions often being intertwined. It also concerns the participants in these forums. Hybrid forums are deliberative arenas hosting a heterogeneous set of actors: farmers, consumers, government bodies, scientists, and so on, but also the natural entities, such as the soil, climate, and ecosystems, for which these actors are the spokespeople.

The range of actors in the participatory workshops of our experiment was not that broad. The grain and livestock farmers close to GAB Bio 82 were the only ones to be invited and take part in the discussion with the researchers. In particular, there were no plans to have agricultural cooperatives, the firms that supplied agricultural inputs, and hauliers—all entities that might have proposed other organizational solutions—to take part. Still, one must keep in mind that the farmers and researchers did not come to these workshops alone and their sociotechnical networks affected the choice of concerns on which they settled. The researchers brought information that came out of their sociotechnical network, i.e., scientific publications that highlighted the benefits of crop-livestock integration. The farmers, for their part, regalvanized some concerns that were highly valued in the national network to which GAB Bio 82 belongs, that is, the French National Federation of Organic Agriculture (FNAB). The will to create an alternative network conducive to crop diversification, territorial embeddedness, and the farms' self-sufficiency corresponded to the introduction of concerns that were closely tied to this affiliation. Let's stress the farms' self-sufficiency. It was important for both groups of farmers to increase their independence from both input suppliers and the cooperatives that distributed their produce via this experiment. They reproached these suppliers and distributors for forcing commodity chains over which they had no influence on them, especially when it came to the prices practiced. We thus understand better why these actors were not invited to take part in the "hybrid forum."

A hybrid "qualculation" between alternative and conventional

Hybrid forums do not aim solely to establish a dialog about the actors' concerns. The idea is indeed to include these concerns in the market framings. This perspective leads one to consider that operations making it possible to renew the areas of qualculation are core operations for creating alternative agri-food networks. Being attentive to these processes enables one to take a relatively clear-eyed view of things. This in turn makes it possible to establish the points on which the new agencement generates alternativeness but also the points on which it maintains existing assessment modes. So, in focusing attention on the modalities of qualculation, the sociology of market agencements gives an original boost to the traditional issue of alternative-conventional hybridization (Hinrichs 2003; Ilbery and Maye 2005; Kneafsey et al. 2008). In alternative agri-food networks, qualculation is hybrid: it combines innovative-alternative references and existing-conventional references. This hybridization is clearly visible in Steps 2 and 4 of the participatory design work concerning the scenario and ex-ante evaluation of performance, respectively.

The scenario drawn up in Step 2 thus strove to include new concerns, such as the farms' territorial embeddedness or autonomy (preferring feed based on local alfalfa rather than South American soybeans). However, it also strove to maintain some existing performance levels. So, the scenario defined the production and exchange levels that would maintain both the nutritional quality of the livestock's diet and the fertility of the grain farmers' land. For the first point, the initial and projected inputs were evaluated in energy and protein equivalents according to references produced by the French National Institute of Agricultural Research. For the second point, the calculations concerned the nitrogen inputs taking the inputs provided by the introduction of alfalfa and legumes and the inputs from the composted manure into account.

The indicators used in Step 4 confirm this (see Table 1). On the one hand, indicators of dependence on outside inputs and marketing structures were added to reflect the new concerns. For the production of ecosystem services they concerned the degree of crop diversification, lengths of crop rotations, etc. Similarly, indicators of dependence on outside inputs and marketing structures were created to assess the concern for autonomy or self-sufficiency. However, on the other hand, some classic indicators of conventional agri-food networks were maintained. This was the case for soil nitrogen fertilization levels, the profit margins per hectare (in the case of vegetable crops), and feed costs (in the case of livestock). What is more, to evaluate the more quantitative indicators, the researchers relied on mean sectoral references. As Donald MacKenzie also pointed out in his study of carbon emissions market creation, agencing a new market can be done without re-opening all the "black boxes," *i.e.*, contesting all the conventions required to construct the computation and equivalences (MacKenzie 2009). Some are modified or added, with the aim of generating alternativeness, while others are maintained.

Tricky nature pacification operations

The farmers' choice to exchange alfalfa hay for composted manure is also highly significant. It attests to the core nature of the goods quality adjustments to be made. An alternative agri-food network is not created around goods whose qualities, *i.e., characteristics*, are known beforehand; on the contrary, the process entails coming gradually to an agreement on the qualities (Callon et al. 2002). The decisions to dry the alfalfa and compost the manure both affect the conditions under which the goods are produced with the aim of making them easier to store and to transport, but also to make them suitable for the user farms. Several stakes ride on this: guaranteeing nutritional inputs for the livestock and nutrient inputs for the soil, but also preventing the manure from disseminating weeds to the grain fields.

The fact that manure and alfalfa are living things or come from living things explains part of the difficulties encountered. Manure and alfalfa must be "domesticated" (Callon 1986) so that they do not act contrary to the aims of the agencement. They must be "pacified" to become merchandise. The weeds that manure can introduce when it is spread on fields come from the seeds present in the cattle's excrement. These seeds, which were initially in the grass or hay in their rations, are not destroyed by digestion. They are destroyed by the composting step, where the temperature usually reaches 60 °C. Alfalfa can be sun-dried out in the field after it is cut or dried in a barn with the possible addition of a heat source. Drying is necessary to prevent fermentation and rotting once the alfalfa is stored. However, this in-field drying has the disadvantage of losing the leaves, which contain the bulk of the plant's nutrient value. Producing good quality alfalfa hay is thus no mean feat: You have to know how to measure out the drying or have a barn with drying facilities.

The difficult recreation of market encounters and price setting

Paying attention to the processes making it possible to match supply and demand and set prices is also useful for analyzing alternative agri-food network creation. These processes are far from ancillary. The problems of identifying the available supply, storage, transportation, and price setting must be solved for the trade to take place. Having farmers willing to sell and consumers willing to buy is not enough for an alternative agri-food network to take shape. All the market mediation conditions must be met, too. When a market exists and operates stably, these elements may appear to be minor at first glance. However, as our experiment shows, when a new market agencement is being established, they are quickly seen to be core elements.

To show this, we can start by going back to Step 3 in designing the experiment. Once the production and trade scenario was established, the doctoral student and GAB Bio 82's technical adviser conducted a workshop to arbitrate among three models for organizing the exchanges: (i) a multirelational model along the lines of "classifieds," with a website for posting offers and requests and bilateral exchanges between livestock and grain farmers; (ii) a completely centralized model including investment in shared packaging, warehousing, and transportation facilities and equipment; and (iii) a polycentric model involving small groups of farmers who were geographically close to each other and would decide on and refine the ways they were organized locally (*Cf.* Fig. 1).

To help them choose, each farmer was asked to write down the five strengths and five weaknesses of these three scenarios on post-its. The compilation of these strengths and weaknesses and following debate resulted in the group's opting for the third formula, which was deemed to be a good compromise between the long-term continuation of the exchanges and development of relations of trust (difficult to create in the "classifieds" model), on the one hand, and ease of implementation (as the centralized model required excessively high investments and commitment) on the other hand.

Nevertheless, many questions remained hanging at the end of this workshop. In theory, each group was supposed to find practical solutions for the transport, storage, and pricing issues, but also for composting the manure and drying the alfalfa. In actual fact, the failure to clarify all of these points led to these exchanges' not taking place during the first attempt to implement the scheme. Transportation was a crucial factor here. Some of the farmers envisioned doing the job themselves, but they failed to reach agreement with their colleagues on the amount of financial compensation that they would get in exchange. Others found that the professional hauliers refused to transport manure, even when it was composted, simply because it was dirty merchandise requiring the trucks to be cleaned afterwards. This last example shows that the scope of actors whose behaviors must adjust in a new market agencement can be quite broad. In the case at hand, the farmers were unable to "enroll" the hauliers (Callon 1986), thereby rendering the otherwise so conscientiously developed scenario null and void.

Stressing the importance of these marketization processes forces one more generally to take the intermediaries' roles in both alternative and conventional agri-food networks seriously. The movements that promote alternative networks regularly accuse intermediaries of being useless parasites that enrich themselves at the farmers' expense. There is no denying that the intermediaries in many cases do capture a

Fig. 1 Schematic representation of the three considered models for organizing market encounters

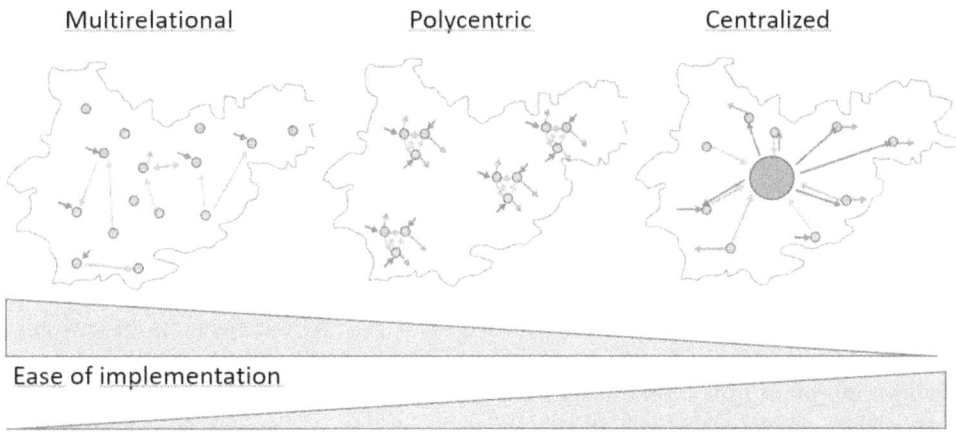

large share of the value chain. On the other hand, the sociology of market agencements reveals that these intermediaries often have central roles in creating market relations. Through their work, they make complex adjustments between production and consumption, for example to allow for seasonality. They are also very often the ones that define and guarantee the quality of produce in order to make it "qualculable" for the other operators. The ease of using the long market chains that is the daily practice of the majority of farmers and consumers in the industrialized world is the result of the mediation by middlemen. So, one should not be astonished to see that it is sometimes harder to sell or buy in short supply chains than in long ones. The former, like those studied in this article, often must be forged, whereas the latter are already established, broken in, and stabilized.

No new agencements without new attachments

Starting with the list of marketization processes, we can finally focus on the dynamics of detachment and attachment. The creation of an alternative food network assumes that detachments and attachments take place: Some actors must detach themselves from certain entities of conventional networks and attach themselves to new entities. According to Callon (2017), participatory design processes such as the one that was used in our case are particularly effective for generating such movement. By permitting each party to express their expectations and constraints, they foster the learning and adjustments that ultimately allow detachments to take place and new attachments to stabilize. Yet in the case that we studied, that clearly was not enough.

On paper the farmers agreed to take the leap. What is more, during the second attempt to implement the experiment (with only six farmers), a host of organizational details had, in principle, been well defined. Yet, the experiment failed when it came time to put things in practice. The failure of one of the grain farmers to deliver grain to one of the livestock farmers in the project sufficed to capsize this second attempt. At the time, this farmer had spoken of "forgetting" and we unfortunately know nothing more about his motivations. Such an observation nevertheless enables us to assert, in an echo of the conclusions drawn in several recent publications, that "agencing and controlling markets are far from being synonymous" (Cochoy et al. 2016, p. 11; Geiger and Gross 2018). Building alternative agri-food markets comes up against the autonomy of the players, who may ultimately refuse the new attachments that they are supposed to adopt (see also Hébert 2014).

The existing attachments that form conventional agri-food networks are robust. They offer many advantages, even if they are not completely satisfying. In our case study, turning toward their cooperatives to market their harvests or to their input or feed suppliers to enrich their soil or feed their stock

were simple routines that gave the farmers guarantees as to what would happen. The professional suppliers of inputs also gave the farmers advice about their cropping or animal husbandry practices, something that a neighboring stock or grain farmer would not be able to do. In a nutshell, detaching the farmers from their supply or sales chains is never a simple matter (Le Velly and Goulet 2015).

To come back to our case study, the attachment process's failure was equivalent to the failure to create a new agencement. Ending the analysis by bringing up the matter of agencement again is not trivial. Saying that the ability to act is distributed is tantamount to saying that the inability to act may come from any piece missing from a complex puzzle, a puzzle that is so complex that the pieces adopt their shapes as the puzzle is put together. If the outputs of one or the other party are not modified as planned; if, at the end of the day, quality, storage, or transportation is not guaranteed; or if the purchases and sales do not come about, then the alternative agri-food network does not see the light of day. Reasoning this way prevents one from overestimating the importance of one or the other factor in the creation process. Attitudes and motivations are vital for alternative agri-food networks to function; studies of the subject refer abundantly to this (Lockie and Halpin 2005; Best 2008; Raynolds 2009). Yet, for all that, the sociology of market agencements reminds us that all the technical and material characteristics of the network, e.g., the logistic schemes that are set up, quality adjustments, price-setting rules, etc., are essential for the successful creation of an alternative agri-food network.

Discussion and conclusions

The sociology of market agencements is a theoretical framework that is particularly well suited for analyzing the creation of alternative agri-food networks. Its principal message can be summed up as follows: Agencing new agri-food networks is a complex process that combines the completion of a series of operations that have already been done in the existing agri-food networks. To create an alternative agri-food network, which must function in ways that are aligned with the network's new purposes, there is a need to (i) agree on new definitions of quality, (ii) redefine the ways that goods and services are assessed, (iii) restore functional trading circuits, (iv) agree on prices, and (v) detach the actors from their customary networks. All of these operations are probably not always present in the agencing of new agri-food networks, but it is worthwhile to keep them in mind in order to be aware of the magnitude of the work required. We thus understand why there is nothing simple about creating alternative agri-food networks and why conventional agri-food networks are particularly resistant, even when their critics show the many problems that they generate. The sociology of market agencements tells us that creating alternative agri-food networks is a dicey operation that calls for settling a considerable number of issues. It therefore is not surprising that their creation, one that culminates in a new, stabilized sociotechnical arrangement, is seldom successful.

To conclude, we can finally come back to the multilevel perspective. This discussion is a must, given that this analytical frame is tending to become an unavoidable reference for analyzing transition processes in agriculture and food (for a review of the literature, see López-García et al. 2019). What is more, Callon and Latour's sociology of innovation is one of the sources of inspiration of the multilevel perspective, notably through its assertion of the co-determination of the various components of sociotechnical networks (Geels 2004).

In comparison, does the sociology of market agencements contribute anything different? In our opinion, it makes a contribution in the form of a different theoretical conception of the process of innovation and sociotechnical lock-ins. The researchers inspired by the multilevel perspective thus differentiate niches, in which the alternative agri-food networks take shape, from sociotechnical regimes, which are vectors of lock-ins that maintain conventional agri-food networks, and landscapes, where exogenous modifications that can reinforce or weaken the niches and regimes arise. The sociology of market agencements recognizes the existence of lock-ins fully (Callon 1998, 2017). Yet for all that, like ANT, it requires an analysis in which "nothing is left outside" (Çalişkan and Callon 2010, p. 9): In other words, the explanation must not be sought outside the agencements. As Geels points out quite clearly, it rests upon a "flat ontology" as opposed to the "hierarchical" analysis used for the different levels of niche, regime, and landscape (Geels 2011). For the sociology of market agencements, the alternative is not on the side of niches and the conventional is not on the side of the regime.

Just as ANT makes it possible to think non-dualistically, in a way that mixes the local and global (Murdoch 1998), nature and society (Goodman 1999), or production and consumption (Lockie and Kitto 2000), the sociology of market agencements makes it possible to understand that conventional agri-food networks are present in all the steps of an alternative agri-food network's creation. Whether they are contested and replaced when it comes to certain points or kept in part, with some of their entities serving as foundations in the alternative network, conventional networks are part of alternative networks. The alternative networks may even remain in thrall to their efficiency and not be able to detach themselves from them. Finally, the sociology of market agencements combines the niche and regime, innovation and stability, and the alternative and conventional in one and the same process.

Acknowledgements The authors thank Frédéric Goulet, Jérémy Forney and the participants of the New directions in agri-environmental governance workshop in Neuchatel, and the two AHV reviewers for their comments on an earlier version and Gabrielle Leyden for her translation. This research has been funded by the French National Research Agency (Grant ANR-15-CE21-0006'Institutionnalisation des agroécologies').

References

Akrich, M., M. Callon, and B. Latour. 2002. The key to success in innovation part II: The art of choosing good spokespersons. *International Journal of Innovation Management* 6 (2): 207–225.

Allen, P., and J. Guthman. 2006. From "old school" to "farm-to-school": Neoliberalization from the ground up. *Agriculture and Human Values* 23 (4): 401–415.

Asai, M., M. Moraine, J. Ryschawy, J. de Wit, A.K. Hoshide, and G. Martin. 2018. Critical factors for crop-livestock integration beyond the farm level: A cross-analysis of worldwide case studies. *Land Use Policy* 73: 184–194.

Best, H. 2008. Organic agriculture and the conventionalization hypothesis: A case study from West Germany. *Agriculture and Human Values* 25 (1): 95–106.

Bui, S., A. Cardona, C. Lamine, and M. Cerf. 2016. Sustainability transitions: Insights on processes of niche-regime interaction and regime reconfiguration in agri-food systems. *Journal of Rural Studies* 48: 92–103.

Buller, H., and E. Roe. 2014. Modifying and commodifying farm animal welfare: The economisation of layer chickens. *Journal of Rural Studies* 33 (1): 141–149.

Busch, L., and A. Juska. 1997. Beyond political economy: Actor networks and the globalization of agriculture. *Review of International Political Economy* 4 (4): 688–708.

Çalişkan, K., and M. Callon. 2010. Economization, Part 2: A research programme for the study of markets. *Economy and Society* 39 (1): 1–32.

Callon, M. 1986. Some elements of a sociology of translation. Domestication of the Scallops and the Fishermen of St. Brieuc Bay. In *Power, action and belief. A new sociology of knowledge?*, ed. J. Law, 196–223. London: Routledge.

Callon, M. 1987. Society in the making: the study of technology as a tool for sociological analysis. In *The social construction of technological systems: New directions in the sociology and history of technology*, ed. W.E. Bijker, T.P. Hughes, and T. Pinch, 83–103. Cambridge MA: The MIT Press.

Callon, M. 1998. Introduction: The embeddedness of economic markets in economics. *The Sociological Review* 46 (S1): 1–57.

Callon, M. 2009. Civilizing markets: Carbon trading between in vitro and in vivo experiments. *Accounting, Organizations and Society* 34 (3–4): 535–548.

Callon, M. 2017. *L'emprise des marchés. Comprendre leur fonctionnement pour pouvoir les changer*. Paris: La découverte.

Callon, M., C. Méadel, and V. Rabeharisoa. 2002. The economy of qualities. *Economy and Society* 31 (2): 194–217.

Chatellier, V., and C. Gaigné. 2012. Les logiques économiques de la spécialisation productive du territoire agricole français. *Innovations Agronomiques* 22: 185–203.

Clark, J.K., and S.M. Inwood. 2016. Scaling-up regional fruit and vegetable distribution: Potential for adaptive change in the food system. *Agriculture and Human Values* 33 (3): 503–519.

Cochoy, F. 2008. Calculation, qualculation, calqulation: Shopping cart arithmetic, equipped cognition and the clustered consumer. *Marketing Theory* 8 (1): 15–44.

Cochoy, F., J. Deville, and L. McFall (eds.). 2015. *Markets and the arts of attachment*. London: Routledge.

Cochoy, F., and S. Dubuisson-Quellier. 2000. Les professionnels du marché: Vers une sociologie du travail marchand. *Sociologie du Travail* 42 (3): 359–368.

Cochoy, F., P. Trompette, and L. Araujo. 2016. From market agencements to market agencing: An introduction. *Consumption Markets & Culture* 19 (1): 3–16.

de Raymond, A.B., L. Bonnaud, and M. Plessz. 2013. Les fruits et légumes dans tous leurs états. La variabilité, la périssabilité et la saisonnalité au coeur des pratiques sociales. *Revue d'études en agriculture et environnement* 94 (1): 3–12.

Doganova, L., and P. Karnøe. 2015. Building markets for clean technologies: Controversies, environmental concerns and economic worth. *Industrial Marketing Management* 44 (1): 22–31.

Dubuisson-Quellier, S. 2013. A market mediation strategy: How social movements seek to change firms' practices by promoting new principles of product valuation. *Organization Studies* 34 (5–6): 683–703.

DuPuis, E.M., and S. Gillon. 2009. Alternative modes of governance: Organic as civic engagement. *Agriculture and Human Values* 26 (1–2): 43–56.

Dwiartama, A., and C. Rosin. 2014. Exploring agency beyond humans: The compatibility of Actor-Network Theory (ANT) and resilience thinking. *Ecology and Society* 19 (3): 28.

Fouilleux, E., and A. Loconto. 2017. Voluntary standards, certification, and accreditation in the global organic agriculture field: A tripartite model of techno-politics. *Agriculture and Human Values* 34 (1): 1–14.

Geels, F.W. 2004. From sectoral systems of innovation to socio-technical systems: Insights about dynamics and change from sociology and institutional theory. *Research Policy* 33 (6): 897–920.

Geels, F.W. 2011. The multi-level perspective on sustainability transitions: Responses to seven criticisms. *Environmental Innovation and Societal Transitions* 1 (1): 24–40.

Geiger, S., and N. Gross. 2018. Market failures and market framings: Can a market be transformed from the inside? *Organization Studies* 39 (10): 1357–1376.

Geiger, S., D. Harrison, H. Kjellberg, and A. Mallard (eds.). 2014. *Concerned markets. Economic ordering for multiple values*. Cheltenham: Edward Elgar.

Goodman, D. 1999. Agro-food studies in the 'age of ecology': Nature, corporeality, bio-politics. *Sociologia Ruralis* 39 (1): 17–38.

Goodman, D., M. DuPuis, and M. Goodman. 2012. *Alternative food networks. Knowledge, practice and politics*. London: Routledge.

Guthman, J. 2004. *Agrarian dreams: The paradox of organic farming in California*. Berkeley: University of California Press.

Hall, A., and V. Mogyorody. 2001. Organic farmers in Ontario: An examination of the conventionalization argument. *Sociologia Ruralis* 41 (4): 322–399.

Hébert, K. 2014. The matter of market devices: Economic transformation in a southwest Alaskan salmon fishery. *Geoforum* 53: 21–30.

Henry, M. 2017. Meat, metrics and market devices: Commensuration infrastructures and the assemblage of 'the schedule' in New Zealand's red meat sector. *Journal of Rural Studies* 52: 100–109.

Henry, M., and R. Prince. 2018. Agriculturalizing finance? Data assemblages and derivatives markets in small-town New Zealand. *Environment and Planning A: Economy and Space* 50 (5): 989–1007.

Hinrichs, C.C. 2003. The practice and politics of food system localization. *Journal of Rural Studies* 19 (1): 33–45.

Ilbery, B., and D. Maye. 2005. Alternative (shorter) food supply chains and specialist livestock products in the Scottish–English borders. *Environment and Planning A* 37 (5): 823–844.

Jaffee, D., and P.H. Howard. 2010. Corporate cooptation of organic and fair trade standards. *Agriculture and Human Values* 27 (4): 387–399.

Karpik, L. 1996. Dispositifs de confiance et engagements crédibles. *Sociologie du Travail* 38 (4): 527–550.

Karpik, L. 2000. Le Guide rouge Michelin. *Sociologie du Travail* 42 (3): 369–389.

Kjellberg, H., F. Azimont, and E. Reid. 2015. Market innovation processes: Balancing stability and change. *Industrial Marketing Management* 44 (1): 4–12.

Kjellberg, H., and C.-F. Helgesson. 2007. The mode of exchange and shaping of markets: Distributor influence in the Swedish post-war food industry. *Industrial Marketing Management* 36 (7): 861–878.

Klein, K. 2015. Values-based food procurement in hospitals: The role of health care group purchasing organizations. *Agriculture and Human Values* 32 (4): 635–648.

Kloppenburg Jr., J., J. Hendrickson, and G.W. Stevenson. 1996. Coming in to the foodshed. *Agriculture and Human Values* 13 (3): 33–42.

Kneafsey, M., L. Holloway, L. Venn, E. Dowler, R. Cox, and H. Tuomainen. 2008. *Reconnecting consumers, producers and food: Exploring alternatives*. Oxford: Berg Publishers.

Kristensen, D.K., and C. Kjeldsen. 2016. Imagining and doing agrofood futures otherwise: Exploring the Pig City experiment in the foodscape of Denmark. *Journal of Rural Studies* 43: 40–48.

Laforge, J.M.L., C.R. Anderson, and S.M. McLachlan. 2017. Governments, grassroots, and the struggle for local food systems: Containing, coopting, contesting and collaborating. *Agriculture and Human Values* 34 (3): 663–681.

Latour, B. 1987. *Science in action: How to follow scientists and engineers through society*. Cambridge.: Harvard University Press.

Latour, B. 2005. *Reassembling the social: An introduction to actor-network-theory*. Oxford: Oxford University Press.

Le Velly, R., and I. Dufeu. 2016. Alternative food networks as "market agencements": Exploring their multiple hybridities. *Journal of Rural Studies* 43: 173–182.

Le Velly, R., and F. Goulet. 2015. Revisiting the importance of detachment in the dynamics of competition. *Journal of Cultural Economy* 8 (6): 689–704.

Lockie, S., and D. Halpin. 2005. The 'Conventionalisation' thesis reconsidered: Structural and ideological transformation of australian organic agriculture. *Sociologia Ruralis* 45 (4): 284–307.

Lockie, S., and S. Kitto. 2000. Beyond the farm gate: Production-consumption networks and agri-food research. *Sociologia Ruralis* 40 (1): 3–19.

López-García, D., L. Calvet-Mir, M. Di Masso, and J. Espluga. 2019. Multi-actor networks and innovation niches: University training for local Agroecological Dynamization. *Agriculture and Human Values* 36 (3): 567–579.

MacKenzie, D. 2009. Making things the same: Gases, emission rights and the politics of carbon markets. *Accounting, Organizations and Society* 34 (3): 440–455.

MacKenzie, D.A., F. Muniesa, and L. Siu (eds.). 2007. *Do economists make markets? On the performativity of economics*. Princeton: Princeton University Press.

Martin, G., M. Moraine, J. Ryschawy, M.-A. Magne, M. Asai, J.-P. Sarthou, M. Duru, and O. Therond. 2016. Crop–livestock integration beyond the farm level: A review. *Agronomy for Sustainable Development* 36 (3): 53.

Mason, K., M. Friesl, and C.J. Ford. 2017. Managing to make markets: Marketization and the conceptualization work of strategic nets in the life science sector. *Industrial Marketing Management* 67: 52–69.

Meynard, J.-M., A. Messéan, A. Charlier, F. Charrier, M.H. Fares, M. Le Bail, M.-B. Magrini, and I. Savini. 2013. Freins et leviers à la diversification des cultures: Étude au niveau des exploitations agricoles et des filières. *OCL* 20 (4): D403.

Miele, M., and J. Lever. 2013. Civilizing the market for welfare friendly products in Europe? The techno-ethics of the Welfare Quality® assessment. *Geoforum* 48: 63–72.

Moraine, M., P. Melac, J. Ryschawy, M. Duru, and O. Therond. 2017. A participatory method for the design and integrated assessment of crop-livestock systems in farmers' groups. *Ecological Indicators* 72: 340–351.

Muniesa, F., Y. Millo, and M. Callon (eds.). 2007. *Market devices*. Oxford: Blackwell Publishers.

Murdoch, J. 1998. The spaces of actor-network theory. *Geoforum* 29 (4): 357–374.

Onyas, W.I., M.G. McEachern, and A. Ryan. 2018. Co-constructing sustainability: Agencing sustainable coffee farmers in Uganda. *Journal of Rural Studies* 61: 12–21.

Onyas, W.I., and A. Ryan. 2015. Agencing markets: Actualizing ongoing market innovation. *Industrial Marketing Management* 44 (1): 13–21.

Ouma, S. 2015. *Assembling export markets. The making and unmaking of global food connections in West Africa*. Oxford: Wiley.

Ouma, S., M. Boeckler, and P. Lindner. 2013. Extending the margins of marketization: Frontier regions and the making of agro-export markets in northern Ghana. *Geoforum* 48: 225–235.

Peyraud, J.-L., M. Taboada, and L. Delaby. 2014. Integrated crop and livestock systems in Western Europe and South America: A review. *European Journal of Agronomy* 57: 31–42.

Phillips, C. 2016. Alternative food distribution and plastic devices: Performances, valuations, and experimentations. *Journal of Rural Studies* 44 (1): 208–216.

Raynolds, L.T. 2009. Mainstreaming fair trade coffee: From partnership to traceability. *World Development* 37 (6): 1083–1093.

Raynolds, L.T., D.L. Murray, and J. Wilkinson (eds.). 2007. *Fair trade. The challenges of transforming globalization*. New York: Routledge.

Renard, M.-C. 1999. The interstices of globalization: The example of fair coffee. *Sociologia Ruralis* 39 (4): 484–500.

Renting, H., T.K. Marsden, and J. Banks. 2003. Understanding alternative food networks: Exploring the role of short food supply chains in rural development. *Environment and Planning A* 35 (3): 393–412.

Ryschawy, J., G. Martin, M. Moraine, M. Duru, and O. Therond. 2017. Designing crop–livestock integration at different levels: Toward new agroecological models? *Nutrient Cycling in Agroecosystems* 108 (1): 5–20.

van der Ploeg, J.D., H. Renting, G. Brunori, K. Knickel, J. Mannion, T. Marsden, K. De Roest, E. Sevilla-Guzmán, and F. Ventura. 2000. Rural development: From practices and policies towards theory. *Sociologia Ruralis* 40 (4): 391–408.

Wang, C.-M. 2018. Assembling lettuce export markets in East Asia: Agrarian warriors, climate change and kinship. *Sociologia Ruralis* 58 (4): 909–927.

Wang, C.-M. 2019. Performing and counter-performing organic food markets in East Asia: The role of ahimsa, scientific knowledge and faith groups. *The Geographical Journal*.

Wiskerke, J.S.C. 2003. On promising niches and constraining socio-technical regimes: The case of Dutch wheat and bread. *Environment and Planning A* 35 (3): 429–448.

Publisher's Note Springer Nature remains neutral with regard to jurisdictional claims in published maps and institutional affiliations.

Ronan Le Velly is professor of sociology in Montpellier SupAgro and member of the UMR Innovation. His research is at the crossroads of market sociology and rural studies and focuses on alternative agri-food networks such as fair trade, short food supply chains, and organic agriculture.

Marc Moraine is researcher in agronomy in the French National Research Institute for Agriculture, Food and Environment (INRAE) and member of the UMR Innovation. His research in agronomy and agroecology focuses on crop-livestock integration and farming systems' diversity and complementarity at the territorial level.

Agriculture and Human Values (2020) 37:1013–1025
https://doi.org/10.1007/s10460-020-10034-8

Competing food sovereignties: GMO-free activism, democracy and state preemptive laws in Southern Oregon

Rebecka Daye[1]

Accepted: 11 April 2020 / Published online: 30 April 2020
© Springer Nature B.V. 2020

Abstract

Indicators of food sovereignty and food democracy center on people having the right and ability to define their food polices and strategies with respect to food culture, food security, sustainability and use of natural resources. Yet food sovereignty, like democracy, exists on multiple and competing scales, and policymakers and citizens often have different agendas and priorities. In passing a ban on the use of genetically-modified (GMO) seeds in agriculture, Jackson County, Oregon has obtained some measure of food sovereignty. Between 2016 and 2017 ethnographic research was undertaken in rural Southern Oregon where local community and State of Oregon priorities regarding the use of GMO crops are in conflict. This article presents ethnographic research findings about the expression and negotiation of multiple food sovereignties by civil society in rural southern Oregon and the State of Oregon via democratic processes. In particular, these findings illustrate the effects of socio-political power dynamics on local and state acts of food sovereignty, democracy and agrifood policy by analyzing what the different expressions of food sovereignty reveal for its implementation at the local level.

Keywords Food sovereignty · Food democracy · Genetically-modified organisms (GMOs) · GMO-free activism · North america · Oregon

Abbreviations

GMO	Genetically-modified organisms
GMOFJC	GMO-free Jackson County
GMOFJoCo	GMO-free Josephine County

Introduction

Food sovereignty, a policy framework first proposed by the peasant organization La Via Campesina in 1996 and increasingly recognized by governments worldwide, refers to the rights of communities to define and shape the structure of their food systems through the democratic process (Alonso-Fradejas et al. 2015; Conti 2016; FIAN 2005; McMichael 2014; Nyéléni 2007). In a recent study on the food sovereignty agenda of the UN, Conti (2016) asserts that the term "food sovereignty" can be assumed to mean the general will of the people expressed in the concept of participatory democracy. Both food sovereignty and food democracy center on people having the right and ability to define their food polices and strategies with respect to food culture, food security, sustainability and use of natural resources. Yet food sovereignty, like democracy, exists on multiple and competing scales (local, regional and global) and policymakers and citizens often have different agendas and priorities (Conti 2016; Iles and Montenegro de Wit 2015; Shattuck et al. 2015). This research asks what the different expressions of food sovereignty reveal about the weakness in its implementation. This is important because it is the value of local diversity and experiences that have the potential to shape different meanings and acts of food sovereignty (Shattuck et al. 2015). This article presents ethnographic research findings about the expression and negotiation of multiple food sovereignties by civil society in rural southern Oregon and the State of Oregon where the use of genetically modified organisms (GMO/GEs) in agriculture is in conflict. In particular, these findings illustrate that the socio-political power dynamics between local communities and state acts of food sovereignty, as well as the vague nature of the sovereign, continue to pose practical and ethical dilemmas for implementing food sovereignty at the local level.

✉ Rebecka Daye
    rebeckadaye@gmail.com; dayer@oregonstate.edu

1   Department of Anthropology, Oregon State University, 203
    Waldo Hall, 2250 SW Jefferson Way, Corvallis, OR 97331,
    USA

Food sovereignty and the vague nature of the sovereign

Food sovereignty is undoubtedly coming of age—as a movement and a set of ideas about how to democratize both access to [productive] resources and political power (Shattuck et al. 2015: 422).

The food sovereignty strategies developed by the international peasant movement La Via Campesina in the mid-1990s aspire towards the democratic governance of all stages of human interaction with food, from production to consumption, and seeks to guarantee the human right to food (La Via Campesina 2007). The original concept of food sovereignty as proposed by La Via Campesina was intended to facilitate the transformation of oppressive trade relations, replacing them with socially embedded markets and democratic governance over the agrifood system in support of small-scale indigenous farmers and food producers in the global south (Agarwal 2014; Edelman 2014; Fairbairn 2012; McMichael 2014; MacRae 2016; Shattuck et al. 2015). By challenging the political and economic power of multinational agrifood complexes, food sovereignty has materialized as a strong alternative to the neoliberal model of agriculture (Wittman et al. 2010). Ultimately proponents of food sovereignty seek to create a socially just and environmentally sustainable food system via democratic control over food and food-producing resources while rejecting such contemporary agrifood policies as the prioritization of export agriculture, the privatization of natural resources and the use of corporate-owned biotechnology, commonly referred to as genetically-modified organisms or genetic engineering (GMO/GEs) (Shawki 2015; see also McMichael 2014; Patel 2010).

Schiavoni notes that although food sovereignty emerged as a way for peasant farmers and nation-states to reclaim some control over domestic food production in the face of expanding neoliberal policies, it also inspired a radical agrarian populism and "a local community-led political project independent of the state" (2015, p. 468). The spread of the food sovereignty movement from the global south to the global north has been seen as an attempt to challenge the current global food paradigm by reasserting people's right to participate in the food system at multiple levels (Alkon and Mares 2012; Ayres and Bosia 2011; Brent et al 2015; Clendenning et al 2016; Dekeyser et al. 2018). Food sovereignty thought and practice is also shifting beyond its initial agrarian focus to embrace the role of consumers and urban areas (Dekeyser et al. 2018; García-Sempere et al. 2018; Navin and Dieterle 2018). Although food sovereignty proponents have made substantial claims about the potential of food sovereignty to transform the global food paradigm, the framework has often been criticized for its lack of implementation (Alonso-Fradejas et al 2015; Hospes 2014; Iles and Montenegro de Wit 2015; McMichael 2014; Shattuck et al. 2015; Shawki 2015). Scholars and policymakers have argued that the vague nature of the sovereign has inhibited the implementation of the food sovereignty framework (see Edelman 2014; Hospes 2014; Iles and Montenegro de Wit 2015; Schiavoni 2015, 2016; Shattuck et al. 2015). The lingering problematic is thus: who is the sovereign in food sovereignty?

A modern definition of the term "sovereign" as referenced in food sovereignty suggests a level of autonomy and power in controlling food policies; the distribution of resources; and local, national and international decision-making processes for those directly affected by these policies (Dekeyser et al. 2018). The term therefore has a strong connotation of democracy and participatory development (FIAN 2005). Yet sovereignty is built and contested at multiple levels, including popular culture, civil society, the market and governmental institutions. Shattuck et al. (2015) assert that there is an existing duality of sovereignty: state food sovereignty refers to national control over a country's food supply and productive resources, while people's sovereignty refers to a more populist vision of internal control. Shattuck et al. suggest that rather than looking at food sovereignty as being either of the state or of the people, food sovereignty frameworks should leave room for "different sovereign actors to coexist, with the terms of their engagement under ongoing negotiation" (2015, p. 425). Thus, food sovereignty should involve understanding sovereignty as a relational form and a process that creates and sustains multiple sovereignties (Conti 2016; Iles and Montenegro de Wit 2015; Schiavoni 2015; 2016; Shattuck et al. 2015). Schiavoni (2016) notes that the food sovereignty agenda also calls for interactions between the nation-state and civil society to create a mutually reinforcing 'state-society' synergy capable of creating systemic change or reform. This research thus asks "What do the different expressions of food sovereignty reveal about the weakness in its implementation"?

Food sovereignty and food democracy

Most contemporary struggles over the future of the agricultural and food system—such as debates over genetic engineering…—seem to be fundamentally about democracy. (Hassanein 2008)

Food democracy is a core aspect of the food sovereignty model and research participants in this study largely framed their activities in terms of food democracy rather than food sovereignty. Although there are discursive and tactical differences between the two approaches, their ultimate goal is similar. As with food sovereignty, food democracy can best be understood as a process of continued development toward a democratic, socially just, economically viable and environmentally sustainable food system (Carlson and Chappell 2015; Dekeyser et al. 2018; Hassanein 2008; Prost et al. 2018). However, as Prost et al. assert, "Radical models, such as food democracy [also] stand for a redistribution of power and a democratization of the food system, i.e. a shift toward increased control by civil society over state regulation and market forces", and thus can be seen as a threat to the state (2018, p. 71; see also Carlson and Chappell 2015; Dekeyser et al. 2018; Hassanein 2008; Porter 2013).

Food sovereignty in the United States

The food sovereignty model has spread from its origins in the global south to locales as diverse as industrialized France, rural Vermont, and urban Chicago in the United States (see Alkon and Mares 2012; Ayres and Bosia 2011; Clendenning et al. 2016). This demonstrates the strength of the food sovereignty model, as well as its appeal and potential for use in varied localized settings. Many of the case studies in the United States and the global north have discussed efforts for food sovereignty as existing within, or being constrained by, the forces of neoliberalism, with its focus on market opportunities and individual choice rather than state responsibility (see Alkon and Mares 2012; Brent et al. 2015; Clendenning et al. 2016; Fairbairn 2012). In her review of food sovereignty initiatives in the US, Fairbairn notes that two aspects lacking from the US-based food sovereignty movement are "challenging the dominant economic system and combating social injustice" (2012, p. 222). She posits, however, that while food sovereignty advocates endorse a mix of entrepreneurial projects and local policy changes, thereby ultimately shifting the right to food from state to personal responsibility, the promotion of policy change still constitutes food sovereignty as critical of neoliberal agrifood policies. Nonetheless, food sovereignty movements in the US continue to be criticized for the lack of transformative potential, although very few actual case studies have been carried out in local communities.

GMO-free activism

One highly contested function of civil society and democratic processes in the exercise of food sovereignty is the rejection of GMOs via GMO/GE-free activism. In recent years a plethora of resistance movements and legal battles over the use of GMO/GEs in agriculture have surfaced in many locales and countries around the world (see Buiatti et al. 2012; Falkner and Gupta 2009; Fitting 2006; Glover 2010; Godfray et al. 2010; Kloppenburg 2014; Levinson

2014; Lieberman and Gray 2008; McCauley 2015; Newell and Glover 2003; Pearson 2012; Pechlaner and Renting 2012; Spinelli 2013; Stone 2002, 2010; Tiberghien 2012; Walsh-Dilley 2008). Because the corporate agrifood complex is constituted through local/global dynamics located in various sectors and levels of the socio-political economy, movements such as those for food sovereignty need to be ready to collaborate and to resist these processes at multiple scales (Iles and Montenegro de Wit 2015). Iles and Montenegro de Wit (2015) posit that in order for this to occur, food sovereignty movements need to be recognized internally (locally) as well as externally (at various state, national and international levels). With multiple and contested sovereignties, this can be extremely difficult to achieve. This challenge can be witnessed by the introduction and corporate control of GMO crops and the ensuing path dependency for farmers, which have led to structural constraint on the part of farmers not wanting to grow GMOs (Iles and Montenegro de Wit 2015).

As of 2013, at least 19 districts in the United States had enacted ordinances at the municipal and county levels restricting the propagation of GMO crops (Levinson 2014), and currently more than 28 countries have enacted regulation prohibiting the propagation, cultivation or import of GMOs (Alonso-Fradejas et al 2015; Library of Congress 2014). Similarly to the US, Costa Rica has several municipalities that have adopted GMO-free ordinances (Pearson 2012). Although the movement failed to establish a moratorium on the propagation of GMOs at the level of the nation-state, organizers have continued to utilize grassroots strategies and connection with other activist networks to effectively pass ordinances in eight municipalities (Pearson 2012). Other countries, like Bolivia and Ecuador, have total bans on the propagation of GMOs, while many European countries have regional GMO bans (GMO-free Europe 2019). Additionally, at least 15 countries have adopted and incorporated food sovereignty legislation into their constitutions (Dekeyser et al. 2018; Schiavoni 2016). These ordinances and legislation are the result of civil society networks coming together to assert their right to food sovereignty and to determine the design of the local agricultural landscape through the democratic process. This is evidenced in the growth of local actors, communities and institutions seeking their own power over seeds, knowledge, farming practices and food systems. In this context, sovereignty itself can be seen as renegotiated at multiple levels (Schiavoni 2015). However, efforts toward local food sovereignty continue to be contested at multiple levels. For example, if an initiative to ban the propagation of GMOs is successful in one county, proponents of the ban may connect with other similar movements locally and globally. This could result in the obstruction of commercial GMO crop production, which thrives in a supportive regulatory environment, thereby triggering pushback from industry (Pechlaner and Renting 2012).

In 2004, Mendocino County, California was the first county in the United States to pass legislation that prohibited the propagation of GMO crops. Walsh-Dilley's 2009 case study on the GMO ban in Mendocino County found that the ballot measure was successful primarily because the issue was reframed as a broader social problem that citizens were able to solve collectively. The success of the ballot was deemed to be a direct result of locally embedded structures, networks and trust in community leaders, rather than a blanket rejection of GMOs. Walsh-Dilley claims this embeddedness was also the movement's weakness because such site-specific localism could not be scaled up to connect with the wider anti-GMO movement. Nonetheless, there is considerable reason to think that civil society movements for food sovereignty/food democracy can have deep and lasting impacts at multiple levels, including local, regional, state and global (Levinson 2014; Pechlaner and Renting 2012; Walsh-Dilley 2008). Levinson's (2014) brief comparative study of the GMO-free movements in Mendocino and Jackson Counties found that many people thought county ordinances would have impacts beyond their respective boundaries.

GMO-free activism in Southern Oregon

In 2014, the voters of Jackson and Josephine Counties in southern Oregon each passed a law prohibiting the cultivation of genetically-engineered crops in the county. The GMO/GE-free campaigns in both Jackson and Josephine counties employed grassroots participatory democracy, which emphasizes the participation of constituents in a liberal democracy and assumes the existence of shared ethical and moral values within a community (Levinson 2014; Nugent 2008; Paley 2002; Walsh-Dilley 2008). Neither campaign framed their activities in terms of a food sovereignty agenda, choosing instead to focus on food democracy via citizens' initiative and democratic process. Nonetheless, in passing a GMO-free law Jackson and Josephine Counties each obtained some measure of food sovereignty. Gaining food sovereignty is a dynamic process not a state of being (Schiavoni 2015).

Jackson and Josephine Counties are part of the growing, yet highly contested, global phenomenon of adopting GMO/GE-free legislation via democratic processes. The gradual inclusion of GMO-free legislation at the local and national levels has been both a major achievement and a challenge. In 2013, when several counties in Oregon were proposing local GMO-free ballot initiatives, the State of Oregon passed a preemptive law banning counties from regulating GMOs or setting their own agricultural seed policy (State of Oregon 2013). Because Josephine County was preempted from enacting their ban while Jackson County was grandfathered in (discussed further below), Jackson and Josephine Counties provide a striking example of the complex negotiations of multiple and competing sovereignties. Oregon is not alone in passing preemptive laws in relation to the propagation or cultivation of GMOs. GMO preemptive laws in the US go back as far as 2005, shortly after the Mendocino County ban when at least nine states enacted preemptive laws restricting local counties from regulating the use of GMOs (Organic Consumers Association 2005). In 2014, three counties in Hawai'i passed GMO bans via democratic process prompting the state to enact a preemptive law banning counties from restricting the propagation or cultivation of GMOs (Herbers 2016). A similar strategy has been used by the federal government to preempt state and local communities from enacting GMO labelling laws (Corbett 2016).

Methodology

Structure of the research

The data presented here are the result of an ethnographic study conducted on GMO-free activism and participatory food democracy in southern Oregon between 2016 and 2017. The following sections present research findings on the relationship between corporate agriculture, the State of Oregon and citizens' efforts for local control of agricultural laws pertaining to the use of GMO/GEs in southern Oregon.

Research site

Jackson County is situated in southern Oregon near the California border (Fig. 1). It is the sixth most populated county in the state. With an estimated population of 212,567 it contains 2783 square miles of land and 214,069 acres of farmland. The average size of farms is 124 acres, though the majority of farms are between 10–49 acres. Farmland usage includes 58.9% pastureland, 19.5% woodland, 15.3% cropland and 6.3% other. Crop sales account for 58% of market value of products sold, and fruits, tree nuts and berries are the number one crop category in terms of value by commodity group (USDA 2012; US Census Bureau 2015). Josephine County borders Jackson County to the north. While this research was focused primarily in Jackson County, Jackson and Josephine Counties together comprise the narrow windy southern Oregon Rogue Valley. Josephine County has a population of 84,063. In 2012 it contained 28,256 acres of farmland with an average farm size of 46 acres. Farmland use includes 33.1% pastureland, 29.6% cropland, 27% woodland and 10.3% other.

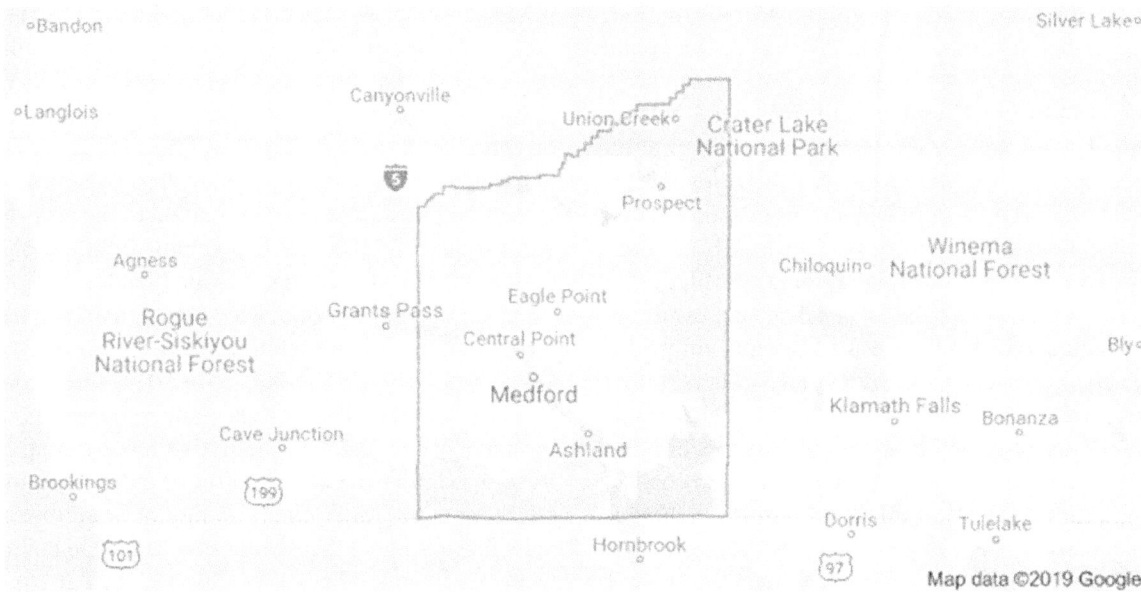

Fig. 1 Southern Oregon Map highlighting Jackson County

Sample

Ethnographic field research was undertaken between January and December 2017 (3 years after the campaign) during which time I conducted forty-four in-depth, semi-structured interviews and engaged in seven participation observation activities with Our Family Farms (OFF) at various events such as State of Oregon legislative hearings and presentations on regenerative agriculture and emerging types of genetic engineering (e.g. CRISPR and gene-editing techniques). Limitations to the study include participants' memory recall of campaign events and the timeframe in which those events occurred. Participant recall was at times imprecise because the movement began in February 2012, the official campaign began in January 2014 and this study was conducted in 2017, 3 years after the campaign.

Sampling strategy

Research questions presented in this paper were designed to elicit feedback about participants' involvement in the campaign and their understanding of food sovereignty and democratic processes in shaping the agrifood system. Research participants were selected via a targeted snowball sample (Bernard 2006). The sample was targeted in the sense that the parameters for research participants were defined as those who were actively engaged in the GMO-free Jackson County campaign. It was necessary to interview research participants who were active with the GMO-free Jackson County movement in some capacity because the research questions involve this population specifically. An effort was

made to talk with local residents who opposed the ban, however ultimately only two local representatives of organizations opposing the ban returned my calls and agreed to be interviewed.

Individuals were recruited for the study via participant referral from preliminary research, Our Family Farms outreach and respondent-driven referrals. Research participants are individuals who were involved with the GMO-free Jackson County movement (Table 1) specifically GMO-free Jackson County (GMOFJC) and Our Family Farms Coalition (OFFC) volunteers (n = 36) and members of the Josephine County GMO-free movement who volunteered with GMOFJC and OFFC (n = 4), as well as representatives of organizations that either supported (n = 2) or opposed (n = 2) the GMO ban.

GMO-free Jackson County is the group of local farmers and citizens that started the movement in 2012. Our Family Farms Coalition (OFFC) is the local farmer-led organization that led the campaign for a GMO-free Jackson County in

Table 1 Interview demographics

	N	% of 44
Our Family Farm Coalition volunteers	40	91
Non-farmers	29	66
Farmers	11	25
Organizations in favor	2	4.5
Opposition	2	4.5

Non-farmers and farmers are categories of volunteers who worked with Our Family Farms Coalition

2014. Founding members are primarily farmers but membership is inclusive of those interested in working to safeguard local biodiversity and protect the family farm (personal communication 2016). Our Family Farms (OFF) is a non-profit educational organization established in 2015 to continue the outreach and promotion of a GMO-free Seed Sanctuary in southern Oregon. Our Family Farms stated primary goal is to protect and promote family farms that grow traditional crops by strengthening local communities to create GMO-free zones in the United States and around the world (Our Family Farms 2016).

Secondary sources

Additional research included an examination of GMO-free Jackson County (GMOFJC) and Our Family Farms Coalition (OFFC) records and promotional materials, newspaper clippings and online news sources related to the GMO-free Jackson County campaign and State of Oregon legislative records. These materials allowed for the examination and documentation of the different ways GMOFJC and OFFC accessed, used and shared information relating to GMO-free activism, food sovereignty and democratic action.

Participant observation

In my role as participant-observer, I attended and participated in Our Family Farms events (2017) including Seed Sanctuary celebrations (n = 2), Farm to Table fundraising dinners (n = 2), a public lecture on the Future of GMOs (n = 1), and the State of Oregon Legislative Hearings (n = 2). Participant-observation of OFF events allowed me to experience and document the different ways OFF/OFFC members and attendees talked about, accessed, used and shared information related to past and present GMO-free activism, food sovereignty and democratic action.

Research findings

Based on analysis of the data collected, two central themes were identified that highlight the difficulties of enacting local food sovereignty via democratic processes. The first theme captures tensions between local organic, conventional and non-GMO seed growers and farmers and concerned citizens on the one hand, and GMO/GE farmers, multinational agrifood corporations and governmental institutions on the other hand, regarding the use of GMO/GE seeds in the Rogue Valley. This theme highlights the socio-political power dynamics that mediated between the local community's efforts toward food sovereignty and the multinational agrifood corporation Syngenta, which was supported by the Oregon Farm Bureau and Cattleman's Association, among

others. The second theme captures tensions between civil society and State of Oregon acts of food sovereignty, including the complexities of implementing local agrifood policy via democratic processes. This theme exemplifies the challenges of competing sovereignties—state food sovereignty versus people's sovereignty—and the limitations of food democracy when the state enacts a statewide agricultural seed preemptive law.

Theme 1: Farmers versus corporate agriculture

The first theme captures tensions between local farmers and corporate agrifood policy regarding agricultural co-existence of genetically-engineered crops with organic and non-GMO/GE crops at the local level.

In February of 2012, a local organic seed farmer discovered that Syngenta's GMO sugar beets were being grown in Oregon's Rogue Valley near his Ashland farm site. Within weeks, a core group of local farmers and concerned citizens began working on a GMO/GE-free citizen's initiative ballot measure and GMO-free Jackson County (GMOFJC) was born. Interview participants unanimously agreed that the primary reason for the proposed ban was concern over transgenic trespass (i.e., cross-contamination) of GMO/GEs with organic and non-GMO seed and vegetable crops, which are the principal economic livelihood for many small-scale family farms in the area that sell both retail and wholesale. On January 3, 2013, GMOFJC submitted over 6700 signatures to qualify the measure for the ballot, nearly double the requisite number (BallotPedia 2015; Wilce 2014). GMO-free Josephine County (GMOFJoCo) was also formed in late 2012 and ran its own ballot measure campaign separately from GMOFJC, although the two groups were in regular communication.

Upon learning that GMO/GE sugar beets were being grown in the Rogue Valley, the Southern Oregon Seed Growers Association (SOSGA) was formed in urgency to protect seed farmers in the region and to facilitate discussion on the issue of cross-contamination. When SOSGA was formed, there was no legal mechanism to discern who was growing GMO/GE crops or where (sugar beets were deregulated in 2012). Furthermore, Syngenta refused to disclose the location of sugar beet test plots. However, they were willing to disclose whether the locations of other farm plots were within the USDA recommended 4-mile radius (8-mile diameter) buffer zone for GMO sugar beet crops. Based upon Syngenta's limited disclosure, local farmers developed a pinning map which highlighted suspected GMO test plots in Jackson County (see Fig. 2). In response to the threat of cross-contamination, several organic vegetable and seed growers in the Rogue Valley ripped up their own crops.

The GMO-free movement in the Rogue Valley had been well underway for a year before Syngenta began meeting

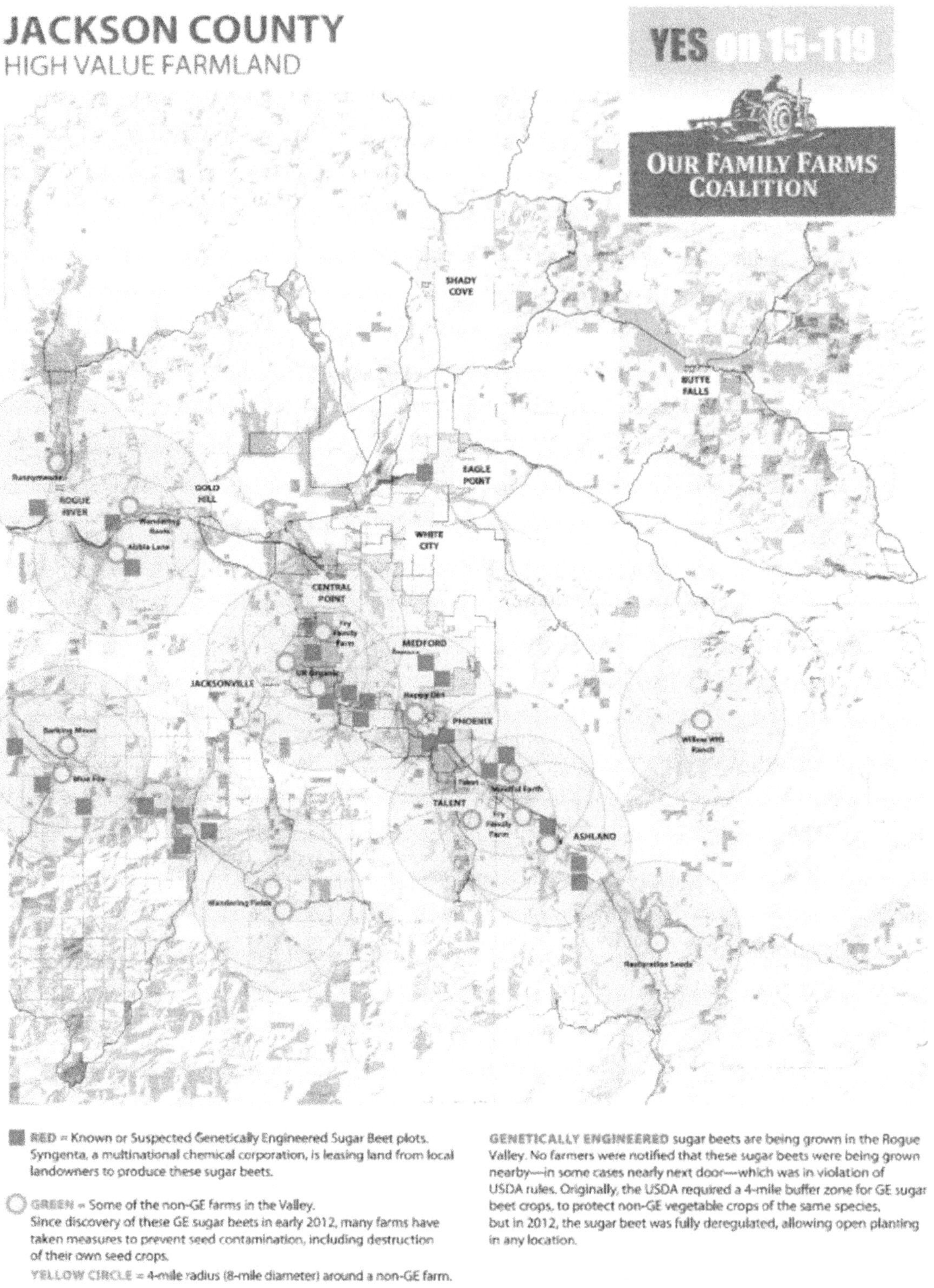
JACKSON COUNTY
HIGH VALUE FARMLAND
YES on 15-119
OUR FAMILY FARMS COALITION
SHADY COVE
BUTTE FALLS
EAGLE POINT
GOLD HILL
ROGUE RIVER
WHITE CITY
CENTRAL POINT
MEDFORD
JACKSONVILLE
PHOENIX
TALENT
ASHLAND
RED = Known or Suspected Genetically Engineered Sugar Beet plots.
Syngenta, a multinational chemical corporation, is leasing land from local landowners to produce these sugar beets.
GREEN = Some of the non-GE farms in the Valley.
Since discovery of these GE sugar beets in early 2012, many farms have taken measures to prevent seed contamination, including destruction of their own seed crops.
YELLOW CIRCLE = 4-mile radius (8-mile diameter) around a non-GE farm.
GENETICALLY ENGINEERED sugar beets are being grown in the Rogue Valley. No farmers were notified that these sugar beets were being grown nearby—in some cases nearly next door—which was in violation of USDA rules. Originally, the USDA required a 4-mile buffer zone for GE sugar beet crops, to protect non-GE vegetable crops of the same species, but in 2012, the sugar beet was fully deregulated, allowing open planting in any location.

Fig. 2 Jackson County Pinning Map 2014

with SOSGA in 2013. According to participants, Syngenta's business model in the area was to rent a plot of land from land owners and pay them a small annual stipend. By 2012, Syngenta had accumulated over 20 GMO test plots in the narrow Rogue Valley, which could at best accommodate 8–10 small GMO test plots. Several farmers noted that Syngenta is one of the largest agrochemical seed companies in the world and could have planted the GMO test

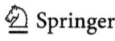

crops anywhere. A few farmers shared that they believed Syngenta consciously knew they were coming into a valley of small organic farms, perhaps with the intention to contaminate every organic and non-GMO farm in the valley. Nonetheless, the Oregon Farm Bureau, the Cattleman's Association, Syngenta and GMO farmers were arguing for co-existence. The USDA claims that co-existence of genetically-engineered crops with organic and non-GMO/GE crops is vital for US agriculture (USDA 2019). However, research participants stated that they didn't think co-existence of GMOs and non-GMOs was possible in the narrow, windy Rogue Valley. They believed co-existence meant the eventually destruction of small family farms due to the threat of cross-contamination.

A sentiment shared by research participants regarding the corporate use of co-existence in the valley was well-expressed by one organic seed grower who explained the problem of co-existence in terms of trying to negotiate with his neighbor over where to plant GMO beet seeds so that cross-contamination with his chard seeds was less likely to occur:

> The [Oregon] Farm Bureau President said, 'Can't we all just get along and … talk this deal out between neighbors?'… So that season I started to grow my chard seed and I talked to my neighbor [farmer]. We talked about co-existence and he says, 'Yeah, you're right… I won't plant any [GMOs]… here. I'll plant them somewhere else.' The following week I drive by their farm… and there they are… planting beet seeds. So I go in and talk to him. He says, 'Well, corporate said I have to plant my beet seeds here'. (Steve 2017)

Contamination from cross-pollination was seen as a real threat in the area for several reasons (1) transgenic sugar beets are open pollinated and can easily cross-pollinate with other members of Beta vulgaris such as chard as well as with other beets that local seed farmers were growing in the area; (2) opening the door to GMOs could mean that other transgenic crops would also be introduced in the area (GMO sugar beets and GMO alfalfa were both being grown in the Rogue Valley at that time); and (3) GMOs are patented and owned by multinational corporations that have been known to file patent infringement claims on farmers who have any trace of the patented seed on their farms, whether the farmer intentionally planted the seed or it was the result of cross-contamination via wind or insects.

Research participants were unanimously confident that the physical characteristics of the narrow Rogue Valley prohibited co-existence of organic and GMO crops. This was commonly understood as real evidence of intentional corporate override in the area. Furthermore, there was no evidence of sincere farmer-to-farmer collaboration toward co-existence between Syngenta and non-GMO farmers. This

was confirmed in mid-2013 when Syngenta sent an attorney to negotiate with SOSGA for co-existence, but negotiations ended when Syngenta's attorney and representatives got up and walked out of the meeting, claiming it did not fit with their business model. Ultimately the failure of the co-exist meetings to reach an agreement on the use of GMO/GE seeds in the Rogue Valley highlights co-existence as a point of tension that was unsuccessfully negotiated at the local level. This would suggest that local food sovereignty could only operate if farmers voted to transform the region from small family farms to GMO/GE farms if the majority so desired. This point of tension is directly linked to the State of Oregon and local community actors involved in the campaigns for a GMO-free Jackson County and a GMO-free Josephine County, discussed in the next section.

Theme 2: Farmers versus the state (food democracy and state preemptive laws)

The second theme captures the tensions between local democratic efforts for a GMO-free Rogue Valley and the State of Oregon statewide agricultural seed preemptive law:
OR SB863—The Preemptive Seed Law.

> Makes legislative finding and declaration that regulation of agricultural seed, flower seed, nursery seed and vegetable seed and products of agricultural seed, flower seed, nursery seed and vegetable seed be reserved to state. (State of Oregon 2013)

To elucidate how multiple food sovereignties were understood and negotiated between civil society and the State of Oregon, research participants in this study were asked what they thought the role of citizens and government should be in shaping the food system and whether that was in accordance with current agrifood policies. It should be noted that respondents generally did not care for the term food sovereignty noting that the meaning was hard to define. Participants preferred the terms food democracy with a focus on local control. Sixty-five percent of respondents thought citizens primarily shape the food system via activism, and fifty-five percent cited education as a primary component of activism and citizen responsibility. Respondents generally agreed that if citizens want to have rights, they need to work to secure them:

> Citizens have to work for [their] rights, for democracy: food sovereignty and food justice is one of many different aspects in which we have to be vigilant (Deb 2017). The foundation of democracy is an educated population. And I think that's where we really stepped up. People were out there day after day, door to door, events, tabling, just constantly making information available to people. In order to be able to make a good

decision you have to have good information. And we felt like we really did an excellent job of providing not just media information but… human being to a human being (Anna 2017).

While many participants felt that activism was a crucial characteristic of food democracy, many also believed that democratic processes were under threat from the power and influence of multinational corporations. Sixty-four percent of respondents thought that the role of government in shaping the food system should be via regulation and legislation that promotes the best interest of the people, not corporations, and fifty percent thought government's primary role is to protect citizens. However, respondents unanimously agreed that because of the influence of corporate money and lobbying, governmental agencies have not represented the interest of the communities or the farms by developing clear and effective agrifood policy:

> The government should not be favoring any corporation or individual over others by creating legislation or policy or other means of influence (Eli 2017).
>
> It didn't have to be this way, but because of corporate greed, there needs to be protection of the people in the businesses, the farmers who are growing the food, making the food, whatever food we're talking about, so that it's safe for us to eat. In the case of the State of Oregon all that has been done is to protect the corporations (Barbara 2017).
>
> [The preemptive seed law] SB863 took the power away from every county across the State of Oregon to be able to protect their local agricultural interests from the multinational chemical companies that came here to… stifle democracy, take our voice away, and to confuse the issue. They meddled in our local democratic process. That is wrong. Fundamentally wrong (Chris 2017).

In January 2013, both GMO-free Jackson County (GMOFJC) and GMO-free Josephine County (GMOFJoCo) submitted signatures to qualify the local GMO-free initiatives for the ballot in May 2014. In February of 2013, Oregonians for Food and Shelter, under the direction of lobbyist John DiLorenzo, drafted a piece of legislation to preempt local jurisdictions from having any say over agricultural seed or seed products and to move those decisions to the state level (Center For Food Safety 2017; Friends of Family Farmers 2017; Wilce 2014). The proposed State of Oregon legislative bill, SB633, also known as the Preemptive Seed Law (commonly referred to as the Monsanto Protection Act), would have effectively banned counties in the state from enacting a local GMO ban (several counties in the state had versions of a GMO ban underway at the time). Wilce (2014) notes that many citizens saw the bill as a corporate-backed

reaction to the local ballot initiatives. Furthermore, legislators in Salem were lobbied by the largest agrochemical companies to pass the bill, with support from the American Legislators Exchange Council (Wilce 2014).

In response to the proposed SB633, a group of Rogue Valley farmers and concerned citizens involved with GMOFJC and GMOFJoCo (together with the Center for Food Safety, Friends of Family Famers and Food Democracy Now) organized against the bill and attended the legislative sessions in Salem to speak on the topic. Senator Bates and Representative Buckley from Jackson County also testified on behalf of their constituents and against the proposed preemptive law. Efforts against the bill were framed very broadly as having to do with local choice, democratic processes and the diversity of agriculture in different areas across the state. The main argument against the bill was that something that works in one county is not necessarily going to work in other places. Moreover, the Center for Food Safety (2017) stated that regulations at the county level are important not just for the GMO/GE issue but more generally for other agriculture and forestry issues that require local control. Ultimately, SB633 did not pass in the regular session, but it was brought back as SB863 by Governor Kitzhaber in the special session of the same year (Levinson 2014; Wilce 2014). SB863 was passed in August 2013 as a part of grand bargain having to do with funding for the State of Oregon Public Employees Retirement System (PERS). Senator Bates and Representative Buckley successfully negotiated an exemption for Jackson County because it had qualified for the ballot before the January 31st deadline. Josephine County, on the other hand, was not exempted because it had not qualified for the ballot before the deadline.

Research participants saw SB863 as a direct action against the principles and foundations of democracy because state preemption laws deny local residents their right to democratic processes. One key informant shared that the preemptive seed law was one of the main reasons she and others joined the campaign:

> Wow. Here we have this opportunity in Jackson County… [because] we were grandfathered in… we're one of the few counties in Oregon that have the ability to pass this ban right now. And if we don't do it now, we don't know that there will be an opportunity in the future. It was definitely a driving force for quite a few people in Jackson County to work on the campaign (Elise 2017).

Both Jackson and Josephine Counties moved forward with their local ballot measures in the spring of 2014. GMO-free Jackson County ballot measure 15–119 was passed by citizens on May 20, 2014 with 65.79% of the vote. GMO-free Josephine County ballot measure 17–58 was voted on and passed by citizens on May 20, 2014, with 58% voting in favor of the ballot measure. Although a full discussion of the ensuing lawsuits is beyond the scope of this article, a brief summary is important to note. In 2014, a lawsuit was filed by two GMO alfalfa farmers in Jackson County. The court ultimately ruled against the plaintiffs and Measure 15–119 was upheld in June 2015. In 2015, the day before Josephine County's ban was to take effect, a lawsuit was filed by a GMO sugar beet farmer. The court ruled in favor of the state preemptive law and Measure 17–58 was never enacted. Nonetheless, Oregonians for Safe Farms and Families (who ran the Josephine County campaign) remains committed to the original vision of a GMO/GE-free zone for Josephine County to protect the integrity of the food supply and farmers in the region:

> OSFF continues to work for the will of the people and implementation of the ordinance restricting the growth of genetically engineered seed in Josephine County to be enforced (Oregonians for Safe Farms and Families, Executive Director 2017).

The preemptive seed law highlights a point of tension between local and State of Oregon acts of food sovereignty that was unsuccessfully negotiated. Despite the inescapable curtailing of food democracy discussed above, democratic values of community participation and collaboration between local communities and the State of Oregon were still envisioned by research participants as routes to long-term structural change. In January 2015, Our Family Farms established the educational arm of the organization to continue the outreach and promotion of a GMO/GE-free Seed Sanctuary in southern Oregon. The Seed Sanctuary is seen as a mechanism to protect and promote seed diversity and traditional food supply through public outreach and agricultural community workshops (Our Family Farms 2017). Our Family Farms and Oregonians for Safe Farms and Families remain committed to the vision of a GMO/GE-free Rogue Valley to protect the integrity of the food supply and farmers in the region. Both OFF and OSFF continue to actively work on GMO/GE-free legislation at the state and national levels. In 2017, two legislative bills were put forth in the State of Oregon. The GMO Liability Law, HB2739, would hold GMO/GE patent holders financially responsible for damage caused by contamination of non-GMO/GE crops. The Local Control Law, HB2469, would overturn SB863, the Seed Preemptive Law, and allow local governments to determine their own agricultural policy. Although neither bill was passed in 2017, informants believed the process raised awareness among the legislators. Efforts to overturn the preemptive law and to create a seed sanctuary in the Rogue Valley highlight a point of ongoing tension between local and state acts of food democracy.

Discussion and conclusion

> Who controls the seed gains a substantial measure of control over the shape of the entire food system. (Kloppenburg 2014)

One finding of this study is that from its inception, the GMO/GE-free movement in southern Oregon was in the process of negotiating multiple sovereignties, including concerned citizens in the community, local conventional and organic farmers and seed growers, GMO/GE farmers and the State of Oregon. Indicators of food sovereignty center on people having the right and ability to define their food polices and strategies with respect to food culture, food security, sustainability and use of natural resources. In passing a ban on the use of genetically-modified (GMO) seeds in agriculture, Jackson County, Oregon has obtained some measure of food sovereignty. However, the term food sovereignty was generally rejected by research participants because the nature of the sovereign was too broad and too vague. This is important to note as many scholars also argue that the vague nature of the sovereign has inhibited and prevented the implementation of the food sovereignty framework (see Edelman 2014; Hospes 2014; Iles and Montenegro de Wit 2015; Schiavoni 2015, 2016; Shattuck et al. 2015). Marc Edelman (2014) illustrates this in his article on food sovereignty regulatory challenges which highlights the ambiguity of the sovereign in both definition and application, as well as the fact that many food sovereignty movements are engaged in adversarial relationships with the governments with which they are interacting. To combat the dichotomy of locality versus the nation-state, Shattuck et al. suggest that rather than looking at food sovereignty as being either of the nation or of the people, food sovereignty frameworks should leave room for "different sovereign actors to coexist, with the terms of their engagement under ongoing negotiation" (2015, p. 425). However, this research shows that the coexistence of different sovereign actors is exceptionally challenging to implement as governments and civil society often have opposing viewpoints and agendas.

Food democracy, a core component of food sovereignty, is one possible way to deal with the ambiguity of the sovereign in food sovereignty. Food democracy is also the term used and the strategic actions employed by research participants. However, a secondary finding of this study is that food democracy is under threat when governments enact laws that deny citizens their right to democratic process. This was evidenced in the agricultural seed preemptive law enacted by the State of Oregon in 2013 that denied any county in Oregon (other than Jackson County, which was grandfathered in) from passing a ban on GMOs. Small-scale farmers and local communities are the most affected by decisions regarding the seed system and therefore within a food democracy framework should be given support to define their own food systems, rather than being undermined by government and corporate agrifood interests. Preemptive seed laws not only deny local residents the right to democratic processes and the right to determine what their food system will look like, they also hinder and curtail efforts for a true food democracy and stifle the advancement of food sovereignty.

The central tenets of food sovereignty and food democracy are in direct opposition to the concentration of power and decision-making processes by large corporations and the influence those corporations exert on governments and policy. The food sovereignty movement specifically seeks to return decision-making control to producers and consumers in the food system while promoting agrifood policies that favor small-scale producers over multinational corporations. As such, the role of law as it pertains to food systems should encompass multiple actors—including citizens, institutions and government—that influence and frame decision-making processes and policies (Kennedy and Liljeblad 2016). However, food policy is currently dominated by corporate interests and governmental regulations that favor multinational corporations with limited civil society involvement (Hospes and Brons 2016). In fact, in the US the predominant regulatory approach to GMO/GE applications in agriculture has been, and continues to be, industry self-regulation of biotechnology (Falkner and Gupta 2009; Hospes and Brons 2016; Newell and Glover 2003; Perry 2016). Regulatory capture occurs when governmental regulations favor corporate interests as a result of industry actions, such as lobbying and campaign donations. An example of regulatory capture occurred in 2013 during the GMO-free ballot initiatives in southern Oregon when Monsanto, Syngenta and other agribusiness corporations contributed more than $127, 000 to legislators' campaign committees following the announcement of the agricultural seed preemptive law that was voted in during the special session of the same year (Loew 2017).

Newell and Glover note that the privileged position afforded to the biotechnology industry in policy debates stems from "their superior access to scientific expertise and capital," as well as "the strategic importance attached to biotechnology by governments" (2003, p. 3). This affords multinational biotechnology corporations the power to influence policy by proposing solutions to policymakers who struggle to understand and respond to biotechnology with appropriate legislation (Newell and Glover 2003). The emergence of civil society movements in the US to contest the ongoing commercialization of GMO/GE crops and to influence policy stems in part from the barriers to appropriate public regulation in this arena, as well as a lack of citizen involvement in setting policy (Alkon and Mares 2012; Allison and Varone 2009; Ayres and Bosia 2011; Brent et al 2015; Clendenning et al 2016; Dekeyser et al. 2018; Newell and Glover 2003; Schiavoni 2015).

In conclusion, a major objective of this research was to analyze what the different expressions of food sovereignty reveal about the weakness in its implementation. The enactment of food sovereignty at a relational scale that creates and sustains multiple sovereignties has to date largely been a theoretical concept (Conti 2016; Iles and Montenegro de Wit 2015; Schiavoni 2015, 2016; Shattuck et al. 2015). This study provides empirical data on the challenges of enacting multiple food sovereignties when citizens' sovereignty is trumped by acts of state sovereignty that appear to be influenced by multinational corporations. Schiavoni (2016) highlights that the food sovereignty agenda calls for interactions between the nation-state and civil society to create a mutually reinforcing 'nationstate-society' synergy capable of creating systemic change or reform. As such, 'nation-state-society' synergy must include more involvement from citizens in decision-making processes and more control at the local level. The same is true for the local state and communities within its borders. In a democratic country the government has a mandate to protect the people and to do the will of the people, but its actions often favor agrifood corporations. Democracy is one way to help people navigate the tensions between citizen and governmental acts of sovereignty through elected representatives and a legislative system. Democracy hinges on the idea that the perspectives of all citizens are valuable and essential to decision-making processes, and thus it runs deep in citizen responses to corporate overreach (Carlson and Chappell 2015). However, when unsuccessful legislative bills are brought back as backdoor riders in a special session without adequate time for citizen response and preempting citizens' ballot initiatives is a result, neither food sovereignty at a relational scale nor food democracy are possible.

Scholars and food sovereignty activists generally recognize that food sovereignty cannot be disseminated identically across cultures but needs to adapt to rhythms of local peoples, including their social, political and cultural ideology and practices (Alkon and Mares 2012; Alonso-Fradejas et al. 2015; Ayres and Bosia 2011; Iles and Montenegro de Wit 2015; Shattuck et al. 2015). In other words, it is the value of local diversity and experiences that have the potential to shape different meanings and acts of food sovereignty (Shattuck et al. 2015). As such, although the food sovereignty framework offers a vehicle for change, the socio-political power dynamics between local communities and state acts of food sovereignty, as well as the vague nature of the sovereign, continue to pose practical and ethical dilemmas for implementing food sovereignty at the local level.

Acknowledgements I would like to thank my advisor, Dr. Lisa Price, for her ongoing support, encouragement and feedback throughout all aspects of the research project. I would also like to give special thanks to Our Family Farms Coalition who granted access to study participants and archival records which made this study possible.

Compliance with ethical standards

Ethical approval All procedures performed in studies involving human participants were in accordance with the ethical standards of the Oregon State University Institutional Review Board (*7476*) and with the 1964 Helsinki declaration and its later amendments or comparable ethical standards.

Informed consent Informed consent was obtained from all individual participants included in the study.

References

Agarwal, Bina. 2014. *Food sovereignty, food security and democratic choice: Critical contradictions, difficult conciliations. Paper presented at the food sovereignty: A critical dialogue*. New Haven: Yale University.

Alkon, Alison Hope, and Teresa Marie Mares. 2012. Food sovereignty in US food movements: Radical visions and neoliberal constraints. *Agriculture and Human Values* 29 (3): 347–359.

Allison, Christine Rothmayr, and Frederic Varone. 2009. Direct legislation in North America and Europe: Promoting or restricting biotechnology. *Journal of Comparative Policy Analysis: Research and Practice* 11 (4): 425–449.

Alonso-Fradejas, Alberto, Saturnino M. Borras, Todd Holmes, Eric Holt-Giménez, and Martha Jane Robbins. 2015. Food sovereignty: Convergence and contradictions, conditions and challenges. *Third World Quarterly* 36 (3): 431–448.

Ayres, Jeffrey, and Michael J. Bosia. 2011. Beyond global summitry: Food sovereignty as localized resistance to globalization. *Globalizations* 8 (1): 47–63.

BallotPedia. 2015. Local GMO on the Ballot. In Local GMO on the Ballot: The Encyclopedia of American Politics. https://ballotpedia.org/Local_GMO_on_the_ballot. Accessed 16 Mar 2016

Bernard, H.Russell. 2006. *Research Methods in anthropology: Qualitative and quantitative approaches*, 4th ed. Lanham: AltaMira Press.

Brent, Zoe W., Christina M. Schiavoni, and Alberto Alonso-Fradejas. 2015. Contextualising food sovereignty: The politics of convergence among movements in the USA. *Third World Quarterly* 36 (3): 618–635.

Buiatti, M., P. Christou, and G. Pastore. 2012. The application of GMOs in agriculture and in food production for a better nutrition: Two different scientific points of view. *Genes Nutrition* 8: 255–270.

Carlson, Jill, and M. Jahi Chappell. 2015. *Deepening food democracy*. Geneva: Institute for Agriculture and Trade Policy.

Clendenning, Jessica, Wolfram H. Dressler, and Carol Richards. 2016. Food justice or food sovereignty? Understanding the rise of urban food movements in the USA. *Agriculture and Human Values* 33 (1): 165–177.

Conti, Mauro. 2016. Food sovereignty agenda of transnational social movements in the UN global goverance. Paper presented at the Global governance/politics, climate justice & agrarian justice: Linkages and challenges, The Hague, Netherlands February 2016.

Corbett, Alicia. 2016. Preemption—Lessons from the Federal GMO Disclosure Law. The Network for Public Health 2017.

Dekeyser, Koen, Lise Korsten, and Lorenzo Fioramonti. 2018. Food sovereignty: Shifting debates on democratic food governance. *Food Security* 10: 223–233.

Edelman, Marc. 2014. *Food sovereignty: Forgotten genealogies and future regulatory challenges. Paper presented at the food sovereignty: A critical dialogue*. New Haven: Yale University.

Fairbairn, Madeleine. 2012. Framing transformation: The counter-hegemonic potential of food sovereignty in the US context. *Agricultural and Human Values* 29 (2): 217–230.

Falkner, Robert, and Aarti Gupta. 2009. The limits of regulatory convergence: Globalization and GMO politics in the south. *International Environmental Agreements* 9: 113–133.

FIAN. 2005. Food sovereignty framework: Concept and historical context. *Food Sovereignty Framework*, 1–23.

Fitting, Elizabeth. 2006. Importing corn, exporting labor: The neoliberal corn regime, GMOs, and the erosion of Mexican biodiversity. *Agriculture and Human Values* 23 (1): 15–26.

García-Sempere, Ana, Moisés Hidalgo, Helda Morales, Bruce G. Ferguson, Austreberta Nazar-Beutelspacher, and Peter Rosset. 2018. Urban transition toward food sovereignty. *Globalizations* 15 (3): 390–406.

Glover, Dominic. 2010. The corporate shaping of GM crops as a technology for the poor. *Journal of Peasant Studies* 37 (1): 67–90.

GMO-free Europe. 2019. European GMO-free regions conference. https://www.gmo-free-regions.org/. Accessed 12 June 2019.

Godfray, H.Charles, et al. 2010. Food security: The challenge of feeding 9 billion people. *Science* 327: 812–818.

Hassanein, Neva. 2008. Locating food democracy: Theoretical and practical ingredients. *Journal of Hunger & Environmental Nutrition* 3 (2–3): 286–308. https://doi.org/10.1080/19320240802244215.

Herbers, Greg. 2016. Ninth circuit holds anti-GMO regulations in Hawaii preempted by federal and state law. https://www.forbes.com/sites/wlf/people/gregherbers/. Accessed 2017.

Hospes, Otto. 2014. Food sovereignty: The debate, the deadlock, and a suggested detour. *Agricultural and Human Values* 31 (1): 119–130.

Hospes, Otto, and Anke Brons. 2016. Food system governance: A systematic literature review. In *Food systems governance: Challenges for justice, equality and human rights*, ed. Amanda Kennedy and Jonathan Liljeblad, 13–42. New York: Routledge.

Iles, Alastair, and Maywa Montenegro de Wit. 2015. Sovereignty at what scale? An inquiry into multiple dimensions of food sovereignty. *Globalizations* 12 (4): 481–497.

Kennedy, Amanda, and Jonathan Liljeblad. 2016. Introduction. In *Food systems governance: Challenges for justice, equality and human rights*, ed. Amanda Kennedy and Jonathan Liljeblad, 1–12. New York: Routledge.

Kloppenburg, Jack. 2014. *Re-purposing the master's tools: The open source seed initiative and the struggle for seed sovereignty. Paper presented at the food sovereignty: A critical dialogue*. New Haven: Yale University.

LaVia Campesina. 2007. Declaracion de Nyeleni. https://www.nyeleni2007. Accessed 2014.

Levinson, L. R. 2014. Taking back the commons: Motivating factors for the local control of GMOs. Master of community and regional planning graduate theses and dissertations. Paper 14065, Iowa State University.

Library of Congress. 2014. Restrictions on genetically modified organisms. Global Legal Research Center.

Lieberman, Sarah, and Tim Gray. 2008. GMOs and the developing world: A precautionary interpretation of biotechnology. *British Journal of Politics and International Relations* 10 (3): 395–411.

Loew, Tracy. 2017. Oregon bill would restore local control over GMO crops. Statesman Journal. https://www.statesmanjournal.com/story/news/politics/2017/02/08/oregon-bill-would-restore-local-control-over-gmo-crops/97643898/. Accessed 15 May 2019

MacRae, Graeme. 2016. Food sovereignty and the anthropology of food: Ethnographic approaches to policy and practice. *Anthropological Forum* 26 (3): 227–232.

McCauley, Darren. 2015. Protest, politics and produce: A resource account of anti-genetically modified organism activism. *Local Environment* 20 (1): 34–49.

McMichael, Philip. 2014. *Historicizing food sovereignty: A food regime perspective. Paper presented at the food sovereignty: A critical dialogue.* New Haven: Yale University.

Navin, Mark Christopher, and J.M. Dieterle. 2018. Cooptation or solidarity: Food sovereignty in the developed world. *Agricultural and Human Values* 35: 319–329.

Newell, Peter, and Dominique Glover. 2003. Business and biotechnology: Regulation and the politics of influence. In *Biotechnology Policy Series*. Sussex, England: Institute for Development Studies.

Nugent, David. 2008. Democracy otherwise: Struggles over popular rule in the Northern Andes. In *Democracy: Anthropological approaches*, ed. Julia Paley, 21–62. Santa Fe: The School for Advanced Research.

Nyéléni. 2007. Declaration of the Forum for Food Sovereignty, Nyéléni 2007. In Declaration of the Forum for Food Sovereignty, Nyéléni 2007. https://www.nyeleni.org/spip.php?article290. Accessed 29 June 2017.

Organic Consumers Association. 2005. US States Passing Laws to Block Local GMO-Free Ordinances. *Organic Consumers Association*.

Our Family Farms Coalition. https://www.ourfamilyfarms.org/. Accessed 2016.

Paley, Julia. 2002. Toward an anthropology of democracy. *Annual Review of Anthropology* 31: 469–496.

Patel, Raj. 2010. What does food sovereignty look like? In *Food sovereignty: Reconnecting food, nature and community*, ed. Hannah Wittman, AnnetteAurelie Desmarais, and Nettie Wiebe. Oakland: Food First.

Pearson, Thomas W. 2012. Transgenic-free territories in Costa Rica: Networks, place, and the politics of life. *American Ethnologist* 39 (1): 90–105.

Pechlaner, Gabriela, and H. Renting. 2012. GMO-free America? Mendocino County and the impact of local level resistance to the agricultural biotechnology paradigm. *International Journal of Sociology of Agriculture* 19 (3): 445–464.

Perry, Mark. 2016. Sustaining food production in the Anthropocene: Influences by regulation of crop biotechnology. In *Food systems governance: Challenges for justice, equality and human rights*, ed. Amanda Kennedy and Jonathan Liljeblad, 127–142. New York: Routledge.

Porter, Matthew. 2013. State preemption law: The battle for local control of democracy. *Pesticides and You* 33 (3): 5.

Prost, Sebastian, Clara Crivellaro, Andy Haddon, and Rob Comber. 2018. Food democracy in the making: Designing with local food networks. Paper presented at the proceedings of the 2018 CHI conference on human factors in computing systems, Montreal QC, Canada.

Schiavoni, Christina M. 2015. Competing sovereignties, contested processes: Insights from the Venezuelan food sovereignty experiment. *Globalizations* 4: 466–480.

Schiavoni, Christina M. 2016. The contested terrain of food sovereignty construction: Toward a historical, relational and interactive approach. *The Journal of Peasant Studies* 44: 1–32.

Shattuck, Annie, Christina M. Schiavoni, and Zoe VanGelder. 2015. Translating the politics of food sovereignty: Digging into contradictions, uncovering new dimensions. *Globalizations* 12 (4): 421–433.

Shawki, Noha. 2015. Transnationalism and diffusion: A study of the food sovereignty movements in the UK and Canada. *Globalizations* 12 (5): 758–773.

Spinelli, Margherita. 2013. The Policy Controversy of GMOs in Ecuador: Mechanisms of framing and polarization in coping with a wicked problem. Masters Thesis, Wageningen University.

State of Oregon. 2013. 77th Oregon Legislative Assembly 2013 1st Special Session Oct 2 2013. https://www.oregonlegislature.gov/.

Stone, Glenn Davis. 2002. Both sides now: Fallacies in the genetic-modification wars, implications for developing countries, and anthropological perspectives. *Current Anthropology* 43 (4): 611–630. https://doi.org/10.1086/341532.

Stone, Glenn Davis. 2010. The Anthropology of genetically modified crops. *Annual Review of Anthropology* 39: 381–400.

Tiberghien, Yves. 2012. The global battle over the governance of agricultural biotechnology. In *Regulating next generation agri-food biotechnologies: Lessons from European, North American, and Asian experiences*, ed. Michael Howlett and David Laycock. New York: Routledge.

US Census Bureau. 2015. QuickFacts, Jackson County, Oregon. https://www.census.gov/quickfacts/table/PST045215/41029. Accessed 29 June 2017.

USDA. 2012. Census of Agriculture: Jackson County Profile, ed. United States Census Bureau: https://www.agcensus.usda.gov/Publications/2012/Online_Resources/County_Profiles/Oregon/cp41029.pdf. Accessed 29 June 2017.

USDA. 2019. Agricultural Coexistence. https://www.usda.gov/topics/farming/coexistence. Accessed 3 July 2019.

Walsh-Dilley, Marygold. 2008. Localizing control: Mendocino County and the ban on GMOs. vol 26.

Wilce, Rebekah. 2014. Oregon's GMO Sellout. The progressive. https://progressive.org/magazine/oregon-s-gmo-sellout/. Accessed 24 Feb 2019.

Wittman, Hannah, Annette Aurelie Desmarais, and Nettie Wiebe (eds.). 2010. *Food sovereignty: Reconnecting food, nature and community*. Food First: Oakland.

Publisher's Note Springer Nature remains neutral with regard to jurisdictional claims in published maps and institutional affiliations.

Rebecka Daye has a Ph.D. in Applied Anthropology from Oregon State University. Dr. Daye is a socioeconomic anthropologist with a focus on morality and economics, social ecology, the political economy, local/global agrifood systems, Latin American studies and sociolinguistics. Her research has focused on various aspects of alternative agrifood systems including food sovereignty, social ecology, farming practices, GMO-free activism, environmental ethics and well-being in Ecuador and the United States.

Agriculture and Human Values (2020) 37:1027–1040
https://doi.org/10.1007/s10460-020-10031-x

Feeding the melting pot: inclusive strategies for the multi-ethnic city

Anke Brons[1,2] · Peter Oosterveer[2] · Sigrid Wertheim-Heck[1,2]

Accepted: 4 April 2020 / Published online: 16 April 2020
© The Author(s) 2020

Abstract

The need for a shift toward healthier and more sustainable diets is evident and is supported by universalized standards for a "planetary health diet" as recommended in the recent EAT-Lancet report. At the same time, differences exist in tastes, preferences and food practices among diverse ethnic groups, which becomes progressively relevant in light of Europe's increasingly multi-ethnic cities. There is a growing tension between current sustainable diets standards and how diverse ethnic resident groups relate to it within their 'culturally appropriate' foodways, raising questions around inclusion. What are dynamics of inclusiveness in migrant food practices? And what does this mean towards the transition to healthy and sustainable food? We study this question among Syrian migrants with different lengths of stay in the Netherlands. Our theoretical framework is based on practice theories, which emphasize the importance of socio-material context and of bodily routines and competences. We use qualitative methods, combining in-depth semi-structured life-history interviews with participant observation. Our findings indicate that inclusiveness takes different forms as migrants' food practices and the food environment change. Regarding health and sustainability in food practices, understandings and competences around particularly fresh food change over time among both short- and long-term migrants, replacing making things from scratch with seasonal products with buying more processed products and out-of-season vegetables and fruits. We conclude that the performances of food practices and their configurations in food environments and lifestyles are dynamic and cannot unequivocally be interpreted as in- or exclusive, but that a more nuanced understanding is required.

Keywords Food consumption · Healthy and sustainable food · Food environment · Multi-ethnic city · Migrants · Inclusiveness · Practice theories

Abbreviations

ST	Short-term [migrants]
LT	Long-term [migrants]
NGO	Non-governmental organisation

✉ Anke Brons
a.brons@aeres.nl

Peter Oosterveer
peter.oosterveer@wur.nl

Sigrid Wertheim-Heck
s.wertheim-heck@aeres.nl

[1] Food and Healthy Living, Aeres University of Applied Sciences Almere, Stadhuisstraat 18, Almere 1315 HC, The Netherlands

[2] Environmental Policy Group, Wageningen University, Hollandseweg 1, Wageningen 6706 KN, The Netherlands

Introduction

In light of the serious threats from global climate change and the increasing world population, the need for a shift toward healthier and more sustainable diets is evident (Burlingame and Dernini 2012; Garnett 2014; Lang 2017; EAT-Lancet Commission 2019). Precisely what constitutes a healthy and sustainable diet is, however, a highly debated issue, with new scientific evidence constantly being developed (Tilman and Clark 2014; Nelson et al. 2016; Mason and Lang 2017; Springmann et al. 2018; Béné et al. 2019). A growing body of literature integrates the two aspects of health and sustainability into an overall 'sustainable diet' (Hallström et al. 2018), which Lang (2017) defines as a diet that is health-enhancing, has low environmental impact, is culturally appropriate and economically viable (see also Burlingame and Dernini 2012).

Within this ongoing debate, there is a trend toward universalizing, uniform standards of healthy and sustainable diets that are valid across a nation or even the entire globe.

Reprinted from the journal

The recent EAT-Lancet report on Food, Planet and Health recommends that everyone adopts a 'planetary health diet': a universal diet rich in plant-based, fresh or minimally processed food (EAT-Lancet Commission 2019). On a national level, many countries have a similar approach to integrating health and sustainability into one set of dietary guidelines. In the Netherlands, the Dutch Nutrition Centre promotes the 'Wheel of Five', consisting of five main food groups that make up the recommended plate of an average Dutch consumer, which is based on traditional Dutch foods (Brink et al. 2017). The latest version of the Wheel of Five (2016) for the first time takes into account sustainability, by putting a limit on the amounts of meat and fish recommended per week and advising to consume legumes and nuts.

How can such standardized norms for a sustainable and healthy diet be combined with the central element of 'cultural acceptability' of diets, which requires taking into account dietary tastes and preferences of different cultural groups (Burlingame and Dernini 2012; Lang 2017)? This question is critical in light of the increasingly multi-ethnic cities in Europe (BCFN and MacroGeo 2018; Crul 2016). With a growing diversity in cultural groups moving to cities in larger numbers, there is a rise of so-called 'majority-minority cities' like Amsterdam or Brussels, in which the majority of the urban population consists of cultural minorities, each with their own food practices (Crul 2016).

The required scale of the transformation towards more sustainable and healthy food practices means all citizens need to be on board for it to be effective. Yet, in light of our multi-ethnic societies, generic standards might have limited reach, as they tend to lack cultural sensitivity while food consumption patterns are highly culturally defined (Nicolaou et al. 2009). Additionally, migrants and their home-country food practices may culturally clash with the host culture's definition of what constitutes healthy and sustainable food (Guthman 2008; Johnston et al. 2011; Rice 2015). Moreover, it is crucial to understand cultural acceptability itself as a process rather than as a static goal (Hammelman and Hayes-Conroy 2015). As Hammelman and Hayes-Conroy (2015) argue, the focus of current urban policy is often just on availability of culturally appropriate food, whereas food is much more than just a 'nutrient vessel': it comes with important cultural values and identity. This is particularly relevant in light of increasing calls for more inclusive food system transformations (Bui et al. 2019; Dubbeling et al. 2017; Raja et al. 2017): what exactly does this inclusiveness mean?

This question is also important because much of the current literature on in- and exclusion regarding access to 'good food' focuses primarily on the supply side to explain exclusion, looking at the influence of retail availability and product range on consumption patterns of poor ethnic minorities (Walker et al. 2010). However, as authors such as Alkon et al. (2013), Shannon (2014), Bedore (2014) and more recently also Allcott et al. (2019) argue, the relationship between consumption and provision is not as unidirectional as often thought. Allcott et al. (2019)'s study shows that exposing poorer households to the same products and prices available to higher income households only reduced nutritional inequality by roughly ten percent. Moreover, most literature on in- and exclusion regarding food access is geographically oriented towards the US, where the focus is on black and poor minorities who live in strongly segregated urban environments (Mata 2013; Raja et al. 2008; Walker et al. 2010). These spatial settings generally differ from the European context in their food environments, with a lower prevalence of 'food deserts' and less segregated ethnic neighbourhoods (see for instance Helbich et al. 2017 on Amsterdam).

In short, there is an urgency to understand how inclusiveness regarding healthy and sustainable food works within a multi-ethnic urban context. What are dynamics of inclusiveness in migrant food practices? And what does this mean for the transition toward healthy and sustainable food? In this paper we start by referring to inclusiveness primarily in relation to culture, as being respectful of cultural tastes and preferences (Sustainable Development Goals 2015), further exploring the meaning of this concept through our empirical work.

In answering our research questions, we conducted a comparative analysis between short- and long-term Syrian migrant residents in the Netherlands, investigating how food practices and understandings of health and sustainability develop over time within a changing food environment. We focus on Syrian migrants because this allows for the short-/long-term migrant comparison as the Netherlands had two Syrian migration flows. One flow dates back several decades and relates to religious persecution of Syrian-orthodox populations in Syria. The other regards the recent influx of Syrians due to the civil war. Within and across these two groups, we study how and to what extent food practices change over the course of migration. We look at how inclusiveness works differently for short- and long-term migrants, also taking into account differences and changes in the food environment and in lifestyles over time. We use a practice theories approach because this is instrumental for uncovering dynamics and arriving at a nuanced understanding of the complexity of inclusiveness.

Below we elaborate on our practices theoretical perspective on inclusiveness and present the conceptual framework of our study. This is followed by an exposition of our methodological approach and a description of the population under study. We then proceed with presenting our empirical results in two main parts, following the main elements of our conceptual model, and end with a discussion and conclusion on our results in light of our research questions.

Fig. 1 Conceptual framework as inspired by Spaargaren and Van Vliet (2000)

Theoretical and conceptual framework

In obtaining more nuanced and contextualised understandings of inclusiveness within the dynamics of healthy and sustainable food consumption among short- and long-term (ST-LT) migrant groups, social practice theories that focus on the habitual nature of consumption appear especially suitable. When migrants arrive into a new food environment with existing country-of-origin routines and competences related to food, de- and reroutinization takes place. This dis- and re-embedding means some practices persist, others disappear and new practices may appear. A practice theories perspective highlights these dynamics by examining how food practices are dynamically co-constituted by their material (food) environment and changing lifestyles.

Theories of social practice focus on daily lives by means of identifying routinized behaviour, shared norms, knowledge and competences within a material context (Nicolini 2013). Practice theories aim to bypass both individualist and holistic social ontologies by conceptualizing social reality as made up of 'practices'. Practice theories are an aggregate of theories which emphasize different elements of practice, but key aspects across theorists are embodied routines, skills and knowledge, shared (social) meanings, norms or understandings, and a material infrastructure (Halkier et al. 2011; Schatzki 2002; Shove et al. 2012; Spaargaren and Van Vliet 2000; Warde 2005; Reckwitz 2002).

Studying practices requires the double move of zooming in and out (Nicolini 2013). This two-step approach allows for both a concrete (zoomed-in) and abstract (zoomed-out) understanding of daily practices. Zooming in entails closely examining how practices are actually performed in everyday life, focusing on competences, tastes and preferences and how they may change over time within these food practices. Specifically, we pay attention to people's understandings of health and sustainability.

Subsequently, zooming out means taking a step back, to see how these situated practices relate to other practices in space and over time. In zooming out, practices are studied relationally, comparing and contrasting different instances of the performance of one particular practice over time and space, within the material context of the food environment, as well as in relation to bundles of practice (see Fig. 1). The food environment is defined by Turner et al. (2018) as "the interface that mediates people's food acquisition and consumption within the wider food system" (p. 95) and includes both market-based sources and home growing. This food environment contains many cues and clues for action that inform the performance of food practices, while changing practice elements or changing bundles of practice can in turn also transform the food environment (Warde 2016). Lastly, bundles of practices are sets of practices that are loosely linked based on co-existence in time or space (Schatzki 2011; Shove et al. 2012). Studying bundled practices means understanding how practices connect, either through restricting, enabling or conditioning each other (Shove et al. 2012).

Our conceptual model (see Fig. 1) illustrates how food practices are located in the interaction of the food environment or system of provision with the wider bundles of practices that together constitute daily life. We look at the practices of food acquisitioning and preparing food at home, paying particular attention to meanings or understandings associated with health and sustainability. These food practices are subject to change over time and space, and connect with the food environment and bundles of practices in different ways, corresponding to different degrees of de- and reroutinization among short- and long-term migrants. Sometimes novel practices emerge which link to existing food system practices, while other (elements of) migrant food practices may disappear or transform by integrating with locally dominant practices, as will be illustrated in more detail below.

These processes of forging connections between migrant food practices and locally dominant food habits can take different forms, which have implications regarding their inclusiveness. In considering inclusiveness, we start from the definition of social inclusion by Hinrichs and Kremer (2002) as "an ongoing and reflexive process of full and engaged participation by all interested or affected social actors, regardless of their socio-economic or cultural resources"

(p. 68). Although we are aware that inclusiveness is a broad concept which is determined by multiple social, economic and cultural factors, in this paper, we understand inclusiveness primarily in relation to the latter aspect of culture, i.e. as being culturally appropriate or acceptable (Hammelman and Hayes-Conroy 2015) or respectful of cultural tastes and preferences (Sustainable Development Goals 2015). We treat inclusiveness as an emerging and dynamic concept and study it inductively, identifying different dynamics of inclusiveness over time and space.

Methodology

Methods

Exploring the notions and dynamics of inclusiveness, our study focused on the lived experiences of migrants in their daily food routines and understandings of health and sustainability. Given its exploratory nature, we used qualitative methods to study practices. Given the short-term/long-term comparative nature of this study we combined in-depth semi-structured life history interviews (accounts of performative action) with food practice observations (direct access to performative action). We applied these methods to study the practices of acquisitioning food and preparing food at home. Within these food practices, we looked at the dynamics in tastes and preferences, and skills and knowledge, in relation to the experiences with the changing food environment. We specifically focused on uncovering (shifting) understandings of health and sustainability.

To inform the interview guide, the study started with the consultation of a dietician from the Arabic region currently residing in the Netherlands. She volunteered with a Dutch NGO called Pharos—a center of expertise that strives to reduce population health disparities and has dedicated programs for migrants. This expert consultation aided an initial understanding of Syrian food culture. Next, the interview guide was tested, after which interviews were conducted to understand current and past performances of the practices of acquisitioning and preparing food. Within the interviews, a life histories approach (Perez 2017) was applied in which people were asked to highlight food-related life events (i.e. favourite childhood food, first time cooking, first meal in the Netherlands) in relation to the two practices under study, to take a historical perspective and understand changing food practices over the course of migration.

The interviews were semi-structured and were conducted between summer 2018 and spring 2019. A total of 26 people were interviewed over a total of 23 sessions (some people were interviewed together, either as couples or friends). Most interviews were conducted in Dutch and on occasion an Arabic translator was used, a Libyan woman who was a native Arab speaker. She was trained by the first author to conduct interviews and did so on a voluntary basis, together with the first author. The interviews were transcribed in Dutch and coded through the open source coding programme QDA Miner Lite. General code categories were drawn up a priori, based on the interview guide, and more specific sub codes were added inductively. Quotes used in this article were translated by the first author who is a native Dutch speaker and have occasionally been edited for grammatical mistakes to ease comprehension. Finally, to get a better understanding of the interaction between food practices and the food environment, the practice of acquisitioning food was observed with five participants in Almere, by accompanying participants in their grocery shopping trips in various stores.

Sampling and recruitment

Participants were recruited based on their length of stay in the Netherlands, to arrive at a balanced sample of short-term (ST < 5 years) and long-term (LT > 5 years) Syrian migrants (see Table 1). Short-term migrants were recruited from the city of Almere, where Syrians have arrived only over the past five years and the food environment is still actively changing. Long-term migrants were recruited from other Dutch cities where there has been a longer presence of Syrians, among which most prominently the city of Enschede. Convenience and snowball sampling was used to recruit participants for both groups, who were approached in various ways. In Almere, the local language education center was approached to recruit participants. A retired volunteer at the Almere asylum seeker center who had a network among the Syrian population was also contacted to recruit participants. The translator was also asked to recruit among her own network among Syrians in Almere. For long-term residents, a primary school in Enschede was approached through a personal contact who put the researchers in touch with mothers from a Dutch language practice class and social hub. Other long-term residents were recruited through the researchers' personal networks.

In general, most Syrians face large differences in socio-economic status between Syria and the Netherlands, as a recent report on Syrians in the Netherlands outlines (Dagevos et al. 2018). Whereas almost everyone was employed in Syria, currently only 22% of the Syrians in the Netherlands hold a job, of whom almost half work below their educational level. Roughly a quarter of women used to be employed in Syria, while 42% indicated managing the household as their primary activity in Syria. Financially, most Syrians struggle to make ends meet in the Netherlands. The large majority of Syrians is religious: 76% identify as Muslim and 8% as Christian. In terms of health, the rate of overweight and obesity is significantly higher among

Table 1 General characteristics of short vs. long term research participants

	Short-term (N = 14)	Long-term (N = 12)
Average length of stay	3 years (range 10 months–4,5 years)	20 years (range 5–32 years)
Average age	41 (22–63)	39 (31–52)
Average age of arrival in the Netherlands	38 years old (20–61)	18 years old (0–47)
Origin in Syria	Urban (Damascus, Aleppo, Homs, Qamishli)	10 rural, 2 urban (all 12 from North-West Syria)
Religion	Muslim, Syrian-Orthodox	Syrian-Orthodox
Reasons for migrating	Civil war	Religious persecution
Place of residence	Almere	Enschede, Zwolle, Rotterdam, Amsterdam
Educational level	5 university 3 vocational school 6 high school	3 university 4 vocational school 5 high school
Occupation outside the home	2 employed 2 volunteering job 5 taking language classes	4 employed 1 volunteering job
Housing condition	4 house (with garden) 10 apartment (with balconies)	House (with garden)

Syrians than among the average Dutch population, with 26% of youngsters (ages 15–24) and up to 75% of people over 45 being diagnosed as overweight (Dagevos et al. 2018).

Regarding our study population, there were some general differences between short- and long-term migrants (see Table 1). In terms of the total study population, 19 were female and 7 were male. All except one was married and had children, varying between babies and adult children who had already moved out or still lived in Syria. Almost everyone had come to the Netherlands with at least some relatives (often parents or in-laws).

Results and analysis

In this section we present our empirical results by making Nicolini (2013)'s two moves of zooming in and out. We zoom in on the practices of acquisitioning and preparing food, with specific attention to meanings of health and sustainability. Zooming out, we observe how these practices relate to the changing food environment and to changing bundles of practices or lifestyles, and elaborate how these interactions illustrate different dynamics of inclusiveness.

Zooming in: practice as performances

Acquisitioning food

Based on our empirical results we identify two types of acquisitioning practices: purchasing (through market-based sources, Turner et al. 2018) and home-growing. Among long-term migrants, food purchasing primarily takes place at regular supermarkets. Ethnic stores, either specifically Syrian or Turkish/Moroccan are also an important source.

Some specific products (milk, cheese and meat) are purchased directly at farms or slaughterhouses. Fresh market shopping is not very prevalent. Participants mentioned the restricted opening times, limited offer and the quality of produce as reasons for not frequenting the Dutch market:

> I do not go to the markets very much because the products on the market are not of good quality, especially in terms of freshness. In the Netherlands the market is not really fresh (M, age 45, LT (19yr))

Many long-term migrants are engaged in the practice of home-growing, which is often a continuation of habits from back in Syria, where most people lived in rural areas with gardens:

> Almost every house has a vegetable garden, we had one back home and now here. All my sisters-in-law have a garden. We eat fresh, we just pick and eat: Grapes, all kinds of fruit, cucumber, salad, a lot, everything (F, age 31, LT (8yr))

All long-term migrants in our study live in a house with a garden, which enables them to easily engage in home-growing. Some grow distinct varieties (particularly smaller sized zucchinis and eggplants), as a coping strategy for addressing their needs and preferences for culturally specific food that they could not buy anywhere. Besides more extensive home growing practices, almost all long-term migrants have at least a grapevine in their backyard to be able to make dolma or yaprak. This was almost everyone's favourite dish from childhood and consists of grape leaves filled with rice and vegetables or meat.

By contrast, almost none of the short-term migrants are engaged in home-growing. Only two used to have a garden back in Syria. Material housing conditions of short-term migrants also prevent them from growing their own food, as most people live in apartments rather than houses with gardens. Still, only one participant expressed his desire to grow his own vegetables if he would have a garden. As home-growing was not part of the daily routines of most short-term migrants in the past, this practice is not common now either. The regular Dutch supermarket and ethnic store figure prominently in their food purchasing practices. Moreover, the practice of fresh market shopping is integrated into the rhythms of daily life of short-term migrants, where they shop for fruits, vegetables and fish. Back in Syria, it was common to go food shopping almost daily, either at the fresh market or at small shops. These habits are continued in the Netherlands among the recently arrived migrants, many of whom shop almost daily for fresh products. This rhythm of daily shopping also fits well with shopping at ethnic stores, as in ethnic stores the price as well as the quality of fruits and vegetables is lower, meaning these products wither more quickly. This is however less of an issue for these migrants, as they buy fresh fruits and vegetables almost daily.

Among short-term migrants, online resources are also used in acquisitioning practices. This involves finding specific Syrian products through digital networks (Facebook, WhatsApp) in Arabic.[1] On these online platforms, information is exchanged about where to find a specific kind of Syrian vegetable; what to do with unknown, typically Dutch vegetables; which retail outlet offers the best food quality; or about a new Syrian business in the area. By being available in their native language, these digital tools provide easy access to culturally appropriate food. Both young and old short-term migrants engage in these online platforms, as this married couple notes:

> M: 'On Facebook, they always talk, what does that person make, where can you buy that? (…) Someone will say, 'I found a store. It's located in Amsterdam; it sells the small zucchinis. So who wants to go to Amsterdam the next day?' F: 'Yes, usually it is with older people. They cannot get used to life here easily. They still have left their heart and everything in Syria' (M+F, age 32, ST (4.5yr))

This digital coping strategy is not present among long-term migrants, neither young nor old. Some long-term migrants did recall other coping strategies to get specific food items, i.e. asking a relative to bring products from Syria on their travels.

These different coping strategies illustrate the flexibility and creativity of migrants over time in performing their cultural food needs and preferences in their acquisitioning

practices. Both short-term and long-migrants generally aim to continue their habits, in which there are some differences among short- and long-term migrants. Adjustment strategies to sustain cultural food practices are accordingly also different among short- and long-term migrants, which is related to changes in the food environment. The recent rise in online shopping means continuing cultural food practices allows for different coping strategies now than when entering the food environment over a decade ago, illustrating the dynamic interaction between food practices and the food environment which will be elaborated below when zooming out. In any case, these differences over time and between groups of migrants illustrate how understanding what is inclusive here and what is not is hard to distinguish from an outsider's view, with migrants sometimes happily taking up their own role in getting what they want and need, and at other times suffering from not being able to eat their preferred cultural food.

Preparing food

Moving to the practice of preparing food, there are differences in the types of food prepared by short- and long-term migrants. Long-term migrants have gained knowledge about Dutch cuisine and are skilled at cooking typical Dutch dishes, which enables them to regularly prepare these meals. By contrast, short-term migrants are constrained by their lack of competences and know-how about what actually constitutes Dutch cuisine and how to prepare it, although they are curious about it—on several occasions during interviews the first author was asked about Dutch food. Short-term migrants almost exclusively prepare the Syrian dishes that they are more familiar with, which take significantly more time and skills to prepare.

A recurring theme among all participants is the understanding that fresh is best. Fresh is associated with healthfulness and tastiness, a good and natural rather than a bad and chemical taste, and with being rich in vitamins. Fresh food is also of cultural value: being able to prepare food from scratch is seen as a sign of being a good Syrian woman. For long-term migrants, fresh is associated with home-growing:

> You have to eat fresh. I'd never eat a ready-made meal. I think it's because we grew up with fresh, my father's vegetable garden, we are used to it (F, age 32, LT (28 yr))

Although the importance of fresh perpetuates over the course of migration, the actual performances around fresh food change upon coming to the Netherlands, among both short- and long-term migrants. In Syria, doing 'fresh' food involved making food from scratch with products from the season that would be stored to last throughout the winter. Fruits and vegetables were bought in bulk (50–100 kg) at

---

[1] The first author got access to the online platforms by looking at them together with the translator.

low prices in the season when it was actually fresh. The practice of preserving was performed by women, who would together engage in canning or drying fruits and vegetables, making tomato paste, all kinds of jam or 'makdous' (stuffed eggplant). The older female research participants were all engaged in this practice back in Syria, and the younger women who migrated before coming of age also recall their mothers and grandmothers doing it in Syria:

> If you see the somewhat older Syrian women, they can make so many things. Those jars, the readymade things, they just make it themselves, they make everything themselves (F, age 30, LT (9 yr))

After moving to the Netherlands, fresh remains a central element in participants' food practices but is performed differently. Long-term migrants who engaged in home-growing still consume fresh vegetables from their gardens in summer, but in winter purchase out-of-season vegetables, while those not engaging in home-growing started buying fresh fruits and vegetables year-round regardless of the season. In the Dutch food environment, most fruits and vegetables are always available with much smaller price differences between in and out of season than in Syria. Rather than making from scratch and preserving food, participants now purchase ready-made tomato paste and jam in an ethnic store or a regular supermarket:

> R: I always have to have cucumber, fruit and (…) I have to have it. I: And did you eat that in Syria? Could you buy it all year? R: Yes, you can get it. But it's very expensive (F, age 45, LT (28 yr))

> So here, all the Syrian women take their habits with them. So they are also busy making 'makdous'. But they stopped doing it here. It's difficult here, because everything is available. Everything is cheap. So why do I do it and then I am tired. I can just buy it at the same price (F, age 32, ST (4.5 yr))

This shift towards buying out of season is mostly driven by convenience and financial incentives, and also occurs among the older women interviewed who used to engage in food preserving until recently in Syria, but who do not see the need to continue in the Netherlands as everything is now always available. Consequently, tensions or contradictions between 'doings and sayings' (Schatzki 1996) around fresh food occur. In terms of 'sayings', participants repeatedly emphasize the importance of fresh, seasonal food. Complaints about the taste of out of season vegetables and fruits are also abundant:

> Everything comes from the fridge. Nothing is fresh. (…) For example there are the cucumbers, I saw them in Emmen [a city in the Netherlands], all in greenhouses. And if you get these big cucumbers or these

tomatoes and peppers and they all grow so quickly, well, then you know that it's not really fresh, that it doesn't grow by itself. It's all pumped up with needles, with water (M, age 35, LT (30 yr))

However, in terms of actual 'doings', this participant and others still buy these cucumbers and tomatoes all year long, regardless of the season. Convenience seems to trump convictions and taste in the new food environment where everything is always available.

There are two exceptions to changing performances of practices around fresh, which involve the products of grape leaves and labneh (a kind of strained yoghurt or fresh cheese). As mentioned before, grape leaves are essential for preparing the popular dish of dolma or yaprak. However, some long-term migrants note that fresh grape leaves were hard if not impossible to buy in the Netherlands, so many resorted to drying the grape leaves from their backyard grapevines in order to be able to also make yaprak in winter. By contrast, neither growing nor drying grape leaves is common practice among short-term residents, who instead buy dried grape leaves at ethnic stores.

The second example concerns the practice of making labneh. This practice perpetuates across migration among some, both long- and short-term, older and younger practitioners, although they are engaged in it for different reasons. Two relatively young interviewees (ages 34/35, ST (3.5/4 years) were recruited to the practice only upon coming to the Netherlands. They enrolled because they could not find the labneh they wanted in the existing food environment (similar to the grape leaves). They acquired the skills and competences to make labneh online, through Facebook and YouTube. However, this performance of the practice disappeared as soon as it had to compete with another, new means of acquiring labneh (i.e. online shopping), which was more convenient. This practice enrolment and engagement differs from other both long- and short-term migrants who already learned the required skills from their mothers when growing up, either in the Netherlands or in Syria, and are in the habit of doing it from a younger age. This example illustrates how the performances of a practice may look similar but that meanings, recruitment and engagement within it can take different forms, contributing to different dynamics of inclusiveness.

Health and sustainability

Within the practices of acquisitioning and preparing food, we zoom in further on understandings of health and sustainability, and their possible change over the course of migration. In being de-routinized upon coming to a new country with a different food environment and food habits, interviewees had often started to reflect more on their own food practices, with changing ideas about health and about the healthfulness of Syrian cuisine. In explaining what meanings they associate with health, many participants describe health in terms of what a healthy diet should *not* contain: a healthy diet consists of less fat, less sugar and less salt. This is often referred to as the opposite of the Syrian cuisine, which contains a lot of sugar and animal fat (in particular ghee):

> I don't think Syrian food is healthy. No, we use a lot of fat. Sometimes they make salad with olive oil, then it's OK, but most use butter very often, or ghee. And that has a lot of fat in it, and that's really not healthy I think. Because my mother also has issues with her cholesterol, and the doctor says, there's so much fat in Syrian food, take it easy and don't eat that too much (F, age 34, LT (born in NL))

This focus on what is not healthy and specifically participants' reflexivity towards Syrian cuisine came about in different ways by cues in the socio-material environment, either back in Syria or in the Netherlands. Some interviewees recall Syrian information campaigns on television in recent years on reducing fat and oil consumption. Yet, most had started reflecting on health in relation to their food practices after coming to the Netherlands. As one interviewee (F, age 32, ST (4.5 years)) notes, "Whereas in Syria it was my own choice to eat healthily or just meat and rice, in the Netherlands you read and hear about healthy food all the time".

In response, the family's food practices shifted to consuming more vegetables and fruits, less sugar and to having more diversity in their meals. The dietician also notes changing performances around health, illustrating this with a traditional Arabic dish called 'musakhan' that is typically served with Arabic flatbread which contains a lot of fat. Now, instead of the Turkish bread with all the oils and fats, she notes that people start using thin bread, tortilla etc.—"musakhan 2018". Others became aware of food-related health issues through personal experience, and recall changing their food practices (towards less fat, less sugar and less salt) after going to the doctor in the Netherlands for obesity, type 2 diabetes or another food-related health issue.

Sometimes there are explicit tensions between Dutch dietary guidelines and cultural food practices, leading to challenges for migrants. For instance, one interviewee (F, age 33, LT (20 years)) recalls that when she was pregnant and suffered from iron deficiency, the doctor prescribed her to eat more rye bread and apple spread, which is typically Dutch food. This required some cultural know-how and competences that this interviewee did not possess at the time, which was constraining for her. She felt a mismatch between the doctor's advice and her own food habits: "And at the beginning I was, how I do that because I am used to eating something different at home. (...) And she told me, eat that and eat that. And yeah, it's very difficult to eat differently".

Recognizing these constraints experienced by migrants in receiving food-related health recommendations, the dietician indicates that the NGO she volunteered with was actually in the process of releasing an Arabic version of the Wheel of Five, which indicates a kind of mutual reflexivity. In this version, the NGO will translate not only the texts but also the kinds of foods included in the guidelines into culturally appropriate foods, in light of the current discrepancies with the food practices of the many different ethnicities of refugees and migrants they work with. This serves as one empirical example of how healthy and sustainable dietary guidelines could become more inclusive by taking into account more culturally acceptable foods.

Another dominant understanding of health among participants is related to consuming fruits and vegetables. Some refer to the Syrian cuisine as rather healthy because it commonly includes fresh vegetables at all meals as well as many fruits. The need to incorporate sufficient amounts of vegetables into the diet is particularly brought about by having children. This life event sparks an increased motivation for eating more vegetables among many participants. The issue of sugar is also raised in relation to children as well as in a broader sense, where the Dutch food environment is sometimes blamed for its wide availability of processed sugary foods:

> These days you really have to pay attention to those things. (...) There's sugar in everything. And in Syria, three quarters of the food came from the land. Candy doesn't come from a tree, I always say. It's all natural wat you consume there. And here it isn't (F, age 33, LT (20 yr))

Many others do however note that traditional Syrian cuisine includes many sugary snacks such as baklava, which also poses health risks.

Moving to understandings of sustainability, the concept of sustainability is hard to translate in Arabic: the equivalent is not commonly known among participants. Only one of the short-term migrants knows the meaning of the word itself, because she took Dutch language classes—but only in relation to mobility, not food. Among long-term migrants, the term is recognized more frequently but does not figure in their daily food practices. Two participants associate sustainability with eating seasonal food. However, although there is much awareness, know-how and appreciation of consuming seasonal food, it does not figure strongly in their current food practices after migration.

Like seasonality, there are more aspects within interviewees' food practices that could be earmarked as sustainable, when relying on the abovementioned current guidelines for sustainable consumption practices (EAT-Lancet Commission 2019; Brink et al. 2017). Such practices or practice elements are 'inconspicuously sustainable': not intentionally sustainable but nonetheless having positive environmental effects (following Dubuisson-Quellier and Gojard 2016). For instance, religious fasting was common among many long-term migrants who were Syrian-Orthodox, which entailed consuming no land animal products on Wednesdays and Fridays nor during the 40 days of Lent before Easter—essentially eating a plant-based diet for two days a week. Consuming local food is another example: back in Syria, in particular long-term migrants who left more than a decade ago were in the habit of acquisitioning only locally produced food (i.e. produced in Syria) in stores and markets, or engaged in home growing. All of these routines and relationships with food (being connected to local and seasonal food consumption and/or production and having a plant-based diet) are an integral part of the cultural identity of these migrants but are also significant in terms of sustainability. These habits continue only among some long-term migrants after migration, by home-growing and shopping at local farms and slaughterhouses. In short, the process of migration changes practical understandings of health and sustainability as well as cultural relationships with food, with increased reflexivity in terms of health sometimes leading to healthier food practices, but potentially sustainable practices often change towards less sustainable practices over the course of migration.

Zooming out

Changing food practices, lifestyles and bundles of practices

In zooming out, we first look at how changes in food practices can happen through changing lifestyles or bundles of practices. Upon migration, many daily practices become deroutinized, as one interviewee who volunteers to help out newly arrived Syrians notes:

> The rhythm, (…) they really miss the system. They cannot live systematically, with everything being a routine. So for instance, having to wake up early in the morning, school, children, making appointments, being on time, they are not used to that (M, age 44, LT (18 yr))

This de- and re-routinization also (in)directly affects food practices, as "common patterns of adjustment reported [are] often a result of changes in practices other than those directly associated with eating" (Warde 2016, p. 133/4). We illustrate this by highlighting how within changing occupational household dynamics three practices bundle with and change food practices: working outside the home, caring for children or parents and going to school.

First, through enrolment in the practice of working, changes occur within eating practices. When participants start working, this almost immediately affects meal timing among both short- and long-term migrants. In Syria, the main meal took place between two and three in the afternoon, and a light dinner (similar to breakfast) was consumed around 7 or 8. In the Netherlands, as soon as one adult in the household starts working, these times shift to having the main hot meal around 5 p.m. or 6 p.m., as is common practice in the Netherlands: "I have to eat like Dutch people because I go to school, I work. The break time is like Dutch people so you have to change" (F, age 33, LT (20 years)). By bundling with working in this way, the temporality of eating practices thus changes and adapts more to local practice rhythms. This contrasts with households in which no one works, where meal timing continues according to Syrian rhythms, also among long-term migrants.

The practice of working also bundles with the practice of food preparing, marked by competition for the same resource: time. These practices are connected through changing gender roles which influence the temporality of both practices. Upon coming to the Netherlands, some women start to work or work at different times than in Syria. This constrains them by limiting their available time for cooking, leaving less time for the typically elaborate Syrian dishes and for cooking from scratch:

> Because my mother had more time at home, I work and before that I was studying and it's not like I always have a lot of time to make a big meal like Syrians (…). They cook a big meal almost every day. I can't do that every day. It's more like in the evening some soup, spaghetti, macaroni, some easier meals (F, age 33, LT (20 yr))

Some women now only prepare more elaborate Syrian dishes in the weekend, when they frequently have family and friends over for breakfast or lunch, as was common in Syria. Participants also express that caring for family (children or elderly parents) takes up more time in the Netherlands and similarly limits their time for food preparing.

Finally, enrolment in the practice of going to school leads to adjustments in the practice of eating, in terms of meal timing and meal content. Dutch eating routines at schools consist of bringing sandwiches to school which are consumed around noon. Adapting to these new practice rhythms is challenging at first for many Syrians. One interviewee (M, age 44, LT (18 years)) noted that he often hears about Syrian children not bringing sandwiches to school for lunch, as they are still in the habit of eating after school, around 3 p.m. However, after a while, Syrian children also start bringing sandwiches to school. This habit in turn influences eating practices at home by introducing Dutch bread to the breakfast, which is common even among short-term migrants who had an otherwise predominantly Syrian diet. In short, through changing lifestyles and enrolment into locally common practices with their rhythms, competences, materials and meanings, the performance of cultural food practices is sometimes constrained and sometimes enabled, with cultural identity changing concurrently. This illustrates how inclusiveness itself is dynamic, whereby what is culturally acceptable or not is changing over time and varies between different people and practices.

Changing food practices in a changing food environment

Migrant food practices are not only affected by changing lifestyles but are also embedded within a changing food environment. Building on the food-based life histories, changes in the food environment became apparent when comparing experiences of long- and short-term migrants. This goes both for the Syrian food environment migrants left and for the Dutch food environment they entered. When long-term migrants left, there was little 'multicultural' or 'globalised' food such as kiwis on offer in Syria, which was more common when short-term migrants left Syria. Moreover, the country was not at war yet when long-term migrants left, and the availability of food was not an issue, while for some short-term migrants, buying sufficient food in the Syrian food environment was a challenge during the war. This translated into different expectations of the food environment upon migration: some short-term migrants just aim for having enough food, which is more important than for instance consuming healthy food.

When the first long-term migrants we interviewed arrived in the Netherlands around thirty years ago, the offer of ethnic food was also limited. Ethnic stores in the Netherlands did exist but were Turkish or Moroccan rather than Syrian, and were not as omnipresent as today, which rendered the process of acquiring culturally appropriate food more challenging (see also Huizinga and van Hoven 2018). Over the past decade or so, ethnic food entered the mainstream food system, with Dutch supermarkets increasingly offering ethnic, including Syrian, food products:

> Now everything can be found. After the arrival of the Syrian migrants, the goods are everywhere. They can even deliver Syrian food to your home for free. Before there wasn't pomegranate syrup or tahini. Now if you need them you can call and it arrives within 24 hours. When I arrived here in 1992, it was not there because there was no demand. Eggplants were not there, basil, nobody knew it (M, age 45, LT (26yr))

With the increase in ethnic food products on supermarket shelves, short-term migrants face a different food environment than long-term migrants did in their days upon migration. This corresponds with different practice dynamics, rhythms and adaptation strategies among long- and short-term migrants in interaction with the changing food environment. For instance, one interviewee recalls having to drive 15 km to the next Turkish store:

> We went once every two months or so, but now I can find it everywhere. I told the new people, you have a lot more luck, you can find it anytime you want. You can even see it on Facebook (F, age 33, LT (20 yr))

This example illustrates how the food environment is actively changing, in a relationship of co-creation with migrant food practices. In Enschede, where there has been a continuous and growing community of Syrian migrants since at least two decades, there are multiple Syrian shops which have been present for over a decade. By contrast, in Almere with its relatively new Syrian community, the first Syrian store appeared only around two or three years ago, although many Turkish and Moroccan ethnic stores had been present in the food environment. The Almere food environment is also still actively changing, with for instance a new Syrian bakery opening during the interviewing period.

This interaction between food environment and migrant food practices demonstrates a variety in practice configurations of emergence, integration and transformation over different times and places, illustrating once more how cultural inclusiveness is dynamic rather than static as both the food environment and food practices change. We will illustrate this again through the case of labneh. In Enschede, Syrian migrants who wanted to make labneh started going to Dutch farmers to buy milk, who gradually ended up selling labneh themselves. Here, ethnic food was integrated into an existing local food provisioning practice, ran by Dutch entrepreneurs, implying a transformation of an existing practice. By contrast, in Almere a novel, separate practice emerged to provide for labneh: businesses initiated and run by Syrian migrants themselves, with labneh being ordered online and home-delivered. Similarly, for meat, participants from Almere shopped at ethnic stores which offer halal meat. In Enschede, participants instead went to the Dutch butcher who possessed the required skills to prepare meat for typically Syrian dishes such as kibbeh:

> I order meat from the butcher. I will say, 2 kilos for kibbeh, 3 kilos for kebab. He knows. Yes, sometimes he asks, for kibbeh or for kebab? (…) He definitely doesn't know the taste, but he knows the name (F, age 31, LT (8 yr))

These changing food environments indicate that there is interaction between food environment and food practices, and that this interaction does not evolve in the same way over time and space, indicating diversity between different practices and people also in terms of what might be inclusive or not.

Dynamics of inclusiveness in relation to health and sustainability

As our results indicate, food practices, their performances and configurations in food environments and lifestyles are dynamic and cannot unequivocally be interpreted as in- or exclusive. Returning to our understanding of inclusiveness as "full and reflexive participation" from Hinrichs and Kremer (2002) while being respectful of cultural dietary needs and preferences, our findings illustrate how this concept of inclusiveness can take different forms within specific contexts. As performances of practices within a food environment change over time and space, with an older versus a newer migrant community, so do the dynamics around inclusiveness change. We identify three main ways in which these dynamics work, which coincide with the three central elements of the conceptual model presented in Fig. 1: (1) the interaction between food practices and lifestyles or bundles of practices is dynamic; (2) performativity of food practices is dynamic; and (3) the interaction between food practices and the food environment is dynamic.

First, lifestyles or bundles of practices are not static. Changes in geographic and occupational household dynamics contribute to changes in cultural identity and related food practices. Dietary tastes and preferences change over the course of migration, which can be characterized by a spectrum running from maintaining one's cultural identity (multiculturalism) to adapting to the local cultural habits (homogenization), with many hybrid forms in between. Change along the spectrum does not always happen voluntarily: participation in some practices dictates the rhythm or content of others bundled to it. This makes it difficult to deem one particular practice or either end of the spectrum more or less inclusive than the other. For instance, by going to school, know-how about Dutch food habits increased and practices adapted accordingly to include Dutch bread into breakfast. At the same time, by starting to work, meal times had to change towards local practice rhythms. Both are examples of shifts toward cultural homogenization, but they are not necessarily similarly in- or exclusive. This also underlines the need to understand the cultural acceptability of food as a dynamic process rather than as static, as previously argued by Hammelman and Hayes-Conroy (2015).

Secondly, food practices and their performances change. Sometimes new practices emerged that became linked to existing food system practices (e.g. Syrian migrants starting a Syrian cheese business). Other times new practice elements were integrated into existing local provisioning practices (a Dutch farmer including Syrian labneh in their offer), thereby transforming local food system practices. Additionally, for migrants, acquiring food according to cultural tastes and preferences in a new food environment sometimes meant enrolment into new practices, as happened in the case of home growing, food preserving and online shopping practices. However, although the performance of some of these practices may look similar from the outside, they contained different meanings and modes of engagements. In the case of making labneh, some practitioners performed this practice because they were already used to doing it, whereas others started because they experienced a lack of availability of labneh in their new food environment. Unravelling these dynamics of practice shows how complex the issue of inclusiveness is: simply deeming the practice of making labneh as such to be in- or exclusive is difficult and does not do justice to the complexity of people's experiences.

Thirdly, food practices also have an interactive relationship with the food environment. Migrant food practices influence the food environment and vice versa, which works out differently in different times and spaces. Short-term migrants arriving into the current food environment that offers opportunities for digital communities and platforms develop different coping strategies to fulfil their cultural dietary needs and preferences than long-term migrants did back in their day, which translates into different practices.

Finally, relating these dynamics of inclusiveness to health and sustainability, we identify potential for latching onto existing elements of migrant food practices for transitioning towards a healthier and more sustainable food system. In terms of health, being de-routinized after migration stimulates reflection, foregrounding (unhealthy) elements of people's food practices, which has potential for making practices healthier. Furthermore, a tool like the Arabic food groups-based Wheel of Five encourages diversity in dietary standards, providing a better match with migrant food practices and making healthy food recommendations more culturally appropriate.

Regarding sustainability, the identified know-how on fresh, seasonal and local food among the migrants studied offers potential for the transition towards a more sustainable food system through a focus on such fresh, local and seasonal food. This focus fits well with some elements of migrant food practices, who are routinized in buying such food. Consuming—and to some extent also growing—local, seasonal and fresh food is thus essentially part of their cultural identity, and this cultural knowledge should be appreciated (Hammelman and Hayes-Conroy 2015). Instead of losing these cultural habits upon migration, these routines should be facilitated and encouraged—although perhaps not explicitly in the name of sustainability, as this framing did not resonate. Rather, a meticulous approach is needed to make the connection between sustainability and migrants' everyday food practices. For instance, as many migrants frequently shop at ethnic stores and/or fresh markets, offering more fresh, local and seasonal food that is also culturally acceptable in these places could be an interesting option for stimulating healthy and sustainable consumption.

Discussion and conclusion

This paper aimed to understand what inclusiveness means in light of our increasingly multi-ethnic cities. Appeals for a more inclusive food system suggest a current state of exclusion for vulnerable groups like migrants. However, our results indicate that rather than working from a normative frame that is imposed top-down on a given population, a more nuanced understanding of what constitutes in- and exclusion is required. Inclusiveness is a dynamic process, in which migrants can be capable of including themselves, demonstrating creativity in sustaining their cultural practices and developing coping strategies in interaction with a changing food environment. Our practice theories approach has been instrumental in identifying these dynamics. We see a number of ways in which our findings can contribute to shaping more robust pathways to a healthier and more sustainable food system that is also inclusive: (1) moving beyond a supply side only-perspective on in/exclusion; (2) taking a critical look at nationally or even globally defined dietary guidelines; (3) emphasizing citizens' creativity in organizing their food practices; (4) acknowledging that migrants are also consumers driven by 'lifestyle' needs like convenience; and (5) identifying health as an interesting access point for dietary change. At the core of these recommendations lies the observation that inclusive transitions to sustainable diets should be informed by how migrants actually engage with food in their daily lives—rather than making assumptions about their food habits and values from a distance—and that these food practices are dynamic and change over time and space.

First of all, in response to food desert thinking we referred to in the introduction, our findings highlight a more multidirectional and dynamic interaction between food environment and food practices. We fully support Hammelman and Hayes-Conroy (2015)'s call to look beyond food availability only, and start paying attention to "how cultural acceptability develops through complex relationships between people and food systems" (p. 44) in order to effectively understand cultural inclusiveness. Food culture, practices and the food environment are dynamic in multiple senses, with changes occurring in both the country of residence and country of origin. This requires appreciating the complexity of everyday life of how migrants negotiate their food practices, as our study has aimed to illustrate through a practice theories lens.

Secondly, in line with this complexity of daily life, dynamics and variation between groups of migrants, our findings illustrate how there are limits to the extent to which nationally defined dietary guidelines can be effective. For instance, as our findings demonstrated, seasonal food consumption may be prevalent in the home country but is affected and diminished by migrating to a new food environment. Even though migrants' 'home-country' practices of seasonal consumption might fit with the Dutch dietary guidelines in terms of consuming fresh foods, the specific foods associated with home-country seasonal consumption are not necessarily available as such in the Netherlands. Drawing up one set of guidelines that is culturally appropriate or inclusive is therefore complicated. Rather, adaptive and reflexive capacity is key, where migrants themselves are being involved, for instance following the initiative of the NGO Pharos to draw up an Arabic foods-based Wheel of Five in cooperation with health experts and consumers from the region.

Thirdly, for local governments working on inclusive food system transformations, our research suggests that policymakers should recognize citizens' flexibility and creativity when tailoring interventions. Inclusiveness is hard to measure based on simple socio-economic parameters but is rather diverse in form and subject to change over time and space. Migrants themselves are not passively waiting to be included but actively shape their food environment and develop creative coping strategies, as "knowledgeable and capable agents" (Giddens 1984), with some transformative capacity to actively interact with and change elements in the food environment to fulfil their dietary needs and preferences. Rather than using quantitative parameters such as only measuring the availability of certain "culturally appropriate" food items to indicate in-or exclusion, our qualitative approach shows the complexity of what inclusiveness means in practice, illustrating different dynamics between short- and long-term migrants and in interaction with changing food environments and bundles of practices.

Fourthly, convenience also played a role in the changing lifestyles and food practices of migrants. Inclusiveness means taking into account that also migrants seek convenience, in acquiring and preparing foods to their 'culturally appropriate' foods, which was illustrated by the example of preparing or buying labneh. This means that solutions such as proposing urban agriculture for cultural inclusion might not be appropriate when migrants' changing lifestyles increasingly require convenience. The supposition of migrants having the time and interest needed to grow their own food, practiced by many short-term migrants, might not uphold when lifestyles change, like women working out-of-home. This research only lightly touched upon convenience and further research into the role of convenience in relation to inclusiveness is required, which might differ between groups of people.

Finally, when defining an inclusive food system from a health and sustainability perspective, our findings indicate that health is an easier access point than sustainability. De- and reroutinization upon migration often includes moments of reflection that help transition to healthier food practices. For sustainability, the connection with migrant food practices is less obvious and more attention should be paid to how sustainability can be integrated within food practices. Moreover, sometimes there were trade-offs, as in the case of fresh vegetables, where migrants changed towards buying fresh year-round rather than eating preserved food, which was beneficial in terms of health but less so in terms of sustainability.

The present study has zoomed in on one particular group of migrants to conduct a comparison over time, among different lengths of stay. This allowed for an in-depth understanding of the dynamics of inclusiveness and change among this population group. While there are most likely similarities to be found in migrant groups from the same region with similarities in food culture, our sample also shows how diverse the dynamics of inclusiveness already are within one cultural group over time. In reflecting upon the cultural diversity of most current metropolises, further research is therefore required among other cultural groups, to explore to what extent the identified dynamics transpire among other migrant populations. This calls for a careful consideration of cultural food practices among different groups of migrants in a city in order to achieve truly inclusive strategies to feeding the multi-ethnic city.

Acknowledgements This research has been funded by Gemeente Almere (Almere 2.0) in cooperation with Flevo Campus. We would like to thank all participants who shared their stories with us and the editor and three anonymous reviewers for their constructive and helpful comments on earlier versions of the paper.

Author contributions All authors conceived and designed the experiments; the first author collected and analyzed the data; and all authors wrote the paper.

Open Access This article is licensed under a Creative Commons Attribution 4.0 International License, which permits use, sharing, adaptation, distribution and reproduction in any medium or format, as long as you give appropriate credit to the original author(s) and the source, provide a link to the Creative Commons licence, and indicate if changes were made. The images or other third party material in this article are included in the article's Creative Commons licence, unless indicated otherwise in a credit line to the material. If material is not included in the article's Creative Commons licence and your intended use is not permitted by statutory regulation or exceeds the permitted use, you will need to obtain permission directly from the copyright holder. To view a copy of this licence, visit http://creativecommons.org/licenses/by/4.0/.

References

Alkon, A.H., D. Block, K. Moore, C. Gillis, N. DiNuccio, and N. Chavez. 2013. Foodways of the urban poor. *Geoforum* 48: 126–135.

Allcott, H., R. Diamond, J.P. Dubé, J. Handbury, I. Rahkovsky, and M. Schnell. 2019. Food deserts and the causes of nutritional inequality. *The Quarterly Journal of Economics* 134 (4): 1793–1844.

BCFN and MacroGeo. 2018. Food & migration: Understanding the geopolitical nexus in the Euro-Mediterranean. BCFN and MacroGeo. https://paper.foodandmigration.com. Accessed 23 Mar 2020.

Bedore, M. 2014. Food desertification: Situating choice and class relations within an urban political economy of declining food access. *Studies in Social Justice* 8 (2): 207–228.

Béné, C., P. Oosterveer, L. Lamotte, I.D. Brouwer, S. de Haan, S.D. Prager, E.F. Talsma, and C.K. Khoury. 2019. When food systems meet sustainability—Current narratives and implications for actions. *World Development* 113: 116–130.

Brink, E.J., A. Postma-Smeets, A. Stafleu, and D. Wolvers. 2017. The wheel of five factsheet. Voedingscentrum website. https://mobiel.voedingscentrum.nl/Assets/Uploads/voedingscentrum/Documents/Professionals/Pers/Factsheets/English/Fact%20sheet%20The%20Wheel%20of%20Five.pdf. Accessed 23 Mar 2020.

Bui, S., I. Costa, O. De Schutter, T. Dedeurwaerdere, M. Hudon, and M. Feyereisen. 2019. Systemic ethics and inclusive governance: two key prerequisites for sustainability transitions of agri-food systems. *Agriculture and Human Values* 36 (2): 277–288.

Burlingame, B., and Dernini, S. 2012. Sustainable diets and biodiversity: Directions and solutions for policy, research and action. Food and Agriculture Organization of the United Nations (FAO) website. https://www.fao.org/3/i3004e/i3004e.pdf. Accessed 23 Mar 2020.

Crul, M. 2016. Super-diversity vs. assimilation: how complex diversity in majority–minority cities challenges the assumptions of assimilation. *Journal of Ethnic and Migration Studies* 42 (1): 54–68.

Dagevos, J., W. Huijnk, M. Maliepaard, and E. Miltenburg. 2018. Syriërs in Nederland. Sociaal Cultureel Planbureau website. https://www.scp.nl/dsresource?objectid=7ad7406c-723d-4954-a601-239b4ae10008&type=org. Accessed 23 Mar 2020.

Dubbeling, M., G. Santini, H. Renting, M. Taguchi, L. Lançon, J. Zuluaga, L. de Paoli, A. Rodriguez, and V. Andino. 2017. Assessing and planning sustainable city region food systems: insights from two Latin American cities. *Sustainability* 9 (8): 1455.

Dubuisson-Quellier, S., and S. Gojard. 2016. Why are food practices not (more) environmentally friendly in France? The role of collective standards and symbolic boundaries in food practices. *Environmental Policy and Governance* 26 (2): 89–100.

EAT-Lancet Commission. 2019. Food, planet, health: Healthy diets from sustainable food systems. The Lancet website. https://www.thelancet.com/journals/lancet/article/PIIS0140-6736(18)31788-4/fulltext. Accessed 23 Mar 2020.

Garnett, T. 2014. Changing what we eat: A call for research & action on widespread adoption of sustainable healthy eating. Food Climate Research Network website. https://www.fcrn.org.uk/sites/default/files/fcrn_wellcome_gfs_changing_consumption_report_final.pdf. Accessed 23 Mar 2020.

Giddens, A. 1984. *The constitution of society: Outline of the theory of structuration*. Berkeley: University of California Press.

Guthman, J. 2008. 'If they only knew': Color blindness and universalism in California alternative food institutions. *The Professional Geographer* 60 (3): 387–397.

Halkier, B., T. Katz-Gerro, and L. Martens. 2011. Applying practice theory to the study of consumption: Theoretical and methodological considerations. *Journal of Consumer Culture* 11 (1): 3–13.

Hallström, E., J. Davis, A. Woodhouse, and U. Sonesson. 2018. Using dietary quality scores to assess sustainability of food products and human diets: A systematic review. *Ecological Indicators* 93: 219–230.

Hammelman, C., and A. Hayes-Conroy. 2015. Understanding cultural acceptability for urban food policy. *Journal of Planning Literature* 30 (1): 37–48.

Helbich, M., B. Schadenberg, J. Hagenauer, and M. Poelman. 2017. Food deserts? Healthy food access in Amsterdam. *Applied Geography* 83: 1–12.

Hinrichs, C., and K.S. Kremer. 2002. Social inclusion in a Midwest local food system project. *Journal of Poverty* 6 (1): 65–90.

Huizinga, R.P., and B. van Hoven. 2018. Everyday geographies of belonging: Syrian refugee experiences in the Northern Netherlands. *Geoforum* 96: 309–317.

Johnston, J., M. Szabo, and A. Rodney. 2011. Good food, good people: Understanding the cultural repertoire of ethical eating. *Journal of Consumer Culture* 11 (3): 293–318.

Lang, T. 2017. Re-fashioning food systems with sustainable diet guidelines: Towards a SDG2 strategy. Food Research Collaboration website. https://foodresearch.org.uk/publications/re-fashioning-food-systems-with-sustainable-diet-guidelines/. Accessed 23 Mar 2020.

Mason, P., and T. Lang. 2017. *Sustainable diets: How ecological nutrition can transform consumption and the food system*. London: Routledge, Taylor & Francis Group.

Mata, C.T. 2013. *Marginalizing access to the sustainable food system: An examination of Oakland's minority districts*. Lanham, MD: University Press of America.

Nelson, M.E., M.W. Hamm, F.B. Hu, S.A. Abrams, and T.S. Griffin. 2016. Alignment of healthy dietary patterns and environmental sustainability: A systematic review. *Advances in Nutrition* 7 (6): 1005–1025.

Nicolaou, M., C.M. Doak, R.M. van Dam, J. Brug, K. Stronks, and J.C. Seidell. 2009. Cultural and social influences on food consumption in Dutch residents of Turkish and Moroccan origin: A qualitative study. *Journal of Nutrition Education and Behavior* 41 (4): 232–241.

Nicolini, D. 2013. *Practice theory, work, and organization: An introduction*. Oxford: Oxford University Press.

Perez, R.L. 2017. Interviewing epistemologies: From life history to kitchen table ethnography. In *Food culture: Anthropology, linguistics, and food studies*, ed. J. Chrzan and J. Brett, vol. 2, 47–57. New York: Berghahn.2

Raja, S., C. Ma, and P. Yadav. 2008. Beyond food deserts: Measuring and mapping racial disparities in neighborhood food environments. *Journal of Planning Education and Research* 27 (4): 469–482.

Raja, S., K. Morgan, and E. Hall. 2017. Planning for equitable urban and regional food systems. *Built Environment* 43 (3): 309–314.

Reckwitz, A. 2002. Toward a theory of social practices: A development in culturalist theorizing. *European Journal of Social Theory* 5 (2): 243–263.

Rice, J.S. 2015. Privilege and exclusion at the farmers market: Findings from a survey of shoppers. *Agriculture and Human Values* 32 (1): 21–29.

Schatzki, T.R. 1996. *Social practices: A Wittgensteinian approach to human activity and the social*. Cambridge, UK: Cambridge University Press.

Schatzki, T.R. 2002. *The site of the social: A philosophical account of the constitution of social life and change*. University Park, PA: Pennsylvania State University Press.

Schatzki, T.R. 2011. Where the action is (on large social phenomena such as sociotechnical regimes). Working Paper 1, Sustainable Practices Research Group.

Shannon, J. 2014. Food deserts: Governing obesity in the neoliberal city. *Progress in Human Geography* 38 (2): 248–266.

Shove, E., M. Pantzar, and M. Watson. 2012. *The dynamics of social practice: Everyday life and how it changes*. Thousand Oaks, CA: Sage.

Spaargaren, G., and B. Van Vliet. 2000. Lifestyles, consumption and the environment: The ecological modernization of domestic consumption. *Environmental Politics* 9 (1): 50–76.

Springmann, M., M. Clark, D. Mason-D'Croz, K. Wiebe, B.L. Bodirsky, L. Lassaletta, W. de Vries, S.J. Vermeulen, M. Herrero, K.M. Carlson, M. Jonell, M. Troell, F. DeClerck, L.J. Gordon, R. Zurayk, P. Scarborough, M. Rayner, B. Loken, J. Fanzo, H.C.J. Godfray, D. Tilman, J. Rockström, and W. Willett. 2018. Options for keeping the food system within environmental limits. *Nature* 562 (7728): 519–525.

Sustainable Development Goals. 2015. Transforming our world: The 2030 agenda for sustainable development. Sustainable Development Goals Knowledge Platform. https://sustainabledevelopment.un.org/post2015/transformingourworld. Accessed 23 Mar 2020.

Tilman, D., and M. Clark. 2014. Global diets link environmental sustainability and human health. *Nature* 515 (7528): 518–522.

Turner, C., A. Aggarwal, H. Walls, A. Herforth, A. Drewnowski, J. Coates, S. Kalamatianou, and S. Kadiyala. 2018. Concepts and critical perspectives for food environment research: A global framework with implications for action in low-and middle-income countries. *Glob Food Secur* 18: 93–101.

Walker, R.E., C.R. Keane, and J.G. Burke. 2010. Disparities and access to healthy food in the United States: A review of food deserts literature. *Health & Place* 16 (5): 876–884.

Warde, A. 2005. Consumption and theories of practice. *Journal of Consumer Culture* 5 (2): 131–153.

Warde, A. 2016. *The practice of eating*. Cambridge, UK: Polity Press.

Publisher's Note Springer Nature remains neutral with regard to jurisdictional claims in published maps and institutional affiliations.

Anke Brons is a Ph.D. candidate at Aeres University of Applied Sciences and at the Environmental Policy Group at Wageningen University, both in the Netherlands. Her Ph.D. project focuses on questions of inclusiveness around access to healthy and sustainable food in a Western urban context, from a sociological perspective. Her research interests include food consumption, food systems, social equity and social practice theories.

Peter Oosterveer is a professor at the Environmental Policy Group at Wageningen University, the Netherlands. His research interests are in global public and private food governance arrangements and innovative institutional developments in sustainable food production and consumption. He is studying food consumption practices from a sociological perspective and is particularly interested in how consumers access sufficient, sustainable and healthy food.

Sigrid Wertheim-Heck is a professor of Food and Healthy Living at Aeres University of Applied Sciences and a senior research fellow at the Environmental Policy Group at Wageningen University, both in the Netherlands. Her interest in global urban food security informs her research on the relationship between metropolitan development, food provisioning and food consumption, focusing on equitable access to sustainable, safe and healthy foods.

Agriculture and Human Values (2020) 37:1041–1053
https://doi.org/10.1007/s10460-020-10032-w

Acting like an algorithm: digital farming platforms and the trajectories they (need not) lock-in

Michael Carolan[1]

Accepted: 4 April 2020 / Published online: 13 April 2020
© Springer Nature B.V. 2020

Abstract

This paper contributes to our understanding of farm data value chains with assistance from 54 semi-structured interviews and field notes from participant observations. Methodologically, it includes individuals, such as farmers, who hold well-known positionalities within digital agriculture spaces—platforms that include precision farming techniques, farm equipment built on machine learning architecture and algorithms, and robotics—while also including less visible elements and practices. The actors interviewed and materialities and performances observed thus came from spaces and places inhabited by, for example, farmers, crop scientists, statisticians, programmers, and senior leadership in firms located in the U.S. and Canada. The stability of "the" artifacts followed for this project proved challenging, which led to me rethinking how to approach the subject conceptually. The paper is animated by a posthumanist commitment, drawing heavily from assemblage thinking and critical data scholarship coming out of Science and Technology Studies. The argument's understanding of "chains" therefore lies on an alternative conceptual plane relative to most commodity chain scholarship. To speak of a data value *chain* is to foreground an orchestrating set of relations among humans, non-humans, products, spaces, places, and practices. The paper's principle contribution involves interrogating lock-in tendencies at different "points" along the digital farm platform assemblage while pushing for a varied understanding of governance depending on the roles of the actors and actants involved.

Keywords Digital agriculture · Precision agriculture · Big data · Dependency · Algorithms · Knowledge · Data cleaning

Abbreviations

AI	Artificial Intelligence
GNSS	Global Navigation Satellite Systems
IoT	Internet of Things
LiDAR	Light Detection and Ranging
STS	Science and Technologies Studies
USDA	United States Department of Agriculture

Introduction

Smart farming—also known as digital agriculture, e-agriculture, precision agriculture, agriculture 4.0, etc.—is an umbrella term referencing data- and software-intensive platforms widely said to be transforming agriculture (DeClercq et al. 2018; Walter et al. 2017). Farm managers, ranchers, and producers, in conjunction with industry and government

actors, are leveraging the capabilities of the Internet of Things (IoT), from sensors to GNSS (Global Navigation Satellite Systems), cloud computing, weather and climate modeling, high-speed internet, historical yield data, and LiDAR (Light Detection and Ranging) systems to usher in what has been described as a digital revolution in agriculture (e.g., Claver 2018; DeBoar 2015). This disruption has been met with a mix of optimism and anxiety: the former, as producers are enticed with promises of increased profits and efficiencies with the adoption of these platforms; the latter, as farmers, peasants, and food justice activists worry about the uncertainties, dependences, and digital divides that these innovations have been known to create.

Most of the academic literature on smart farming focuses on improving economic and technological efficiencies while grappling with the question of how to better incentivize adoption (for a review of this scholarship, see Finger et al. 2019). Consequently, there remains a lot we do not know about these artifacts, especially in terms of their impacts—social, political and otherwise. Fortunately, the available critical social science scholarship on the subject is growing. This literature ranges from qualitative studies examining the

✉ Michael Carolan
  Michael.Carolan@colostate.edu

1   Department of Sociology, Colorado State University, B241
    Clark, Fort Collins, CO 80523, USA

ways that smart farming platforms are accommodated within (and alter) the everyday lives, practices, social networks, and identities of farmers (e.g., Carolan 2017a, 2017b, 2018a, 2020b; Higgins et al. 2017; Jakku et al. 2018; Jayashankar et al. 2019) to surveys on farmer attitudes toward "smart" technologies and the organizations that promote and manufacture them (e.g., Gardezi and Bronson 2019; Wiseman et al. 2019). There is also a suite of studies examining actors linked to digital farming innovation, design, and investment (Bronson 2019; Fielke et al. 2019; Carolan 2020a, 2018a, b) and other "key governance actors" (Regan 2019).

This paper contributes to the literature by looking *at* as well as *beyond* actors already known for their positionality within the digital agriculture space. Rather than study pre-determined groups, which to date has tended to focus overwhelmingly on those generating data and using platform outputs (i.e., farmers), or to a lesser extent on those identified as playing a major role in the design phase (i.e., investors), I attempted to cast a wider methodological net. This net was defined less by what I thought I knew about data values chains and oriented more toward discovery. This put me into contact with who I was expecting: farmers, senior leadership of tech firms, and the like. But it also led me to actors engaged in important acts of tinkering; activities that might not rise to the celebratory status of "investor" and/or "inventor" (i.e., the Elon Musk's of the digital agriculture world), as observed, for instance, among those experimenting with self-learning machines and algorithms.

This brings me to what I am calling farm data value chains. While inspired by earlier agrifood scholarship directed at following the commodity (e.g., Friedland et al. 1981; Wells 1996), the paper diverges from this framework in a number of important ways. My use of the term "chain" differs from how it might otherwise be used when talking about food and agriculture, such as when those in business logistics talk about supply chains (e.g., Bosona and Gebresenbet 2013; Nicholson et al. 2011) or when political economists interrogate global commodity chains (e.g., Challies 2008; Selwyn 2012). This argument draws from the likes of Bruno Latour and the field of Science and Technology Studies (STS) broadly. Latour (1999), for instance, advances the concept "chain of translation." This speaks to an orchestrating set of relations among humans, non-humans, products, spaces, places, practices, and times producing assemblages as opposed to hierarchies or linearly-defined chains.

The analysis draws from 54 semi-structured interviews with actors identified as having a role in farm data value chains in North America. This sample was derived from a hybrid methodological approach, combining snowball sampling inspired by a follow-the-commodity (e.g., de Sousa and Busch 1998; Friedland et al. 1981; Wells 1996) spirit alluded to earlier. Giving respondents' space to direct me to other potential participants helped draw me to less visible

elements of smart farming assemblages, including those yet to be fully analyzed by agrifood scholars—like practices linked to the creation and use of algorithms in agriculture.

The paper's principle finding is that these platforms are linked to types of lock-in—a term used to describe when seemingly small alterations produce immensely consequential pathways that become calcified and resistant to change. These tendencies are identified as taking shape in the design phase as well as among farmers, though how they become expressed across chains is shown to differ. The paper concludes situating these findings within a discussion of governance. In rejecting essentialism, the paper's posthuman lens leaves space to talk about how lock-in tendencies are not destiny. How we *do* governance, I argue, ought to reflect the types of lock-in confronted and the ways in which actors and actants are involved.

Smart farming landscapes

The digital agriculture sector has changed radically in the last decade. Monsanto (now owned by Bayer) invested early and aggressively in this space. In 2012, the company spent US$210 million to acquire Precision Planting, which at the time was one of the largest precision planting equipment manufacturers in the U.S. In 2013, the firm invested close to US$1 billion to buy Climate Corporation, the world's largest digital agriculture platform. Estimates from 2014 reported the platform was used on as much as one-third of all U.S. farmland (McDonnell 2014). In 2014, Monsanto purchased the soil analysis division of Solum, and, in a separate deal, 640 Labs, a mobile technology and cloud-computing firm specializing in smart farming applications. In 2016, the company acquired the European farm management software company VitalFields, while 2017 they added to their portfolio the irrigation-specific data analytics firm Hydrobio (for further details about these acquisitions see, e.g., Bassetti et al. 2017; McDonnell 2014; Pletz 2014).

John Deere has also made significant investments in this space, especially since 2017 with the purchase of the artificial intelligence startup Blue River Technology (Kolodny 2017). Weeds cost farmers an estimated $11 billion a year. Blue River was one of the first to innovate into the "See & Spray" space, creating a weed control cotton machine capable of scanning the ground and distinguishing between crops and the weeds, down to a level of millimeters. It is designed to target weeds with herbicides and kill them (Gagliordi 2018).

Smart farming platforms are not only interested in planting and growing periods. Diagnostic tools have also been designed for harvest. Yield monitors keep detailed records of within-field yield variability. In addition to crop quantity, quality can also be monitored. In cereals, traits like protein and moisture content are measured in combines equipped with near-infrared spectroscopy. Yield quality monitoring is of particular relevance for high-value horticultural crops (Finger et al. 2019).

Case IH and New Holland introduced their autonomous tractors at the 2016 Farm Progress Show. The former can be controlled from a PC, tablet, or smart phone. The French company Naio Technologies engineered a tractor that uses laser and camera guidance to navigate between rows of fruits and vegetables while its sensors scan individual plants as it hunts for weeds.

Automation has also made inroads in dairy farming with automated milking parlors. Though less than 5% of U.S. dairy farms utilize robots, the figure is expected to increase by 20 to 30% annually "for the foreseeable future" (Mulvany 2018). Robotics in dairy farming is already a US$1.6 billion global industry (Mulvany 2018). A related suite of technologies draws from the facial recognition. Dairy and other livestock operations are now using repurposed facial recognition algorithms to detect lameness or variability in animals. Estimates place these A.I.-assisted detections as occurring a full two days prior to when a trained technician can see problems (Griffiths 2020).

The agricultural drone market is likewise experiencing exponential growth, with analysts expecting its valuation to hit US$2 billion by 2026 due to the technology's versatility regardless of a farm's size (Ag Daily 2018). A 2016 report calculated the value of drone-powered solutions in all applicable industries to be more than US$127 billion, noting that the most promising applications are likely to take place in agriculture (Mazur and Wisniewski 2016). Drones are part of digitization trends in agriculture not only because they assist in data collection but when coupled with machine-learning platforms they become an important part of a larger sensing assemblage. Drones are used for applications ranging from crop assessment, where they generate spectral data and reveal crops that have been injured by the drifting pesticide dicamba and can spot herbicide-resistant weeds growing between rows. Other uses include counting livestock, monitoring for disease, monitoring water (e.g., checking irrigation cannels), and even pollination—a New York–based start-up has developed pollinating drones for almond, cherry and apple orchards (Ehrenberg 2018).

Much of this technology is rooted in processes captured by such terms as "machine learning," "Artificial Intelligence" (AI), "deep learning," and "algorithms," which essentially speak to the idea that computers can be taught without explicit programming. Specific examples of these techniques and practices are discussed in the findings. Providing a few definitions at this point in the paper might help better situate those empirics.

Machine learning involves drawing conclusions from large datasets. In an agricultural context, it "can take a decade of field data—insights about how crops have performed

in various climates and inherited certain characteristics—and use this data to develop a probability model" (Kharkovyna 2019). The description continues,

With all this information, far more than any single human can grasp, machine learning can predict which genes will most likely contribute a beneficial trait to a plant. Of the millions of combinations, advanced software greatly narrows the search (Kharkovyna 2019).

Meanwhile, an algorithm, to quote directly from Wikipedia (n.d.), "is a finite sequence of well-defined, computer-implementable instructions, typically to solve a class of problems or to perform a computation." The term therefore refers to, drawing now from critical data studies scholars, any "abstract, formalized description of a computational procedure" (Dourish 2016, p. 3). This procedure, in turn, "autonomously makes decisions based on statistical models or decision rules without explicit human intervention" (Lee 2018, p. 3). While imaginations tend to go immediately to Google, Netflix's recommender, or predictive policing systems when hearing the term, algorithms can carry out both "simple calculations or highly complex reasoning tasks" (Janssen and Kuk 2016, p. 372), realizing that an algorithm like bubble sort can be described in a sentence (Seaver 2017).

"Chains" and "lock-in": situating the argument conceptually

In *Pandora's Hope*, Latour (1999) recounts his time with French and Brazilian scientists in the Amazon as they studied the boundary between forest and savannah; an activity that included collecting soil samples. Latour used the experience to follow the samples of soil from the rainforest to peer-reviewed academic journals—the "passage from a clump of earth to a sign" or scientific fact (Latour 1999, p. 79). Bruno Latour used the term "chain of translation" to talk about this process, where material objects, judgments, and practices produce ontological multiplicity in the context of representational coherence—e.g., *peer-reviewed facts* about soil is not ontologically equivalent to *Amazon soil* but each eventually comes to represent the other. I also want to introduce Latour's (Latour 1999, p. 40) concept "circulating reference," as it introduces ways of thinking that lend themselves well to the empirics of this paper. The term speaks to the idea that while processes "along" the chain are heterogenous and multiple they are not entirely unconnected. Each sequence always, to some degree, refers "back" to prior objects and activities, though Latour's thick description made it clear that countless judgements are always made along the way.

Others have appropriated Latour's understanding of a "chain" to understand issues central to data studies scholarship, such as Marieke de Goede's (2018, p. 1) application to interrogate "illicit" financial transactions, "whereby commercial transactions are collected, stored, transferred, and analysed in order to arrive at security facts (including for example frozen assets, closed accounts, and court convictions)." As she writes,

> If we liken a transaction to the soil sample in the Latourian schema, we can render visible the practices that transform—for example—a financial record from simple bank registration, to suspicious transaction to (in some cases) court evidence. As in Latour's chain of soil analysis, translation is key to the movement and sequencing of transactions data across the security chain. When transactions are reported from one professional domain to another, they are not simply moved but also modified: they acquire new meanings, new combinations with other data, and new capabilities (de Goede 2018, p. 6).

These modifications take any number of forms within farm data value chains, such as when machine-sensing platforms convert observations into 1s and 0s or when software engineers undertake the process of "cleaning" data.

Data chains are therefore presented below as neither discrete nor linear; the same also pertains to the material (e.g., data, code, algorithms) said to be moving through them. Rather, the chains discussed are recursive, multiple, contested, and heterogeneous. *Folds* supplant *flows* for the verb of choice in this approach as phenomenon like "commodity," "infrastructure," and "actors" are shown to be distributed throughout the network as opposed to representing parts within it (Lee et al. 2019). This also has implications for how we think about issues of governance, realizing that "governing the world on the basis of the politics of modernity (top-down, cause-and-effect understandings) is dangerous, false and hubristic" (Chandler 2015, p. 850).

The term "lock-in," traditionally defined, speaks to how even relatively small changes in institutional arrangements can become self-reinforcing and result in very different paths than if those changes had not been implemented (Arthur 1989; Bui et al. 2019; David 1985; Frank 2007). North (1990), for example, famously used this reasoning to help explain cross-national income disparities. The phenomenon of lock-in helps explain how practices and ideas become entrenched even though they produce lower social utility than the alternatives. Some examples in the literature include high-levels of red meat consumption (Frank 2007), the standard keyboard arrangement (David 1985), and the carbon economy (Unruh 2002).

The automobile's rise as the dominant form of transportation provides a well-observed example of technological and institutional lock-in. Its "effectiveness" as a form of personal mobility cannot be separated from the co-emergence of numerous supporting technologies and industries and the core competencies these activities require, involving, for instance, the production and distribution of petroleum, glass, rubber, concrete, asphalt, and steel. In addition, we cannot ignore the role played by investments in an extensive road network, the building of service stations and motels, the development of car-based disciplines (e.g., automotive engineering), "free" parking, and the like. In assemblage language, it is impossible to separate what an automobile *is* from this heterogenous network. When needing to get from Point A to B, what form of transportation typically comes to mind? What future mobilities are we more likely to envision? (*Personal* transportation, think X-wings and landspeeders, often crowds out *public* forms of transportation in most futuristic depictions.) Societal answers to these questions are shaped deeply by the fact that we, too, are part of the automobile-assemblage. We are locked-in to this architecture—this car-chitecture.[1]

The term lock-in might be perceived by some as ill-suited for a posthuman approach, especially among those reading the concept as embracing a nondynamic and linear—deterministic, even—understanding of history. This would explain attempts to propose complimentary concepts, like path generation (Djelic and Quack 2007), in an effort to provide balance. I see the term, rather, as offering a way to talk about assemblages while moving beyond a purely flat ontology, where agency is distributed equally among actors and actants. It is worth pointing out that Deleuze and Guattari (1988), and scholars inspired by their thinking (e.g., Adkins 2015), routinely deploy the noun "calcified" and the verb "calcification" to talk about when phenomenon, from political forms to language, harden and become resistant to change, which I would argue is "lock-in" by another name.

Lock-in, fundamentally, recognizes that power is situated across heterogenous networks. Power is not "held" according to this concept and linearity is abandoned. A "path" supported by lock-in is not destiny, just as it is not a path in the sense that we tend to think of the term, unless what comes to mind is something that folds onto itself, veers, and multiplies. This also speaks to what I like about an STS-inspired reading of lock-in, as it rejects dichotomist (i.e., either/or) outlooks. This is to say: assemblages can both/and—they can *both* harden *and* multiply.

---

[1] "Yellow No. 2" field corn is a USDA (United State Department of Agriculture) grade designation with a minimum test weight of 54 lb per bushel at 15.5 percent moisture. It has a maximum limit of percent broken kernels and foreign material and cannot exceed 5 percent total damaged kernels (Extension 2009). It is also often used to speak generally of conventional corn seed.

Table 1 Categorizing respondents (n = 54) by job titles (capital letters designate code identifiers)

Job title	n
Programmer (P)	2
Engineer (E)	6
Firm Leadership (FL): E.g., CEO/Vice President/Director	7
Crop Specialist (CS)	5
Agronomist (A)	6
Computational Biologist (CB)	2
Statistician (S)	1
Farmer[a] (F)	
Annual gross revenue: less than US$350,000	15
Annual gross revenue: US$350,000 or greater	10

[a]According to the United States Department of Agriculture (USDA), "small" farms as those with an annual gross revenue below US$350,000

Table 2 Digital agriculture platform-types reported among farmers and adoption frequencies

Platform-type	Frequency
Mobile devices for (often) real-time monitoring	25
Global Navigation Satellite Systems	20
Variable rate seeding	0
Variable-rate fertilizer/nitrogen modeling/etc	20
Aerial vehicle technology (drones)	13
Robotics/automation	5
Remote sensing	5
Precision irrigation	3

Methods

Rather than fixate on pre-identifed groups (e.g., software engineers with Company X) and/or roles (e.g., farmers), I entered the field inspired by a sense of exploration. My goal was to cast a broad methodological net and not be hemmed in by preconceived notions about who and what are part of data value chains. Some of this inspiration came from a well-traveled path within agrifood scholarship, known generally as the "follow the commodity" approach (e.g., de Sousa and Busch 1998; Friedland et al. 1981; Wells 1996). Initially, the idea was to follow big agricultural data; to start by talking with farmers, who both produce and consume 1s and 0s, and work my way "out." The approach quickly broke down, however, as I learned there was no-*thing* to follow, only heterogeneous assemblages. Regardless, the approach helped expand my awareness of the actors and actants involved in the making of these messy networks.

Fifty-four face-to-face semi-structured interviews were conducted between July 2017 and October 2019. Respondents consisted of individuals from across North America who were involved in various ways with farm data value chains, including programmers, software engineers, computational biologists, crop specialists, and farmers. Farmers came from the U.S. states of Iowa (n = 12), Kansas (n = 10), Nebraska (n = 8), Illinois (n = 5) and the Canadian province of Saskatchewan (n = 10). See Table 1 for a breakdown of respondents' roles and identities as they were related to agricultural data. Smart farming platform-types reported among farmers interviewed, and their adoption frequencies, are listed on Table 2. Thirty-nine respondents resided in the U.S. The other fifteen participants lived in Canada.

Participants were also followed and observed engaging in situ with the "chain." Some observations took the better part of a day (like when a farmer brought me along to plant corn); others were folded into the interview itself, like when some participants (e.g., engineers) were asked to explain "their" algorithms. Extensive field notes were taken during these observations. All interviews were recorded, transcribed, and ultimately coded. A research assistant and I independently coded the same three randomly selected transcripts and any inconsistencies in coding were reviewed until consensus was reached. I then coded the remaining transcripts, using those initial codes as a guide. Only adults were interviewed. Pseudonyms are used to protect the anonymity of respondents.

A convenience sample was sought as it can be difficult to convince actors in these sectors to agree to an interview. Personal connections were used to enroll respondents. A snowball sampling method took shape by starting with those known to myself. In the spirit of following the commodity, respondents guided me to actors/roles who they thought I could learn from to better understand the data valorization process. Names were obtained in most instances by asking respondents about "who I should talk to in order to better understand how these platforms get created." This was especially useful for an outsider—disciplinarily- and organizationally-speaking—such as myself as it helped identify actors and practices. For instance, in eleven cases I was directed outside of the corporate world to independent consultants, which included a few university professors; individuals who would have been missed had the focus been on, say, a particular firm or firms. As I learned, the expertise and know-how related to these platforms is highly distributed and therefore hard to pinpoint by means other than, for lack of a better phrase, following the commodity.

The fact that my study was multinational in scope—it included actors within the U.S. and Canada—reflects the extent to which I tried to stay true to that "following" ethos. The geographic scope of this study reflects both intent, of wanting to follow as best I could, and material realities; after all, following costs money. While I would have welcomed the opportunity to fly wherever and talk to and observe whomever, qualitative research is an exercise in what is, among other things, financially feasible.

Seaver (2017) makes some important methodological points about studying objects like algorithms, which are ontologically opaque in part because of access issues. Those involved with these artifacts worry about revealing trade secrets and breeching intellectual property protections. These barriers often encourage approaches to data collection that can appear highly "undisciplined" (Seaver 2017, p. 6) and that look to be "a departure from the idealized image of a fieldworker embedded long-term in a bounded society" (Seaver 2017, p. 6). In these instances, Seaver encourages becoming a methodological "scavenger." As he argues:

> Ethnographers have always gleaned information from diverse sources, even when our objects of study appear publicly accessible. Moreover, the scavenger replicates the partiality of ordinary conditions of knowing—everyone is figuring out their world by piecing together heterogeneous clues—but expands on them by tracing cultural practices across multiple locations (Marcus 1995) and through loosely connected networks (Burrell 2009).

"Scavenging" nicely captures what I was attempting to accomplish with the above-methods. I was turned down for interviews and access to some spaces alluded me, like the data/server farm I tried to visit. Methods sections for qualitative research tend to read as success stories, in the sense of mentioning only those efforts that resulted in an interview. They also rarely admit to the role played by money in the research design—as admitted earlier, I would have welcomed flying to interview and observe more people if my budget had allowed for it. Such shortcomings associated with this research are fully recognized. But I also realize that qualitative research is ultimately, to quote Hannerz (2003, p. 213), "an art of the possible." The below represents my best attempt to do what I could with what I had and with whom I had access to.

Findings: lock-ins along the assemblage

The methodological technique of "following" meant I also followed those interviewed as they engaged with data and code. For farmers, this involved being given demonstrations of what their smart platforms did. Those responsible for creating these platforms, alternatively, showed me why they did what they did, explanations that gave me an inside look to see, as one engineer described it, "how the algorithm-sausage is made" (E2). During these encounters, I was especially interested in how farm-level data were aggregated and turned into something deemed valuable.

I was told repeatedly about the importance of making "dirty" farm data useful by way of a "cleaned" process. Cass was a software engineer. She described how part of her job was to get data, in her words, "talking the same language" (E6). She used a wood analogy to describe the cleaning process.

> Take a pile of wood, which in itself doesn't have much if any value, and turn it into a bunch of interchangeable pieces that connect up and make something beautiful and functional. That's basically what I do with data.

I was immediately struck by this analogy. To liken what she does to the craft of woodworking and to allude to aesthetics, by describing her job as involving the creation of "something beautiful," is a radical departure from the position described in conventional scientific chains of translations—as described in Latour (1999), for example. I asked Cass about this, wanting to hear more about why she chose to talk about what she did in those terms.

> "Could you say more about this," I asked. "The way you describe it," I continued, "makes what you do sound as much like an art as a science, which isn't how it's promoted and sold to the public."

Cass smiled at the question and did not immediately respond. She breathed deeply before hissed softy through her teeth, as if contemplating just how much to say. She went on to admit that "data cleaning does involve judgement calls," though, echoing language from my question, she pointed out that "just because I make artistic judgments those judgments are heavily inform by facts."

Data were deemed "dirty" by a number of respondents by nature of not sharing the same format or structure. Numerous reasons given for this. Data could have been collected by different tractors from any variety of companies using their own propriety software and systems. Or perhaps the data were from one company's tractors but from different models. I learned that even the same tractor model can generate different data depending on the version of the operating system used. Sometimes the problem lied in user-error, as when data are coded improperly even prior to being collected. One programmer mentioned how she "routinely comes across 200 bushel per acre soybean fields" (P1)—signifying a combine that had not been reset after being put to work in a corn field.

Data cleaning has received surprisingly little attention, either among STS scholars or the larger scientific community, even though it involves making changes to data (it is sometimes referred to as data editing) (Leahey 2008). Another way to refer to the process is to talk about creating conforming data, which does a better job of calling out the practice's normative nature. To clean data is really to create data that conform to a particular, predetermined moral architecture that specifies what *is* and what *ought* to be. When

112

thought of in these terms, cleaning can be viewed as a prime example of the chain of translation at work if value is to be generated from the (dirty) data coming from farm implements and other sources.

I interviewed an engineer, Hadi, who told me he specialized in "cyber-agriculture." For one project, he created algorithms designed to help spot disease in fields. In his words:

Diagnosing for plant disease is all about pattern recognition. Usually, we—humans—spot disease after it is too late. [...] Imagine feeding millions of imagines into a computer and have it do pattern recognition, down to the cellular level even, before it becomes visible to the naked eye. [...] *It*—the computer—learns these patterns and finds them for us. [...] Now you can take a drone, fly it over a field, and have it look at the cellular structure of every plant and see disease when it is still easily treatable. (E3)

Or take the following quote from Faye, a computational biologist, taken in the context of discussing machine learning as applied to plant breeding:

You take a couple decades of field data and develop a probability model that predicts the genes most likely to contribute a beneficial trait to a plant.[...]. You train the program to look for patterns [in the genes] to see if there are any interactional effects; this is stuff that even a hundred years of field trail data wouldn't tell us.[...] Basically, the program learns to understand why certain traits get expressed.[...] We exploit that knowledge and create new varieties. (CB1).

Both Hadi and Faye talked as if the data inputted into these self-learning systems were of the God's Eye variety. For example, Faye's reference to taking "a couple decades of field data" glossed over the fact that she was talking about decades of field data from *very specific fields*—namely, those of the monocrop, industrial-scaled variety. Respondents involved in the making of algorithms were continuously making judgments about what data ought to be included when instructing these self-learning machines; judgements driven by assumptions concerning what patterns were worthy of recognition.

Mark was a V.P. of a division of a smart farming firm. At one point in the interview he talked about how his company was trying to perfect the sensor accuracy on combines and drones for detecting water and protein content of corn.

We feed the algorithm endless pictures. [...] I can input images of kernels from any farm. But then if I went one farm over the kernels might look different, even if the same variety. The difference could be due to soil type or because that field recorded different weather events for the year. [...] The algorithm gets

better every year we collect more and more field data. (FL6)

"But your algorithms are only seeing Yellow No. 2," I pointed out, adding, "which means they are only learning to be valuable to a very distinct subset of farmers, right?".[2]

Hearing this produced a blank look on his face, as if my question were somehow unfair or that its answer obvious.

Yes, but that's what Iowa corn farmers grow. This [technology] is what farmers want. They don't grow broomcorn, or Mexican blue corn, or sweet corn. If they did, we'd develop predictive analytics for those crops. [...] We're in the business of improving the efficiencies of large-scale corn growers.

The above speaks to one form of lock-in, which resembles the technological and institutional lock-in described in prior scholarship (e.g., Cowan 1990; Unruh 2002). Not surprisingly, because these platforms are taught to only "see" certain variations across monocultures within monocultures—e.g., a corn field containing genetically identical corn—their value proposition lies, first, in encouraging producers to transform their operations in order to have them conform to the "needs" of these platforms. This form of lock-in is a continuation of what has been happening in agriculture since at least the advent of agricultural mechanization. Kloppenburg (2005), for instance, details how corn hybridization was a response to the invention of the mechanical corn picker, which introduced new "needs" into the assemblage, such as ears of uniform height and stalks that remained standing at harvest.

I was able to observe this movement-toward-increased-conformity firsthand. Take Josh's experience, a farmer in Illinois who a few years prior grew soybeans, field corn, and sweetcorn, the latter for markets in and around Chicago. He spoke positively when asked whether he found value in adopting smart farming technologies.

I honestly don't know how I lived without them [smart farming technologies]. In addition to plant counts, I use drone mapping to assess stand establishment; to tell me if there's are any place [in the field] that needs replanting. [...] I've saved hundreds of hours not having to manually count plants. [...] I probably make tens of thousands of dollars more every growing season because I'm now putting in the best crop possible. (F5)

Yet Josh was also clear about how those smart platforms are variety specific. He noted a strong incentive to focus on growing field corn and soybeans, even though at one point

---

[2] I cannot take credit for this neologism, as I heard it from Michael Bell some twenty years ago.

he contemplated "further diversification"—more varieties of sweetcorn "so I could sell it over the span of a month or so" to generate additional household income.

"You'd think corn is corn, but no. The results aren't as reliable," he answered when asked if he was able to use the drone and field corn assessment software on his sweetcorn. Later he confessed to having "shrunk my acreage in sweetcorn after adopting the [aforementioned] technology" in order to "take full advantage" of the technology.

Also observed were a number of examples of *knowledge lock-in*, which was most acutely observed among farmers. Prior scholarship notes, on the one hand, how technology increases farmer knowledge as they learn to use and accommodate new technology into daily routines (Arthur 1989). On the other hand, knowledge-gained has also been shown to come at the expense of knowledge-lost, creating another type of lock-in as technologies risk eroding local analogue knowledge thereby further encouraging even more intensive use of a given platform (e.g., Stone and Flachs 2018).

All the farmers interviewed believed they were more technologically savvy than prior generations; sentiments that, if true, speak to knowledge gained among farmers. Yet a number of farmers also talked about knowing their farms and fields differently than their parents on account of these platforms—e.g., "I don't think it's a matter of one generating knowing more or less than another generation but knowing *differently*" (F11). Digging further to understand what precisely this meant revealed "knowing differently" to be a euphemism for "knowing less" about certain elements, namely, those elements dealing with local analogue knowledge.

Analogue knowledge and the digital knowledge generated by so-called smart, precision platforms are both local. The latter are in fact marketed on that very basis, pledging to generate highly local (if applying a Euclidean, geospatial understanding of the term) knowledge. The distinction rests heavily on digital knowledge's promotion of seeing like an algorithm, where what is knowable and actionable are dictated by an exogenous sensing infrastructure (Lee 2020) created by individuals acting at a distance with their own assumptions and value judgments and animated by a you-can't-manage-what-you-can't-measure ethos. Conversely, what I am calling local analogue knowledge does not travel as well in large part because of its representational exuberance—it cannot be "contained" by a number or dashboard.

James' case is representative of this group. James farmed approximately 2,000 acres and has come to rely, in his words, "heavily on these digital platforms for knowing what's going on in my fields" (F8). He was quick to show me on his phone a variety of different dashboards and interfaces that told me almost anything a conventional farmer would ever want to know about their fields, from predictive weather software to plant counts, end-of-season

yield estimates, and satellite soil maps. With a few taps of his screen he could bring up soil composition, crop residue levels from prior years, etc. The farm had been in his family for multiple generations. I therefore asked how previous generations knew the land, without the aid of satellites, predictive analytics, drones, and broadband internet.

"My dad knew his ground like the back of his hand," James explained, later adding that "while he didn't have the detailed knowledge that I have—it's called 'precision agriculture' for a reason—he learned to read the land better than I ever could." When asked to say more about how his knowing of the farm differed from how his parents knew the land, he explained:

I read screens and dashboards. That's where I get my info from. My parents read fields. They made judgements by the look of the plants. I even remember dad when he used to smell and taste the soil.

I do not have data to support the thesis that smart platforms were directly responsible for the loss of local analogue knowledge among adopters—knowledge associated with, for instance, how rich, organic soil ought to *taste* and *smell* and how healthy plants ought to *look*. A number of farmers interviewed did, however, acknowledge that prior generations were better at, as James put it, reading their fields, whereas smart farmers' expertise today rested in reading screens, charts, and maps.

Lyle, a farmer from Kansas, explained the difference between what he knows and what, in his words, "old timers" know this way, pointing specifically to his neighbor as someone illustrative of the latter category.

I've got a neighbor like that. That's not to say he's a bad farmer. He's a damn good one, thanks in part to having been on his ground for the entire 80 years of his life. [...] His is old school precision agriculture. He knows every inch of that property because he's walked it hundreds of time. [...] God bless him. I'll never get to know my land that well; not *that* way at least. I'm [my farm's] too big. But I know it as well as anyone can at this scale thanks to these technologies. (F17)

Lyle's comment brings up another theme common among farmers interviewed: the importance of smart platforms for knowing large expanses of land "as well as anyone can at this scale." A lot has been written about how farmers have lost important local knowledge of their land and livestock thanks to intensification and scale-expansion (e.g., Flora 1992; Nazarea et al. 2013; Shiva 2016). Smart farming promises to correct this knowledge deficit, allowing farmers the ability to make management decisions down to the square-inch of their fields (Nebraska Corn Board 2019). At one level, then, techniques like precision agriculture make adopters as knowledgeable about their land as Lyle's neighbor, with all his intimate local knowledge.

Yet at another level this knowledge is different; a difference that induces dependencies on these platforms once farmers lose the ability to read fields and animals in that way that Lyle's neighbor does and James' parents did. Learning to "see like an algorithm" is also predicated on the fact that those making algorithms are enacting futures in the image of their assumptions, thus further constraining what (future) farmers see and thus know. (The optic language is intentional as digital interfaces in smart farming tend to be visual in nature.) So: when actors like Mark, the earlier-introduce V.P., decide farmers do not what platforms knowing anything about "broomcorn, or Mexican blue corn, or sweet corn," to use his own examples, he was making a decision about how knowledgeable farmers ought to be in their quest to practice "smart farming."

Governance: "going under" and "staying above" an algorithm's hood

To stimulate meta-reflection, Lee and Björklund Larsen (2019, p. 2) "draw playfully on the metaphor of the engine hood" when they "ask what positions, other than 'going under the hood' to uncover the hidden normativities of the algorithm, are there?" In addition to this position, where "the politics, effects, and normativities that are designed into algorithms become foregrounded" (Lee and Björklund Larsen 2019, p. 2), they talk about "staying above" the hood.

> Here, on the other side of our constructed spectrum of ideal types, we could place ethnomethodological analyses of the achievement of social order. In this analytical position, algorithms would emerge as 'contingent upshot of practices, rather than [as] a bedrock reality' (Woolgar and Lezaun 2013, p. 326). (Lee and Björklund Larsen 2019, p. 2–3)

Finally, Lee and Björklund Larsen (2019, p. 3) introduce "a middle road between going under the hood and staying above it"—an "ideal type that approaches algorithms, and technology, through an analysis of nonhuman agency and relationality."

This meta-reflection helps me analytically categorize the above findings. I went under the hood when interrogating the politics and normativities of digital farming platforms. At other moments I was above it, such as when emphasizing how such technologies cannot be abstracted from their larger assemblages. And I was operating somewhere in the middle when talking about knowledge lock-in and the enfolding of values, materialities, and practices as engendering novel ways of knowing, anticipating, and doing foodscapes while potentially closing off other pathways.

These analytic categories also help link the above findings with Campbell-Verduyn et al.'s (2017) paper where they distinguish between governance *with*, *through*, and *by* algorithms in order to speak to how big data-related innovations "have important but varied implications for accountability, effectiveness and power" (p. 230). To talk about governance *through* algorithms is to place impacted humans in the driver's seat. Governance *by* algorithms, conversely, speaks to scenarios were human agency is greatly reduced. Finally, to talk of governing *with* algorithms represents a mixed form of governance, which lies at some point between the prior extremes.

To go under the hood and point out the normativities of these platforms is not to critique that fact that these technologies are value-laden—no socio-technical assemblage exists without having politics (Winner 1980). Those who critique smart farming platforms on the basis that they are biased are engaging in the practice of tail chasing—you cannot avoid creating these platforms absent of value judgments. The act of pointing out normativities is more about interrogating the *types* of values at work and the *processes of design* that afford access to some voices but not others (Bronson 2019). To talk about governance *through* algorithms is to target "points" along assemblages where humans do have a say in what the platforms look like, and what they look at (e.g., monocultures within monocultures versus, say, polycultures), so as to ensure those politics are truly participatory.

Alternatively, agency, as conventionally conceived, is highly debatable when situated with an algorithmic context. As Mau (2019, p. 3–4) argues,

> If everything we do and every step we take in life are tracked, registered and fed into evaluation systems, then we lose the freedom to act independently of the behavioural and performance expectations embodied in those systems.

The ability to affect or influence someone else's social and psychological characteristics is a capability had regardless of the physical or mental architecture (digital, mechanical, or biological) in question (Schwitzgebel and Garza 2015; Vladeck 2014). Instances of governance *by* algorithms therefore needs to be addressed through additional governance *of* algorithms—this is where the state can play a role. Such is especially important in light of the distinctly private character of algorithmic governance regimes, where shareholders represent the entirety of the *dêmos* (Kalpokas 2019). To not include in our suite of options governance *of* algorithms is to abdicate even greater influence to those assemblages assisted by the affordances of capital.

Governance *with* algorithms, finally, pushes conventional understandings of what it means to govern given the human and non-human heterogeneity involved. How does one approach the subject of governance if not through a humanist lens? What does a participatory politics look like that includes "representatives" from the heterogenous assemblage that *is* digital agriculture? Questions like this often assume governance depends entirely on just, inclusive processes.

Yet political theorists have not reached a consensus on this point. What if we were to reconsider whether political process is the be-all and end-all solution to what it means to engage in good governance? Rothstein (2009, p. 311), for instance, argues that "electoral democracy is highly over-rated when it comes to creating legitimacy," adding that "even in the successful and stable Nordic democracies, there is scant evidence that legitimacy is created on the input side of the political system." Muirhead and Rosenblum (2019, p. 33) believe it important to distinguish between philosophic and sociological legitimacy when talking about political systems of governance: the former "asks what kind of regime, in principle, would be worthy of support," while the latter focuses on "whether citizens in fact view their political order as worthy of their support" (see also Kalpokas 2019, p. 103).

I am not suggesting we abandon the rich scholarship arguing on behalf of democratizing socio-technical transitions (e.g., Brunori et al. 2011) and design (e.g., Koskinen and Norros 2018) in agriculture. This is not about creating competing frameworks. The aim here is to think about complimenting what we already know about governing *through* algorithms. The following brief example may prove generative for such ends.

Plantix is a free mobile app that was launched in 2016. By 2017, the app. had been downloaded more than a million times in over 155 counties. By late-2019, total downloads exceeded 10 million in 16 languages (Willmer 2019). Farmers from around the world upload an average of 50,000 images daily, using the app. to identify plant (and weed) type, possible disease, and pest or nutrient deficiency. With more than 15 million images on file, the platform represents the world's largest plant database (Martyn-Hemphill 2019). This real-time data stream means the platform can give users alerts about the spread and direction of diseases and pests. To quote Monti, a university-employed crop specialist, while describing Plantix and other platforms like it:

> The tool is learning every day, not based on what some investors or engineers what it to know but based on what farmers of any scale want it to know. […] People are even starting to teach the app. about houseplants. […] This isn't about making money. It's about making a tool for and by the people. (CS1)

The penetration of smartphones into the rural areas of India, countries throughout Africa, and Pakistan has helped give millions access to this database. This, in turn, has given the self-learning system access to imagines and data that it otherwise would not have "known"—a type of double-movement democratization, mutually benefiting and informing actors (e.g., pheasant farmers) as well as actants (e.g., Plantix algorithums).

Two farmers interviewed reported using Plantix. (I did not seek out users so these data are largely antidotal, which is why this discussion was not included in the Findings section.) Both spoke positively of the platform. To quote one, a smaller-scale grower actively involved with the National Farmers Union of Canada, an organization with close ties to the international food sovereignty movement *La Via Campesina*:

> [Plantix is] an example of how we *should* be building these technologies, to learn from farmers of all sizes and to be of value to them all, too; but also, the proof is in the pudding. It's useful. Enough said. (F12)

This quote also aligns with both forms of political legitimacy mentioned by Muirhead and Rosenblum (2019): philosophical and sociological—the *principles* of the platform engender legitimacy as do the *outcomes* it affords, respectively.

This concludes my discussion of Plantix. It was not my intent to study the platform, having only learned about it through the interviews. Plantix does, however, seem good to think with when contemplating what it means to govern *with* algorithms, as an activity that grapples with the heterogeneous enfoldings common to digital farming platforms.

Conclusion

Drawing from 54 semi-structured interviews and many hours of observations, this paper adds to our understanding of farm data value chains. Before talking about the paper's principle conceptual and empirical contributions, it is worth reviewing its methods and theoretical approach. Methodologically, I entered the field wanting to follow data from farm-to-farm, taking cues from the farm-to-fork approaches used in more conventional agrifood "follow the commodity" projects (e.g., de Sousa and Busch 1998; Friedland et al. 1981; Wells 1996). I did this to cast a broad empirical net, realizing, too, some of the limits and tradeoffs of this approach. Theoretically, the paper is animated by a posthumanist commitment, drawing heavily on assemblage thinking. This orientation positions its understanding of "chains" on a different conceptual plane relative to most commodity chain scholarship. As opposed to those linear applications, chains are understood here as referring to an orchestrating set of relations among humans, non-humans, products, spaces, places, and practices. This brings me to the paper's principle contribution, which involves interrogating lock-in tendencies at different "points" of the assemblage while pushing for a varied understanding of governance depending on the roles of the actors and actants involved.

In future research, I would like to see additional empirical work targeting the heterogeneity of data value chains, involving both qualitative interviews and ethnographic methods. I would also encourage scholars to think more like a scavenger (Seaver 2017) when studying these platforms given their wildly distributed nature. For example, every farmer interviewed stored her data in the Cloud, meaning their data could have been anywhere in the world. (Cloud providers typically store your data in different locations for reliability; only a few give users the ability to choose which countries to store data in.) Assuming no social scientist has access to unlimited resources (including time), no one can be expected to faithfully trace this heterogeneity in its entirety. To say nothing about access: some of these spaces and practices are as guarded as nuclear codes. As more of us scavenge the digital foodscape, however, the clearer these routines, actors, and actants become, which can afford us insight into how best to work with these platforms and not be governed by them.

Acknowledgements This research was supported in part by the Ministry of Education of the Republic of Korea and the National Research Foundation of Korea (NRF-2016S1A3A2924243), the National Institute of Food and Agriculture (NIFA-COL00725), and the Office for the Vice President for Research, College of Liberal Arts, and Office of Engagement at Colorado State University.

References

Adkins, B. 2015. *Deleuze and Guattari's a thousand plateaus.* Edinburgh: Edinburgh University Press.

Ag Daily. 2018. Ag drone market to near $2 billion in value by 2026. *Ag Daily*, February 26. https://www.agdaily.com/technology/ag-drone-market-to-near-2-billion-in-value-by-2026/.

Arthur, W.B. 1989. Competing technologies, increasing returns, and lock-in by historical events. *The Economic Journal* 99 (394): 116–131.

Bassetti, V., J. Davidson, and T. Finck-Haynes. 2017. Bayer-Monsanto merger: Big data, big agriculture, big problems, Report published by Friends of the Earth, Open Markets, and SumOfUs, November, Friends of the Earth, Amsterdam, The Netherlands. https://1bps6437gg8c169i0y1drtgz-wpengine.netdna-ssl.com/wp-content/uploads/2017/11/Bayer-Monsanto-merger-report-Nov-2017.pdf.

Bosona, T., and G. Gebresenbet. 2013. Food traceability as an integral part of logistics management in food and agricultural supply chain. *Food Control* 33 (1): 32–48.

Bronson, K. 2019. Looking through a responsible innovation lens at uneven engagements with digital farming. *NJAS-Wageningen Journal of Life Sciences.* https://doi.org/10.1016/j.njas.2019.03.001.

Brunori, G., A. Rossi, and V. Malandrin. 2011. Co-producing transition: Innovation processes in farms adhering to solidarity-based purchase groups (GAS) in Tuscany, Italy. *International Journal of Sociology of Agriculture & Food* 18 (1): 28–53.

Bui, S., I. Costa, O. De Schutter, T. Dedeurwaerdere, M. Hudon, and M. Feyereisen. 2019. Systemic ethics and inclusive governance: Two key prerequisites for sustainability transitions of agri-food systems. *Agriculture and Human Values* 36 (2): 277–288.

Burrell, J. 2009. The field site as a network: A strategy for locating ethnographic research. *Field Methods* 21 (2): 181–199.

Campbell-Verduyn, M., M. Goguen, and T. Porter. 2017. Big Data and algorithmic governance: The case of financial practices. *New Political Economy* 22 (2): 219–236.

Carolan, M. 2020a. "Urban farming is going high tech": Digital urban agriculture's links to gentrification and land use. *Journal of the American Planning Association* 86 (1): 47–59.

Carolan, M. 2020b. Automated agrifood futures: Robotics, labor and the distributive politics of digital agriculture. *Journal of Peasant Studies* 47 (1): 184–207.

Carolan, M. 2018a. "Smart" farming techniques as political ontology: Access, sovereignty and the performance of neoliberal and not-so-neoliberal worlds. *Sociologia Ruralis* 58 (4): 745–764.

Carolan, M. 2018b. Big data and food retail: Nudging out citizens by creating dependent consumers. *Geoforum* 90: 142–150.

Carolan, M. 2017a. Agro-digital governance and life itself: Food politics at the intersection of code and affect. *Sociologia Ruralis* 57 (51): 816–835.

Carolan, M. 2017b. Publicising food: Big data, precision agriculture, and co-experimental techniques of addition. *Sociologia Ruralis.* 57 (2): 135–154.

Challies, E.R. 2008. Commodity chains, rural development and the global agri-food system. *Geography Compass* 2 (2): 375–394.

Chandler, D. 2015. A world without causation: Big data and the coming of age of posthumanism. *Millennium: Journal of International Studies* 43 (3): 833–851.

Claver, H. 2018. Farmer at the core of precision farming revolution. *Future Farming*, November 7. https://www.futurefarming.com/Smart-farmers/Articles/2018/11/Farmer-at-the-core-of-precision-farming-revolution-356631E/.

Cowan, R. 1990. Nuclear power reactors: Aa study in technological lock-in. *The Journal of Economic History* 50 (3): 541–567.

David, P.A. 1985. Clio and the economics of QWERTY. *American Economic Review* 75: 332–337.

De Goede, M. 2018. The chain of security. *Review of International Studies* 44 (1): 24–42.

Deleuze, G., and F. Guattari. 1988. *A thousand plateaus: Capitalism and schizophrenia.* London: Bloomsbury Publishing.

Dourish, P. 2016. Algorithms and their others: Algorithmic culture in context. *Big Data & Society.* https://doi.org/10.1177/2053951716665128.

Extension. 2009. What is No. 2 field corn. *National Extension*, May 19. https://articles.extension.org/pages/39109/what-is-no-2-field-corn.

DeBoar, J. (2015, May/June). The precision agriculture revolution. *Foreign Affairs.* https://www.foreignaffairs.com/articles/united-states/2015-04-20/precision-agriculture-revolution.

DeClercq, M., A. Vats, and A. Biel. 2018. Agriculture 4.0: The future of farm technology. World Government Summit, Dubai, United Arab Emirates, February. https://www.worldgovernmentsummit.org/api/publications/document?id=95df8ac4-e97c-6578-b2f8-ff0000a7ddb6.

Djelic, M.L., and S. Quack. 2007. Overcoming path dependency: path generation in open systems. *Theory and Society* 36 (2): 161–186.

Ehrenberg, R. 2018. Eyes in the sky: 5 ways drones will change agriculture. *Knowable Magazine*, October 11. https://www.knowablemagazine.org/article/technology/2018/eyes-sky-5-ways-drones-will-change-agriculture.

Lee, F. and L. Björklund Larsen. 2019. How should we theorize algorithms? Five ideal types in analyzing algorithmic normativities. *Big Data and Society.* https://doi.org/10.1177/2053951719867349

Fielke, S.J., R. Garrard, E. Jakku, A. Fleming, L. Wiseman, and B.M. Taylor. 2019. Conceptualising the DAIS: Implications of the 'digitalisation of agricultural innovation systems' on technology and policy at multiple levels. *NJAS-Wageningen Journal of Life Sciences*. https://doi.org/10.1016/j.njas.2019.04.002.

Finger, R., S.M. Swinton, N. El Benni, and A. Walter. 2019. Precision farming at the nexus of agricultural production and the environment. *Annual Review of Resource Economics*. https://doi.org/10.1146/annurev-resource-100518-093929.

Flora, C.B. 1992. Reconstructing agriculture: The case for local knowledge. *Rural Sociology* 57 (1): 92–97.

Frank, J. 2007. Meat as a bad habit: A case for positive feedback in consumption preferences leading to lock-in. *Review of Social Economy* 65 (3): 319–348.

Friedland, W.H., A.E. Barton, and R.J. Thomas. 1981. *Manufacturing green gold: Capital, labor, and technology in the lettuce industry*. New York: Cambridge University Press.

Gagliordi, N. 2018. How self-driving tractors, AI, and precision agriculture will save us from the impending food crisis. *Tech Republic*, December 12. https://www.techrepublic.com/article/how-self-driving-tractors-ai-and-precision-agriculture-will-save-us-from-the-impending-food-crisis/.

Gardezi, M., and K. Bronson. 2019. Examining the social and biophysical determinants of US Midwestern corn farmers' adoption of precision agriculture. *Precision Agriculture*. https://doi.org/10.1007/s11119-019-09681-7.

Griffiths, C. 2020. 20 Mega trends for 2020 and beyond. *AgroProfessional*, January 29. https://www.agprofessional.com/article/20-mega-trends-2020-and-beyond.

Hannerz, U. 2003. Being there… and there… and there! *Ethnography* 4 (2): 201–216.

Higgins, V., M. Bryant, A. Howell, and J. Battersby. 2017. Ordering adoption: Materiality, knowledge and farmer engagement with precision agriculture technologies. *Journal of Rural Studies* 55: 193–202.

Jakku, E., B. Taylor, A. Fleming, C. Mason, S. Fielke, C. Sounness, and P. Thorburn. 2018. "If they don't tell us what they do with it, why would we trust them?" Trust, transparency and benefit-sharing in Smart Farming. *NJAS-Wageningen Journal of Life Sciences*. https://doi.org/10.1016/j.njas.2018.11.002.

Janssen, M., and G. Kuk. 2016. The challenges and limits of big data algorithms in technocratic governance. *Government Information Quarterly* 33: 371–377.

Jayashankar, P., W.J. Johnston, S. Nilakanta, and R. Burres. 2019. Co-creation of value-in-use through big data technology-a B2B agricultural perspective. *Journal of Business & Industrial Marketing*. https://doi.org/10.1108/JBIM-12-2018-0411.

Kalpokas, I. 2019. *Algorithmic governance: Politics and law in the post-human era*. Cham: Springer Nature.

Kloppenburg, J.R. 2005. *First the seed: The political economy of plant biotechnology*. Madison: University of Wisconsin Press.

Kolodny, L. 2017. Deere is paying over $300 million for a start-up that makes "see-and-spray" robots. *CNBC*, September 6. https://www.cnbc.com/2017/09/06/deere-is-acquiring-blue-river-technology-for-305-million.html.

Koskinen, H., and L. Norros. 2018. The participatory design of tools: Foreseeing the potential of future internet-enabled farming. *Interaction Design and Architectures* 37: 175–205.

Kharkovyna, O. 2019. 7 Reasons why machine learning is a game changer for agriculture. *Towards Data Science*, July 4. https://towardsdatascience.com/7-reasons-why-machine-learning-is-a-game-changer-for-agriculture-1753dc56e310.

Latour, B. 1999. *Pandora's hope: essays on the reality of science studies*. Cambridge: Harvard University Press.

Leahey, E. 2008. Overseeing research practice: The case of data editing. *Science, Technology, & Human Values* 33 (5): 605–630.

Lee, F. 2020. Sensing Salmonella: Modes of sensing and the politics of sensing infrastructures. In *Sensing security*, ed. N. Witjes, N. Pöchhacker, and G. Bowker. Manchester: Mattering Press.

Lee, M.K. 2018. Understanding perception of algorithmic decisions: Fairness, trust, and emotion in response to algorithmic management. *Big Data & Society*. https://doi.org/10.1177/2053951718756684.

Lee, F., J. Bier, J. Christensen, L. Engelmann, C.F. Helgesson, and R. Williams. 2019. Algorithms as folding: Reframing the analytical focus. *Big Data & Society*. https://doi.org/10.1177/2053951719863819.

Marcus, G. 1995. Ethnography in/of the World System: The emergence of multi-sited ethnography. *Annual Review of Anthropology* 24: 95–117.

Martyn-Hemphill, R. (2019, November 21). Crop disease recognition app Plantix raises €6.6m Series A led by RTP Global. *AgFunderNews*. https://agfundernews.com/breaking-crop-disease-recognition-app-plantix-raises-e6-6-million-series-a-led-by-rtp-global.html.

Mau, S. 2019. *The metric society: On the quantification of the social*. Cambridge: Polity Press.

Mazur, M., and A. Wisniewski. 2016. Clarity from above: PwC Global Report on the Commercial Applications of Drone Technology. *PwC Poland*, May. https://www.pwc.pl/clarityfromabove.

McDonnell, T. 2014. Monsanto is using big data totakeover the World. *Mother Jones*, November 19. https://www.motherjones.com/environment/2014/11/monsanto-big-data-gmo-climate-change/.

Mulvany, L. 2018. Robots coming to a dairy farm near you. *Farm Futures*, January 30. https://www.farmfutures.com/dairy/robots-coming-dairy-farm-near-you. Accessed 10 June 2019.

Muirhead, R., and N.L. Rosenblum. 2019. *A lot of people are saying: The new conspiracism and the assault ondemocracy*. Princeton: Princeton University Press.

Nebraska Corn Board. 2019. Corn production: Kernels of knowledge. Nebraska Corn Board, Lincoln, Nebraska, https://nebraskacorn.gov/todays-farm/.

Nazarea, V.D., R.E. Rhoades, and J. Andrews-Swann (eds.). 2013. *Seeds of resistance, seeds of hope: Place and agency in the conservation of biodiversity*. Tucson, AZ: University of Arizona Press.

Nicholson, C.F., M.I. Gómez, and O.H. Gao. 2011. The costs of increased localization for a multiple-product food supply chain: Dairy in the United States. *Food Policy* 36 (2): 300–310.

North, D.C. 1990. *Institutions, institutional change and economic performance*. Cambridge: Cambridge University Press.

Pletz, J. 2014. Monsanto nabs Chicago big-data startup 640 labs. *Crain's Chicago Business*, December 9. https://www.chicagobusiness.com/article/20141209/BLOGS11/141209758/monsanto-nabs-chicago-big-data-startup-640-labs.

Regan, Á. 2019. "Smart farming" in Ireland: A risk perception study with key governance actors. *NJAS-Wageningen Journal of Life Sciences*. https://doi.org/10.1016/j.njas.2019.02.003.

Rothstein, B. 2009. Creating political legitimacy: Electoral democracy versus quality of government. *American Behavioral Scientist* 53 (3): 311–330.

Schwitzgebel, E., and M. Garza. 2015. A defense of the rights of artificial intelligences. *Midwest Studies in Philosophy* 39 (1): 89–119.

Seaver, N. 2017. Algorithms as culture: Some tactics for the ethnography of algorithmic systems. *Big Data & Society* 4 (2): 2053951717738104.

Selwyn, B. 2012. Beyond firm-centrism: Re-integrating labour and capitalism into global commodity chain analysis. *Journal of Economic Geography* 12 (1): 205–226.

Shiva, V. 2016. *The violence of the green revolution: Third world agriculture, ecology, and politics*. Lexington, KY: University Press of Kentucky.

de Sousa, I., and L. Busch. 1998. Networks and agricultural development: the case of soybean production and consumption in Brazil. *Rural Sociology* 63 (3): 349–371.

118

Stone, G.D., and A. Flachs. 2018. The ox fall down: Path-breaking and technology treadmills in Indian cotton agriculture. *The Journal of Peasant Studies* 45 (7): 1272–1296.

Unruh, G.C. 2002. Escaping carbon lock-in. *Energy Policy* 30 (4): 317–325.

Vladeck, D.C. 2014. Machines without principals: Liability rules and artificial intelligence. *Washington Law Review* 89 (1): 117–150.

Walter, A., R. Finger, R. Huber, and N. Buchmann. 2017. Opinion: Smart farming is key to developing sustainable agriculture. *Proceedings of the National Academy of Sciences.* https://doi.org/10.1073/pnas.1707462114.

Wells, M.J. 1996. *Strawberry fields: Politics, class, and work in California agriculture.* Ithaca: Cornell University Press.

Wikipedia (n.d.). Algorithm. https://en.wikipedia.org/wiki/Algorithm.

Willmer, G. 2019. Tailored, targeted AI apps pave way for smart farming. *Sci Dev Net*, October 16. https://www.scidev.net/global/agriculture/feature/tailored-targeted-ai-apps-pave-way-for-smart-farming.html.

Winner, L. 1980. Do artifacts have politics? *Daedaleus* 109: 121–136.

Wiseman, L., J. Sanderson, A. Zhang, and E. Jakku. 2019. Farmers and their data: An examination of farmers' reluctance to share their data through the lens of the laws impacting smart farming. *NJAS-Wageningen Journal of Life Sciences.* https://doi.org/10.1016/j.njas.2019.04.007.

Woolgar, S., and J. Lezaun. 2013. The wrong bin bag: A turn to ontology in science and technology studies. *Social Studies of Science* 43: 321–340.

Publisher's Note Springer Nature remains neutral with regard to jurisdictional claims in published maps and institutional affiliations.

Michael Carolan is a Professor of Sociology and Associate Dean of Research and Graduate Affairs for the College of Liberal Arts. Other appointments include: Distinguished Fulbright Research Chair, University of Ottawa, Ottawa, Canada; Professor of Political Science (Status-Only), University of Toronto, Toronto, Canada; Visiting Professor, Ruralis Research Institute, Trondheim, Norway; and Research Affiliate, Centre for Sustainability, University of Otago, Dunedin, New Zealand.

Agriculture and Human Values (2020) 37:1055–1071
https://doi.org/10.1007/s10460-020-10137-2

Sustainability transitions in agri-food systems: insights from South Korea's universal free, eco-friendly school lunch program

Jennifer E. Gaddis[1] · June Jeon[2]

Accepted: 22 July 2020 / Published online: 3 August 2020
© Springer Nature B.V. 2020

Abstract

Government-sponsored school lunch programs have garnered attention from activists and policymakers for their potential to promote public health, sustainable diets, and food sovereignty. However, across country contexts, these programs often fall far short of their transformative potential. It is vital, then, to identify policies and organizing strategies that enable school lunch programs to be redesigned at the national scale. In this article, we use document analysis of historical newspapers and government data to examine the motivating factors and underlying conditions that allowed South Korea's universal free, eco-friendly (UFEF) school lunch program to become a tool for advancing social justice and ecological goals at the national scale. We analyze the socio-historical evolution and current status of the Korean school lunch program, combining the multi-level perspective with insights from environmental sociology and critical food studies, in order to shed light on the factors that enabled the program to become an innovative niche and articulate the opportunities and challenges it now faces. We identify the state-sponsored creation of what we call "precautionary infrastructure" as a key anchoring mechanism between the school food niche and agri-food regime. Precautionary infrastructure includes new supply chains, certification standards, and sourcing policies that provide a stable market for eco-friendly farms and small-scale producers, while minimizing the environmental health risks of school lunch by delivering organic and pesticide-free ingredients to on-site kitchens that serve free lunches to all students. This analysis offers insight into how public school-lunch programs can become protected niches that help drive sustainability transitions within agri-food systems.

Keywords School lunch · Sustainability transitions · Precautionary consumption · Corporeal citizenship · Korean school lunch policy · Food sovereignty

Abbreviations

UFEF   Universal free, eco-friendly
PNAE   Programa Nacional de Alimentação Escolar
MLP   Multi-level perspective
KINDS   Korea integrated news database system
KWPA   Korean Women Peasants Association
NAQS   The National Agricultural Products Quality Management Services

✉ Jennifer E. Gaddis
   jgaddis@wisc.edu

   June Jeon
   June.Jeon@tufts.edu

[1] School of Human Ecology, University of Wisconsin-Madison, 1300 Linden Drive, Madison, Wisconsin, USA

[2] Jonathan M. Tisch College of Civic Life, Tufts University, 163 Packard Ave., Medford, Massachusetts, USA

Introduction

School lunch has long been a contested political arena shaped by government agencies, civil society activists, and powerful agri-food companies concerned with what and how children are fed (Morgan and Sonnino 2013; Robert and Weaver-Hightower 2011; Gaddis and Coplen 2018). They are a public form of care (Gaddis 2019), which is best thought of as a "species activity that includes everything we do to maintain, continue, and repair our world so that we may live in it as well as possible" (Fisher and Tronto 1990, p. 40). Thus, in designing school lunch programs, governments must grapple with what political theorist Joan Tronto (2013, p. 139) describes as "the larger structural questions of thinking about which institutions, people and practices should be used to accomplish concrete and real caring tasks."

The prevailing ideology that both food and care should be cheap has kept public school-lunch programs around the world locked into highly industrialized systems of production and consumption (Sonnino et al. 2014). Multiple factors—including global climate change, rising obesity rates, the decline of family-scale agriculture, continuing rural–urban migration, and the Westernization of food cultures—have motivated national governments and civil society organizations to leverage their public school-lunch programs to support *sustainability transitions,* or "long-term, multi-dimensional and fundamental transformation processes through which established socio-technical systems shift to more sustainable modes of production and consumption" (Markard et al. 2012, p. 956).

As complex socio-technical systems of care provisioning, government-sponsored school lunch programs are not only an outcome of political processes, but also a historical achievement, established and negotiated in relation to local and global contexts. They have undergone three distinct phases of development in the global north (Oostindjer et al. 2016). From the 1850s–1950s, programs were created to reduce hunger and malnutrition and, in some cases, to redistribute surplus agricultural commodities. In the 1970s, some countries improved the nutritional quality of school lunches, while others sought to reduce costs by outsourcing food preparation, program management, or both. The third and contemporary phase encompasses a wide range of reforms motivated by concerns about public health, environmental sustainability, economic development, and social justice. It includes the United Nations' efforts to support sustainable and equitable development in the global south through the Home Grown School Meals initiative (World Food Program 2016), which is grounded in the principles of food sovereignty, or "the right of peoples to healthy and culturally appropriate food produced through ecologically sound and sustainable methods, and their right to define their own food and agriculture systems" (Declaration of Nyéléni 2007).

Policymakers and development organizations have identified school lunch as a key arena for promoting *sustainable diets*, or "diets with low environmental impacts which contribute to food and nutrition security and to healthy life for present and future generations" (Food and Agriculture Organization 2010). Yet the bulk of scholarship on school food, which occurs within the fields of nutrition and public health, does not offer guidance for planned transitions or engage with questions of how school food policy changes over time. Recently, however, scholars have begun to fill this gap by examining the relationship between school food and food sovereignty (Kleine and Brightwell 2015; Wittman and Blesh 2017; Powell and Wittman 2018; Stapleton 2019), documenting how farm-to-school programs can be designed in ways that prioritize equity and support cooperative regional economies (Lakind et al. 2016), and identifying viable transition pathways (Lehtinen 2012; Morgan and

Sonnino 2013; Galli et al. 2014; Bui et al. 2016; Ilieva and Hernandez 2018; Gaddis and Coplen 2018; Gilbert et al. 2018; Gaddis 2019).

Within this literature, Brazil's Programa Nacional de Alimentação Escolar (PNAE) has emerged as a high-profile case study, with scholarship focusing on the program's potential to strengthen food security/sovereignty (Sidaner et al. 2013; Sonnino et al. 2014; Wittman and Blesh 2017) and scale-up ethical consumption (Kleine and Brightwell 2015). The PNAE has undergone several programmatic redesigns over the past 70 years, the most recent of which occurred in the early 2000s with the enactment of a multisectoral food and nutrition security strategy designed to support sustainable rural development and food sovereignty (Sidaner et al. 2013). Since 2009, in response to demands from social movements and civil society activists, PNAE schools have been required to source at least 30% of their ingredients from local, small-scale, family farmers and/or organic producers (Kleine and Brightwell 2015). At present, the Brazilian constitution guarantees free school meals to the country's 43 million public-school students through the PNAE.

South Korea's universal free, eco-friendly (UFEF)[1] school lunch program is an equally compelling case, yet there is minimal English-language scholarship on the program besides Kang's (2011) account of the partisan political debates surrounding its creation. Both the Brazilian and Korean programs provide free meals to all students and prioritize serving traditional dishes prepared from scratch with ingredients sourced from alternative food networks. However, in Brazil, a new 20-year budget severely caps public spending, which may have negative consequences for the PNAE and its broader impact.

Conversely, public support for the Korean UFEF school lunch program has increased since its creation in 2011. Korea's capital, Seoul, announced it would triple the city's school lunch budget by 2021 to include all elementary, middle, and high schools (Ilbo 2018). Moreover, the national government plans to replicate the model of procurement developed for public schools in other public institutions–including hospitals, social welfare facilities, and correctional facilities–which feed roughly 25% of the Korean population (Korea Agro-Fisheries & Food Trade Corporation 2017). Thus, it is especially important to understand the motivating factors and underlying conditions that allowed Korea's UFEF school lunch program to become an innovative niche for sustainable agri-food policy development.

In the remainder of this article, we examine these factors and conditions, drawing from the multi-level perspective

---

[1] In the Korean context "eco-friendly" refers to healthy and traditional foods that are either organic or pesticide-free.

(MLP) (Geels 2002, 2011) and insights from environmental sociology and critical food studies. We trace the transition pathway and current status of what we call "precautionary infrastructure," which includes new procurement policies, distribution hubs, certification standards, food preparation methods, and serving requirements underwritten by municipal, provincial, and national actors. The state-led development of precautionary infrastructure for the UFEF program is intended to protect children from the health risks of consuming agro-industrial chemicals, while providing a stable market for eco-friendly farms and small-scale producers. It represents one way that the precautionary principle can be embedded "across institutions at all levels of government, as well as in science and corporate research and development" (MacKendrick 2018, p. 168).

In this paper, we analyze key moments in which notions of risk, responsibility, and sovereignty within the Korean school food system were destabilized and renegotiated. After presenting our theoretical framework, we provide a review of the literature on agri-food sustainability transitions, explain our research methods, and contextualize our case study. Next, we analyze the historical evolution of Korean school lunch policy and develop the concept of precautionary infrastructure. We identify several factors that motivated the state to develop precautionary infrastructure, including: (1) policy demands at the niche level of the public school-lunch program, (2) challenges to the social legitimacy of the neoliberal agri-food regime posed by civil society activists and the philosophy of *sint'o puri* (body and earth are one), and (3) shifts within the larger landscape related to international trade relations, Korean political-economic conditions, and a change in consumer consciousness about the origins of risk within the food system. Lastly, we discuss implications for sustainability transitions within public school-lunch programs and offer suggestions for future research.

Sustainability transitions in agri-food systems

Sustainability transitions involve changes not only in technology, but also in government policies, consumer habits, business practices, cultural values, and infrastructure (Geels 2011). Such processes are often multidimensional, context-dependent, and subject to power dynamics (Kern and Mackard 2016). In the case of agri-food systems, El Bilali et al. (2019) have identified three main strategies for accomplishing sustainability transitions: (1) increasing the efficiency of food systems, i.e., sustainable intensification, (2) reshaping consumer demand, i.e., supporting sustainable diets, and (3) transforming food systems, i.e., fostering alternative food networks. Yet questions remain about the range of viable transition

pathways (Pitt and Jones 2016) and the conditions that enable transitions to occur in one place and not another (Hansen and Coenen 2015).

Scholars have increasingly studied agri-food transition pathways (Lawhon and Murphy 2012; Spaargaren et al. 2013; Kirwan et al. 2013; Hinrichs 2014; Bui et al. 2016; Ilieva and Hernandez 2018; Rut and Davies 2018; Rossi et al. 2019; O'Neill et al. 2019) using approaches that account for both power relations and place-specificity in determining the processes and outcomes (Hansen and Coenen 2015). It is vital to understand, for example, how different path dependencies (e.g., environmental, infrastructural, institutional, cultural, and economic) affect the viability of transitions across a range of places and scales.

The most prominent sustainability transitions framework within the agri-food systems literature is the MLP (El Bilali 2019), which conceptualizes sustainability transitions as systems changes that emerge dynamically from interactions and changes across the three analytical levels of niche, landscape, and regime. A niche is the space in which radical innovations are initially developed (e.g., municipal policy or direct-contracting with agricultural cooperatives) (Geels and Schot 2007). They can take a variety of forms, including technologies, new rules and legislation, programs, and organizations. The regime is where socio-technical structures are stabilized and become dominant systems that rarely undergo transformation or reconfiguration. Regimes are constituted of informal and formal rules, technologies, institutions, actors and social groups (Geels 2011). Their elements can be tangible (e.g., laws, procurement standards, certifications) or intangible (e.g., culture, social norms, and policy paradigms) (Geels 2011). Lastly, in the MLP the landscape plays a key role in structuring the relationship between niche and regime. It includes the economic, political, and cultural context beyond the influence of niche and regime actors (e.g., demographic trends, political-economic developments, international trade agreements, global climate change) and cannot be easily changed in the short-term (Lachman 2013).

Within the agri-food literature, scholars using the MLP typically conceptualize alternative agri-food systems (e.g., organic, fair trade, local) as niches (El Biali 2019) that coexist with, and at times challenge, dominant food regimes. Niches enable innovation by providing some level of protection from the dominant rules of the regime, however they must be robust and mature enough to challenge the regime in order for a transition to occur. The MLP suggests that niche development in itself is not enough to cause a regime shift; theoretically, transitions can only occur when there is strong niche-regime-landscape alignment, however recent empirical work on niche-regime linkages within agri-food systems challenges this notion (Bui et al. 2016).

Changes within the socio-technical landscape that cause regime destabilization open up opportunities for niche-innovations to be integrated into or reconfigure the existing regime. To better articulate how niche-regime linkages begin, Elzen et al. (2011) introduced the concept of "anchoring," which they define as an interaction that leads to a durable link between these two levels of the MLP. Niches can anchor to regimes by: (1) establishing new rules or institutions (institutional anchoring), (2) fostering new technical systems (technological anchoring), and (3) building new networks and social groups (network anchoring). There is some concern, however, that the "radicalness" of niche-innovations will be diluted as they scale up and out (El Biali 2019), which is particularly relevant given the Korean government's plan to replicate and scale the UFEF school lunch program.

Notably, while the MLP doesn't preclude discussions of social justice, it is underemphasized in the literature relative to considerations of economic and environmental goals. Our case allows for an integrated analysis of the social, economic, and ecological dimensions of sustainability since the UFEF school lunch program is the outcome of two parallel transition processes (i.e., making meals universally free *and* eco-friendly). We conceptualize the UFEF school lunch program as a protected niche within the mainstream Korean agri-food regime and consider both niche formation and development in our analysis, alongside changes at the regime and landscape levels. Agri-food transitions scholars have often devoted more analytical attention to understanding niche innovations, leaving processes occurring at the regime and landscape levels comparatively understudied. Thus, when tracing the historical evolution of Korean school food policy prior to the enactment of the UFEF school food policy, we pay careful attention to elements at both the regime and landscape levels that shaped this niche-in-the-making.

Korea's universal free, eco-friendly school lunch program

South Korea offers an ideal site to examine potential pathways through which national school lunch programs can become protected policy niches that foster the development of eco-friendly agri-food systems and universal social welfare provisioning. During the country's first direct election of superintendents of education and local boards of education in June 2010, progressive candidates campaigned on a platform that included a promise to convert the existing school lunch program—which provided free lunches only to basic livelihood security recipients—into a universal program that would serve free, safe, eco-friendly meals to all students. The progressives won in most provinces and refused to compromise on their vision for a universal free program, even though fiscal conservatives severely criticized this aspect of their proposal.

Since then, the proportion of students participating in Korea's UFEF school lunch program has increased both within and across municipalities, from 56.8% in 2012 to 76.2% in 2017 (Ministry of Education 2018). In 2017, 4.28 million students participated in the program, which was initially implemented in elementary and middle schools and has subsequently been introduced to high schools in four provinces (Monthly Nutriand 2018). With a total budget of ₩2942 billion (roughly 2.6 billion USD), the program provides elementary- and middle-school students with a standardized meal (Fig. 1) consisting of rice, soup, kimchi, vegetables, and fish or meat (Ministry of Education 2018).

The Korean government finances the program through a cost-sharing agreement that relies on national, district, and local contributions, but lunch prices vary by municipality and type of school. In Seoul, for example, the price of lunch during the 2018–2019 school year was ₩3628 (3.13 USD) in public elementary schools, ₩4649 (4.01 USD) in private

Fig. 1 Examples of Korean school lunch consisting of rice, soup, kimchi, vegetable, and protein

elementary schools, and ₩5406 (4.66 USD) for middle and high schools. The Korean Office of Education pays 50% of the cost, while the District Office of Education contributes 20% and the City Office of Education covers the remaining 30% (Ministry of Education 2018).

Social movement activism in support of food sovereignty and food safety pushed the Korean government to align school food policy with the "Food from Somewhere" regime (McMichael 2002; Campbell 2005, 2009; Friedmann 2005), which includes both local food networks and the corporate environmental food regime. This emergent regime operates in complex opposition to the "Food from Nowhere" regime (McMichael 2002) that has dominated global agriculture and food policy since the mid 1990s. In Korea, social movement opposition to globalized, industrial food production was instrumental in creating the enabling conditions for the UFEF school lunch policy to pass in the late the 2000s. Likewise, consumer concern about the human health risks of the Food from Nowhere regime put pressure on regime actors to change school food policy, as Koreans increasingly recognized the limitations of "precautionary consumption" (MacKendrick 2014, 2018) as an individualized, consumer-based approach to risk management.

Precautionary consumption entails a variety of individualized actions to address both real and perceived risks, including but not limited to: researching potential risks, comparing available products, purchasing items that may be sold only in specialty stores or farmers markets, and preparing a greater percentage of food and other household items from scratch. People turn to precautionary consumption, a form of gendered care-work (Cairns et al. 2013), in order to compensate for weak regulatory systems that fail to adequately mitigate chemical body burdens (i.e., the total accumulation of environmental chemicals like pesticide residues and plasticizers in an individual's body). In addition to requiring more time and money than some individuals have, precautionary consumption may elicit a false sense of security that inadvertently undercuts public support for government action, while allowing companies to profit from the sale of "safe" products (Szasz 2007).

Moreover, precautionary consumption is "likely to exacerbate health disparities along social modalities of race/ethnicity, education levels, and socio-economic status" (Scott et al. 2016, p. 327), while adding to women's already disproportionate share of household reproductive labor (MacKendrick 2014; Castellano 2015, 2016). These multiple shortcomings led MacKendrick (2018, p. 156) to suggest that societies reject precautionary consumption in favor of environmental justice, which requires "a shift from seeing the government as a stumbling block to innovation and progress to seeing it as a democratic institution that can and should provide collective protection from environmental health risks."

In Korea, niche actors have successfully managed to enroll the state in creating *precautionary infrastructure* that incentivizes Korean farmers and food companies to adopt eco-friendly production practices in order to access the multi-billion-dollar school lunch market. Precautionary consumption within the school food environment is no longer a class-based project since the Korean government guarantees all students the right to a free, eco-friendly school lunch. As such, the UFEF school lunch program is an especially fruitful case study for understanding how sustainability transitions can address environmental and social justice concerns in tandem.

Methods

The MLP is a widely used framework within sustainability transitions studies, but it does not adequately account for the role of power relations or human agency in motivating sustainability transitions within agri-food systems (Hargreaves et al. 2013). Scholars have responded to such criticisms by enriching the MLP with other frameworks (Bui et al. 2016; Elzen et al. 2011; Ilieva and Hernandez 2018; Kuokkanen et al. 2018). We enhance the MLP with insights from environmental sociology and critical food studies—a combination that allows us to take advantage of the analytical perspective of the MLP, while addressing some of its weaknesses, specifically in relation to the role of political economy and civil society activism in transition processes.

The data for this paper is drawn from a content analysis of primary and secondary sources. A Korean member of the research team conducted a literature review of Korean- and English-language peer-reviewed articles and books on the Korean school lunch program. Factual information contained in Korean articles was cross-referenced with the original primary sources in order to establish a comprehensive history of school lunch in Korea and an accurate description of the current UFEF program. Next, we compiled and analyzed primary documents pertaining to school lunch, including newspaper articles and governmental documents, from online archives such as the Korea Integrated News Database System (KINDS) and Naver news library (dna.naver.com/search/searchByDate.nhn).

We collected and reviewed: (1) all newspaper articles in the KINDS database published between 1990–2018 that mentioned the search term "school lunch program" (and variations thereof), (2) all newspaper articles published between 1953–2000 in South Korea's four major presses (Kyunghyang Shinmoon, Donga Daily, Maeil Economy, and Hangyurae Shinmoon) and archived by the Naver news library, and (3) government documents accessed electronically via the online archive of the National Archives of Korea (archives.go.kr) and Seoul Metropolitan

Government's public information disclosure portal (opengov.seoul.go.kr).

School lunch as a niche-in-the-making

South Korea's school lunch program began in the early 1950s after the Korean War as a mechanism for distributing international food aid to poor, rural children (Gyunghyang Newspaper 1957, 1963). At the regime and landscape levels, economic instability and the potential for widespread famine were viewed as the major sources of risk within the Korean food system, until 1977 when a massive outbreak of food poisoning alerted the public to the need for a more robust set of food hygiene laws and practices. In one of the most publicized episodes impacting the school lunch program, 7872 students experienced vomiting and diarrhea, while another 948 required hospitalization after eating a cream bread made from donated milk powder and wheat flour. The Korean government supplied this bread free-of-charge to 13,000 of Seoul's poorest students and at a discounted price to the remainder of the city's 97,000 elementary and secondary students (Gyunghyang Newspaper 1977).

Many parents had opted to purchase the government-supplied bread because they thought it was safer than what they could purchase on the open market, but the mass food poisoning incidents shattered this illusion. The school lunch program was not yet operating as a protected niche. Lack of food safety infrastructure at the regime level meant schools had little ability to gauge the bread's freshness. The bread company did not print the manufacturing date on the packaging and inadequate temperature control during distribution and storage could cause cream bread to spoil. Without the political-economic resources to fix the problem, Korean officials discontinued the bread program in late 1977 (Gyunghyang Newspaper 1977).

Regime-level constraints nearly destroyed the viability of the niche program as parents increasingly reverted to packing their children's lunches. In 1981, over 97% of Korean students brought their own lunches to school (Kim 2013). However, public health experts argued that the public school-lunch program should be expanded and strengthened, not disbanded. "Food poisoning in 1977 was the problem of unethical suppliers and corrupted school officers, not the problem of the policy itself," one such advocate wrote in a national newspaper (Dong-Ah Daily 1979). Supporters used the rhetoric of nation-building and international competition, pointing to high-income countries with robust public school-lunch programs, to argue that school lunch would improve public health while simultaneously bolstering economic development

and innovation within the agriculture and food sectors (Dong-Ah Daily 1979).

The Korean government didn't disagree. As early as 1975, policymakers had begun debating the merits of creating a national school lunch program that would ensure Korean children were as healthy as their counterparts in high-income countries. In 1981, the Korean government passed the School Lunch Act of 1981, to support the "healthy growth of students' minds and bodies" and the widespread "improvement of people's dietary life" (Ministry of Education 1981). Ultimately, however, the legislation provided more in rhetoric than resources.

The timing of the 1977 food poisoning incidents and a lack of financial resources for public projects under President Doohwan Jeon's harsh military regime (1980–1987) restricted the expansion of Korea's school lunch program. At the same time, landscape-level factors like the rapid Westernization of the Korean diet in the 1980s— marked by a dramatic increase in the consumption of meat (especially beef) and fast food—sparked a consumer movement to protect "traditional foods," along with the economic and cultural vitality of the country's rural agricultural communities (Yang 2010).

Food sovereignty activism and landscape shifts: 1970s–1990s

Civil society organizations (as new niche actors) leveraged the food safety scares of the 1970s, which illustrated the state's inability to provide safe food through the dominant agri-food regime, as an opportunity to propose an alternative model of school food provisioning. Under the banner of the Women Catholic Peasants Association, and later the Korean Women Peasants Association (KWPA), small-scale women farmers urged the Korean government to purchase their locally grown produce and locally raised livestock for use in school lunches. They believed food safety was about more than hygiene and argued that women peasant-farmers could be trusted to provide schools with a reliable supply chain of safe, high quality ingredients produced using traditional agricultural practices (Park and Jeong 2010). Their proposal was rejected, but it indicates an important shift in the country's agri-food landscape.

For much of Korea's post-war history, peasant farmers had advocated with little success for protectionist trade policies and financial support for domestic agriculture. In the 1970s, a social movement of peasant farmers rallied against state economic programs that threatened the economic viability of Korean agriculture and livestock producers. Some farmers committed suicide by ingesting agricultural chemicals, while thousands more engaged in public demonstrations against global trade agreements

(Abelmann 1996). During this time, the KWPA organized against both the global food regime and the disproportionate workload that patriarchal rural culture placed on their shoulders. As early as 1977, local women's organizations (which later merged under the national umbrella of the KWPA) had begun to advance the interests of "independent women peasants" by bringing a sophisticated class and gender analysis to their diagnosis of the multiple problems facing Korean farmers in general, and women farmers in particular (Park and Jeong 2010, p. 109). They believed that true independence for women farmers could only be achieved by breaking free of the imperial global agricultural system (i.e., achieving food sovereignty), upending the patriarchal structure of Korean society, and erasing the social boundaries that restricted women's empowerment (Jung 2005; Park 1995).

As the food sovereignty movement gained strength in the early 1990s, the KWPA once again petitioned the Korean government to purchase locally produced agricultural products for the country's school lunch program. However, the KWPA and their allies, including the National Farmers Association (*Jeon-nong*), were never able to overcome the hurdles posed by limited finances and lack of investment, particularly at the local level (Park 2008). At the time, local offices of education administered lunch programs using funding supplied by the central government and revenue from children's fees. It was illegal for provincial governments to provide supplemental funding because of how the governance and budgetary systems were structured. And even if schools had been able to increase their lunch budgets, the Uruguay Round multilateral trade negotiations (1986 to 1993) prevented the government from incentivizing public schools to give preference to local producers (Kim 2009).

Paradoxically, as the Korean government was entering into this new international trade agreement and the Korean population was becoming more dependent on food imports, the National Agricultural Cooperative Federation (Nonghyop Chung'ang hoe) and the Ministry of Agriculture, Forestry, and Fisheries (Nong lim Susanbu) began to incorporate nationalistic slogans like "healthy eating—just eat our rice" and *sint'o puri* (body and earth are one) in food advertisements and official dietary advice (Cwiertka 2013). This publicity campaign helped to cement a new public belief that food from Korean soil (i.e., Food from Somewhere) is best for Korean bodies because it is better tasting and healthier than imported food. Regime-actors encouraged urban Koreans, who made up nearly 80% of the country's population by the end of the twentieth century, to purchase food grown by their farmer-compatriots in the countryside.

In sum, food sovereignty activism and s*int'o puri* ideology generated pressure for regime change. However, the niche school food program was not yet robust enough to take advantage of the opportunity.

Niche developments and constraints: *wei-tak egupsik* and austerity politics

School lunch was a hot-button political issue in the early 1990s, especially among mothers employed outside the home who found preparing school lunch boxes to be a burdensome task (Lee et al. 1994). During Korea's 1992 election, every presidential candidate pledged to expand the school lunch program to all Korean elementary schools (Kang 2011). Soon after, the national government passed the 1996 School Lunch Act Amendment, which partially subsidized the cost of building new kitchens (Ministry of Education, Science, and Technology 2011) and passed a controversial policy that allowed schools to outsource food preparation to for-profit catering firms (*wei-tak geupsik*).

This public–private partnership helped to rapidly expand the national school lunch program from 11.3% of Korean elementary schools in 1992 to 99.2% in 1998 (Kang 2011). By 2004, 99.9% of elementary schools, 97.8% of middle schools, and 98.7% of high schools provided school lunch (Kang 2011). During the late 1990s and early 2000s, catering companies competed for market share within the two trillion-won (roughly 2 billion USD) market for prepared lunch boxes (Maeil Economy 1997). Most students paid about ₩2500 per meal (roughly 2 USD) (Gyunghyang Newspaper 1997). Only ten thousand students qualified for the government's free lunch subsidy.

Schools lagged behind other sectors of the Korean agri-food economy in reducing the risk of food poisoning. School lunches were responsible for 19.4% of Korean food poisoning cases in 1996, 30.3% in 1998, and 70% in 2001 (Lee, E. H. et al. 2016). Students and parents both considered lunches prepared in on-site school kitchens to be far superior to catered lunches in terms of hygiene, nutrition, and taste (Park et al. 1997). In one 1997 survey of 541 middle school parents, 89% said they only let their children eat *wei-tak geupsik* because it reduced the amount of cooking they had to do at home; 70% of these parents said meal quality should be improved (Park et al. 1997). Only about 11% of students who ate *wei-tak geupsik* said their lunches were delicious, in contrast to the 70% of students who ate lunches prepared in on-site school kitchens. Yet the majority of Korean schools lacked on-site kitchens for meal preparation.

Teachers and parents told the media that they could not trust the safety and quality of school lunches delivered through this system (Munhwa Newspaper 2002). The option to feed students attending *wei-tak geupsik* schools "better" homemade lunches put pressure on parents (especially mothers) to practice precautionary consumption. Students eating *wei-tak geupsik* often complained about finding hairs and bugs in their soup and said they would rather purchase cheap ramen from the school market than eat the catered meals

(Kyung-In Daily 2002). There were nutritional concerns, however, associated with giving children pocket money to purchase instant-cup ramen or other convenience foods that caused some mothers to experience guilt and stigma (Kyung-nam Newspaper 2002). As one mother explained: "I cannot simply look over my kids eating ramen every day, instead of eating warm lunch boxes" (Munhwa Newspaper 2002).

Deep reforms were needed, but no major policy changes happened until 2006. The massive economic crisis that impacted East Asia in the late 1990s put a significant strain on the Korean government's ability to respond. The International Monetary Fund (IMF) lent South Korea 58 billion USD in 1997 and placed the country under a structural adjustment program that catalyzed major neoliberal reforms to Korea's financial institutions, labor markets, and public service sectors (Rodier 2014). Many of the country's nationalized companies were privatized in favor of small government.

The continued reliance on the privatized model of school lunch provisioning using *wei-tak geupsik* aligned with neoliberal free market logic. So, too, did the government's response to concerns about food safety. In place of a robust regulatory and inspection regime, the Ministry of Education published in 2000 the "Guidelines of Hygiene Control for School Lunch," a manual designed to instruct catering firms in hygiene control and general food safety protocol (Kang 2011).

Civil society activists wanted the government to provide more than a set of guidelines. They wanted an entirely different school lunch program—one that would rely on eco-friendly agriculture, promote socio-cultural and economic exchanges between rural and urban areas, and advance democracy by preserving local autonomy (Choi 2006). At the provincial level, the Chonnam Christian Peasants Association convinced local officials to pass a 2003 ordinance that included a 5-year plan for purchasing eco-friendly agricultural products from Chonnam producers using a subsidy provided by the provincial government (Kim 2009; Yoon 2018).

At the national level, a coalition of leftist political parties, labor unions, and roughly 650 civil society organizations pushed for change (Kang 2011). School Lunch Network Nationwide, one of the coalition's leaders, put forward a proposal for a universal free lunch program that would use locally grown, organic foods. They had reason to hope that the government would act on their demands since the IMF loan was fully repaid in August 2001 and public support was on their side. However, the Korean economy was still in recovery. During structural adjustment, jobs had become increasingly precarious and government social services had been rolled back. This macroeconomic context prevented the Korean government from investing in a new model of school lunch provisioning. Instead, policymakers issued a stricter

set of regulations for private caterers in 2004, requiring all *wei-tak geupsik* factories to implement a Hazard Analysis Critical Control Point (HACCP) management system.

A stronger regulatory response was necessary. All of the food poisoning cases that took place in Seoul schools between 2003–2006 were linked to *wei-tak geupsik* caterers, some of whom had even bribed teachers and school administrators in order to renew their contracts and avoid health inspections (Kang 2011). In the most high-profile incident, fifteen-hundred students were sickened at 23 Seoul-area schools after consuming meals supplied by one of Korea's largest catering companies (Hankook Economy 2006). Parents and civil society organizations demanded an end to the widespread practice of contracting out meal preparation to for-profit *wei-tak geupsik* caterers—a change that was subsequently codified into national legislation with the 2006 amendment to the 1981 Korean School Lunch Act.

The 2006 amendment dramatically expanded the government's role in financing and managing the school lunch program. It removed a prior rule that capped the government subsidy for poor students at 50% of the total meal cost, required schools to assume responsibility for managing their own lunch programs, and earmarked money to help schools build their own kitchen and cafeteria facilities. The central government subsequently provided funding to employ school nutritionists (in a position equivalent with teachers) to work in each of the nation's schools (Ministry of Education 2014). Nutritionists were expected to design menus that: (1) relied on various cooking methods and ingredients, (2) promoted traditional Korean food culture, (3) maximized the use of seasonal and natural ingredients, and (4) reduced the prevalence of salt, fat, monosaccharides, and food additives in school lunches.

In sum, the 2006 amendment switched the Korean school lunch program onto a new policy track—one of increased social welfare spending and tighter regulatory controls—that departed from the neoliberal laissez-faire approach favored by Korean policymakers throughout the 1980s and 1990s (see Table 1). High-profile food poisoning accidents alerted the public to the limits and hidden dangers of cheapness and privatization as strategies for organizing government programs and managing risk in the food system. This led to an investment and strengthening of the school food niche, via the widespread development of kitchen infrastructure and managerial expertise in schools that had previously contracted with *wei-tak geupsik* companies, as overall public interest in eco-friendly food and food sovereignty continued to grow.

Table 1 Characteristics of the School Lunch Act of 1981, 1996 amendment, 2006 amendment, and 2011 UFEF school lunch policy

	School Lunch Act of 1981	1996 Amendment	2006 Amendment	Post-2011
Purpose	To mobilize the healthy growth of students' minds and bodies and improve people's dietary life	To mobilize the healthy growth of students' minds and bodies and improve people's dietary life	Improve the quality of school meals, mobilize the healthy growth of students' minds and bodies, and improve people's dietary life	Improve the quality of school meals, mobilize the healthy growth of students' minds and bodies, and improve people's dietary life
Law restricting management of school lunch program	Not stated	External catering service providers or direct management by schools allowed	Only direct management by schools (with minor exceptions) allowed	Only direct management by schools (with minor exceptions) allowed
Managers of lunch program	School principals	External suppliers or school committee	Committee on school meals (led by local superintendent of education) and schools' own meal preparation staff	Committee on school meals (led by local superintendent of education) and schools' own meal preparation staff
Government subsidy for lunch program	Commodity donations to elementary schools in rural areas	50% financial subsidy for poor students	100% financial subsidy for poor students	Universal free meals in most provinces
Funding mechanism	Facilities and infrastructure: paid by parents Other expenditures: paid by parents	Facilities and infrastructure: paid by parents Ingredients paid for by parents or national/local government Other costs: paid by schools or parents	Facilities and infrastructure: paid by schools or national/local governments Ingredients: paid by parents or national/local government Other costs: paid by schools or parents	All costs paid by either national or local governments on behalf of caregivers* *The term "parents" was changed to "caregivers" in a 2008 amendment

Corporeal citizenship and the early stages of alignment

A well-being craze spread through Korean society in the early 2000s, leading to an increase in producer–consumer activism in support of agri-food regime change (Yang 2010). The volume of domestic agricultural products (excluding livestock) that the Korean government certified as eco-friendly jumped from 87,279 metric tons in 2001 to 2,188,311 metric tons in 2008. This change in producer activity was accompanied by an evolving consumer consciousness. By the end of the decade, producer–consumer activism began to take on the characteristics of what Gabrielson and Parady (2010) call "corporeal citizenship." Precautionary consumption rests on the assumption that human bodies can be isolated from their natural environments and protected from exposure to chemical risks. Corporeal citizenship instead acknowledges the permeable boundaries between bodies and the environment, pushing individuals to expand their sphere of responsibility to encompass care for others (human and nonhuman) both in proximity and at a distance (Scott et al. 2016).

This more sophisticated understanding of risk and responsibility—a key change in landscape conditions—is apparent in the coordinated social and political resistance of Korean agri-food producers and consumers against American beef imports in 2008 (Chang 2010). By 2006, sixty-five nations, including Korea, had adopted restrictions on importing American beef products due to concerns about the neurological risks associated with mad cow disease (bovine spongiform encephalopathy, or BSE). Yet in 2008, the Korean government committed to importing American beef as a pre-condition of the Korea-US Free Trade Agreement.

Korean mass media depicted American beef, along with the Food from Nowhere regime, as inherently risky. For example, *PD Notebook,* a popular liberal-leaning television news program, broadcast a segment entitled "American beef, is it safe from BSE?," which included a clip of downer cows. *PD Notebook* told its viewers that the trade agreement would allow specified risk materials—tissues in cattle that are considered to be of high risk for prion contamination, such as brains, eyes, spinal cord, and skull (United States Department of Agriculture 2019)—to enter the Korean food system (MBC 2008). Conservative pundits argued that anti-free trade liberals were using graphic imagery to intentionally manipulate public support for the trade agreement (Chae 2009). Regardless, Koreans took to the streets in a series of 2398 candlelight protests between 2008 and 2009, culminating in a gathering of one million protesters on June 10, 2009. Their direct action continued until President Myung-Bak Lee apologized and promised to renegotiate the trade agreement (Bak 2012).

BSE was a focal point for Korean's concerns about food safety and globalization, but it was not the only food product or process subjected to heightened public scrutiny. According to the 2011 Korean general social survey, 80% of Korean adults were concerned about residual pesticide contamination on imported produce and 75% were worried about the effects of consuming genetically modified organisms (Korean Social Statistics 2011). However, less than 6% of Korean agricultural products were grown using eco-friendly methods in 2006 and only 0.4% with organic methods (Kim and Lee 2011). This mismatch between supply and demand posed a problem for the many Koreans who believed it was healthier for their families to consume fewer agricultural chemicals.

At the provincial level, some policymakers recognized the need to take a more active role in supporting eco-friendly agriculture and identified the school lunch program as a viable focus of their efforts. In April 2009, Sanggon Kim, a progressive candidate for the Superintendent of Education of Gyunggi, Korea's largest province, launched a campaign for a "free for all, organic school lunch system." Roughly 60% of Gyunggi citizens approved of the policy, which attracted strong support from the Korean Teachers and Educational Workers Union, but conservative members of the Gyunggi provincial school board blocked Kim's proposal due to budgetary limitations. Gyunggi Governor Moonsu Kim, a radical conservative, argued that Kim's proposal was nothing more than shallow populism. However, Kim's proposal motivated similar policy debates in other political jurisdictions, the largest and most controversial of which took place in Seoul, a city with 1,162,000 schoolchildren (Seoul Metropolitan Office of Education 2013).

UFEF school lunch policy and precautionary infrastructure in Seoul

On December 1, 2010, a group of leftist city counselors backed by Seoul's Superintendent of Education enacted a local "free for all, organic school lunch" ordinance through the Seoul city parliament. Sehoon Oh, the city's conservative mayor, immediately vetoed the ordinance (Maeil Economy 2010). Citing recent events in Greece, Mayor Oh insisted that such "politically-motivated populism" could ruin the country's economy since the Korean pension fund and welfare budget were already operating in crisis mode (Oh My News 2011). He suggested that Seoul citizens vote directly on the matter and vowed to resign if the election results upheld the ordinance. Turnout for the August 2011 special election was so low—only 25.7% of eligible voters—that quorum was not reached. The ordinance was therefore upheld and Mayor Oh resigned his post. Wonsoon Park, a former social movement activist, human-rights lawyer, and co-founder of People's

Solidarity for Participatory Democracy (one of Korea's largest NGOs), won the election to replace Oh and immediately enacted the school lunch ordinance (Ju 2016).

At the time, a majority of Koreans (62%) believed social welfare should be prioritized over economic growth (Korean Gallup 2014). Even so, the proposal to convert the existing means-tested school lunch program into a universal social service was much more controversial than the proposal to source pricier, eco-friendly ingredients. The tax burden of the program was widely discussed, but the Korean Democratic Party successfully argued in favor of both measures, thereby practicing corporeal citizenship by extending care to all students and making the program fully public.

Seoul Mayor Wonsoon Park believed that investing in UFEF school lunches would have positive social, economic, ecological, and community impacts (Kang 2016). He envisioned using school lunch funds to contract directly with farmers and incentive them to use eco-friendly practices. This aligned with broader government attempts to revitalize the communal character of rural areas and enhance the profitability of small-scale farming (Choi and Kim 2015). Accomplishing Mayor Park's vision, and satisfying the city's 2010 ordinance for schools to purchase only certified eco-friendly and organic food, meant using *Orbon*—an aggregation, certification, and distribution center built in 2009 by Seoul Agro-Fisheries & Food Corporation (a public enterprise funded by the Seoul Metropolitan Government)—to develop a new supply chain.

The National Agricultural Products Quality Management Service (NAQS) coordinates a nationwide eco-friendly certification system, encompassing multiple criteria, including: where the food is produced, antibiotic usage, HACCP certification, and the presence of pesticide residues (Gyunggi Province 2018). In Seoul, *Orbon* workers oversee the NAQS certification process and conduct their own independent testing to ensure the government's food safety standards are met. As of 2018, 67% of the foods served in Seoul schools were domestic products with NAQS eco-friendly certification (Seoul Metropolitan Office of Education 2018).

During the early years of the UFEF school lunch program, procurement policies privileged price over a more holistic set of social, cultural, or ethical values (Kim 2013; Lang et al. 2009). Until 2015, schools in Seoul were prevented from direct contracting with small-scale farmers, unless they offered the cheapest prices through the Electronic Agriculture Trade procurement system that schools are required to use.[2] This policy disadvantaged small-scale farmers with high land and labor costs (Korean Rural Economic News 2019). In 2015, the Electronic Agriculture

Trade procurement system was redesigned to penalize vendors whose prices are too far below the average of other firms. In addition, Orbon now allows schools to contract directly with small-scale farmers (for purchases up to 20,000 USD) instead of requiring them to take the lowest competitive bid.

Despite its limitations, the rapid development of this precautionary infrastructure is impressive. Early signs suggest the UFEF school lunch policy is helping to facilitate a sustainability transition since it encourages Korean farmers to reduce their use of antibiotics and pesticides in order to become NAQS certified (Kim et al. 2014). The number of newly certified organic farms has increased consistently, from an 11.6% annual increase in 2014 to a 15.5% increase in 2018 (Jung et al. 2019). Likewise, domestic production of organic food has grown in market size from 170 million USD in 2007 to 380 million USD in 2018.

However, it is too soon to tell whether recent changes will fully integrate Korea's most marginalized farmers into the precautionary infrastructure that Seoul and other municipalities are building for their public-school-lunch programs. The government could facilitate this process by underwriting the expansion of the KWPA's *toet bat* (kitchen garden) initiative to schools. This initiative helps women farmers sell seasonal produce and traditional processed foods such as tofu and red pepper paste to urban households (Burmeister and Choi 2012). Alternatively, the government could align school lunch procurement criteria with the KWPA's 2010 policy platform, which includes a rice-price guarantee, gender equality on rural farms, the realization of women peasants' rights to protect seeds, and an increase in farmers' participation in the production, processing, and distribution of agricultural goods (Park and Jeong 2010).

Another primary shortcoming of the UFEF school lunch program has to do with the outsourcing of culinary labor to for-profit food companies that rely on part-time workers and the subsequent deskilling of the country's 74,079 school kitchen and cafeteria workers. In 2016, these workers prepared approximately seven million lunches per day in the country's 11,389 school kitchens. While labor efficiency is already quite high in Korean schools—with each worker providing lunches for over one hundred students (Maeil Economy 2019)—it is not high enough to satisfy the tight fiscal constraints that government-employed nutritionists are expected to work within. Schools have less than 4.66 USD to spend per lunch, which makes it difficult to pay for both the higher cost of eco-friendly ingredients and the on-site labor necessary to transform minimally processed ingredients into ready-to-eat lunches. As a result, nutritionists are increasingly choosing NAQS-certified semi-prepared foods over basic ingredients that require additional on-site processing and preparation.

---

[2] The electronic system was established in 2010 to reduce the opportunity for corruption within the school lunch procurement system.

The economy of scale provided by private-sector factories helps schools serve NAQS-certified lunches. However, in prioritizing the public health and ecological dimensions of eco-friendly food, this certification deflects attention away from the poor job quality in Korean school kitchens that makes it challenging to recruit and retain enough workers. Recent surveys of frontline kitchen and cafeteria workers show the physical and emotional demands of the job are contributing to high rates of emotional exhaustion, job burn-out, and workplace injury (Lee, O. et al. 2014; Lee. D. et al. 2016).

While most nutritionists are directly employed by the government in full-time positions, the majority of frontline kitchen and cafeteria staff are part-time workers. Only 2100 out of 74,079 kitchen and cafeteria workers were full-time employees in 2018. These workers are part of the precautionary infrastructure that Korean schools are developing to provide children with safe, eco-friendly lunches, yet they receive little compensation for the mental, manual, and emotional labor they perform. In protest of their job conditions, non-permanent cafeteria workers staged a nationwide strike in July 2019 and at least 4601 schools stopped serving lunches for several days (BBC Korea 2019).

To date, social equity concerns related to the receivers of public care (i.e., children) have been much more strongly integrated into UFEF school lunch policy than social welfare concerns related to the providers of care (i.e., kitchen and cafeteria workers). Efforts to localize Seoul's school-lunch supply chain have operated largely within a market-based system that is slow to incorporate labor and social justice concerns, much like farm-to-school programs in the United States, which Allen and Guthman (2006) have criticized for reproducing neoliberalism and inadvertently restricting a politics of the possible (Harris 2009).

Thus, the next step in advancing food sovereignty and corporeal citizenship via Korea's UFEF school lunch program would be to extend the sphere of ethical and political responsibility to attend to the lives and livelihoods of food- and farm-workers across Korea's global and domestic school-food supply chains. The 2015 UFEF school food procurement policy reforms and the 2019 nationwide cafeteria worker strike suggest that such social justice concerns will continue to be raised and potentially integrated into the rules governing the niche, which, may, in turn, have a larger impact on both the regime and landscape.

Discussion and Conclusion

In this article, we have analyzed the socio-historical context of Korea's UFEF school lunch program, combining the MLP with perspectives drawn from environmental sociology and critical food studies, in order to equip scholars, policymakers, and civil society activists with fresh insights about how public school-lunch programs can become protected niches that help drive sustainability transitions within agri-food systems.

First, our analysis of Korea's UFEF school lunch program underscores the importance of alignment for the transition process (see Table 2). It was only after the Korean school lunch program became fully public in 2011, with the government assuming fiscal and administrative responsibility for providing free lunches to all children, that it became a protected space (partially removed from the market-based economy) conducive to the development of precautionary infrastructure. This confirms and extends existing theories of how and when sustainability transitions occur by placing niche-regime interactions within a country-specific socio-historical context and demonstrating that structural conditions can be both enabling and constraining (Slingerland and Schut 2014).

In the Korean case, some factors (e.g., consumer perception of risk within the food system, *sint'o puri* ideology, and food sovereignty activism) helped launch the school lunch program along a viable transition pathway, while others (e.g., structural adjustment, trade liberalization, and neoliberal social policy) prevented the national school lunch program from operating as a radical niche. The process of decoupling Korea's school food procurement from neoliberal market logic appears to have only begun after the government stopped the widely used practice of outsourcing meal preparation to for-profit catering firms and began providing free meals to all children. Thus, it seems that addressing the social equity dimensions of the school food program helped to create space for ecological and economic development goals to be more aggressively pursued at the niche level.

Second, this paper responds to criticisms of the MLP's lack of attention to agency and power dynamics by providing a full account of how consumer consciousness, social movement activism, and direct action in opposition to the Food from Nowhere regime eventually brought the niche, regime, and landscape into sufficient alignment to enable the UFEF school food policy to be implemented. Women's social movement activism, through organizations such as the KWPA and during critical moments of social movement mobilization (e.g., to expand the national school lunch program in the 1990s and later to prevent the import of American beef in 2008), played an especially important role in enabling Korea's UFEF school lunch program to develop as a radical niche. This confirms findings from Ilieva and Hernandez (2018) about the importance of women's groups in bringing about sustainability transitions in agri-food systems.

From 2008 to 2011, civil society activists, who were later joined by progressive politicians (many with social movement backgrounds), used mass street protests and national

Table 2 Summary of the MLP analysis of Korean school lunch policies from the 1960s–2010s

		1960s-80s	1990s	2000s	Post 2011
Niche level		Niche-in-the-making	Niche-in-the-making	Niche developments	
		International aid distribution and creation of the national program	Growth of the national program through private-public partnership	Deprivatization of the national program and investment in school-based infrastructure	Precautionary Infrastructure
Regime level	Enabling	Food sovereignty activism	Food From Somewhere regime	Growth of organic food market and the movement for food sovereignty	Public health, social welfare, educational, economic, and ecological goals are simultaneously pursued through the UFEF school lunch program operating as a protected niche
	Constraining	Poor risk management	Food from Nowhere regime		
Landscape level	Enabling			Economic prosperity Corporeal citizenship Welfare state expansion	
	Constraining	Military dictatorship Economic hardship	Financial crisis Neoliberal trade policy Precautionary consumption		

media coverage to shift the narrative about risk and responsibility. Policymakers subsequently reshaped market conditions and regulations for the public school-lunch program to align with the Food from Somewhere regime and leveraged tax dollars to build precautionary infrastructure for Seoul and other municipalities/provinces.

Institutional, technological, and network anchoring (Elzen et al. 2011) occurred in Seoul as the Korean government, civil society activists, and private sector actors developed precautionary infrastructure for the UFEF school lunch program. Evidence of this niche-regime linkage is visible in changes to municipal school food policy, the creation of new supply chains and certification schemes, and collaborations with a wider range of stakeholders including farmers using (or willing to use) eco-friendly production practices. These niche activities are leading to a gradual reconfiguration of the agri-food regime through a two-fold process (Bui et al. 2016): first, by establishing a shared vision (i.e., for public food programs to act as drivers of food sovereignty, sustainable diets, and social welfare) and second, by embedding this vision into public policy (i.e., the plan to develop precautionary infrastructure for cafeterias serving hospitals, correctional facilities, and government workers).

Third, this paper adds nuance to theories of niche development—specifically in relation to public school-lunch programs—and their potential to catalyze regime-change by documenting how regime- and landscape-level factors affected the niche based on its maturity and ability to function as a protected space. Other scholars have examined niche development of school food policy at multiple scales using the MLP with examples from Brazil, New York, and Senegal (Ilieva and Hernandez 2018). They identified a number of factors that can act as potential levers for sustainability transitions, which we also see in Korea. These include the ability of niche actors to: (1) respond to environmental

pressures, (2) frame their innovations as political tools, (3) remain open to experimentation, (4) create new markets, (5) engage in partnerships and coalition building, (6) build and maintain autonomy while working with public institutions, (7) mobilize women's groups, and (8) impact and participate in the policy process. Notably, we also found that provincial- and municipal-level innovation and policymaking played a critical role in scaling up the niche innovation of a municipal-level UFEF school lunch program to the national level.

Lastly, scholars using the MLP to analyze sustainability transitions within agrifood systems have warned about the dilution of niche-innovations (El Biali 2019). There is evidence to suggest that Korea's UFEF school lunch policy is helping to establish new markets for domestically grown eco-friendly food and reducing children's overall consumption of agricultural chemicals. However, the UFEF school lunch program now operates at a much higher standard of environmental and social justice than the Korean agri-food regime. Therefore, applying the MLP framework to this case suggests that future multi-regime interactions (e.g., labor market policies, public education budgets, sustainable rural development) and landscape-level changes (e.g., gender dynamics and neoliberal ideology) will be necessary in order to overcome the current shortcomings of the UFEF school lunch program and further allow this niche-innovation to support a society-wide sustainability transition in the agri-food sector.

Future Research

Results from this paper are not generalizable beyond the specific socio-historical conditions of Korea, however this does not preclude the possibility that findings may be relevant to other countries' public school-lunch programs or agri-food transition pathways. Future research on how national government-sponsored school-lunch programs may or may not contribute to sustainability transitions within agri-food systems would benefit from delving deeper into how transition processes unfold within the subcomponents that constitute the niche (e.g., school building, municipality, region, state). For instance, while Seoul's Orbon is constantly expanding its partnership with small-scale organic farms, other municipal provinces are still compromising their UFEF agendas due to budget constraints (Korean Agricultural Policy News 2020).

Thus far, the government-led creation of precautionary infrastructure and the continuous strengthening of social policy have bolstered the nationwide implementation of the UFEF school lunch program. However, it is unclear whether the radical potential of this protected niche will be maintained as UFEF policies are scaled up and out to additional public food programs. The niche-regime-landscape alignment present in the formative years of the UFEF school lunch program may become de-aligned and potentially realigned at a lower or higher level of sustainability. Future research is needed in order to understand this process and guard against the dilution of the program's radical potential. Likewise, research that supports cross-country comparison of public school-lunch programs at all scales (e.g., school building, municipality, state, nation) would help clarify best practices for program design and shed light on factors that make sustainability transitions more or less likely to occur. Pursuing such a research agenda has the potential to equip governments and civil society activists with a multiplicity of ideas and approaches for ensuring that school food becomes increasingly safe, healthy, eco-friendly, and fair.

Acknowledgements The authors would like to express their sincere gratitude to the peer reviewers, journal editors, Jane Collins, Alfonso Morales, and Seulgi Son for their helpful comments on earlier versions of this manuscript.

Funding Support for this research was provided by the University of Wisconsin-Madison, Office of the Vice Chancellor for Research and Graduate Education with funding from the Wisconsin Alumni Research Foundation.

Compliance with ethical standards

Conflict of interest The authors declare that they have no competing interest.

References

Abelmann, N. 1996. *Echoes of the past, epics of dissent: A South Korean social movement.* Berkeley: University of California Press.

Allen, P., and J. Guthman. 2006. From "old school" to "farm-to-school": Neoliberalization from the ground up. *Agriculture and Human Values* 23 (4): 401–415.

Bak, H.J. 2012. Public perceptions of the risk of BSE and the risk-avoidance behavior in Korea. *The Journal of Rural Society* 22 (1): 311–341.

BBC Korea. 2019. Why are non-permanent school workers striking?. https://www.bbc.com/korean/news-48849581 Accessed May 2020.

Bui, S., A. Cardona, C. Lamine, and M. Cerf. 2016. Sustainability transitions: Insights on processes of niche-regime interaction and regime reconfiguration in agri-food systems. *Journal of Rural Studies* 48: 92–103.

Burmeister, L.L., and Y. Choi. 2012. Food sovereignty movement activism in South Korea: National policy impacts? *Agriculture and Human Values* 29 (2): 247–258.

Cairns, K., J. Johnston, and N. MacKendrick. 2013. Feeding the 'organic child': Mothering through ethical consumption. *Journal of Consumer Culture* 13 (2): 97–118.

Campbell, H. 2005. The rise and rise of EurepGAP: European (re)invention of colonial food relations. *International Journal of Sociology of Agriculture and Food* 13 (1): 1–19.

Campbell, H. 2009. Breaking new ground in food regime theory: Corporate environmentalism, ecological feedbacks and the 'food from somewhere' regime? *Agriculture and Human Values* 26 (4): 309.

Castellano, R.L.S. 2015. Alternative food networks and food provisioning as a gendered act. *Agriculture and Human Values* 32 (3): 461–474.

Castellano, R.L.S. 2016. Alternative food networks and the labor of food provisioning: A third shift? *Rural Sociology* 81 (3): 445–469.

Chae, J. 2009. The conservative counter discourses on "candlelight protest". *Korean Political Science Review* 43 (3): 129–150.

Chang, D. 2010. Politicization of risk in the 2008 candlelight protests. In *Risk society and risk politics*, ed. J. Jung et al., 159–203. Seoul: Seoul National University Press.

Choi, K. 2006. *Problems of school lunch and directions of school lunch movements by civil organizations* (Unpublished master's thesis). Graduate School of NGO Policies, Hanil Jangsin University, Seoul, Korea.

Choi, Y., and H. Kim. 2015. Success factors of the local food movement and their implications: The case of Wanju-Gun, Republic of Korea. *Procedia Economics and Finance* 23: 1168–1189.

Cwiertka, J.J. 2013. *Cuisine, colonialism and cold war: Food in twentieth-century Korea.* Islington: Reaktion Books.

Declaration of Nyéléni. 2007. Selingue, Mali. https://nyeleni.org/spip.php?article290. Accessed December 2019.

Dong-Ah Daily. 1979. Restart the school lunch. https://dna.naver.com/viewer/index.nhn?articleId=1979112000209204002&editNo=2&printCount=1&publishDate=1979-11-20&officeId=00020&pageNo=4&printNo=17885&publishType=00020. Accessed March 2019.

El Bilali, H. 2019. The multi-level perspective in research on sustainability transitions in agriculture and food systems: A systematic review. *Agriculture* 9 (4): 74.

El Bilali, H., C. Callenius, C. Strassner, and L. Probst. 2019. Food and nutrition security and sustainability transitions in food systems. *Food and Energy Security* 8 (2): e00154.

Elzen, B., F.W. Geels, C. Leeuwis, and B. Van Mierlo. 2011. Normative contestation in transitions 'in the making': Animal welfare concerns and system innovation in pig husbandry. *Research Policy* 40 (2): 263–275.

Fisher, B., and J. Tronto. 1990. Toward a feminist theory of care. In *Circles of care: Work and identity in women's lives*, ed. E.K.

Abel, and M.K. Nelson, 36–54. Albany: State University of New York Press.

Food and Agriculture Organization. 2010. Biodiversity and sustainable diets. https://www.fao.org/3/a-i3004e.pdf. Accessed May 2019.

Friedmann, H. 2005. From colonialism to green capitalism: Social movements and emergence of food regimes. In *New directions in the sociology of gobal development*, 227–64. Emerald Group Publishing Limited.

Gabrielson, T., and K. Parady. 2010. Corporeal citizenship: Rethinking green citizenship through the body. *Environmental Politics* 19 (3): 374–391.

Gaddis, J.E. 2019. *The labor of lunch: Why we need real food and real jobs in American public schools*. Berkeley: University of California Press.

Gaddis, J.E., and A.K. Coplen. 2018. Reorganizing school lunch for a more just and sustainable food system in the US. *Feminist Economics* 24 (3): 89–112.

Galli, F., G. Brunori, F. Di Iacovo, and S. Innocenti. 2014. Co-producing sustainability: Involving parents and civil society in the governance of school meal services, a case study from Pisa, Italy. *Sustainability* 6 (4): 1643–1666.

Geels, F.W. 2002. Technological transitions as evolutionary reconfiguration processes: A multi-level perspective and a case-study. *Research Policy* 31 (8–9): 1257–1274.

Geels, F.W. 2011. The multi-level perspective on sustainability transitions: Responses to seven criticisms. *Environmental Innovation and Societal Transitions* 1 (1): 24–40.

Geels, F.W., and J. Schot. 2007. Typology of sociotechnical transition pathways. *Research Policy* 36: 399–417.

Gilbert, J.L., A.E. Schindel, and S.A. Robert. 2018. Just transition in a public school food system: The case of Buffalo, New York. *Journal of Agriculture, Food Systems, and Community Development* 8: 95–113.

Gyunggi Province. 2018. *Proceedings of committee on ecofriendly school meal support*.

Gyunghyang Newspaper 1957. Hundred million sacks of wheat have been sent from the United States. https://dna.naver.com/viewer/index.nhn?articleId=1957041500329203001&editNo=1&printCount=1&publishDate=1957-04-15&officeId=00032&pageNo=3&printNo=3580&publishType=00020. Accessed May 2019.

Gyunghyang Newspaper. 1963. Hundred-thousand children are starving. https://dna.naver.com/viewer/index.nhn?articleId=1963013000329206001&editNo=6&printCount=1&publishDate=1963-01-30&officeId=00032&pageNo=6&printNo=5310&publishType=00020. Accessed March 2019.

Gyunghyang Newspaper. 1977. Poor students are skipping lunches due to the sudden termination of the bread program. https://dna.naver.com/viewer/index.nhn?articleId=1977092100329207019&editNo=2&printCount=1&publishDate=1977-09-21&officeId=00032&pageNo=7&printNo=9841&publishType=00020. Accessed March 2019.

Gyunghyang Newspaper. 1997. Free lunches for 10,000 students. https://dna.naver.com/viewer/index.nhn?articleId=1997011800329102004&editNo=45&printCount=1&publishDate=1997-01-18&officeId=00032&pageNo=2&printNo=15998&publishType=00010. Accessed February 2019.

Hankook Economy. 2006. The worst food poisoning in school food. https://www.hankyung.com/society/article/2006062215771. Accessed January 2020.

Hansen, T., and L. Coenen. 2015. The geography of sustainability transitions: Review, synthesis, and reflections on an emergent research field. *Environmental Innovation and Societal Transitions* 17: 92–109.

Hargreaves, T., N. Longhurst, and G. Seyfang. 2013. Up, down, round and round: Connecting regimes and practices in innovation for sustainability. *Environment and Planning A* 45 (2): 402–420.

Harris, E. 2009. Neoliberal subjectivities or a politics of the possible? Reading for difference in alternative food networks. *Area* 41 (1): 55–63.

Hinrichs, C.C. 2014. Transitions to sustainability: A change in thinking about food systems change? *Agriculture and Human Values* 31 (1): 143–155.

Ilieva, R., and A. Hernandez. 2018. Scaling-Up sustainable development initiatives: A comparative case study of agri-food system innovations in Brazil, New York, and Senegal. *Sustainability* 10 (11): 4057.

Ju, E.H. 2016. Analysis on free school meal policy in Seoul: Focusing on diagnostics of public value failure. *The Korean Administration for Policy Studies* 25 (1): 269–297.

Jung, K. 2005. A case study on the women's peasant movement in Gyeongbuk areas: Female activists and their activities. *The Journal of Rural Society* 15 (1): 59–101.

Jung, H., J. Sung, and H. Lee. 2019. *Domestic eco-friendly agricultural goods: Demands and prospects*. Naju: Korean Rural Economic Institute.

Kang, M. 2011. Free for all, organic school lunch programs in South Korea. In *School food politics: The complex ecology of hunger and feeding in schools around the world*, ed. S. Robert and M.B. Weaver-Hightower, 120–139. New York: Peter Lang.

Kang, S. 2016. 5 years of school meals in Seoul. *Hankook-Nongjung*.

Kern, F., and J. Markard. 2016. Analysing energy transitions: Combining insights from transition studies and international political economy. In *Palgrave handbook of the international political economy of energy*, ed. T. Van de Graf et al., 291–318. London: Palgrave Macmillan.

Kim, H.J. 2009. Building local food system through school foods safety movement: A case study of Naju City in GeonNam privnce, Korea. *The Journal of Rural Society* 19 (2): 63–92.

Kim, H.J. 2013. School food and local food: A comparative study of Korea and Japan. *The Journal of Rural Society* 23 (1): 87–139.

Kim, H.J., H.J. Lee, and S. Kim. 2014. A study on the social characteristics and types of environment-friendly farmers. *Korean Research on Environmental Sociology (ECO)* 18 (2): 45–82.

Kim, I.J., and J.H. Lee. 2011. The housewives' purchase behaviors on environment-friendly agricultural products in Daejeon area. *Korean Journal of Community Nutrition* 16 (3): 386–397.

Kirwan, J., B. Ilbery, D. Maye, and J. Carey. 2013. Grassroots social innovations and food localisation: An investigation of the local food programme in England. *Global Environmental Change* 23 (5): 830–837.

Kleine, D., and M. das Graças Brightwell. 2015. Repoliticising and scaling-up ethical consumption: Lessons from public procurement for school meals in Brazil. *Geoforum* 67: 135–147.

Korean Agricultural Policy News. 2020. Is Choongnam abandoning eco-friendly school lunch?. https://www.ikpnews.net/news/articleView.html?idxno=40805. Accessed May 2020.

Korea Agro-Fisheries & Food Trade Corporation. 2017. *Survey for the expansion of public meal plan*. https://edu.at.or.kr/cmm/fms/FileDown.do?atchFileId=FILE_000000000003233&fileSn=0. Accessed December 2019.

Korean Gallup. 2014. https://www.gallup.co.kr/gallupdb/reportDownload.asp?seqNo=580. Accessed April 2019.

Korean Rural Economic News. 2019. Bidding system is modified in school lunch. Accessed in May 2020.

Korean Social Statistics. 2011. *A status of Korean society*. Korea: Seoul.

Kuokkanen, A., A. Nurmi, M. Mikkilä, M. Kuisma, H. Kahiluoto, and L. Linnanen. 2018. Agency in regime destabilization through the selection environment: The finnish food system's sustainability transition. *Research Policy* 47 (8): 1513–1522.

Kyung-In Daily. 2002. Discussion on school lunch hygiene management. Accessed May 2020.

Kyungnam Newspaper. 2002. For tasty and nutrient school lunch. Accessed May 2020.

Lachman, D.A. 2013. A survey and review of approaches to study transitions. *Energy Policy* 58: 269–276.

Lakind, A., L. Skipper, and A. Morales. 2016. Fostering multiple goals in farm to school. *Gastronomica: The Journal of Critical Food Studies* 16 (4): 58–65.

Lang, T., D. Barling, and M. Caraher. 2009. *Food policy: Integrating health, environment and society*. Oxford: Oxford University Press.

Lawhon, M., and J.T. Murphy. 2012. Socio-technical regimes and sustainability transitions: Insights from political ecology. *Progress in Human Geography* (3): 354–378.

Lee, D., S. Ju, and J. Han. 2016a. A study of communal feeding establishment employees' work-family conflict level family burnout, and job burnout. *International Journal of Tourism and Hospitality Research* 30 (6): 197–211.

Lee, E.H., G.Y. Yun, and S.H. An. 2016. An analysis of the policy change in free school meals using multiple streams framework. *The Korea Educational Review* 22 (1): 77–104.

Lee, K., Y. Jang, and W. Kim. 1994. A study on the state of lunchbox preparation and the opinion of school lunch program of mothers with elementary school children in Seoul. *Family and Environment Research* 32 (5): 135–142.

Lee, O., M. Cho, and H. Chang. 2014. The organization commitment and perception of human resource management by employment types of school foodservice employees. *Journal of the Korean Society of Food Science and Nutrition* 43 (1): 162–171.

Lehtinen, U. 2012. Sustainability and local food procurement: A case study of Finnish public catering. *British Food Journal* 114 (8): 1053–1071.

MacKendrick, N. 2014. More work for mother: Chemical body burdens as a maternal responsibility. *Gender & Society* 28 (5): 705–728.

MacKendrick, N. 2018. *Better safe than sorry: How consumers navigate exposure to everyday toxics*. Berkeley: University of California Press.

Maeil Economy. 1997. Catch the two-trillion won market. https://www.mk.co.kr/news/home/view/1997/01/1034/. Accessed March 2019.

Maeil Economy. 2010. Mayor Oh rejects catastrophic populism. https://www.mk.co.kr/news/society/view/2010/12/668382/. Accessed December 2018.

Maeil Economy. 2019. Each school lunch worker is in charge of more than 100 students, twice other public organizations. https://www.mk.co.kr/news/society/view/2019/01/22594/. Accessed April 2019.

Markard, J., R. Raven, and B. Truffer. 2012. Sustainability transitions: An emerging field of research and its prospects. *Research Policy* 41 (6): 955–967.

McMichael, P. 2002. The global restructuring of agro-food systems. *Mondes En Développement* 1: 45–53.

Ministry of Education. 1981. *The school lunch act*. Seoul Korea.

Ministry of Education. 2014. *The school lunch act*. Seoul Korea.

Ministry of Education. 2018. The expansion of universal free eco-friendly school lunch program. https://opengov.seoul.go.kr/sanction/14564268. Accessed April 2019.

Ministry of Education, Science, and Technology. 2011. *Evaluation of supports for school lunch*. Seoul Korea.

Monthly Nutriand. 2018. Universal-free school lunch policy at 17 local governments. https://m.blog.naver.com/nutriand/221241014026. Accessed December 2019.

Morgan, K., and R. Sonnino. 2013. *The school food revolution: Public food and the challenge of sustainable development*. London: Routledge.

Munhwa Broadcasting Corporation (MBC). 2008. American beef, is it safe from BSE? *PD Notebook*.

Munhwa Newspaper. 2002. Hygiene, nutrition… I can't trust school lunches. parents and teachers are complaining. Accessed May 2020.

Oh My News. 2011. Mayor Oh and superintendent of education debated. https://www.ohmynews.com/NWS_Web/View/at_pg.aspx?CNTN_CD=A0001610363&PAGE_CD=N0000&BLCK_NO=3&CMPT_CD=M0001. Accessed April 2019.

O'Neill, K.J., A.K. Clear, A. Friday, and M. Hazas. 2019. 'Fractures' in food practices: Exploring transitions towards sustainable food. *Agriculture and Human Values* 36 (2): 1–15.

Oostindjer, M., J. Aschemann-Witzel, Q. Wang, S.E. Skuland, B. Egelandsdal, G.V. Amdam, A. Schjøll, M.C. Pachucki, P. Rozin, J. Stein, V.L. Almli, and E.A. Kleef. 2016. Are school meals a viable and sustainable tool to improve the healthiness and sustainability of children's diet and food consumption?: A cross-national comparative perspective. *Critical Reviews in Food Science and Nutrition* 57 (18): 3942–3958.

Park, H. 2008. The political process and the effect of the participatory democracy in Korea: The comparative study of "the initial movement for child care ordinance amendment of gwacheon" and "the initiative movement for school lunches ordinance enactment". *Memory and Prospect* 18 (18): 307–344.

Park, S. 1995. *A study of the Korean women peasant movement: Experience of organization and individuals*. Yonsei: Yonsei University Press.

Park, S., and E. Jeong. 2010. Formation of social identity of women peasants and development of women peasants movement: Focusing on the Korean women peasant association. *The Journal of Rural Society* 20 (1): 89–129.

Park, Y., J. Lee, and M. Lee. 1997. Comparisons of students' and their parents' satisfaction of school lunch program in middle school by foodservice management. *Korean Journal of Community Nutrition* 2 (2): 218–231.

Pitt, H., and M. Jones. 2016. Scaling up and out as a pathway for food system transitions. *Sustainability* 8 (10): 1025.

Powell, L.J., and H. Wittman. 2018. Farm to school in British Columbia: Mobilizing food literacy for food sovereignty. *Agriculture and Human Values* 35 (1): 193–206.

Robert, S.A., and M.B. Weaver-Hightower. 2011. *School food politics: The complex ecology of hunger and feeding in schools around the world*. Bern: Peter Lang.

Rodier, L. 2014. Assessing the role of the IMF in South Korea during the Asian financial crisis. *Journal of Economics* 2 (2): 107–113.

Rossi, A., S. Bui, and T. Marsden. 2019. Redefining power relations in agrifood systems. *Journal of Rural Studies*. 68: 147–158.

Rut, M., and A.R. Davies. 2018. Transitioning without confrontation?: Shared food growing niches and sustainable food transitions in Singapore. *Geoforum* 96: 278–288.

Scott, D.N., J. Haw, and R. Lee. 2016. 'Wannabe toxic-free?': From precautionary consumption to corporeal citizenship. *Environmental Politics* 26 (2): 322–342.

Segye Ilbo. 2018. By 2021, all Seoul high schools will serve universal-free, eco friendly school lunches. https://www.segye.com/newsView/20181029005096. Accessed December 2019.

Seoul Metropolitan Office of Education (SMOE). 2013. Seoul student population has been decreased into half since 1990, https://stat.seoul.go.kr/pdf/e-webzine68.pdf. Accessed April 2019.

Seoul Metropolitan Office of Education (SMOE). 2018. *Seoul educational statistics*. Seoul: Seoul Metropolitan Office of Education.

Sidaner, E., D. Balaban, and L. Burlandy. 2013. The Brazilian school feeding programme: An example of an integrated programme in support of food and nutrition security. *Public Health Nutrition* 16 (6): 989–994.

Slingerland, M., and M. Schut. 2014. Jatropha developments in Mozambique: Analysis of structural conditions influencing niche-regime interactions. *Sustainability* 6 (11): 7541–7563.

Sonnino, R., C.L. Torres, and S. Schneider. 2014. Reflexive governance for food security: The example of school feeding in Brazil. *Journal of Rural Studies* 36: 1–12.

Spaargaren, G., P. Oosterveer, and A. Loeber (eds.). 2013. *Food practices in transition: Changing food consumption, retail and production in the age of reflexive modernity*. London: Routledge.

Stapleton, S.R. 2019. Parent activists versus the corporation: A fight for school food sovereignty. *Agriculture and Human Values* 36 (4): 805–817.

Szasz, A. 2007. *Shopping our say to safety: How we changed from protecting the environment to protecting ourselves*. Minneapolis: University of Minnesota Press.

Tronto, J.C. 2013. *Caring democracy: Markets, equality, and justice*. New York: New York University Press.

United States Department of Agriculture. 2019. *Bovine spongiform encephalopathy (BSE) and specified risk materials (SRM) guidance materials and resources*, https://www.fsis.usda.gov/wps/portal/fsis/topics/regulatory-compliance/specified-risk-material/specified-risk-materials. Accessed April 2019.

Yang, Y. 2010. Well-being discourse and Chinese food in Korean society. *Korea Journal* 50 (1): 85–109.

Yoon, S.J. 2018. On the organization and practice of Christian peasants' association in Chonnam region. *Journal of Democracy and Human Rights* 18 (4): 225–284.

Wittman, H., and J. Blesh. 2017. Food sovereignty and fome zero: Connecting public food procurement programmes to sustainable rural development in Brazil. *Journal of Agrarian Change* 17 (1): 81–105.

World Food Program. 2016. Home grown school feeding. https://www.wfp.org/home-grown-school-feeding. Accessed in December 2019.

Publisher's Note Springer Nature remains neutral with regard to jurisdictional claims in published maps and institutional affiliations.

Jennifer E. Gaddis is an assistant professor of Civil Society and Community Studies at the University of Wisconsin-Madison and the author of *The Labor of Lunch: Why We Need Real Food and Real Jobs in American Public Schools* (University of California Press, 2019). She received a Ph.D. in environmental studies from Yale University in 2014. Her research on school lunch programs has appeared in numerous journals, including *Feminist Economics* and the *Journal of Agriculture, Food Systems, and Community Development*, and in popular media outlets such as the *New York Times, Washington Post, USA Today, and Teen Vogue*.

June Jeon is a Postdoctoral Fellow in Civic Science at Tufts University. He received Ph.D. in Sociology and Environmental Studies at the University of Wisconsin-Madison. He studies how social powers operate in the scientific field to engage with subsequent social and environmental consequences. His works have been published in journals, such as *Social Studies of Science, and Engaging Science, Technology, and Society*.

Agriculture and Human Values (2020) 37:1073–1081
https://doi.org/10.1007/s10460-020-10036-6

The real meal deal: assessing student preferences for "real food" at Fort Lewis College

Kathleen Hilimire[1] · Carl Schnitker[1]

Accepted: 18 April 2020 / Published online: 28 April 2020
© Springer Nature B.V. 2020

Abstract

Fort Lewis College committed to purchasing 20% real food by 2020 as part of a national campaign called the Real Food Challenge, an initiative on college campuses that aims to shift food procurement toward real food, defined as ecologically sound, humane, fair, or local. Our research explored student preferences regarding food at Fort Lewis College. We analyzed students' willingness-to-pay for 20% real food and the characteristics that predicted this willingness-to-pay. We also examined food preference parameters outside of the Real Food Challenge categories, considering how factors such as taste, health, price, and convenience influenced students when choosing what to eat. We found that all four Real Food Challenge categories were important to respondents, and two-thirds of respondents were willing to pay more for real food. Yet overall, taste, health, and price were statistically more important than any of the Real Food Challenge categories to respondents. Statistically significant positive parameters driving willingness-to-pay more for real food were that a respondent considered environment an important factor when deciding what to eat, that a respondent agreed with the statement "It is important that my food reflect my values," that sustainability was a factor for the respondent in deciding what school to attend, and that a respondent was from Colorado (in-state). Convenience was a negative parameter. In order to match these student preferences, campus dining services should emphasize benefits to taste and health when creating real food meals and should attempt to meet Real Food Challenge objectives with minimal price increases.

Keywords Environmental decision-making · Contingent valuation · Willingness-to-pay · Campus sustainability · Real Food Challenge

Abbreviations

ANOVA	Analysis of variance
FLC	Fort Lewis College
HEI	Higher education institutions
RFC	Real Food Challenge
SPSS	Statistical Package for Social Sciences
STARS	Sustainability tracking, assessment, and rating system
USDA	United States Department of Agriculture

✉ Kathleen Hilimire
kehilimire@fortlewis.edu

Carl Schnitker
cjschnitker@fortlewis.edu

[1] Environment & Sustainability, Fort Lewis College, 1000 Rim Drive, Durango, CO 81301, USA

Introduction

Sustainability is increasingly a part of the ethos of higher education institutions (HEIs) (Casarejos et al. 2017; Lozano et al. 2015; Wright 2002). Students demand and expect sustainability, with many sustainability efforts originating from student campaigns (Murray 2018). Further, students have now come to expect sustainability at HEIs. Research on prospective students over the past decade shows that a school's commitment to "green" issues influences the choice of school for over 60% of students (PR 2019). Learning sustainability also fits with the longstanding tradition of HEIs as places for students to learn social responsibility and leadership (Filho et al. 2019; O'Connor 2006; Smiley Smith 2017).

Concurrent with this uptick in interest for HEI sustainability has been a rise in formal HEI commitments and reporting for sustainability (Ceulemans et al. 2015; Dyer and Dyer 2017). For example, Second Nature is an organization dedicated to accelerating climate action in HEIs. The organization began in 2006 with just twelve signatories to its American College & University Presidents' Climate Commitment. By 2018, Second Nature recognized 486 HEI signatories committed to carbon neutral campuses (Dyer and Dyer 2017; SN 2018, nd). Similarly, the Association for the Advancement of Sustainability in Higher Education reported that 477 institutions participated in its 2018 sustainability tracking, assessment, and rating system (STARS) (AASHE 2015, 2018). STARS allows HEIs to track and quantify efforts to reduce greenhouse gas emissions and waste, increase sustainability in academics and research, improve campus dining, and more. Second Nature and STARS both offer HEIs the ability to track and benchmark their sustainability efforts against other HEIs, practices that have helped to create shared language and transparency in the HEI sustainability community of practice. These entities have also catalyzed action at HEIs. By naming sustainability objectives and creating measurable targets, HEIs have been moved toward real progress for sustainability. This relationship between commitment and action has spawned numerous other sustainability-oriented commitments among HEIs, including the Real Food Challenge.

The Real Food Challenge

The Real Food Challenge aims to create a just and sustainable food system (RFC nd-b). Founded in 2007 with the goal of shifting $1 billion of HEI dining budgets toward "real" food by 2020, the Real Food Challenge provides a platform for student leaders to make change at their HEIs. Over 80 institutions participate in the Real Food Challenge by committing to shift 20% of food procurement dollars to "real" food by 2020. "Real" food is defined carefully in the organization's Real Food Standards and includes broad food categories of ecologically sound, humane, fair, or local and community based (RFC nd-c). For the ecologically sound, humane, and fair categories, classification is determined primarily on certification from a third-party group. The local and community based category does not require third-party certification, instead relying upon individual research by Real Food Challenge signatory institutions to determine if food is produced within a specific set of criteria.

Guiding principles for ecologically sound food set by the Real Food Challenge include ecologically sound pest management, soil conservation, biodiversity and habitat protection, water conservation, sustainable waste management, and energy conservation (RFC nd-c). Examples of foods that would meet the Real Food Challenge standards as ecologically sound include USDA certified organic, Smithsonian certified Bird Friendly, and Rainforest Alliance certified. Humane food, under the Real Food Challenge, includes animals that are provided nutritious feed free of non-therapeutic antibiotics and hormones, a low stress environment, limited physical alterations, careful handling, minimized

transportation, and humane slaughter (RFC nd-c). Examples of foods that would meet the Real Food Challenge standards as humane include Animal Welfare Approved/Certified by a Greener World and Biodynamic Certified by Demeter. For fair food, the Real Food Challenge highlights concepts of working with dignity, such as workers having access to potable water and clean sanitary facilities; worker bargaining and advocacy; fair compensation; a safe working environment; non-discrimination, gender equity, and gender justice; and job security (RFC nd-c). Examples of international foods that would be Real Food Challenge fair include Ecocert Fair Trade Certified, Fair for Life Certified by Institute for Marketecology, and Fairtrade America. For domestic foods, examples include Food Justice Certified by Agricultural Justice Project and certain unionized farms.

The criteria for local and community based set by the Real Food Challenge are multi-faceted in recognition of the fact that distance is an imperfect metaphor for the actual aims of local and community based food movements. This is a concept known as the "local trap," an acknowledgement of the reality that every farm is geographically close to some consumer and that proximity implies nothing about values-based food production (Born and Purcell 2006). Another argument that is often used to justify distance as an important measure for food production is that local food reduces greenhouse gas emissions due to shorter transport distances between producer and consumer. This too turns out to be a trap, since the bulk of greenhouse gas emissions associated with agriculture come from production, not transportation (Garnett 2011; Weber and Matthews 2008). As such, the Real Food Challenge requires that food in this category meet not just a distance metric but also criteria for ownership and size (RFC nd-c). For example, a produce farm in this category must be privately or cooperatively owned, gross less than $5 million/year, and be located within a 250-mile radius of the institution. In this, the Real Food Challenge aims to shift HEI food procurement toward small- and mid-size farms and food businesses, diversified ownership and control, reduced distance between producers and consumers, and traceability (RFC nd-c).

The Real Food Challenge ultimately aims to transform the food system by leveraging the purchasing power of HEIs and investing in alternative food production, labor, and sales practices. Inherent in this effort is a critique of conventional "industrial" food systems (Barlett 2011). Industrial agriculture is based on the paradigm of the farm as a sort of factory, comprised of inputs and outputs, that is designed to yield maximum profits with minimum costs (Horrigan et al. 2002). This intensive form of agriculture is a critical component of the modern food systems that sustain a significant proportion of the world's population. However, major concerns have been raised about industrial agriculture's reliance on monocultures and attendant dependence on high levels of nutrient inputs and chemical pest controls (Horrigan et al. 2002; Tilman 1999). Fertilizers can cause hypoxia in aquatic systems (Smith et al. 2008). Pesticides have been linked to public health concerns for farmworkers and consumers (Horrigan et al. 2002; Mascarelli 2013), surface water and aquatic life contamination (Stehle and Schulz 2015), and a treadmill-like effect whereby the pesticide itself is no longer effective (Altieri and Nicholls 1997). Further concerns about industrial agriculture include consolidation, exploitation of worker's rights, erosion of biodiversity, and inhumane treatment of animals (Chittapur Doddabasawa and Umesh 2017; Heise and Theuvsen 2017; Ruben et al. 2009). The Real Food Challenge aims to minimize, or even reverse, these impacts by shifting institutional purchasing toward foods that are produced in ways that are ecologically sound, humane, fair, or local and community based.

Theoretical context

Food preferences are multi-faceted and hierarchical. Understanding the attributes that students seek in food can help HEIs achieve success with a Real Food Challenge effort. For example, taste is often identified by individuals as more important than foods with instrumental value such as Real Food Challenge foods (Satter 2007). This does not mean that HEIs should only consider taste. Rather, HEIs need to understand the complex, hierarchical preferences of their students with regards to food and integrate real food offerings with these. For example, if students report valuing health and convenience in the form of grab-and-go foods, a Fair Trade banana could be a good offering to introduce fair food in a way that appeals to students. Contingent valuation is a method of measuring consumers' demand for an intangible item by offering them a hypothetical set of prices for a product that contains the attributes being studied; in this case, real food. Porter et al. (2017) demonstrated that contingent valuation, as measured by willingness-to-pay, can be used as an effective proxy measure for real food preference in a collegiate setting. Willingness-to-pay refers to the amount that an individual would be willing to pay for a good or service and is commonly used to study values-based food consumption (Belcher et al. 2007; Costanigro et al. 2016; Porter et al. 2017). The specific maximum price level that an individual would be willing to pay is the willingness-to-pay level. For example, an individual may purchase an apple if it costs one dollar but may not purchase the apple if it costs two dollars. Given the two options, the individual's willingness-to-pay level is one dollar. In the case of assessing willingness-to-pay for real food, what is being assessed are perceived attributes of the food. Given that a fairly produced apple may look and taste the same as a conventional apple, a higher willingness-to-pay for the fair apple indicates a perceived attribute of the food (Ulvila et al. 2009).

Porter et al. (2017) evaluated student willingness-to-pay for real food at the University of Vermont. They surveyed students and asked whether respondents would pay more for a meal plan with 20% real food. They found that a majority of students were willing to pay more for real food. However, the actual willingness-to-pay increase was small, at a median of 1%. They then conducted binary logistic regression with a suite of independent variables to investigate the characteristics of respondents who were willing to pay more. Being from Vermont ("in-state"), being female, being from an environmental school within the university, and valuing the origin of food all correlated positively with willingness-to-pay. Finding the price of food to be important correlated negatively.

The study

We evaluated student food preferences and willingness-to-pay for real food at Fort Lewis College, closely following the framework and methods of Porter et al. (2017). Fort Lewis College is a small public liberal arts college with a student body of approximately 3400 located in Durango, Colorado (FLC nd). The student body at the College is very diverse. Approximately one-third of students are Native American. Many students are first generation college students. Nearly 1300 students live on campus, and these on-campus residents must purchase a meal plan (FLC nd). Dining services at Fort Lewis College are provided by Sodexo, a French multinational corporation that is a major player in the North American collegiate food service sector (Sodexo nd). There are five options for students to purchase food at Fort Lewis College: San Juan Dining Hall and four retail outlets. At the time of data collection during February and March 2018, individual entries to the dining hall (not including tax) were $6.28 for breakfast, $9.20 for lunch (and weekend brunch), and $9.86 for dinner.

Fort Lewis College committed to the Real Food Challenge in 2013. As of the most recent inventory, 10% of food purchasing dollars were spent on real food (RFC nd-a). As the 2020 deadline for 20% real food approaches, efforts to shift more purchasing dollars to real food are intensifying. Student buy-in is a critical component of the Real Food Challenge at Fort Lewis College and at other institutions, especially considering potential increases in financial cost for students. Our research addressed important questions associated with this transition: (1) What are the most important factors for students in deciding what to eat? (2) Would students be willing to pay more for real food? If so, how much? (3) What characteristics predict whether a student is willing to pay more for real food?

Methods

Survey instrument and administration

We used a paper survey to collect data from Fort Lewis College students eating in San Juan dining hall, the only cafeteria on campus. The survey instrument was aimed at assessing food preferences, willingness-to-pay for real food, and drivers of willingness-to-pay. It was comprised of twelve questions with scale, multiple choice, and open answer response formats. The survey began with questions about food preference using a five-point Likert scale that assessed importance of the following variables in deciding what to eat: taste, health, price, convenience, appearance of food, and the four Real Food Challenge categories. Next, there were multiple choice questions asking students how important values and peer influence were to their food choices and how important Fort Lewis College's sustainability commitments were to them. Following this, the survey provided information about the definition of real food. Then, respondents were asked to select one of four willingness-to-pay levels for 20% real food at Fort Lewis College dining facilities. Finally, demographic data were collected. All responses were kept anonymous. The Institutional Review Board at Fort Lewis College approved the research on 12/21/2017 (IRB-2017-176).

Students in the Environmental Research Methods and Design course administered the surveys in February and March 2018. Student surveyors used two specific, defined methods for administering surveys for voluntary completion. One method was to enter the San Juan dining hall during a meal, pick a table, and count five tables to the right. Students would then ask every student seated at that table to complete a survey. They would then count five more tables to the right, administer the survey again, and so on. The other method was for student surveyors to stand outside the dining hall during mealtimes and ask every third student leaving the dining hall to complete a survey. Verbal screening questions were asked before survey administration to ensure that no individual completed the survey more than once and all respondents were students. Data from paper surveys were then manually input into an online data collector.

Student food preferences data analysis

To conduct a quantitative comparison of responses about food preferences, we converted all Likert scale responses into quantitative scores from 1 to 5. A score of 1 represented the least impact on student food preference, or "not at all important," and a score of 5 represented the most

impact on student food preference, or "very important." Food preferences included in this analysis were taste, health, price, convenience, appearance of food, and the four Real Food Challenge categories. We then determined the average quantitative response for each food preference type and used one-way Analysis of Variance (ANOVA) with a Tukey's Honesty Significant Difference (HSD) post-hoc test to compare means. Statistical tests were conducted using Statistical Package for Social Sciences (SPSS) Version 24 with an alpha level of ≤ 0.05.

Willingness-to-pay data analysis

To assess willingness-to-pay, we first converted the dependent variable into a binary format of 1 (willing to pay more for real food) or 0 (not willing to pay more for real food). Then, we used binary logistic regression to analyze the role of multiple independent variables in determining willingness-to-pay (Table 1). We conducted binary logistic regression using SPSS Version 26 with an alpha level of ≤ 0.10. We used this higher alpha level in keeping with the Porter el al. (2017) paper to ensure the best comparison between results.

Results

Respondent characteristics

Out of 309 respondents, 52% were first years, 25% were sophomores, 14% were juniors, and 9% were seniors. There were more females than males, 52.4% and 46.3% respectively and 1% identifying as other. The most prominent majors were Biology (9.7%), Business Administration (9.7%), Psychology (8.1%), Public Health (4.8%), Environmental Studies (4.8%) and Engineering (4.5%). The average age of respondents was 19.8, and the median age was 19. Respondents were primarily from Colorado (38%), New Mexico (14%), Arizona (8%), and California (7%). Most respondents had campus dining meal plans; 21% had a 19 meals/week plan, 23% had a 14 meals/week plan, 32% had a 10 meals/week plan, 11% had other meal plans, and 14% did not have a meal plan.

Student food preferences

We found that the Real Food Challenge categories of foods were important to a majority of students. Specifically, 79% of students identified animal welfare as "moderately important" or higher when choosing what to eat. 80% of students said the same of the working conditions of farmworkers, 79% said so of environmentally friendly, and 67% for local. We then compared student preferences for

Table 1 Independent variables in binary logistic regression model

Variable	Criteria for variable
Local	1 = student considers that food was grown locally an important or very important factor when deciding what to eat; 0 = otherwise
Environment	1 = student considers that food was produced in an environmentally-friendly way an important or very important factor when deciding what to eat; 0 = otherwise
Farmworkers	1 = student considers working conditions of farmworkers an important or very important factor when deciding what to eat; 0 = otherwise
Animal welfare	1 = student considers animal welfare an important or very important factor when deciding what to eat; 0 = otherwise
Taste	1 = student considers taste an important or very important factor when deciding what to eat; 0 = otherwise
Health	1 = student considers health an important or very important factor when deciding what to eat; 0 = otherwise
Price	1 = student considers price an important or very important factor when deciding what to eat; 0 = otherwise
Convenience	1 = student considers convenience an important or very important factor when deciding what to eat; 0 = otherwise
Appearance	1 = student considers appearance of food an important or very important factor when deciding what to eat; 0 = otherwise
Values	1 = student agrees or strongly agrees with the statement "It is important that my food reflect my values"; 0 = otherwise
College's sustainability	1 = student agrees or strongly agrees with the statement "FLC's sustainability commitments are part of the reason I chose this school"; 0 = otherwise
Social influence	1 = student agrees or strongly agrees with the statement "My food choices are influences by what my friends eat"; 0 = otherwise
In-state	1 = student is from Colorado; 0 = otherwise
Gender	1 = student is female; 0 = otherwise
Year in college (sophomore)	1 = student is a sophomore; 0 = otherwise
Year in college (junior)	1 = student is a junior; 0 = otherwise
Year in college (senior)	1 = student is a senior; 0 = otherwise

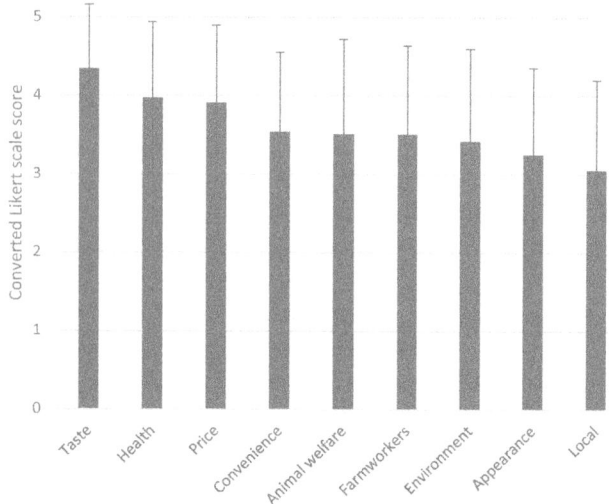

Fig. 1 Average Likert scores (±1 standard deviation) for nine food preference factors (N = 309)

taste, health, price, convenience, appearance of food, and the four Real Food Challenge categories using one-way ANOVA ($F(8,2742) = 44.223$, $p = 0.000$). All four Real Food Challenge categories were important to respondents when deciding what to eat; average scores were from 3.05 to 3.51 out of 5 (Fig. 1). Of the four Real Food Challenge categories, there was no statistically significant difference between animal welfare (3.51/5 ± 1.21), farmworkers' working conditions (3.51/5 ± 1.13), and environmentally-friendly food production (3.42/5 ± 1.18) (p values ranging from 0.98 to 1.00). All three of these categories were more important to respondents than local sourcing of food (3.05/5 ± 1.15) (p values ranging from 0.000 to 0.001).

Overall, taste was the most important factor for respondents when deciding what to eat (4.34/5 ± 0.82) compared to all other factors (p values ranging from 0.000 to 0.001). Health (3.97/5 ± 0.97) and price (3.91 ± 0.99) did not differ significantly from one another and were the next most important factors compared to all other factors (p values ranging from 0.000 to 0.001). Convenience (3.54/5 ± 1.01) was more important than appearance (p = 0.026) and local (p = 0.000) but not statistically more important than the other Real Food Challenge categories of environment, animal welfare, and farmworkers' working conditions. Appearance of food (3.25 ± 1.10) was not more important than any of the real food challenge categories.

Understanding willingness-to-pay for real food

To answer the question of whether students would be willing to pay more for real food, we asked the multiple-choice question "Consider the resources you and your parents/guardians have to pay for your meals at college. How much more would you be willing to pay for food at FLC if over 20% of the food was defined as real?" Two-thirds of respondents reported that they would be willing to pay more for real food at Fort Lewis College. Of the three cost increase options, the most popular choice (27%) was an increase of 50 cents per meal (Fig. 2).

We then used binary logistic regression to analyze what predicted respondent likelihood of willingness-to-pay more for real food. The overall logistic regression model was statistically significant $\chi^2(17) = 47.432$, $p = 0.000$.

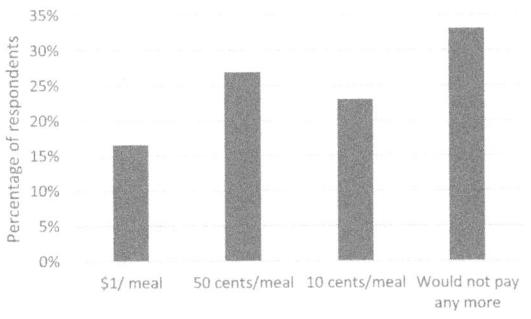

Fig. 2 Respondent willingness-to-pay for real food levels (N = 309)

The Nagelkerke R^2 value was 0.198, and the model correctly assigned 69.9% of cases. According to the model, statistically significant parameters for predicting whether a respondent would pay more for real food were environment, convenience, values, college's sustainability, and in-state (Table 2). Of those, environment, values, college's sustainability, and in-state had a positive effect on willingness-to-pay, while convenience had a negative effect. Specifically, respondents who reported that it was important or very important to them that food was produced in an environmentally-friendly when deciding what to eat were significantly more likely to be willing to pay more for real food (odds ratio = 2.012). In terms of values, respondents who agreed or strongly agreed with the statement "It is important that my food reflect my values" were significantly more likely to be willing to pay more for real food (odds ratio = 1.879). For college's sustainability, respondents who agreed or strongly agreed with the statement "FLC's sustainability commitments are part of the reason I chose this school" were significantly more likely to be willing to pay more for real food (odds ratio = 2.260). For in-state, respondents from Colorado were significantly more likely to be willing to pay more for real food (odds ratio = 1.872) than respondents from out of state. In terms of convenience, respondents who reported that convenience was an important or very important factor when deciding what to eat were significantly less likely to be willing to pay more for real food (odds ratio = 0. 0.521).

Table 2 Logistic regression results predicting respondents' willingness-to-pay more for real food

Predictor variables	B	S.E. of B	Wald	D.F	P value	EB (0dd ratio)
Local	0.528	0.369	2.044	1	0.153	1.695
Environment	**0.699**	**0.383**	**3.339**	**1**	**0.068[a]**	**2.012**
Farmworkers	−0.246	0.377	0.425	1	0.515	0.782
Animal welfare	0.394	0.340	1.345	1	0.246	1.483
Taste	0.624	0.418	2.223	1	0.136	1.866
Health	0.049	0.313	0.024	1	0.876	1.050
Price	0.027	0.310	0.007	1	0.931	1.027
Convenience	**−0.652**	**0.308**	**4.488**	**1**	**0.034[a]**	**0.521**
Appearance	−0.201	0.289	0.484	1	0.486	0.818
Values	**0.631**	**0.304**	**4.313**	**1**	**0.038[a]**	**1.879**
College's sustainability	**0.815**	**0.426**	**3.656**	**1**	**0.056[a]**	**2.260**
Social influence	0.099	0.356	0.076	1	0.782	1.104
In-state	**0.627**	**0.287**	**4.777**	**1**	**0.029[a]**	**1.872**
Gender	−0.369	0.284	1.691	1	0.194	0.691
Year in college (sophomore)	0.044	0.328	0.018	1	0.893	1.045
Year in college (junior)	−0.282	0.397	0.503	1	0.478	0.755
Year in college (senior)	−0.203	0.486	0.175	1	0.676	0.816
Constant	−0.328	0.436	0.564	1	0.453	0.721

Variables indicated in bold are statistically significant

[a]Significant at alpha level ≤ 0.10

Discussion

We found that real food was important to our respondents but not as important as other food attributes. In our study, taste was the most important food preference factor, followed closely by health and price. Although not as important as taste, health, and price, the four Real Food Challenge categories were, on average, important to respondents. Of these, organic, humane, and fair trade scored higher with respondents than local. Most of the local food at the Fort Lewis College dining hall comes from the Old Fort at Hesperus, a farm operated by the school primarily as an educational facility. Food from the Old Fort meets the Real Food Challenge because it is locally produced; however, it also uses organic, humane, and fair practices. The produce is raised without synthetic inputs, the cattle are grass fed and humanely treated and slaughtered, and employees are paid a living wage (Beth LaShell, personal communication), but the Old Fort does not carry certifications related to organic, humane, and fair. The findings of this research suggest that students would appreciate knowing about the multi-faceted real food attributes of food from this source.

Similar to other studies (e.g., Feenstra et al. 2011; Porter et al. 2017), we found that respondents in this survey were willing to pay more for real food. However, it is important to note that there was not a consensus among respondents about the amount of cost increase they would accept, likely an indication of varying capacities to pay, differing levels of value for real food, and different definitions of sustainability in food (O'Neill et al. 2019). Fewer than one-fifth of respondents supported a cost increase of one dollar per meal. In addition, approximately one-third of respondents said that they would not be willing to pay any more for 20% real food. This indicates that while there was some student support for a cost increase in exchange for more real food, campus dining leadership should be wary of raising prices by any significant amount.

The predictor variables that correlated most strongly with a positive willingness-to-pay were the college's sustainability, environment, and values. More specifically, respondents who reported that they had considered Fort Lewis College's sustainability commitments when choosing where to go to college, students who take into consideration the environmental conditions of food production when deciding what to eat, and students who reported that it was important that their food reflect their values were all significantly more likely to report a positive willingness-to-pay for more real food. This likely indicates a role for personal environmentalist identity in real food commitment and is consistent with much of the environmental behavior literature on identity (e.g., Fielding and Hornsey 2016; Freed 2018; McGuire et al. 2013). People who identify strongly with their values about environmental issues are much more likely to take genuine actions in support of these values.

We also found that students from Colorado were significantly more likely to be willing to pay more for real food than out of state students. Porter et al. (2017) found the same result in Vermont and attributed this to the conceptual heft of local food and to the familiarity of in-state students with local foods produced in that state. These were likely also driving factors in our study. For example, a student from Colorado may have a favorable association with a product like the state's famous Palisade peaches that an out of state student would not find familiar.

Convenience correlated negatively with willingness-to-pay. Students who valued convenience were less likely to be willing to pay more for real food. This could be due to a perception that increasing the proportion of real food at Fort Lewis College might make wait times longer or reduce food options, both of which could be perceived as reducing convenience. Another explanation is that students who reported that they value convenience perceive food as simply 'fuel' and want it to be both convenient and cost-effective.

In summary, this research found many similar findings to the study on which it was based. Like Porter et al. (2017), we found in-state to be a statistically significant variable in predicting willingness-to-pay more for real food. Other variables that Porter et al. (2017) found to be statistically significant were gender, with women willing to pay more for real food; origin, with students who considered origin of food to be important willing to pay more; and school, with students in environmental schools within the university willing to pay more. We did not find a significant effect for gender in our study; 68% of women reported a positive willingness-to-pay level compared to 66% of men. Local was our proxy for the origin variable in the Porter et al. (2017) study. Of students who said that local sourcing of food was important to them, 81% were willing to pay more, and 19% were not willing to pay more, but this finding was not statistically significant. Similarly, we did find a potential role for major, a proxy for the school variable in the Porter et al. (2017) study. Every Environmental Studies major (100%) that was surveyed reported being willing to pay more for real food, while only 58% of students in all other majors reported a positive willingness-to-pay, but we were not able to include this parameter in the model due to the small sample size of Environmental Studies students (n = 15).

This research can be used by dining services at Fort Lewis College to meet the Real Food Challenge commitment to 20% real food by next year, 2020. Our research indicates further implementation of the Real Food Challenge at Fort Lewis College should aim to purchase real foods that are tasty and healthy without increasing price. Given that real food often costs more than conventional foods, it would be worthwhile to pursue creative menu changes rather than simply substituting conventional food items for more expensive real substitutes. This might mean substituting a five-ounce 'conventional' burger patty for three ounces of local and humane steak served in a stir-fry. Using this method, it might be possible to substitute several items for real foods without a noticeable increase in price.

Acknowledgements Thank you to Maggie Magierski for contributing to this research. A huge thank you to students in the spring semester of Environmental Studies 320 for collecting data. Thank you to the Undergraduate Research in the Humanities support from the Mellon Foundation for funding this research. And thank you to the two anonymous reviewers whose insights improved this manuscript.

References

AASHE. 2015. 2015 Sustainable Campus Index. https://www.aashe.org/wp-content/uploads/2017/10/aashe_2015_sustainable_campus_index.pdf. Accessed 22 April 2019.

AASHE. 2018. 2018 Sustainable Campus Index. https://www.aashe.org/wp-content/uploads/2018/08/SCI-2018.pdf. Accessed 22 April 2019.

Altieri, M.A., and C.I. Nicholls. 1997. Conventional agricultural development models and the persistence of the pesticide treadmill in Latin America. *International Journal of Sustainable Development & World Ecology* 4 (2): 93–111.

Barlett, P.F. 2011. Campus sustainable food projects: Critique and engagement. *American Anthropologist* 113 (1): 101–115.

Belcher, K.W., A.E. Germann, and J.K. Schmutz. 2007. Beef with environmental and quality attributes: Preferences of environmental group and general population consumers in Saskatchewan, Canada. *Agriculture and Human Values* 24 (3): 333–342.

Born, B., and M. Purcell. 2006. Avoiding the local trap: Scale and food systems in planning research. *Journal of Planning Education and Research* 26 (2): 195–207.

Casarejos, F., L.M. Gustavson, and M.N. Frota. 2017. Higher Education Institutions in the United States: Commitment and coherency to sustainability vis-a`-vis dimensions of the institutional environment. *Journal of Cleaner Production* 159: 74–84.

Ceulemans, K., I. Molderez, and L. Van Liedekerke. 2015. Sustainability reporting in higher education: A comprehensive review of the recent literature and paths for further research. *Journal of Cleaner Production* 106: 127–143.

Chittapur Doddabasawa, B.M., and M.R. Umesh. 2017. On-farm crop diversity for sustainability and resilience in farming: A review. *Agricultural Reviews* 38 (3): 191–200.

Costanigro, M., O. Deselnicu, and D. McFadden. 2016. Product differentiation via corporate social responsibility: Consumer priorities and the mediating role of food labels. *Agriculture and Human Values* 33 (3): 597–609.

Dyer, G., and M. Dyer. 2017. Strategic leadership for sustainability by higher education: The American College & University Presidents' Climate Commitment. *Journal of Cleaner Production* 140: 111–116.

Feenstra, G., P. Allen, S. Hardesty, J. Ohmart, and J. Perez. 2011. Using a supply chain analysis to assess the sustainability of farm-to institution programs. *Journal of Agriculture, Food Systems, and Community Development* 1 (4): 69–84.

Fielding, K.S., and M.J. Hornsey. 2016. A social identity analysis of climate change and environmental attitudes and behaviors: Insights and opportunities. *Frontiers in Psychology* 7: 1–12.

Filho, W.L., F. Doni, V.R. Vargas, T. Wall, A. Hindley, L. Rayman-Bacchus, K. Emblen-Perry, J. Boddy, and L.V. Avila. 2019. The integration of social responsibility and sustainability in practice: Exploring attitudes and practices in Higher Education Institutions. *Journal of Cleaner Production* 220: 152–166.

FLC. nd (no date). About Fort Lewis College. https://www.fortlewis.edu/Home/About/AboutFortLewisCollege.aspx. Accessed 2 May 2019.

Freed, A. 2018. The relationship between university students' environmental identity, decision-making process, and behavior. *Environmental Education Research* 24 (3): 474–475.

Garnett, T. 2011. Where are the best opportunities for reducing greenhouse gas emissions in the food system (including the food chain)? *Food Policy* 36 (Supplement 1): S23–S32.

Heise, H., and L. Theuvsen. 2017. What do consumers think about farm animal welfare in modern agriculture? Attitudes and shopping behaviour. *International Food & Agribusiness Management Review* 20 (3): 379–399.

Horrigan, L., R.S. Lawrence, and P. Walker. 2002. How sustainable agriculture can address the environmental and human health harms of industrial agriculture. *Environmental Health Perspectives* 110 (5): 445–446.

Lozano, R., K. Ceulemans, M. Alonso Almeida, D. Huisingh, F.J. Lozano, T. Waas, W. Lambrechts, R. Lukman, and J. Hugé. 2015. A review of commitment and implementation of sustainable development in higher education: Results from a worldwide survey. *Journal of Cleaner Production* 108: 1–18.

Mascarelli, A. 2013. Growing up with pesticides. *Science* 341 (6147): 740–741.

McGuire, J., L. Morton, and A. Cast. 2013. Reconstructing the good farmer identity: Shifts in farmer identities and farm management practices to improve water quality. *Agriculture and Human Values* 30 (1): 57–69.

Murray, J. 2018. Student-led action for sustainability in higher education: A literature review. *International Journal of Sustainability in Higher Education* 19 (6): 1095–1110.

O'Connor, J.S. 2006. Resource review: Civic engagement in higher education. *Change: The Magazine of Higher Learning.* 38 (5): 52–58.

O'Neill, K.J., A.K. Clear, A. Friday, and M. Hazas. 2019. 'Fractures' in food practices: Exploring transitions towards sustainable food. *Agriculture and Human Values* 36 (2): 225–239.

Porter, J., D. Conner, and J. Kolodinsky. 2017. Get real: An analysis of student preference for real food. *Agriculture and Human Values* 34 (4): 921–932.

PR. 2019. The Princeton Review 2019 college hopes & worries survey report. https://www.princetonreview.com/college-rankings/college-hopes-worries. Accessed 22 April 2019.

RFC. nd (no date)-a. Real Food Calculator: Fort Lewis College. https://calculator.realfoodchallenge.org/institutions/67/profile. Accessed 1 May 2019.

RFC. nd (no date)-b. Real Food Challenge. https://www.realfoodchallenge.org/. Accessed 22 April 2019.

RFC. nd (no date)-c. Real Food Standards 2.1. https://www.realfoodchallenge.org/resources/real-food-resources/real-food-standards-20/. Accessed 1 May 2019.

Ruben, R., R. Fort, and G. Zúñiga Arias. 2009. Measuring the impact of Fair Trade on development. *Development in Practice* 19 (6): 777.

Satter, E. 2007. Hierarchy of food needs. *Journal of Nutrition Education and Behavior* 39 (5): S187–S188.

Smiley Smith, S.E. 2017. Seeking equilibrium: Exploring environmental sustainability and decision making in higher education. PhD diss., School of Forestry and Environmental Studies. New Haven, CT: Yale University.

Smith, D.R., S.J. Livingston, B.W. Zuercher, M. Larose, G.C. Heathman, and C. Huang. 2008. Nutrient losses from row crop agriculture in Indiana. *Journal of Soil & Water Conservation* 63 (6): 396–409.

SN [Second Nature]. 2018. Second Nature 2017–2018 impact report. https://secondnature.org/wp-content/uploads/2017-18_SeconddNature_ImpactReport.pdf. Accessed 22 April 2019.

SN [Second Nature]. nd (no date). Second Nature history. https://secondnature.org/history/. Accessed 22 April 2019.

Sodexo. nd (no date). About Sodexo. www.sodexo.com/about-us. Accessed 1 May 2019.

Stehle, S., and R. Schulz. 2015. Agricultural insecticides threaten surface waters at the global scale. *Proceedings of the National Academy of Sciences of the United States of America* 112 (18): 5750.

Tilman, D. 1999. Global environmental impacts of agricultural expansion: The need for sustainable and efficient practices. *Proceedings of the National Academy of Sciences of the United States of America* 96 (11): 5995.

Ulvila, K.M., A. Paloviita, and A. Puupponen. 2009. Consumers' perceptions of sustainably produced food: A focus group study. *Progress in Industrial Ecology, An International Journal* 6 (4): 355–370.

Weber, C.L., and H.S. Matthews. 2008. Food-miles and the relative climate impacts of food choices in the United States. *Environmental Science & Technology* 42 (10): 3508–3513.

Wright, T.S.A. 2002. Definitions and frameworks for environmental sustainability in higher education. *Higher Education Policy* 15 (2): 105–120.

Publisher's Note Springer Nature remains neutral with regard to jurisdictional claims in published maps and institutional affiliations.

Kathleen Hilimire PhD, is an Associate Professor in the Department of Environmental & Sustainability at Fort Lewis College. She is also co-founder of the Regenerative Food Systems Certificate at FLC. Dr. Hilimire teaches courses in political ecology of food, agroecology, and research methods. Her research interests explore environmental decision-making and conservation agriculture.

Carl Schnitker is an Environmental Studies and Adventure Education major at Fort Lewis College

Agriculture and Human Values (2020) 37:1083–1094
https://doi.org/10.1007/s10460-020-10039-3

From texts to enacting practices: defining fair and equitable research principles for plant genetic resources in West Africa

F. Jankowski[1] · S. Louafi[2] · N. A. Kane[3] · M. Diol[4] · A. Diao Camara[5] · J.-L. Pham[6,7] · C. Berthouly-Salazar[6] · A. Barnaud[6]

Accepted: 28 April 2020 / Published online: 9 May 2020
© Springer Nature B.V. 2020

Abstract

Collaborative research practices in the field of plant genetic resources must follow the principles of fairness and equity as defined in the Convention on Biological Diversity (CBD) and in the International Treaty on Plant Genetic Resources for Food and Agriculture (ITPGRFA). In this context the concepts of fairness and equity generally refer to the substantive and procedural dimensions associated with sharing the benefits of this research. But neither term is clearly defined by these international treaties, and the meanings attributed to the concepts vary among different societies. This paper looks at the question of how to account for the diversity among value systems when conducting research that implicates diverse stakeholders and respects the requirements of fairness and equity. We incorporated an auto-ethnography method developed as part of a multi-stakeholder network involved in research projects on plant genetic resources in West Africa. A theatrical device was used as a framework for testing the principles of fairness and equity, helping us to collectively identify feelings of injustice, and explore the conditions for making collaborative research practices more ethical in ways that respect the perspectives of different stakeholders. In an environment of extensive political and socio-cognitive inequality, this approach makes it possible to explain the criteria relating to interactional justice and expectations in terms of socio-political and socio-emotional benefits. It also invites us to consider the principles of fairness and equity in a framework of ethical competence that goes beyond international directives.

Keywords Access and benefit sharing (ABS) · Ethics · Interactional justice · Transdisciplinary research · Plant genetic resources · Forum theater

Abbreviations

ITPGRFA	International treaty on plant genetic resources for food and agriculture
NP	Nagoya protocol
CBD	Convention on biological diversity
PIC	Prior informed consent
ABS	Access and benefit sharing

✉ F. Jankowski
frederique.jankowski@cirad.fr

S. Louafi
selim.louafi@cirad.fr

N. A. Kane
ndjidokane@gmail.com

M. Diol
diol6@yahoo.fr

A. Diao Camara
astoudiaocamara@gmail.com

J.-L. Pham
jean-louis.pham@ird.fr

C. Berthouly-Salazar
cecile.berthouly@ird.fr

A. Barnaud
adeline.barnaud@ird.fr

1   GREEN, CIRAD, University of Montpellier, Bâtiment F - Bureau 105, TA C-47 / F, Campus international de Baillarguet, 34398 Montpellier, Cedex 5, France

2   AGAP, CIRAD, University of Montpellier, 34398 Montpellier, France

3   CERAAS, ISRA, Thiès, Sénégal – LMI LAPSE, Campus de Bel Air, route des Hydrocarbures, Dakar, Sénégal

4   Kaddu Yaraax, Dakar, Sénégal

5   BAME, ISRA, Dakar, Sénégal

6   DIADE, University of Montpellier, IRD, Montpellier, France

7   Agropolis Fondation, Montpellier, France

FT   Forum theater
ASPSP   Association Sénégalaise de Producteurs de
        Semences Paysannes (Senegalese association
        of peasant seed producers

Introduction

The International Treaty on Plant Genetic Resources for Food and Agriculture (ITPGRFA) and the Nagoya Protocol (NP) of the Convention on Biological Diversity (CBD) have contributed to a renewed form of socialization and politicization of research practices in the field of agrobiodiversity. Although access to genetic resources and the fair and equitable sharing of the benefits stemming from their use occupy a central place in international texts, these ideas are seldom precisely defined. It is often noted that the concepts of fairness and equity refer to substantive (contents of the agreement) and procedural (legitimacy of the process to reach the agreement) dimensions (Vermeylen 2007; Parks 2018; Morgera et al. 2018). However, the determination of what is fair and equitable does not rest on any pre-established principle or criterion. It falls to the contractual frameworks as defined in the texts, either multilaterally as in the case of the ITPGRFA (an agreement negotiated between States and imposed on stakeholders) or bilaterally as in the case of the CBD.

In both frameworks, the operationalization of the principles of fairness and equity is confronted with two challenges: one related to the diversity in the types of relationships that are potentially affected by benefit-sharing, and another related to the diversity in the types of benefits themselves. There are numerous types of relationships (Morgera et al. 2018): relationships between States, between government and communities within a State, between private and public entities in the same jurisdiction, between different jurisdictions, or between different entities in the same community. Therefore, the goal is to include the wide diversity of stakeholders involved in using or conserving plant genetic resources. In terms of benefits, the NP and the ITPGRFA insist on including both monetary and non-monetary advantages such as access to technology and training, the sharing of results from the use of genetic resources, the reinforcement of institutional capacity, and the creation of employment opportunities.

Access and benefit-sharing agreements need to provide for the diversity of value systems at work among stakeholders who often have different visions and objectives for the entities exchanged. Prior Informed Consent (PIC), community protocols in the framework of the NP, and farmers rights in the context of the ITPGRFA are all mechanisms aimed at achieving this operationalization of the principles of fairness. Nevertheless, several authors have noted that

unbalanced power relationships and the unilateral conception of tools sometimes prevent these instruments from fully accomplishing their goals (Parks 2018). In addition, some studies describe how the concepts of sharing, justice, and benefits have very different meanings in many communities when compared to those in Western cultures (Vermeylen 2007; De Jonge 2011). The incompatibilities exposed when comparing certain local values or cosmovisions with modern economic and political value systems highlight the respect required for the former in order to truly respond to the principles of fairness and equity in international treaties (Byström et al. 1999; Graddy 2013).

This paper is aimed at contributing to discussions about how the requirements of fairness and equity can be respected while identifying and taking account of the diversity in value systems encountered in a context of research partnerships involving diverse actors. The legal instruments found in the international texts are important vehicles for raising awareness of the ethical risks associated with procedures for accessing and using genetic resources in research and development (Lavery and IJsselmuiden 2018). However, their impact on research culture is difficult to measure in the specific context of research partnerships where it is difficult to ensure the recognition and accounting of all values relevant to what can, or should be, a fair and equitable sharing of benefits. To be able to define practices that can be considered fair and equitable from the perspective of the different parties involved, researchers have taken a relation-based approach. Notably, this relational approach involves an accounting of the socio-historical context that links the different protagonists and the socio-political context of inequality (Whiteman and Mamen 2002; O'Faircheallaigh 1998). It demands a contextual understanding of the concepts of fairness and equity, in concrete situations of their implementation rather than simply in a substantial, formal way.[1] It is from this perspective that we propose a pragmatic approach to the concepts of fairness and equity in the collaborative research practices employed in the field of plant genetic resources. It offers an observation of situations of ethical deliberation, of the ways in which the ideas of fairness and equity are treated by the actors and stakeholders in interactional contexts. We revisit a collective, auto-ethnographic experiment developed in the framework of a multi-stakeholder network (researchers, farmers, gene bank directors, civil society actors, etc.) involved in collaborative projects concerning plant genetic resources in West Africa. First, we describe the way that we collectively created a theater device for testing the concepts of fairness

---

[1] In other words, extrinsic to the subject who only needs to use pre-defined rules and apply them to the situation through a deductive and decontextualized process of analysis.

and equity in a participatory format. This device enabled (i) an analysis of the feelings of injustice in collaborative research practices, (ii) the expression of the principles that the different stakeholders use to evaluate the fairness of their respective behaviors, and (iii) a collective exploration of the conditions that make it possible for collaborative research practices to be more ethical from the point of view of different stakeholders. The principles of interactional justice that were employed to judge the actions of researchers as well as those of farmers provide direction for the examination of the socio-emotional and socio-political benefits for stakeholders interested in collaborative research.

Creating a theater device to test the concepts of fairness and equity

Context and posture of collective auto-ethnography

The context is a multi-stakeholder network composed of French and West African researchers, breeders, farmer organizations, actors from civil society, state representatives, farmer organizations and managers of gene banks. Over the course of the collaborative research within this network, and within the framework of the application of the international texts, ethical questions have emerged concerning the way that research can truly take into account the institutional constraints and the interests and objectives of the various parties involved. Two issues have appeared. The first is cognitive and relates to the explanation of a shared vision of the exchanged objects and resources. The second is operational and concerns the definition of a collaborative charter defining the principles of ethical practices coherent with the socio-cultural context in which they are applied. Neither the international texts, nor the various national legislatures or guides edited by the research organisms offer much in the way of elements that respond to these operational issues (Engels et al. 2011). With the aim of improving their collaborative practices, some researchers decided to organize a participatory workshop that brought together members of this international network to discuss the concepts of fairness and equity in a contextualized manner, i.e. by considering the way they are lived and defined by the various stakeholders in the network.

Previous research highlight the way some value and knowledge systems are discredited by the predominance of a participation format used in international arenas (Richard-Ferroudji 2011; Cheyns 2014). In order to avoid this, the workshop organizing committee chose to mobilize a singular format of participation through a theatrical device. This device has been designed as an approach of collaborative auto-ethnography through the theatrical staging of collaborative research situations familiar to the various stakeholders.

The use of collaborative, theatrical forms is drawn from participatory science and performance or art-based ethnography (Banks 2001; Colleyn 2011; Conrad 2004; Denis 2009; Müller et al. 2017). All of these approaches are based on the analysis of self-representations, considered to be representations of the reality created by the concerned actors. The interpretation of these self-representations is part of a process of negotiating meanings; the role of interpreter is shared by the researcher and the different participants in a collective analytical framework. In these types of approaches, the goal is also to give the actors "tools" that they can use in their particular environments. These approaches themselves respond to the ethical and political issues of the research. They involve different levels of participation: (i) the personal experiences of the participants may form the basis of the theatrical creation, or (ii) a play may be created by a team of facilitators and then performed in front of a target audience who then participate in a forum about the play. Depending on the objectives, the use of theatrical forms can be developed during different phases of research, from data collection to analysis and representation, to convey meanings that could be otherwise unavailable (Heras and Tabara 2014).[2] In our case, the approach was not part of a research project but was deployed at the end of collaborations and with the perspective of future projects. It was therefore an issue of learning from our collaborative experiences in order to define fairer future practices.

The participatory theatrical form that we chose is forum theater (FT). FT was conceived specifically to discuss situations of tension, conscious or not, between a diversity of actors (Boal 1985). The orchestration of a forum theater is characterized by four principal stages: (i) the creation of a play inspired by real facts, representing the tensions between different characters; (ii) the presentation of the play in front of a "concerned" audience; (iii) the addition of a moderator who invites the audience to share their feelings and interpretations of the play and the actions of the different characters; (iv) an invitation for the "spect-actors" to participate in the play, replacing certain characters to test possible solutions and discuss them collectively. The last step thus incorporates the collective experience of new alternatives, and together, the different stages offer a veritable collective investigation (Dewey 1938).[3] The forum puts the audience members in a reflective posture, where they are called upon to judge the different characters of the play. Members of the audience are asked to justify their conclusion and clarify the criteria

---

[2] For a description of the advantages and limitations of using applied theatre, according to participation levels in the theatrical approach and duration of the process, see Heras and Tabara (2014).

[3] Dewey proposes a model of five stages: (i) awareness of the problem, (ii) understanding its construction, (iii) suggestions for possible solutions, (iv) examination of the suggestions and their consequences, (v) testing the hypotheses.

used to evaluate the character's behavior. In this context, the spect-actors submit their qualifications and interpretations in terms of the elements of tension dramatized in the play. The audience is therefore obligated to participate in the explanation of the justification criteria. Thus, the forum theater offers a methodological framework that responds to the theoretical postulates of the French pragmatic sociology (Bréviglieri and Stavo-Debauge 1999; Barthe et al. 2013). A key notion of this sociology is that of *test*[4] (Latour 1988; Thévenot 2001; Boltanski and Chiapello 2007; Hennion 2017). A *test* is any situation in which actors experience the vulnerability of the social order, because they have doubts about what reality is. The *test* constitutes a moment that engenders an activity of judgement and a dynamic of reaffirmation or readjustment, of institutions and initial conceptions. The pragmatic sociology favors the analysis of situated action where uncertainty and the forms of adjustment of actors among themselves can be observed. It is more particularly concerned with the justification practices of social actors to denounce an offence or injustice (Boltanski and Thevenot 2006). It underlines the way in which operations of justification mobilize a plurality of principles of justice according to the situation. By offering participants the opportunity to express themselves from concrete interactional situations and to engage in the exploration for overcoming feelings of injustice, forum theater provides spaces for testing participants' representations. It allows to understand the notions of justice as an ongoing practical process in which individuals participate through their justified actions rather than as an ideology or a shared belief.

Identify elements of tension and feelings of injustice

Researchers of the workshop organizing committee were in charge to define the script for staging a piece of collaborative practices. The group was characterized by diversity in disciplines (genetics, anthropology, political science), in international collaboration experience, in fieldwork experience (practices), in institutional responsibilities (institutional ABS referents) and in nationalities (French and Senegalese). Two questions were posed to this group: (i) In your experience, what are the principal problems related to intellectual property in genetic resources? and (ii) What, in your opinion, are the principal problems related to the sharing of benefits in multi-partner practices? We made the choice to begin with specific situations that might provoke feelings of

injustice. Each of the researchers had to respond to the questions from the perspective of actual fieldwork experience. This elicited clarifications about experiential knowledge, and also the associated feelings such as being powerless, or constrained, or the sense that issues are known but not "talked about". The researchers' responses were regarded as narratives about experiences of injustice, situations that were not fair[5] from their point of view.

From these experiences, certain situations–defined as an ensemble of characters, or a place, or subject–were identified. The selection of situations was made according to a tested criterion of meaning: the situation should recall those experienced by the different researchers, without being identical to those that they could have experienced alone.

On the basis of these situations a diversity of issues or elements of tension were more generically identified. These included four issues or elements of tension: (i) the diversity of issues generated by the different types of stakeholders involved, particularly between researchers and farmers; (ii) the gap between practices and legal frameworks for circulation and access to seeds (genetic resources); (iii) between institutional frameworks (research centers)[6]; and (iv) between production and knowledge systems at the heart of the same discipline (research posture and relationship to transdisciplinarity). The co-definition of the forum-theater play enabled the clarification of certain dimension that constrain, from the researchers' point of view, the definition of fair and equitable practices. Through this piece, the idea was to test these points of view in interaction with the experiences of the other members of the network.[7]

Representing collaborative research practices

A script was drafted from scenes representing the different elements of tension, and then given to a professional theater troupe from Dakar named Kaddu Yaraax. The actors assumed control of the script, introducing an ensemble of artistic expertise and using the devices inherent in forum theater (humor, caricature, plurality of space and time, etc.)

---

[4] *Test* is the English translation of *épreuve*. In the Francophone world, indeed, the term *épreuve* has a more complex meaning, referring also to "trial", "ordeal", and "proof".

[5] Fair in the sense here of being "right", morally and ethically correct in the situation and in the response to the situation.

[6] Defining the script opened the discussion to subjects that are rarely, if ever, broached in the framework of international research partnerships. International exchanges and the reception of researchers are governed by conventions at the scale of the research institutes, that never take into account the divergence in mandated authority that exists among the different institutes.

[7] When only a small group defines the initial script, it is important to evaluate the effect of the framing of the play on the debates. The appropriateness of the situations is crucial. In our case, this was possible because of the long-standing collaboration between the different stakeholders. Moreover, during the forums, the spectators have had the possibility to modify the script of the play by adding scenes or characters that they feel are missing from the situations represented.

to give life to the situations. Different proposals for staging presented opportunities for exchanges between the actors and the researchers in order to make adjustments to the play. The play itself was composed of five scenes which described various characters: a Brazilian researcher hosted for several years, a Senegalese researcher, farmers from the village of Ndiayenne, and a Brazilian gene bank with an international conservation mandate. The first scene covered the modes of communication between researcher and farmer, and the sharing of stakes related to the research, including the objects of exchange (the seeds) which also represent a component of a larger system of resources for each of these stakeholders. The second scene placed multiple values on the stage, as well as the identifying dimensions linked to the seeds for the farmers and the associated issues at stake. The third scene questioned the national and international framework governing seed circulation in relation to the collaborative practices between researchers and farmers. The fourth scene introduced the fact that the stakeholders in the scene do not act solely according to their own motivations, that they are part of broader frameworks such as collaborative networks, institutions, and regulations. These frameworks can favor the emergence of ethical dilemmas, as we see for example when the different partnered research institutes are not subject to the same rules concerning the recognition of farmers' knowledge. The fifth and last scene brings the question of intellectual property rights for research results to both the individual scale and that of collectives.

The play composed of these five scenes has been subjected to scrutiny in a variety of arenas. It was presented at a 2016 workshop that brought together stakeholders from eight West African countries. At the end of this workshop, a Senegalese farmers' organization decided to program the forum theater in the framework of two national and international fairs devoted in 2017 to the seeds of West African farmers. Similarly, actors from civil society have mobilized this play during the first Dakar AlimenTerre Festival in 2018 and it was performed on the campus of international research in Dakar. For each of these representations, the objective was to discuss the principles of justice and equity relevant to the diversity of experiences involved. All performances and forums in the play were filmed. In the following parts, we present an analysis of the practices of valuation and justification, and collective explorations for overcoming feelings of injustice.

Principles of interactional justice

At the end of each performance, during the first part of the forum, the troupe Kaddu Yaraax usually conducts what she calls a "trial." Each character's behavior is subjected to this judgment. If he is deemed to be fair, the

character is placed in the shade of an imaginary baobab. If, on the contrary, a majority of the audience thinks that he has acted improperly, the character is placed under the imagined burning sun of the Sahel. Each member of the audience is systematically asked to justify his conclusion and clarify the criteria used to evaluate the character's behavior. We can see that considerably different criteria are used to evaluate fairness in the behavior of farmers, local researchers, and international researchers.

The criteria for evaluating fairness in farmers' actions

The specific case of a farmer who has given seeds to a foreign researcher was judged by the audience, who then clearly described several criteria for evaluating the fairness of this farmer's attitude.

The first criterion identified for judging the attitude is his loyalty to his community. This is understood in the sense of respect for values and tradition. Some audience members emphasized the farmer's positive reception of the foreign researcher. Ly (1966), in his analysis of Senegalese moral philosophy, underlined the way that their society is structured by an ensemble of values. Even if these values have to be put into current social perspectives, they provide an idea of the normative framework in which social relationships would ideally develop, helping us understand the impact that such values can have on the perception of good and evil. One of the first values is that of *teranga*, the value of hospitality. In spite of ethnic and religious differences, the foreigner must be made to feel welcome in Senegal. He must always be welcomed in both words and acts, even if this requires certain sacrifices. Even if *teranga* does not translate into some sort of gift, it is embodied in an unwavering attitude of courtesy towards others. Some spectators emphasized the way in which the farmer responded to the demands of the foreigner even though the motivations of the latter did not seem clear to him. He thus inscribes his actions in another key value of Senegalese society: *kersa*. This value reflects a form of modesty that corresponds to the respect of oneself, but also of the other. In respecting the value of *kersa*, the individual acts in the best interest of the social relationship in any social situation. *Kersa* is the foundation of the virtue of a diplomacy that forbids upsetting others or placing them in an uncomfortable situation.

The reference to tradition when judging the farmers' actions is consistent with the notion of a justification of the domestic order as defined by Boltanski and Thévenot (2006). According to the authors, the reference to the domestic world appears "each time that research on fairness emphasizes personal relations between people...human beings are immediately qualified by the relationship that they maintain with their own kind" (Boltanski and Thévenot 2006, p. 206). The criteria mobilized to judge the correctness of the farmer's behavior are primarily based on his respect for the norms and values defined by his community. Some emphasize that he only did his duty when he gave the seeds demanded by the foreign researcher, because seeds have to be given to anyone who asks for them. *"You asked him for seeds, he gave them to you. That's what was done, that's a normal act."* But for others who think of the seeds as property of the community, this farmer acted disrespectfully when he didn't first consult with the traditional authorities and other farmers who depend on the seeds collected. In the domestic world, it is through the reference to generation, tradition, and hierarchy that order can be established between domestic beings (Boltanski and Thévenot 2006): according to some audience members, an order that this farmer would not have respected. From the perspective of these spectators, the farmer reappropriated a common good when he gave seeds to the Brazilian researcher. In this sense, he would have placed his own interests and those of the foreigner above those of the members of his community.

The farmer's behavior is therefore judged by criteria such as loyalty to tradition, hierarchy, and respect for the values of, and honesty towards, the community. Acting appropriately with the researchers means, above all, acting in accordance with those principles and values.

Criteria for evaluating the fairness of researchers' actions vis-à-vis the farmers

The criteria of loyalty and honesty are also called upon to evaluate the researchers' actions. Several spectators thought the attitude of the foreign researcher was not right in the sense that he did not respect his commitment to the farmer. When the researcher presents the project to the farmer, he promises to link the latter's name to the results. But in the end, there is no recognition for the contribution of the farmer in the valorization (in this case, a patent) of the research. It should be noted that the piece's scenario specifies that it is a local researcher, not his foreign colleague, who initiates the patent. However, from the point of view of farmers present in the audience, the latter is also held responsible for the situation since he alone made the commitment with the farmer in the piece. This situation can be considered as unjust from the perspective of the foreign researchers. The group of scientists emphasize certain effects related to the fact that the countries involved in the collaborative research projects are not subject to the same institutional obligations and regulations.

The researchers' actions are also judged with regard to their mandate. In the evaluating arguments it becomes clear that what is seen as fair or legitimate for the foreign researcher is not systematically applied to the national researcher who is considered to be in the service of his country and thus its farmers:

> These two researchers do not have the same job. The first is a national researcher. He is in the service of his country and its people. If he selects a gene, he has an obligation to see that local farmers are aware of, and understand, the vital importance of this gene. He should be aware that he works in the interest of the country.
>
> (Farmer at the seed fair in Fatick)

The local researcher is evaluated in the context of being a representative of the state. According to the analytical framework of Boltanski and Thévenot's justification theory, the evaluation falls within the civic world; here the representative has certain qualities that enable him to accomplish his mission (Boltanski and Thévenot 2006). In addition, the aspect of legality defines a form of greatness particularly appreciated in this world. In this spirit, a local researcher justifies his actions to the farmers by emphasizing that *"the State supports what we are doing here...for everything that we do, clear rules and procedures are followed. There isn't any deception or equivocation".* Moreover, as a full member of the local society, the behavior of the local researcher is also evaluated according to local hierarchies and values. Wagué (2012) describes the way in which the local researcher fits into the local power networks and how his family and ethnic links can create the perception among certain stakeholders that he is constrained by a "social record" and risks being compromised by those social relationships and obligations. Farmers and researchers repeatedly draw attention to the difference between mandates and responsibilities towards the society when comparing local and foreign researchers. The latter can be judged negatively by some spectators because they are considered to ultimately serve foreign interests. African researchers also emphasize the lack of information generally received about their foreign colleagues' activities within the local communities. Appearing implicitly in these interventions is the balance of power between local and foreign researchers that is manifested in both the definition of research issues and the ways they are implemented in the postcolonial context of West Africa (Ela 2007; Fall 2011).

One criterion used to evaluate both local and foreign researchers is based on the attitude that they adopt in the presence of the farmer. The ethical evaluation of the researchers' behavior considers the characteristics of modesty, humility, and respectfulness in their interactions with the farmers. Farmers and representatives from farmers' organizations mentioned the fact that researchers often have preconceived impressions of the farmers: they consider farmers to be naive and without pertinent knowledge or competence. Several members of the audience particularly noted that the researchers in the play never show any concern about the needs or interests of the farmers. The president of an international group of farmers in West Africa mentions that the researcher should have first addressed the farmer in the form of a "*question*" and that he should have "*asked the farmer if he wanted the seeds to be preserved or kept*" in order to establish their relationship as a collaboration. According to the president and the farmers in the audience, this is rarely the case.

Kindness and well-meaning are also often cited as criteria for judging attitudes. When researchers adopt the perspective of farmers and implement actions that enable the latter to respond to their interests, that behavior is considered to be fair-minded. This sense is reflected in the positive or "indulgent" judgment of the foreign researcher who tried to recover the seeds from the gene bank when the farmers notified him of the extinction of their seed supply.

Kind and respectful postures are strongly associated with whether or not the researchers communicate successfully with the farmers. Several farmers insisted on the need to restore a relationship of confidence weakened by the past collaborative experiences with researchers. The farmers' comments about their relationship with researchers depends largely on the degree of deception: "*Ever since our ancestors, this has been the way we are deceived*". Many farmers speak about the lack of trust related to the regular stream of researchers coming to meet with them and collect seeds, only to never be seen again, not even to inform them about the progress of their work. In the forum, it is noted that the foreign researcher has explained the project to a certain degree, but that he has never clearly informed the farmers of the risks and limits of the research. He is criticized for failing to explain to the farmer that part of the seeds will be sent out of the country and will no longer be available.

Redressing feelings of injustice

Ensure mutual and informed engagement

Diverse criteria were mentioned for evaluating the behavior of the various characters, farmers and researchers, in the forum-theater play. The theatrical performance, with a specific focus on the flow of interactions, offers a collective analytical scale that illuminates the importance that individuals place on the way in which they are considered and the way that this consideration participates in their feelings of injustice. In this context, the principal criteria used by both the researchers and the farmers are consistent with interactional justice (Bies and Moag 1986; Bies 1987, 2005; Collie et al. 2002; Ando and Matsuda 2010; Dai and Xie 2016). Bies defines interactional justice as "people's concern about the quality of interpersonal treatment they receive during the enactment of organizational procedures" (Bies and Moag 1986, p. 44). These works emphasize that even though the individuals are preoccupied with the official procedures used in a decision process, they are equally preoccupied with the interpersonal treatment that they experience from other people. An ensemble of work demonstrates that interactional justice makes it possible to understand the processes of confidence, the legitimacy of authority, and the engagement of the actors in the organizations (Barling and Phillips 1993).

Bies (1987) identifies four determining criteria in the perception of fair treatment: honesty, courtesy, respect, and appropriate feedback. Five criteria are also considered key in the perception of fair treatment within an organization: adequate consideration for opinions, suppression of personal bias, uniform application of decision making among people and over time, and availability of precise information and adequate explanation concerning decisions (Tyler and Bies 1990). We see the ensemble of the criteria expressed by the different audiences when explaining their judgment of the characters' behavior in the play. Beginning with these criteria, Tyler and Bies (1990) demonstrate that the perception of interactional justice is particularly influenced by two important factors: the individual's perception of the interpersonal treatment and the appropriateness of the explanations of formal procedures. These two factors define both the interpersonal dimension and the informational dimension of interactional justice. Greenberg (1993) thus proposes to consider two dimensions of interactional justice: informational justice based on the accuracy and quality of information received, and interpersonal justice corresponding to the quality of interpersonal exchanges (honesty, respect, etc.).

It is specifically these two dimensions of interactional justice that form the basis of the collective experiments for overcoming or reducing the injustices perceived in the play. One group of propositions involved the farmers' access to information and the researchers' attention to the interests of the farmers. Several spectators discussed the use of the PIC (prior informed consent) and its capacity to fully inform, and take into account, the farmers' potential stakes in the research. This engendered further debate on the practical questions posed by the implementation of this type of tool. In the forum framework, these efforts clarifying the PIC brought to light the difficulty of actually sharing the different parties' interests when using the typical formulations found in PIC document. The forum theater has also revealed a feeling of uneasiness of the individuals seeking the signature of PIC. Beyond the often opaque formulations, this instrument generally clarifies the project's global objectives but seldom makes explicit the objectives of the project's other partners. In its typical form, it does not establish a mutual engagement among the parties, according to their respective interests. To truly develop confidence and create a relationship between researchers and farmers, it seems essential that potential issues and expectations among the different stakeholders be fully clarified in the document, and that they are formulated in terms of commitments. In and of itself, such a tool does not enable the recognition of a role for farmers in research procedures: it remains a passive agent from whom a simple agreement of principle is required for a project that in the end has little to do with him. The need to manage the discrepancy between PIC as process and its implementation as a formal administrative procedure is in line with existing international guidelines[8] and echoes with a burgeoning literature. In particular, Marchegiani et al (2020) in a different context (mining in South America) have showed the importance of the reliability, accessibility and framing of information and the need for continuous, good-faith dialogue on free and prior informed consent and benefit-sharing.

Several propositions for overcoming the sense of injustice were focused on access to information about current procedures and regulations. In the propositions, this information is seen as a socio-political resource that offers farmers some understanding of the procedures that concern them but for which they had no part in defining. As noted by Marchegiani et al (2020), the power to shape the relationship lies overwhelmingly with the ones who have the resources to define which communities they will consult with, for how long and in what terms. Considering informational justice as a dimension of interactional justice highlights its relational aspect. It makes it possible to reintroduce the question of inequity in access to information and the way it involves stakeholders, such as the researcher's access in relation to that of other stakeholders.

Recognizing cognitive plurality

Some spectators also point to the central role of the researcher in recognizing the farmers' contribution in terms of creating and conserving crop varieties. Many exchanges over the course of the forums involved the traceability of seeds and farmers' access to them in the gene banks. The process of redefining seeds as genetic resources in order to stock them in a gene bank is a product of the anonymization of those seeds and thus a loss in the value and knowledge that they represent. Boltanski (2009) emphasizes that

defining a relationship as equitable or inequitable supposes, first, a definition of what gives things and people value. For things being paired to individuals, the values associated with one are equated to the other. By the same process, subtracting the socio-cultural dimension of the seeds, their origin and thus their identity, engenders a negation of the farmers' cognitive and experiential value. It is in this sense that the situation is considered unfair and inequitable, even beyond the issue of the farmers' access to the resources of the gene banks. Some farmers repeatedly emphasized the role that seeds play as a living entity, a full member of the community:

> We think of the seeds as living matter, and that is the problem. By virtue of the fact that they are thought of as living matter, they become part of our society, of our lives, our daily routine; they are an integral part of our tradition, they enable us to live.
> (Head of the association of farm seeds)

The traceability of seeds in the gene banks is intimately linked to the idea of cognitive justice (Leach and Scoones 2006; van der Velden 2009). By recognizing the plurality of knowledge systems, this principle aims to insure equal treatment of the different ways of understanding the world. As such, cognitive justice emphasizes the inherent right of different knowledge systems to exist in the framework of discussion and debate (Visvanathan 2005, p. 92 in De Jonge 2011). In addition to its explicative role in the discussion of knowledge-system plurality, the traceability question reflects the issue of a true accounting of the ontological plurality associated with the entity that circulates among stakeholders and spaces (from the village to the gene bank by way of the laboratory). Seen as a living, cultural entity, the seeds transform into a biological entity, isolable as a genetic resource. Minimally, the seeds' traceability would involve the reinsertion into ex situ conservation spaces of the socio-cultural dimensions associated with the seed and therefore being constitutive of genetic resources. To define the modalities of seed traceability in the forum-theater play, a researcher takes on the role of a West African gene bank director speaking to a researcher.

> You would need to give me the name, but also where you collected them…It will be necessary to note the geographically referenced coordinates related to the place where you collected them and then the way they were used, the type of usage for which these resources are used locally…it would be necessary to know more to be able to reference them correctly and if ever they ask us, to have the ability to find them, to then be able to restore the material
> (Researcher in political science playing the role of a gene bank director speaking to a researcher)

---

[8] CBD, Mo'otz Kutal Voluntary Guidelines for the Developments of Mechanisms, Legislation or Other Appropriate Initiatives to Ensure the 'Prior and Informed Consent', 'Free, Prior and Informed Consent' or 'Approval and Involvement', Depending on National Circumstances, of Indigenous Peoples and Local Communities for Accessing their Knowledge, Innovations and Practices Relevant for the Conservation and Sustainable Use of Biological Diversity, and for Reporting and Preventing Unlawful Appropriation of Traditional Knowledge, CBD decision XIII/18 (2016), https://www.cbd.int/decisions/cop/13/18/6

A spectator noted that the absence of an international legal framework could now constitute an opportunity for gene banks to define their own internal regulations. These regulations could stipulate that the seed placed in the bank remain accessible to the concerned researchers and farmers.

The recognition of knowledge and the respect for farmers' contributions to the creation of varieties are submitted as essential elements in the definition of fair and equitable situations. These dimensions appear in the farmers' remarks as one of the principal benefits expected from the collaboration with researchers. Having been asked a question about the advantages that a farmer can anticipate when collaborating with a researcher, a Senegalese farmer who is president of an association of farm seeds responded:

> It is fairly complex and also difficult to determine the benefits, for me, the most important is to first discuss the coexistence, how we are going to coexist together in a process and how we are going to define the process from beginning to end, it will be necessary to define things in advance and those involved must each try to see what his commitment is, and in the end, what will be the contribution of the others, that is what is more important for me right now and once that is defined one can really think about the benefits, but in my opinion, once the partnership framework is defined at the level of consistency, for me this is the first benefit...
> Yes, for me, it is very important because very often this is not defined enough, and also in the definition it is necessary that each of the parties respect the vis-à-vis, and above all, in an ethical framework, that is very important, and there, that we at least get away from a dominator and dominated structure, in my opinion, once this is defined and clear, I don't see what would keep the two from working together.

This farmer thus defines certain expected benefits in terms of respect for commitments, for differences, and for a well-intentioned relationship between the researcher and the farmer. The issue here in terms of respect is one of overcoming the relationships of power between the different stakeholders implicated, and particularly between the researchers and farmers. We can see how distributive justice can be linked to the dimensions of interactional justice in collaborative research projects. The notion of benefits appears in the exchanges in terms of socio-political resources (such as access to information, increasing competence, and the power of action) and socio-emotional resources (such as fulfillment of needs for respect and independence, the possibility to influence, and the perpetuation of values).

Conclusion

This research was motivated by the need to move from a theoretical and normative reasoning for the concepts of fairness and equity to a practical reasoning that accounts for the diversity of knowledge systems and the constituent relational asymmetries in the context of collaborative research on plant genetic resources.

To do this, we used the forum theater as a device to examine these principles of fairness and equity. This theater device made it possible to highlight an ensemble of criteria relative to interactional justice and to clarify expectations in terms of socio-political and socio-emotional benefits. In a context characterized by a high level of political and socio-cognitive inequality (in terms of access to decision-making authorities, global knowledge, power relationships between knowledge systems, etc.), these resources appear essential to the definition of fair and equitable situations from the perspective of the different parties involved. It therefore seems that the different instruments developed by the international treaties for operationalizing these principles cannot suffice in themselves. Being reduced to simple generic codes of good practices, they are not able to guarantee collaborative practices considered to be fair and equitable by the different parties involved. The operationalization of the ethical principles requires the consideration of the cultural, social, historical and political contexts in which the collaborative research projects are conducted. The explorations resulting from the play contributed to the formalization within another participatory research project[9] in 2019, of a handbook on ethical collaborative practices between farmers' and research organizations. This handbook has also benefitted from the experiences of actual collaborative agreements defined by the French and West African research centers involved in the project. Such initiative allowed bridging with institutional practices through exchanges with the administrative support and contract services of these research centers and identification of shortcomings and gaps in existing collaborative practices. The operationalization of the ethical principles calls also for the principles of fairness and equity to be conceived not only as formalized techniques, but as the embodied practices and competencies of the various stakeholders involved. To this end, the play is still mobilized by members of the network in varied instances. From the point of view of farmers' organizations, the medium not only gives farmers a voice, but also strengthens their awareness of institutional issues

---

[9] Research project of the Agropolis Foundation *"Adaptive Governance of Coexistence of Crop Diversity Management Strategies"* (CoEx) (ANR-10-LABX-001–01).

related to research and their ability to express their needs to research partners. Likewise, researchers use the video of the forum theater in post graduate school and teaching for students working on genetic resources. At the end of screening, learners are invited to experiment with role-playing situations. It thus participates in the development of a reflective posture for young researchers. Besides, it offers them a methodological framework for the co-definition of practical ethical conditions in participatory research context.

The arenas reveal by the forum theater sessions are related to spaces of joint investigation, and equally important, of collective learning. Using theatrical staging as a representative language, these spaces help reduce the inequalities in cognitive resources between researchers and other stakeholders who are less accustomed to mobilizing these resources in conventional deliberation. This language offers a non-discriminatory space for listening, and for recognizing knowledge that has not necessarily found an explicit formulation in an academic or political context. The forum theater experiences also encourage new perspectives: through the inquiry process, participants focus their attention to new issues and explore them from different points of view. This posture enables the collective identification of sources of injustice, and the clarification of shared responsibilities vis-à-vis these situations in order to explore new collaborative modalities. The situation's definition in terms of shared responsibility requires each party to consider the situation from the perspective of the other participants, and to take part in an effort to improve it. This means assuming a responsibility in relation to what is observed, what is felt, and act on it with the intention of responding to the needs identified. These collective investigations thus participate in a more thoughtful, dialogical and collaborative practice of ethics (Aiguiers and Cobbaut 2016), in other words, a shift from a prescribed duty to act to an ethical knowledge of how to act.

Acknowledgements The authors gratefully acknowledge Claire Billot and Diegane Diouf for their contributions to the construction of the play. The involvement of all the actors of the Kaddu Yaraax theater company contributed to the quality of the collective debates. We also thank the organizers of farmers' seed fairs, such as the ASPSP (Association Sénégalaise de Semences Paysannes), who have made it possible to involve a wide variety of stakeholders in the forums. We would like to thank all the participants in the various forum theater performances, that made it possible to carry out a collective analysis on collaborative research practices in the field of plant genetic resources. This work has benefited from the support of the West Africa Agricultural Productivity Program (WAAPP/PPAAO 2A) through the Fonio project (CERA58ID06 SE), the CERAO projects (ANR-13-AGRO-0002), and the CoEx project that was publicly funded through ANR (the French National Research Agency) under the "Investissements d'avenir" programme with the reference ANR-10-LABX-001–01 Labex Agro and coordinated by Agropolis Foundation.

References

Aiguiers, G., and J.-P. Cobbaut. 2016. Le tournant pragmatique de l'éthique en santé : enjeux et perspectives pour la formation. *Journal International de Bioethique et d'ethique des Sciences* 27 (1): 17–40.

Ando, N., and S. Matsuda. 2010. How employees see their roles: The effect of interactional justice and gender. *Journal of Service Science and Management* 3: 281–286.

Banks, M. 2001. *Visual methods in social research.* London, UK: Sage.

Barling, J., and M. Phillips. 1993. Interactional, formal, and distributive justice in the workplace: An exploratory study. *The Journal of Psychology Interdisciplinary and Applied* 127 (6): 649–656.

Barthe, Y., D. de Blic, J.-P. Heurtin, É. Lagneau, C. Lemieux, D. Linhardt, C. Moreau de Bellaing, C. Rémy, and D. Trom. 2013. Pragmatic sociology : A user's guide. *Politix* 103 (3): 175–204.

Bies, R.J. 2005. Are procedural justice and interactional justice conceptually distinct? In *Handbook of organizational justice*, ed. J. Greenberg and J.A. Colquitt, 85–112. Mahwah, NJ, US: Lawrence Erlbaum Associates Publishers.

Bies, R.J. 1987. The predicament of injustice: The management of moral outrage. *Research in Organizational Behavior* 9: 289–319.

Bies, R.J., and J.S. Moag. 1986. Interactional justice: communication criteria of fairness. *Research on Negotiation in Organizations* 1: 43–55.

Boal, A. 1985. *Theatre of the oppressed.* New York, NY: Theatre Communications Group.

Boltanski, L. 2009. *De la critique. Précis de sociologie de l'émancipation.* Paris: Gallimard NRF Essais.

Boltanski, L., and E. Chiapello. 2007. *The new spirit of capitalism.* London, UK: Verso Books.

Boltanski, L., and L. Thevenot. 2006. *On justification: Economies of worth.* NJ, Princeton University Press: Princeton.

Bréviglieri, M., and J. Stavo-Debauge. 1999. Le geste pragmatique de la sociologie française. *Antropolítica* 7: 7–22.

Byström, M., P. Einarsson, and G. A. Nycander. 1999. *Fair and equitable. sharing the benefits from use of genetic resources and traditional knowledge.* Technical Report, ed. Swedish Scientific Council on Biological Diversity. Stockholm.

Cheyns, E. 2014. Making "minority voices" heard in transnational roundtables: the role of local NGOs in reintroducing justice and attachments. *Agriculture and Human Values* 31: 439–453.

Colleyn, J-P. 2011. *De l'anthropologie visuelle.* L'HOMME no 198 and 199.

Collie, T., G. Bradley, and B.A. Sparks. 2002. Fair process revisited: differential effects of interactional and procedural justice in the presence of social comparison information. *Journal of Experimental Social Psychology* 38: 545–555.

Conrad, D. 2004. Exploring risky youth experiences: popular theatre as a participatory, performative research method. *International Journal of Qualitative Methods* 3 (1): 12–25.

Dai, L., and H. Xie. 2016. Review and prospect on interactional justice. *Open Journal of Social Sciences* 4: 55–61.

De Jonge, B. 2011. What is fair and equitable benefit-sharing? *Journal of Agricultural and Environmental Ethics* 24 (2): 127–146.

Denis, B. 2009. Acting up: Theater of the oppressed as critical ethnography. *International Journal of Qualitative Methods* 8 (2): 65–96.

Dewey, J. 1938. *Logic: The Theory of Inquiry.* New York, NY: Henry Holt and Company.

Ela, J.-M. 2007. *Les Cultures africaines dans le champ de la rationalité scientifique.* Paris, FR: L'Harmattan.

Engels, J.M.M., H. Dempewolf, and V. Hensons-Appollonio. 2011. Ethical considerations in agrobiodiversity research, collecting, and use. *Journal of Agricultural and Environmental Ethics* 24 (2): 107–126.

Fall, M.A. 2011. Décoloniser les sciences sociales en Afrique. *Journal des Anthropologues* 124 (125): 313–330.

Graddy, G. 2013. Regarding biocultural heritage: In situ political ecology of agricultural biodiversity in the Peruvian Andes. *Agriculture and Human Values* 30 (4): 587–604.

Greenberg, J. 1993. Stealing in the name of justice: Informational and interpersonal moderators of theft reactions to underpayment inequity. *Organizational Behavior and Human Decision Processes* 54: 81–103.

Hennion, A. 2017. Attachments, you say? How a concept collectively emerges in one research group. *Journal of Cultural Economy* 10 (1): 112–121.

Heras, M., and J.D. Tabara. 2014. Let's play transformations! performative methods for sustainability. *Sustainability Science* 9: 379–398.

Latour, B. 1988. *The pasteurization of France.* Cambridge, MA: Harvard Univ. Press.

Lavery, J.V., and C. IJsselmuiden. 2018. The research fairness initiative: Filling a critical gap in global research ethics. *Gates Open Research* 2: 58–68.

Leach, M., and I. Scoones. 2006. *The slow race.* London, UK: DEMOS.

Ly, B. 1966. *L'honneur et les valeurs morales dans les sociétés ouolof et toucouleur du Sénégal.* Thèse de doctorat: Paris, Université de Paris I.

Marchegiani, P., E. Morgera, and L. Parks. 2020. Indigenous peoples' rights to natural resources in Argentina: the challenges of impact assessment, consent and fair and equitable benefit-sharing in cases of lithium mining. *The International Journal of Human Rights* 24 (2–3): 224–240.

Morgera, E., L. Kramer, and E. Orlando, eds. 2018. Fair and equitable benefit-sharing. In: *Principles of Environmental Law.* Cheltenham, UK: Edward Elgar Publishing Limited.

Müller, B., C. Pasqualino, and A. Schneider. 2017. *Le terrain comme mise en scène.* Lyon, FR: Presses Universitaires de Lyon.

O'Faircheallaigh, C. 1998. Resource development and inequality in indigenous societies. *World Development* 26 (3): 381–394.

Parks, L. 2018. Challenging power from the bottom up? Community protocols, benefit-sharing, and the challenge of dominant discourses. *Geoforum* 88: 87–95.

Richard-Ferroudji, A. 2011. Limites du modèle délibératif : Composer avec différents formats de participation. *Politix* 4 (96): 161–181.

Thévenot, L. 2001. Pragmatic regimes governing the engagement with the world. In *The practice turn in contemporary theory*, ed. K. Knorr-Cetina, T. Schatzki, and V.S. Eike, 56–73. London, UK: Routledge.

Tyler, T.R., and R.J. Bies. 1990. Beyond formal procedures: The interpersonal context of procedural justice. In *Applied social psychology and organizational settings*, ed. J.S. Carroll, 77–98. Hillsdale, NJ: Erlbaum.

Van der Velden, M. 2009. Design for a common world: On ethical agency and cognitive justice. *Ethics and Information Technology* 11 (1): 35–47.

Vermeylen, S. 2007. Contextualizing "fair" and "equitable": The San's reflections on the hoodia benefit-sharing agreement. *Local Environment* 12 (4): 423–436.

Visvanathan, S. 2005. Knowledge, justice and democracy. In *Science and citizens: Globalization and the challenge of engagement*, ed. M. Leach and I. Scoones, 83–94. London, UK: Zed Books.

Wagué, C. 2012. La fabrique d'un savoir scientifique sur sa communauté : témoignage d'une relation au terrain. In *L'Afrique des savoirs au Sud du Sahara (XVI-XXI siècle) – Acteurs, supports, pratiques*, ed. D. Gary-Tounkara and D. Nativel, 313–331. Paris, FR: Karthala.

Whiteman, G., and K. Mamen. 2002. Examining justice and conflict between mining companies and indigenous peoples: Cerro Colorado and the Ngabe-Bugle. *Journal of Business and Management* 8 (3): 293–310.

Publisher's Note Springer Nature remains neutral with regard to jurisdictional claims in published maps and institutional affiliations.

Frédérique Jankowski socio-anthropologist, is a Senior Research Fellow at the Centre International de Recherche Agronomique pour le Développement (CIRAD, Montpellier, France) where she is part of an interdisciplinary team working on the collective decision-making processes that structure the socio-ecological systems. Her work focuses on the socio-cognitive and political dimensions of participatory management of renewable resources and territories.

Sélim Louafi is a Senior Research Fellow at the Centre International de Recherche Agronomique pour le Développement (CIRAD, Montpellier, France) where he is part of a team of biologists and geneticists to work on science and policy interface in the field of agricultural biodiversity. From 2007 to 2009, he served as Senior Officer at the Secretariat of the International Treaty on Plant Genetic Resources for Food and Agriculture (FAO) where he was in charge, amongst other things, of the implementation of the Multilateral System of Access and Benefit Sharing.

Ndjido Ardo Kane is a plant geneticist, researcher at the Senegalese Institute for Agricultural Research (ISRA), Director of the Regional Center for the Improvement of Adaptation to Drought (CERAAS), Co-director of the international laboratory for the adaptation of plants and associated microorganisms to environmental stresses. Expert in genetic resources management and biosafety, he coordinated from 2013 to 2018 the national program "Management of agrobiodiversity and biotechnologies".

Mouhamadou Diol is the director of the forum theater troupe Kaddu Yaraax created in 1994 in Dakar. The company offers performances and forum theater workshops on social and environmental themes in Senegal. Her interventions can accompany community mobilizations, integrate into development or research projects. The company offers forum theater training and has been organizing the Senegalese forum theater festival for the past fourteen years, which welcomes theater groups from all continents.

Astou Diao Camara is a socio-pastoralist, researcher at Senegalese Institute for Agricultural Research (ISRA) and Coordinator of Pastoralist Pole and Dry Areas (PPZS). She capitalizes ten years of experience of research on family farms and support of agricultural civil society. Over the last five years, she has coordinated several research operations and expertise, particularly on participatory approaches to support actors from various categories.

Jean-Louis Pham plant population geneticist, is a Senior Research Fellows at the French National Research Institute for Sustainable Development (IRD, Montpellier, France). He coordinated the development of the Agropolis Resource Centre for Crop Conservation, Adaptation and Diversity (ARCAD). He leads the IRD (Nagoya Committee) on Access to genetic resources and Benefit-Sharing.

Cécile Berthouly-Salazar plant population geneticist, is a Senior Research Fellows at the French National Research Institute for Sustainable Development (IRD, Montpellier, France). She is interested in pearl millet adaptation to ongoing climate changes.

Adeline Barnaud plant population geneticist, is a Senior Research Fellows at the French National Research Institute for Sustainable Development (IRD, Montpellier, France). She explores crop evolutionary history and adaptation in response to both social and environmental drivers.

Agriculture and Human Values (2020) 37:1095–1109
https://doi.org/10.1007/s10460-020-10113-w

A carrot isn't a carrot isn't a carrot: tracing value in alternative practices of food exchange

Galina Kallio[1]

Accepted: 8 May 2020 / Published online: 15 May 2020
© The Author(s) 2020

Abstract

Questions of value are central to understanding alternative practices of food exchange. This study introduces a practice-based approach to value that challenges the dominant views, which capture value as either an input for or an outcome of practices of exchange (value as values, standards, or prices). Building on a longitudinal ethnographic study on food collectives, I show how value, rather than residing in something that people share, or in something that objects have, is an ideal target that continuously unfolds and evolves in action. I found that people organized their food collectives around pursuing three kinds of *value-ideals,* namely good food, good price and good community. These value-ideals became reproduced in food collectives through what I identified as *valuing modes,* by which people evaluated the goodness of food, prices and community. My analysis revealed that, while participating in food collectives in order to pursue their value-ideals, people were likely to have differing reasons for pursuing them and tended to attach different meanings to the same value-ideal. I argue that understanding how value as an ideal target is reproduced through assessing and assigning value (valuing modes) is essential in further explorations of the formation of value and in better understanding the dynamics of organizing alternative practices of food exchange.

Keywords Value · Alternative food practices · Practice-based approach · Ethnography · Food collectives

Introduction

The past two decades have shown a rapid increase in alternative ways of organizing food supply from farm to fork that challenge the industrial food system, which is sustained by practices of mass production, distribution and consumption (Goodman et al. 2012; Holt Giménez and Shattuck 2011). Simultaneously, grassroots food movements have called into question conventional quality standards established predominantly by a small number of experts, industry representatives or political authorities, and introduced alternative approaches to valuing "good food" (Goodman 2003; Pollan 2010). These have included, for instance, promoting new standards and pricing mechanisms (Raynolds 2000; Reinecke et al. 2012), developing new discourses and labels (Pratt 2007; Van Bommel and Spicer 2011), and establishing

new practices of local food exchange (Brunori et al. 2012; Hinrichs 2000; Werkheiser and Noll 2014).

Questions of value are essential in all market exchanges (Helgesson and Kjellberg 2013), but appear particularly interesting in the context of alternative food practices (Dahlberg 1988; Forssell and Lankoski 2015). In moving away from globalized impersonal markets towards direct exchanges, people need to establish new quality standards and ways of valuing (Weber et al. 2008). On the one hand, studies have shown how formation of value is contingent on *values,* such as cultural frames, norms and discourses that motivate action (Doran 2009; Thompson and McDonald 2013). On the other hand, scholars suggest that value is formed through the properties of an object (such as its qualities or characteristics) as captured by labels and standards, or through a price reflecting multiple attributes or values attached to an object (Miller 2008; Reinecke and Ansari 2015; Van Bommel and Spicer 2011). But while these studies have broadly explored the formation of value and shed light on various valuation practices (for reviews, see e.g. Kjellberg et al. 2013; Lamont 2012), scholars have tended

✉ Galina Kallio
  galina.kallio@helsinki.fi

[1] Ruralia Institute, Faculty of Agriculture and Forestry, University of Helsinki, Lönnrotinkatu 7, 50100 Mikkeli, Finland

to treat value as either an *input for*, or an *outcome of* human practice (Muniesa 2011; Orlikowski and Scott 2013).

Motivated by examining value not as a fixed variable, but as collectively enacted accomplishment (Gherardi 2009; Graeber 2001), I mobilize a practice-based approach (Gherardi 2012; Sandberg and Tsoukas 2011) to examine how value is formed in alternative food practices. Due to the exploratory nature of the study, this paper is based on an open-ended and emergent research design (Wiedner and Ansari 2017) with insights arising from a longitudinal ethnographic study on food collective organizations in Finland. Food collectives are groups of households who procure local and organic food directly from small-scale farmers and other types of food suppliers and distribute it among the participating members. Food collectives provide an excellent setting for studying the formation of value, because they have emerged in a situation in which commonly agreed standards for local food as an object of exchange, or regulations for food collectives as an alternative practice for food exchange, have not existed. Therefore, food collective members have needed to work out for themselves what to value and how.

This study found that food collectives were organized around three *value-ideals*: good food, good price and good community. These value-ideals are contextual and continuously unfolding signifiers that guide action. In food collectives, value was formed through what I identified as *valuing modes* that engaged people in continual evaluation of the goodness of the value-ideals. While enabling people to evaluate the goodness of food, prices and community, valuing modes simultaneously functioned as ways of re-producing these value-ideals. I argue that this dynamic movement between evaluating and re-producing value-ideals is essential in understanding the formation of value. This study contributes to a better understanding of value as constituted in action (Graeber 2001; Heuts and Mol 2013; Hutter and Stark 2015; MacIntyre 2008) and adds to the previous research by problematizing the general belief that value resides in something that people share, or in something that objects have.

Formation of value in practices of exchange

Questions of value have attracted attention within a plethora of different academic fields (Aspers and Beckert 2011; Lamont 2012; Otto and Willerslev 2013). Research within the fields of economic sociology and anthropology question the dominance of economic approaches that equate (exchange) value with price (Graeber 2001; Muniesa 2011) and argue that in the economic sphere, like in other spheres of life, the formation of value should be understood as a culturally and materially mediated process. In this vein, an increasing amount of research labelled as valuation studies (Helgesson and Muniesa 2013; Lamont 2012), and research

within economic sociology and anthropology more broadly, has suggested that scholars should look into how various practices enable people to assign or assess value (Fourcade 2011; Graeber 2001; Kjellberg et al. 2013).

Understanding how value is formed is central to understanding alternative practices of exchange (Dahlberg 1988; Hughes 2005; Parker et al. 2014). Prior research has, for instance, found that prices do not merely measure the value of the product but mediate various moral, environmental and ethical values (Fourcade 2011; Zelizer 1978) and thereby lead to action. A good example is Fair Trade pricing, which signals valuing the work of farmers, but has also been suggested to reflect the values of those buying the products (Doran 2009; Murray and Raynolds 2007). Similarly, other types of valuing schemes such as eco-labels and quality rankings enable the production and categorization of value (Brunsson et al. 2012; Karpik 2010; Lamont 2012). At the same time, however, higher prices for specialty products like organic and fair-trade food may signal "elite" values and inhibit action despite initial motivations that people have towards buying "ethical" food (Johnston et al. 2011). In this sense, while standards and pricing enable the creation of value through certifying and framing activities, they may also come to represent unwanted, controversial or "false" values (DeLind 2011; Pratt 2007).

While the existing literature has shown that values play a significant role in alternative food practices by motivating and guiding action (Brunori et al. 2012; Diekmann and Theuvsen 2019; Seyfang 2007; Weber et al. 2008), there seems to be a tendency to try to capture value in the form of shared values (ethical, environmental, economic, etc.) or in the value of an object (price, quality ranking, eco-labels) and thereby see "value" as more or less stable. This is perhaps because many scholars build on either economic or sociological premises that treat value as the product of, or the starting point for, valuing practices (Aspers 2008; Aspers and Beckert 2011; Muniesa 2011; Orlikowski and Scott 2013). This, argues Graeber (2001, p. 49), makes it difficult to imagine "a theory of value starting from the assumption that what is ultimately being evaluated are not things, but actions". Along with anthropological scholarship (Graeber 2001, 2013; Otto and Willerslev 2013), sociologists have also called for a more comprehensive understanding of value and proposed studying value as a verb ("to value") rather than as a noun (Hutter and Stark 2015; Kjellberg et al. 2013).

Through empirical investigations, anthropological research has provided important insights into how valuing occurs in different practices of exchange (Malinowski 2002; Mauss 1954). Understanding these practices is crucial since, as several scholars suggest, they reveal that it is not the object per se but the reproductive action (rituals, practices) around these objects that is valued (Graeber 2001; Lambek 2013; Mauss 1954). Thus, by showing how objects like food ultimately end up constituting social relations, scholars have problematized the common assumption that it is the value of these objects, referred to as exchange or use value, that is produced in and through practices of exchange. In other words, understanding the formation of value requires us to look beyond *what* is being exchanged (the object of exchange), and explore *how* exchange happens in practice.

A good example of the importance of understanding the performative nature of value is the concept of local that appears central to understanding value in alternative practices of food exchange (Feagan 2007; Hinrichs 2003; Werkheiser and Noll 2014). Several studies suggest that "local" is not a mere quality of a product, or an attribute of a community, but commonly refers to the nature of relationships formed with the food and between the people (Albrecht and Smithers 2018; Delind 2006; Hinrichs 2000; Trivette 2017; Weber et al. 2008). These studies show how, through the formation of relationships, a sense(s) of closeness, connectivity, and trust local is produced in practice.

In order to better understand the performative nature of value (Graeber 2001; Hutter and Stark 2015; Lambek 2013), I adopt a practice-based approach emphasizing everyday action as the primary source of knowing and theorizing (Sandberg and Tsoukas 2011; Schatzki 2001a). Practices can be understood as recurrent patterns of socially sustained and materially mediated action (Schatzki 2001a). They form what Nicolini (2009b) refers to as the "sites of knowing" that provide a context for people to make judgments on what is to be held as good or bad, appropriate or inappropriate, beautiful or ugly, valuable or worthless—and how these assessments should happen in practice (Gherardi 2009; Heuts and Mol 2013). Several practice scholars point towards a definition of practices which emphasizes that practices are normatively shared and assume a common understanding among the participants of the ends (aims, values) of the activity that constitutes a practice (Gherardi 2011; Schatzki 2001b). Exploring value from a practice-based approach, then, involves examining what ends people pursue through their practice and how they evaluate that what is collectively held as good, valuable, and worth pursuing (Hutter and Stark 2015; MacIntyre 2008; Thévenot 2001).

Methodology and the context of the study

Along with providing theoretical tools, the practice-based approach extends to methodological considerations on how to study practices (Miettinen et al. 2009). This study draws on a 6-year ethnographic study among 22 food collectives in Finland conducted over the years 2010–2016. Food collectives are groups of households who procure local and/or organic food directly from farmers and other small-scale food suppliers such as hunters, gardeners, fishermen, or

"mushroom grannies" and distribute it among the members of the collective. Many food collectives also supplement their supply with purchases from organic wholesale distributors. Food collectives operate on a non-profit basis and are located in specific neighbourhoods, regions, or cities. Legally, the majority of these organizations are not formal in the sense of being registered as non-governmental organizations or co-operatives.

Food collectives can be found across the country, with the densest concentration in the capital region. Furthermore, their sizes vary, accounting for anywhere from a dozen to a couple of hundred households. Food collectives bring together several interconnected practices like farming, ordering, transporting, distributing, and cooking, among others that need to be coordinated in order to enable the exchange of food on a regular basis. To do this, the members of each food collective establish and maintain relationships with suppliers, negotiate terms of delivery, create ordering systems and manage orders, distribute the food, and communicate among the participating members and with the farmers, among other tasks.

Unlike many other practices of local and organic food exchange, such as farmer's markets, food co-ops or food box deliveries (CSA shares), food collectives are initiatives organized primarily by households, not farmers. Moreover, as orders are placed collectively, it is not individuals who engage in exchange but the collective as a whole. The households thus organize the entire web of interconnected practices from production to consumption and are in charge of the whole value chain. This requires a lot of work and forces people to reflect on what is held valuable in food collectives, how, and why. These factors make food collectives very interesting sites for studying the formation of value.

Fieldwork and data sources

My fieldwork included participant observation, site observation and shadowing, open-ended in-depth interviews and informal conversations, and following social media discussions. I supplemented these sources with information from food collectives' web pages and member-surveys conducted by the collectives. These methods allowed me to gain deep insight into the practices of food collectives (Gherardi 2012; Nicolini 2011).

Participation, observation and shadowing

Participating in and observing the everyday activities of food collectives provides the basis for the analysis. As a participant in three different food collectives, varying in sizes and locations, I was able to actively follow how the procurement and distribution of food as well as communication and coordination were organized in the collectives. I began observing the first food collective in spring 2010. As a member of the collective, I participated in the weekly orders and pick-ups of food, and voluntary work, such as receiving food deliveries, sorting and (re-) packing food, guiding people at the pick-up events, and sanitation work. I got to know the coordinators and had the opportunity to shadow (Czarniawska 2014) private moments of planning that dealt with the food collective's organizational tasks.

After 6 months, I joined a much smaller food collective located in my own neighborhood. This collective was "closed" to people outside the neighborhood and could only be joined by invitation from one of the members. This collective required more intensive volunteering that made me switch to the third collective, which I studied as a participant observer for 5 years. At this stage, I also began to search for new farmers and help with (re-) organizing tasks related to changes in distribution places or product selection. Additionally, I conducted on-site observations 1–2 days long in five other food collectives.

Interviews and informal conversations

In addition to several informal discussions, I conducted 25 open-ended interviews among 22 food collectives with food collective founders, coordinators, and members (see Table 1). These interviews lasted from half an hour to three hours. With the aim of gaining better understanding of what people valued in food collectives and how their particular food collective functioned in regard to what they valued, I began each interview with an open inquiry about their reasons for participating in a food collective. I then asked the respondents to reflect concretely on how their food collective functioned and how they personally engaged in the collective, and what were the "goods" and "bads" of procuring food this way. I soon discovered that the respondents brought up similar themes (reasons for participating in food collectives, vivid stories about encounters with the food, challenges in the practicalities related to organizing a collective) and paid close attention to asking follow-up questions on these and guiding the respondents to describe the related activities in concrete terms. In some cases, I followed-up the interviews and discussions by e-mail to gather additional information.

Table 1 Data on food collectives

Food Collective (type)	Location	Members	Distribution	Available data
FC 1 (organic/local)	Capital region	300	1/week	Observation 6 months, shadowing 2 days, 2 interviews, informal conversations, emails, photos, web page
FC 2 (local/organic)	Capital region	330	1/week	Observation 5 years, 3 interviews, informal conversations, Facebook, emails, 3 surveys, photos, web page
FC 3 (local/organic)	Capital region	25	1/week	Observation 6 months, 1 interview, informal discussions, emails, photos
FC 4 (local/organic)	Southern Finland	31	1/month	Observation (distr. event), 1 interview, informal conversations, Facebook, photos
FC 5 (local/organic)	Southern Finland	30	1/month	Observation (distr. event), 1 interview, informal conversations, photos
FC 6 (local/organic)	Capital region	80	1/week	Observation (distr. event), 1 interview, informal conversations, emails, phone calls, Facebook, photos, 2 surveys, web page
FC 7 (organic/local)	Southern Finland	100	1/month	1 interview, email conversations, photos, web page
FC 8 (local/organic)	Eastern Finland	70	1/month	1 interview, photos, web page
FC 9 (local/organic)	Eastern Finland	250	1/month	1 interview, photos, web page
FC 10 (local/organic)	Central Finland	40	1/month	1 interview, web page
FC 11 (local/organic)	Capital region	60	1/month	1 interview, photos, Facebook, web page
FC 12 (local/organic)	Capital region	92	1/week	1 interview
FC 13 (local/organic)	Capital region	15	1/week	1 interview (2 interviewees), photos
FC 14 (organic/local)	Northern Finland	12	1/month	Observation (distr. event), group interview, informal conversations, photos
FC 15 (local/organic)	Eastern Finland	100	1/month	1 interview, Facebook
FC 16 (local/organic)	Northern Finland	10	1/month	1 interview
FC 17 (organic/local)	Southern Finland	60	2/month	1 interview
FC 18 (organic/local)	Western Finland	120	1/month	1 interview, photos, web page
FC 19 (local/organic)	Southern Finland	60	1/month	1 interview, emails, phone calls, photos, web page
FC 20 (local/organic)	Southern Finland	18	1/month	1 interview, emails, photos
FC 21 (local/organic)	Capital region	15	1/month	1 interview, emails
FC 22 (local/organic)	Eastern Finland	25	1/month	1 email interview

Social media discussions and online data

During my fieldwork, some of the food collectives began using Facebook for communicating and organizing their everyday activities. I gained access to six closed groups, including members from various food collectives around the country and collected social media discussions from these group conversations. In these discussions, the participants of food collectives, sometimes including the farmers, exchanged information regarding the food and the suppliers, coordinated ordering and distribution, and shared information on various other practicalities. Additionally, I collected the information available on food collectives' websites, which some collectives maintained, in order to obtain an understanding of how a food collective is described to a broader audience.

Data analysis

My fieldwork was driven by a broad question: how and why food collectives were organized around the exchange of organic and local food. At an early stage, it became evident that understanding questions of value was essential in understanding the functioning of food collectives. This was an observation that I brought to the centre of my analysis. In an abductive analytical process (Timmermans and Tavory 2012), I iterated between the data, the existing literature and different theoretical concepts. In studying the formation of value in food collectives' practices, I focused my analysis specifically on both the doings and sayings (Nicolini 2009a) of the participating household members.

In the first stage, I went through all the data looking into what "things", including material, abstract and activity-based things, were valued in food collectives. This task produced 78 first-order codes (coding in the language of the original data) from which I identified three broader categories, namely food-related, price-related, and community-related "things". What sparked my interest was that pursuing "goodness" (of food, prices, and community) seemed to unite all these three categories, which I then named the "core value-ideals in food collectives".

This led to the second phase of the analysis, in which I looked more specifically into individual food collectives, and analyzed *how* people pursued good food, good price and good community through the different practices of food collectives. By identifying the different practices in which people participated (see Appendix 1), I was able to see that people encountered what Hutter and Stark (2015) call "moments of valuation" that engaged them in assessing the goodness of food, prices and community throughout different situations in different practices.

In the third phase of the analysis, I looked specifically into the moments of valuation, which I used to analyze the relationship between *what* people valued (value-ideals) and *how* they assessed their goodness. I identified these as *valuing modes*. Valuing modes refer to actions incorporating both sayings and doings that emerge in and evolve throughout the different interlinked, materially mediated practices of food collectives.

Valuing modes and formation of value in food collectives

Due to the lack of local food standards on the one hand, and benchmarks for exchanging food through a food collective on the other, households and farmers needed to establish for themselves what to value and how. I found that food collectives were organized around pursuing three value-ideals, namely good food, good price and good community. While separated for analytical purposes, in reality these value-ideals were entwined and interdependent. Moreover, not all value-ideals appeared significant in similar ways. In performing the core activities of the collectives, individual people as well as a particular food collective as a group could prioritize one value-ideal over another—with some emphasizing the nature of food, some being more conscious about pricing, and some being driven primarily by communal aspects.

In pursuing good food, good price and good community, people engaged in evaluating their goodness through what I identified as *valuing modes* (see Fig. 1). Valuing modes refer to embodied and materially mediated activities comprising sayings and doings by which people engaged in assessing the value-ideals that were pursued in food collectives.

I found that good food was assessed through the valuing modes of *sensing quality* and *knowing the origin*, good price through *making comparisons* and *reflecting on profits*, and good community through *encouraging participation* and *sharing work*. In the following, I draw on the analysis and show how people assessed the value-ideals throughout the different practices of food collectives through these valuing modes. Each chapter begins with a quote referring to the value-ideal in question, after which the findings on the valuing modes are presented.

Valuing good food—from feeling bad to feeling good

I: This idea that there are these good products here [pointing at the kitchen where distribution is taking place], local food and organic food...and then, good food.
Q: What is good food?
I: Well, pure food.
Q: And what kind of food is that?

Fig. 1 Value-ideals and valuing modes in food collectives

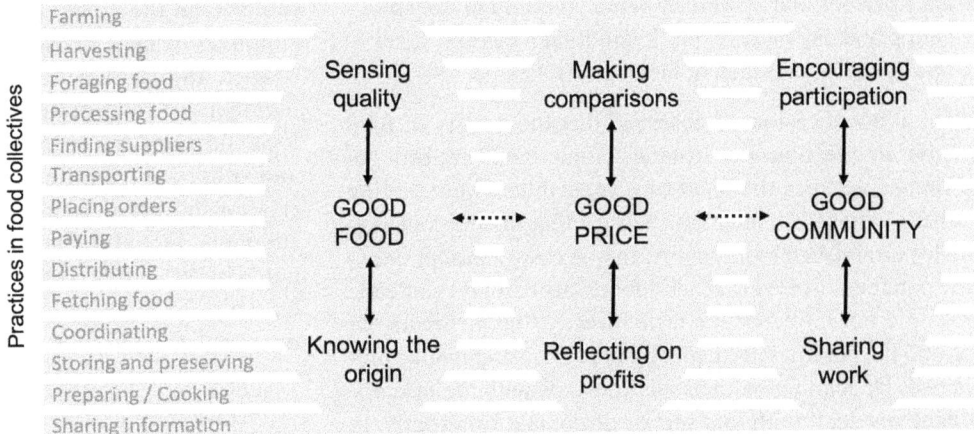

I: Well, that you know where it came from and how it was produced, and then these pure and raw ingredients from which you can prepare good food at home. (field-notes, coordinator, FC5)

Members in all food collectives studied claimed to be pursuing *good food*. They most commonly described good food as being pure, fresh, and tasty—and the kind of food whose origins one knew. Practically, this meant procuring locally-grown and organically-produced food. However, local and organic did not always go hand-in-hand in practice, as an organic label did not guarantee proximate production, and locally grown food was not necessarily certified as organic. Since standards for *local* did not exist, the members could not be reliant merely on external evaluation such as the common (industrial) food quality standards or eco-labels. Instead, in order to assess the goodness of their food, the members used their senses to assess its quality and built connections with the origins of the food.

Sensing quality

The great majority of food entering food collective exchange was unprocessed or very little processed. Generally, the basic supply included (root) vegetables, such as potatoes, carrots, onions, and other row crops, and often also eggs and grain, which formed the basis for farmers' regular deliveries. On top of these, each collective worked out its unique assortment, including specialty items like dairy and meat products, fruit and berries, and other foodstuff that the members wanted. Since most of this food came directly from local farms, its appearance varied greatly in terms of shape and size, variety, and packaging. Additionally, some products were only available through food collectives and could not be found in grocery stores, like certain parts of animals, or raw milk. The non-standardized nature of food and originality of the assortment came to signify good food and thus formed an essential part of its evaluation.

As many of the core practices of food collectives, such as ordering, coordinating and distributing were shared, the quality of food was often collectively checked and explored. Each food collective received the food in bulk, requiring volunteer members to sort and repack the foodstuffs before distributing them among the members who had placed orders. This meant unpacking sacks and boxes of root vegetables covered in soil, using tools to sort the food into smaller packaging, and encountering oddly-shaped vegetables or feathers stuck onto eggshells while re-packing eggs into smaller pods. Through continual exchange of information with farmers and through dealing with the raw nature of food, soil on vegetables came to signify purity, the smell of grain freshness, and feathers on eggs naturalness. Carrots often functioned as the quality proxy for good food; besides their taste, people assessed their colour, size, shape, and the texture of soil they were covered in.

Subsequently, this enabled the members of the collective to share their personal experiences of "dealing" with the food when they were preparing and processing, cooking, and consuming it at home. In food collectives, all the observed and interviewed participants cooked regularly. Dealing with raw ingredients as well as collectively learning about the quality of food enabled people to describe their senses and the qualities of food in very concrete ways, as my conversation with one of the active members revealed:

Q: Do you notice the quality in other [things] than taste? You have told me a lot about how good this [food] tastes…
I: Well, flour for example. I did not believe it when they [the other members] told me that this flour that we order from [name of the farmer] is so good. It has no additives. Grains have been milled fresh and so on, and so people have told me that baking is totally different with it. And at first I thought, "this is nonsense". But now I have become a believer. We have a gas oven, and it is extremely difficult to bake in it. But now, it still

isn't perfect, but it's much better, the dough doesn't slump and dry in the oven. There [when baking] I can feel it. (fieldnotes, active member, FC12)

On a few occasions, I observed that the quality of food was put in question. For instance, some members brought up that sometimes the local root vegetables "were getting leathery" after the winter and claimed that storing root vegetables diminished their quality. In one case, a farmer delivered potatoes covered in black spots. Surprisingly, while the looks did not influence the taste directly, the members did not find the potatoes aesthetically appealing and many complained. Depending on their relationships with the farmers, collectives dealt with this sort of discontent differently. In some collectives, assessments led to changing the suppliers, while in others communicating with farmers led to mutual learning about the why's, what's and how's of evaluating the quality of food.

Finally, evaluating the goodness of food through sensing was as much connected to experiences with the food as to concrete feelings. Many claimed to "have more energy", to "feel good", and some people brought up that they didn't have any allergic reactions, unlike with the majority of vegetables that were sold in grocery stores. However, sensing food was inseparable from knowing its origin.

"Knowing" the origin

It was important for the members of food collectives to know where the food came from and that it had been produced ethically and using organic methods. But assessing these issues was not always straightforward. While sorting out foodstuffs at a distribution, I had the following conversation with another volunteer, revealing the key challenge of evaluating the origin of food:

I: That [this] food is organic is important to me.
Q: Why is organic important?
I: It makes you feel better; you feel good.
Q: Right. So how about local food, do you find it somehow…? [interruption by the volunteer]
I: If it is not organic then I don't think it is any better. But if there is Finnish organic food available, I prefer that.
Q: But what about this [showing a bag of onions] farmer then, he does not have an organic certificate…
I: The certificate is not that important if the food is pure. That I know it [is]. But if I don't know the farmer personally, then I can't know for sure. (fieldnotes, volunteering member, FC1)

In food collectives, assessing organic, which commonly meant "pure", was often a question of trusting the farmer rather than the label. Thus, finding good farms and food suppliers formed an essential part of assessing the origins of food. In practice, the founders, leaders, and other active members of food collectives searched for and contacted local farmers, inquiring about their production methods and other information about the farm. Forming direct relationships with a unique set of farmers and other food suppliers, including "mushroom grannies", fishermen and home bakers, allowed the members in charge of these relationships to repeatedly exchange information and talk to the suppliers. Often, when suppliers delivered their products to the collective's distribution point, they provided information about the harvest and shared details of the production process, like the influence of the weather on the growing season, or characteristics of specific varieties. Most importantly, from time to time, the farmers shared stories about farming and through these opened a window onto their farming philosophy.

Interestingly, however, most of the members of the collectives never met any farmers because the farmers almost never stayed for the actual distribution. This was also the case with the person quoted in the conversation above—a piece of information that appeared contradictory to me. It seemed that, in food collectives, knowing and trusting were inseparable: "knowing" the origin meant on the one hand *trusting that someone within the collective knew*, and on the other hand it meant *having a possibility of knowing*—that is, the opportunity to talk to the farmer, or to visit the farm, as articulated by one of the founders:

Last year we only visited [name of the farm], but many of us are very interested in knowing what the farm is like. Visits give you a totally different picture, or information about farming that you could not transmit in any other way, really. Of course, you can read from the homepage, but there [at a farm], you can see with your own eyes, and you can ask questions and talk with them [farmers]. (interview, founder, FC6)

Traces of information about the origin of food could be found throughout the practices of food collectives. When ordering the food, the members saw the names and contact information of the farms and the suppliers, and received information about varieties and occasionally also about the harvest days. In addition to farmers' stories circulating in food collectives, many were active in following the media and acquiring information about food production from multiple sources. This information was shared during distribution events, in Facebook conversations, and in some cases, when meeting another member on the street. The increased awareness about the origin of food also made people more conscious about its pricing.

Valuing good price—from profits to what is reasonable

Well, the economic side, it's clear that the farmer gets a better price and the household gets a better price. Because there are no margins for the middlemen. And then, prices here [in food collectives] do not reflect the volatility of global market prices. Our prices have been the same all the time. There has been no rise or fall. And we don't expect that either, there is no reason to demand that from the local producer. Middlemen of course buy where it is cheapest and refer to global prices. But that's such a perverted and unfair business practice. (interview, founder, FC10)

Another shared value-ideal in food collectives was *good price*. From the households' perspective, good price meant affordable prices for organic and local food and fair compensation for the work of producers. But cutting out the middlemen and procuring food directly from farmers or through wholesale distributors required extra work from the members of the collectives. Because volunteer labour that enabled lower prices was not measured in monetary terms and because there existed no uniform standard for good food, assessing the goodness of price was not at all straightforward. I found that making comparisons about the prices and qualities of food and striving for transparency regarding the non-profit nature of food collectives appeared central in evaluating the price.

Making comparisons

People participating in food collectives were generally quite conscious about food prices, and had a good idea about what organic and local products cost in supermarkets, marketplaces or specialty stores. Members often tried to use these sales channels as proxies when they compared "regular" prices with the prices they paid in a collective. There existed a common understanding that the prices for organic and local food in supermarkets and in other stores were high and it seemed important for many to emphasize that this was not the case in their collective. Some people calculated, for example, that in a food collective they paid the same amount for organically produced food as they would pay for conventionally produced food in a supermarket. Others suggested that they paid half of what they would have paid if they had bought the same amount of organic/local food in a supermarket.

However, attempting to compare prices disconcerted many people, because in the course of comparing they often ended up concluding that one couldn't find "this food" elsewhere. While some explained that the quality was totally incomparable, others simply referred to the unavailability of organically produced local food in supermarkets:

[…] you can't compare the prices of many of these products because they do not exist in supermarkets. Like back then [when we started] and even now…well, now you might be able to find some sort of organic meat, but not really… And eggs for example, specifically these local ones you can't find… (interview, founder, FC16)

In some cases, people appeared upset about the low prices of food, or consumption habits in general, and seemed to need to justify that what they were doing was actually investing in good food. As the following conversations at a food collective's distribution reveals, comparing prices included broader reflection and evaluation of one's priorities in consuming:

Q: So, how about the price then?
I-1: Well, it's a matter of importance. When something is important, you invest in it. Some consider a car important. For me, this [food] is important, and I am ready to put effort into it, to invest in it.
I-2: Why do people think that food should be so cheap? When beer and booze can cost whatever. Tobacco can cost whatever. But food needs to be cheap. (fieldnotes, FC14)

Due to the lack of uniform quality standards for good food and the very limited availability of organically produced local food in supermarkets, comparing prices was not easy. Additionally, and surprisingly, people's perceptions of good prices extended beyond food to reflect broader understandings of what was important in life in general and worth investing in. As people in food collectives often eschewed for-profit businesses, reflecting on profits and assessing the non-profit nature of collectives was equally important when evaluating the price.

Reflecting on profits

In food collectives, along with pursuing affordable local and organic food, good price meant that farmers would have fair compensation for their work. All the observed and interviewed members wanted to support small-scale farmers rather than have their money go to the profits of the middlemen. Direct exchange relationships meant that people knew that their money went directly to the farmers. Thus, and contrary to market logic, negotiating prices with producers was not common in food collectives. Rather, the members paid what the farmers asked, as expressed by one of the coordinators:

I think it is important not to think that [we buy] cheap at all costs, or that we would run after the producers who sell at the lowest prices. Not like that. (interview, coordinator, FC18)

While most collectives operated without any internal fees, some charged a yearly membership fee or took a small margin in order to cover operating costs. For this reason, and contrary to the operating logic of food collective organizations, some outsiders mistook them as middlemen distributors. In order to justify internal fees and to emphasize their non-profit nature, food collective founders and coordinators talked openly about their financial situations and, as the following conversation reveals, sought to communicate that they sought no profits:

Q: So, do you take some percentage, if you need buy, for example, a scale or if you have other expenditures? A: Yes, we take 2–5% margins to cover some costs. But we don't make any profits by any means. (interview, coordinator, FC20)

Due to transparency regarding fees and margins, the members were able to learn about the income of the producers, and also to reflect on how the income of the collective was spent. Transparent practices also enabled people to trust that no one was cheating or pursuing profits, which in turn influenced the perception that prices were "good". But being dependent on investing not only money but also time in organizing and participating in a food collective made people assess the communal aspects as well.

Valuing good community—from being a consumer to being a member of a collective

For many, communality is important. But how do you get people to participate? That would be important. Even though the main idea of a food collective and the way we operate unites people, it is totally different if people really engage [as volunteers]. The question is, does one want to experience oneself in other ways than just [being] a consumer or a customer? (interview, founder, FC11)

Alongside good food and good price, *good community* was one of the core value-ideals in food collectives. In practice, however, community signified different things for different people. But while the members had varying reasons for participating in a collective, many of the observed collectives were founded in pursuit of communality. Furthermore, the community enabled access to local and organic food at a certain price level and the opportunity to influence one's own food choices, which appeared important for many people participating in a collective. Many felt happy about food collectives uniting likeminded people, bringing together

mothers in the same life situation, or connecting people living in the same neighborhood. Simultaneously, being part of a food collective community required both commitment and work. Community was then evaluated by how it encouraged participation, and how successfully work was shared among the members.

Encouraging participation

Because people seemed to pursue many different things as members of food collectives, how and how much they participated in creating and sustaining the community differed. Due to the nature of collectives, there was great potential to form relationships and catch up on a regular basis. As food collectives operated very locally, people could quite easily get to know others living in the same area. In fact, for some, this appeared to be one of the main reasons for founding a food collective:

One of my main motivations to start a food collective was absolutely communality, to get to know people in my neighborhood. In almost a year that we had lived here [name of the neighborhood] I hadn't actually gotten to know any "neighbors". Now, with this food collective, I have gotten to know other people living here. Even within one week I have been able to greet a person at the bus stop, or when doing groceries. This neighborhood has started to become more my neighborhood. (Facebook, founder of a food collective)

In food collectives, interaction and engaging in a deeper exchange at a more personal level was encouraged. I observed that a sense of community was easier to achieve and maintain in smaller than in larger collectives, in which close relationships were usually formed among a relatively small core group of members. In many collectives, the members were encouraged to participate not only by regularly ordering the food but also by taking an active part in volunteering. But although it was the household members who were responsible for organization, in many collectives the ideal community included not only the member households, but also the farmers:

And then we organize farm visits this summer so that the relationships between these two [parties], which the conventional shops have kept as far apart as possible, eventually form and get established. (interview, coordinator, FC10)

Being a member of a food collective also meant having a greater ability to influence one's own food choices. This was important for many members as they felt that, as consumers, they had no choice but to purchase what the supermarkets offered. Unlike food collectives, chain stores appeared to discourage participation, as the following conversation with one of the coordinators at a distribution event revealed to me:

> I: I have also very strong feelings [...] when someone decides what I eat...then I decided that I will stop going to these chain stores.
> Q: So, do you feel that your ability to influence is small?
> I: Yes, it is... It is useless [shaking head], so absolutely useless to say anything. No one there [in the chain stores] listens. And believe me, I've tried.[...] It [not responding to my feedback] is so unscrupulous [...] But now my ability to influence is so great. That we collectively search for these good products and have real choices. (fieldnotes, coordinator, FC2)

Some people emphasized that, in food collectives, making conscious, informed, and ethical choices was a norm rather than an exception. Often, being able to influence one's own food choices was not merely related to having limited choice with respect to the quality or the assortment of food, but equally important was the ability to choose to support the farmers. However, despite all the benefits of the communality of food collectives, not all members sought to form relationships with each other or wanted to participate actively in community efforts. Specifically, in larger collectives, many members acted more like regular consumers, which could easily become a problem, since in order to function, a food collective required more engagement than what was expected from a regular supermarket customer. Hence, evaluating community was also closely related to how well work was shared among the members.

Sharing work

As the members of food collectives were responsible for organizing the exchange, there was a lot of work that needed to be shared. Group size often influenced how this happened within a collective. In smaller food collectives it was easier to share work more equally than in larger collectives, where there were more "free riders" and work tended to pile up on a few people's shoulders. Thus, figuring out fair ways to share work among the members was crucial.

On a weekly basis, the members of the collectives needed to accomplish several tasks. These included taking care of orders, organizing deliveries, communicating with the farmers and the participating member households, distributing the food, paying the bills, and keeping the books. Usually, certain key people took responsibility for coordinating these core tasks for an agreed period. Distribution was the most laborious and time-consuming task, and each collective needed to come up with a workable solution for sharing this work among all the members. Overall, the amount of work depended on how the food collective's distribution was organized in practice and on how many members had placed an order at a particular time.

Small food collectives with up to 35 members relied on what could be characterized as *obligatory volunteering*, as explained by one of the active members:

> I: This is based completely on volunteering, but with the principle that everyone needs to do something.
> Q: And do you then recruit volunteers, or?
> I: The distribution crew is always formed of those people who have ordered. And then there is this solidarity principle that if you have been a couple of times in a row, I can come this time. And from the beginning we have had this [principle] that everyone does something. (fieldnotes, active member, FC14)

In smaller food collectives, where everyone knew each other, it was easier to divide the workload and circulate different tasks more equally among all the members. Small group sizes also created social pressure to fulfill one's volunteering duty. In larger collectives (with more than 35 members), however, it was more difficult to oblige people to do volunteer work and to keep track of the members' volunteering. This put a lot of pressure on the few most active members. Larger collectives relied on what I characterize as *voluntary obligation*. This meant that the coordinators faced situations in which the continuity of the collective was at risk and they had to communicate that volunteering was an obligation for keeping the collective alive:

> At times it is somewhat challenging. And at times you find people [to volunteer] quite easily. Then like last winter we had to send this email reminding people that this needs to be done collectively, that we are not a store and that we hope that everyone participates at least in some ways or we won't be able to organize the distributions and continue [to exist]. (interview, coordinator, FC9)

Sharing different work tasks was one of the most important ways of taking part in the community—and of evaluating how good the community was. Sharing work was also a way of enforcing communal feeling, as people could meet each other, talk and exchange on a more personal level, or take part in decision making and thereby enforce the feeling of being part of a collective. Thus, goodness of community was evaluated through how well people in a food collective succeeded in sharing work and in encouraging participation among both households and farmers.

In this chapter, I have offered an account of how valuing happened in food collectives. My analysis showed how people engaged in continuous re-evaluation of the value-ideals around which their food collectives were organized. Instead of providing uniform criteria, valuing modes denoted ways of assessing the goodness of food, price and community in food collectives. It is important not to confuse valuing modes with a scheme or instrument for defining value, but see them as action that is valuable in itself; action that contributes to and is inseparable from the formation of the pursued value-ideals. Conceptualizing value as something one *does* creates an opportunity to reflect on value(s) as something intrinsically connected to action and relationships (Graeber 2001) rather than to things or people. In the following, I elaborate on this idea further as I discuss the implications of the findings and the contributions they make to the existing literature.

Towards a practice-based understanding of value

The study of food collectives reveals how formation of value is a dynamic, relational and continuous process that engages people not only in the pursuit of what they find valuable, but requires them to continually assess what they are pursuing. Developing a practice-based understanding of value is important, because it directs one to zoom into what people do and what they say while doing it (Nicolini 2009a), and thereby allows us to theorize value as intrinsic to and emerging from action (Graeber 2001; MacIntyre 2008). Building on the findings presented above, I wish to discuss two particularly important contributions made by this study.

First, this study introduces an understanding of value through the notion of good, suggesting that *value* is ultimately not only contextual but also an ideal target that is impossible to fully reach. Hence, I coin the term *value-ideals*. Furthermore, contrary to what both practice theory and prior research in alternative food movements imply (Schatzki 2001b; Weber et al. 2008; Werkheiser and Noll 2014), my analysis suggests that, despite pursuing the same value-ideals, people are likely to understand them and relate to them differently from each other. This, in turn, may invoke dissonance within the practices. Second, the concept of *valuing modes* draws attention to the movement between assessing and assigning value and thereby captures the formation of value as a socially dynamic, materially mediated and continually unfolding process. In the following, I discuss these contributions in more detail.

Value-ideals—understanding value through the notion of *good*

The economic and sociological traditions seem to have created a divide between value and values by enforcing definitions of what, and by trying to figure out where (a) value is (Miller 2008). Claiming that value is an ideal target that guides action, and that it does it without a guarantee of ever

being achieved, requires an ontological turn to practice. My ethnographic study made visible how value is to be found in the pursuit of *the good*, supporting an argument that some practice-oriented scholars have made (Boltanski and Thevenot 2006; Dewey 1939; Hutter and Stark 2015; MacIntyre 2008; Thévenot 2001), that people's actions are guided by their concern for the good. The question then revolves around *how* something is made good, rather what makes something good. Thus, like the concept of value, the concept of *good* should not be understood as a quality of "a thing" (e.g. of food, price, community), but as something being continually (re-) produced in a particular practice.

In food collectives, this was visible in the lack of standards and common metrics that would explicitly define what comprised good food, good price and good community. The non-standard quality of the food, direct relationships with farmers and buying as a collective effort produced incommensurability and created a sense of uniqueness that people did not directly attach to a specific food but to the food procured through a food collective in particular. Thus, the practice of exchange—the ways in which food collectives operated and engaged people to participate—functioned as a guarantee of the goodness of the value-ideals. But as the same value-ideals could signify various things to different people—even within the same collective—it is here that the belief of *shared* value(s) needs to be questioned.

As several practice scholars note (Nicolini 2011; Pantzar and Shove 2010), practices cannot simply be copied, but need to be adapted to local contexts. The findings of this study reveal how the locations and sizes of food collectives led to local adaptations and specific characteristics, making individual collectives sensitive to particular conditions and resources. This, in turn, produced various practical understandings of "the same" value-ideals. Some people valued primarily locally and seasonally sourced food, others organically produced food; some were after vegetarian food, others after ethical meat; some participated because of affordable prices, others did not mind paying extra; some tried to build a community, and others eschewed it. But sharing the same action qualified all these different interpretations of value-ideals as *good* in the context of food collectives. In support of what some authors have previously acknowledged (DeLind 2011; Kloppenburg Jr et al. 2000; Werkheiser and Noll 2014), my findings suggest that "the same" value(s) may encompass various meanings within the same practice, and thus while people share particular kinds of action and pursue similar value-ideals, they may be guided by variable "goods" within the same value-ideal. This, I believe, is a very important insight as it reveals how dissonance may exist in alternative practices of food exchange along with aspiring to the same value-ideals.

Further, the findings imply that the so-called two sides of direct food exchanges, producers and consumers, each of which is often treated as a monolithic group sharing "the same" values and goals (Albrecht and Smithers 2018; Trivette 2017), should be more closely investigated before being considered as such. It may well be, as the case of food collectives reveals, that the disharmony within a group of producers or of consumers becomes more visible. Understanding how, despite differing meanings, the practical, embodied and materially mediated ways of sharing action re-produce the (perception of) goodness of value-ideals is therefore important. Value(s) thus cannot simply be treated as harmonious triggers or as unanimous outcomes of practices of exchange—in this case, food collectives—upon which conclusions can be based.

Valuing modes—formation of value through assessing and assigning

I have previously argued that value should not be taken as a static entity, but that its formation is a continuous process in which people actively (though not necessarily reflexively) participate in and through different practices. Building on what Lambek (2013) and Vatin (2013) have noted, namely that in examining the formation of value there appears to be a tendency to make a distinction between evaluation (assessment) and production (assigning) of value, the findings of this study suggest that both are equally essential and inseparable features in the formation of value. To capture this movement between assessing and assigning value happening as part of the very quotidian participation in the practices of food collectives, I coined the term of *valuing modes*.

By participating in the different practices of individual food collectives, people encountered several situations in which food, price and community became collectively assessed. Valuing modes produced very practical and embodied understandings of what was considered *good* in the context of food collectives: they enabled people to judge "wrong" actions and justify "right" ones by expressing and reflecting on their bodily feelings and attitudes. Depending on the food collective, some valuing modes came to occupy a more central role than others. Specifically, in larger food collectives that included not only many members but also numerous different producers, some of which were not local, the community ideal was often put in question as valuing modes of *encouraging participation* and *sharing work* were likely to be in conflict. In this sense, trying to be inclusive and to encourage the participation of various members and farmers with multiple needs reduced the sense of connectedness and intimacy, which appear crucial for building relationships and local embeddedness (Feagan 2007; Feagan and Morris 2009; Hinrichs 2000), and distanced the collectives from achieving this goal. Also, needing to continuously balance between sharing and daring—that is, to find ways to share work in an egalitarian way while acknowledging the

different situations that people were in—made people reflect on and sometimes even question the goodness of the different value-ideals.

However, despite the challenge of not being able to fully realize the value-ideals, my findings reveal that the very acts of assessing the goodness of the value-ideals concurrently also assigned (produced) value. In other words, as Thévenot (2001) notes, evaluating performance simultaneously produces *the good* through the ways in which people learn within their practices to feel, understand and make judgements (Gherardi 2009; MacIntyre 2008). Using one's senses is a good example of how assessing and assigning are intrinsically entwined (Hennion 2004; Mann et al. 2011; Mol 2009). In food collectives, while sensing made it possible to evaluate the quality of food, at the same time people developed their sensing (c.f. senses)—tasting, smelling, touching, observing and handling food—through collectively shared and reflexive activity influenced by different material intermediators (such as the oven, the lack of wrapping, diverse appearance) and this, in turn, produced understandings and experiences of quality. Similarly, the valuing modes by which people assessed the goodness of price and community became to be considered as *good* in themselves, which, in turn, contributed to (re-) producing the value-ideals. It is therefore important to understand that different kinds of valuing modes (standards, labels and rankings) are not static and objective instruments that merely help people to make choices, but that they also assign value and thereby re-produce the actions and qualities that people end up considering good.

The findings raise further interesting questions on the nature of trust in practices of exchange. There seems to be a common understanding that alternative practices of food exchange are based on and bring about trust because of interconnected relationships between farmers and consumers (Feagan 2007; Hinrichs 2000; Trivette 2017). However, despite the existence of direct exchange relationships, in food collectives it was not common for each member to know the farmer personally, nor the other way around. In food collectives, trust was formed through the different valuing modes that increased transparency and created *the sense of knowing*. Thus the participants did not feel that it was necessary to know the farmer or the farming practices personally. These findings suggest that creating trust in alternative practices of food exchange does not necessarily require direct personal relationships. In this sense, valuing modes also made visible how a particular practice of exchange can in itself become trustworthy and thereby inform value.

Acknowledgements I thank the founders and members of food collectives for opening their homes and sharing their experiences for this research. I also wish to thank Nina Granqvist and Keijo Räsänen, and the members of the Organization and Management group in Aalto University School of Business, Neil Fligstein and the members of the Center for Culture, Organization and Politics at the University of California, Berkeley and Silvia Gherardi for their constructive, caring and critical comments on previous versions of this manuscript. Open access funding provided by University of Helsinki including Helsinki University Central Hospital.

Open Access This article is licensed under a Creative Commons Attribution 4.0 International License, which permits use, sharing, adaptation, distribution and reproduction in any medium or format, as long as you give appropriate credit to the original author(s) and the source, provide a link to the Creative Commons licence, and indicate if changes were made. The images or other third party material in this article are included in the article's Creative Commons licence, unless indicated otherwise in a credit line to the material. If material is not included in the article's Creative Commons licence and your intended use is not permitted by statutory regulation or exceeds the permitted use, you will need to obtain permission directly from the copyright holder. To view a copy of this licence, visit http://creativecommons.org/licenses/by/4.0/.

Appendix 1

Practices in food collectives

Practice	Description
Farming	Producing food by cultivating crops and growing vegetables, growing animals
Harvesting	Harvesting crops, vegetables, and berries from the fields, collecting eggs, etc.
Foraging food	Catching fish, picking berries or mushrooms in the forest, hunting
Processing food	Processing harvested food, e.g. milling the grains, butchering, producing juice out of berries
Finding suppliers	Member households engage in seeking small-scale farmers and other food suppliers to procure food from. In order to ensure variability in selection, food collectives order from several suppliers
Transporting	Food needs to be transported from farms and processing units to the food collective's distribution point
Placing orders	Orders must be placed in advance. Member households place orders via some technological medium, such as Excel sheets, Google docs, email, webstores, or Facebook
Paying	Member households pay for the food either in cash when picking up the food, or via bank transfer. Each member can either pay individually directly to the producer, or the food collective may act as a medium for payment

Practice	Description
Distributing	Food is delivered in large quantities and variable qualities, and needs to be distributed among the member households according to their orders. Food is sorted by volunteer members and put into piles or bags for members to pick up
Fetching food	Member households need to come and pick up the food from a particular place at a specific time. Depending on how distribution is organized, a person may be picking up a ready-packed bag, or placing orders into their own bags
Coordinating	Food collectives need to assign volunteers, communicate with producers and among member households, negotiate terms of delivery, place orders with suppliers, manage monetary transfers between member households and farmers, and keep the books
Storing and preserving	Households need to store or preserve the food due to the large quantities and variable qualities of foodstuff (e.g. by making jam, freezing food in small portions)
Preparing/cooking	Unprocessed or very little processed food needs to be handled, processed and prepared in order to make them ready for cooking and eating
Sharing information	Food collective members exchange recipes, and share information on food quality and practices of food production; they also organize events around local food

References

Albrecht, C., and J. Smithers. 2018. Reconnecting through local food initiatives? Purpose, practice and conceptions of 'value'. *Agriculture and Human Values* 35 (1): 67–81.

Aspers, P. 2008. Analyzing order: Social structure and value in the economic sphere. *International Review of Sociology-Revue Internationale de Sociologie* 18 (2): 301–316.

Aspers, P., and J. Beckert. 2011. Value in markets. In *The worth of goods: Valuation and pricing in the economy*, ed. P. Aspers and J. Beckert, 3–38. Oxford: Oxford University Press.

Boltanski, L., and L. Thevenot. 2006. *On justification: Economies of worth*. Princeton: Princeton University Press.

Brunori, G., A. Rossi, and F. Guidi. 2012. On the new social relations around and beyond food. Analysing consumers' role and action in Gruppi di Acquisto Solidale (Solidarity Purchasing Groups). *Sociologia Ruralis* 52 (1): 1–30.

Brunsson, N., A. Rasche, and D. Seidl. 2012. The dynamics of standardization: Three perspectives on standards in organization studies. *Organization Studies* 33 (5–6): 613–632.

Czarniawska, B. 2014. Why I think shadowing is the best field technique in management and organization studies. *Qualitative Research in Organizations and Management: An International Journal* 9 (1): 90–93.

Dahlberg, K.A. 1988. Ethical and value issues in international agricultural research. *Agriculture and Human Values* 5 (1–2): 101–111.

Delind, L.B. 2006. Of bodies, place, and culture: Re-situating local food. *Journal of Agricultural and Environmental Ethics* 19 (2): 121–146.

DeLind, L.B. 2011. Are local food and the local food movement taking us where we want to go? Or are we hitching our wagons to the wrong stars? *Agriculture and Human Values* 28 (2): 273–283.

Dewey, J. 1939. Theory of valuation. *International encyclopedia of unified science*, vol. 2. Chicago, IL: The University of Chicago Press.

Diekmann, M., and L. Theuvsen. 2019. Value structures determining community supported agriculture: Insights from Germany. *Agriculture and Human Values* 36 (4): 733–746.

Doran, C.J. 2009. The role of personal values in fair trade consumption. *Journal of Business Ethics* 84 (4): 549–563.

Feagan, R. 2007. The place of food: Mapping out the 'local' in local food systems. *Progress in Human Geography* 31 (1): 23–42.

Feagan, R.B., and D. Morris. 2009. Consumer quest for embeddedness: A case study of the Brantford Farmers' Market. *International Journal of Consumer Studies* 33 (3): 235–243.

Forssell, S., and L. Lankoski. 2015. The sustainability promise of alternative food networks: An examination through "alternative" characteristics. *Agriculture and Human Values* 32 (1): 63–75.

Fourcade, M. 2011. Cents and sensibility: Economic valuation and the nature of "nature". *American Journal of Sociology* 116 (6): 1721–1777.

Gherardi, S. 2011. Organizational learning: The sociology of practice. In *Handbook of organizational learning and knowledge management*, 2nd ed, ed. M. Easterby-Smith and M.A. Lyles, 43–65. Chichester: Wiley.

Gherardi, S. 2012. *How to Conduct a Practice-Based Study: Problems and methods*. Cheltenham: Edward Elgar Publishing.

Gherardi, S. 2009. Practice? It's a matter of taste! *Management Learning* 40 (5): 535–550.

Goodman, D. 2003. The quality 'turn' and alternative food practices: Reflections and agenda. *Journal of Rural Studies* 1 (19): 1–7.

Goodman, D., M. DuPuis, and M. Goodman. 2012. *Alternative food networks: Knowledge, practice, and politics*. London: Routledge.

Graeber, D. 2001. *Toward an anthropological theory of value: The false coin of our own dreams*. New York, NY: Palgrave.

Graeber, D. 2013. It is value that brings universes into being. *HAU: Journal of Ethnographic Theory* 3 (2): 219–243.

Helgesson, C.-F., and H. Kjellberg. 2013. Introduction: Values and valuations in market practice. *Journal of Cultural Economy* 6 (4): 361–369.

Helgesson, C.-F., and F. Muniesa. 2013. For what it's worth: An introduction to valuation studies. *Valuation Studies* 1 (1): 1–10.

Hennion, A. 2004. Pragmatics of taste. In *The Blackwell companion to the sociology of culture*, ed. M. Jacobs and N. Hanrahan, 131–144. Oxford: Blackwell.

Heuts, F., and A. Mol. 2013. What is a good tomato? A case of valuing in practice. *Valuation Studies* 1 (2): 125–146.

Hinrichs, C.C. 2000. Embeddedness and local food systems: Notes on two types of direct agricultural market. *Journal of Rural Studies* 16 (3): 295–303.

Hinrichs, C.C. 2003. The practice and politics of food system localization. *Journal of Rural Studies* 19 (1): 33–45.

Holt Giménez, E., and A. Shattuck. 2011. Food crises, food regimes and food movements: Rumblings of reform or tides of transformation? *The Journal of Peasant Studies* 38 (1): 109–144.

Hughes, A. 2005. Geographies of exchange and circulation: Alternative trading spaces. *Progress in Human Geography* 29 (4): 496–504.

Hutter, M., and D. Stark. 2015. Pragmatist perspectives on valuation: An introduction. In *Moments of valuation: Exploring sites of dissonance*, ed. A.B. Antal, M. Hutter, and D. Stark, 1–12. Oxford: Oxford University Press.

Johnston, J., M. Szabo, and A. Rodney. 2011. Good food, good people: Understanding the cultural repertoire of ethical eating. *Journal of Consumer Culture* 11 (3): 293–318.

Karpik, L. 2010. *Valuing the unique: The economics of singularities*. Princeton, NJ: Princeton University Press.

Kjellberg, H., A. Mallard, D.-L. Arjaliès, P. Aspers, S. Beljean, A. Bidet, A. Corsin, E. Didier, M. Fourcade, and S. Geiger. 2013. Valuation studies? Our collective two cents. *Valuation Studies* 1 (1): 11–30.

Kloppenburg Jr., J., S. Lezberg, K. De Master, G.W. Stevenson, and J. Hendrickson. 2000. Tasting food, tasting sustainability: Defining the attributes of an alternative food system with competent, ordinary people. *Human Organization* 59 (2): 177–186.

Lambek, M. 2013. The value of (performative) acts. *HAU: Journal of Ethnographic Theory* 3 (2): 141–160.

Lamont, M. 2012. Toward a comparative sociology of valuation and evaluation. *Annual Review of Sociology* 38: 201–221.

MacIntyre, A. 2008. *After virtue: A study in moral theory*. Indiana, IN: University of Notre Dame Press.

Malinowski, B. 2002. *Argonauts of the Western Pacific: An account of native enterprise and adventure in the archipelagoes of Melanesian New Guinea*. London: Routledge.

Mann, A., A. Mol, P. Satalkar, A. Savirani, N. Selim, M. Sur, and E. Yates-Doerr. 2011. Mixing methods, tasting fingers: Notes on an ethnographic experiment. *HAU: Journal of Ethnographic Theory* 1 (1): 221–243.

Mauss, M. 1954. *The gift: Forms and functions of exchange in archaic societies*. Illinois, IL: The Free Press.

Miettinen, R., D. Samra-Fredericks, and D. Yanow. 2009. Re-turn to practice: An introductory essay. *Organization Studies* 30 (12): 1309–1327.

Miller, D. 2008. The uses of value. *Geoforum* 39 (3): 1122–1132.

Mol, A. 2009. Good taste: The embodied normativity of the consumer-citizen. *Journal of Cultural Economy* 2 (3): 269–283.

Muniesa, F. 2011. A flank movement in the understanding of valuation. *The Sociological Review* 59 (2): 24–38.

Murray, D.L., and L.T. Raynolds. 2007. Globalization and its antinomies: Negotiating a Fair Trade movement. In *Fair Trade: The challenges of transforming globalization*, ed. L.T. Raynolds, D.L. Murray, and J. Wilkinson, 19–30. London: Routledge.

Nicolini, D. 2009a. Zooming in and out: Studying practices by switching theoretical lenses and trailing connections. *Organization Studies* 30 (12): 1391–1418.

Nicolini, D. 2009b. Articulating practice through the interview to the double. *Management Learning* 40 (2): 195–212.

Nicolini, D. 2011. Practice as the site of knowing: Insights from the field of telemedicine. *Organization Science* 22 (3): 602–620.

Orlikowski, W., and S. Scott. 2013. What happens when evaluation goes online? Exploring apparatuses of valuation in the travel sector. *Organization Science* 25 (3): 868–891.

Otto, T., and R. Willerslev. 2013. Introduction: "Value as theory" Comparison, cultural critique, and guerilla ethnographic theory. *HAU: Journal of Ethnographic Theory* 3 (1): 1–20.

Pantzar, M., and E. Shove. 2010. Understanding innovation in practice: A discussion of the production and re-production of Nordic Walking. *Technology Analysis & Strategic Management* 22 (4): 447–461.

174

Parker, M., G. Cheney, V. Fournier, and C. Land (eds.). 2014. *The Routledge companion to alternative organization*. London: Routledge.

Pollan, M. 2010. The food movement, rising. New York Review of Books. https://www.nybooks.com/articles/archives/2010/jun/10/food-movement-rising/. Accessed 30 Mar 2020.

Pratt, J. 2007. Food values: The local and the authentic. *Critique of Anthropology* 27 (3): 285–300.

Raynolds, L.T. 2000. Re-embedding global agriculture: The international organic and fair trade movements. *Agriculture and Human Values* 17 (3): 297–309.

Reinecke, J., and S. Ansari. 2015. What is a "fair" price? Ethics as sensemaking. *Organization Science* 26 (3): 867–888.

Reinecke, J., S. Manning, and O. Von Hagen. 2012. The emergence of a standards market: Multiplicity of sustainability standards in the global coffee industry. *Organization Studies* 33 (5–6): 791–814.

Sandberg, J., and H. Tsoukas. 2011. Grasping the logic of practice: Theorizing through practical rationality. *Academy of Management Review* 36 (2): 338–360.

Schatzki, T. 2001a. Introduction: Practice theory. In *The practice turn in contemporary theory*, ed. T. Schatzki, K. Knorr Cetina, and E. von Savigny, 1–14. London: Routledge.

Schatzki, T. 2001b. Practice mind-ed orders. In *The practice turn in contemporary theory*, ed. T. Schatzki, K. Knorr Cetina, and E. von Savigny, 42–55. London: Routledge.

Seyfang, G. 2007. Cultivating carrots and community: Local organic food and sustainable consumption. *Environmental Values* 16: 105–123.

Thévenot, L. 2001. Pragmatic regimes governing the engagement with the world. In *The practice turn in contemporary theory*, ed. T. Schatzki, K. Knorr Cetina, and E. von Savigny, 64–82. London: Routledge.

Thompson, D.B., and B. McDonald. 2013. What food is "Good" for you? Toward a pragmatic consideration of multiple values domains. *Journal of Agricultural and Environmental Ethics* 26 (1): 137–163.

Timmermans, S., and I. Tavory. 2012. Theory construction in qualitative research: From grounded theory to abductive analysis. *Sociological Theory* 30 (3): 167–186.

Trivette, S.A. 2017. Invoices on scraps of paper: Trust and reciprocity in local food systems. *Agriculture and Human Values* 34 (3): 529–542.

Van Bommel, K., and A. Spicer. 2011. Hail the snail: Hegemonic struggles in the slow food movement. *Organization Studies* 32 (12): 1717–1744.

Vatin, F. 2013. Valuation as evaluating and valorizing. *Valuation Studies* 1 (1): 31–50.

Weber, K., K.L. Heinze, and M. DeSoucey. 2008. Forage for thought: Mobilizing codes in the movement for grass-fed meat and dairy products. *Administrative Science Quarterly* 53 (3): 529–567.

Werkheiser, I., and S. Noll. 2014. From food justice to a tool of the status quo: Three sub-movements within local food. *Journal of Agricultural and Environmental Ethics* 27 (2): 201–210.

Wiedner, R., and S. Ansari. 2017. Resisting the Urge to Follow Set Plans. In *The Routledge companion to qualitative research in organization studies*, ed. R. Mir and S. Jain, 343–358. New York, NY: Routledge.

Zelizer, V. 1978. Human values and the market: The case of life insurance and death in 19th-century America. *American Journal of Sociology* 84 (3): 591–610.

Publisher's Note Springer Nature remains neutral with regard to jurisdictional claims in published maps and institutional affiliations.

Galina Kallio is a postdoctoral researcher in Ruralia Institute at the University of Helsinki. Her research focuses on alternative food economies, regenerative agriculture, ecocentric epistemologies, and non-capitalist conceptualizations of value and work.

Agriculture and Human Values (2020) 37:1111–1123
https://doi.org/10.1007/s10460-020-10114-9

From left behind to leader: gender, agency, and food sovereignty in China

Li Zhang[1]

Accepted: 13 May 2020 / Published online: 20 May 2020
© Springer Nature B.V. 2020

Abstract

Capitalist reforms usually drive outmigration of peasants to cities, while elders, children, and women responsible for their care are "left behind" in the countryside. The plight of these "left behind" populations is a major focus of recent agrarian studies in China. However, rural women are not merely passive victims of these transformations. Building on ethnographic research in Guangxi and Henan provinces from 2013 to 2017, and drawing on critical gender studies and feminist political ecology, I show how the food safety crisis in China creates conditions for peasant women to increase control and income from organic food production, often establishing alternative food networks with the support of female scholars and NGO organizers. Thus, I shift focus of scholarship on rural women from "left behind" to leaders in struggles for justice and food sovereignty.

Keywords China · Left-behind populations · Gender · Agency · Alternative food networks · Food sovereignty

Abbreviations

AFN	Alternative Food Network
BOFM	Beijing Organic Farmers' Market
CCP	Chinese Communist Party
COHD	College of Humanities and Development Studies, China Agricultural University
CSA	Community Supported Agriculture

Introduction

A central characteristic of China's recent market-oriented reforms has been the massive outmigration of peasants to the cities, where they take up temporary jobs as migrant workers in industry, construction, and various service sectors. This results from an urban-focused export-oriented industrial policy, price differentials for agricultural and manufactured products, uneven incomes from agriculture and manufacturing/services, and an urban bias in cultural attitudes and the provision of social services (Wen 2001; Yan 2003). Moreover, China's household registration system (*hukou*)

generally curtails the permanent settlement of rural populations in major cities, excluding them and their families from essential social services such as education (Wen 2001; Yan 2003). So as the working-age rural population migrates out for temporary urban employment, elders, children, and women responsible for their care are "left behind". The characteristics and plight of these "left behind" populations have become focus of much scholarship in development studies, agrarian studies, and various social sciences (Ye and Wu 2008; Wu and Ye 2016; Ye et al. 2016), and these have contributed to promoting various government policies to address the predicament of these people and the "hollow villages" where they remain. This scholarship and the political mobilization around it are commendable for bringing much needed governmental policies and resources to address the social (economic, cultural, ecological, etc.) problems that come about through increasing rural–urban inequality. However, this scholarship and much of the policy recommendations it provides also faces important limitations. My purpose in this paper is to build upon this literature and advance it further through stronger and deeper engagement with feminist political ecology and critical gender studies.

Two aspects of this literature are particularly useful for expanding and deepening this scholarship. First, there is recognition that women are the pillar of "left behind" populations, as they are "left behind" precisely because they are considered to be responsible for taking care of children

✉ Li Zhang
  li.zhang@uci.edu

[1] Department of Global and International Studies, University of California, Irvine, 3151 Social Science Plaza, Irvine, CA 92697-5100, USA

who cannot advance their education at the urban centers, and elders who are not capable of migrating to work in new factories and social services. Moreover, since women generally live longer than men, most of the "left-behind elderly" are also women. Second, there is also growing recognition that the tidal waves of migrant workers results in the "feminization" of agriculture, that is, the fact that much agricultural labor and other rural work is being done increasingly by women (Zuo and Song 2002; Chang et al. 2011). This feminization of agriculture had already been widely recognized across India, Africa, Latin America, and much of the rest of the world (Fortmann and Rocheleau 1985; Deere 2005; FAO 2010; de Schutter 2013; Lahiri-Dutt and Adhikari 2016). In China, however, there were powerful voices utilizing neoliberal discourses and patriarchal assumptions (mainly in economics, political science, and sociology) to question the prevalence of feminization and challenge those who argued this was taking place extensively (e.g. Zhang et al. 2004; de Brauw et al. 2008), since much of the female work in agriculture focused on household subsistence, and encompassed as well various other forms of unpaid, non-cash "household" economy (cf. Barker 2005; de Schutter 2013). As extensive documentation of feminization of agriculture continues to emerge through rigorous, extensive, and in-depth fieldwork-based research (mainly in critical agrarian studies, development studies, anthropology, and to a certain extent sociology as well) critics were forced to revise their previous statements (e.g. de Brauw et al. 2013). This growing recognition of the feminization of agriculture in China, therefore, is an important accomplishment in its own right.

In this paper, I argue that we must advance from merely describing the characteristics of women as "left behind", and demonstrating the feminization of agriculture, to pay more attention to the manner that rural women are not merely passive victims during these transformations. Maintaining this currently limited perspective and purpose in the literature can even risk aggravating the condition of these women, reproducing a discourse of *victimization* that makes their agency invisible and their initiatives unimportant, and may even coopt their self-empowerment efforts (cf. Sangtin Writers Collective 2010; Gilson 2016). This critique is not new in gender studies, including the argument that certain "burdens" may also be opportunities for greater female agency (e.g. Schneider 1993; Chung et al. 2019). However, the growing attention to women and gender issues among overseas development practice, international scholarship, and policy since the 1990s (FAO 1996, 2010; UNDP 2003), including the promotion of women's rural cooperatives and contract farming schemes (Dolan and Sorby 2003; de Schutter 2011), has generated a powerful new wave of scholarship on this topic, particularly in the way that NGOs and "participatory rural development" initiatives that were designed to "empower women" often failed to do so, and sometimes even have the opposite effect (e.g. Sangtin Writers Collective 2010; Jacka 2013). Albeit focused on empirical cases and literature about China, therefore, my article does not rest upon nor suggest any exceptionalism about this country, but rather it engages with key debates in the international and interdisciplinary field of "agriculture and human values" worldwide.

In short, I argue we must shift focus of scholarship on rural women from "left behind" to leaders in various forms of resistance to displacement, marginalization, and discrimination. Discussion of feminization of agriculture in feminist political ecology, after all, has often indicated this can become an opportunity for female empowerment (Carney and Watts 1990; Rocheleau et al. 1996; Schroeder 1996; Vaz-Jones 2018). With this argument, moreover, we can also begin to deconstruct the dichotomies that separate "left behind" rural women from others in non-rural spaces where they exercise their agency, contributing to new analytic frameworks that recognize "translocal family reproduction" as key to understanding contemporary agrarian change (Jacka 2018), and female-led "rooted networks" as central to rural and environmental social movements (Escobar et al. 2002; Rocheleau and Roth 2007).

Theoretically, I build upon critical agrarian studies, development studies, gender studies, and feminist political ecology, particularly the feminist critique of Fraser (2003, 2009), Tamara Jacka (1997, 2010, 2013, 2018), Judith Butler (2004), and Erinn Gilson (2016), and both classic and new works of feminist political ecology (Fortmann and Rocheleau 1985; Carney and Watts 1990; Rocheleau et al. 1996; Jarosz 2011; Ge et al. 2011; Elmhirst 2011). Methodologically, I utilized ethnographic methods of semi-structured interviews, qualitative surveys, and participant observation during several months of in-depth fieldwork in Guangxi and Henan provinces from 2014 to 2017, which I supplemented with a critical review of media and government reports.

The paper is organized as follows. In the second section, I review the literature and outline my theoretical framework. In the third section, I present my methods and field sites. Then in the fourth section, I discuss various findings from my fieldwork to highlight how the feminization of agriculture and the ongoing food safety crisis in China are creating conditions for peasant women to increase control over food production and increase their income through sales of safer, organic food through "alternative food networks" (AFNs).[1]

---

[1] AFNs contrast with mainstream agri-food commercial channels (such as major agribusiness companies, wholesalers, supermarkets, institutional canteens and restaurants), and include community-supported agriculture (CSA) initiatives, farmers' markets, buying clubs, peasant cooperatives and even informal (e.g. family) producer-consumer connections that embed agri-food distribution in stronger social and ecological relations. For more details see Si and Scott (2019).

In the fifth section, I briefly discuss the ongoing challenges and obstacles faced by these female leaders, who are still subjected to the multiple burdens of advancing their agricultural, community, and/or political work alongside extensive unpaid domestic labor, and pervasive sexism and discrimination. In the conclusion, I revisit feminist debates in agrarian studies to argue that shifting our focus to women's role as leaders contributes to a better understanding of the complex manner in which the feminization of agriculture constitutes both a disproportionate burden for rural women and an important opportunity for female empowerment. This generates conceptualizations that better reflect these women's subjective understandings of their own condition and experiences, but also more productive grounds for scholarship that does not simply describe their plight, but also recognizes and contributes to the advancement of their struggles.

Literature review and theoretical frameworks

Critical agrarian studies and development studies

Scholarship from and about China has been formative to the international and interdisciplinary fields of agrarian and development studies. Arguably, Mao Zedong himself introduced the idea of the revolutionary leadership of the peasantry to the communist movement through his studies of the conditions of the peasantry in his native Hunan province (Mao [1926–1927] 1971). Liang Shuming and Yan Yangchu also led the creation of a non-communist "rural construction movement", advancing both social science scholarship on agrarian societies and a broader social movement for peasant cooperatives (Si and Scott 2019). In addition, Fei Xiaotong is widely considered the founder of Chinese sociology through his ethnographic studies of the rural foundations of Chinese modernizing society (Fei [1948] 1992). Across all their scholarly and political differences, however, there is a theoretical commitment to researching the *agency* of peasants, a basic but fundamental insight that should orient critical agrarian studies worldwide.

Chinese agrarian studies then transformed radically from the socialist period, when the peasantry was discussed (at least officially among scholars and government officials) in very high regard, into the period of "reform and opening up", when "members of the urban educated elite [began] seeking to reclaim a positive status and future for both themselves and the Chinese nation in the aftermath of late Maoist zealotry, in part by emphasizing the 'backwardness' of the peasantry" (Jacka 2013, p. 986; Schneider 2015). The peasantry began to be seen as "low quality" people whose numbers had to be contained through the one-child policy, and "backward" people who needed to be "modernized"

(Jacka 2013; Schneider 2015). An anti-Marxist and anti-Maoist neoliberal consensus began to emerge that agricultural development takes place through "technological modernization", reducing the need for labor in the countryside while increasing "economic efficiency" and "productivity" of agriculture (e.g. Zhang et al. 2004; Huang et al. 2008; de Brauw et al. 2008, 2013). Such neoliberal agrarian studies became mainstream during the 1990s and 2000s, informing and supporting capitalist reforms, and removing the agency of the peasants from theoretical discussion.

In opposition to such neoliberal agrarian studies, there has been an increasingly strong current of what we call critical agrarian studies. These are largely driven by scholars who refer back to the non-communist currents of agrarian studies and "rural construction movement", particularly Wen Tiejun (2001) and He X. (2007), as well as new Marxist scholarship in anthropology and sociology (Yan 2003, 2008; Yan and Chen 2013; Zhang 2015), and critical development studies (Ye and Wu 2008; Ye 2010). These scholars criticize the capitalist reforms in the Chinese countryside and offer alternative visions for Chinese development. They call attention to the historical and ongoing contributions of the peasantry to the wellbeing and advancement of society, and the need for continued and/or renewed labor-intensive agro-ecological production to reverse the socio-ecological crisis that China is facing.

A central aspect of this crisis turns on food safety, as became widely recognized in 2008 when adulterated milk formula caused the death of many infants. Major incidents of food contamination have continued to cause national public health scares each year. This crisis results from the commodification of food and farming, which enables and incentivizes overuse of toxic agrochemicals and adulteration of agri-food products (Zhang 2017; Zhang and Qi 2019). Consequently, Chinese society has a growing concern to access safer and organic food, creating conditions for peasant women to increase control and income from organic food production by establishing AFNs in collaboration with female scholars and NGO organizers. In this context, the struggles of peasants and urban food consumers to network for the provision of safer foods is fundamentally about regaining control (sovereignty) over food by re-embedding agri-food markets into social relations (as in a Polanyian countermovement). My contribution to this literature, therefore, simultaneously expands the empirical and theoretical scope of food sovereignty, and interlinks critical agrarian studies with broader debates about development studies.

Development studies emerged as a distinct field in China following upon the expansion of overseas development aid during the 1980s and 1990s. Since that time, overseas development agencies began funding not only development projects directly, but also an increasingly large number of development research initiatives, and training in development project implementation and research (Ye 2010; Jacka 2013). This led to the creation of China's first College of Rural Development at the China Agricultural University in 1998, which later became the College of Humanities and Development Studies (COHD). The emergence of development studies, and its close association with critical agrarian studies, "reflected a broad shift in scholarly approaches to rural issues, away from a predominant focus on achieving increases in agricultural productivity toward a broader, more holistic conceptualization of rural social and economic development" (Jacka 2013, p. 988; cf. Ye 2010), including most prominently the need to recognize and address the plight of the "left behind" populations.

The terms "left behind" used to describe rural people who (mostly) remain in the countryside while others migrate for temporary employment in urban areas first emerged in short local news articles in the mid 1990s (Shangguan 1994; Yi 1994; Lu 1996), and the first scholars to discuss the topic academically began publishing in 2004 (Du 2004; Luo and Chai 2004). This issue of "left behind" populations continued gathering academic attention during the late 2000s, and received even more academic attention when Ye Jingzhong's team at the COHD gathered substantial resources to conduct national-level quantitative and qualitative surveys of "left behind" populations, triggering a larger wave of publications and even government attention to the topic (Ye and Wu 2008; Wu and Ye 2016; Ye et al. 2016).

These efforts have produced very empirically rich scholarship on the topic, demonstrating in very vivid terms the plight and suffering of "left behind" women, children, and elders, and critiquing this as a serious problem of contemporary Chinese development. These include mainly examination of the economic hardship faced by these individuals (low income, heavy workloads in agricultural production and care work, limited financial and other contributions from family members who migrated for temporary urban employment, and limited access to good quality social services, particularly healthcare and education), and their personal and psychological suffering (loneliness, depression, anxiety, problems with self-esteem, etc.). These challenges are especially difficult for women who suffer multiple layers of these problems, who are described as "burdened" with agricultural work to maintain the family's fields in addition to all the care work for elders and children, while receiving the least economic and social recognition, and facing the worst exclusion and marginalization among the family clans and villages of their husbands, as women traditionally "marry out" of their own family to go live *and work for* the husband's family (Zhang 2009).

Yet this literature has come under increasingly more sustained criticism in recent years for remaining limited to a description of the negative experiences of these victimized individuals, without theoretical advancements about their condition or recognition of their agency.[2] Indeed, the most explicit attempt by members of the COHD team to advance this scholarship continue to frame the issue in terms of "burdens" and victimization (Ye 2018, 2019). When Ye Jingzhong (2018) wrote most explicitly about "left behind women's contribution to development", for example, he still regarded this contribution as the passive "sacrifice" of these women so that men can migrate to work in the cities, sustaining household reproduction and cheap labor for export-oriented industrialization. Evidently, this scholarship continues to neglect longstanding feminist debates regarding female agency, subjective interpretations of burden/care, and the opportunities that feminization of agriculture may generate for female empowerment (Carney and Watts 1990; Schneider 1993; Schroeder 1996; Chung et al. 2019), sidestepping the feminist arguments of female scholars, even when produced and/or presented at the COHD (Jacka 2012; Zhang 2016, 2018).

Gender studies and feminist political ecology

To build upon and advance this scholarship, I turn to critical gender studies and feminist political ecology. In particular, I draw upon Nancy Fraser for a feminist theory of justice that is especially attentive to the post-socialist condition and everyday capitalist relations (Fraser 2003, 2009), and build upon Tamara Jacka's feminist critique of the "rural reconstruction movement" and participatory development scholarship and practice in China (Jacka 2013). This feminist scholarship has shown that justice and injustice have multiple dimensions that go beyond economic exploitation and political oppression, and I focus particularly on what they call "cultural injustice", which includes not only cultural imposition or appropriation, but broader forms of disrespect, marginalization, and "non-recognition", that is, the rendering of a person as "invisible" (Jacka 2013, p. 984; Fraser 2003). Critical agrarian studies and development studies literature on "left behind" populations have explicitly sought to make these persons "visible" in a context where neoliberal agrarian studies, mainstream culture, and government policy was making them "invisible", and in this regard this literature has contributed to overcoming this cultural injustice.[3]

---

[2] Remarks made by Luo Cheng, professor at the Shaanxi Academy of Social Sciences, prefacing his presentation "Current situation and recommendations of support for poor rural left behind families", at the seminar on Rural Left Behind Populations: New Questions, New Characters, New Actions, China Agricultural University, College of Humanities and Development Studies (COHD), Beijing, March 23, 2019.

[3] Since the publication of the Chinese Central Government Document Number 1 of 2008, for example, the government utilizes the explicit terms of this scholarship in its rural development efforts.

However, this scholarship would be problematic if it remains limited to this discourse of "the plight of the left behind", since it is a reductionist approach to understanding a complex social problem that (1) does not necessarily identify concrete and constructive solutions to this crisis, (2) generates a discourse of victimization that makes the agency of these people invisible, and (3) may even aggravate their condition by undermining their initiatives, agency, self-esteem, or even coopt their self-empowerment efforts. In other words, this scholarship can "potentially help to address economic and cultural injustice by shifting understandings of 'development' and how it is achieved, and by changing perceptions of rural citizens and rural culture", but since it also reproduces a discourse of victimization at the same time, the emancipatory potential of this scholarship could possibly be "undermined by a failure to develop effective strategies for overcoming gender injustice" and may even "contribute to the reproduction of injustice" (Jacka 2013, p. 985).

A common challenge to this critique has been that victimization and stigmatization are not actually *created* or *imposed* by scholars who research it, and in fact this concern amounts merely to a "misunderstanding" that results from the "shallow imagination" of society and "one-sided" presentation of information in the media (Ye 2019, p. 24). In order to advance this debate, I follow Butler (2004) and Gilson (2016) in theorizing stigmatization, victimization, vulnerability, and precarity in relation to various expressions of human identity and agency. Even when these women *become victims* of increased exploitation and oppression, "being a victim" is not their personal identity as the discourse of "left behind women" appears to suggest. Therefore, our own scholarship must shift theoretical focus to follow these women in their own agency, recognizing how their choices—albeit from precarious positions of vulnerability—still reveal daily-life struggles against displacement, marginalization, and discrimination. This includes their work in agriculture, rural cooperatives, and rural livelihoods, but also other work in non-rural spaces where they exercise agency, particularly their efforts to restructure gender, class, and rural-urban relations in the first place, and implicitly, address the greater social injustices engendered by these inequalities.

In this way, my feminist critique also builds on the theoretical advances of female Chinese scholars who already deconstructed similar victimization discourses about women who *did* migrate for temporary work in urban industries (e.g. Lee 1998; Ngai 2005; Yan 2008), and post-colonial studies of "quiet social movements" and "everyday life" resistance among the poor and marginalized elsewhere in the Global South (Bayat 2000, 2013; Roy 2015; Vaz-Jones 2018). In particular, I theorize food sovereignty initiatives among rural women as a *feminist movement* in China, which is unlike the high-profile account of middle-class liberal feminism

that is gaining attention recently (Milwertz 2002; Fincher 2016), as those accounts are almost entirely disconnected from the deeper social, political, economic, and ecological analysis present in critical agrarian studies and feminist political ecology.

Thus, I bring Fraser's (2003) and Jacka's (2013) feminist theory of justice to bear upon the broader fields of critical agrarian studies and feminist political ecology, enabling us to recognize the limitations of the existing literature on "left behind" women and promote distinct frameworks in the following manner. The mere description of rural women as simply "left behind" with the "burdens" of farming and social reproduction—which isolates rural women from their translocal family reproduction and political networks, and reduces them to a homogeneous and isolated group of victims—constitutes what Fraser (2003) and Jacka (2013) call an "affirmative" conception and strategy of justice. Affirmative approaches pivot on "inclusion". They seek to address injustices by "identifying" and "including" victims of injustice in social, political, and economic structures, yet they do not call attention to or challenge the underlying structures of power that produce "invisibility" and "burdens" in the first place, nor do they reflect the agency of those who actively struggle against these conditions. In part as a result of scholarship on "left behind" populations, for example, the Chinese government is now superficially including women and other "left behind" populations in government policies for "poverty alleviation" and "rural vitalization" without challenging the capitalist reforms that generate this condition, or supporting the rooted networks and bottom-up initiatives of these vulnerable persons themselves to overcome this injustice.

Transitional conceptions and strategies of justice, on the other hand, do not simply rest upon the "inclusion" of the marginalized, but pivot upon their own agency to "alter the terrain" upon which struggles are waged in ways that may ultimately transform the underlying structures that generate injustice in the first place (Fraser 2003, p. 74; Jacka 2013, p. 985). Developing our conceptual framework from "left behind" to "leaders" harnesses the commitment to scholarship and engagement with people's agency, and enables recognition of their quiet struggles in everyday life as a form of transitional approach to justice. Women across China's villages, townships, and cities—including women who migrate from rural to urban spaces for higher education—are engaged in various forms of agroecological production to satisfy their household's basic needs for food, especially safer, organic food in face of an aggravating food safety crisis (Zhang and Qi 2019). They are also collaborating in the creation of AFNs designed to cultivate and support the livelihoods of women, children, the elderly, the disabled, and other vulnerable persons, particularly through alliances between female scholars who may have their own roots in the countryside, and now partner with those who remain engaged in agricultural production. As women take up leadership roles and positions of power and authority in rural cooperatives, local governments, universities, and AFNs, they *alter the terrain of struggle* and open the possibility for additional claims and forms of recognition that can *transform the underlying structures of power* that cause injustice. While both "affirmative" and "transitional" conceptions of justice have been features of critical agrarian studies and feminist political ecology, this explicit analysis of their different approaches enables us to recognize the limitations of the former and the need to expand the latter.

This work resonates with earlier critiques of liberal feminism and capitalist development from more radical perspectives (Carney and Watts 1990; Schroeder 1996), including feminist political ecology arguments that dismissed the "myths" that women do not engage in agricultural production and leadership of political struggles (Fortmann and Rocheleau 1985; Rocheleau et al. 1996), and that call attention to the "rooted networks" of female-led social movements (Escobar et al. 2002; Rocheleau and Roth 2007). The merits of this approach includes a refusal of simple binary thinking (such as rural/urban, producer/consumer, passive victim/active organizer, etc.), an attention to entanglements of power within networks, and a recognition that networks both shape and are shaped by territories (Escobar et al. 2002; Rocheleau and Roth 2007). Thus, the agency of peasant women in AFNs can be theorized as a form of "self-organization from below", which reveals their "power of mobility and connectivity in horizontal and vertical dimensions" (Rocheleau and Roth 2007, p. 436) in ways that transcend the static imagined territoriality of the "rural left behind", and the powerless condition of passive victim this discourse engenders.

Methods and field sites

I draw upon ethnographic research methods, including participant observation (of AFNs and government regulations of food safety), semi-structured interviews, and qualitative surveys. These methods have been widely used in interdisciplinary social sciences, and proved to be particularly useful in identifying the nuances of gender injustices in everyday life situations such as the plight and agency of female migrant workers in China (e.g. Lee 1998; Ngai 2005; Yan 2008) and female leaders and critics of rural development initiatives in China, India, Latin America, and beyond (Escobar et al. 2002; Rocheleau and Roth 2007; Jacka 2010, 2013, 2018; Deere 2005; Sangting Writers Collective 2010; de Schutter 2013; Ge et al. 2011; Elmhirst 2011).

Most of my fieldwork was undertaken in Gu[4] village in Guangxi Zhuang Autonomous Region and Bian[5] village in Henan province during the Summer of 2014, Spring of 2015, and Spring of 2017. I undertook 126 semi-structured interviews in Gu and Bian villages with peasant households, which included interviews with ten childless elders, disabled and orphans (五保户), six school teachers, and four spiritual leaders, the majority of whom were all female. I also undertook 86 semi-structured interviews with other key informants, including rural cooperative leaders, county and township officials, food vendors and brokers, agricultural input vendors and brokers, local food market and restaurant managers, and urban representatives of food safety-oriented NGOs, community supported agriculture (CSA) initiatives, and buyers' groups.

The main focus of my research at the time was the establishment of new top-down government laws and regulations on food safety, and the AFNs among peasants and between peasants and urban consumers to produce and distribute safer, organic food (Zhang 2017; Zhang and Qi 2019). But one of my key findings was that the articulation of gender, ethnic, and class identity among peasants and rural cooperative leaders appears to influence how much they prioritize the production of safer organic food, as the female-led cooperative was doing in the ethnic minority village of Gu, or the scaling-up and commercial success of agricultural production, as was taking place in the male-led cooperative in Bian village (Zhang 2016, 2018). This two-case comparison may not be sufficient to draw clear conclusions about gender as a determinant factor, which requires not only more case studies but also clearer analysis of the way gender, class, ethnic identity, and other factors articulate in each situation. But it certainly enables us to pose questions about gender, agrarian studies, and rural development politics as undertaken in this present article.

Female leadership in food sovereignty

Many believe that China does not have "social movements" because of the authoritarian nature of its state, and the limited space for "civil society" to coordinate nationwide protests and organize openly, independently, and especially in opposition to the Communist Party (Ho and Edmonds 2007). Yet I argue bottom-up initiatives for self-protection in face of China's ongoing food safety crisis (i.e. the establishment of AFNs) constitutes a key aspect of the global social movement for *food sovereignty*. This fits the theoretical foundations of "food sovereignty" as a political struggle for greater

---

[4] Pseudonym.

[5] Pseudonym.

control and autonomy over food production and consumption, contrasting it with commercial and distributive frameworks of "food security" that have not prioritized issues of quality or the agency of food producers (Wittman et al. 2010; McMichael 2013; Bezner Kerr 2013). Moreover, the literature on food sovereignty has become increasingly attentive to household-level power and gender dynamics (Wittman et al. 2010; McMichael 2013; Bezner Kerr 2013), and by reflecting upon female-led AFNs as part of the global food sovereignty movement, my work also expands upon the role of women in this struggle. In order to sustain this argument, I call attention to the growing literature on "everyday life" resistance among the poor and marginalized in the Global South as a form of "quiet social movement" (Zhang and Qi 2019; cf. Bayat 2000, 2013; Roy 2015; Vaz-Jones 2018). In this context, I discuss female leadership not merely in rural cooperatives, agrarian studies, development initiatives, CSAs and other AFNs, but collectively as female leadership in the food sovereignty movement in China, echoing the work of Diane Rocheleau and other feminist political ecologists on rooted networks of environmental and rural social movements worldwide (Rocheleau et al. 1996; Escobar et al. 2002; Rocheleau and Roth 2007; Jarosz 2011; Nyantakyi-Frimpong 2017). Identifying these AFNs as female-led rooted networks for food sovereignty accomplishes two theoretical and empirical purposes: first, it delineates the translocal connections through which so-called "left behind" women exercise agency and power, and second, it enriches formulations of the global food sovereignty movement with attention to these less confrontational everyday life practices and the centrality of food safety concerns for such struggles in places under more authoritarian regimes.

The role of female scholars and educated young women

First, it is worth highlighting that even though male scholars like Wen Tiejun usually get credit for leading the "new rural reconstruction movement" and several of their associated initiatives, very often there are younger people, and particularly younger women, who actually do the hard work of organizing, implementing, and cultivating these initiatives. This is particularly evident in some of the most famous AFNs emerging in China. One example is the Little Donkey Farm, a peri-urban farm in Beijing where urban intellectuals and young volunteers have been establishing a CSA and organic farming initiative. Its core founder was Shi Yan, a young woman who was a PhD student of Wen Tiejun, and spent some time as exchange student in a US university, where she learned the CSA model and practice. Upon her return to China, Shi Yan became one of the founders the Little Donkey Farm in 2008, and continues to play a leading role in promoting organic food production in China as

founder of Shared Harvest, another high-profile CSA-turned-agribusiness in Beijing.

The situation is similar with the Beijing Organic Farmers' Market (BOFM), another very high-profile AFN in China. Chang Tianle was a young female social activist among the first group of volunteers of the BOFM, joining it upon her return from studying abroad in the US in 2010, while working in the Institute for Agriculture and Trade Policy think tank. The BOFM was originally founded by a foreign couple, but it was Chang Tianle's initiative to create an online presence for the BOFM. Her online promotion was extremely successful, and as the market grew, Chang Tianle became increasingly involved, eventually leaving her other work to assume full-time management of the BOFM, and networking even more to expand China's organic food social movement.

But the majority of female scholars and educated women cultivating their own and other female leadership in food sovereignty in China obtained their education and remain firmly rooted within China itself. Their input as critical agrarian scholars has been instrumental for the development of multiple other AFNs and food sovereignty initiatives in China. Tamara Jacka was supportive of the transformative potential of their work, yet apprehensive and critical of their limitations in addressing gender justice explicitly (Jacka 2013). My research findings support some of her critique, but also reveal more positive and optimistic trajectories. One of the female scholars somewhat critiqued by Tamara Jacka was He Huili, a professor of development studies at COHD who was very involved in the creation of the Bian rural cooperative and CSA in Henan province. As Jacka correctly points out, He Huili's community engagement did attempt to empower women and improve the condition of the most marginalized persons—the "left behind" women, elders, and children—more clearly than Wen Tiejun or He Xuefeng, yet her publications never addressed gender issues directly (e.g. He H. 2007). After Tamara Jacka's publication, I began my own fieldwork in Bian village, and examined the development of the Bian village cooperative that He Huili helped create.

The Bian village cooperative was created in 2004 with 39 households adapting the CSA model, in a village highly controlled by five men from the leading family clans. The only exception was He Huili herself, who was not only responsible for academic support for the project, but also politically responsible as deputy governor of the county in which Bian village is located. Their original aim was to produce organic rice for members who paid in advance to assist the cooperative with production. However, the cooperative was not able to fully abandon the use of synthetic fertilizers, and they failed to obtain government certification as "organic", so they marketed it instead as "pollution free" (Zhang and Qi 2019). In 2009, the Bian village case received national attention, as their cooperative was showcased by then-vice president Xi Jinping as an example to be followed for rural development. He Huili was a key organizer of this political and publicity stunt.

However, as expected from Jacka's (2013) critical assessment of the "new rural reconstruction" movement and other critical scholarship of the limitations and cooptation of China's new cooperatives (e.g. Yan and Chen 2013; Zhang 2015), the efforts of the Bian village cooperative to produce "pollution free" rice were largely coopted by male local cadres for their own personal gains. This happened especially after severe droughts affected the cooperative's own rice production in 2014, threatening the economic viability of the project. After all, the cooperative was contracting over 300 households to provide an ever growing amount of rice, even selling beyond their own CSA members, especially after their case gained national-level attention. And the male cooperative leaders feared their customers would not accept the CSA terms of shared cost, shared risk, and shared results. Therefore, the male leadership of the cooperative began buying up regular (i.e. not "pollution free") rice from neighboring villages, processing and repackaging them with the cooperative's brand, and selling it as if it was their own "pollution free" production. In other words, focusing on branding and sales instead of production. This was never admitted publically, but it was an open secret among residents of Bian village and in the surrounding area at the time I conducted my fieldwork from 2014 to 2017.[6] He Huili herself became frustrated with this outcome (and other complicated issues beyond the scope of this article), withdrew her leadership role in the Bian village project, and shifted instead to new collaborative research in her own home village, where she is placing culture and gender issues more prominently in her research and development agenda, as she indicated to me in a personal conversation in 2017.

He Huili's new collaborative research project in her own village (in Lingbao, Henan province) started around 2013, when she was growing distant from the Bian village cooperative to which she devoted her work for ten years. She realized it was not enough to promote economic production alone, and it was necessary to refocus on cultural and gender issues in their own right, as she perceived women to be "more active" in such initiatives already. Therefore, she combined various existing peasant cooperatives to established the Peasant Grassroots College (弘农书院), focusing on cultivating the traditional Chinese agricultural practice and spirit. All key leaders of the Peasant Grassroots College are female, after the only young man who participated its core group gave up the project. When I first met one of the

young leaders at Grassroots College in 2014, she was still very shy and nervous. With He Huili's support and encouragement, and especially after they transitioned to a female-only core leadership group, she was transformed. When I met her again, she was a strong and confident leader, even acting as the main organizer of the province-wide Grassroots College Forum in 2017. Despite the shift away from He Huili's intellectual leadership in the Bian village case, therefore, her own leadership role and attention to gender issues continues to grow with transitional approaches to justice, creating conditions for transformation of the structural conditions that negatively affect rural women.

In other words, when women advance in their own education, they can lead food sovereignty initiatives like Shi Yan and Chang Tianle have done, thus contributing to a change in the social and economic terrain upon which female peasants are marginalized (i.e. mainstream food networks), creating new markets and discourses that can empower broader counter-movements to the capitalist reforms that are aggravating women's exploitation and marginalization in Chinese society (Zhang 2016, 2018). When women take up leadership in academia and local government, as He Huili has done, they may even undertake efforts to alter more directly the structural conditions that preclude or enable other women to empower themselves, as illustrated by the case of this young female leader who emerged from the Grassroots College. These are not merely affirmative strategies of justice, but rather transitional strategies, since they alter the terrain upon which justice is conceived and grappled. Rather than merely affirming the existence of such women, or pivoting on their "inclusion" in agroecological initiatives, we can only fully grasp their significance when emphasizing their agency in a struggle for transitional justice.

Female leaders in local government, cooperatives, and AFNs

As implicit in the sub-section above, the role of female scholars and young social activists requires networking with the rooted leadership of cooperatives and CSAs. A clearer example of women networking in leadership across all these roles was evident in my second case study in Gu village, Guangxi province.[7] In 2001, a female scholar and proponent of participatory rural development from the Chinese Academy of Sciences, Song Yiqing, went to Gu village to launch a development project focused on breeding local maize

---

[6] Field site observations in 2014, 2015, and 2017, and various surveys and interviews with peasant households and key informants in and around Bian village, Henan.

[7] The information in this and the following paragraphs comes from my field site observations in 2014, 2015, and 2017, and various interviews with peasant households and key informants in and around Gu village, Guangxi.

varieties and sustaining local culture. There she collaborated primarily with Lu Yanyan[8], a female cadre who joined the village committee in 1991, and was vice-director of the village since 1999. Lu Yanyan was also among the most well-educated people in the village, having completed high school in a remote mountainous region where most ethnic minority children abandon school much earlier to work in the fields or migrate out to work in the factories and social services of neighboring Guangdong province. In addition, she is a committed CCP member, and received several awards from the CCP for the work I describe below.

Song Yiqing went to work in Gu village because Lu Yanyan had already established a cultural cooperative—which was composed almost exclusively of "left behind" women and female elders—to sustain Zhuang and Yao ethnic minority dances and traditions since 1998. Like He Huili, Song Yiqing believed they could develop from these cultural initiatives to economic projects (Zuo and Song 2002). Her initial efforts were limited to a traditional participatory rural development approach involving participatory mapping, rapid rural appraisal through surveying, and provision of seed varieties and short-term extension of breeding assistance. As was also found in several other similar cases (cf. Cahn and Liu 2008; Jacka 2010, 2013; Zhao 2011; Ge et al. 2011), these efforts themselves failed to produce any significant transformation of Gu village's difficult social and economic condition. On the other hand, Song Yiqing's intervention did serve a *transitional* function for Lu Yanyan and her female partners in the village to advance *their own* initiatives afterwards, shifting the conditions of the terrain of struggle and opening new opportunities for mobilization (cf. Fraser 2003). In particular, it enabled Lu Yanyan to lead the transformation of the cultural cooperative into bottom-up construction of AFNs of their own.

Lu Yanyan and her whole cultural cooperative were invited to the COHD in Beijing to give a show and participate in workshops, and connected with other peasant and ethnic minority leaders to cultivate a network of solidarity, especially in the practice of saving and reproducing local seed varieties. She continues to be frequently invited to national and even international workshops and meetings organized by critical agrarian studies scholars, but she almost always politely declines these invitations. Lu Yanyan explained to me in one of our many personal conversations

> I don't have time for all those meetings… my work needs to continue to focus on our cooperative, our village government, our own problems at home. All that training and experience-sharing are not really applicable to our village and cooperative, so instead of spend-

ing time on that, now I am more and more focused on our own things and experiences.[9]

Indeed, Lu Yanyan herself deserves credit for the most successful advancements in her village. In 2006, she led the development of their cooperative from merely cultural activities to the organization of organic vegetable production.[10] Her efforts were directed primarily at improving the economic condition of "left behind" women, and particularly elderly women. As she explained to me in an interview:

> Only the poorest villagers have the willingness to join the ecological cooperative to produce pollution-free vegetables and raise pigs and chickens. This is because they are old, and cannot migrate out of the village to earn cash. So this is a source of sustainable livelihood for them.[11]

At first their production focused on distribution among the "left behind" households. But due to Lu Yanyan's efforts, the cooperative grew from an initial 11 members to over 57 by 2010, renamed Yangshan Yanyan Ecological Planting and Breeding Cooperative[12], and expanded distribution to a NGO-operated farm-to-table restaurant in the provincial capital of Nanning (the Farmer's Friend restaurant). In an even more illustrative contrast with the case of Bian village cooperative in Henan, when drastic floods destroyed much of the production of the cooperative in Gu village, Lu Yanyan and her female partners in the cooperative preferred to sustain organic production for their own household consumption, rather than scale-up production with a greenwashed alternative that could maintain their commercial supplies to the farm-to-table restaurant in Nanning. As Lu Yanyan explained to me, this required active leadership by her and the other elder women in the cooperative:

> The ecological planting and breeding cooperative is facing a problematic issue: the younger peasants want to use a modern way to produce with hybrid seeds and fertilizer to sell to the ordinary market. However, the elderly members and I insist on using the ecological

---

[9] Personal interview with Lu Yanyan, Gu village, Guangxi, January 15, 2017.

[10] Their production was not certified organic, because the cost of obtaining and renewing government certification was beyond their capacity, so it was marketed as "green food" instead. But the cooperative members and its CSA consumers both recognized it as "organic" (绿色有机的). I verified through field site visits in 2014, 2015, and 2017 that in fact they do not use chemical pesticides and fertilizers, and so in this article I follow their convention in calling it "organic".

[11] Personal interview with Lu Yanyan, Gu village, Guangxi, January 15, 2017.

[12] Pseudonym.

---

[8] Pseudonym.

way to produce less but safer and good food to sell to those who think it is worthy to buy.[13]

The restaurant was also pushing down prices by purchasing from various other villages, and complaining that Gu's cooperative could not scale-up and guarantee a steady supply of all the vegetables they needed, so Lu Yanyan led efforts to establish new marketing channels at farmers' markets in their own county. Through her leadership, Lu Yanyan not only improved the economic conditions of the "left behind" women in her village, but also gained further political power for herself, becoming village director and Communist Party secretary since 2008.

We can conclude, therefore, the case of Gu village demonstrates precisely a successful case of the strategy of developing from cultural initiatives to economic cooperatives, and although external support was important, the determinant factor was essentially the strong female leadership by Lu Yanyan. Her bottom-up initiatives effectively transformed the "burden" of agricultural production faced by women labeled as "left behind" into a more fundamentally transitional strategy that is enabling vulnerable women, particularly elderly women, to transform structural conditions through self-empowerment by cooperation in agricultural production, self-governance, and food sovereignty. Affirmative approaches to justice, such as the simple identification of these women as "left behind" and their inclusion in externally-organized development projects, are not sufficient to recognize and leverage their self-empowerment initiatives. These initiatives rest upon their own agency and leadership, which improves both their livelihoods and their self-esteem, as they do not identify themselves as "left behind" victims, but women leading efforts in the production of safe and organic food for themselves and their own alternative markets.

Continuing challenges and obstacles

Despite the advancement of all these transitional approaches to gender justice across China's countryside and their networking with urban-based scholars and consumers, with their potential for transforming the structural conditions of power that make women more vulnerable to exploitation and oppression, there are still various challenges and obstacles to be overcome. As documented in several other cases across the "developing world", the "economic inclusion" of women in capitalist societies and the inclusion of "gender issues" in new governmental initiatives may actually

strengthen hierarchical power relations of female subordination to fathers, husbands, their families and clans, and even patriarchal states themselves (Ge et al. 2011; Lyon et al. 2017). Moreover, "female empowerment" initiatives may even become coopted to sustain neoliberal discourses and practices that ultimately undermine gender justice even further (e.g. World Bank and IFPRI 2010; cf. Sangtin Writers Collective 2010; Fraser 2009). Real transformations of society ultimately require radical shifts in social norms and institutional organizations.

Social and political conditions in China, however, remain very challenging for transitional strategies for gender justice. These range from social norms that discriminate against women in educational and employment opportunities, differential incomes and advancement trajectories for women in the workplace, gender bias in the recruitment and advancement in political offices, and social practices in both domestic and political spaces that are "both unappealing and risky for women" (Howell 2008, p. 76). Recognizing and encouraging women's leadership is therefore necessary, but not sufficient. In addition, it is also necessary to simultaneously redistribute unpaid care work and other *domestic* labor from women to men, and alter the social norms and institutional structures of *political* life. Otherwise, these new roles and responsibilities of female leadership may compound burdens rather than become a means for empowerment, as has been widely acknowledged in feminist literature (de Schutter 2013; Lyon et al. 2017).

The case of Lu Yanyan can be used once again for illustration, yet the narrative below is representative of virtually every single female leader who I have encountered through the course of this research. Many of the key challenges and obstacles she identified ultimately arise from the patriarchal relations with her husband, his family clan, and their children:

> I got married when I was only 17 years old, and came from another poorer village. I am not a local person here and have different family name. I have to be very cautious to do anything, even to be excellent, because my husband belongs to the biggest clan in this village. There are more eyes on my behavior. When I began to engage in the village management affairs, my husband and his relatives did not believe that I could do well as a woman who came from outside [the village]. So I had to try very hard to convince them that I can take charge the village even though I am an "outside woman". Now I became very busy with my work, so I do not have time to cook for the family and to take care of my grandchildren. So sometimes my husband, my sons, especially my two daughters-in-law complain with me about this. I have no choice now because I

---

[13] Personal interview with Lu Yanyan, Gu village, Guangxi, May 12, 2015.

have to sacrifice the time with them to help more the others.[14]

Additional challenges include women's systematic disenfranchisement from property ownership in both urban and rural areas, particularly in cases of divorce and displacement (Li and Bruce 2005; Sargeson and Song 2010; Fincher 2016), the destabilization of peasant knowledge for agroecological production (Bezner Kerr 2013), the limited understanding of middle-class consumers about the nature of the food safety crisis and the challenges of peasant production, and the sustainability of networks of mutual trust between peasant producers and urban consumers (Zhang and Qi 2019), all of which are especially serious obstacles for China's food sovereignty movement, and consequently women's leadership within it. Further empirical evidence of what is often termed the "multiple burdens" faced by women engaged in economic and political leadership seems hardly necessary in this article, as this finding is widespread among scholars who examine this topic in China (e.g. Howell 2008; Jacka 1997, 2018; Ge et al. 2011). Rather, it is more important to relate feminist theories of justice to the complexity of burden and opportunity for empowerment that results from the feminization of agriculture, and how this approach enables us to move beyond the victimization of supposedly isolated and homogeneous "left behind" rural women.

Conclusion

I have argued that the scholarship and advocacy on "left behind" populations, particularly women, needs to advance through deeper engagement with feminist theories of justice and feminist political ecology. Therefore, I developed a theoretical framework and illustrated it with my empirical research on how we can and must pay more attention to the manner that rural women are not merely passive victims during recent social transformations associated with rapid rural-to-urban migration and new dynamics of translocal family reproduction (Jacka 2018). In fact, these women are becoming leaders in agricultural production initiatives, particularly for safer and organic foods to address China's ongoing food safety crisis. This constitutes a "quiet" social movement for feminism and food sovereignty, as it addresses various forms of resistance to displacement, marginalization, and discrimination.

My theoretical contribution and empirical findings thus contribute to broader debates about capitalist transformation, rural activism and the "politics of possibility" in China

(Ho and Edmonds 2007; Day and Schneider 2018), and broader feminist debates about the feminization of agriculture as a burden involving socio-economic exploitation on the one hand, and opportunities for female leadership and empowerment on the other. In problematizing state efforts and academic scholarship that focus on "left behind" populations merely as victims, I contribute to the advancement of a collective argument that simply including women and other vulnerable populations in affirmative approaches to justice may still aggravate social relations of production that exclude, marginalize, and exploit women (Ge et al. 2011; Jacka 2013; Day and Schneider 2018). Moreover, such "affirmation" of rural women as a supposedly homogenous and isolated group of victims overlooks their agency, their heterogeneity in terms of socio-economic and geographical mobility, and their rooted networks that constitute a quiet social movement for food sovereignty. The significance and implications of my research are the following: shifting focus to women's role as leaders—rooted and networked peasant women, local cadre, scholars and NGO organizers—identifies a more productive path for research in critical agrarian studies and development studies in China, recognizing and supporting female-led transitional strategies that may transform the basic conditions of struggle for social justice, the reproduction of livelihoods, and food sovereignty in China.

Acknowledgements For their constructive criticism, support, and feedback on earlier versions of this paper, I would like to thank Qi Gubo, Philip McMichael, Ewan Robinson, Hilary Faxon, Gustavo Oliveira, the anonymous reviewers, and colleagues at the 2016 conference of the Sociology of Development section of the American Sociology Association (ASA), the 2018 International Conference of the BRICS Initiative for Critical Agrarian Studies (BICAS), and the 2019 Scaling Out Agroecology Practices in China conference at Nanjing University. I would also like to thank Lu Yanyan, Song Yiqing, He Huili, and all others who agreed to participate in my research.

References

Barker D.K. 2005. Beyond women and economics: Rereading "women's work." *Signs: Journal of Women in Culture and Society* 30(4):2189–2209

Bayat, A. 2000. From "dangerous classes" to "quiet rebels": Politics of the urban subaltern in the Global South. *International Sociology* 15 (3): 533–557.

Bayat, A. 2013. *Life as politics: How ordinary people change the Middle East*, 2nd ed. Palo Alto: Stanford University Press.

Bezner Kerr, R. 2013. Seed struggles and food sovereignty in northern Malawi. *Journal of Peasant Studies* 40 (5): 867–897.

Butler, J. 2004. *Precarious life: The powers of mourning and violence.* New York: Verso.

Cahn, M., and M. Liu. 2008. Women and rural livelihood training: A case study from Papua New Guinea. *Gender & Development* 16 (1): 133–146.

Carney, J., and M. Watts. 1990. Manufacturing dissent: Work, gender and the politics of meaning in a peasant society. *Africa* 60 (2): 207–241.

---

[14] Personal interview with Lu Yanyan, Gu village, Guangxi, June 5, 2015.

Chang, H.Q., F. MacPhail, and X.Y. Dong. 2011. The feminization of labor and the time-use gender gap in rural China. *Feminist Economics* 17 (4): 93–124.

Chung, Y.B., S.L. Young, and R. Bezner Kerr. 2019. Rethinking the value of unpaid care work: Lessons from participatory visual research in central Tanzania. *Gender, Place & Culture* 26 (11): 1544–1569.

Day, A., and M. Schneider. 2018. The end of alternatives? Capitalist transformation, rural activism and the politics of possibility in China. *Journal of Peasant Studies* 45 (7): 1221–1246.

de Brauw, A., Q. Li, C.F. Liu, S. Rozelle, and L.X. Zhang. 2008. Feminization of agriculture in China? Myths surrounding women's participation in farming. *China Quarterly* 194: 327–348.

de Brauw, A., J.K. Huang, L.X. Zhang, and S. Rozelle. 2013. The feminization of agriculture with Chinese characteristics. *Journal of Development Studies* 49 (5): 689–704.

de Schutter, O. 2011. *Contract farming and the right to food*. Report of the Special Rapporteur on the right to food to the sixty-sixth session of the General Assembly, UN doc. A/66/262.

de Schutter, O. 2013. The agrarian transition and the "feminization of agriculture". *Food Sovereignty: A Critical Dialogue. International Conference at Yale University, Paper Series #37*. New Haven, CT: Yale University.

Deere, C.D. 2005. The feminisation of agriculture? Economic restructuring in rural Latin America. *Occasional Paper 1*. Geneva: United Nations Research Institute for Social Development.

Dolan, C.S. and K. Sorby. 2003. Gender and employment in high-value agriculture industries. *Agriculture and Rural Development Working Paper* 7, Washington, D.C.: World Bank.

Du, P. 2004. 聚焦"386199"现象关注农村留守家庭 (The phenomenon of "386199", focusing on rural left-behind families). *Population Research* 04: 25–36.

Elmhirst, R. 2011. Introducing new feminist political ecologies. *Geoforum* 42 (2): 129–132.

Escobar, A., D. Rocheleau, and S. Kothari. 2002. Environmental social movements and the politics of place. *Development* 45 (1): 28–36.

Fei, X.T. [1948] 1992. *From the soil: The foundations of Chinese society*. Berkeley: University of California Press.

Fincher, L.H. 2016. *Leftover women: The resurgence of gender inequality in China*. New York: Zed Books.

Food and Agriculture Organization (FAO). 1996. *FAO focus: Women and food security: Women hold the key to food security*. Rome: FAO.

Food and Agriculture Organization (FAO). 2010. *The state of food and agriculture 2010–11. Women in agriculture: Closing the gender gap for development*. Rome: FAO.

Fortmann, L., and D. Rocheleau. 1985. Women and agroforestry: Four myths and three case studies. *Agroforestry Systems* 2 (4): 253–272.

Fraser N. 2003. Social justice in the age of identity politics: Redistribution, recognition, and participation. In: Fraser N, Honneth A (eds) *Redistribution or recognition? A political–philosophical exchange*. Verso, London, pp 7–109

Fraser, N. 2009. Feminism, capitalism and the cunning of history. *New Left Review* 56: 98–117.

Ge, J., B. Resurreccion, and R. Elmhirst. 2011. Return migration and the reiteration of gender norms in water management politics: Insights from a Chinese village. *Geoforum* 42 (2): 133–142.

Gilson E. 2016. Vulnerability and victimization: Rethinking key concepts in feminist discourses on sexual violence. *Signs: Journal of Women in Culture and Society* 42(1):71–98

He, H.L. 2007. Experiments of new rural reconstruction in Lankao. *Chinese Sociology and Anthropology* 39 (4): 50–79.

He X.F. 2007. New rural construction and the Chinese path. *Chinese Sociology and Anthropology* 39 (4):26–38

Ho P, Edmonds R. 2007. *China's embedded activism: Opportunities and constraints of a social movement*. Routledge, London and New York

Howell, J. 2008. Gender and rural governance in China. In *Women's political participation and representation in Asia: Obstacles and challenges*, ed. K. Iwanaga, 55–80. Copenhagen: NIAS Press.

Huang, J.K., K. Otsuka, and S. Rozelle. 2008. Agriculture in China's development: Past disappointments, recent successes, and future challenges. In *China's great economic transformation*, ed. L. Brandt and T. Rawski, 467–505. Cambridge: Cambridge University Press.

Jacka, T. 1997. Women's work in rural China: Change and continuity in an era of reform. Cambridge: Cambridge University Press.

Jacka, T. 2010. Women's activism, overseas funded participatory development, and governance: A case study from China. *Women's Studies International Forum* 33: 99–112.

Jacka, T. 2012. An alternative framework for understanding the situation of the "left-behind". Paper presented at the COHD Seminar Series: Critical Issues in Agrarian and Development Studies, n. 4. China Agricultural University, College of Humanities and Development Studies (COHD), September 27, 2012.

Jacka, T. 2013. Chinese discourses on rurality, gender and development: A feminist critique. *Journal of Peasant Studies* 40 (6): 983–1007.

Jacka, T. 2018. Translocal family reproduction and agrarian change in China: A new analytical framework. *Journal of Peasant Studies* 45 (7): 1341–1359.

Jarosz, L. 2011. Nourishing women: Toward a feminist political ecology of community supported agriculture in the United States. *Gender, Place & Culture* 18 (3): 307–326.

Lahiri-Dutt, K., and M. Adhikari. 2016. From sharecropping to croprent: Women farmers changing agricultural production relations in rural South Asia. *Agriculture and Human Values* 33 (4): 997–1010.

Lee, C.K. 1998. *Gender and the South China miracle: Two worlds of factory women*. Berkeley: University of California Press.

Li, Z.M., and J. Bruce. 2005. Gender, landlessness and equity in rural China. In *Development dilemmas: Land reform and institutional change in China*, ed. P. Ho, 308–337. London and New York: Routledge.

Lu, B. 1996. 农村"留守女"呼唤关注 (Rural "left-behind women" call attention). *Professional Households* 07: 56.

Luo, Y.Y., and D.H. Chai. 2004. 半流动家庭中留守妇女的家庭和婚姻状况探析 (Analysis of family and marital status of left-behind women in semi-mobile families). *Theoretical Monthly* 03: 103–104.

Lyon, S., T. Mutersbaugh, and H. Worthen. 2017. The triple burden: The impact of time poverty on women's participation in coffee producer organizational governance in Mexico. *Agriculture and Human Values* 34 (2): 317–331.

Mao, Z.D. [1926-7] 1971. Analysis of the classes in Chinese society. Report on an investigation of the peasant movement in Hunan. In *Selected works from Mao Zedong*, 11-39. Beijing: Foreign Languages Press.

McMichael, P. 2013. *Food regime and agrarian questions*. Halifax: Fernwood Publishing.

Milwertz, C. 2002. *Beijing women organizing for change: A new wave of the Chinese women's movement*. Copenhagen: NIAS Press.

Ngai, P. 2005. *Made in China: Women factory workers in a global workplace*. Durham and London: Duke University Press.

Nyantakyi-Frimpong, H. 2017. Agricultural diversification and dietary diversity: A feminist political ecology of the everyday experiences of landless and smallholder households in northern Ghana. *Geoforum* 86: 63–75.

Rocheleau, D., and R. Roth. 2007. Rooted networks, relational webs and powers of connection: Rethinking human and political ecologies. *Geoforum* 38 (3): 433–438.

Rocheleau, D., B. Thomas-Slayter, and E. Wangari. 1996. *Feminist political ecology: Global issues and local experiences.* New York: Routledge.

Roy, A. 2015. Introduction: The aporias of poverty. In *Territories of poverty: Rethinking North and South*, ed. A. Roy and E. Crane, 1–38. Athens, GA: University of Georgia Press.

Sangtin Writers Collective. 2010. Still playing with fire: Intersectionality, activism and NGOized feminism. In *Critical Transnational Feminist Praxis*, ed. A. Swarr and R. Nagar, 403–419. Albany: SUNY Press.

Sargeson, S., and Y. Song. 2010. Land expropriation and the gender politics of citizenship in the urban frontier. *The China Journal* 64: 19–45.

Schneider, E. 1993. Feminism and the false dichotomy of victimization and agency. *New York Law School Law Review* 38: 387–400.

Schneider, M. 2015. What, then, is a Chinese peasant? *Nongmin* discourses and agroindustrialization in contemporary China. *Agriculture and Human Values* 32 (2): 331–346.

Schroeder, R. 1996. "Gone to their second husbands": marital metaphors and conjugal contracts in the Gambia's female garden sector. *Canadian Journal of African Studies* 30 (1): 69–87.

Shangguan, M.Z. 1994. "留守儿童"问题应引起重视 (More attention should be paid to "left behind children"). *China Scholars Abroad* 6: 39.

Si, Z.Z., and S. Scott. 2019. China's changing food system: Top-down and bottom-up forces in food system transformations. *Canadian Journal of Development Studies* 40 (1): 1–11.

United Nations Development Programme (UNDP). 2003. *China's accession to WTO: Challenges for women in the agricultural and industrial Sector—Overall report.* Collaborative research report by UNDP, UNIFEM, ACWF, NDRC, and CCAP. Beijing: United Nations Development Programme in China.

Vaz-Jones, L. 2018. Struggles over land, livelihood, and future possibilities: Reframing displacement through feminist political ecology. *Signs: Journal of Women in Culture and Society* 43(3): 711-735.

Wen Tiejun, J. 2001. Centenary reflections on the "three dimensional problem" of rural China. *Inter-Asia Cultural Studies* 2 (2): 287–295.

Wittman, H., A. Desmarais, and N. Wiebe. 2010. *Food sovereignty: Reconnecting food, nature, and community.* Halifax & Winnipeg, Nova Scotia: Fernwood Publishing.

World Bank and IFPRI. 2010. *Gender and governance in rural services: Insights from India, Ghana, and Ethiopia.* Washington, D.C.: World Bank and IFPRI.

Wu, H.F., and J.Z. Ye. 2016. Hollow lives: Women left behind in rural China. *Journal of Agrarian Change* 16 (1): 50–69.

Yan, H.R. 2003. Spectralization of the rural: Reinterpreting the labor mobility of rural young women in post-Mao China. *American Ethnologist* 30 (4): 1–19.

Yan, H.R. 2008. *New masters, new servants: Migration, development, and women workers in China.* Durham: Duke University Press.

Yan, H.R., and Y.Y. Chen. 2013. Debating rural cooperative movements in China, the past and the present. *Journal of Peasant Studies* 40 (6): 955–981.

Ye, J.Z. 2010. 再论"参与式发展"与"发展研究". 序 ("Participatory development" and "development studies" revisited. Foreword). In 参与式发展研究与实践方法 (*Methodologies of participatory development studies and practices*), ed. Li Ou, 1-8. Beijing: Social Science Academic Press.

Ye, J.Z. 2018. 留守女性的发展贡献与新时代成果共享 (Left behind women's contribution to development and achievement sharing in a new era). *Journal of Chinese Women's Study* 1: 11–13.

Ye J.Z. 2019. 农村留守人口研究:基本立场、认识误区与理论转向 (Rural left-behind population research: Fundamental standpoint, misunderstandings and theoretical turn.) *Population Research* 43(2):21-31.

Ye, J.Z., and H.F. Wu. 2008. 阡陌独舞: 中国农村留守妇女 (*Dancing solo: Women left behind in rural China*). Beijing: China Social Sciences Academic Press.

Ye, J.Z., H.F. Wu, J. Rao, B.Y. Ding, and K.Y. Zhang. 2016. Left-behind women: Gender exclusion and inequality in rural-urban migration in China. *Journal of Peasant Studies* 43 (4): 910–941.

Yi, Zhang. 1994. "留守儿童" (Children left behind). *Outlook Weekly* 45: 37.

Zhang, L. 2016. The role and plight of female leaders in rural development in China. Paper presented at the 5th Annual Conference of the American Sociological Association (ASA) Section on the Sociology of Development, Cornell University, Ithaca, USA, October, 2016.

Zhang, L. 2017. *The politics and governance of food and farming system change in China: Case studies of Bian Village in Henan and Gu Village in Guangxi.* PhD dissertation, Department of Development Studies, COHD, Beijing: China Agricultural University.

Zhang, L. 2018. From "left behind" to leader: Female leaders in food sovereignty and local governance in China. *BICAS Working paper no. 50*, BRICS Initiative for Critical Agrarian Studies.

Zhang, L., and G. Qi. 2019. Bottom-up self-protection responses to China's food safety crisis. *Canadian Journal of Development Studies* 40 (1): 113–130.

Zhang, L.X., A. de Brauw, and S. Rozelle. 2004. China's rural labor market development and its gender implications. *China Economic Review* 15: 230–247.

Zhang, Q.F. 2015. Class differentiation in rural China: Dynamics of accumulation, commodification and state intervention. *Journal of Agrarian Change* 15 (3): 338–365.

Zhang, W.G. 2009. "A married out daughter is like spilt water"? Women's increasing contacts and enhanced ties with their natal families in post-reform rural North China. *Modern China* 35: 256–283.

Zhao, J. 2011. Developing Yunnan's rural and ethnic minority women: A development practitioner's self-reflections. In *Women, gender and rural development in China*, ed. T. Jacka and S. Sargeson, 171–189. Cheltenham, UK and Northampton, MA: Edward Elgar.

Zuo, J.P., and Y.Q. Song. 2002. 农业女性化与夫妻平等: 性别与发展研究的一次本土化尝试及其政策思考 (Feminization of agriculture and equality of couples: A localization attempt of gender and development studies and its policy thinking). *Tsinghua Sociology Review* 2002: 40–46.

Publisher's Note Springer Nature remains neutral with regard to jurisdictional claims in published maps and institutional affiliations.

Li Zhang is visiting assistant professor of global and international studies at the University of California, Irvine. She was visiting fellow in the Department of Development Sociology at Cornell University in 2015–2016, and remained research fellow at the Cornell Contemporary China Initiative. Dr. Zhang was assistant professor of sociology at Henan Agricultural University, and research fellow at the China Agricultural University, COHD, where she obtained her PhD in 2017. Dr. Zhang has published on democracy and socialist theory, ecological agriculture, urban farming, and China's food safety crisis.

Agriculture and Human Values (2020) 37:1125–1138
https://doi.org/10.1007/s10460-020-10071-3

Effects of institutional pressures on the governance of food safety in emerging food supply chains: a case of Lebanese food processors

Gumataw Kifle Abebe[1] 🄞

Accepted: 6 May 2020 / Published online: 11 May 2020
© Springer Nature B.V. 2020

Abstract

Food safety has become a major development challenge and a key influence on the strategic behavior of food companies. The study seeks to analyze the effect of perceived institutional pressures on the governance of food safety and the effect this may have on food safety performance in emerging food supply chains. The research develops a conceptual framework that links perceived institutional pressures, degree of food manufacturer-supplier relationships, food safety practices, and food safety output. The hypothesized relationships were tested in the Middle Eastern context, where food safety concerns are rising. Accordingly, a survey was carried out to collect data from food quality/safety managers representing 94 food processors across Lebanon. The study finds that perceived institutional pressures have a direct and strong effect on the degree of integration in the agro-food supply chain (i.e., in terms of long-term relationships, strategic information sharing, information technology connection, and logistic integration), and such an integration, in turn, increases the intensity of food safety practices and food safety performance. However, in the absence of strong manufacturer-supplier relationships, perceived institutional pressures do not lead to improved food safety performance. The study suggests that long-term food manufacturer-supplier relationship is necessary if agro-food chain actors are to respond to established regulatory demands, industry practices, and social norms.

Keywords Institutional pressures · Food safety governance · Food safety performance · Supply chain integration

Abbreviations

GVC	Global value chain
TCE	Transaction costs economics
IT	Information technology
FSMS	Food Safety Management Systems
HACCP	Hazard analysis critical control point
BRC	British Retail Consortium
ISO	International Organization for Standardization
SQF	Safe quality food
IFS	International Food Standard
FSSC	Food Safety System Certification
IRB	Institutional Review Board
AUB	American University of Beirut
PLS-SEM	Partial least squares structural equation modeling
CR	Composite reliability
VIF	Variance inflation factor
HTMT	Heterotrait–Monotrait ratio
AVE	Average variance extracted
R^2	Coefficient of determination
Q^2	Cross-validated redundancy
IPMA	Importance-performance map analysis

✉ Gumataw Kifle Abebe
  gumataw@gmail.com

[1] Faculty of Agricultural and Food Sciences, American University of Beirut, Riad El Solh, Beirut 1107 2020, Lebanon

Introduction

In the highly globalized agro-food supply chains, food safety has become a societal concern and critical factor in influencing the strategic behavior of food companies (Diekmann and Theuvsen 2019; Ait Hou et al. 2015). Agro-food supply chains are increasingly exposed to various food safety hazards (Esteki et al. 2019; Nayak and Waterson 2019). Past and recent high profile food scares such as the *E. coli* contamination from a sprout farm in Germany (2011), the *salmonella* outbreak from peanut butter paste in the US and Canada (2008–2009), the Irish pork products contamination with dioxin (2008), and the adulteration of Chinese milk products with melamine (2008) have exerted unprecedented scrutiny on food companies to guarantee food safety. Failure to address food safety concerns can have far-reaching consequences, including economic losses, reputational damages, and business closures (Esteki et al. 2019) and a loss of livelihoods for farming communities and workers (Davidson et al. 2016). Firms may respond to real and perceived food safety hazards by strengthening food safety cultures within the organization and adopting appropriate food safety governance regimes in their relationships with other supply chain actors (Tan et al. 2017; van Ruth et al. 2017; Baur et al. 2017; Nyarugwe et al. 2020). In today's market environment, the effect of institutional pressures on the governance of agro-food supply chains is evident as new food safety-oriented agro-food supply chains (e.g., organic farming, short food supply chains, etc.) have become embedded in the food safety regimes dominated by the large-scale agro-food companies (Laforge et al. 2017; Dubois 2019; Gale and Hu 2011).

The present study focuses on food manufacturer-supplier relationships in the context of emerging food supply chains and analyzes the effect of perceived institutional pressures on the food safety culture and performance of food manufacturers. Following Schuster and Maertens (2013, p. 292), emerging food supply chains can be defined as, in this study, "supply chains with increasing food safety requirements to take advantage of the potential opportunities the modern chains offer." The "global value chain (GVC)" concept (Gereffi 2005), distinguishes global supply chains as "buyer-driven" and "producer-driven" to explain the transformations in the production and distribution of goods at the global level. Increasingly, the organization of global food chains is being reshaped by powerful buyers and thus has become buyer-driven (Gereffi and Lee 2012; Hattersley et al. 2013). Accordingly, powerful buyers ("lead-firms") are influencing the organization of agro-food supply chains and governance of food safety. Access to high-value markets by economic actors in the developing countries relies on their ability to improve competitiveness in the global supply chains coordinated by the lead firms (Gereffi and Lee 2012) and establish strategic responses to changes in the institutional environment (Mercado et al. 2018). Consequently, firms in emerging agro-food supply chains are under increasing pressures to participate in high-value modern chains and to implement food safety governance regimes to improve their position in the global market. The theory of institutions predicts that firms tend to respond to institutional pressures by choosing a governance regime that can conform to established regulatory demands, industry practices, and social norms.

The present research assesses the effect of perceived institutional pressures on the governance of food safety and the impact this may have on food safety performance. The study introduces a model that links the principal elements of supply chain integration and food safety output (performance) to test the effect of perceived institutional pressures on the governance of food safety. More specifically, the study analyzes the effect of institutional pressures on the degree of food manufacturer-supplier relationships and intensity of food safety practices and food safety output. The perishability of food products (Akkerman et al. 2010), the extended supply relationship and interconnectedness of transactions (Trienekens and Zuurbier 2008), and the recurrent food fraud (Esteki et al. 2019) have increased the vulnerability of food supply chains to various food safety risks. The current study intends to expand our understanding as to how firms, in the context of emerging food supply chains, are responding to food safety concerns while enhancing their operational efficiency and market competitiveness the modern food supply chain offers.

The empirical study was carried out among food processors in Lebanon. Lebanon is a compelling case as food safety issues are attracting increasing public attention (e.g., Bou-Mitri et al. 2018; Abebe et al. 2017; Abebe et al. 2020), following the uncovering of several food safety scandals. In 2014, for example, the Lebanese Ministry of Health launched a series of inspection campaigns targeting around 1,005 companies that included big slaughterhouses, factories, restaurants, bakeries, and supermarkets across the country. The public pressure had led the Lebanese parliament to approve its first and long-awaited food safety law in 2016 (Law No. 35).

The conceptual framework: key concepts and hypotheses

The conceptual framework presented in Fig. 1 builds on Prajogo and Olhager (2012) and links the principal elements of manufacturer-supplier relationships (hereafter called supply chain integration), food safety practices, and food safety output (performance). The current research aims to extend the work of Prajogo and Olhager (2012) by incorporating perceived institutional pressures as drivers of potential transformations in the governance of food safety in emerging supply chains. In the conceptual model developed by Prajogo and Olhager (2012), long-term relationship is viewed as a response to market competitiveness and operational efficiency; this may be useful in the context of non-food industries. However, in food supply chains, organizations must conform to established regulatory demands, industry practices, and social norms to prevent food safety hazards (Hoffman 2001). In fact, food safety is found to have played an important role in the transformations of food supply chains in developed countries (Trienekens and Wognum 2013; Trienekens et al. 2012).

The present study is informed by the theory of institutions (North 1994) and supply chain integration. According

Fig. 1 Relationships of institutional pressures, supply chain integration, and food safety output

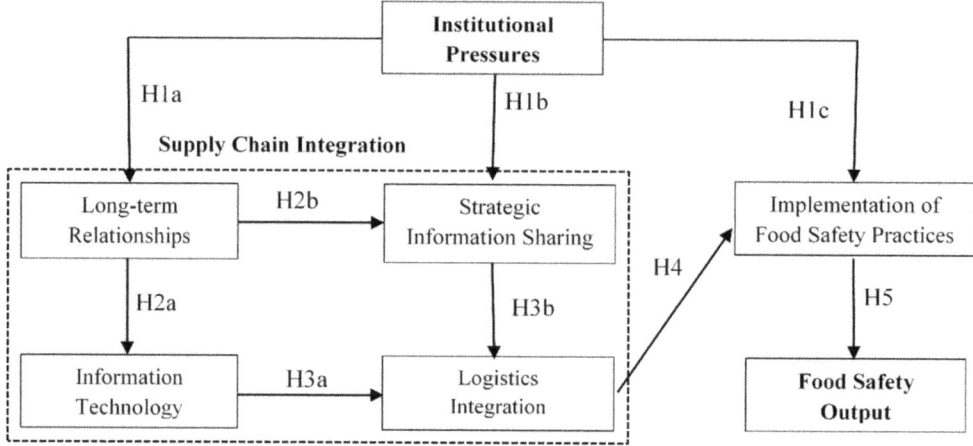

to North (1994, p.360), institutions are "humanly devised constraints that structure human interaction; they [include] formal constraints (rules, laws, constitutions), informal constraints (norms of behavior, conventions, and self-imposed codes of conduct), and their enforcement characteristics." In supply chains, relationships, interests, and values among different actors converge in institutions (Hoffman 2001). Generally, two (alternative) forms of food safety governance regimes can be identified in agro-food supply chains (1) chain-based—i.e., supply chain integration (i.e., long-term relationships, information, and logistics integration) and (2) firm-specific—i.e., food safety practices/ food safety cultures (Trienekens and Wognum 2013). The conceptual model presented in Fig. 1 tests the effect of (perceived) institutional pressures on the governance of food safety and the effect this may have on food safety output (i.e., in terms of microbiological food safety complaints, hygiene-related complaints, sampling techniques used to confirm the microbiological performance of raw materials and final products, and hygiene and pathogen conformities). The model incorporates the principal elements of integrated supply chains, namely long-term relationships, information, and logistics integration and food safety activities (i.e., food safety culture) and microbiological safety output (Luning et al. 2011).

Institutional pressures

The literature distinguishes institutional pressures at three levels (DiMaggio and Powell 1983): coercive, normative, and mimetic pressures. Coercive pressure arises from the presence of strict government rules and regulations and the organization's attempt to comply with established regulatory demands (Wu et al. 2013). Powerful trading partners (e.g., retailers) can also exert coercive pressure on other supply chain actors (e.g., manufacturers) to enforce compliance with specific (e.g., food safety) requirements (Hattersley et al. 2013). Normative pressure is exerted by professional

networks or (trade) associations as organizations try to establish legitimacy within their professional associations (Bhakoo and Choi 2013); also, it may arise from social obligations as organizations attempt to do the right thing for societies (March and Olsen 1983). Mimetic pressure relates to uncertainties in the market as organizations try to emulate other organizations that they believe are successful (John et al. 2001). In practice, coercive, normative, and mimetic pressures are interdependent; thus, a single practice can prompt the three pressures at a time (DiMaggio and Powell 1983).

Consumers' demand for food safety may affect not only direct suppliers but also other upstream actors in the food supply chain (Wever et al. 2012; Stranieri et al. 2017). Food supply chains face multiple transaction risks (Trienekens and Wognum 2013). In the context of the study, food manufacturers are exposed to both supply- and demand-side risks. For example, supply-side risks associated with the sourcing of raw materials can expose the manufacturer to demand-side risks, such as product recalls, negative media image, and reduced market competitiveness. Firms may seek a governance regime that can both optimize transaction risks and enhance competitiveness in the supply chain (Trienekens and Wognum 2013; Wever et al. 2012). Manufacturers may respond to supply- and demand-side food safety risks by transforming their supply management from arm's length (transactional) to long-term relationships (Tan et al. 2017). Also, firms can be exposed to food safety risks related to their production processes. They may respond to such risks by implementing robust food safety standards and procedures (Wever et al. 2012; Stranieri et al. 2017; Trienekens et al. 2012).

Recent studies have documented a positive relationship between institutional pressures and sustainable supply chain practices such as green initiatives (Zhu and Sarkis 2007), halal standards (Tan et al. 2017; Kurth and Glasbergen 2017), and corporate social responsibility (Awaysheh and

Klassen 2010). The food industry is one of the highly regulated sectors; firms must respond to such legislative demands by implementing various food safety practices. Thus, institutional pressure is expected to influence the integration of food supply chains, as hypothesized below:

H1a Institutional pressure is positively correlated with the degree of long-term relationships between manufacturers and suppliers.

H1b Institutional pressure is positively correlated with the degree of strategic information sharing between manufacturers and suppliers.

H1c Institutional pressure is positively correlated with the degree of food safety practices implementation by food manufacturers.

Supply chain integration

The present study builds on Prajogo and Olhager (2012), who conceptualize supply chain integration as a function of long-term relationships, strategic information sharing, information technology connection, and logistics integration.

Long-term relationship

The Transaction Costs Economics (TCE) framework (Williamson 1991) provides three generic forms to govern supply chain relationships; they include market, hybrid, and hierarchal, which may vary in terms of incentives, uncertainty reduction, and transaction costs. In a market type of governance, price is the sole mechanism to coordinate transactions, often, between anonymous parties, and such relationships are short-term. On the other extreme is a hierarchical form of governance where all the production, distribution, and marketing activities are under the control of one entity. The hybrid form of governance lies between the classical market and hierarchy forms of governance (Williamson 1991). According to TCE, the behavior of the trading partners (e.g., opportunism) and the nature of transactions (e.g., uncertainty and specificity of investments to safeguard food safety) may provide useful information to determine the type of governance regime that is needed to instill order in the relationships. The hybrid form of governance has different forms, including long-term relationships, strategic alliances, and joint ventures (Zhang and Aramyan 2009). According to Menard and Valceschini (2005, p. 424), the hybrid form is more common and attractive than the other two regimes to govern complex relationships of food supply chains. They argue that the hybrid form allows supply chain actors to keep property and decision rights distinct while providing each actor the flexibility to select reliable

partners for joint planning and investments (in key supply chain resources such as information technology and logistics) and employ strategic information exchanges with other actors in the chain.

Prior studies (e.g., Chen and Paulraj 2004; Prajogo and Olhager 2012) found a positive correlation between long-term relationships and information and logistics integration. In this era of complex and rapidly evolving food supply chains, long-term partnerships can be paramount to deal with food safety hazards (Trienekens and Zuurbier 2008; Trienekens et al. 2012). The present study extends the work of Prajogo and Olhager (2012) by testing this relationship in the context of the food industry, which is under increasing pressure and scrutiny to ensure food safety. Also, in the present study, long-term relationship is conceptualized as an organizational response to food safety risks (see H1a) and the basis for information and logistics integration. Accordingly, the second hypothesis follows:

H2a Long-term relationship is positively correlated with the intensity of information technology connection between food manufacturers and key suppliers.

H2b Long-term relationship is positively correlated with the degree of strategic information sharing between food manufacturers and key suppliers.

Information integration

Several studies have applied the notion of information integration to refer to either the use of information technologies or information sharing practices between trading partners (Jacobs et al. 2016). Prajogo and Olhager (2012) argue that both the technological and social aspects of information integration are necessary for supply chain integration. On the technological side, the internet, electronic data interchange, and advanced information technology (IT) systems can provide interactive ways to coordinate activities and manage relationships in food supply chains (Kittipanya-Ngam and Tan 2020). IT systems such as warehouse management systems (for the control of quality and safety in storage and distribution processes), laboratory information management systems (for the control of products or animals), and enterprise resource planning systems (for collaborative planning and forecasting programs) are crucial for food supply chain transparency and integration (Trienekens et al. 2012, p.62). IT connections allow supply chain actors to manage complex information exchanges in real-time (see Prajogo and Olhager 2012; Kittipanya-Ngam and Tan 2020). Studies such as Vanpoucke et al. (2017), using global manufacturing data, and Zhang et al. (2016), based on evidence of Chinese manufacturing firms, have shown a direct (positive) effect of IT connection on supply integration. Thus, a third hypothesis is presented:

H3a Information technology connection is positively correlated with logistics integration between food manufacturers and key suppliers.

Although IT connection is necessary for information integration, its intensive use may depend on the quantity and quality of information and the willingness of other supply chain actors to share strategic information (i.e., the social aspect, Prajogo and Olhager 2012). In food supply chains, information related to process and product characteristics is critical to implement traceability systems and for logistics integration (Trienekens and Wognum 2013). Following Prajogo and Olhager (2012) but applied in a different context (food supply chains), this study intends to test the effect of information integration on logistics integration from both the technological and social aspects by formulating the following hypothesis:

H3b Strategic information sharing is positively correlated with logistics integration between food manufacturers and key suppliers.

Logistics integration

Logistics integration aims to provide a seamless connection for the flow of materials from suppliers to processors (Prajogo and Olhager 2012). There is, however, a lack of consensus regarding the relationship between logistic integration and supply chain performance. According to Prajogo et al. (2016), logistics integration and performance are indirectly related via other operations such as lean production processes (e.g., total quality control principles). In the food supply chain, logistics integration involves the management of microbial growth and food waste as raw materials (or primary) products move from suppliers to manufacturers. This is necessary but not sufficient to ensure the quality and safety of the final product. In the present study, food safety is the performance indicator, which may also depend on intermediate processes such as the implementation of prerequisite programs, critical controlling and monitoring points, and compliance procedures (Luning et al. 2011). Logistics integration can influence the intensity of food safety practices and thereby food safety output. This relationship is hypothesized as follows:

H4 Logistics integration is positively correlated with the implementation of food safety practices by food manufacturers.

Implementation of food safety practices

For decades, supply chain performance has been associated with cost, quality, price, and delivery conditions and applied in the context of non-food industries (Prajogo and Olhager 2012). However, the nature of products/production and distribution are significantly different in food supply chains. In the present study, the intensity of food safety practices (i.e., activities necessary to control microbiological safety hazards and assure product safety) is measured (1) by assessing the organizational structures supporting food safety (Luning et al. 2011)—i.e., administrative conditions and information systems—and implementation of food safety management systems. In the latter, a plethora of Food Safety Management Systems (FSMS) are present to govern food safety in supply chains, including Hazard Analysis Critical Control Point (HACCP), British Retail Consortium (BRC), International Organization for Standardization (ISO) 22,000, Safe Quality Food (SQF), International Food Standard (IFS), and Food Safety System Certification (FSSC) 22,000 (Akkerman et al. 2010). Strong organizational structures supporting food safety (Luning et al. 2011) and the implementation of one or more FSMSs can enhance food safety output (Trienekens et al. 2012). Thus, the final hypothesis is stated as follows:

H5 Implementation of food safety practices by food manufacturers is positively correlated with food safety output.

Method

Data collection and study sample

A survey questionnaire, approved by the Institutional Review Board (IRB) at the American University of Beirut (AUB), was used to collect data from food manufacturers in Lebanon. The survey was executed through face-to-face interviews and online with the quality/food safety managers of each company. For the online survey, AUB-supported Limesurvey was applied. First, 342 food companies were invited for the study, which 124 of them participated. However, responses from 30 food companies were incomplete and excluded from further consideration. The study finally included the responses of 94 food processors, which constituted a response rate of 27.5%. In fact, this can be considered a success given the sensitivity of food safety issues in Lebanon. A response rate of above 20% is generally considered desirable in studies involving business organizations (Malhotra and Grover 1998). The companies included in this study were engaged in the production of baked goods, meat, dairy, fruits and vegetables, confectionery, cereals, and oils, and represented all the governorates of Lebanon.

A description of food companies participated in the study showed that 68% (64) were family-owned. Only 29% (27) manufacturers had less than 20 years of operation, while the majority of the food manufacturers (71%) were in the business for 20 years or more. Approximately 30% (28) of the manufactures in the study were predominantly export-oriented. The majority of manufactures (63%) were in the business of baked goods, confectionery, cereals, and oils processing, while the remaining 37% were primarily engaged in the meat, dairy, fruits, and vegetable processing businesses. Around 51% (48) of the manufactures were small size (i.e., 50 or less full-time equivalent employees), while 49% of them were medium (51 to 250 employees) and large-scale (more than 250 employees) enterprises.

Measurement and structural model assessment

A structured questionnaire was used to gather information needed to test the hypothesized relationships in Fig. 1. The survey questionnaire included five reflectively and two formatively measured multi-item constructs. Distinguishing measurement modes into reflective and formative is necessary to establish the content validity and structural relations of unobserved (latent) constructs (Coltman et al. 2008); in reflectively measured constructs, causality flows from the latent construct to the indicator and the opposite is true informatively measured constructs. The reflectively measured constructs are: perceived institutional pressures, degree of supply chain integration, food safety practices, and food safety output. A 7-point-Likert scale, ranging from 1 ("very unimportant") to 7 ("very important"), was applied to measure (perceived) institutional pressures and the four supply chain integration constructs. Twenty-nine items were included in the questionnaire, and all the multi-item scales were based on (or adapted from) an extensive review of prior studies (e.g., Qijun and Batt 2016; Prajogo and Olhager 2012) to ensure their content validity.

For the analysis of food safety practices and food safety output, the study has adapted the formative indicators from Luning et al. (2015). The extent of food safety practices implementation is measured as follows. First, food safety managers were asked to indicate the status of ISO 22,000 implementation in their respective organization: "Already implemented and certified by a third party" (= 1), "Already implemented but not certified by a third party" (= 2), "In-progress" (= 3), "Intended to implement in the next two years" (= 4), and "Do not intend to implement in the near future" (= 5). The study focused on ISO 22,000 mainly because it is an international, science-based, risk management approach to food safety (Codex Alimentarius Commission 2003) and has gained increasing attention due to its ability to address food safety concerns across the whole food supply chain. Second, food safety managers were asked to

characterize the food safety cultures of their organization as supportive (= 1), constrained (restricted = 2), or poor (= 3). Following Luning et al. (2015), seven items were included in the initial analysis: technological staff, the variability of workforce composition, operator competences, management commitment, employee involvement, formalization of food safety decisions, and information systems supporting food safety decisions. For the construct measuring food safety output (performance), a total of seven items were included in the initial analysis: advancedness of product sampling to confirm microbiological performance, the character of food safety evaluation, nature of microbiological food safety complaints, type of hygiene and pathogen non-conformities, the seriousness of FSMS evaluation remarks, type of hygiene-related complaints, and the extensiveness of judgment criteria. Food safety managers were asked to evaluate each item based on four situational descriptions: a low (= 1), basic (= 2), average (= 3), and advanced (= 4) performance level (Luning et al. 2015 provide a detailed description of each level).

Finally, the direct and indirect influence of perceived institutional pressures on food safety output (performance), via the mediating constructs of supply chain integration and food safety practices, was measured using Partial Least Squares Structural Equation Modeling (PLS-SEM). PLS-SEM allows the inclusion of the (unobservable) constructs discussed above (i.e., perceived institutional pressures, long-term relationships, information technology, strategic information sharing, logistics integration, implementation of food safety practices, and food safety output); several indicators (observed variables) measured each construct (see Tables 1 and 2). PLS-SEM has received increased attention in business and social science research (Sarstedt et al. 2014) as it combines optimal prediction efficiency and robustness with different scale types and small sample size.[1] Furthermore, PLS-SEM does not have distributional restrictions on measured variables and thus allows the incorporation of nominal-, ordinal- and interval-scaled indicators (Henseler et al. 2016). As discussed above, all the indicators measuring the seven constructs are ordinal-scaled. The study applied the latest algorithm (SmartPLS 3 software, Ringle et al. 2015) to investigate the strength and direction of relationships between the multi-item constructs of perceived institutional pressures, supply chain integration, food safety practices, and food safety output.

---

[1] The rule of thumb to determine the sample size is 10 times the number of maximum arrowheads pointing on a single latent variable.

Table 1 Reflective construct indicators

Indicators (sources)	Loadings	AVE	CR
Institutional pressures (Qijun and Batt 2016; Prajogo and Olhager 2012)		0.671	0.934
There are internal and external pressures to implement food safety practices to			
Enhance our export competitiveness	0.817		
Increase our competitive advantage	0.884		
Improve our company image	0.823		
Avoid negative media attention	0.705		
Improve our relationship with suppliers	0.758		
Meet customer requirements	0.871		
Improve our relationship with customers	0.861		
Long term relationship (Prajogo and Olhager 2012)		0.731	0.915
We expect our relationship with key suppliers to last a long time	0.886		
We collaborate with key suppliers to improve their quality in the long run	0.876		
The suppliers see our relationship as a long-term alliance	0.893		
We view our suppliers as an extension of our company	0.757		
Information technology connection (Prajogo and Olhager 2012)		0.639	0.898
Inter-organizational coordination with our key suppliers is achieved using electronic links	0.719		
We use information technology-enabled transaction processing	0.855		
We have electronic mailing capabilities with our key suppliers	0.763		
We use electronic transfer of purchase orders, invoices, and/or funds with our key suppliers	0.857		
We use advanced information systems to track and/or expedite shipments/deliveries with our key suppliers	0.795		
Strategic information exchange (Prajogo and Olhager 2012)		0.703	0.904
Suppliers are provided with any information that might help them	0.726		
Exchange of information with our key suppliers takes place frequently, informally, and timely	0.846		
We keep each other informed about changes that may affect the other party	0.908		
We have frequent face-to-face planning/communication with our suppliers	0.864		
Logistic integration (Prajogo and Olhager 2012)		0.710	0.907
Our inter-organizational logistics activities are closely coordinated	0.850		
Our logistics activities are well integrated with suppliers' logistics activities	0.867		
We have a seamless integration of logistics activities with our key suppliers	0.862		
Our logistics integration is characterized by excellent distribution, transportation, and warehousing facilities	0.790		

Table 2 Formative construct indicators

Formative construct indicators (source)	Outer weights	Outer loadings	VIF
Food safety practices[a] (Luning et al. 2015)			
Extent of ISO 22,000 implementation	0.558***	0.812***	1.1196
Sufficiency of information system toward food safety	0.212	0.703***	1.885
Extent of management commitment to food safety	0.472**	0.842***	2.048
Food safety performance (output)[b] (Luning et al. 2015)			
Advancedness of product sampling to confirm microbiological performance	0.551***	0.668***	1.502
Character of food safety evaluation	0.650***	0.736***	1.052
Nature of microbiological food safety complaints	− 0.088	0.516***	1.711
Type of hygiene and pathogen non-conformities	0.359**	0.554***	1.200

***p<0.01, **p<0.05, *p<0.1

[a]Excluded indicators due to low factor loadings: The degree of employee involvement, presence of technological staff, the variability of workforce composition, and sufficiency of operator competences. [b] Excluded indicators due to low factor loadings: The seriousness of FSMS evaluation remarks, type of hygiene-related complaints, and the extensiveness of judgment criteria

Table 3 Assessment of discriminant validity

	(1)	(2)	(3)	(4)	(5)	(6)	(7)
Fornell–Larcker criterion							
Food safety output (1)	Formative						
Info exchange (2)	0.150	**0.839**					
Info technology (3)	0.362	0.434	**0.800**				
Logistic integration (4)	0.137	0.711	0.564	**0.843**			
Long term relationship (5)	0.200	0.653	0.593	0.708	**0.855**		
Institutional pressure (6)	0.165	0.447	0.331	0.460	0.494	**0.819**	
Food safety practices (7)	0.576	0.179	0.304	0.191	0.216	0.016	Formative
Heterotrait–Monotrait ratio	(1)	(2)	(3)	(4)	(5)		
Info exchange (1)	–						
Info technology (2)	0.492						
Logistic integration (3)	0.812	0.649					
Long term relationship (4)	0.739	0.672	0.806				
Institutional pressure (5)	0.505	0.380	0.503	0.543	–		

Results

Reliability and validity checks

The five reflective constructs (i.e., perceived institutional pressures, long-term relationships, information technology, strategic information sharing, and logistics integration) reported in Table 1 were measured using a 7-point Likert scale, and thus an internal consistency reliability test is required to assess the interrelatedness and unidimensionality of the set of items measuring each construct. Cronbach's alpha' is traditionally used for estimating internal consistency reliability in social science research (Wong 2013). However, it tends to underestimate the true reliability value (Sijtsma 2009). "Composite Reliability (CR)" has become a preferred approach for PLS-SEM analysis (Sarstedt et al. 2014), where a higher CR value corresponds to a higher level of reliability. According to Sarstedt et al. (2014, p.108), CR values between 0.60 and 0.70 are "acceptable" while CR values between 0.70 and 0.95 are "satisfactory to good". CR values above 0.95 may indicate items that are redundant and may lead to undesirable response patterns. In the study, five indicators were removed from the final analysis as they exhibited loadings below 0.70. All five reflectively measured constructs have a value greater than 0.890 (Table 1), which is greater than the 0.7 threshold for composite reliability. Next, the convergent validity of all items associated with the five reflectively measured constructs was performed using the Average Variance Extracted (AVE). The five constructs have a value higher than the minimum acceptable value of 0.5 (i.e., 0.639 to 0.731); meaning, on average, each construct explains over 50% of the variance of its items.

The inclusion of the formative construct indicators in the model follows the guideline provided by Wong (2013, p.28), Hair et al. (2016), and Sarstedt et al. (2014). Internal consistency reliability is not appropriate for the formatively measured constructs; they are assessed based on the statistical significance and relevance of the indicator weights and loadings and collinearity. In the study, only indicators with insignificant outer weights and low outer loadings (i.e., below 0.50) are dropped. As shown in Table 2, seven indicators are retained for the two formative constructs because either their outer weights (relative importance) are significant or outer loadings (absolute importance) are greater than 0.5. Multicollinearity may be a problem among the formative constructs, and thus each item's variance inflation factor (VIF) was calculated (Sarstedt et al. 2014). All the VIF values are below 2.1, and hence multicollinearity is not a concern (Table 2).

As shown in Tables 1 and 2, the reliability and convergent validity of the constructs are well established; this leads to the next step—the assessment of discriminant validity. Table 3 assess discriminant validity using Fornell–Larcker Criterion and Heterotrait–Monotrait Ratio (HTMT). The AVE values for all the latent variables are greater than the minimum value of 0.5, confirming convergent validity. The square root of each latent variable's AVE is suggested to check for discriminant validity, and a value larger than other correlation values of the latent variables indicates that discriminant validity is satisfied (Fornell and Larcker 1981). Also, the recently proposed HTMT method was applied (Henseler et al. 2015) to assess the discriminant validity of reflectively measured constructs; a value of less than one is considered acceptable to establish discriminant validity (Henseler et al. 2016). Based on a comprehensive literature review and Monte Carlo simulation, Voorhees et al. (2016), published in the *Journal of the Academy of Marketing Science*, conclude HTMT as a "method [that] offers the best balance between high detection and low arbitrary violation" (p.119). They propose a

Fig. 2 Path model and PLS-SEM results

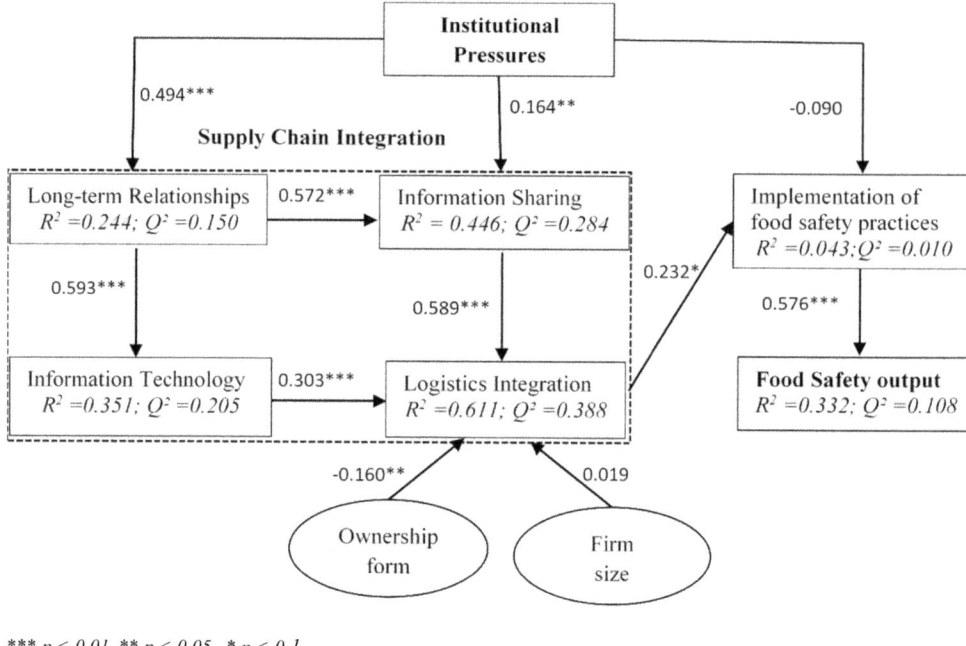

*** p < 0.01, ** p < 0.05, * p < 0.1

cutoff point of 0.85 as "the best assessment" to establish discriminant validity and as a standard for publication. The results in Table 3 confirm that the values are below this cutoff point, and hence discriminant validity is well established.

Assessing the PLS-SEM output

A bootstrapping procedure (based on 94 cases, 5000 samples, no sign change option) was performed using the latest version of SmartPLS 3 (version 3.2.8) to calculate bootstrap standard errors and check the statistical significance of various PLS-SEM results such as path coefficients (Ringle et al. 2015). PLS-SEM does not provide a standard goodness-of-fit statistic, and thus its output quality is assessed using the coefficient of determination (R^2), cross-validated redundancy (Q^2) and the path coefficients (Sarstedt et al. 2014). Figure 2 presents the R^2 and Q^2. A high R^2 value suggests greater predictive accuracy. Hair et al. (2011) noted that the interpretation of R^2 depends on the discipline, and in some business studies, a value of 0.20 is still considered high. In the study, five of the six constructs have an R^2 value greater than 0.20. Also, the "blindfolding" procedure available in SmartPLS was applied to calculate Q^2 and assess the predictive relevance of the endogenous constructs, with an omission distance of seven. The values of all the endogenous constructs are substantially greater than zero (Fig. 2), indicating the acceptability of the PLS path model (Sarstedt et al. 2014).

Path model and PLS-SEM results

As shown in Fig. 2, the empirical evidence from the Lebanese food manufacturers supported eight of the nine hypotheses (H1a, H1b, H2a, H2b, H3a, H3b, H4, and H5). Only the relationship between institutional pressures and the implementation of food safety practices (H1c) was not supported.

Discussion

The empirical study of the Lebanese food manufacturers supports the appropriateness of the developed conceptual model. Perceived institutional pressures were expected to influence food safety performance by strengthening the manufacturer-supplier relationships and improving the food safety cultures in the organizations. The study confirms the influence of institutional forces on the degree of supply chain integration, i.e., in terms of enhancing long-term relationships (H1a) and strategic information sharing (H1b) between food manufacturers and key suppliers, in keeping with earlier studies (Aung and Chang 2014; van Ruth et al. 2017; Gale and Hu 2011). However, contrary to expectation, perceived institutional pressures did not directly influence the implementation of food safety practices in the organizations but rather indirectly through the strengthening of food manufacturer-supplier relationships. The findings suggest that firms can improve food safety performance directly by increasing the intensity of food safety practices (food safety cultures) in the organization or indirectly by strengthening

Table 4 Total effects of the structural model relationships

	Total indirect effects	Total direct effects	Total effects (direct + indirect)
Institutional pressures → Long term relationship	–	0.494***	0.494***
Institutional pressures → Information sharing	0.282***	0.164**	0.447***
Institutional pressures → Information technology	0.293***	–	0.293***
Institutional pressures → Logistic integration	0.352***	–	0.352***
Institutional pressures → Food safety practices	0.082	– 0.090	– 0.009
Institutional pressures → Food safety output	– 0.005	–	– 0.005
Long term relationship → Information sharing	–	0.572***	0.572***
Long term relationship → Information technology	–	0.593***	0.593***
Long term relationship → Logistic integration	0.516***	–	0.516***
Long term relationship → Food safety practices	0.120*	–	0.120*
Long term relationship → Food safety output	0.069*	–	0.069*
Strategic information sharing → Logistics integration	–	0.589***	0.589***
Strategic information sharing → Food safety practices	0.137*	–	0.137*
Strategic information sharing → Food safety output	0.079*	–	0.079*
Information technology → Logistic integration	–	0.303***	0.303***
Information technology → Food safety practices	0.070	–	0.070
Information technology → Food safety output	0.041	–	0.041
Logistic integration → Food safety practices	–	0.232*	0.232*
Logistic integration → Food safety output	0.134*	–	0.134*
Food safety practices → Food safety output	–	0.576***	0.576***

***p < 0.01, **p < 0.05, *p < 0.1

the relationships with key suppliers. In keeping with the TCE theory, food companies have opted for a supply chain approach to respond to food safety risks and economize on transaction costs. Due to the globalization in food trade and the increasing distance, where food is produced and consumed, a supply chain approach has increasingly become a strategic response by food manufacturers in emerging supply chains (Aung and Chang 2014; Ait Hou et al. 2015).

Zooming in the effects of institutional pressures on the food manufacturer-supplier relationships, long-term relationships have strong effects on information technology connection (H2a) and strategic information sharing (H2b) between food manufacturers and their suppliers. Information technology connection (H3a) and strategic information sharing (H3b), in turn, have a strong influence on logistics integration, which is positively associated with the implementation of food safety practices (food safety culture) within the organizations (H4). These relationships are in keeping with prior studies (e.g., Chen and Paulraj 2004; Prajogo and Olhager 2012). The added value of the present study is that such relationships are tested in the context of food supply chains and food safety.

The implementation of food safety practices is strongly and positively correlated with food safety output (H5), as expected. Past studies have indicated the importance of

organizational structures supporting food safety such as technological staff, workforce composition, operator competences, technological expertise and laboratory facilities (e.g., Luning et al. 2011; Luning et al. 2015) and food safety culture such as the implementation of food safety standards—ISO 22,000 and HACCP and pre-requisite programs (Psomas and Kafetzopoulos 2015)—to enhance food safety performance in the food industry (Nyarugwe et al. 2020).

PLS-SEM method also allows analyzing the total effects (i.e., the direct and indirect effects in the structural model) and offers a richer picture of the relationships in the PLS-SEM path model (Sarstedt et al. 2014). Accordingly, the PLS-SEM algorithm has been applied to assess the indirect influence of institutional pressures on the target constructs (i.e., logistics integration and food safety output). The analysis revealed interesting perspectives (Table 4). Perceived institutional pressures have stronger indirect effects on strategic information sharing (i.e., 0.282, which is higher than its direct effects, 0.164) and information technology (0.293) and logistic integration (0.347). Strategic information sharing, in turn, has indirect effects on the intensity of food safety practices (0.137) and food safety output (0.079). These results suggest that supply chain integration mediates the relationship between perceived institutional pressures and food safety performance.

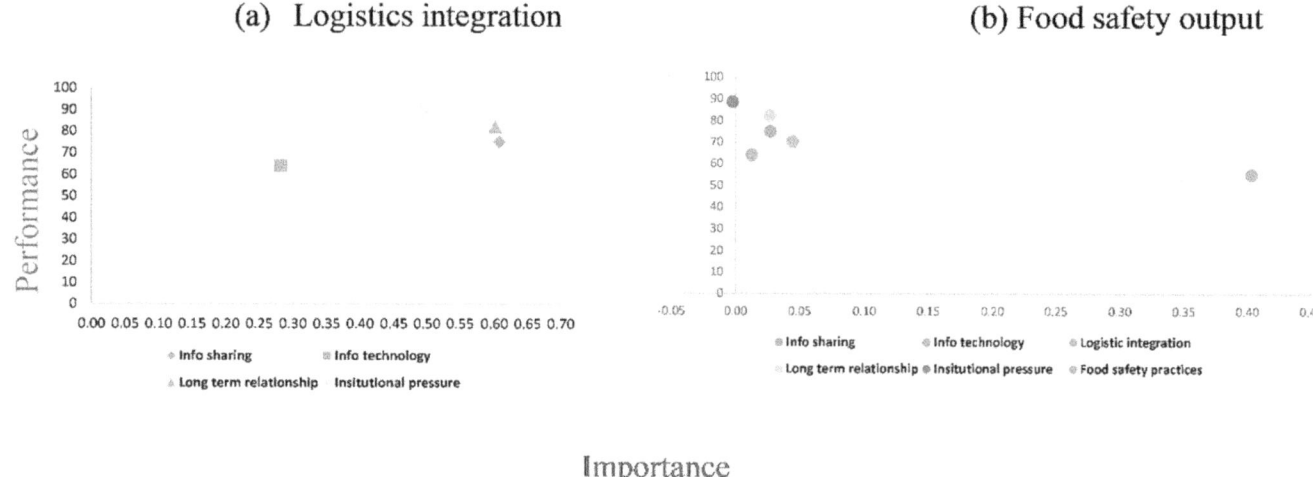

Fig. 3 Importance-performance map of target constructs: **a** logistics integration and **b** food safety output

Finally, PLS-SEM offers the importance-performance map analysis (IPMA), which may be a useful tool to provide practical recommendations for managers on how to enhance their supply relationships and food safety performance. By extending the results of the standard path coefficient estimates in Smart PLS, IPMA helps to identify factors that have a relatively strong influence and potential for improvement on the target constructs (Ringle and Sarstedt 2016). The goal is to identify predecessors with relatively strong total effects (high importance) for the target construct but low average latent variable scores (low performance).

Figure 3 displays the (unstandardized) total effects (importance) on the x-axis and the latent variable scores (performance) on the y-axis. Long-term relationships and strategic information sharing have relatively high importance on the target construct 'logistics integration' (Fig. 3a); food manufacturers may still strengthen the performance of logistics integration by improving their long-term relationships and strategic information sharing with suppliers. More importantly, in Fig. 3b, the latent variable implementation of 'food safety practices' has high importance (0.40) but low performance (55%) for the final target construct 'food safety output'. Meaning, food manufacturers have substantial room to enhance food safety performance by improving food safety practices (food safety culture) within the organizations. All other things remain equal, a unit increase in the intensity of food safety practices would enhance food safety performance by 40%.

Conclusion

Food safety is a public health concern and a major development challenge. Empirical evidence is scarce to guide public policy in the governance of food supply chains. The study has explored the effect of perceived institutional pressures on the degree of food manufacturer-supplier relationships and food safety practices and food safety performance in the context of emerging food supply chains, by applying the latest version of PLS-SEM to handle a complex model structure with a relatively small sample size (see Voorhees et al. 2016; Henseler et al. 2016). The study finds that perceived institutional pressures have strong effects on supply chain integration—i.e., long-term relationships (0.494), strategic information sharing (0.447), information technology connection (0.293), and logistic integration (0.349). However, in the absence of strong manufacturer-supplier relationships, perceived institutional pressures may not lead to improved food safety performance.

The study provides several implications for food manufacturers and policy guiding food safety governance in the context of emerging agro-food supply chains. First, the findings indicate long-term relationships as immediate responses to perceived institutional pressures (on food safety) and the basis for building integrated agro-food supply chains. Thus, food manufacturers should maintain and strengthen their relationships with key suppliers by investing in information technology and strategic information sharing. Second, the findings suggest that food manufacturers cannot improve the performance of food safety without strong food safety cultures within the organization. Although 50% of the food manufactures reported to have implemented ISO 22,000, those companies with own technical staff (i.e., food microbiologists, food quality management expert, etc.) and research laboratories for microbial analyses are only 30% and those that have detailed food safety strategy and food quality (safety) team are only 38%. Thus, food companies have substantial room to enhance the food safety culture and food safety performance of their organization by strengthening their food safety staff (in terms of quantity and competences) and facility and increasing the involvement of employees in the design and modifications of food safety control systems. Third, the findings encourage food manufacturers to establish long-term supply relationships, invest in information technology, strengthen strategic information sharing and logistics integration with supply chain partners to respond to established regulatory demands (coercive pressures), industry competition (mimetic pressures) and social norms (normative pressures). As agro-food companies strengthen their supply chain relationships, they tend to increase the intensity of food safety practices (food safety cultures), which, in turn, are necessary to enhance food safety performance. Fourth, the study establishes perceived institutional pressures transforming the governance of food safety in emerging agro-food supply chains. Public policy may support such transformations by taking (chain-wide rather than firm-specific) measures to enhance the integrity and transparency of agro-food supply chains. Finally, the findings are based on food manufacturers within Lebanon. Future studies may expand the scope of analysis to other agro-food supply chain actors and across geographies.

The study addresses important gaps in the agro-food supply chain literature by exploring the effect of perceived institutional pressures on the governance and food safety performance of agro-food supply chains. First, the study was able to develop and test a conceptual framework that links the principal elements of supply chain integration (i.e., long-term relationships, strategic information sharing, information technology connection, and logistics integration) and food safety management systems to advance our understanding of firm strategic responses to food safety concerns. Second, due to varying levels of regulatory demands, customer requirements, and cultural factors, context-specific studies are paramount to understand the strategic behavior and food safety culture of firms operating in a different set of institutions (Mercado et al. 2018). This study provides empirical evidence on the restructuring of agro-food supply chains in the Middle Eastern context (Lebanon) where food safety concerns are rising (e.g., Bou-Mitri et al. 2018; Abebe et al. 2017; Abebe et al. 2020). Finally, the study establishes a direct link between institutional pressures and the governance of food safety in emerging food supply chains.

Acknowledgements This work was supported by the Lebanese National Council for Scientific Research (Award No. 103256). There is no conflict of interest to report.

References

Abebe, Gumataw Kifle, Ali Chalak, and Mohamad G. Abiad. 2017. The effect of governance mechanisms on food safety in the supply chain: Evidence from the Lebanese dairy sector. *Journal of the Science of Food and Agriculture* 97 (9): 2908–2918. https://doi.org/10.1002/jsfa.8128.

Abebe, Gumataw Kifle, Rachel Anne Bahn, Ali Chalak, and Abed Al Kareem Yehya. 2020. Drivers for the implementation of market-based food safety management systems: Evidence from Lebanon. *Food Science & Nutrition* 8 (2): 1082–1092. https://doi.org/10.1002/fsn3.1394.

Ait Hou, Mohamed, Cristina Grazia, and Giulio Malorgio. 2015. Food safety standards and international supply chain organization: A case study of the Moroccan fruit and vegetable exports. *Food Control* 55: 190–199. https://doi.org/10.1016/j.foodcont.2015.02.023.

Akkerman, Renzo, Poorya Farahani, and Martin Grunow. 2010. Quality, safety and sustainability in food distribution: a review of quantitative operations management approaches and challenges. *OR Spectrum* 32 (4): 863–904.

Aung, Myo Min, and Yoon Seok Chang. 2014. Traceability in a food supply chain: Safety and quality perspectives. *Food Control* 39: 172–184.

Awaysheh, Amrou, and Robert D. Klassen. 2010. The impact of supply chain structure on the use of supplier socially responsible practices. *International Journal of Operations & Production Management* 30 (12): 1246–1268.

Baur, Patrick, Christy Getz, and Jennifer Sowerwine. 2017. Contradictions, consequences and the human toll of food safety culture. *Agriculture and Human Values* 34 (3): 713–728. https://doi.org/10.1007/s10460-017-9772-1.

Bhakoo, Vikram, and Thomas Choi. 2013. The iron cage exposed: Institutional pressures and heterogeneity across the healthcare supply chain. *Journal of Operations Management* 31 (6): 432–449.

Bou-Mitri, Christelle, Darine Mahmoud, Najwa El Gerges, and Maya Abou Jaoude. 2018. Food safety knowledge, attitudes and practices of food handlers in Lebanese hospitals: A cross-sectional study. *Food Control* 94: 78–84.

Chen, Injazz J., and Antony Paulraj. 2004. Towards a theory of supply chain management: The constructs and measurements. *Journal of Operations Management* 22 (2): 119–150.

Codex Alimentarius Commission. 2003. Hazard analysis and critical control point (HACCP) system and guidelines for its application. ANNEX to recommended international code of practice/general principles of food hygiene. CAC/RCP 1–1969, Rev 4. FAO/WHO Codex Alimentarius Commission.

Coltman, Tim, Timothy M. Devinney, David F. Midgley, and Sunil Venaik. 2008. Formative versus reflective measurement models: Two applications of formative measurement. *Journal of Business Research* 61 (12): 1250–1262. https://doi.org/10.1016/j.jbusres.2008.01.013.

Davidson, Debra J., Kevin E. Jones, and John R. Parkins. 2016. Food safety risks, disruptive events and alternative beef production: A case study of agricultural transition in Alberta. *Agriculture and Human Values* 33 (2): 359–371. https://doi.org/10.1007/s10460-015-9609-8.

Diekmann, Marie, and Ludwig Theuvsen. 2019. Value structures determining community supported agriculture: Insights from Germany. *Agriculture and Human Values* 36 (4): 733–746. https://doi.org/10.1007/s10460-019-09950-1.

DiMaggio, Paul J., and Walter W. Powell. 1983. The iron cage revisited: Institutional isomorphism and collective rationality in organizational fields. *American Sociological Review* 48: 147–160.

Dubois, Alexandre. 2019. Translocal practices and proximities in short quality food chains at the periphery: The case of North Swedish farmers. *Agriculture and Human Values* 36 (4): 763–778. https://doi.org/10.1007/s10460-019-09953-y.

Esteki, M., J. Regueiro, and J. Simal-Gándara. 2019. Tackling fraudsters with global strategies to expose fraud in the food chain.

Comprehensive Reviews in Food Science and Food Safety 18 (2): 425–440. https://doi.org/10.1111/1541-4337.12419.

Fornell, Claes, and David F. Larcker. 1981. Structural equation models with unobservable variables and measurement error: Algebra and statistics. *Journal of Marketing Research.* https://doi.org/10.1177/002224378101800313.

Gale, H.Frederick, and Dinghuan Hu. 2011. Food safety pressures push integration in China's agricultural sector. *American Journal of Agricultural Economics* 94 (2): 483–488. https://doi.org/10.1093/ajae/aar069.

Gereffi, Gary. 2005. The global economy: Organization, governance, and development. *The Handbook of Economic Sociology* 2: 160–182.

Gereffi, Gary, and Joonkoo Lee. 2012. Why the world suddenly cares about global supply chains. *Journal of Supply Chain Management* 48 (3): 24–32.

Hair, Joe F., Christian M. Ringle, and Marko Sarstedt. 2011. PLS-SEM: Indeed a silver bullet. *Journal of Marketing Theory and Practice* 19 (2): 139–152.

Hair, Joseph F., G. Tomas, M. Hult, Christian Ringle, and Marko Sarstedt. 2016. *A primer on partial least squares structural equation modeling (PLS-SEM).* Thousand Oaks: Sage Publications.

Hattersley, Libby, Bronwyn Isaacs, and David Burch. 2013. Supermarket power, own-labels, and manufacturer counterstrategies: International relations of cooperation and competition in the fruit canning industry. *Agriculture and Human Values* 30 (2): 225–233. https://doi.org/10.1007/s10460-012-9407-5.

Henseler, Jörg, Geoffrey Hubona, and Pauline Ash Ray. 2016. Using PLS path modeling in new technology research: Updated guidelines. *Industrial Management & Data Systems* 116 (1): 2–20.

Henseler, Jörg, Christian M. Ringle, and Marko Sarstedt. 2015. A new criterion for assessing discriminant validity in variance-based structural equation modeling. *Journal of the Academy of Marketing Science* 43 (1): 115–135. https://doi.org/10.1007/s11747-014-0403-8.

Hoffman, Andrew J. 2001. *From heresy to dogma: An institutional history of corporate environmentalism.* Stanford: Stanford University Press.

Jacobs, Mark A., Yu Wantao, and Roberto Chavez. 2016. The effect of internal communication and employee satisfaction on supply chain integration. *International Journal of Production Economics* 171: 60–70.

John, Caron H.St, Alan R. Cannon, and Richard W. Pouder. 2001. Change drivers in the new millennium: Implications for manufacturing strategy research. *Journal of Operations Management* 19 (2): 143–160.

Kittipanya-Ngam, Pichawadee, and Kim Hua Tan. 2020. A framework for food supply chain digitalization: Lessons from Thailand. *Production Planning & Control* 31 (2–3): 158–172.

Kurth, Laura, and Pieter Glasbergen. 2017. Serving a heterogeneous Muslim identity? Private governance arrangements of halal food in the Netherlands. *Agriculture and Human Values* 34 (1): 103–118.

Laforge, Julia M.L., Colin R. Anderson, and Stéphane M. McLachlan. 2017. Governments, grassroots, and the struggle for local food systems: Containing, coopting, contesting and collaborating. *Agriculture and Human Values* 34 (3): 663–681. https://doi.org/10.1007/s10460-016-9765-5.

Luning, P.A., W.J. Marcelis, J. Rovira, M.A.J.S. van Boekel, M. Uyttendaele, and L. Jacxsens. 2011. A tool to diagnose context riskiness in view of food safety activities and microbiological safety output. *Trends in Food Science & Technology* 22: S67–S79. https://doi.org/10.1016/j.tifs.2010.09.009.

Luning, P.A., K. Kirezieva, G. Hagelaar, J. Rovira, Mieke Uyttendaele, and Liesbeth Jacxsens. 2015. Performance assessment of food safety management systems in animal-based food

companies in view of their context characteristics: A European study. *Food Control* 49: 11–22.

Malhotra, Manoj K., and Varun Grover. 1998. An assessment of survey research in POM: From constructs to theory. *Journal of Operations Management* 16 (4): 407–425. https://doi.org/10.1016/S0272-6963(98)00021-7.

March, James G., and Johan P. Olsen. 1983. The new institutionalism: Organizational factors in political life. *American Political Science Review* 78 (3): 734–749.

Menard, C., and E. Valceschini. 2005. New institutions for governing the agri-food industry. *European Review of Agricultural Economics* 32 (3): 421–440. https://doi.org/10.1093/j.eurrag/jbi013.

Mercado, Geovana, Carsten Nico Hjortsø, and Benson Honig. 2018. Decoupling from international food safety standards: How small-scale indigenous farmers cope with conflicting institutions to ensure market participation. *Agriculture and Human Values* 35 (3): 651–669. https://doi.org/10.1007/s10460-018-9860-x.

Nayak, Rounaq, and Patrick Waterson. 2019. Global food safety as a complex adaptive system: Key concepts and future prospects. *Trends in Food Science & Technology* 91: 409–425. https://doi.org/10.1016/j.tifs.2019.07.040.

North, Douglass C. 1994. Economic performance through time. *The American Economic Review* 84 (3): 359–368.

Nyarugwe, Shingai P., Anita R. Linnemann, Yingxue Ren, Evert-Jan Bakker, Jamal B. Kussaga, Derek Watson, Vincenzo Fogliano, and Pieternel A. Luning. 2020. An intercontinental analysis of food safety culture in view of food safety governance and national values. *Food Control* 111: 107075. https://doi.org/10.1016/j.foodcont.2019.107075.

Prajogo, Daniel, Adegoke Oke, and Jan Olhager. 2016. Supply chain processes: Linking supply logistics integration, supply performance, lean processes and competitive performance. *International Journal of Operations & Production Management* 36 (2): 220–238.

Prajogo, Daniel, and Jan Olhager. 2012. Supply chain integration and performance: The effects of long-term relationships, information technology and sharing, and logistics integration. *International Journal of Production Economics* 135 (1): 514–522. https://doi.org/10.1016/j.ijpe.2011.09.001.

Psomas, Evangelos L., and Dimitrios P. Kafetzopoulos. 2015. HACCP effectiveness between ISO 22000 certified and non-certified dairy companies. *Food Control* 53: 134–139. https://doi.org/10.1016/j.foodcont.2015.01.023.

Qijun, Jiang, and Peter J. Batt. 2016. Barriers and benefits to the adoption of a third party certified food safety management system in the food processing sector in Shanghai, China. *Food Control* 62: 89–96.

Ringle, C. M, S. Wende, and J.-M. Becker. 2015. SmartPLS 3. Boenningstedt: SmartPLS GmbH. https://www.smartpls.com. Accessed 07 June 2019.

Ringle, Christian M., and Marko Sarstedt. 2016. Gain more insight from your PLS-SEM results: The importance-performance map analysis. *Industrial Management & Data Systems* 116 (9): 1865–1886.

Sarstedt, Marko, Christian M. Ringle, Donna Smith, Russell Reams, and Joseph F. Hair. 2014. Partial least squares structural equation modeling (PLS-SEM): A useful tool for family business researchers. *Journal of Family Business Strategy* 5 (1): 105–115. https://doi.org/10.1016/j.jfbs.2014.01.002.

Schuster, Monica, and Miet Maertens. 2013. Do private standards create exclusive supply chains? New evidence from the *Peruvian asparagus* export sector. *Food Policy* 43: 291–305. https://doi.org/10.1016/j.foodpol.2013.10.004.

Sijtsma, Klaas. 2009. On the use, the misuse, and the very limited usefulness of Cronbach's alpha. *Psychometrika* 74 (1): 107.

Stranieri, Stefanella, Luigi Orsi, and Alessandro Banterle. 2017. Traceability and risks: an extended transaction cost perspective. *Supply Chain Management: An International Journal* 22 (2): 145–159.

Tan, Kim Hua, Mohd Helmi Ali, Zafir Mohd Makhbul, and Azman Ismail. 2017. The impact of external integration on halal food integrity. *Supply Chain Management: An International Journal* 22 (2): 186–199.

Trienekens, Jacques, and Nel Wognum. 2013. Requirements of supply chain management in differentiating European pork chains. *Meat Science* 95 (3): 719–726.

Trienekens, Jacques, and Peter Zuurbier. 2008. Quality and safety standards in the food industry, developments and challenges. *International Journal of Production Economics* 113 (1): 107–122.

Trienekens, Jacques, P.M. Wognum, Adrie J.M. Beulens, and Jack G.A.J. van der Vorst. 2012. Transparency in complex dynamic food supply chains. *Advanced Engineering Informatics* 26 (1): 55–65.

van Ruth, Saskia M., Wim Huisman, and Pieternel A. Luning. 2017. Food fraud vulnerability and its key factors. *Trends in Food Science & Technology* 67: 70–75.

Vanpoucke, Evelyne, Ann Vereecke, and Steve Muylle. 2017. Leveraging the impact of supply chain integration through information technology. *International Journal of Operations & Production Management* 37 (4): 510–530.

Voorhees, Clay M., Michael K. Brady, Roger Calantone, and Edward Ramirez. 2016. Discriminant validity testing in marketing: An analysis, causes for concern, and proposed remedies. *Journal of the Academy of Marketing Science* 44 (1): 119–134.

Wever, Mark, Petronella Maria Wognum, Jacques H. Trienekens, and Simon Willem Frederik Omta. 2012. Supply chain-wide consequences of transaction risks and their contractual solutions: Towards an extended transaction cost economics framework. *Journal of Supply Chain Management* 48 (1): 73–91.

Williamson, Oliver E. 1991. Comparative economic organization: The analysis of discrete structural alternatives. *Administrative Science Quarterly* 36: 269.

Wong, Ken Kwong-Kay. 2013. Partial least squares structural equation modeling (PLS-SEM) techniques using SmartPLS. *Marketing Bulletin* 24 (1): 1–32.

Wu, Ting, Elizabeth M. Daniel, Matt Hinton, and Paul Quintas. 2013. Isomorphic mechanisms in manufacturing supply chains: A comparison of indigenous Chinese firms and foreign-owned MNCs. *Supply Chain Management: An International Journal* 18 (2): 161–177.

Zhang, X., and L.H. Aramyan. 2009. A conceptual framework for supply chain governance An application to agri-food chains in china. *China Agricultural Economic Review* 1 (2): 136–154. https://doi.org/10.1108/17561370910927408.

Zhang, Xuan, Dirk Pieter Van Donk, and Taco van der Vaart. 2016. The different impact of inter-organizational and intra-organizational ICT on supply chain performance. *International Journal of Operations & Production Management* 36 (7): 803–824.

Zhu, Qinghua, and Joseph Sarkis. 2007. The moderating effects of institutional pressures on emergent green supply chain practices and performance. *International Journal of Production Research* 45 (18–19): 4333–4355.

Publisher's Note Springer Nature remains neutral with regard to jurisdictional claims in published maps and institutional affiliations.

Gumataw Kifle Abebe is an assistant professor of Agribusiness Marketing and Management at the Faculty of Agricultural and Food Sciences, American University of Beirut, Lebanon. He has several years of experience in interdisciplinary research related to agri-food supply chains in sub-Saharan Africa and Middle East and North Africa. His research focuses on exploring the efficiency and effectiveness of agri-food supply chains in response to recent trends for food safety; food retail expansion, its drivers and impact on food security; and economics of agricultural technology adoption in arid and semi-arid regions.

Agriculture and Human Values (2020) 37:1139–1154
https://doi.org/10.1007/s10460-020-10115-8

Farmer field schools and the co-creation of knowledge and innovation: the mediating role of social capital

Chrysanthi Charatsari[1] · Evagelos D. Lioutas[2] · Alex Koutsouris[3]

Accepted: 13 May 2020 / Published online: 15 May 2020
© Springer Nature B.V. 2020

Abstract

Research has repeatedly confirmed that farmer field schools (FFS) can serve as a bridge between science and farm practice, enhancing simultaneously rural social energy. However, even though social capital is a burgeoning topic in FFS research, it is not clear whether and how it mediates FFS performance. In this mixed-methods study, using data from two FFS projects conducted in Greece, we examined if social capital among trainees facilitates the co-creation of knowledge and the co-development of agricultural innovations by farmers. A thematic analysis was performed to analyse qualitative data, whereas regression models were employed for the quantitative strand of analysis. Results revealed that social capital evolves progressively during FFS, enabling the achievement of the project's aims by enhancing in-group communication, establishing affective ties, instilling a sense of community, and triggering motivational contagion among participants. Statistical analyses confirmed that the "softest" sides of social capital (bonding and connection) significantly contribute to knowledge and innovation co-production. Although this work was based on data derived only from two FFS projects that took place in Greece, our findings underline the importance of social capital for the success of any FFS project and emphasise the need for identifying routes to nurture social capital within FFS.

Keywords Farmer field schools · Social capital · Innovation · Knowledge · Extension

Abbreviations

FFS   Farmer field schools
ICM   Integrated crop management

✉ Chrysanthi Charatsari
chcharat@agro.auth.gr

Evagelos D. Lioutas
evagelos@agro.auth.gr

Alex Koutsouris
koutsouris@aua.gr

1   Department of Agricultural Economics, School of Agriculture, Aristotle University of Thessaloniki, University Campus, 54124 Thessaloniki, Greece

2   Department of Supply Chain Management, International Hellenic University, Kanellopoulou 2, 60100 Katerini, Greece

3   Department of Agricultural Economics and Rural Development, Agricultural University of Athens, Iera Odos 75, 11855 Athens, Greece

Introduction

Conventional agricultural knowledge diffusion approaches face considerable difficulties in meeting farmers' real needs (Lioutas et al. 2019), in building upon their experience (Calo 2018), and in overcoming their culturally rooted disbeliefs about farming (Salite 2019). Farmer Field Schools (FFS) represent a response to the limited ability of these traditional, linear practices of knowledge transfer, to supply farmers with systemic and experiential knowledge. Starting in the late 1980s as an FAO-supported initiative in Indonesia (Van de Fliert et al. 1995; Winarto 1995), FFS continue to help farmers—especially in the developing countries—to enhance their knowledge potential (Davis et al. 2012).

Built on the core principles of adult education which emphasise the cultivation of learners' ability to use their creative thinking in order to respond to varying problems and situations (Gallagher 1999), FFS conceive of learning as a holistic process of adaptation to the challenges imposed by the changing nature of agriculture. Three different types of adult learning converge to set the ground of the FFS approach: experiential learning, social learning, and transformative learning. First, FFS aim at offering farmers a wide array of experiential learning activities (Röling and de Jong 1998), thus creating opportunities for acting and reflecting. Second, top-down knowledge flow gives its position to activities that promote social learning (Pimbert 2002; Tripp et al. 2005). Hence the learning emerges within a social context, through a process of social interaction and mutual production of knowledge (Pretty and Buck 2002). Third, this collective process has a transformative character (Taylor et al. 2012), since it can help trainees to transform their worldviews and attitudes, thus helping them to empower their perspectives.

Several studies support that within the framework of FFS farmers can expand a system thinking ability (Yang et al. 2008) by developing a logic-based understanding of the complex ways farm practices interrelate with the agroecosystem producing diverse crop responses (Dalton et al. 2014). By focusing on context-specific rather than generic agricultural knowledge (Ortiz et al. 2004), FFS strengthen farmers' problem-solving skills (Dzeco et al. 2010) and help them make better decisions (Yang et al. 2005). On the other hand, by actively engaging trainees in the process of co-evolution of site- and time-specific innovations (Charatsari 2015), FFS improve farmers' innovation capacity (Mfitumukiza et al. 2017).

Some studies suggest that FFS graduates achieve higher yields (Cai et al. 2016; Ortiz et al. 2019) and enjoy higher incomes (Chhay et al. 2017). However, FFS participation is also associated with benefits that go beyond the economic dimension of farming. For example, Tripp et al. (2005) found that FFS attendance leads to a reduction of agrochemicals use, whereas Chandra et al. (2017) noted that FFS have the potential to promote the adoption of climate resilience practices by farmers. In addition, Clausen et al. (2017) concluded that farmers who took part in an FFS project were more likely to use safety practices during the application of pesticides than non-participants. In other cases, FFS have been used as a vehicle for raising awareness of community health issues (Van den Berg and Knols 2006) or human rights (Friis-Hansen et al. 2012). Hence, FFS can be applied as a knowledge-building instrument in a broad range of farm-related problems and non-farming topics as well.

Yet, a fundamentally important attribute of FFS is their ability to generate social value. Much more than traditional adult education approaches, FFS emphasise group dynamics, non-hierarchical relationships, and rural networking to elevate their performance (Tripp et al. 2005; Friis-Hansen et al. 2012; Taylor et al. 2012). A consistent finding in the relevant literature is that the participatory process of knowledge construction and the collaborative evolution of innovations within the framework of FFS on the one hand help attendees to sharpen their social skills (Feder et al. 2004; Van den Berg and Jiggins 2007; Braun and Duveskog 2008) and on the other hand facilitate the development of social capital among participants (Settle and Garba 2011). In the present study, building upon the classical work of Putnam (1995) and Fukuyama (2000), we operationally define social capital as the development of connections and informal norms among individuals that are based on and promote reciprocity and trust, thus facilitating cooperation for mutual benefit. In this vein, as Coleman (1988) postulates, social capital serves a major function within social groups: it produces positive outcomes which cannot be easily achieved when social capital is absent.

Nevertheless, although social capital is a dominant issue in FFS discourse, relevant research tends to see social capital as a product of FFS participation (David 2007; David and Asamoah 2011). However, it is well supported by the literature that in educational settings the cultivation of social capital among learners facilitates their integration into the educational system and paves the way for higher learning achievements (Anderson 2008). On the other hand, work on social psychology (Ryan and Deci 2000) posits that the development of a sense of relatedness among learners—which may be the first step in the process of social capital formation—generates the conditions for a positive educational climate, fostering in parallel their motivation to learn. Therefore, social capital is not only an outcome of any educational process but also a catalyst for the process. In this study, we sought to delineate this role of social capital in FFS, by pursuing two research questions: First, how does social capital among farmers evolve in FFS? Second, which aspects of social capital contribute to FFS performance (i.e., the degree to which FFS attendance supplies farmers with knowledge and contributes to the participatory development of innovations)?

By answering these questions we attempt to help designers of FFS interventions understand if and how social capital can enhance the performance of their projects, and to provide an insight into the ways social capital can be triggered within the framework of FFS. By doing so, the present study contributes to the ongoing discussion on the role of rural social capital in facilitating farmers' access to spaces of knowledge and innovation (Leonardo et al. 2020; Saint Ville et al. 2020), and in helping them adapt to different stresses and shocks that affect the productivity of their enterprises (Roy et al. 2019). However, as King et al. (2019) explain, the presence of social capital in any project aimed at leveraging knowledge creation opportunities and at developing farmers' innovation capacity should not be taken for granted, whereas the ways different forms of social capital affect projects' success are still underinvestigated. Our work responds to this call for further research on the ways social capital unfolds and affects the performance of FFS projects.

Conceptual framework

Social capital was and remains a widely used concept in all disciplines that study human behaviour within social contexts. From educational (Croninger and Lee 2001) and workplace settings (Carmeli et al. 2009) to the context of the modern family (Dufur et al. 2015), the positive/negative influence of present/absent social capital in optimal individual and social functioning is more than well-documented. By definition, social capital has the potential to enhance social solidarity (Bourdieu 1980), to raise action within a social structure (Coleman 1988) and to promote cooperation among actors (Putnam 1995). Civic participation (Onyx and Bullen 2000) and action (Larsen et al. 2004), reciprocity (Fukuyama 2001), social trust (Whiteley 2000), and psychological involvement with a social group (Brehm and Rahn 1997) represent different facets of social capital.

In rural settings, a broad diversity of research continues to point out that social assets are far from equally distributed (Bock 2016). For instance, research findings reveal that rural women (Meinzen-Dick et al. 2014), homosexuals (Kuhar and Švab 2014), small scale farmers (Shortall 2008), older people (Winterton et al. 2014), and new members of rural communities (Charatsari and Papadaki-Klavdianou 2017) might have limited access to social capital. The relevant literature examines social capital both as a prerequisite for communal activities aimed at rural development and as an output of collective action (Bisung et al. 2014; Cummings et al. 2019).

Within the strand of research dedicated to the FFS, the issue of social capital holds a predominant position, since—by default—this alternative agricultural extension paradigm was built around the idea that the collective character of this approach potentiates both knowledge co-creation and innovation development. Several indications support this contention by showing that FFS attendees form social bonds (Palis 2006; Phillips et al. 2014), construct networks (David and Asamoah 2011; Mancini et al. 2007), establish collaboration platforms (David 2007), develop peer-confidence and trust (Pretty and Buck 2002; Mancini and Jiggins 2008), and easily engage in the logic of mutual support (Dzeco et al. 2010) and solidarity (Settle et al. 2014). Such a social capital-rich environment provides a fertile ground for the redefinition of deeply rooted prejudices (Najjar et al. 2013) and the development of a collective mindset among trainees (Friis-Hansen and Duveskog 2012).

However, most of the published work in this area conceptualises social capital as a product of FFS participation. On the contrary, little is known about the mediating role of social capital in FFS performance. Research in both formal (Brouwer et al. 2016) and informal educational contexts (Bartsch et al. 2013) supplies evidence that social capital is not only an outcome of participation in a group of learners but also an important mechanism amplifying the educational outcomes of a project. Social capital enhances motivation (Palmer and Gasman 2008), increases knowledge sharing among individuals (Bakker et al. 2006), improves learning accomplishments (Yli-Renko et al. 2001), and, finally, leads to higher levels of knowledge acquisition (Valenzuela and Dornbusch 1994). Consequently, it is expected that the development of social capital among FFS participants would facilitate the success of an FFS project as expressed through the amounts of knowledge constructed by farmers and the

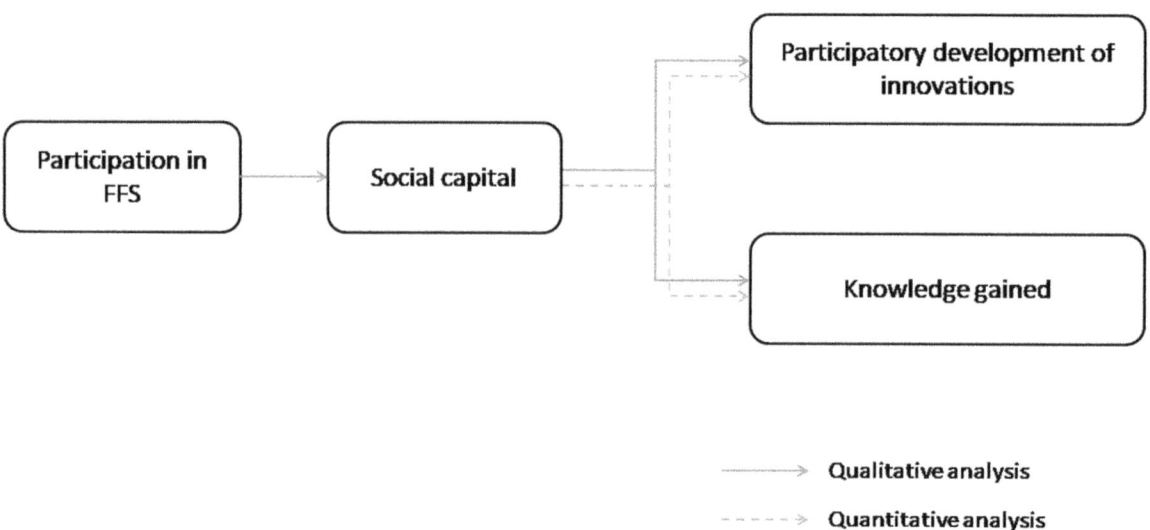

Fig. 1 Conceptual framework and data analysis methods used for this study

degree to which attendees relationally co-develop innovations. Our conceptual framework (Fig. 1) summarises the expected relationships between social capital and the two key-outcomes of an FFS project: participatory development of innovations and knowledge gained by farmers.

Nevertheless, as Portes (1998) argues, sometimes social capital might turn into an obstacle to individual expression and success. In this vein, high levels of group closure within an FFS project might restrict individual initiative and creativity. This "dark" side of social capital (Gargiulo and Benassi 1999; King et al. 2019) might negatively influence the levels of knowledge that farmers gain and their active involvement in the co-evolution of innovations within the field school. Hence, high levels of specific dimensions of social capital may lead to negative outcomes. Given that social capital is a multi-component construct encompassing different layers, it is important to illustrate how these layers unfold throughout FFS and to examine their contribution to FFS success. Therefore, our analysis also focuses on the ways social capital takes different forms during FFS projects.

The present study

To address our research questions we followed a mixed-methods approach. A qualitative analysis was carried out to answer the first research question (How does social capital among farmers evolve in FFS?). Both qualitative and quantitative analyses were used to respond to the second question (Which aspects of social capital contribute to FFS performance?). Data were obtained from two FFS projects (a cotton field school and a rice field school) conducted in Thessaly (Greece) during the growing season of 2015 (April to October). The projects were advertised for about three weeks in local newspapers, farmers' blogs, and websites. For both projects, participation was free of charge and without any financial compensation. Farmers could express their willingness to participate by call or by filling and emailing an application form. After collecting 94 applications, we informed farmers about the details (location, aims, and duration of the two projects). Some of the applicants did not finally participate in the projects, indicating as the main reason the geographical distance between their place of residence and the places where FFS were conducted. From the 74 farmers who participated in the first meetings, five left the project for personal reasons or due to workload. Among the remaining 69 trainees, the average attendance rate was 83.3%, whereas 23 participants attended all the meetings. Both field schools were centred around the aims of familiarising farmers with the principles and practice of integrated crop management (ICM), supplying them with knowledge and skills on issues related to farm management, and enhancing their knowledge on occupational safety issues.

In each FFS meeting, a team of three to five extensionists undertook the guidance and facilitation of farmers' learning process. Facilitators were Greek extensionists, with more than five years of professional experience, recruited from both the public and the private sector. All facilitators had a degree in agronomy, but they also received an eight-hour training programme on the principles and aims of the FFS approach, with a special focus on facilitation techniques. The training programme included four sections, referring to the philosophy of the FFS approach, the use of participatory techniques, the application of knowledge-discovery processes, and the use of group learning activities.

To collect our data we followed two different paths. Participant observation was used to gather in-depth information on the ways FFS participants interacted within the framework of each FFS project and behaved under the various social situations that emerged across different cases. The first two authors collected independently systematic observations throughout the two field schools, while unstructured interviews and informal discussions with farmers and extensionists were also used to increase the richness of these data. All the participants in the two projects (both farmers and trainers) offered data for the qualitative part of the study. To prevent observer-expectancy bias or Hawthorne effects, non-specific treatment was given to individuals or sub-groups (Gale 2004), whereas no performance demands or aspirations were imposed (Campbell et al. 1995; Hansson and Wigblad 2006). At the end of each meeting, observers transcribed their data and after the end of FFS, these data were integrated into a common analytical framework.

However, to enhance the inferential value of the study, as well as to obtain standardised comparable data, we also used questionnaires including closed-ended questions. Multi-item quantitative peer- and self-assessment instruments were designed to assess different facets of social capital, the levels of knowledge gained by farmers, and the degree to which trainees actively took part in the process of mutual development of innovations. These instruments are presented in the section "Measures". Quantitative data were provided by 69 farmers and 11 extensionists who participated in the cotton and rice FFS. These data were analysed to uncover statistically significant associations and differences between the main variables of interest.

By combining these two research paradigms we attempted to gain a deeper understanding of the ways social capital among farmers takes different forms during FFS, as well as of the degree to which different aspects of social capital affect the performance of FFS projects. Such a mixed research design permits the integration of "how" and "why" questions into a common analytical procedure, whereas it reduces the disadvantages of single-method approaches. As Johnson and Onwuegbuzie (2004) explain, in this type of design qualitative data are used to provide meaning to numbers while quantitative findings add precision to qualitative analyses.

Methods

Participants and procedure

Participants in the first FFS were 36 cotton producers (34 men; mean age = 40.53 years, S.D. = 14.72), and six trainers/facilitators (5 men; mean age = 44.83 years, S.D. = 14.22). Given that participants resided in nine different communities in the region of Thessaly, it is not surprising that only twelve of them reported having social relationships (i.e., they already had social interactions) with other trainees before the starting day of the FFS project (mean number of social relationships with other trainees = 0.56, S.D. = 0.91). Cotton producers had an average annual income of €13,750 (S.D. = 2941). Most of them (44.4%) reported having completed secondary education, 25% had a diploma from vocational or technical tertiary education, 13.9% had a university degree, and 16.7% had primary education or had not completed primary school. In the second field school project participated 33 rice farmers (31 men; mean age = 41.30 years, S.D. = 12.07) and five trainers/facilitators (4 men; mean age = 48.40 years, S.D. = 16.06). In the case of the rice field school, producers resided in six different communities in the region of Thessaloniki. Ten attendees reported previous social relationships with some of their co-trainees (mean number of social relationships with other trainees = 0.76, S.D. = 0.56). Most of the rice farmers had completed secondary education (51.5%), whereas 24.3% of the participants had primary education or reported not having completed primary school, 9.1% had a tertiary vocational or technical education degree, and 15.1% of them had a university education. The average income for this group was €14,121 per year (S.D. = 3,018).

As we noted above, a combination of participant observation techniques, informal discussions and unstructured interviews throughout the two projects was used to obtain information on the main research questions. Moreover, farmers and extensionists completed a battery of quantitative instruments, presented in the next section. For both strands of the study, questions were posed in plain language, to avoid confusing participants. A paper-and-pencil questionnaire, administered by the two first authors was used for the quantitative part of the study. The questionnaire required around 10 min to be completed.

Measures

Social Capital Scale

To assess the levels of social capital among trainees we used a multi-item measure. After a thorough literature review, we generated 20 items to capture several aspects of social capital already identified in previous work on human social and in-group behaviour. To extend our conceptualisation of social capital we expanded our review to include disciplines beyond sociology, such as social psychology and cognitive science. To determine whether these items can capture important dimensions of social capital the list was sent to four experts in rural sociology for the assessment of scale's content and face validity, as suggested by some scholars (Flood and Carson 1993; Teddlie and Tashakkori 2009). Each assessor rated the quality of the items as "poor", "fair" or "good". Only items gathering more than 75% "good" ratings were retained in the final list. After this procedure, the administered questionnaire included 14 items referring to different aspects of social capital (Table 1). After this stage, the original items (developed in the English language) were translated into Greek by the two first authors, and then they were translated back to English by a professional translator. The back-translated English version was compared with the original measure by the authors. The process revealed no inaccuracies or gaps in the meaning conveyed.

Trainees completed the Greek version of the instrument in the final meeting of FFS by using a scale anchored by 1 (strongly disagree) and 7 (strongly agree). Such a Likert-type scale can be used to assess opinions, attitudes, beliefs, and characteristics, having as main advantages its simplicity, as well as its ability to adequately capture the assessed attribute (Maurer and Pierce 1998) and to create scale variance (Netemeyer et al. 2003). Similar ways have been used in different fields of study to assess various dimensions of social capital (e.g., Onyx and Bullen 2000; Yip et al. 2007).

To examine the dimensionality of the measure, data were subjected to exploratory factor analysis using varimax rotation. The analysis yielded four factors that represent different dimensions of social capital (Table 2). The first factor corresponded to "Social bonding" (Cronbach's alpha = 0.89) and included four items referring to the development of trust, support, homogeneity, and the cultivation of a sense of belongingness. The second factor was termed "Social cohesion" (Cronbach's alpha = 0.89) as it reflected the mutuality of shared aims, problems, understandings, and the satisfaction that emerges from the group membership. The third factor (Cronbach's alpha = 0.91) comprised items about group centrality, group commitment, and group importance to the self, and therefore was labelled "Group identification". Finally, the factor "Social connection" (Cronbach's alpha = 0.86) was composed of three items concerning

Table 1 Items used to measure social capital

Domain	Source	Item
Connectedness between group mates	Putnam (1995)	I feel connected with the other members of the group, even with those who I don't know well
Development of common values and a civic culture within the group	Forrest and Kearns (2001)	I feel that I belong to a group that shares a common aim
Build-up of a sense of in-group homogeneity	Putnam (1995)	I feel that with these people we are a homogeneous group
Enhancement of a sense of common fate	Jansen et al. (2006)	I feel that with my co-learners we face the same problems
Perceived importance of the group	Luhtanen and Crocker (1992)	To participate in this group of people is really important for me
Group commitment	Ellemers et al. (1997)	I don't feel that I have any special commitment to this group (reverse-worded item)
Satisfaction of belongingness need through group participation	Baumeister and Leary (1995)	It is really important for me to know that I belong to this group of people
Strength of group attachment	Epley et al. (2008)	Sometimes I feel isolated within the group (reverse-worded item)
Mutual understanding among group members	Kearns and Forrest (2000)	With the other farmers, we can understand each other
Presence of in-group support	Turner (1999)	I like to offer support to the other participants
Establishment of in-group trust	Adler and Kwon (2002)	I really feel that I can trust my co-trainees
Emotional value attributed to group membership	Friedkin (2004)	I really like the sense of being a member of that group
Engagement in joint actions with the group-mates	Marsh et al. (2009)	I take part in every join action in the group
Centrality of the group for the farmer's self-concept	Leach et al. (2008)	To be a member of that group is an integral part of my life

Table 2 Social Capital Scale: Factors, loadings, eigenvalues and explained variance

Subscale/item[a]	Loading
Social bonding (Eigenvalue: 5.73; Explained variance: 40.95%)	
I really feel that I can trust my co-trainees	0.94
I like to offer support to the other participants	0.83
I feel that we are a homogeneous group with these people	0.76
It is really important for me to know that I belong to this group of people	0.63
Social cohesion (Eigenvalue: 2.94; Explained variance: 21.01%)	
I feel that I belong to a group that shares a common aim	0.77
With the other farmers, we can understand each other	0.76
I feel that we face the same problems with my co-learners	0.75
I really like the sense of being a member of that group	0.71
Group identification (Eigenvalue: 1.38; Explained variance: 9.89%)	
To be a member of that group is an integral part of my life	0.83
I don't feel that I have any special commitment to this group*	−0.81
To participate in this group of people is really important for me	0.77
Social connection (Eigenvalue: 1.03; Explained variance: 7.33%)	
I take part in every join action in the group	0.82
Sometimes I feel isolated within the group*	−0.79
I feel connected with the other members of the group, even those who I don't know well	0.70

*Negatively worded item

in-group connectedness, engagement in common actions, and in-group inclusion.

Pearson's correlations were computed between the resulting subscales to examine bivariate associations. The strongest correlations were found between social bonding and connection, connection and cohesion, as well as between cohesion and identification ($r > 0.30$, $p < 0.01$ in all cases). This pattern of correlations suggests that the four dimensions of social capital can be seen on a continuum ranging from social bonding (the "softest" and easily achieved aspect of social capital) to social connection, to cohesion and finally to identification (the "hardest" and most difficult to be reached aspect of social capital).

Knowledge gained during FFS

To evaluate farmers' levels of knowledge we used three self-assessment instruments. The first one included 11 items referring to knowledge on integrated crop management (e.g., "Integrated disease management"). The second comprised four items designed to assess participants' knowledge of occupational safety (e.g., "Use of protective equipment"). Finally, the third measure consisted of five items related to different facets of farm management (e.g., "Cultivation practices"). Trainees were asked to complete these instruments before and after the attendance of FFS by using a scale ranging from 1 (very low level) to 5 (very high level). A knowledge score for each category was calculated as the difference between the final score and the pre-participation score. In all cases, reliability coefficients were satisfactory (Cronbach's alphas > 0.70), suggesting substantial internal consistency.

Participatory development of innovations

Three items were generated to assess trainees' involvement in the process of co-evolution of innovations. These items were: "Participant X actively participated in the collective processes of discovering gaps and proposing new ways to overcome them", "Participant X proposed to and discussed with the other members of the group innovative ways to solve problems", and "Participant X helped other trainees to make sense of the experiences they have encountered during FFS and to generate ideas collaboratively". Items were addressed to the facilitators who worked in the two FFS projects. Response options ranged from 1 (not at all) to 5 (very much). After data collection, items were averaged and a new variable was calculated for each trainee. The estimate for the internal consistency of the three items was quite high (Cronbach's alpha = 0.79).

Data analysis procedures

As we explained above, the analysis for this study was based on two different types of data. Qualitative data, collected through participant observation, unstructured interviews and discussions, were subjected to thematic analysis. As Braun and Clarke (2006) suggest, we first transcribed data in chronological order and then we coded data items. In the next step, we collated relevant codes to produce sub-themes that then were sorted into the main themes. Quantitative

data, collected using the instruments reported in the section "Measures", were analysed using the appropriate bivariate tests, as well as regression procedures. Linear hierarchical regressions were performed to assess the contribution of each independent variable to the outcome variables (participatory development of innovations and knowledge gained). Instead of simultaneous regression models, this hierarchical strategy performs a series of regression analyses, calculating the predictive influence of every variable or set of variables in the equation (Warner 2008).

In the first model, the variable "participatory development of innovations" was used as the dependent measure. Farmers' age and education were entered in the first step, as control variables. We then entered the four dimensions of social capital. In the third step, a binary variable indicating the FFS project each farmer attended (1: Cotton FFS, 2: Rice FFS) was entered. For the next models, using the same sets of independent variables, we examined if and how they were associated with the knowledge gained by farmers in ICM, farm management, and occupational safety. This strategy permitted us to examine the hypothesis that social capital is associated with trainees' performance in the FFS, i.e. in the degree to which they participated in the co-evolution of innovations and in the knowledge they gained. By working in this way, we controlled for potential effects of age and education (Gliner and Morgan 2000), whereas we examined whether the variable referring to the FFS to which farmers participated in explains the variance of the dependent measures not explained by the other two sets of predictors (Meyers et al. 2013).

Results

Thematic analysis

The thematic analysis of our qualitative data provided a list of different codes. Codes sharing the same underlying meaning were combined and sorted into 15 sub-themes (Table 3). These sub-themes were then collated under main themes referring to the different forms of social capital among FFS participants and the ways these social capital dimensions affected the co-production of knowledge and innovation. Below we present the analysis of each one of these themes.

Tying and group formation

The first theme was labelled "Tying and group formation" because it reflects the degree to which trainees in each FFS project formed social relationships and were integrated into the team. In both FFS projects, farmers were characterised by a great variance in personality traits. Some of them appeared to have higher levels of openness to social

Table 3 Sub-themes and themes emerged from the thematic analysis

Theme	Definition	Sub-themes
Tying and group formation	Development of social relationships among trainees and integration into the team	1. Building of social relationships 2. Intra-group communication 3. Development of collective responsibility for project success 4. Positive/negative group attitudes 5. Sense of homogeneity
Connectedness and embeddedness	Development of connections among trainees, creation of a sense of group membership and attachment	1. Feelings of group attachment 2. Positive emotional states 3. Involvement in group activities 4. Sense of community 5. Pursuit of common aims 6. Motivation to share knowledge
Coherence and group importance	Perceived importance of the group for trainees	1. Affiliation with the group 2. Psychological commitment to the group 3. Sense of belongingness 4. Emergence of collective selves

experiences and more energetic group behaviour while others tended to be more introverted, spending limited time in discussing with their team-mates. A variety of knowledge-generating activities were designed by the FFS planner in collaboration with facilitators, to instigate the involvement of such individuals in the processes of knowledge creation and innovation co-evolution.

In the cotton FFS, farmers spent much time discussing problems and proposing solutions. Our qualitative data confirmed that this pattern of finding problems and suggesting solutions was fostered by the six facilitators who worked in the project. Although in the first meeting trainers seemed to feel that farmers' involvement in the process of knowledge co-production restricted their role as "teachers"—and, perhaps, put in doubt their expertise and experience—they gradually accepted their new role as "co-evolvers" of knowledge. On the other hand, in the rice FFS, trainers seemed to prefer conventional knowledge delivery practices, emphasising more the instruction and paying limited attention to the knowledge discovery process. Hence, rice producers had fewer opportunities to interact with each other and to use their experience to translate knowledge into practice. This feature eliminated the opportunity to boost the integration of more introverted farmers into the team spirit.

Moreover, in the cotton FFS, farmers already from the second meeting developed a sense of collective responsibility for the success of the project. Despite their disagreements on several issues, cotton farmers easily expressed a positive group attitude, encouraging the more inhibited or wary trainees to actively engage in the group activities. Hence, all participants were involved in the group interaction and communication process. Within this climate, new activities which helped farmers on the one hand to sharpen their knowledge and on the other hand to develop camaraderie were proposed by some trainees and were adopted by the

group. For example, in the "cycle of knowledge creation"—a new knowledge-sharing game initiated by some trainees in the second FFS meeting—all the farmers had to form a circle and each one of them should explain a symptom she/he observed in the cotton field to the trainee who stayed next to her/him. Interestingly, when this strategy was introduced in the rice FFS some of the participants appeared reluctant to join the circle. Despite the efforts taken by facilitators, the process of knowledge creation and sharing through the game of circle had limited success compared to the cotton field school.

Our data suggested that this difference in the quality of communication between the two FFS projects can be attributed to a climate of competition that developed between some of the rice farmers. Indeed, some of the rice producers stated during the interviews that before the beginning or at the first stages of the project they viewed other participants more as competitors than as co-trainees. Trying to explain this attitude, we organised a series of informal discussions with farmers and facilitators. This procedure revealed that, although cotton producers usually sell their yields in cotton-ginning factories at standard prices, rice farmers sell their production to large private traders and/or processors, who use to bargain the price down. Thus, different rice producers may enjoy significantly different prices and, consequently, incomes. In the first two meetings, we observed that the issue of price fairness was stressed many times by some trainees. To help farmers re-conceptualise their role in price formation, in the third meeting we initiated a joint discussion on the issue. Facilitators induced farmers to propose ways to overcome price fluctuations and to collectively negotiate rice prices. This discussion has shown to encourage the involvement of farmers in the development of consensus around the topic, expediting in parallel the shaping of a sense of homogeneity between participants, which further increased in the next meetings. As a rice producer stated: "We all have the same problems, we have to fight the same enemies; we need to understand our common destiny".

Connectedness and embeddedness

The second theme that emerged from our analysis refers to the extent to which farmers were gradually embedded in the social group of FFS, i.e., they were connected to each other, felt attached to the group, and acted as group members rather than as distinct individuals. Interestingly, our data showed that apart from the pursuit of a common aim, there were also affective forces (positive emotional responses associated with the feeling of being a member of the group) which bound trainees to one another, thus facilitating the cultivation of a sense of community within the FFS projects. In its turn, this feeling of community contributed to the engagement of attendees in a gradually developed web of relationships.

Although the influence of farmers' personality traits on the levels of their group inclusion cannot be overlooked, our observations indicated that the congruence between a trainee's interests and the project's foci was an important antecedent of her/his active participation. Hence, farmers who felt that the project corresponded to their needs shaped connections with their co-trainees and displayed high levels of active participation already from the first meetings. On the contrary, a lack of fit between one's conceptualisation of the ideal project and her/his perception of the FFS she/he attained seemed to impede embeddedness. For instance, the development of connections with other group members was slower for trainees who mainly interested in the choice of fungicides and other agrochemicals for the treatment of fungal diseases (while facilitators aimed at helping them understand the ways such diseases emerge and the prevention strategies that can be used).

Arguably, the development of affective connections appeared to generate a domino-like effect, facilitating farmers' embeddedness, which increased their self-confidence and strengthened their feelings of competence. At the other end of the spectrum, the increase of both self-confidence and self-efficacy encouraged farmers' active involvement in the process of co-evolution of innovations. What we noticed was that affective connections created a comfort zone, which facilitated the expression of individual opinions in front of the group. In the cotton FFS, where interpersonal connections were based on more positive emotions, the formation of positive intra-group attitudes was easier. Within this context, even the more introverted farmers contributed their ideas to the reflection process, thus facilitating the development of complex and radical managerial or organisational innovations. For instance, in

the fourth FFS meeting, cotton producers jointly developed a new framework for the management of irrigation water.

Another remarkable finding was that the development of affective ties led also to the build-up of positive psychological states within the two groups of trainees, increasing the amount of optimism among them, their belief that they will become better farmers, and the exchange of motivational energy within the group. In particular, our data indicated that farmers' motivation not only to learn but also to share knowledge had a contagious quality. For instance, as some facilitators noted, willingness to actively participate in group discussions and volition to engage in FFS's activities were spreading from the more active farmers to their more passive colleagues. It is noteworthy that facilitators seemed also to follow this motivational stream. Farmers' increasing motivation to learn during the course of FFS many times led extensionists to go the extra mile, that is to work beyond the standard tasks and their prescribed duties, by providing special support to some trainees, spending effort to motivate farmers, and devoting more time to explaining complicated concepts related to the agroecosystem. As one of the facilitators in the cotton FFS commented:

> You can't stop when farmers want to continue. It's a self-respect thing, you know. But, it is also something inner, something like a spontaneous reaction. You feel that they really want to continue, you should continue but you also want to continue. Nobody looks at his watch, you don't look either. Time goes fast.

Notably, in about half of the meetings of the cotton FFS, facilitators worked in the field beyond the scheduled time, continuing to guide the process of knowledge construction or offering directions for further reflection. Another interesting observation was that after each meeting most of the participants (trainees and trainers) used to get together at local cafés discussing farming issues but also talking about everyday life. On the contrary, this pattern of extracurricular social interaction was observed only after the last meeting of rice FFS, suggesting that the development of strong social connections among participants followed a slower time course. In parallel, the process of motivational contagion evolved along a similar time course: the conveyance of motivation from the more motivated farmers to their co-trainees as well as to facilitators was slower and weaker in valence. Our observations in both FFS projects revealed that the strength of motivational arousal within each group affected the number of topics covered during FFS meetings. This can explain the difference in the number of topics discussed between cotton and rice FFS (43 in cotton field school versus 33 in rice field school).

Coherence and group importance

The third theme of our analysis refers to the importance attributed by farmers to the groups they belonged to. As we have already mentioned, during the FFS journey participants had the opportunity to know each other, develop social relationships, and form sentimental bonds. Nevertheless, not all trainees developed a sense of group centrality for their lives. Despite the positive social climate created and maintained in both FFS projects—and especially in the cotton field school—the levels of felt affinity to the group varied among participants. Hence, some trainees exhibited greater levels of affiliation with the group and a strong psychological commitment to their group-mates, while others, although seemed to voluntarily follow the prevailing group norms and participated in group activities, did not develop an important level of group favouritism.

Our data showed that in the more coherent group of cotton farmers the expression of a collective self was more frequent than in the group of rice producers. A plausible explanation is that in the cotton field school facilitators used different techniques (direct questions to participants, assignment of group activities) to encourage the collective analytical reflection among farmers. Such a framework enhanced trainees' feeling of belongingness to the group, impelling in parallel their pro-group behaviour. In both projects, we noticed many expressions of this kind of behaviour, ranging from actions associated with field school's well functioning (e.g., spending time to explain topics to other farmers, encouraging their co-trainees' involvement in the reflection process) to behaviours associated with personal problems of some attendees (e.g., lending money to another participant).

On the other hand, the engagement in the knowledge co-production process evoked trainees' sense of enjoyment from their participation in FFS, which encouraged their further involvement in the activities performed, thus leading to a strong adherence to the group. However, not all farmers developed a sense of adherence or felt immersed in their groups. This holds true especially for those farmers who did not manage to appreciate the process of co-evolution of knowledge and innovation through teamwork, and they showed a preference towards more traditional ways of knowledge and innovation diffusion.

Quantitative analysis

Preliminary analyses

To provide a first overview of our quantitative data we performed a series of bivariate analyses aimed at identifying significant associations and differences between the main study variables. As a first step, we examined differences in demographic characteristics between the two groups of farmers. Results indicated no significant differences in age ($t = -0.24$, $p > 0.05$), education ($\chi^2 = 1.21$, $p > 0.05$), and income ($t = 0.52$, $p > 0.05$). Kruskal–Wallis tests showed that participants' educational level was not associated with the participants' scores on the four sub-scales of social capital or the degree to which trainees participated in the process of co-evolution of innovations ($p > 0.05$ in all cases). Concerning age, Pearson's correlation coefficients indicated that younger farmers experienced a higher sense of bonding with their co-trainees ($r = -0.340$, $p < 0.01$), whereas they also gained higher levels of knowledge on farm management issues ($r = -0.260$, $p < 0.05$).

Paired samples t-tests were conducted to examine for differences between cotton and rice farmers on the variables of interest. As Table 4 illustrates, trainees in the cotton field school participated in a higher degree to the process of innovation co-creation ($t = 6.26$, $p < 0.01$), whereas they gained significantly higher levels of knowledge on ICM practices ($t = 7.89$, $p < 0.01$), farm management techniques ($t = 5.19$, $p < 0.01$) and occupational safety issues ($t = 2.30$, $p < 0.05$). Interestingly, we also observed that the scores for all four dimensions of social capital were significantly higher for cotton farmers compared to those for rice producers. In all cases, p-values were lower than 0.01.

Bivariate correlational analyses using Pearson's r showed that the four dimensions of social capital were positively associated with the degree to which trainees participated in the process of innovation co-development, as well as with the knowledge they gained throughout FFS. In particular, social bonding and social connection were found to correlate highly with all outcome variables ($r > 0.40$ and $r > 0.46$, respectively), social cohesion appeared to correlate strongly with these variables ($r > 0.37$), whereas moderate to high

Table 4 Mean scores of key variables and differences between FFS projects

Variable	Group		Difference
	Cotton farmers	Rice farmers	
Participatory development of innovations	3.77	2.36	1.41**
Knowledge scores			
ICM	0.99	0.41	0.58**
Farm management	0.78	0.23	0.55**
Occupational safety	0.69	0.41	0.28*
Dimensions of social capital			
Social bonding	4.31	3.45	0.86**
Social cohesion	3.82	2.39	1.43**
Group identification	3.26	1.82	1.44**
Social connection	4.04	3.14	0.90**

$*p < 0.05$, $**p < 0.01$

correlation coefficients were also obtained for group identification ($r > 0.25$).

Social capital and participatory development of innovations

To test for associations between the different facets of social capital and trainees' involvement in the process of co-evolution of innovations during the FFS projects we performed a hierarchical regression analysis, as it is explained in the section "Data analysis procedures". Regression results (Table 5) showed that farmers' demographics had no significant effect on the degree to which they participated in the process of innovation co-production ($\Delta R^2 = 0.01$, $p > 0.05$). On the other hand, the aspects of social capital were found to contribute significant variance to the regression ($\Delta R^2 = 0.56$, $p < 0.01$). Among the four subscales, the development of connectedness in the FFS setting and the strength of social bonding were positively associated with the involvement in the innovation co-evolvement process. Finally, the coded

variable for the FFS group accounted for significant additional variance ($\Delta R^2 = 0.04$, $p < 0.05$). The negative sign of the beta coefficient ($\beta = -0.28$) indicates that participation in the cotton FFS (code: 1) was associated with higher levels of involvement in innovation co-creation.

Social capital and knowledge gained

In the regression analyses for knowledge gained we followed the same hierarchical strategy. The three measures referring to the educational objectives of FFS projects (ICM, farm management, and occupational safety) were used as dependent variables (Table 6). In all three regressions, demographic characteristics were not qualified as significant predictors ($\Delta R^2 < 0.08$, $p > 0.05$ in all cases). Again, we found that—as a set—the four dimensions of social capital contribute significantly to the variance of the dependent variables ($\Delta R^2 > 0.33$, $p < 0.01$, in all cases). In particular, social bonding was significantly associated with the knowledge gained on ICM, farm management, and occupational safety, whereas social connection had significant positive effects on farmers' knowledge. Social cohesion was also found to significantly affect the levels of knowledge acquired during the FFS projects on occupational safety issues. Furthermore, the dummy variable representing group membership was negatively associated with the dependent measures in all three regression models ($\Delta R^2 > 0.03$, $\beta < 0$, $p < 0.05$ in all cases), showing that cotton FFS participants gained significantly higher levels of knowledge than rice FFS attendees.

Discussion and conclusions

In the present work, we attempted to investigate whether social capital among FFS participants does catalyse field schools' performance. To this end, we identified different dimensions of social capital, and we examined which of

Table 5 Results of hierarchical regression analysis

	Coefficients		
	R^2 change	β	p
Step 1	0.01		0.710
Age		0.01	0.997
Education		−0.10	0.248
Step 2	**0.56**		0.000
Social bonding		**0.29**	0.005
Social connection		**0.35**	0.001
Social cohesion		−0.04	0.712
Group identification		0.15	0.204
Step 3	**0.04**		0.024
Group		**−0.28**	0.024

Significant coefficients are in boldface

Table 6 Standardized coefficients (β) of regressions used to test the association of social capital with knowledge gained through participation in FFS

	Knowledge score		
	ICM	Farm management	Occupational safety
Step 1			
Age	−0.02	−0.10	0.06
Education	0.04	0.01	0.04
Step 2			
Social bonding	**0.23***	**0.30****	**0.27***
Social connection	**0.29****	**0.30****	**0.32***
Social cohesion	0.09	0.05	**0.41****
Group identification	0.13	−0.06	0.12
Step 3			
Group	**−0.32****	**−0.28***	**−0.33***

Significant coefficients are in boldface; (* $p < 0.05$, ** $p < 0.01$)

these dimensions advance the participatory co-evolution of innovations by trainees and instigate the process of collaborative knowledge construction. In doing so, we conceived of social capital as both an output of FFS participation and an internal mechanism inducing knowledge creation and innovation development. Hence, our theorising suggests that social capital is not a final but an intermediate product of FFS participation, which evolves during an FFS project enlarging the spaces for knowledge and innovation. In line with previous work arguing that the conceptualisation of social capital as a multidimensional construct can shed more light on the ways it affects rural life (Gómez-Limón et al. 2014; McShane et al. 2016; Sseguya et al. 2018), herein we endeavoured to tap distinct dimensions of social capital by integrating two different analytical approaches.

Our qualitative analysis indicated that social capital among trainees unfolds through a gradual process, spanning from the build-up of social ties between farmers to the construction of affective connections that facilitate group embeddedness and motivational contagion, and finally to the development of a feeling of group centrality which enhances group coherence. Both, the quantity and quality of these social capital configurations appeared to improve in-group communication, to foster the development of a sense of community among attendees, and to advance the prioritisation of group goals. Within this framework, social capital serves as a springboard for triggering group reflection and for eliciting farmers' involvement in collective action, thus being instrumental to the success of FFS.

Facilitators, apart from their role in applying participatory methods to guide the learning process (Feder et al. 2004; Taylor et al. 2012), have the challenging task to create and sustain the conditions for the expression of social capital in the framework of FFS. The differences in the performance of the two projects reported in the analysis partly relate to the different facilitation styles used by extensionists. As our qualitative results revealed, the designing of learning experiences that integrate farmers into the spirit of a common purpose and the adoption of their role as co-evolvers of knowledge helped cotton FFS facilitators to cultivate a positive climate that strengthened the coherence of the group. On the other hand, even though the rice FFS project was based on the same principles, the trainer-centred style followed by some facilitators and the competitive stance of some farmers led to lower levels of social capital and lower performance.

Arguably, social capital can act as both an antidote to the antagonistic relationships among trainees and a potentiator of positive psychological connections between FFS participants. In this vein, it can be argued that the creation of a sense of community within the FFS context is expected to open new opportunities for knowledge construction and innovation co-production. Indeed, such a sense can set the ground for individual and collective progress within a community (Dalton et al. 2001), strengthening in parallel community members' energy and motivation (Collins 1993; Parker et al. 2013).

The quantitative analysis provided confirmatory support for these qualitative results, unraveling again a significant influence of social capital on the performance of FFS. The scale we developed to assess social capital offered preliminary evidence that four facets of social capital lie along a continuum ranging from social bonding to connection, to social cohesion and, finally, to group identification. Although the continuum structure of social capital awaits further validation, this finding is consistent with our qualitative analysis which also supports the progressive evolution—and transformation—of social capital among FFS attendees. Our multivariate analyses revealed that the "soft" aspects of social capital (bonding and connection) are significant antecedents of knowledge and innovation co-creation, implying that bonding social capital is a key component for the accomplishment of FFS aims.

Taken together, our quantitative and qualitative findings converge to show that nurturing social capital among trainees should be a high priority for FFS facilitators. Launching group inquiry activities, animating farmers to engage in team reasoning, encouraging their active involvement in the knowledge discovery process, and helping trainees to make sense of their learning experiences through the collective elaboration of knowledge, constitute a spectrum of strategies which—according to our observational data—when used by facilitators can enhance social capital among participants. The qualitative strand of the present analysis showed that in cases when facilitators fail to endorse the importance of promoting social capital generating behaviours—as we observed in the rice field school—the development of social capital is slower and less evident. This limited availability of social capital lessens the potential of the FFS approach to increase farmers' knowledge and innovation capacity. The significant differences between cotton and rice FFS projects, which uncovered through our quantitative analysis, offer strong support to this contention.

Viewed in conjunction with research postulating that farmers' participation in FFS is guided not only by their intent to acquire specialised knowledge but also by their socio-psychological needs (Charatsari et al. 2017, 2018), our results suggest that the social aspects of FFS participation merit further attention by both researchers and FFS planners. Within the framework of field schools, farmers develop and pursue not only learning objectives but also social goals that promote collectivism. Nevertheless, several studies suggest that such goals have psychological antecedents (Perkins et al. 2002; Luthans et al. 2007), which should be taken into consideration in the blueprinting of FFS projects. The uncovering of these factors was beyond the scope of the present study; however, it represents an interesting future research direction.

Although the small sample sizes and the limited proportion of women in our samples were two limitations of our study, we believe that the present work does add new knowledge to the ongoing discussion on rural social capital. From a methodological angle, this research shows that the hybridisation of qualitative and quantitative approaches can provide a nuanced and more in-depth understanding of the ways social capital evolves, takes different forms, spreads over time, and produces positive outcomes. From a practical point of view, the study presented here underscores the prime importance of social capital for FFS success and highlights the need for more empirical research on how to create and augment social capital within FFS contexts.

Acknowledgements This research project is funded under the Project "Research & Technology Development Innovation Projects"-AgroE-TAK, MIS 453350, in the framework of the Operational Programme "Human Resources Development". It is co-funded by the European Social Fund through the National Strategic Reference Framework (Research Funding Program 2007–2013) coordinated by the Hellenic Agricultural Organization-DEMETER. We gratefully acknowledge this support. Finally, we would like to thank the four anonymous reviewers and the editor for their valuable comments and suggestions.

References

Adler, P.S., and S.W. Kwon. 2002. Social capital: Prospects for a new concept. *Academy of Management Review* 27 (1): 17–40.

Anderson, J.B. 2008. Social capital and student learning: Empirical results from Latin American primary schools. *Economics of Education Review* 27 (4): 439–449.

Bakker, M., R.T.A. Leenders, S.M. Gabbay, J. Kratzer, and J.M. Van Engelen. 2006. Is trust really social capital? Knowledge sharing in product development projects. *The Learning Organization* 13 (6): 594–605.

Bartsch, V., M. Ebers, and I. Maurer. 2013. Learning in project-based organizations: The role of project teams' social capital for overcoming barriers to learning. *International Journal of Project Management* 31 (2): 239–251.

Baumeister, R.F., and M.R. Leary. 1995. The need to belong: Desire for interpersonal attachments as a fundamental human motivation. *Psychological Bulletin* 117 (3): 497–529.

Bisung, E., S.J. Elliott, C.J. Schuster-Wallace, D.M. Karanja, and A. Bernard. 2014. Social capital, collective action and access to water in rural Kenya. *Social Science and Medicine* 119: 147–154.

Bock, B.B. 2016. Social and economic equality: A territorial and relational perspective. In *Routledge international handbook of rural studies*, ed. M. Shucksmith and D.L. Brown, 427–432. London: Routledge.

Bourdieu, P. 1980. Le capital social: Notes provisoires. *Actes de la Recherche en Sciences Sociales* 31 (1): 2–3.

Braun, A. and D. Duveskog. 2008. The farmer field school approach: History, global assessment and success stories. *Background Paper for the IFAD Rural Poverty Report*. Rome: IFAD.

Braun, V., and V. Clarke. 2006. Using thematic analysis in psychology. *Qualitative Research in Psychology* 3 (2): 77–101.

Brehm, J., and W. Rahn. 1997. Individual-level evidence for the causes and consequences of social capital. *American Journal of Political Science* 41 (3): 999–1023.

Brouwer, J., E. Jansen, A. Flache, and A. Hofman. 2016. The impact of social capital on self-efficacy and study success among first-year university students. *Learning and Individual Differences* 52: 109–118.

Cai, J., G. Shi, and R. Hu. 2016. An impact analysis of farmer field school in China. *Sustainability* 8 (2): 137.

Calo, A. 2018. How knowledge deficit interventions fail to resolve beginning farmer challenges. *Agriculture and Human Values* 35 (2): 367–381.

Campbell, J.P., V.A. Maxey, and W.A. Watson. 1995. Hawthorne effect: Implications for prehospital research. *Annals of Emergency Medicine* 26 (5): 590–594.

Carmeli, A., B. Ben-Hador, D.A. Waldman, and D.E. Rupp. 2009. How leaders cultivate social capital and nurture employee vigor: Implications for job performance. *Journal of Applied Psychology* 94 (6): 1553–1561.

Chandra, A., P. Dargusch, K.E. McNamara, A.M. Caspe, and D. Dalabajan. 2017. A study of climate-smart farming practices and climate-resiliency field schools in Mindanao, the Philippines. *World Development* 98: 214–230.

Charatsari, C. 2015. *Farmer field schools: a practitioner's guide*. Report No D4.a., Thessaloniki: ELGO-Dimitra (in Greek).

Charatsari, C., and A. Papadaki-Klavdianou. 2017. First be a woman? Rural development, social change and women farmers' lives in Thessaly-Greece. *Journal of Gender Studies* 26 (2): 164–183.

Charatsari, C., A. Koutsouris, E.D. Lioutas, A. Kalivas, and E. Tsaliki. 2018. Promoting lifelong learning and satisfying farmers' social and psychological needs through farmer field schools: Views from rural Greece. *Journal of Agricultural & Food Information* 19 (1): 66–74.

Charatsari, C., E.D. Lioutas, and A. Koutsouris. 2017. Farmers' motivational orientation toward participation in competence development projects: A self-determination theory perspective. *The Journal of Agricultural Education and Extension* 23 (2): 105–120.

Chhay, N., S. Seng, T. Tanaka, A. Yamauchi, E.C. Cedicol, K. Kawakita, and S. Chiba. 2017. Rice productivity improvement in Cambodia through the application of technical recommendation in a farmer field school. *International Journal of Agricultural Sustainability* 15 (1): 54–69.

Clausen, A.S., E. Jørs, A. Atuhaire, and J.F. Thomsen. 2017. Effect of integrated pest management training on Ugandan small-scale farmers. *Environmental Health Insights* 11: 1–10.

Coleman, J.S. 1988. Social capital in the creation of human capital. *American Journal of Sociology* 94 (suppl.): 95–120.

Collins, R. 1993. Emotional energy as the common denominator of rational action. *Rationality and Society* 5 (2): 203–230.

Croninger, R.G., and V.E. Lee. 2001. Social capital and dropping out of high school: Benefits to at-risk students of teachers' support and guidance. *Teachers College Record* 103 (4): 548–581.

Cummings, S., A.A. Seferiadis, J. Maas, J.F. Bunders, and M.B. Zweekhorst. 2019. Knowledge, social capital, and grassroots development: Insights from rural Bangladesh. *The Journal of Development Studies* 55: 161–176.

Dalton, J.H., M.J. Elias, and A. Wandersman. 2001. *Community psychology: Linking individuals and communities*. Belmont: Wadsworth Thomson Learning.

Dalton, T.J., I. Yahaya, and J. Naab. 2014. Perceptions and performance of conservation agriculture practices in northwestern Ghana. *Agriculture, Ecosystems and Environment* 187: 65–71.

David, S. 2007. Learning to think for ourselves: Knowledge improvement and social benefits among farmer field school participants in Cameroon. *Learning* 14 (2): 35–49.

David, S., and C. Asamoah. 2011. The impact of farmer field schools on human and social capital: A case study from Ghana. *The Journal of Agricultural Education and Extension* 17 (3): 239–252.

Davis, K., E. Nkonya, E. Kato, D.A. Mekonnen, M. Odendo, R. Miiro, and J. Nkuba. 2012. Impact of farmer field schools on agricultural productivity and poverty in East Africa. *World Development* 40 (2): 402–413.

Dufur, M.J., J.P. Hoffmann, D.B. Braudt, T.L. Parcel, and K.R. Spence. 2015. Examining the effects of family and school social capital on delinquent behavior. *Deviant Behavior* 36 (7): 511–526.

Dzeco, C., C. Amilai, and A. Cristóvão. 2010. Farm field schools and farmer's empowerment in Mozambique: A pilot study. *Journal of Extension Systems* 26 (2): 1–13.

Ellemers, N., R. Spears, and B. Doosje. 1997. Sticking together or falling apart: In-group identification as a psychological determinant of group commitment versus individual mobility. *Journal of Personality and Social Psychology* 72 (3): 617–626.

Epley, N., S. Akalis, A. Waytz, and J.T. Cacioppo. 2008. Creating social connection through inferential reproduction loneliness and perceived agency in gadgets, gods, and greyhounds. *Psychological Science* 19 (2): 114–120.

Feder, G., R. Murgai, and J.B. Quizon. 2004. Sending farmers back to school: The impact of farmer field schools in Indonesia. *Applied Economic Perspectives and Policy* 26 (1): 45–62.

Flood, R.L., and E.R. Carson. 2013. *Dealing with complexity: An introduction to the theory and application of systems science*, 2nd ed. New York: Plenum Press.

Forrest, R., and A. Kearns. 2001. Social cohesion, social capital and the neighbourhood. *Urban Studies* 38 (12): 2125–2143.

Friedkin, N.E. 2004. Social cohesion. *Annual Review of Sociology* 30: 409–425.

Friis-Hansen, E., and D. Duveskog. 2012. The empowerment route to well-being: An analysis of farmer field schools in East Africa. *World Development* 40 (2): 414–427.

Friis-Hansen, E., D. Duveskog, and E.W. Taylor. 2012. Less noise in the household: The impact of farmer field schools on gender relations. *Journal of Research in Peace, Gender and Development* 2 (2): 44–55.

Fukuyama, F. 2000. *Social capital and civil society*. Working paper WP/00/74. International Monetary Fund, Washington, D.C.

Fukuyama, F. 2001. Social capital, civil society and development. *Third World Quarterly* 22 (1): 7–20.

Gale, E.A. 2004. The Hawthorne studies—A fable for our times? *QJM* 97 (7): 439–449.

Gallagher, K. D. 1999. Farmer education for IPM. *Sustainable Developments International*. Available at https://p2infohouse.org/ref/22/21995.pdf. Assessed 29 February 2020.

Gargiulo, M., and M. Benassi. 1999. The dark side of social capital. In *Corporate social capital and liability*, ed. R.T.A.J. Leenders and S.M. Gabbay, 298–322. Boston: Springer.

Gliner, J.A., and G.A. Morgan. 2000. *Research Methods in Applied Settings: An Integrated Approach to Design and Analysis*. London: Lawrence Erlbaum.

Gómez-Limón, J.A., E. Vera-Toscano, and F.E. Garrido-Fernández. 2014. Farmers' contribution to agricultural social capital: Evidence from southern Spain. *Rural Sociology* 79 (3): 380–410.

Hansson, M., and R. Wigblad. 2006. Recontextualizing the Hawthorne effect. *Scandinavian Journal of Management* 22 (2): 120–137.

Jansen, T., N. Chioncel, and H. Dekkers. 2006. Social cohesion and integration: Learning active citizenship. *British Journal of Sociology of Education* 27 (2): 189–205.

Johnson, R.B., and A.J. Onwuegbuzie. 2004. Mixed methods research: A research paradigm whose time has come. *Educational Researcher* 33 (7): 14–26.

Kearns, A., and R. Forrest. 2000. Social cohesion and multilevel urban governance. *Urban Studies* 37 (5/6): 995–1017.

King, B., S. Fielke, K. Bayne, L. Klerkx, and R. Nettle. 2019. Navigating shades of social capital and trust to leverage opportunities for rural innovation. *Journal of Rural Studies* 68: 123–134.

Kuhar, R., and A. Švab. 2014. The only gay in the village? Everyday life of gays and lesbians in rural Slovenia. *Journal of Homosexuality* 61 (8): 1091–1116.

Larsen, L., S.L. Harlan, B. Bolin, E.J. Hackett, D. Hope, A. Kirby, A. Nelson, T.R. Rex, and S. Wolf. 2004. Bonding and bridging understanding the relationship between social capital and civic action. *Journal of Planning Education and Research* 24 (1): 64–77.

Leach, C.W., M. van Zomeren, S. Zebel, M.L. Vliek, S.F. Pennekamp, B. Doosje, J.W. Ouwerkerk, and R. Spears. 2008. Group-level self-definition and self-investment: A hierarchical (multicomponent) model of in-group identification. *Journal of Personality and Social Psychology* 95 (1): 144–165.

Leonardo, E., P. Dorward, C. Garforth, C. Sutcliffe, and F. Van Hulst. 2020. Conflict-induced displacement as a catalyst for agricultural innovation: Findings from South Sudan. *Land Use Policy* 90: 104272.

Lioutas, E.D., C. Charatsari, M. Černič Istenič, G. La Rocca, and M. De Rosa. 2019. The challenges of setting up the evaluation of extension systems by using a systems approach: The case of Greece, Italy and Slovenia. *The Journal of Agricultural Education and Extension* 25 (2): 139–160.

Luhtanen, R., and J. Crocker. 1992. A collective self-esteem scale: Self-evaluation of one's social identity. *Personality and Social Psychology Bulletin* 18 (3): 302–318.

Luthans, F., B.J. Avolio, J.B. Avey, and S.M. Norman. 2007. Positive psychological capital: Measurement and relationship with performance and satisfaction. *Personnel Psychology* 60 (3): 541–572.

Mancini, F., A.H. Van Bruggen, and J.L. Jiggins. 2007. Evaluating cotton integrated pest management (IPM) farmer field school outcomes using the sustainable livelihoods approach in India. *Experimental Agriculture* 43 (1): 97–112.

Mancini, F., and J. Jiggins. 2008. Appraisal of methods to evaluate farmer field schools. *Development in Practice* 18 (4–5): 539–550.

Marsh, K.L., M.J. Richardson, and R.C. Schmidt. 2009. Social connection through joint action and interpersonal coordination. *Topics in Cognitive Science* 1 (2): 320–339.

Maurer, T.J., and H.R. Pierce. 1998. A comparison of Likert scale and traditional measures of self-efficacy. *Journal of Applied Psychology* 83 (2): 324–329.

McShane, C.J., J. Turnour, M. Thompson, A. Dale, B. Prideaux, and M. Atkinson. 2016. Connections: The contribution of social capital to regional development. *Rural Society* 25 (2): 154–169.

Meinzen-Dick, R., J. Behrman, L. Pandolfelli, A. Peterman, and A.R. Quisumbing. 2014. Gender and social capital for agricultural development. In *gender in Agriculture: Closing the Knowledge Gap*, ed. A.R. Quisumbing, R. Meinzen-Dick, T.L. Raney, A. Croppenstedt, J.A. Behrman, and A. Peterman, 235–266. Dordrecht: Springer.

Meyers, L.S., G.C. Gamst, and A.J. Guarino. 2013. *Performing Data Analysis Using IBM SPSS*. Hoboken, NJ: Wiley.

Mfitumukiza, D., B. Barasa, A.M. Nankya, N. Dorothy, A.H. Owasa, B. Siraj, and K. Gerald. 2017. Assessing the farmer field schools diffusion of knowledge and adaptation to climate change by smallholder farmers in Kiboga District, Uganda. *Journal of Agricultural Extension and Rural Development* 9 (5): 74–83.

Najjar, D., H. Spaling, and A.J. Sinclair. 2013. Learning about sustainability and gender through farmer field schools in the Taita Hills, Kenya. *International Journal of Educational Development* 33 (5): 466–475.

Netemeyer, R.G., W.O. Bearden, and S. Sharma. 2003. *Scaling Procedures: Issues and Applications*. Thousand Oaks, CA: Sage.

218

Onyx, J., and P. Bullen. 2000. Measuring social capital in five communities. *The Journal of Applied Behavioral Science* 36 (1): 23–42.

Ortiz, O., K.A. Garrett, J.J. Health, R. Orrego, and R.J. Nelson. 2004. Management of potato late blight in the Peruvian highlands: Evaluating the benefits of farmer field schools and farmer participatory research. *Plant Disease* 88 (5): 565–571.

Ortiz, O., R. Nelson, M. Olanya, G. Thiele, R. Orrego, W. Pradel, R. Kakuhenzire, G. Woldegiorgis, J. Gabriel, J. Vallejo, and K. Xie. 2019. Human and technical dimensions of potato integrated pest management using farmer field schools: International potato center and partners' experience with potato late blight management. *Journal of Integrated Pest Management* 10 (1): 4.

Palis, F.G. 2006. The role of culture in farmer learning and technology adoption: A case study of farmer field schools among rice farmers in central Luzon, Philippines. *Agriculture and Human Values* 23 (4): 491–500.

Palmer, R., and M. Gasman. 2008. "It takes a village to raise a child": The role of social capital in promoting academic success for African American men at a Black college. *Journal of College Student Development* 49 (1): 52–70.

Parker, A., A. Gerbasi, and C.L. Porath. 2013. The effects of de-energizing ties in organizations and how to manage them. *Organizational Dynamics* 42 (2): 110–118.

Perkins, D.D., J. Hughey, and P.W. Speer. 2002. Community psychology perspectives on social capital theory and community development practice. *Community Development* 33 (1): 33–52.

Phillips, D., H. Waddington, and H. White. 2014. Better targeting of farmers as a channel for poverty reduction: A systematic review of farmer field schools targeting. *Development Studies Research* 1 (1): 113–136.

Pimbert, M.P. 2002. Social learning for ecological literacy and democracy: emerging issues and challenges. In *Farmer Field Schools (FFS): Emerging issues and challenges,* ed. International Potato Center, pp. 21–40. Los Baños: International Potato Center.

Portes, A. 1998. Social capital: Its origins and applications in modern sociology. *Annual Review of Sociology* 24: 1–24.

Pretty, J., and L. Buck. 2002. Social capital and social learning in the process of natural resource management. In *Natural Resources Management in African Agriculture*, ed. C.B. Barret, F. Place, and A.A. Aboud, 23–33. Walingford: CAB International.

Putnam, R. 1995. Bowling alone: America's declining social capital. *Journal of Democracy* 6 (1): 65–78.

Röling, N., and F. de Jong. 1998. Learning: Shifting paradigms in education and extension studies. *The Journal of Agricultural Education and Extension* 5 (3): 143–161.

Roy, R., A.K. Gain, N. Samat, M. Hurlbert, M.L. Tan, and N.W. Chan. 2019. Resilience of coastal agricultural systems in Bangladesh: Assessment for agroecosystem stewardship strategies. *Ecological Indicators* 106: 105525.

Ryan, R.M., and E.L. Deci. 2000. Intrinsic and extrinsic motivations: Classic definitions and new directions. *Contemporary Educational Psychology* 25 (1): 54–67.

Saint Ville, A., G.M. Hickey, U. Locher, and L.E. Phillip. 2020. The role of social capital in influencing knowledge flows and innovation in St. Lucia. In *Food Security in Small Island States*, ed. J. Connell and K. Lowitt, 239–260. Singapore: Springer.

Salite, D. 2019. Explaining the uncertainty: Understanding small-scale farmers' cultural beliefs and reasoning of drought causes in Gaza Province, Southern Mozambique. *Agriculture and Human Values* 36 (3): 427–441.

Settle, W., and M.H. Garba. 2011. Sustainable crop production intensification in the Senegal and Niger River basins of francophone West Africa. *International Journal of Agricultural Sustainability* 9 (1): 171–185.

Settle, W., M. Soumaré, M. Sarr, M.H. Garba, and A.S. Poisot. 2014. Reducing pesticide risks to farming communities: Cotton farmer field schools in Mali. *Philosophical Transactions of the Royal Society of London B: Biological Sciences* 369 (1639): 20120277.

Shortall, S. 2008. Are rural development programmes socially inclusive? Social inclusion, civic engagement, participation, and social capital: exploring the differences. *Journal of Rural Studies* 24 (4): 450–457.

Sseguya, H., R.E. Mazur, and C.B. Flora. 2018. Social capital dimensions in household food security interventions: implications for rural Uganda. *Agriculture and Human Values* 35 (1): 117–129.

Taylor, E.W., D. Duveskog, and E. Friis-Hansen. 2012. Fostering transformative learning in non-formal settings: Farmer-field schools in East Africa. *International Journal of Lifelong Education* 31 (6): 725–742.

Teddlie, C., and A. Tashakkori. 2009. *Foundations of Mixed Methods Research: Integrating Quantitative and Qualitative Approaches in the Social and Behavioral Sciences*. Thousand Oaks, CA: Sage.

Tripp, R., M. Wijeratne, and V.H. Piyadasa. 2005. What should we expect from farmer field schools? A Sri Lanka case study. *World Development* 33 (10): 1705–1720.

Turner, R.J. 1999. Social support and coping. In *A Handbook for the Study of Mental Health: Social Contexts, Theories, and Systems*, ed. A.V. Horwitz and T.L. Scheid, 198–210. New York: Cambridge University Press.

Valenzuela, A., and S.M. Dornbusch. 1994. Familism and social capital in the academic achievement of Mexican origin and Anglo adolescents. *Social Science Quarterly* 75 (1): 18–36.

Van de Fliert, E., J. Pontius, and N. Röling. 1995. Searching for strategies to replicate a successful extension approach: Training of IPM trainers in Indonesia. *European Journal of Agricultural Education and Extension* 1 (4): 41–63.

Van den Berg, H., and B.G. Knols. 2006. The farmer field school: A method for enhancing the role of rural communities in malaria control? *Malaria Journal*. https://doi.org/10.1186/1475-2875-5-3.

Van den Berg, H., and J. Jiggins. 2007. Investing in farmers—The impacts of farmer field schools in relation to integrated pest management. *World Development* 35 (4): 663–686.

Warner, R. 2008. *Applied Statistics: From Bivariate Through Multivariate Techniques*. Thousand Oaks, CA: Sage.

Whiteley, P.F. 2000. Economic growth and social capital. *Political Studies* 48 (3): 443–466.

Winarto, Y.T. 1995. State intervention and farmer creativity: Integrated pest management among rice farmers in Subang, West Java. *Agriculture and Human Values* 12 (4): 47–57.

Winterton, R., J. Warburton, S. Clune, and J. Martin. 2014. Building community and organisational capacity to enable social participation for ageing Australian rural populations: A resource-based perspective. *Ageing International* 39 (2): 163–179.

Yang, P., K. Li, S. Shi, J. Xia, R. Guo, S. Li, and L. Wang. 2005. Impacts of transgenic Bt cotton and integrated pest management education on smallholder cotton farmers. *International Journal of Pest Management* 51 (4): 231–244.

Yang, P., W. Liu, X. Shan, P. Li, J. Zhou, J. Lu, and Y. Li. 2008. Effects of training on acquisition of pest management knowledge and skills by small vegetable farmers. *Crop Protection* 27 (12): 1504–1510.

Yip, W., S.V. Subramanian, A.D. Mitchell, D.T. Lee, J. Wang, and I. Kawachi. 2007. Does social capital enhance health and well-being? Evidence from rural China. *Social Science & Medicine* 64 (1): 35–49.

Yli-Renko, H., E. Autio, and H.J. Sapienza. 2001. Social capital, knowledge acquisition, and knowledge exploitation in young technology-based firms. *Strategic Management Journal* 22 (6–7): 587–613.

Publisher's Note Springer Nature remains neutral with regard to jurisdictional claims in published maps and institutional affiliations.

Chrysanthi Charatsari holds a Ph.D. degree in Agricultural Education and Extension from the Aristotle University of Thessaloniki. She also completed three postdoctoral fellowships at National Agricultural Organization DEMETRA and at the Aristotle University of Thessaloniki where she is currently a research associate. Moreover, she teaches Adult Education at the Hellenic Open University.

Evagelos D. Lioutas has a Ph.D. in Consumer Behaviour from the Aristotle University of Thessaloniki where he also had a postdoctoral fellowship in Consumer Psychology. He currently teaches at the Department of Supply Chain Management (International Hellenic University) and the Hellenic Open University.

Alex Koutsouris is a Professor in the Department of Agricultural Economics & Rural Development, Agricultural University of Athens, Greece. His research interests revolve around sustainable rural development focusing on topics such as innovation, extension & communication, training, and education. Currently, he is Associate Editor of the Journal of Agricultural Education & Extension.

Agriculture and Human Values (2020) 37:1155–1173
https://doi.org/10.1007/s10460-020-10121-w

Virtualizing the 'good life': reworking narratives of agrarianism and the rural idyll in a computer game

Lee-Ann Sutherland[1]

Accepted: 24 June 2020 / Published online: 5 July 2020
© The Author(s) 2020

Abstract

Farming computer games enable the 'desk chair countryside'—millions of people actively engaged in performing farming and rural activities on-line—to co-produce their desired representations of rural life, in line with the parameters set by game creators. In this paper, I critique the narratives and images of farming life expressed in the popular computer game 'Stardew Valley'. Stardew is based on a scenario whereby players leave a [meaningless] urban desk job to revitalize the family farm. Player are given a choice to invest in the Community Center or to support 'JojaMart', a 'big-box' development. The farming narrative demonstrates the hallmarks of classical American agrarianism: farming as the basic profession on which other occupations depend, the virtue of hard work, the 'natural' and moral nature of agricultural life, and the economic independence of the farmer. More recent discourses of critical agrarianism are noticeably absent, particularly in relation to environmental protection. Conflict is centred on urban-based big business, whereas the farm is represented as a 'bolt-hole' or sanctuary from urban life. I argue that embedding issues of big-box development in gameplay enrols players in active reflection and debate on desirable responses, whereas the emphasis on reproducing classical agrarian tropes risks desensitizing game players to contemporary agrarian social and environmental justice issues. However, Stardew Valley gameplay implicitly reinforces the ideal that low input farming is the way that agriculture should be practiced. The success of the game in eliciting on-line debates, and the requirement for active performance and decision-making, demonstrates the specific potential of computer games as mediums for influencing and intervening in ongoing reworking of farming imaginaries, and enabling more critically engagement of the 'desk chair countryside' in important global debates.

Keywords Critical agrarianism · Rural development · Back-to-the-land movements · Big box development · Farming simulator

Abbreviation
RPG   Role playing games

Introduction

You're moving to the Valley...You've inherited your grandfather's old farm plot in Stardew Valley. Armed with hand-me-down tools and a few coins, you set out to begin your new life!
Can you learn to live off the land and turn these overgrown fields into a thriving home? It won't be easy.

Ever since Joja Corporation came to town, the old ways of life have all but disappeared. The community center, once the town's most vibrant hub of activity, now lies in shambles. But the valley seems full of opportunity. With a little dedication, you might just be the one to restore Stardew Valley to greatness!

Thus reads the official advertisement for what GQ magazine dubbed the "unlikeliest independent video game triumph since Minecraft" (White 2018). In stark contrast to Grand Theft Auto and numerous popular first-person shooters, the text and images of Stardew Valley present an opportunity to engage in bucolic farm and community life: the image presented is of rustic housing; a few crops, with a chicken ranging free; travel is by horseback and mine cart; a neighbour comes bearing gifts; trees, wooden fences, mountains and a clear blue sky dominate the landscape. Stardew is pitched as a playground for agricultural production and rural exploration, with a strong 'retro' vibe. It has sold over

✉ Lee-Ann Sutherland
    Lee-Ann.Sutherland@hutton.ac.uk

[1] Social, Economic and Geographical Sciences Department,
     The James Hutton Institute, Aberdeen AB15 8QH, UK

Image 1 Stardew Valley promotional image. *Source* www.stardewval ley.net. Image copyright Eric Barone

10 million copies (Strickland 2020), staying near the top of the Personal Computer (PC) game charts since its launch in 2016. Available in multiple formats (e.g. PlayStation, Xbox, Nintendo Switch) and 12 languages (including Russian, Turkish, Brazilian-Portuguese and several Asian languages), it is a global phenomenon. As a platform for contemporary imaginings and interactions with rurality, I argue that it is an important site of cultural production (Image 1).

In this paper I assess the narratives of rurality and farm life embedded in Stardew Valley gameplay, critiquing the extent to which this portrayal of idyllic rurality reflects tenets of classical and critical agrarianism. I focus the paper particularly on what players of Stardew Valley may learn through gameplay about farming practices, the contemporary imaginaries of rural life which are encouraged or challenged, and how these differ from those found in other forms of media engagement. I thus critically appraise how rurality and farm life are represented within the game. Analysis of Stardew narratives and game constructs offers insights into the recreational appeal of rural landscapes, and the aspects of farming life which are selectively re-configured into recreational experiences. The analysis thus advances thinking on how popular notions of idyllic rurality are constructed for public consumption, and how that public consumes them.

Understanding these representations is important in several respects. Social representations of rurality and agriculture underpin public policy, particularly planning guidelines and agricultural subsidies (Beus and Dunlap 1994; Clark and Jones 1998; Satsangi et al. 2010). Migration—particularly counter urbanization—is also influenced by these representations, both within countries (Halfacree 2011) and between countries (Gaspar 2015). Although these ideals of rural life may be acted upon in only a few cases, it can be a useful indicator of urban understanding and preferences for rural areas (Blekesaune et al. 2010), influencing consumer behaviour (e.g. food, agri-tourism, Flanigan et al. 2015). Critically, players of computer games represent a substantial population which has previously been unrecognized within agrarian and rural studies. I term this cohort the 'deskchair countryside': individuals who primarily experience farming and rural life through their computer screens (see also Sutherland 2020).

In coining the term 'desk chair countryside' I develop work by Bunce (1994) on the 'armchair countryside'—generations of people whose primary experience of rural life is through art, literature, cinema and television (i.e. a step removed from direct experience). Both armchair and deskchair countryside cohorts have highly selective experiences of the countryside, which have been actively edited, produced and marketed to them, and are experienced remotely. For example, as Horton (2008) demonstrates in his analysis of a popular British rural television series, television shows need to be commissioned, scripted and promoted (often by networks of white males), and target specific demographic cohorts. Paul Cloke (2003), writing in the early 2000s, argued that rural life as portrayed in film, television, art, books, toys and magazines is uncritically idyllic—a form of "brainwashing" (p. 1) that underpins public expectations of rurality. These expectations have traditionally omitted less than idyllic features of rural life, such as crime, alcoholism, and homelessness (Yarwood 2001; Jayne et al. 2011; Cloke et al. 2001), focusing instead on representations of farming and rural community life which emphasize the picturesque, recreational and bucolic (Bunce 2003).

More recent media studies reveal a less rosy set of representations, integrating issues of housing and poverty into televisual landscapes for dramatic and comedic effect, but continuing to present rural residents in a positive light (Dickason 2017). Peeren and South (2019) demonstrated the resilience of the 'good life' genre and its association with rurality in their analysis of the multiple genres utilized in the popular Dutch TV show 'The Farmer Wants a Wife'. Early episodes emphasize the farm as a place of difference, an "unsentimental globalized business" (p. 39) utilising modern technology, to which incomers (i.e. prospective wives) must adjust. However, Peeren and Souch argue that by Series 8, the show had evolved to emphasize the romance genre (i.e. dating), with idyllic rurality forming an uncritical backdrop. The agricultural context became incidental, and potential wives—and viewers—were no longer challenged on their idealized conceptions of farming life. Candidates instead engaged in 'romantic' actions of petting livestock and enjoying sunsets over the farm. Peeren and Souch (2019) argue that it is possible to challenge the good life genre through integration with other genres, pointing to spin-off series that more critically present the practicalities of contemporary farming. Although the appeal of the good life genre is highly durable, popular media thus can act as an entry point into more critical representations of rural life.

Computer gameplay has similarities to other forms of popular media engagement, but has some important differences. Like most rural television programs, gameplay is oriented primarily towards entertainment, involves substantive scripted components and evolves over time in response to participant feedback and changes in the market. Consumers choose the programs in which they invest their time. Computer games also include a flow of visual images (Gee 2015)—you can physically see the depiction, the way you would a painting or a television show. Games are more flexible in some respects—unrestricted by prevailing weather or other geophysical realities (see Phillips et al. 2001), but inevitably an animation and thus not bound by logics of gravity or time. However, the major difference is that 'armchair countryside' consumption is largely passive. In contrast, computer and video games require active engagement—players drive the story forward, making choices about how to respond to different scenarios and shape their rural 'worlds'. By changing the story through their own actions, players impact on the outcome and develop their own preferred narratives. Players thus both produce and consume the narratives and experiences embedded in computer games, within the limits set by the game designers.

Murray (2006) argues that contemporary computer games represent a variation on traditional forms of play, where myths and legends are taught and acted out in order to teach children and pass on wisdom. Play involves tangible engagement with objects and practices, implicating multiple senses (e.g. sight, touch, movement). As such, games both influence how the setting is understood, and offer opportunities for experiential learning. The learning potential of games is an opportunity which has been recognized by sociologists like James Coleman since the 1960s (Starr 1994). 'Serious games' are currently utilized for educating children and adults on a range of topics (Wouters et al. 2011; Kaufman and Flanigan 2015). Educational theory argues that people learn better when placed in authentic contexts and are given the opportunity to make decisions and interact in the world of the game (Draefer 2014). However, there is considerable question of the extent to which in-game learning and experiences transfer into 'real life': regular participants in first person shooters do not typically become mass murders. There is one example of a Stardew Valley player who was directly motivated by the game to establish a farm (see Messner 2017), but this is an isolated occurrence. Successful—and enjoyable—gameplay requires learning the 'rules of the game' set by the game developer, but it is unclear to what extent this influences real

world beliefs and practices (Bos 2018). Owing to the specific nature of role playing games like Stardew Valley, I argue that narrative analysis offers insights into what players may consciously and unconsciously 'learn' about rurality and agricultural production through participating in the game.

Agrarianism and the rural idyll

In this paper I critically evaluate the tenets of agrarianism which are—and are not—represented in Stardew Valley gameplay. The basis of agrarianism is the tenet that agriculture is "not an occupation so much as an all-encompassing lifestyle whose purpose was sustaining families and communities in addition to fields and pastures" (Mariola 2005, p. 209). Agrarianism is deeply embedded in Western cultures—Montmarquet (1989) traces cultural scripts describing the positive social and moral value of agriculture and farm work back to Virgil and Cicero. Wolf (1987) describes (state supported) efforts made by scholars and poets in first century Rome to encourage agrarian sentiments, which associated family-led farming with profitability, pleasure and virtue. In the United States, Thomas Jefferson (amongst others) wrote of the moral value specifically associated with agrarian life.

In seminal work on agrarianism, Flinn and Johnson (1974) describe five major elements:

- farming as a basic occupation on which other occupations depend
- the virtue of hard work
- the 'natural' and moral nature of agricultural life
- the economic independence of the farmer
- engagement in agriculture as contributing to the successful maintenance of democracy

For the purposes of this paper, I term these elements 'classical agrarianism': historic ideals of the positive role of agriculture in society. Within classical agrarianism, engaging in farming practices is understood as inherently wholesome and of high moral value—farmers are identified as the 'bedrock' of a successful society. The willingness of farm households to work long hours and the desire for economic independence underpin responses to the 'Agrarian Question'—the persistence of 'family farming' under capitalism (e.g. Kautsky 1988; Chayanov 1927; Friedman 1978). The practices of working the land and caring for livestock are understood as building moral character. As the source of food, fibre, and energy, farming is understood as the only true wealth-generating activity, and thus the basic occupation on which other occupations ultimately depend (Montmarquet 1989).

At root, classical agrarianism tenets idealize and romanticize farming practices, supporting imagery and tropes that feature in historic and contemporary art, literature, and cinema. However, the tenets also reify existing power relations in the countryside (Carlisle 2014), particularly the moral and economic right to private land ownership (and associated inheritance practices i.e. protected rights to transfer land ownership between generations), the right of farmers to pursue profits through intensification, public supports to maintain farm businesses, and the assumption that farmers are producing important public goods and value to society simply by running their farm businesses. Classical agrarian tenets thus underpin contemporary policies to support and protect farmers in the US (Graddy-Lovelace and Diamond 2017) and Australia (Berry et al. 2016).

Classical agrarian representations also reflect differing regional and national histories—American and Australian imagery is lodged in their shared history of colonialism; farmers are portrayed as highly independent, tamers of the wilderness and dominant over their land. Van Keulen and Krijnen (2013) demonstrate how international differences play out in representations of farmers in TV programming: in their cases, Australian ranchers are presented as 'real men'—dominant, rough and muscular outdoor types, whereas the Dutch version of the same program portrays farmers as less physically attractive or socially capable, but more interested in women who will partner with them in operating their farms. In Japan—where Stardew Valley gained its inspiration—geophysical conditions for farming have made it difficult to follow the typical Western agrarian pathway of consolidation and industrialisation; agrarianism remains lodged in small-scale family-style farming, often performed non-commercially and/or on a part-time basis (Hisano et al. 2018).

Resurgence of agrarianism in the mid twentieth century has led to a parallel, more critical agrarian stance, orienting practices and concepts towards addressing social justice and environmental sustainability issues. New forms of 'critical agrarianism' emerged in the 1960s and 1970s, evident in 'back-to-the-land' movements, and literature by writers like Wendell Berry and Rachel Carson, who challenged the growing economic power and environmental impacts of agribusiness, played out against broader social influences such as the Vietnam War (Carlisle 2014). Although these critiques diminished in the 1980s, a resurgence has occurred from the late 1990s onwards (see Halfacree 2011) and agrarian ideals of farming as a simpler, purer life have been taken up by a new generation of newcomers to farming. While both critical and classical agrarianism privilege family farming as a way of life, perspectives on how these ideals should be achieved radically differ (Beus and Dunlap 1994). Whereas classical agrarianism is aligned with industrialisation and modernisation, critical agrarianism proposes radical alternatives to the contemporary food system. A revival of small-scale, arguably 'peasant' style production is presented as essential to countering global food and energy security problems, as well as environmental degradation and climate change (Marsden and Farioli 2015; van der Ploeg 2014).

Contemporary critical agrarianism encompasses a wide scope, bringing together broader societal impetuses for environmental protection and public goods from agriculture (e.g. European and Australian conceptualizations of multifunctionality—see Holmes 2012; Marsden and Sonnino 2008; Wilson 2007) with critical practices of new land holding formations (e.g. community supported agriculture, collective land ownership and co-operatives) (Wittman et al. 2017) and the ecological and financial potential of local food networks (Tregear 2011; Trivette 2012). The opportunities for newcomers to establish farms, enjoy the affordances of farm life (e.g. simplicity, working with nature and animals), to produce healthy food and to protect the environment are vocalized by a growing cohort of activists as environmental and social justice issues, countering intersecting issues of race, gender and socio-economic privilege in the countryside (Carlisle 2014). Establishing a new entrant farm becomes an intentional, critical practice. The arguments of critical agrarianism extend into the global South, where the role of small-scale farmers as efficient, ecological providers of much of the world's food is championed (Ricciardi et al. 2018; Netting 1993).

There is a further cohort of newcomers to agriculture who pursue agrarian ideals. Recent literature has demonstrated the growing cohort of hobby or 'non-commercial' farmers across the global West (Sutherland et al. 2019; Sutherland 2019, 2012; Gosnell and Abrams 2011; Hisano et al. 2018). This approach to farming life as self-actualizing and recreational has roots in both classical and critical agrarianism, but is embedded within broader notions of idyllic rurality. Little and Austin (1996) define the rural idyll simply as a set of myths or images that endure over time, particularly invoking nostalgia and heritage. Halfacree (2010) identifies three 'styles' of consuming the rural, using the metaphors of rural idylls as 'boltholes', 'castles', or 'life rafts'. The rural as bolthole is an escape from toxic urban life, engendering practices of flight and disappearance. The rural as 'castle' involves defensive and protective practices: fortification against urban pressures. The rural is positioned as an escape, but less completely so than the rural as bolthole—as a castle, the rural idyll must be defended against urban threats, including the incursion of other exurbanites. Halfacree's third reading is of 'life rafts'—temporary escapes to the countryside for second homeowners or tourists who engage with the rural idyll on a part-time basis. All three readings of the rural represent critiques of urban life, rather than critiques of agricultural practices.

The key issue for this paper is that idealized representations of rural and farm life selectively emphasize particular aspects; computer games and other media can thus perpetuate, challenge or establish new rural tropes. By engaging in these staged farming practices, game players 'perform' farming and rural life activities, learning in-game skills that may influence in-life perspectives on the desirability of different farming and rural practices.

Stardew Valley: a role playing game

The focus of this paper is on the representations of agrarianism and idyllic rurality produced and consumed within a computer game. In this section I situate Stardew Valley within the broader field of computer and video games.

Contemporary computer games engage millions of players world-wide, representing a larger industry than cinema and music combined (Parsons 2019). 'Role playing games' (RPGs) like Stardew Valley occupy about 11% of the US gaming market (behind 'Shooters'—26% and 'Action'—22% genres, Statistica 2019), representing a type of computer game where the player or 'gamer' constructs a character or 'avatar' that undertakes quests in an imaginary world (Technopedia 2020). Through the avatar, the player pursues storylines and open world opportunities, with varying degrees of scripting. Stardew is based on a scenario whereby the avatar gives up a [monotonous] urban desk job to revitalize the family farm. Freedom is an overarching theme of the game (Lin 2016); players can pursue the main storylines at their own speed or opt to simply explore the game-world. As gameplay progresses, players become embedded in the local community 'Pelican Town', farming, foraging, fishing, and mining to gain the resources necessary to improve the farm, make friends, and (if they choose) restore the Community Center. The game thus embraces familiar tropes of going 'back-to-the-land', family farming and the rural idyll, engaging the player in producing his or her own version of idealized country life.

Stardew Valley was an unexpected hit when it was released in 2016—the product of a single developer, at time when most successful RPGs have hundreds of developers, writers, artists and sound technicians, with budgets in the millions of dollars (White 2018). Creator Eric Barone (also known by his on-line user name 'ConcernedApe') has repeatedly stated that he intentionally modelled Stardew on Harvest Moon, a successful Japanese console-based game (e.g. playable on SuperNintendo) created in the 1990s (Leask 2016). Whereas Harvest Moon was based on its creator (Yasuhiro Wada)'s lived experience of farming in Japan (Wada 2012), Barone's primary experience of farming appears to be through computer games (i.e. he is a member of the desk chair countryside). Barone spent much of his childhood playing near a rural wetland (Grathwohl and Lachausse 2016), and thus has lived experience of rural, outdoor play. Stardew's overtly anti-corporate narrative reflects Barone's personal journey towards creating his own job, on his terms, in his mid twenties. Although he applied (unsuccessfully) for jobs following completion of his computer science degree, he describes his reluctance to become involved in corporate-style employment: *I didn't want to work at a normal job, I wanted to do my own thing, that's kind of the message of Stardew Valley, to follow your heart* (GameInformer 2016). As the sole developer, Barone represents his personal ideals and imaginaries of farming, food consumption, rural space and leisure practices, and those which he believes will appeal to other gamers.

As an RPG, Stardew Valley is an 'open world', where elements of the story become apparent as players interact with different characters or landscape elements. Stardew has two major plot options—to rebuild the Community Center or to purchase a membership in JojaMart, the local branch of a 'big-box' corporate superstore. Rebuilding the Community Center involves donating over 130 different products or resources which the player produces or sources; membership in JojaMart involves substantial donations of 'g' (in the in-game currency, which accumulates largely through produce sales). The two major plot options cannot both be pursued in a single play through the game. RPGs are typically designed to be played multiple times, enabling the player to explore game dynamics and produce different outcomes, thus co-constructing the story to varying degrees. In Stardew Valley players choose whether to build or expand buildings; whether to plant crops and/or have livestock; whether to refurbish the Community Center or join JojaMart; whether to marry, have children and indeed whether to interact with local community members at all (i.e. opting in or out of learning the histories and peculiarities of those people). Players are thus able to influence the story and construct a farm and community life of their choosing.

In Stardew, players learn (the commands to perform) specific skills (e.g. how to cultivate crops, forage, mine etc. and also how to make friends), thus gaining various forms of reward. While these skills in themselves have limited real world value, the principles behind them have meaning. Barone is clear that the critique of corporate practices is intentional:

> Corporations are some of the biggest players in the global arena. They wield extraordinary power over governments, communities, and individuals. Joja Corporation represents that power, taken to a frightening extreme. It's a bit of a caricature, but also disturbingly realistic. I wanted the game to have some

real-world messages, something for modern audiences to relate to. Stardew is mostly just a fun game, but maybe also a plea for individuals and communities to empower themselves. Barone in Leack 2016

Stardew is thus consistent with other contemporary computer games (e.g. Metal Gear Solid, Stamenković et al. 2017), which integrate commentary on contemporary societal issues, while encouraging players to respond in particular ways. However, this commentary does not actively extend to farming practices: although Barone describes how he initially allowed livestock to be butchered in early (pre-release) versions of the game, he ultimately decided that butchering was not in keeping with the peaceful nature of the game (Singal 2016). This is consistent with his vegetarianism, but he has expressed no overt critique of industrialized farming practices in his public interviews.

Stardew Valley begins by requiring players to customize an avatar, select a farm type and name their farm. The avatar is clearly human, male or female, with a wide range of potential features—there are 24 skin colours (including purple and green), 56 hair styles, 20 facial accents (ranging from beards to make-up and jewellery) and 112 shirts from which to personalize the avatar, which the player can also name. The five farm types[1] each offer opportunities for farm development—the 'standard farm' has the largest space for production; others increase access to one of the four major resource extraction opportunities: forestry and foraging on the 'forest farm', fishing on the 'riverland farm', mining on the 'hill-top farm' and nocturnal monsters (which drop sap, slime and the occasional gemstone) on the 'wilderness farm'. Gameplay itself starts with a 'cut scene' (short video) to establish the context to the game; cut scenes appear throughout, as key plot points or milestones in friendship with local community members are reached.

Barone (in Singal 2016) describes initial gameplay as intentionally overwhelming—the numerous possibilities of the farm and village landscapes require players to prioritize which options to pursue. A typical day lasts 15–25 min in real time. There are four, 28-day, 4-week seasons comprising a 'year' of in-game play, 9 annual festivals and 30 local residents, each with their own weekly and seasonal schedules and birthdays. Each season, different crops can be cultivated, forage items (e.g. mushrooms, wildflowers, berries) collected and fish caught. The landscape is malleable: the initial farmhouse, boundaries and community setting is largely stable, but the farm itself can be altered through crop and livestock selection, placement of buildings and fences, as well as afforestation and 'clearing' the land.

Stardew Valley symbolically takes the player back in time, but the specific time period is difficult to determine—the use of hoes and axes to clear and cultivate land are more characteristic of gardening than contemporary Western agricultural production (i.e. there are no tractors, or even horse-drawn equipment). Travel around the landscape is on foot, or later by horse or by mine cart, or magical totems. A steam train runs through the valley, but avatars 'arrive' in Stardew Valley by bus and have access to television. Initial cut scenes show avatars working at desk-top computers. Stardew is thus located outside of conventional time.

The 'end' of Stardew Valley gameplay is largely determined by the player (Moore 2016). Discussion in the forums indicates that some players stop when the Community Center tasks are completed, which can be achieved by early winter of the first year; others consider the return of Grandpa to evaluate player achievements at the beginning of year three to represent the time to stop playing (e.g. Carl's Guide, year unknown). A few players consider exhausting all of the game dynamics (e.g. following through all possible story lines and production options), to be an end point. Even so, gameplay never officially stops: on-line forums and You-Tube videos show farms which have reached up to 20 years in duration, and Barone has released multiple up-dates to the game, increasing the options and experiences available. For the purposes of this paper, the return of Grandpa at the beginning of Year 3 is considered to be the end point. Further detail on gameplay can be found in "Appendix A".

Methodology

There are different schools of thought about the suitability of computer games for narrative analysis. 'Narratologists' argue that video games are story-telling mediums, with game writers holding a similarly expressive position to authors of books; 'ludologists' argue that games are simply games (Mukherjee 2015). Part of the challenge is in the complex array of contemporary computer games. Few would argue that puzzle-based video games (e.g. Tetris) have appreciable narratives. First person shooters and action games may have very limited narratives, whereas 'sandpit' games like Minecraft set players free to create a story in-line with their creations. Within the RPG genre, some RPGs are directly based on substantive works of fiction, such as Lord of the Rings On-line. Others are based on successful movies (e.g. Star Wars), or become successful movies (e.g. Lara Croft Tomb Raider). Stardew Valley has thousands of lines of dialogue and the stated intention of its creator to produce credible in-game interactions: *Ultimately, I wanted the game world to feel like a living place. I wanted you to forget that it was a video game and to feel like these people had a life of*

---

[1] A sixth type enables up to four players to play together, an option that was released after this study began. This option is not considered in the paper.

their own (Barone in White 2018). This depth of narrative development and clear international popularity makes it a suitable candidate for analysis.

This paper is based on over 300 hours of PC gameplay. Stardew Valley is also available for other mediums, with slightly different dynamics (Verret 2017). My initial gameplay was purely recreational; I decided to pursue academic analysis following my first play through the game (approximately 80 hours). I recorded all subsequent gameplay, and selectively transcribed scenes the first time they appeared in gameplay: much of gameplay is rote repetition with few words (e.g. watering plants, feeding livestock) where transcription is unnecessary. I created and played avatars of both genders, and established all five types of farm, in order to assess differences in dialogues and events, discovering that these are very minor. I played both the 'Community Center' and 'support Joja' storylines until all of the associated tasks were completed, in order to elicit the primary game discourses. Data is thus comprised of videos of gameplay, which were transcribed and logged with screen shots for reference. I reviewed 3 YouTube videos and 14 on-line articles of interviews with Eric Barone, reaching saturation in his available statements on game dynamics and his personal intentions for the game. I also reviewed forum discussions on the issues addressed in the findings, in order to provide context and critical support to the narrative analysis. I analysed the data deductively in relation to the tenets of classical agrarianism, critical agrarianism and idyllic rurality. Analysis was thus instrumental in approach (van Vught and Glas 2018): I treated the game as an object of study, following the intended progression set by the game designer.

Analysis of media content inevitably reflect the perspective of the analyst. The positionality of the researcher impacts on every stage of the research process (Coghlan and Brydon-Miller 2014). I bring with me substantial experience in studying agricultural adjustment, which influences how I 'read' the narrative, as does my direct experience with farming practices, gained during my childhood on a commercial livestock farm and engaging in 4H (rural youth) clubs. My personal stance on agrarianism is not strongly towards either critical or classical orientations, whereas some of my colleagues are clearly active proponents of critical agrarianism.[2] Players from other cultures and farming backgrounds may similarly interpret Barone's representations differently. I am a casual gamer with experience of other RPGs; playing Stardew Valley would

be challenging for a non-gamer, as there are no instruction manuals or formal guidelines for gameplay. I found the extensive wiki and forums to be useful both for successfully playing the game and gaining insights into others' experiences, but the observations and analysis presented here are my own unless otherwise identified.

A key challenge in analysing game narratives is differentiating the narrative content from the characteristics of the medium in which it is developed. As a medium, computer games are evolving and difficult to categorize. Written text and language have limitations of expression (e.g. Gkartzios and Remoundou 2018); computer games are also restricted by the capabilities of the code utilized, and the mechanisms of gameplay. Stolnik (2014) argues that conflict is at the root of all good gameplay, setting a challenge for the player to aim to address. This challenge may be artificially exaggerated by game designers to motivate gameplay, as Barone has done in his portrayal of JojaMart. In-game interactions also have limits. In Stardew, statements made by in-game characters are necessarily short, confined to text boxes. Player responses to dialogues are multiple choice, limiting their range. While some text appears to comprise stories to entertain or attract the curiosity of the player, other texts actively encourage the player to explore or do particular tasks (e.g. are game mechanics to encourage players to visit specific locations, learn to fish or explore the mines). RPGs more broadly are based around regular rewards and breaking down the action into small chunks of achievement, which can lead to space/time incongruities (e.g. in Stardew crops typically mature within 4 to 12 days of planting; livestock are bought as young animals but mature to adulthood within days or weeks). These achievements further engage the player in the game (e.g. yielding a profit which enables purchase of further seeds or tool upgrades).

Findings

Findings are organized into a progressive critique of the major narratives identified in the analysis: the rejection of urban life, tenets of classical agrarianism, the critical practices which are (and are not) embedded in game narratives, and the protection of the rural idyll.

Rejection of urban life

The bleakness and futility of urban life is established at the beginning of the game. As twinkly music plays, a cut scene with 'Grandpa'—old and in bed—sees him hand the player's avatar an envelope.

> And for my very special granddaughter, I want you
> to have this sealed envelope ... No, no, don't open

---

[2] This is evident in the papers they write and responses to the content of this paper when presented at conferences—some colleagues were personally outraged at the sanitization of farming practices presented in the game.

Images 2-3 Contrast between the grey monochrome of urban employment and the colourful countryside of "Stardew Valley" Images copyright Eric Barone

it yet… have patience. Now listen please, there will come a day when you feel crushed by the burden of modern life… and your bright spirit will fade before a growing emptiness. When that happens, my dear, you'll be ready for this gift. Now let Grandpa rest.

With this ominous introduction, the scene fades, and a blank screen with *XX Years later* in the middle appears. Cut to scene at the Joja corporate offices, where there are numerous identical cubicles, and overseers watching through glass windows; the 'work' button glows green. The view pans along a row of desks—past one with a skeleton, another with a 'terminated' sign, to one occupied by Lucy (Image 2), my avatar. The office space is grey, with tall barriers separating office workers, and an [ironic] 'Life's better with Joja' moto on the wall.[3] Clearly the time has come to open the envelope from Grandpa, which is found in the desk drawer. It reads:

Dear granddaughter,
If you're reading this, you must be in dire need of a change. The same thing happened to me, long ago. I'd lost sight of what mattered most in life…real connections with other people and nature. So I dropped everything and moved to the place I truly belong. I've enclosed the deed to that place…my pride and joy. It's located in Stardew Valley, on the southern coast. It's the perfect place to start your new life. This was my most precious gift of all, and now it's yours. I know you'll honor the family name, my dear. Good luck.
Love, Grandpa.

The primary themes of the Stardew Valley discourse are thus established: the crushing burden and emptiness of urban life, the importance and authenticity of connecting to people and nature, the pride and joy inherent in agrarian/rural life, and the precious opportunity and 'gift' of engaging in these experiences. The natural and moral nature of rural life—Flinn and Johnson's (1974) third characteristic of agrarianism—is implicit in the words "what matters most in life". Heritage is identified in the reference to family, Grandpa's evident love and forward planning for his grandchild, and in the inheritance of the farm. Grandpa has set the precedent of "dropping everything" and finding a place where he "truly belonged". "Dropping out" is thus not a novel or radical contemporary innovation, but a fulfilment of tradition (and a form of farm succession). Life in Stardew Valley is described as more "real" than urban experiences. There is no mention of why Grandpa left the valley, other family members, why his place has been vacant for a lengthy period of time, or why one would need to experience modern life as a "crushing burden" before moving there. Unlike the characters living in Pelican Town, there will be no opportunity to delve into Grandpa's backstory, beyond the tiny snippets given by his old friend the mayor. The "family name" is unknown, as the player never selects or is identified by a surname—all of the characters in the game have first names only. Notably, Grandpa does not refer to his legacy as a farm, despite multiple opportunities to do so—instead, the word "place" is reiterated three times. The experience of Stardew Valley is thus of the whole valley, rather than the farm alone.

The social and psychological distance between the urban and rural settings are also embodied in landscape differences. The avatar takes an apparently lengthy bus ride into the 'mountains' to reach Stardew Valley. The cheerful palette, wooden fences and dirt paths (Image 3) are in stark contrast to the grey tones and desk-top computers of the Joja offices (Image 2). Cheerful background music plays throughout the game, and non-player characters reinforce the pleasures of farm life in interactions with local residents.

---

[3] The heavy black borders to the scene are characteristic of all indoor scenes.

For example, Leah (the local artist) comments that *The simple things in life are best: a soft summer breeze, majestic clouds, and a goblet full of Stardew Valley red*. Imagery thus emphasizes outdoor experiences, nature and consumption of local produce, although there is no option in the game for local residents to purchase produce directly from the farm. The game thus positions local production as an amenity and opportunity for income generation, rather than a critical practice.

The 'dropping out' proposed by Grandpa to pursue an authentic lifestyle is not 'alternative' in the meaning employed in critical agrarianism. The place is a source of pride, joy and opportunity to connect with people and nature. New friendships are anticipated but they are with existing community members, not other newcomers or members of a social movement. The rejection of urban life is unrelated to alternative discourses around agricultural production (e.g. the meaning of sustainable agricultural practices, 'healthy food', or environmental degradation caused by intensive practices). At no point in the game is there even recognition that there are multiple, competing approaches to farming. The only other agriculturalist in the valley—Marnie, a rancher—has a tiny farm holding, where livestock are kept in small pens. Although sometimes dubbed a 'farming simulator' (e.g. Dieker 2016), the villain is not industrial agriculture, it is a big-box store, unidentified by Grandpa at any stage. This is a new problem in the valley, one he did not face.

Classical agrarianism

The farm is clearly the central location of the game—the player starts each day by waking up at the farmhouse at 6 am and must return every evening or receive a penalty. Gameplay is consistent with the agrarian concept of farming as the central profession, on which all others are dependent (see Flinn and Johnson 1974). The businesses of local community members appear to exist primarily to serve the farmer: local residents are never seen using the blacksmith, carpenter, or fishing hut. They can be found shopping, using the library and drinking in the pub but there is never a queue. The virtue of hard work is implicit in 'grafting'—the daily grind of watering rows of crops, mining ores, foraging or fishing in the lakes and rivers—rewarded with accumulating g. The mayor reinforces this norm by invoking his memories of Grandpa: for example, when the avatar interacts with him in the pub, the mayor states that *Your grandfather always worked himself too hard... I'll have an extra beer in his honor tonight*.

The avatar's primary identity is of farmer—the player is routinely greeted as 'the new farmer' by community residents—although throughout gameplay the player may

choose to spend more time mining or fishing. Farming is thus pluriactive in Stardew, but the farm remains central—crop and livestock production are by far the most lucrative game activities. The multiple reasons for becoming a farmer are raised through an interaction with Leah in Pierre's shop in the second week of play:

> Leah: So why did you become a farmer? [Multiple choice options appear]

- I want to make tons of money.
- It's more "real" than living in the city.
- To follow in grandpa's footsteps.
- I wanted to escape my old life.

The exchange reveals the four primary reasons Stardew's creator sees for moving to the valley. Three of these reasons are already evident in the initial set up of the game, but this is the first mention of the lucrative nature of agricultural production—not economic independence as identified by Flinn and Johnson (1974) but the massive accumulation of g. Although this appears to be an unlikely reason to engage in Stardew Valley, it is consistent with the reward structures embedded in RPGs more generally—it is a clear indicator of progress, always present in the top right of the player's screen, and enables progression through the game's narratives (e.g. accumulating sufficient g to purchase livestock or construct new buildings). Rapid pursuit of wealth is also a reason to choose Joja over the Community Center, and can be linked to a form of freedom—as the author of the guidebook to becoming a Stardew millionaire reasons *once you make enough money, you can play the game however you like without worrying about farming or tedious gameplay* (Verrett 2017, kindle 6%). Notably, there is also no discourse around part-time farming, or the trope of working to build up sufficient capital for financial independence (i.e. to achieve the status of 'full-time farmer'). The avatar's identity as farmer is already secured, and regularly reinforced through interactions with community members.

The second option is consistent with the narrative introduced by Grandpa at the beginning of gameplay, about the "things which matter most" being "real relations" and interacting with nature, but appears somewhat ironic when directly expressed within a cartoonish computer game. If the player chooses this option, Leah will respond *That's pretty much the reason I came here too!* The suggestion that Leah herself has dropped out of urban life is confirmed as the player interacts with Leah (and eventually her ex-boyfriend, who attempts—unsuccessfully—to lure her back to the city). The opportunity to interact with animals may also be implicit in option two. Livestock are individual and personalized: livestock come with a name than can be changed

Image 4 Heart bubbles from happy livestock in a poultry coop (The small green animals are 'dinosaurs', adding a playful and sometimes humorous addition to the standard livestock available: cattle, pigs, sheep, goats, rabbits, chickens and ducks.) Image copyright Eric Barone

upon purchase. 'Happy livestock'—a status achieved by regularly feeding, petting and opening barn doors to enable them to graze outside—produce higher quality products, which earn higher sales values at Pierre's store or through the collection box (Image 4).

Petting an individual animal also raises a heart bubble over it, if the animal has also been fed; a double click will identify the animal's happiness rating (e.g. *Polly looks happy today*! or *Peppa looks grumpy.*). Livestock are thus anthropomorphized—positioned as sources of love (or guilt) for the avatar, with a heart meter (shorter but similar to those of the human residents of Pelican Town).[4] Livestock 'matter', requiring regular care but offering affection to their carer.

Option three, following in Grandpa's footsteps, clearly references the family heritage of farming, although the reasons farming has not been pursued by Grandpa's children are unclear. Throughout the game the avatar's parents will send gifts through the post (ranging from cash to homemade cookies), stating that they are proud of the avatar's progress, but they never appear in person. The 'family' aspect of Grandpa's farm is made possible through game dynamics that encourage courtship, marriage and even having a family with one of over a dozen other community members (selected by the player). Once a spouse is secured (achieved largely through persistent gift giving), the spouse will increase farm labour—occasionally watering and harvesting crops, and feeding livestock. The player is thus encouraged to establish their own family to undertake work on the family

farm, although will typically spend the first 'year' of game time working the farm alone.[5]

Option four reinforces the farm as an escape. Farm life thus represents what Halfacree (2010) terms a 'bolt hole'—a refuge from urban life. The farm is clearly a sanctuary—other community residents rarely venture on to it (and only during cut scenes, typically offering gifts or assistance), unless the player takes a spouse. The farm is a safe environment—there are no farming 'accidents' or events that have the potential to cause physical harm. Although there is considerable work to tidying the farm and producing crops, it is impossible for the avatar to become injured (unless the player has selected the 'wilderness farm' where creatures normally restricted to the mines come out at night). Commodity production is not entirely without challenges: crows will eat individual crops (addressed by installing scarecrows) and single plants occasionally die, but most crops reach maturity and yield a consistent profit. There is no bank: no mortgage or need to borrow; no interest rates. Commodity prices are stable—the economic 'losses' occur only if the advice on the television and letter from Granny Evelyn are ignored and slow-maturing crops are planted at the end of the season, or animals are locked out of the barn at night. By and large g accumulates; the player's success in accumulating it increases in speed over time. The economic independence of the farmer is thus ensured.

---

[4] For further analysis of the role of sentience and affective encounters in Stardew gameplay, see Sutherland (2020).

[5] An update to Stardew Valley in 2018 made it possible for 'multi-player' play on PCs, where a group of players could develop the same farm. This version is not considered here.

Critical practices

Critical agrarianism emphasizes the environmental damage caused by industrialized, large-scale agricultural production, and social justice issues around land access (Carlisle 2014). In Stardew, the farm is inherited; land access is not identified as an issue at any point in the game (i.e. no community members indicate their interest in acquiring a farm, nor is farm expansion an option). Farming practices are clearly low input, involving no mechanization, pesticides or even manure—fertilizers can be made from seashells and tree sap, but are not identified as 'organic' or controversial in anyway. Fertilizer is simply a product to speed plant growth; similar products are not available for livestock. Players are encouraged to clear undergrowth and rocks to use as building materials: library books and the 'Livin' off the Land' television program inform the player that cleared areas are more likely to produce forage items. Trees are attractive and varied, but function as harvestable resources, regenerating within a month. Frogs, squirrels and rabbits scamper away as undergrowth is cleared but appear unharmed. It appears impossible for the farmer to appreciably damage—or improve—the natural environment. There are no rats or other vermin, but it is not possible to remove the litter from areas of the Stardew waterfront or to produce renewable energy. Players can craft a recycling machine, which turns 'trash' collected from fishing into useful resources, representing a nod to contemporary expectations around recycling, but it is never raised in discussions with community members. The opportunity to work in a natural environment is clearly an attraction of farm life but there is no mention of the city as polluted. Moving to Stardew Valley represents a retreat from the monotony and pointlessness of urban life but not an act of environmental activism.

The potential for environmental damage is clearly linked to JojaMart. The player may venture into JojaMart while exploring the landscape, but the first game-directed encounter with Joja does not occur until day 5. The post arrives with the following notice:

> To our valued JojaMart customers: our team members have removed the landslide caused by our drilling operation near the mountain lake. I'd like to remind you that our drilling operation is entirely legal (pursuant to init. L61091, JocaCo Amendment). Responsible stewardship of the local environment is our top priority! We apologize for any inconvenience this accident may have caused. As always, we value your continued support and patronage!—Morris, Joja Customer Satisfaction Representative.

The message is part of a game mechanic that opens the mines for exploration, but it also serves as an expression of the sinister nature of Joja Corp, who are clearly damaging the local environment. It is not an accident in the game writing that Joja Corp was the source of the urban desk-job from which the player has fled. The player's escape from urban life has not led to complete separation from Joja Corp; by positioning JojaMart as the villain, the narrative prompts the player to make that disconnection complete.

The 'evil' of Joja is reinforced by a cut scene in Pierre's shop (the local grocery store), which opens with Joja Manager Morris entering and offering 50% discounts to the customers inside the store. Clearly this is an unethical and socially unacceptable business practice—but Pierre's patrons promptly leave to take up Morris' offer. This lack of support is evidence of the lack of leadership or conviction amongst Pelican Town residents. The action causes Pierre distress: he is unable to compete. However, the opportunity to receive a massive discount at JojaMart is never open to the player. Membership is the only option, requiring a visit to the Joja store, with its sterile rows of products and ice blue interior, where Morris' sales' patter is full of insincere propaganda (e.g. referring to Joja membership as a *joyous experience*). To buy a Joja membership requires 5000 g—a considerable sum at the beginning of the game, when parsnip seeds cost 20 g and ripe parsnips sell for 40 g. The player needs to farm for several weeks to accumulate sufficient g before this decision can be made. In contrast, opportunities to start completing Community Center tasks begin immediately, which involve foregoing income (i.e. 'donating' produce to the Community Center rather than selling it to gain g to invest in the farm). The player is thus clearly positioned to support the Community Center.

JojaMart is juxtaposed against the derelict Stardew Valley Community Center. A cut scene with Mayor Lewis opens up about day 6:

> Lewis: What an eyesore... This is the Pelican Town Community Center... or what's left of it, anyway. It used to be the pride and joy of the town... always bustling with activity. Now... just look at it. It's shameful. These days, the young folk would rather sit in front of the TV than engage with the community. But listen to me, I sound like an old fool.... Joja Corporation has been hounding me to sell them the land so they can turn it into a warehouse... Pelican Town could use the money, but there's something stopping me from selling it...I guess old timers like me get attached to relics of the past... Ah well. If anyone else buys a Joja Co. Membership I'm just gonna go ahead and sell it. * sigh * ... Here, let's go inside...

Lewis thus invokes nostalgia, and the classical trope of disengaged youth. Big-box development is identified as a potential source of revenue but there is concern about the loss of opportunity for community engagement. These are well versed concerns in the rural (and indeed urban) development literatures: Walmart and other superstores typically have negative impacts on rural economic development, offering cheap produce and local jobs, but putting local 'mom and pop' stores out business and reducing local tax revenues (Hernandez 2003; Salkin 2005; Freilich et al. 2010). Notably, competition with Pierre's shop is identified as an issue, but not the benefits to the local communities of jobs provided by Joja. Mayor Lewis instead defines the economic issue purely in relation to the financial gains from selling the Community Center property.

Protecting the rural idyll

If the farm represents a 'bolthole', then the Community Center is a castle, to be fortified and defended from urban influence (following Halfacree 2010). Exploration of the Community Center reveals that it is literally a magical space—populated by 'Junimos': mysterious, harmless spirits who invoke a mystical connection to the environment.

> We the Junimo, are happy to aid you. In return, we ask for gifts of the valley. If you are one with the forest then you will see the true nature of this scroll.

Interacting with the Junimos in the Community Center is embedded in its refurbishment. Completing the Community Center is a mechanic to engage the player in exploring a wide diversity of game activities and nuances. The magic of the Community Center is known only to the player and a local wizard—the other community members do not see it, or indeed contribute in any way to the Community Center's redevelopment. No interactions with local residents are specifically required by the Community Center tasks, although friendship points with several of the residents are identified as rewards. The Community Center is lodged in discourse about nature and the forest but remains part of the magic rather than practical action.

Within the community, Morris is presented as the stereotypical villain: dark hair, dressed in black, and unusual amongst local residents in not having a specified home or friendship meter—it is impossible to give him gifts, gain friendship points and there is no back-story to access. His sole personality trait appears to be his loyalty to Joja. JojaMart itself is notable for not being particularly convenient—players have to travel past Pierre's store and cross a river to reach it. For the player, prices are approximately 20% higher than in Pierre's shop until Joja membership is purchased, at which point they become the same. The products are the same, with the exception of Joja-themed home furnishings and wallpaper, and Joja Cola cans, which can also be fished out of the river as trash. The rewards of completing 'rooms' of the Community Center are identical to the improvements offered for direct payment through JojaMart, but are of dubious 'community value'—the greenhouse solely benefits the player, and although the bus employs a community member, no other characters are seen using the bus, quarry, minecarts or panning for gold options that open up. The primary benefit to the local community (other than a prosperous local farmer) is the refurbished Community Center, which cannot be achieved in collaboration with Joja. Purchasing JojaMart membership is thus a means to more efficiently achieve up-grades which advantage the player, but at the expense of the Community Center, and thus the local community.

The relative merits of supporting the Community Center, and by association Pierre's shop have been substantially debated in the forums.

> Forum Participant A: The bottom line is, the lazy or incompetent citizens of Pelican Town didn't take the initiative to fix their OWN TOWN, even so far as to not have a working transit system for YEARS if the player doesn't fix it themselves, and I'm expected to believe Joja is evil for coming in and simultaneously making a profit and fixing a podunk shantytown back up to a modern standard?

> Forum Participant B: Just like Walmart: minimum wage, little advancement, no benefits, drive the small, local businesses out of business. Good-going JojaMart.

(Steamcommunity 2017)

> So Jojamart came to town to try out their "community revitalization project" and set up a store in a town, population 30. They even want to build a warehouse in an unused abandoned building in town, which would create jobs and help the local economy. They didn't drive their prices absurdly low to put their competition out of business. It would appear that Joja has Pelican Town's best interests at heart except for them sending out a villain in Morris.

(Reddit 2018)

Forum participants point out that Pierre becomes progressively less likeable during gameplay, talking primarily about making money, having an 'illicit stash',[6] and filling players' mailboxes with irritating adverts. On the first play through the game, siding with the Community Center is the obvious choice, but upon repeated plays, the choice becomes more complex, as players seek to explore subplots and make achievements more quickly.

---

[6] There are forum debates on whether his stash is pornography or drugs, but Barone leaves it to players' imaginations.

Image 5 Community members unite against JojaMart. Image copyright Eric Barone

The conflict between Joja and the Community Center is resolved immediately when Joja membership is purchased—Joja employees are seen refurbishing the Community Center into a warehouse that night. Pierre's shop remains open and there appear to be no secondary consequences of the player's choice. If the Community Center option is pursued, victory is more prolonged. The cut scene pans over a range of community members enjoying the different rooms. The major presents the avatar with a 'Stardew hero trophy' for her work in revitalising the Community Center. Morris appears, grumbling that his sales have plummeted and disturbed to see his customers in the Community Center. This leads to an altercation with Pierre. The player is given two options: "Let's be reasonable" and "Let's settle this the old-fashioned way".[7]

The "let's be reasonable" scenario has Pierre giving an impassioned speech on the importance of community, and the specific experiences of community members:

> I remember when I first came to Pelican Town. This building was active and vibrant. We worked together to make the town a better place. There was a real sense of community.[He goes on to reminisce about the idiosyncratic activities of local residents—George with his crossword puzzle, Emily weaving a banner for the fair, Willy's gaff with runaway crabs and Gus's community—building solution.] You see, everyone? Our community is what makes Pelican Town special. When JojaMart came to town we lost sight of that. But now, thanks to Lucy [the avatar], we have a second chance. I'm asking that you join me in boycotting JojaMart! We have the power to reclaim our old way of life. Whose with me?

Pierre thus invokes nostalgia for the 'sense of community' he experienced upon arrival, and how the community 'worked together'. Success for the player is enabling the community to 'regain their old way of life'. The responses are unanimous; Morris recognizes he's *done for* and leaves,

---

[7] The 'old fashioned way' is a fist fight between Morris and Pierre, which Pierre instigates by insulting Joja workers as 'cowards'. They proceed to sling punches and entertaining produce-related insults at each other (e.g. "you're even weaker than your fresh produce selection"). Pierre soon lands a punch that sends Morris literally flying out of the Community Center.

never to be heard from again... The power of 'community' has won out. The Junimos leave but the 'magic' of community is now physically embedded in the Community Center and the unity of the community, brought about by the player as a moral good. The farmer has solidified her position as the pillar and savior of the community. In the coming days, the avatar receives a flood of 'thank yous' and gifts through the mail from grateful community members, due in part to the two heart-point bump granted by Community Center completion (Image 5).

Discussion

The portrayal of farming in Stardew Valley is consistent with classical American tenets of agrarianism (Flinn and Johnson 1974)—the value of hard work, centrality of farming to other occupations, the economic independence of the farmer, and the inherent goodness and moral value of farming as an occupation. The role of farming in preserving democracy is implicit: Pelican town residents 'vote with their wallets', initially by supporting JojaMart, but later follow the leadership of the local farmer, and unite against Joja. The farmer is thus positioned as the moral compass for the community. The Community Center is portrayed as a magical space, of central importance to community life, and the player draws on supernatural assistance to refurbish it.

Stardew Valley players do not have the opportunity to practice critical agrarianism. Although the lifestyle encouraged is pluriactive—involving mining, fishing and foraging, in addition to farming—game dynamics do not allow farming to be undertaken as a political action, and there is no discourse which connects environmental outcomes to farming activities. The environmental preservation discourse is instead associated with magic. Players conquer and subdue their farming landscapes, comfortably deforesting in the secure knowledge that trees will automatically regenerate. Threats to the environment are external (the villain Joja). The sanctity of farming as a moral practice is preserved, but disconnected from debates on meat consumption, animal welfare, pesticide use, pollution and intensification. Instead, long-standing tropes of the rural 'good life' are reinforced—the farm is positioned as a refuge or bolt-hole from the problems of urban life; the local community is a precious castle to be fortified and defended from invaders, particularly big-box development. Players of Stardew arguably consume rurality as a life raft—a temporary escape from urban life (following Halfacree 2010).

My analysis of Stardew Valley is thus consistent with Peeren and Souch (2019) in demonstrating the resilience of the 'good life' genre, and its appeal to contemporary

audiences. Stardew also demonstrates the universality of this appeal, with over 10 million copies sold (Strickland 2020) across 12 languages. The deep connections between people and place inherent in classical American agrarian tenets clearly resounds with a global audience. The distinction is that in Stardew, rather than simply observing, players repeatedly perform farming practices in this sanitized environment, reinforced by rapid accumulation of g and positive representations of farm life (e.g. heart bubbles from happy, productive livestock). Players face none of the risks of fluctuating commodity prices and bank loans, challenges or responsibilities of land ownership, or negative environmental impacts dealt with—or caused—by contemporary farmers. Farming is represented as 'peaceful'—players engage in and reproduce a farm life that is free from moral and ethical dilemmas. Game dynamics mirror the positive aspects of farm life—working outdoors, independence, and being your own boss—identified by Gasson (1973) in seminal work on the goals of contemporary farmers—but offer none of the controversies that might inspire critical sensibilities to arise. Whereas Peeren and Souch identify the television program 'The Farmer Wants a Wife' as a potential gateway to more critical portrayals of farming, this is less clear for Stardew Valley. Although players may be inspired to engage in more realistic games like Farming Simulator, on-line recommendations for other games which players may also enjoy suggest that these games are similarly sanitized and uncritical of farming practices (e.g. Morton 2019; Loveridge 2018).

A key question for this analysis is what players may learn from gameplay. The cartoonish style of Stardew does not preclude experiential learning: studies of recreational computer games have demonstrated that cartoon-based games are more effective than photo-realistic realistic games in facilitating learning (Mayer 2019). Barone has clearly been successful in his aim of engaging players in making decisions about whether to support big-box development, provoking active forum discussions. Although the negative outcomes of big-box developments are well recognized in the academic literature (e.g. Carr and Servon 2008; Vias 2004; Goetz and Swaminathan 2006), the continued popularity of these developments to consumers is evident in their commercial success. The fullness of these issues is not explored in Stardew (e.g. issues around local employment and taxation remain unaddressed) but this is to be expected in a recreational game which is not designed to educate, or to confront. Invoking nostalgia for the Community Center yields an image of rural life which is lodged in the past but may enable a more critical understanding of contemporary rural economic development issues.

It is less clear what players learn about farming practices. Arguably, Stardew Valley reinforces positive normative associations with low input agriculture (i.e. that low input farming is the right or best way to farm). However, by sanitizing farming production, it substantially underplays the importance of contemporary debates on major social justice and environmental degradation issues (ranging from mental health and farm safety, to biodiversity preservation, animal welfare, water pollution and cheap food). I argue that playing Stardew Valley both reinforces classical agrarian tenets and reduces social consciousness of contemporary farming practices, contributing to their invisibility. Many industrialized practices (e.g. intensive pig and poultry production) are located indoors, out of public sight. The distance between Stardew and contemporary practices is also highly varied, making it challenging for players to connect their on-line actions to real farming practices. For example, small-scale egg production and sale at farm gates is common across the global North. This type of poultry may even have names and be treated as pets by their owners (see Kyle and Sutherland 2018), much as they are in Stardew. In contrast, the production of 'free range' eggs sold in supermarkets involves huge barns populated by thousands of nameless poultry with short lifespans—but may evoke a similar image in consumers whose primary experience of farming is from their desk chairs. Gameplay 'teaches' that happy animals produce higher qualities and quantities (e.g. of milk and eggs), which may encourage players to pay premiums for 'organic' or free-range products. However, it may also lead players to believe that all farmers must be committed to high standards of welfare, as these practices yield higher outputs. This connection between perceptions, gaming practices and in-life decision-making is an important topic for future study.

The major contribution of this paper is the recognition and advancement of knowledge about the desk chair countryside. Stardew Valley's creator is clearly a member of this cohort. The consistency of Stardew Valley representations with classical American agrarian tenets is therefore somewhat surprising. Not only has Barone had limited (if any) direct experience of farm life, he was intentionally mimicking a Japanese video game (Harvest Moon) when he designed Stardew Valley. The international popularity of Stardew suggests that Barone's representations are deeply embedded in broader social ideologies which transcend national boundaries. The imagery and positive cultural associations of small, independent family farmers and the simple life in the countryside is shared amongst a wide range of cultures (Netting 1993), not least Japan, where there is a trend towards part-time, rather than industrial farming (Hisano et al. 2018). Although the USA is known for industrial farming practices, part-time farming is also common: the USDA (2015) has identified some 39% of farms as 'off-farm occupation farms' and a further 29% as 'retirement farms'. The recreational appeal of farming life is thus evident amongst farming practitioners as well as participants in the desk chair countryside. What is particularly important about

the desk chair countryside is that they actively engage in farming practices, albeit mediated through their computer systems. Unlike Bunce' (1994) armchair countryside, they are not casual observers but active decision-makers in deciding what to produce, and seeing their production through to sale. How this active engagement is shaping consciously and unconsciously held beliefs and normative associations amongst the desk chair countryside is an important topic for future research.

The reasons behind Barone's omission of critical perspectives on agrarianism also warrants exploration. Clearly, he is not averse to including 'real world messages' and integrated his own vegetarianism into the game. It could be that for the game to be 'fun', he preferred to limit the 'villain' to a single agent which was external to the local community. However, although Barone researched farming production mechanics (White 2018), it could also be that his knowledge of broader agricultural industry issues is primarily informed by computer games. Although he is not comfortable with butchering livestock, he may not be aware of or concerned about the processes and negative externalities of industrialized agricultural production. This may also be true of a substantial cohort within the desk chair countryside. The representations in Stardew Valley thus raise the possibility that its creator—and the people who play it—are disconnected from or disinterested in the debates on critical agrarianism which heavily populate the literature in this journal. If academics are aiming to reach this cohort with their critiques, there is a considerable distance to travel.

Contemporary computer games thus present both challenges and potential solutions for engaging the general public in contemporary agrarian debates. My analysis demonstrates that Barone's game has engaged (some) players in actively debating the merits of big-box development. It follows that there is similar potential to engage players in debates on the practices of contemporary agricultural production, using computer games as mediums. This is already underway, evident in organisations like 'Games for Change' and computer games like Third World Farmer[8] (which challenges players to experience the difficulties of farming amidst military unrest, disease outbreaks and drought). Computer games thus embody not only contemporary representations of rurality, they offer spaces in which a substantial global cohort can be enrolled in actively considering the practices and desired future of the farming sector.

---

[8] www.thirdworldfarmer.org.

Conclusion

In this paper I have analysed a farming computer game as a site of cultural production for and by the 'desk chair countryside': the millions of people who engage in farming and rural life through their computer systems. I see considerable potential in computer game studies for updating our understanding of contemporary representations and classical issues in rural studies, such the reproduction of the rural idyll, ideologies of rural development and critical agrarianism, and associated implications for governance. As highly interactive mediums, computer games offer important opportunities to assess how players interact with and construct rural and agrarian spaces. The precise mechanisms of these encounters—how difference is negotiated and made meaningful by players (Wilson 2017)—are specifically developed in another paper from this dataset (Sutherland 2020), which demonstrates how game players are actively co-constructing and performing 'authentic idylls' through affective encounters in their gaming worlds.

Analysis of Stardew Valley has been particularly useful for demonstrating how recreational game players can be enrolled in debates and personal decision-making on contemporary rural issues. Comparative analysis of other popular games could provide further nuance to the discussion of contemporary renderings of idyllic rurality and their influence on game players. For example, it was not possible to play Farming Simulator (first released in 2011) as a female avatar (farmer) until 2016 (Famularo 2016). Persistent patriarchal relations are well established in the agricultural sociology literature (Shortall 2016; Price and Evans 2009) but it appears that these are being normalized to an enormous global audience—Farming Simulator has sold over 25 million copies across its multiple versions to date (VB Staff 2020). Games like Stardew Valley also offer opportunities for player 'mods' (modifications of game constructs)—altering the underlying code to produce individualized games, which are then made broadly accessible—and open to critique. For example, there was considerable outrage in the forums about a mod that made (mixed race) Maru 'more appealing as a marriage partner'—by lightening her skin tone. Computer games thus offer important opportunities for methodological advances which explore the relationships between on-line and real-life communities and practices, and the materiality and 'more-than-representational' aspects of on-line farming activities. The desk chair countryside is an important cohort with whom to engage, as well as a cohort it may be possible to mobilise, if we are to understand and influence how contemporary ideals of farming and rural life are being remade.

Acknowledgements I wish to thank Tony Craig and Sharon Flanigan of the James Hutton Institute and three anonymous reviewers for their constructive feedback on earlier versions of the paper. A much earlier version was presented at the Transatlantic Rural Research Network (TARRN) Conference in May 2019, where I received helpful reviews from Menelaos Gkartzios and Christian Kelly Scott. I also wish to thank my brother Dennis Small, for gifting me with Stardew Valley and inadvertently starting me on this journey, and my husband Brian, for introducing me to gaming and providing in-house IT support. The writing of this paper was made possible by the James Hutton Institute, and the Rural & Environmental Science & Analytical Services (RESAS) Division of the Scottish Government, through the 2016–2021 RESAS Strategic Research Programme. The views expressed in this paper are my own and not those of the Scottish Government or RESAS.

Open Access This article is licensed under a Creative Commons Attribution 4.0 International License, which permits use, sharing, adaptation, distribution and reproduction in any medium or format, as long as you give appropriate credit to the original author(s) and the source, provide a link to the Creative Commons licence, and indicate if changes were made. The images or other third party material in this article are included in the article's Creative Commons licence, unless indicated otherwise in a credit line to the material. If material is not included in the article's Creative Commons licence and your intended use is not permitted by statutory regulation or exceeds the permitted use, you will need to obtain permission directly from the copyright holder. To view a copy of this licence, visit http://creativecommons.org/licenses/by/4.0/.

Appendix A: Further information on Stardew gameplay

Playing Stardew Valley involves considerable 'grafting'—repetition of activities to gain skills or resources. The screen features a clock, important for identifying progress through the day and following the opening hours of the various businesses, homes and schedules of Pelican Town residents. Under the clock is a statement of the 'g' (Stardew currency) available to spend, and an '!' sign, which links to a 'journal'—a list of current tasks (each of which arrive in the mail or are posted as 'help wanted' at Pierre's shop in the village). Pierre's store window also hosts a monthly calendar which identifies birthdays and the dates of nine annual festivals, which mimic Christmas, Easter, Halloween and assorted village gatherings.

Achievements and skill advances are regularly highlighted through pop-ups on the screen. Speech similarly occurs through pop-ups and occasionally speech bubbles; the soundscape of Stardew is primarily seasonal theme songs, punctuated by seasonal birdsong, although specific locations have their own sounds (e.g. waves at the beach, ominous music in the mines); all can be found in the jukebox at the pub. An energy meter in the bottom right of the screen shows energy consumption, and encourages a varied lifestyle: some activities require more energy than others. Energy can be replaced by eating crops, or after the first month, by soaking in the village spa. A health meter also appears when in the mines, to reflect the damage inflicted by various monsters, but there is no 'death' option (the avatar simply loses consciousness and is rescued). Across the screen is the top line of a 'backpack', which holds the player's tools and portable items. Initially these are farming tools (e.g. hoe, watering can); a basic fishing rod and sword complete the primary set within the first week.

Information on the different characters and items in the game can be found in an extensive on-line wiki[9] and through independent gaming forums. In-game, information comes from four primary sources: the television set, the mail, interactions with other residents of Pelican Town, and a nightly financial reckoning of the returns from the conveniently placed collection box, in which produce is placed for sale during the day. Information is also available from 'books' (short texts) that appear in the library. The television provides daily weather and fortune telling up-dates, which enable a degree of forward planning, as well as programs with farming tips and recipes. Twice weekly 'Livin' off the Land' and Sunday 'Queen of Sauce' programs teach about the opportunities of the landscape (e.g. which berries are in season for foraging) and new recipes (which increase the energy content and occasionally add skills boosts to food produced). Learning also occurs through action (e.g. developing the skill of catching fish, how much energy it takes to do particular tasks).

Most characters are white, with the exception of Demetrius, and his mixed-race daughter Maru. Several community members have useful professions—Robin is a carpenter who can build farm buildings or up-grade the farmhouse; Clint is a blacksmith who can break open geodes found in the mine and up-grade tools; Marnie is a rancher who sells livestock, feed and equipment. Other community members will help to develop skills, following cut scenes about their history: Willy gives you your first fishing pole, and Marlon your first sword. 'Granny' Evelyn will give you gardening tips. All 30 of Stardew's residents will send you recipes, blueprints or resources once you become 'friends' (indicated by a heart score on a friendship bar and achieved largely by giving gifts). As friendship increases, cut scenes will open, showing you a more detailed picture of the character's personality, interests and problems, typically offering multiple choice options to interact. At ten hearts it is possible to marry, and subsequently have 'children' with single characters of either gender. The spouse will water plants and feed livestock, if a high heart score is maintained. Divorce is possible and unwanted children can be turned into doves and set free.

---

[9] https://stardewvalleywiki.com/Stardew_Valley_Wiki.

References

Berry, H.L., L. Courtenay Botterill, G. Cockfield, and N. Ding. 2016. Identifying and measuring agrarian sentiment in regional Australia. *Agriculture and Human Values* 33: 929–941.

Beus, C.E., and R.E. Dunlap. 1994. Endorsement of agrarian ideology and adherence to agricultural paradigms. *Rural Sociology* 59: 462–484.

Blekesaune, A., M.S. Haugen, and M. Villa. 2010. Dreaming of a smallholding. *Sociologia Ruralis* 50: 225–241.

Bos, D. 2018. Answering the call of duty: Everyday encounters with the popular geopolitics of military-themed videogames. *Political Geography* 63: 54–64.

Bunce, M. 1994. *The Countryside Ideal*. London: Routledge.

Bunce, M. 2003. Reproducing rural idylls. In *Country visions*, ed. P. Cloke, 14–30. Harlow, UK: Pearson Education Ltd.

Carlisle, L. 2014. Critical agrarianism. *Renewable Agriculture and Food Systems* 292: 135–145.

Carl's Guide to Stardew Valley. Stardew Valley: Grandpa's Score. https://www.carlsguides.com/stardewvalley/grandpa-end-score.php Accessed 14 March 2019

Carr, J.H., and L.J. Servon. 2008. Vernacular culture and urban economic development: Thinking outside the (big) box. *Journal of the American Planning Association* 75: 28–40.

Chayanov, A. 1927. *Theory of peasant co-operatives*. Translated by David Wedgwood Benn. Columbus, OH: Ohio State University Press.

Clark, J.R.A., and A. Jones. 1998. Agricultural elites, agrarian beliefs, and their impact on the evolution of agri-environment policies: An examination of the British experience, 1981–92. *Environment and Planning A* 30: 2227–2243.

Cloke, P., P. Milbourne, and R. Widdowfield. 2001. Homelessness and rurality: Exploring connections in local spaces of rural England. *Sociologia Ruralis* 41: 438–453.

Cloke, P. 2003. Knowing ruralities? In *Country visions*, ed. P. Cloke, 1–30. Harlow: Pearson Education Ltd.

Coghlan, D., and M. Brydon-Miller. 2014. Positionality. In *The Sage encyclopedia of action research*, ed. D. Coghlan and M. Brydon-Miller. London: Sage.

Dickason, R. 1970s. Visions of rurality in popular British fictional television series from the 1970s to the present day. In *The English countryside: Representations, identities, mutations*, ed. D. Haigron, 83–111. London: Macmillan.

Dieker, N. 2016. The economics of Stardew Valley. Billfold 25 August 2016. https://www.thebillfold.com/2016/08/the-economics-of-stardew-valley/Chayanov, A. 1927. Theory of Peasant Co-operatives. Translated by David Wedgwood Benn. Columbus, OH: Ohio State University Press.

Draefer, N. 2014. 5 Reasons you need to be using games for corporate training. Elearning Industry. https://elearningindustry.com/5-reasons-you-need-to-be-using-games-for-corporate-training Accessed 17 June 2020.

Famularo, J. 2016. Farming Simulator 17 Introducing Female Farmers. Rock Paper Shotgun 16 July 2016. https://www.rockpapershotgun.com/2016/07/16/farming-simulator-17-female-farmers/ Accessed 9 February 2020.

Flanigan, S., K. Blackstock, and C. Hunter. 2015. Generating public and private benefits through understanding what drives different types of agritourism. *Journal of Rural Studies* 41: 129–141.

Flinn, W.L., and D.E. Johnson. 1974. Agrarianism among Wisconsin farmers. *Rural Sociology* 39: 187–204.

Freilich, R.H., R.J. Sitkowski, and S.D. Mennillo. 2010. *From sprawl to sustainability. Smart growth, new urbanism, green development, and renewable energy*, 2nd ed. Chicago: ABA publishing.

Friedmann, H. 1978. World market, state and family farm: Social bases of household production in the era of wage labour. *Comparative Studies in Society and History* 20: 545–586.

GameInformer. 2016. GameInformer GI Show—Gears of War 4 Impressions, The Division, Stardew Valley Interview Published on Mar 10, 2016. https://www.youtube.com/watch?v=krHrxlpca9A&list=RDCMUCK-65DO2oOxxMwphl2tYtcw&start_radio=1#t=4719 Accessed 9 June 2020.

Gaspar, S. 2015. In search of the rural idyll: Lifestyle migrants across the European Union. In *Practicing the good life: Lifestyle migration in practices*, ed. K. Torkington, I. David, and J. Sardinha, 14–32. Newcastle: Cambridge Scholars Publishing.

Gasson, R. 1973. Goals and values of farmers. *Journal of Agricultural Economics* 24: 521–542.

Gee, J.P. 2015. Discourse analysis of games. In *Discourse and digital practices: Doing discourse analysis in the digital age*, ed. R.H. Jones, A. Chik, and C.A. Hafner, 18–27. London: Routledge.

Gkartzios, M., and K. Remoundou. 2018. Language struggles: Representations of the countryside and the city in an era of mobilities. *Geoforum* 93: 1–10.

Goetz, S.J., and H. Swaminathan. 2006. Wal-Mart and county-wide poverty. *Social Science Quarterly* 87: 211–226.

Gosnell, H., and J. Abrams. 2011. Amenity migration: Diverse conceptualizations of drivers, socioeconomic dimensions, and emerging challenges. *GeoJournal* 76: 303–322.

Graddy-Lovelace, G., and A. Diamond. 2017. From supply management to agricultural subsidies—And back again? The U.S. Farm Bill & agrarian (in)viability. *Journal of Rural Studies* 50: 70–83.

Grathwohl, M. and J. Lachausse. 2016. The soloist. An interview with Eric Barone. *Matador Review. A quarterly missive of alternative concern*. https://www.matadorreview.com/eric-barone Accessed 9 June 2020.

Halfacree, K. 2011. Radical spaces of rural gentrification. *Planning Theory and Practice* 12: 618–625.

Halfacree, K. 2010. Reading rural consumption practices for difference: Bolt-holes, castles and life-rafts. *Culture Unbound* 2: 241–263.

Hernandez, T. 2003. The impact of big box internationalization on a national market: A case study of Home Depot Inc. in Canada. *The International Review of Retail, Distribution and Consumer Research* 13: 77–98.

Hisano, S., M. Akitsu, and S.R. McGreevy. 2018. Revitalising rurality under the neoliberal transformation of agriculture: Experiences of re-agrarianisation in Japan. *Journal of Rural Studies* 61: 290–301.

Holmes, J. 2012. Cape York Peninsula, Australia: A frontier region undergoing a multifunctional transition with indigenous engagement. *Journal of Rural Studies* 28: 252–265.

Horton, J. 2008. Producing Postman Pat: The popular cultural construction of idyllic rurality. *Journal of Rural Studies* 24: 389–398.

Jayne, M., G. Valentine, and S.L. Holloway. 2011. *Alcohol, drinking, drunkenness. (Dis)orderly spaces*. London: Routledge.

Kaufman, G., and M. Flanagan. 2015. A psychologically "embedded" approach to designing games for prosocial causes. *Cyberpsychology: Journal of Psychosocial Research on Cyberspace* 9: 5.

Kautsky, K. 1988. *The agrarian question: In two volumes*. London: Zwan Publications.

Kyle, C. and L-A. Sutherland. 2018. Understanding backyard poultry keepers and their attitudes to biosecurity. Final Report. EPIC Centre of Expertise of Animal Disease Outbreaks. https://www.epicscotland.org/resources/reports-by-epic-members/understanding-backyard-poultry-keepers-and-their-attitudes-to-biosecurity-final-report/ Accessed 9 June 2020.

Leack, J. 2016. Interview: Stardew Valley creator on "Surprise" success and favorite game elements Jonathan Leack Friday, March 11, 2016. https://www.gamerevolution.com/features/12324-interview-stardew-valley-creator-on-surprise-success-and-favorite-game-elements Accessed 18 January 2020.

Lin, A. 2016. Stardew Valley pushing the boundaries of farming rpgs. The Cornell Daily Sun February 23, 2016 https://cornellsun.com/2016/02/23/stardew-valley-pushing-the-boundaries-of-farming-rpgs/ Accessed 8 June 2020.

Little, J., and P. Austin. 1996. Women and the rural idyll. *Journal of Rural Studies* 12: 101–111.

Loveridge, S. 2018. 11 games like Stardew Valley that'll keep you farming until the cows come home (literally) July 23, 2018. https://www.gamesradar.com/uk/games-like-stardew-valley/ Accessed 19 March 2020.

Mariola, M.J. 2005. Losing ground: Farmland preservation, economic utilitarianism, and the erosion of the agrarian ideal. *Agriculture and Human Values* 22: 209–223.

Marsden, T., and F. Farioli. 2015. Natural powers: From the bio-economy to the eco-economy and sustainable place-making. *Sustainability Science* 10: 331–344.

Marsden, T., and R. Sonnino. 2008. Rural development and the regional state: Denying multifunctional agriculture in the UK. *Journal of Rural Studies* 24: 422–431.

Mayer, R.E. 2019. Computer games in education. *Annual Review of Psychology* 70: 531–549.

Messner, S. 2017. Meet the man who loved Stardew Valley so much he bought a farm. PC Gamer 17 April 2017. https://www.pcgamer.com/meet-the-man-who-loved-stardew-valley-so-much-he-bought-a-farm/ Accessed 17 September 2019

Montmarquet, J.A. 1989. *The idea of agrarianism: From hunter-gatherer to agrarian radical in western culture*. Idaho: University of Idaho Press.

Moore, B. 2016. What to do in Stardew Valley's endgame. PC Gamer 5 May 2016. https://www.pcgamer.com/uk/what-to-do-in-stardew-valleys-endgame. Accessed 17 September 2019.

Morton, L. 2019. The best games like Stardew Valley on PC. Longing for more games like Stardew Valley? Good news: casual farming sims have taken root on PC. November 15, 2019 https://www.pcgamer.com/uk/games-like-stardew-valley/ Accessed 19 March 2020.

Mukherjee, S. 2015. *Video games and story telling. Reading games and playing books*. Basingstoke: Palgrave Macmillan.

Murray, J.H. 2006. Toward a cultural theory of gaming: Digital games and the co-evolution of media, mind, and culture. *Popular Communication* 4: 185–202.

Netting, R.M. 1993. *Smallholders, householders. Farm families and the ecology of intensive, sustainable agriculture*. Stanford, CA: Stanford University Press.

Parsons. J. 2019. Video games are now bigger than music and movies combined. Metro 3 January 2019. https://metro.co.uk/2019/01/03/video-games-now-popular-music-movies-combined-8304980/ Accessed 17 September 2019.

Peeren, E., and I. Souch. 2019. Romance in the cowshed: Challenging and reaffirming the rural idyll in the Dutch reality TV show Farmer Wants a Wife. *Journal of Rural Studies* 67: 37–45.

Phillips, M., R. Fish, and J. Agg. 2001. Putting together ruralities: Towards a symbolic analysis of rurality in the British mass media. *Journal of Rural Studies* 17: 1–27.

Price, L., and N. Evans. 2009. From stress to distress: Conceptualizing the British family farming patriarchal way of life. *Journal of Rural Studies* 25: 1–11.

Reddit. 2018. Why do people hate Pierre? Stardew Valley Discussion Forum https://www.reddit.com/r/StardewValley/comments/8mx6vv/why_do_people_hate_pierre/ Accessed 9 June 2020.

Ricciardi, V., N. Ramankutty, Z. Mehrabi, L. Jarvis, and B. Chookolingo. 2018. How much of the world's food do smallholders produce? *Global Food Security* 17: 64–72.

Salkin, P.E. 2005. Supersizing small town America: Using regionalism to right-size big box retail. *Vermont Journal of Environmental Law* 6: 48–66.

Satsangi, M., N. Gallent, and M. Bevan. 2010. *The rural housing question*. Bristol: Policy Press.

Shortall, S. 2016. Changing configurations of gender and rural society: Future directions for research. In *Routledge international handbook of rural studies*, ed. M. Shucksmith and D.L. Brown, 349–356. London: Routledge.

Singal, J. 2016. How a first time developer created Stardew Valley, 2016's best game to date. Vulture, Controller Freak14 March https://www.vulture.com/2016/03/first-time-developer-made-stardew-valley.html Accessed 17 September 2019.

Stamenković, D., M. Jaćević, and J. Wildfeuer. 2017. The persuasive aims of Metal Gear Solid: A discourse theoretical approach to the study of argumentation in video games. *Discourse, Context & Media* 15: 11–23.

Starr, P. 1994. Seductions of Sim: Policy as simulation game. The American Prospect Spring 1994. https://prospect.org/environment/seductions-sim-policy-simulation-game/ Accessed 17 September 2019.

Statistica. 2019. Genre breakdown of video game sales in the United States in 2017. https://www.statista.com/statistics/189592/breakdown-of-us-video-game-sales-2009-by-genre/ Accessed 17 September 2019.

SteamCommunity. 2017. Pierre versus Jojamart—Stardew valley discussions 20 Jan 2017: https://steamcommunity.com/app/413150/discussions/0/141136086917151693/?ctp=2 Accessed 9 June 2020.

Stolnik, E. 2014. *Video game storytelling: What every developer needs to know about narrative techniques*. New York, NY: Watson-Guptill Publications.

Strickland, D. 2020. Stardew Valley is a massive success with 10 million copies sold. Tweaktown, 23 January 2020, https://www.tweaktown.com/news/70139/stardew-valley-massive-success-70110-million-copies-sold/index.html Accessed 19 March 2020.

Sutherland, L-A. 2020. The 'Desk-Chair Countryside': Affect, authenticity and the rural idyll in a farming computer game. *Journal of Rural Studies* 78: 350–363.

Sutherland, L-A. 2019. Agriculture and inequalities: Gentrification in a Scottish parish. *Journal of Rural Studies* 68: 240–250.

Sutherland, L-A. 2012. Return of the gentleman farmer?: Conceptualising gentrification in UK agriculture. *Journal of Rural Studies* 28: 568–576.

Sutherland, L-A., C. Barlagne, and A.P. Barnes. 2019. Beyond 'Hobby Farming': Towards a typology of non-commercial farming. *Agriculture and Human Values* 36: 475–493.

Technopedia. 2020. Role Playing Game (RPG) https://www.techopedia.com/definition/27052/role-playing-game-rpg Accessed 9 June 2020.

Tregear, A. 2011. Progressing knowledge in alternative and local food networks: Critical reflections and a research agenda. *Journal of Rural Studies* 27: 419–430.

Trivette, S.A. 2012. Close to home: The drive for local food. *Journal of Agriculture, Food Systems, and Community Development* 3: 161–180.

United States Department of Agriculture (USDA). 2015. 2012 Census of Agriculture: Farm Typology Volume 2 Subject Series Part 10. https://www.agcensus.usda.gov/Publications/2012/Online_Resources/Typology/typology13.pdf. Accessed 15 May 2020.

van der Ploeg, J.D. 2014. Peasant-driven agricultural growth and food sovereignty. *The Journal of Peasant Studies* 4: 999–1030.

van Keulen, J., and T. Krijnen. 2013. The limitations of localization: A cross-cultural comparative study of Farmer Wants a Wife. *International Journal of Cultural Studies* 17: 277–292.

van Vught, J., and R. Glas. 2018. Considering play. From method to analysis. *Transactions of the Digital Games Research Association* 4: 205–244.

VB Staff. 2020. Giants Software on the quiet, surprising success of Farming Simulator. 10 January 2020. https://venturebeat.com/2020/01/10/giants-software-on-the-quiet-surprising-success-of-farming-simulator/ Accessed 19 March 2020.

Verrett, V. 2017. *Stardew PS4 the ultimate unofficial money making guide*. Kindle book.

Vias, A.C. 2004. Bigger stores, more stores, or no stores: Paths of retail restructuring in rural America. *Journal of Rural Studies* 20: 303–318.

Wada, Y. 2012. GDC Vault. Classic Came Postmortem: Harvest Moon. Oral Presentation, accessible on-line at: https://www.gdcvault.com/play/1015842/Classic-Game-Postmortem-Harvest.

White, S. 2018. Valley forged: How one man made the video game sensation Stardew Valley. https://www.gq.com/story/stardew-valley-eric-barone-profile Accessed 19 March 2020.

Wilson, G.A. 2007. *Multifunctional agriculture: A transition theory perspective*. Wallingford: CABI.

Wilson, H.F. 2017. On geography and encounter: Bodies, borders, and difference. *Progress in Human Geography* 41: 451–471.

Wittman, H., J. Dennis, and H. Pritchard. 2017. Beyond the market? New agrarianism and cooperative farmland access in North America. *Journal of Rural Studies* 53: 303–316.

Wolf, A. 1987. Saving the small farm: Agriculture in Roman literature. *Agriculture and Human Values* 4: 65–75.

Wouters, P., H. van Oostendorp, R. Boonekamp, and E. van der Spek. 2011. The role of game discourse analysis and curiosity in creating engaging and effective serious games by implementing a back story and foreshadowing. *Interacting with Computers* 23: 329–336.

Yarwood, R. 2001. Crime and policing in the British countryside: Some agendas for contemporary geographical research. *Sociologia Ruralis* 41: 201–219.

Publisher's Note Springer Nature remains neutral with regard to jurisdictional claims in published maps and institutional affiliations.

Lee-Ann Sutherland Ph.D., is a Research Leader at the James Hutton Institute. She was raised on a family farm in Canada, and has spent her career to date studying the development of farming culture, and how it influences human–environment relations and farm-level decision-making. She is particularly interested in recreational approaches to farming, and her work has recently shifted into studying how perceptions of farming are formed by non-farmers. She holds leadership positions in a number of Scottish Government and European Commission-funded projects focusing on agricultural structural adjustment and the associated knowledge systems, and is a Visiting Professor at the University of Guelph (Canada).

Agriculture and Human Values (2020) 37:1175–1194
https://doi.org/10.1007/s10460-020-10123-8

When farmers are pulled in too many directions: comparing institutional drivers of food safety and environmental sustainability in California agriculture

Patrick Baur[1]

Accepted: 27 June 2020 / Published online: 3 July 2020
© Springer Nature B.V. 2020

Abstract

Aspirations to farm 'better' may fall short in practice due to constraints outside of farmers' control. Yet farmers face proliferating pressures to adopt practices that align with various societal visions of better agriculture. What happens when the accumulation of external pressures overwhelms farm management capacity? Or, worse, when different visions of better agriculture pull farmers toward conflicting management paradigms? This article addresses these questions by comparing the institutional manifestations of two distinct societal obligations placed on California fruit and vegetable farmers: to practice sustainable agriculture and to ensure food safety. Drawing on the concept of constrained choice, I define and utilize a framework for comparison comprising five types of institutions that shape farm management decisions: rules and standards, market and supply chain forces, legal liability, social networks and norms, and scientific knowledge and available technologies. Several insights emerge. One, farmers are expected to meet multiple societal obligations concurrently; when facing a "right-versus-right" choice, farmers are likely to favor the more feasible course within structural constraints. Second, many institutions are designed to pursue narrow or siloed objectives; policy interventions that aim to shift farming practice should thus anticipate and address potential conflicts among institutions with diverging aspirations. Third, farms operating at different scales may face distinct institutional drivers in some cases, but not others, due to differential preferences for universal versus place-specific policies. These insights suggest that policy interventions should engage not just farmers, but also the intersecting institutions that drive or constrain their farm management choices. As my framework demonstrates, complementing the concept of constrained choice with insights from institutional theory can more precisely reveal the dimensions and mechanisms that bound farmer agency and shape farm management paradigms. Improved understanding of these structures, I suggest, may lead to novel opportunities to transform agriculture through institutional designs that empower, rather than constrain, farmer choice.

Keywords California · Institutions · Constrained choice · Farm management · Food safety · Sustainability

Abbreviations

CDC	US Centers for Disease Control
CDFA	California Department of Food and Agriculture
FDA	US Food and Drug Administration
FSMA	US Food Safety Modernization Act
LGMA	Leafy Greens Marketing Agreement
USDA	US Department of Agriculture

✉ Patrick Baur
pbaur@uri.edu

[1] Department of Fisheries, Animal and Veterinary Sciences, University of Rhode Island, Woodward Hall, 9 East Alumni Avenue, Kingston, RI 02881, USA

Introduction

A 2018 article in *Western FarmPress*, a daily news source for farmers, asserted, "[T]he vast majority of farmers… know healthy soil keeps them in business, but more importantly they are committed soil conservationists. They take their land stewardship responsibilities seriously" (Hart 2018). Despite that strong commitment, the article nonetheless proposed: "New idea in soil health: Pay farmers for their actions." Although farmers value soil conservation, the financial costs of implementation may preclude healthy soil practices such as cover cropping or diversifying crop rotations. Even financial concerns represent just one of many considerations that determine whether and how farmers may act on their values (Selinske et al. 2017). California farmers, for instance, have limited options for crops to rotate with baby greens (e.g. spring mix) because mechanical harvesters for baby greens also pick up any detritus left in the field by crops like broccoli or berries, which are otherwise common in the region. This brief example illustrates the concept of constrained choice, a theoretical lens through which farmer values and beliefs are understood as only partly predictive of farming practices (Hendrickson and James 2005; Stuart and Schewe 2016). Many forces beyond the farm level shape what is or is not possible on the farm. In the midst of widespread calls for global agricultural transformation (e.g. Anderson et al. 2019; Oteros-Rozas et al. 2019; Willett et al. 2019), there is a pressing need to better understand precisely how those forces intersect to constrain the extent to which farmers can actually make the farm-level changes that will precipitate such transformation.

Models of agricultural land-use practices often focus on the farmer decision-making space—comprising attitudes, beliefs, and values—and its relation to biophysical and social constraints (Ahnström et al. 2009; Price and Leviston 2014). A practical application of such models has been to increase adoption of socially desirable practices (Reimer et al. 2012a, b) and predict the conditions under which farmers will cooperate with, rather than resist, policy interventions intended to shift farm management (Kaine et al. 2017). Despite relatively sophisticated conceptualization of the mechanisms through which farmers adopt, hold to, and shift attitudes, beliefs, and values (Prokopy et al. 2008; Baumgart-Getz et al. 2012), models of farmer behavior tend to collapse the diverse institutions that shape, constrain, and drive farmer decision-making into simply the 'policy context'.[1] This simplification is especially limiting in the context of proliferating societal visions of better agriculture, in which farmers face accumulating pressure to be more sustainable, safe, healthy, just, efficient, resilient, and so on. Too frequently unaddressed is the question of what happens when the weight of accumulated expectations overwhelms farmers' decision-making and implementation capacity? Worse yet, what happens when different visions of better agriculture pull farmers toward conflicting management paradigms? To encompass these questions, models of farmer behavior and management decisions would benefit from clearer conceptualization of the ways in which cross-scalar cultural, market, legal, and regulatory forces shape and constrain farmers' land-use choices (Reimer et al. 2014).

Toward this end, I conduct an institutional case comparison. I first categorize five types of institutions that constrain (or drive) farmer 'choice': rules and standards; markets and supply chain forces; legal liability; social networks and norms; and scientific knowledge and available technologies. I then use this framework to compare the institutional manifestations of two distinct societal obligations placed on California fruit and vegetable farmers: to practice sustainable agriculture and to ensure food safety. Agricultural sustainability pressures encourage farmers to conserve the farm's natural resource base and preserve its ecological matrix (Perfecto et al. 2019). In general, researchers have found that many California farmers feel they should conserve soil and water, and to a more limited extent protect downstream environmental quality and ambient biodiversity (Stuart 2009; Baur et al. 2016, 2017; Garbach and Long 2017; Kross et al. 2018). Produce farmers also feel they should ensure that the food they grow will not make the people who eat it sick (Stuart 2009; Baur et al. 2016, 2017). On the farm, food safety pressures encourage farmers to minimize potential environmental sources of pathogenic contamination.

In practice, sustainability and food safety diverge (Beretti and Stuart 2008; Stuart 2008, 2009; Karp et al. 2015a, b; Olimpi et al. 2019). In an attempt to reconcile the resulting tension, multi-stakeholder efforts have emerged to promote "co-management" of safety and sustainability (Crohn and Bianchi 2008; Lowell et al. 2010; Wild Farm Alliance 2016). However, such efforts concentrate on how to change farmers' minds (i.e. their attitudes, beliefs, and values) through information transfer, education, ethical appeals, and raising awareness. These efforts fail to address the complex web of legal, regulatory, market, cultural, and agronomic contexts within which California farmers grow vegetables, fruits, and nuts. Various institutions bound the range of possible management options. In other words, explanations and interventions to date have not sufficiently delineated the limits to farmer choice, eliding the discontinuity between aspirations to farm in a socially desirable way and how farmers actually farm. This paper aims to fill that gap by leveraging the contrast between food safety and sustainability obligations as an opportunity to compare the impact of multiple interacting institutions on farmer 'choice'.

The paper proceeds with a review of constrained choice, how institutional theory provides insights into the nature and mechanism of those constraints, and explanation of five types of institutions that shape, constrain, and drive farmer decisions. I then provide background on food safety and agricultural sustainability in California, the national leader in produce farming, before presenting my comparative analysis of these systems. I conclude with applied and theoretical insights drawn from this institutional comparison that can

---

[1] An illustrative example is provided by Liu et al.'s (2018) review on adoption of Best Management Practices. Out of 121 peer reviewed papers, the authors identified only 7 that addressed "macro factors", including just 2 that directly evaluated the "roles of policies, markets, business, or agencies," and concluded that such factors are "rarely investigated.".

advance understanding of why farmers might, or might not, shift their land management practices to align with multiple, overlapping societal aspirations for agriculture.

Dimensions of constrained choice

In their seminal article on constrained choice, Hendrickson and James (2005) argue that the economic and technological consolidation of industrialized agriculture limits the management options available to farmers, for example by reducing the range of possible plants and animals they can use. Over time, the authors warn, these constraints may reduce farmer agency, including their ability to make ethical determinations. Not only might chronic exposure to constrained choice normalize farm practices formerly deemed "unethical", but it might also leave farmers without the knowledge, experience, or strength of moral identity necessary to navigate more complex "right versus right" problems.[2] Several empirical studies have found that farmers' ethical fortitude appears to decline as economic pressures mount (James and Hendrickson 2008; Stuart 2009). In later work, Hendrickson and James (2016) use network exchange theory to define constraints on farmer choice according to the power dynamic created through dependency relationships in agricultural markets—in essence formalizing the degree to which farmers must accept decisions made for them by others to whom they are economically beholden. Such dependencies limit a variety of farmer freedoms, including the "freedom to make ethical decisions" (Hendrickson and James 2016). Constrained choice thus poses a major impediment to policy interventions that hinge on cultivating new farming ethics, for example educational campaigns to encourage farmers to commit to on-farm climate change mitigation and adaptation strategies (Stuart and Schewe 2016).

While this body of literature makes a compelling argument for the *implications* of constrained choice, the theory would benefit from further specification of the dimensions and mechanisms of constrained choice: precisely what kinds of factors impose constraints and create dependencies, and how do they do so? A clear and structured answer to this question would better connect the insights provided by constrained choice theory to the literature on agricultural policies intended to shift farm management practices.

Within that literature, the subject of why farmers do or do not adopt particular management practices that seem socially desirable is the subject of much debate. Warren

et al. (2016) observe, "A disconnection between policy aspirations and the effective delivery of policies at 'ground level' has been a leitmotif of the agrienvironmental policy sphere since the 1980s." Interest in environmentally sustainable agriculture has driven research on adoption of conservation strategies such as erosion control (Knowler and Bradshaw 2007) or biodiversity conservation (Moon and Cocklin 2011), but researchers have also evaluated farmer responses to other social objectives including fair labor practices (Brown and Getz 2008; Harrison and Getz 2015) or animal welfare (Mench 2008; Veissier et al. 2008). However, when translating insights from adoption studies into policy contexts, the disconnect between aspiration and practice is often framed narrowly as a result of insufficient incentives (Barnes et al. 2013) or information/knowledge deficits among farmers (Parker et al. 2016; Calo 2018). Fixation with such perceived absences—i.e. what farmers lack—seems especially misguided given a nagging suspicion that neither researchers nor policymakers fully understand "what it is that makes farmers want to behave in a certain way" (Battershill and Gilg 1997). Other scholars identify political economic constraints that impede farmers who may want to farm in more sustainable or equitable ways (Bacon et al. 2012; Stuart and Gillon 2013; Carlisle et al. 2019a). As Guthman (2016) succinctly observes, "growers are not operating in an entirely voluntary context… in which non-business values might have more salience." When farmers consider adopting conservation or sustainability practices, they "face different financial, agronomic, and environmental contexts" (Reimer et al. 2012a, b), implying that policy interventions must adapt to local conditions in balancing farmer needs against the desire for broader societal benefits (Knowler et al. 2014; Mills et al. 2017).

To more systematically and comprehensively organize this wide-ranging debate, I propose complementing the concept of constrained choice with insights from institutional theory. In the context of this paper, an institution refers broadly to "the humanly devised constraints that shape human interaction" and create stable patterns in social behavior (North 1990). Institutional theory generally divides these patterning forces into regulatory, normative, and cultural-cognitive facets (Scott 2013, pp. 59–74). Regulatory facets of institutions comprise the highly-visible social processes of making rules, monitoring behavior, and implementing incentive structures to encourage compliance; the pressures exerted tend to be coercive. Normative facets of institutions, in contrast, exert 'soft' pressure through social obligation, peer expectations, or standards of appropriate behavior. Lastly, cultural-cognitive facets of institutions are the least visible forces that pattern behavior, as they operate at a "taken-for-granted" level by shaping shared understanding of how the world works and how collective meaning is produced. Whereas transgressing regulatory institutions

---

[2] On the types of ethical problems that may occur in agriculture, see James (2003).

results in punishment (or loss of reward) and transgressing normative institutions results in shame, loss of standing, or social exclusion, transgressing cultural-cognitive constraints results simply in confusion or disorientation. All three institutional facets should be considered when analyzing the pressures that shape farm management.

These three facets of institutions may align in different configurations, conveyed by both symbolic and material "carriers" that span the private–public continuum, operate at different levels (e.g. individual to societal), and are formally organized or emergent from uncoordinated market or social interactions (Scott 2013, pp. 95–107). These configurations can be described as distinct institutional logics (Thornton et al. 2012). As Higgins et al. (2016) warn in their analysis of how responsibility for biosecurity is distributed in the beef industry, "Multiple logics can also cause confusion for individuals who are required to manage competing and possibly misunderstood priorities." In principle, recourse to alternative institutional logics can enhance individual agency to break from dominant logics (Battilana 2006; Battilana and Dorado 2010), for example by equipping practitioners with tools of resistance or resilience. However, when multiple institutional logics become too confused, the resulting muddle can overwhelm farmers and exacerbate their mistrust of expert advice and incentive programs. That mistrust can in turn undercut efforts to align farm management practices with societal goals (Higgins et al. 2016) and may cause farmers to miss out on otherwise beneficial opportunities (Mercado et al. 2018).

Given these conceptual tools, I argue that the structures which constrain farmer choice can be usefully characterized according to their institutional "carriers," which operate along regulatory, normative, and cultural-cognitive dimensions. Moreover, viewing "right versus right" problems (Hendrickson and James 2005) as corresponding to competing institutional logics (Higgins et al. 2016) suggests that constraints do not operate within a single overarching configuration. Rather, references to 'ethical' and 'unethical' behavior overlook the possibility that each choice belongs to complete, but divergent, institutional logics, each with its own set of constitutive institutional carriers. The task then becomes how to classify those carriers so as to best enable comparison of their relative power over farm management.

Toward that end, I define five types of institutional carrier that strike a balance between encompassing the many factors known to shape farm management and maintaining conceptual simplicity: rules and standards, market and supply chain forces, legal liability, social networks and norms, and scientific knowledge and available technologies. To develop this taxonomy, I have drawn upon insights from my previous research in California specialty crop agriculture (documented in Karp et al. 2015a, b; Baur et al. 2016; Baur et al. 2017; Olimpi et al. 2019) and

further developed and refined these grounded insights through a focused literature review. I searched for relevant, well-cited papers that addressed drivers of food safety or environmental sustainability in produce agriculture through diverse disciplinary lenses, prioritizing studies conducted in California or the United States. I followed up by cross-referencing these papers with an interdisciplinary sample of other leading papers that they cite. In the rest of this section, I summarize each institutional carrier separately, but recognize their fundamental interdependence in the comparative analysis that follows.

Rules and standards

Rules, set and enforced by government agencies, may mandate or prohibit the use of certain practices. Standards, set and monitored by public or private entities, likewise limit land use practices to those which will meet desired thresholds or criteria. Comprising a continuum from mandates—including prescriptive 'musts' and prohibitive 'must-nots'—to voluntary metrics, rules and standards influence farming practices by setting benchmarks against which regulators, customers, contracted third-parties, or other actors in a position of authority can measure farm or farmer performance.[3] Though their enforcement methods differ, rules and standards form a hybrid regulatory institutional influence on farm management, increasingly operating across the public–private divide in complementary or mutually-reinforcing relationships (Higgins et al. 2008a, b; Garcia Martinez et al. 2013; Verbruggen and Havinga 2017). Farmers are more likely to comply if they believe that rules and standards do not overly impinge upon their sense of control over their farm (Moon and Cocklin 2011; Price and Leviston 2014). At the same time, farmers more readily accede to rules and standards imposed by powerful social actors whom they believe are beyond their ability to influence (Feola et al. 2015). Together, these findings suggest that multi-level, interlocked systems of rules and standards not only place stronger pressure on farmers to cooperate—especially when backed by enforcement mechanisms such as loss of market access, monetary fines,

---

[3] The literature on the role of standards—and the various systems through which compliance with standards is monitored and approved—in shaping food systems and value chains is both deeper and more wide-ranging than can be fully addressed here. For the purposes of this framework, I focus on standards through a regulatory lens. However, it should be noted that standards may also operate at normative and cultural-cognitive levels, inasmuch as they "reflect much more fundamental social/technical relations that are essential to the establishment and regulation of social and ethical behavior in capitalist markets" (Busch 2000). See also Busch (2020), Hatanaka and Busch (2008), Hatanaka et al. (2012), and Verbruggen and Havinga (2017).

or criminal charges—but are also less likely to be met with attitudes of indifference or defiance (Bartel and Barclay 2011).

Legal liability

The risk that a particular farming practice might generate legal liability can likewise encourage or discourage certain practices. In contrast to the preventive intent of rules and standards, liability represents a reactive regulatory institution that discourages socially undesirable behavior by punishing those responsible after the fact. Farmers may face civil suits or criminal charges brought by government agents if they break a statutory law (Buzby et al. 2001). They may also be liable under common (or tort) law, meaning they could face lawsuits, if their operation harms someone, either through consumption of their farm products (Buzby and Frenzen 1999) or as a result of externalities associated with farming (Bergstrom and Centner 1989). Farmers also experience this risk indirectly through market forces (see above) that are leveraged against farmers by their large, well-recognized customers (e.g. supermarkets) who are more exposed to consumer lawsuits. The level of legal liability risk depends on whether a particular farming practice contravenes existing laws or whether that practice could cause a harm clearly and readily traceable to farmers.

Market and supply chain forces

Certain forms of land use are promoted or limited through the economics of production costs and product prices, as well as through politics of market access and exclusion at play in the supply chain. These market forces are themselves a product of various institutions that give shape and stability to the social interactions that markets comprise. New market incentives may normalize different management paradigms. For example, the promise of a higher price or a competitive advantage to offset additional time, effort, and cost incurred can incentivize farmers to adopt new practices, as has been seen with conversion to organic agriculture (Klonsky and Greene 2005; Uematsu and Mishra 2012). Conversely, market forces can incentivize farmers to abandon practices deemed undesirable by consumers, such as ceasing to use genetically modified seed in order to access expanding markets for non-GMO food (Castellari et al. 2018) or adopting pollinator-friendly agrichemical use patterns (Wollaeger et al. 2015). Other institutional aspects also shape market forces. Labels or certifications are a common form of communicating product attributes—including the farm practices used to grow food—to customers, who may then take that information into account when considering whether to buy the product and at what price (Czarnezki et al. 2018). In this way, labels and certifications operate normatively

by defining what is appropriate or acceptable; however, they also embed shared meanings of what constitutes food and its relevant attributes, and so carry a cultural-cognitive force as well (Busch 2020). On the supply side, production contracts between farmers and their buyers act as regulatory institutions that condition market access. These contracts detail, in legally-binding language, the precise planting, cultivation, and harvesting processes farmers must follow to grow a crop for a buyer (Kelley 1994; Guthman 2017; Rehber 2018). Market and supply chain forces may thus bundle together related institutional constraints or drivers that might otherwise be difficult to parse.

Social networks and norms

Social networks are the primary normative institution through which appeals to farmer values, attitudes, and beliefs attempt to intervene in farm management, and it is critical to recognize that norms emerge, persist, and adapt through webs of interpersonal relationships (Raymond et al. 2016; Duff et al. 2017). Social networks, often diffuse or polycentric and including a range of civic actors such as non-governmental organizations or advocacy groups, regulate information and resource flows to farmers, subtly circumscribing "farmers' opportunity space" (Feola et al. 2015). Peer group norms about acceptable and appropriate farm management emerge from networks (Bacon et al. 2012) and peer relationships (Bartel and Barclay 2011). Moreover, norms shape farmer interpretation of and response to other institutions, and are heavily influenced by information networks, i.e. the sources farmers trust to advise them (Schewe and Stuart 2017). Policies which "nudge" farmers to change their practices in a way that respects existing information networks and social group norms are more effective than policies which try to "budge" farmers out of their familiar relationships of trust and realms of experience (Barnes et al. 2013; Mills et al. 2017).

Scientific knowledge and available technologies

Scientific knowledge and available technologies play a fundamental role in determining which farming practices are considered possible in the first place (and which are unimaginable), setting the cultural-cognitive institutional boundaries for farmer 'choice'. An example can be seen in the context of agroecology, a farming system based on harnessing the benefits of ecological diversity and which is often framed in normative terms (see Altieri 2009; Rosset and Altieri 2017). In discussing the potential for agroecological practices to spread, Montenegro de Wit and Iles (2016) argued that, compared to conventional industrial agriculture, agroecology lacks "thick legitimacy", or the "authority that is woven into the knowledge-making of scientific and political institutions, and embedded in widely practiced social conventions." Farmers are more likely to adopt new farming practices when expert or knowledge-making institutions speak to both the societal importance and feasibility of adoption. Perceptions of uncertainty, conflict, or fraudulence in the underlying institutions that produce scientific knowledge weaken expert authority to guide or regulate farmers' behavior. Doremus (2003) emphasizes the importance of the practical capacity to act, which overlaps with the concept of perceived behavioral control (or perceived available resources) frequently used in social psychological models of farmer behavior (Reimer et al. 2012a, b; Price and Leviston 2014). Moreover, farmers respond positively when they can personally experience the 'success' of a new practice, suggesting that capacity to directly measure desired outcomes is a critical complement to expert assertions (Moon and Cocklin 2011; Price and Leviston 2014). Finally, the degree to which a proposed practice must shift the "momentum" (Hughes 1987) of farm systems matters; farmers are likely to view practices that complicate operations or challenge their "technological beliefs" as impractical (Davies and Hodge 2006).

Factors not included

Beyond these five institutional types, other macro factors also influence farmer 'choice' (Liu et al. 2018). Biophysical aspects of land and climate influence soil fertility and health, pest pressures, cropping decisions, and pathogen loading, setting environmental constraints on what is possible. Property rights, land values, and farm loan conditions impact who owns and has access to farmland; these factors indirectly mediate the relationship between an aspiration to farm 'better' and an actual farming practice through operational characteristics such as land ownership, land tenure, and farm size (Calo and De Master 2016; Calo 2018). Of final note, many farms comprise multiple people acting in decision-making roles, which belies the notion that an individual 'farmer' chooses how to farm. Large-scale farms spread decision-making across many individuals, and even family farms often delegate farm roles and responsibilities among family members. The institutional factors described above likely influence different roles unequally, and the influence of intra-farm dynamics on farm management decisions deserves more in-depth discussion and analysis than is possible here.

Background: safety vs. sustainability in California

Food safety and agricultural sustainability at the farm level parallel societal desires for food that is both safe to eat and sustainably grown. Various economic, cultural, and regulatory forces have emerged to better align agricultural practice with these broad aspirations. However, such forces operate in relative isolation, akin to 'policy silos', which has led to conflict (Broad Leib and Pollans 2019).

Tensions between food safety and sustainability spiked following a deadly 2006 outbreak of *E. coli* O157:H7 linked to spinach grown in California's central coast. Investigators found the outbreak strain in both livestock and wild pigs near the implicated farm field (CDC 2006), catalyzing widespread concern that the farm environment itself might pose a substantial health risk by spreading dangerous pathogens onto growing crops. Fear of human pathogens in the farm environment was renewed after another deadly outbreak of *E. coli* O157:H7 in strawberries was traced back to wild deer (FDA 2011); the same year, an outbreak of listeriosis linked to a cantaloupe farm killed 33 people (CDC 2012). In an effort to prevent such outbreaks at the source, both government officials and produce industry groups developed rules and standards detailing how growers should manage food safety, including provisions to control waterborne, soil-borne, and animal-borne routes of contamination. These policies pressure farmers to suppress potential sources of pathogens in the farm environment, in particular wild animals and their habitat (Olimpi et al. 2019).

Animals can harbor pathogenic strains of *E. coli* and *Salmonella* and spread these pathogens through their feces (Langholz and Jay-Russell 2013). Other potentially pathogenic bacteria, such as *Listeria monocytogenes*, are native to soil (Vivant et al. 2013). Moreover, many enteric pathogens can survive outside of their host environment for extended periods of time, only to be reactivated once they come into contact with a new host (Gutierrez-Rodriguez and Adhikari 2018). Such pathogens pose a high risk of cross-contamination in the open environment of farms: flowing water, wind, other animals, workers, and equipment can move pathogens to new locations and surfaces (Fig. 1).

A survey of growers in California's central coast following the 2006 outbreak revealed that many growers felt pressured by their buyers to remove non-crop vegetation around fields and step-up efforts to trap, poison, or fence-out wild animals (Beretti and Stuart 2008). "Growers are being put in the unfair position," warned the authors, "of choosing between being able to sell their crops or protecting the environment" (p. 72). An aerial-imagery study found that 13% of the riparian habitat in the Salinas Valley was degraded or destroyed in the 5-year period following the 2006 outbreak (Gennet et al. 2013). A follow-up survey in 2014 found that many California produce growers continue to remove habitat around fields, poison and trap wild animals, and install wildlife-deterrent fences due to food safety concerns (Baur et al. 2016). Today, the latest evidence from California indicates that food safety concerns continue

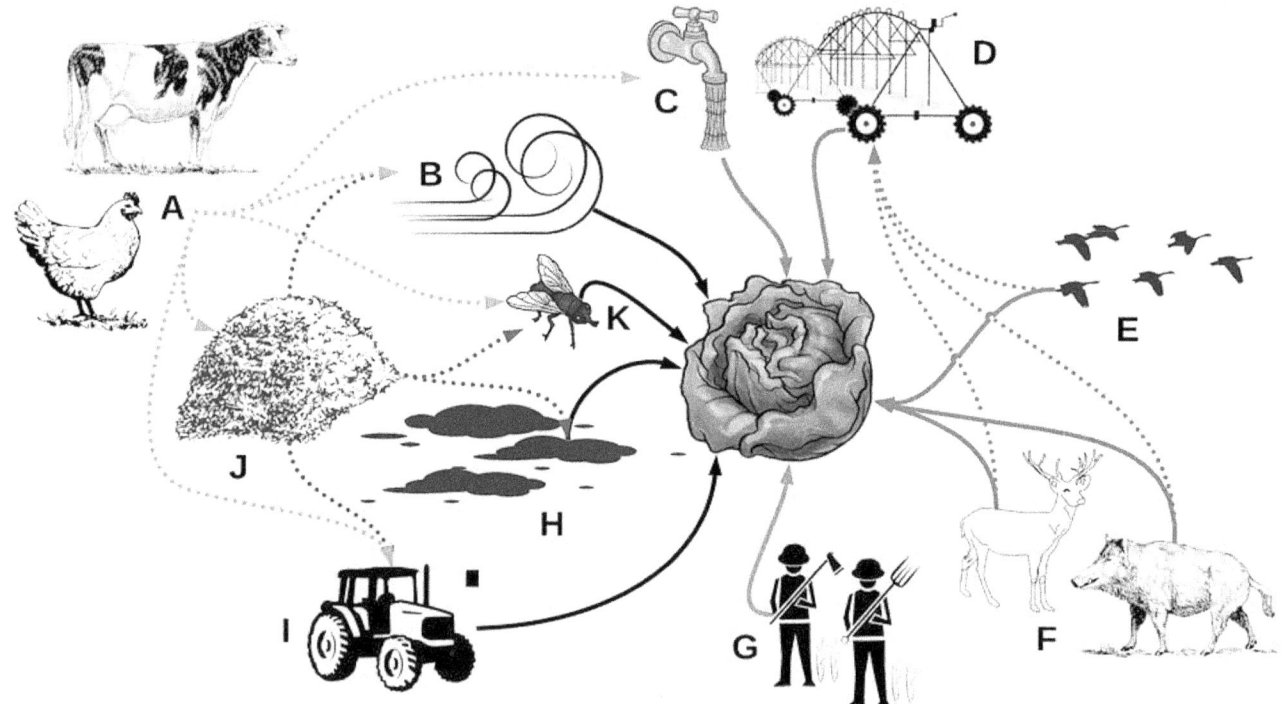

Fig. 1 Example routes of pathogenic contamination in the farm environment. Human pathogens originating in animal fecal matter—from livestock (**a**), terrestrial animals (**f**), birds (**e**) or humans (**g**)—may contaminate crops directly (solid lines) or indirectly by cross-contaminating (dashed lines) irrigation water (**d**), wash water (**c**), or flood water that has been cross-contaminated. Pathogens may also persist in soils (**h**) and soil amendments such as compost and manure (**j**), spreading from there onto crops. Insects (**k**), windborne aerosols (dust) (**b**), and farm equipment (**i**) can carry pathogens from livestock pens or manure piles onto crops

to pose a barrier to environmentally sustainable agriculture (Olimpi et al. 2019).

Such attempts to prevent pathogenic contamination carry an environmental cost. Removing habitat decreases biodiversity, exacerbates erosion, and lessens water quality (Crohn and Bianchi 2008; Lowell et al. 2010; Stuart 2010). Field-edge habitat, such as hedgerows, provides important ecosystem services to farm fields (Long et al. 2017); removing field-edge vegetation not only foregoes these benefits, but also has little to no demonstrated effect on the risk of pathogen contamination (Sellers et al. 2018). Habitat removal may actually increase the risk of pathogen transfer into fields by removing physical barriers to contaminated surface water and increasing the relative abundance of animals known to vector human pathogens (Karp et al. 2015a, b). Food safety pressures have also led some growers to abandon organic soil amendments, in particular compost, in favor of ostensibly 'safer' synthetic or heat-treated, pelletized fertilizers that do not promote ecosystem services such as natural pest predation (Karp et al. 2016; Olimpi et al. 2019). The cumulative effects of these practices has the potential to impact a wide range of ecosystem services in agricultural landscapes across the United States (Karp et al. 2015a, b). Although preventing foodborne illness and

sustaining agroecosystems both represent clear societal obligations, in actual farming practice they often work at cross-purposes.

Analysis: comparing constrained choices

To better understand this tension, I qualitatively compare the institutions constraining or driving food safety and environmental sustainability in California agriculture across the five dimensions of constrained choice defined above. I focus on three primary areas of overlap between safety and sustainability: wildlife/habitat conservation, water conservation/quality, and soil conservation.

Agricultural sustainability

Rules and Standards

Agricultural sustainability comprises agricultural conservation and environmental protection. Agricultural conservation dates to New Deal-era promotion of productivity-enhancing "on-farm benefits", e.g. reducing soil erosion. Rising environmental awareness pushed federal policy, beginning in the 1980s, to pivot toward environmental protection, "managing the 'off-farm' environmental effects of agriculture such as clean water, clean air, biodiversity, and… other ecological goods and services" (Cox 2007). However, the voluntary incentive for farmers to expend their own resources to prevent impacts to other people in other places (i.e. externalities) is low (Knowler and Bradshaw 2007), particularly as "food production environmental externalities are underregulated" (Broad Leib and Pollans 2019). Moreover, US environmental law has largely exempted farms from most environmental regulations (Browne 1988; Ruhl 2000), reflecting a strong cultural commitment to the "liberty of private property" that underpins many limitations on government regulatory authority in the United States (Opie 1987, p. 19). One exception is the Endangered Species Act, which is interpreted to allow direct restrictions on the use of private property in order to protect the habitat of a listed species (Doremus 2010). The Federal Insecticide, Fungicide, and Rodenticide Act also indirectly intervenes by requiring farmers to apply pesticides according to label instructions (Ruhl 2000).

Exempting farmers from command-and-control style regulation extends to state law as well, although California imposes more environmental regulations on its farmers than most states (Pollans 2015, p. 422). The California Water Code, for example, authorizes regional boards to regulate pollutants including nitrates, sediments, and pesticides that farms discharge into surface and ground water. While not every water board uses this authority, the Central Coast Water Board requires growers to use best management practices including vegetated buffers to filter farm runoff; however, the board has limited direct oversight at the farm level (Dowd et al. 2008; Drevno 2016) and generally suffers an antagonistic relationship with growers in the region (Drevno 2018). Since US growers face few regulatory consequences for farming unsustainably, sustainability hinges on 'soft' policy tools, including information gathering, technology transfer, direct payments, cost-sharing, and market-based incentive programs (Heimlich and Claassen 1998; Claassen et al. 2008; Batie 2009; Reimer et al. 2018). An example is the California Healthy Soils Program, which provides state funding for both farmer incentives to improve soil health and demonstrations of best soil practices (CDFA 2020).

Legal liability

In principle, off-farm environmental impacts could pose a legal liability risk to farmers. A downstream party suffering damages from a farm activity, such as pesticide or fertilizer runoff, could bring suit against a polluting farm. However, agricultural externalities generate little true liability (Pollans 2015; Broad Leib and Pollans 2019), especially in cases where state-level "right-to-farm" laws sharply delimit farmer liability (Ruhl 2000). California adopted its Right to Farm Act in 1981 (California Civil Code § 3482.5), and many counties and cities have passed right-to-farm ordinances to protect farmers from nuisance suits brought by neighbors (Wacker et al. 2001). These statutes may also protect farmers from lawsuits brought by other farmers. For example, an organic farmer may not be able to sue a neighboring conventional farm for spill-over damages from pesticide drift (McElwain 2015). Although outside California, a recent legal battle between the city of Des Moines and Iowa farmers over agricultural nitrate pollution underscores the point (Pollans 2016). Although the Des Moines Water Works utility sought to claim damages from upstream agricultural polluters, represented by regional drainage districts, the intervention failed to effect any change—the Iowa Supreme Court dismissed the city's lawsuit on the grounds that "drainage districts are immune from claims for damages or injunctive relief," in effect absolving farmers of responsibility for water pollution from agricultural runoff (Tidgren 2017).

Market and supply chain forces

Economic cost is a primary constraint for sustainable farming, as agricultural conservation and environmental protection take money, time, and often land removed from production. Farms of all types operate on narrow financial margins, and resist unnecessary costs. Moreover, there are few mainstream market incentives for sustainable farming other than the cost of water and fertilizer, which can incentivize growers to conserve these resources (Pollans 2015). While the organic label confers a price premium to certified farmers in exchange for avoiding many pesticides and synthetic fertilizers, the incentive increasingly leads farmers toward an input-substitution approach dubbed "organic lite" that permits certified farms to replicate many of the ecologically-unsustainable aspects of conventional agriculture (Guthman 2004). Competing retailer-driven labels, such as Whole Foods' now-discontinued "Responsibly Grown" program, suffer the same problem, but also lack transparency and legitimacy (Woods and Tropp 2015). Finally, high rates of leased farmland mean that long-term investments to protect a farm's natural resource base offer only a weak incentive for many growers (Lambert et al. 2007; Bigelow et al. 2016; Macaulay and Butsic 2017).

Social networks and norms

Shared cultural norms of land stewardship support agricultural conservation, and to a lesser extent environmental protection from externalities caused by agriculture. In a 2014 survey among California produce growers, over 93% of respondents agreed that they have a responsibility to protect water quality and the environment on their farms (Baur et al. 2016). However, stewardship, conservation, and environmental protection take many forms (Raymond et al. 2016), and farmers are more likely to adopt those that yield direct on-farm benefits, such as soil conservation, than they are to adopt practices that yield "distant" off-farm benefits to society, such as climate change mitigation (Niles et al. 2013, 2015). "Although environmental stewardship is a factor influencing many farmers' decisions," observe Robertson et al. (2014), "sustained profitability is usually the overriding concern." Pragmatic, business-oriented farmers—a dominant group in California's agribusiness-oriented rural economies (Walker 2004; Henke 2008)—are more likely to try agroecological approaches when the resulting tangible benefits to the farm are foregrounded (Warner 2008). Yet California growers espouse widely mixed perspectives on conservation, wildlife, and non-crop vegetation (Brodt et al. 2006). One study among Central Valley growers found an even split in field-edge vegetation management: some farmers preferred "'clean' (weed-free or vegetation-free) edges" while others expressed "esthetic enjoyment in seeing plants and associated vertebrate wildlife... [that] led some to retain naturally occurring trees" (Brodt et al. 2009).

Scientific knowledge and available technologies

Overall, norms in California agriculture may be shifting toward broader emphasis on actively conserving water and building soil health, while attitudes toward wildlife and biodiversity remain mixed. This shift aligns with the increasing legitimacy of scientific evidence pointing to direct on-farm benefits of soil and water conservation practices, evidence which has been matched with experience through cost-sharing, demonstration, and pilot programs provided by conservation extension agencies such as the USDA Natural Resources Conservation Service and the University of California Division of Agriculture and Natural Resources. Many California vegetable growers now use cover crops and practice crop rotation to promote soil health and weed control (Brennan and Smith 2005, 2017; Baur et al. 2016). Likewise for water conservation, many have switched to high-efficiency drip tape and micro-irrigation (Ayars et al. 2015), which Taylor and Zilberman (2017) attribute to effective local cooperative extension advisors who were able to effectively facilitate coevolution of drip irrigation technology with changes in cultivation practices that farmers had to adopt. With the exception of cover crops, which require off-season upkeep and can interfere with tightly-packed planting schedules, these shifts represent relatively simple technological substitutions that require only limited adjustments to the farm operation.

Although most growers support the principle of environmental protection, many resist specific practices that could reduce agricultural externalities because they are seen as unnecessary or impractical. Growers may discount scientific findings that agriculture pollutes water and air or threatens biodiversity (Lubell and Fulton 2007, 2008). High land prices make setting aside field acreage for non-crop vegetation—hedgerows, filter strips, native habitat, riparian buffers—financially burdensome for farmers, especially given the extra costs of upkeep (trimming, irrigating, fencing) and the potential to harbor pests or pathogen vectors (Olimpi et al. 2019). Likewise, farmers may perceive few economically viable alternatives to pesticides for managing insects, fungi, and weeds, especially those who farm conventionally and do not receive organic price premiums. Perceptions that such practices are impractical stem in part from the long-term commitment among California's agricultural research and cooperative extension institutions to facilitating productivity-maximizing industrial agriculture (Henke 2008). Among the casualties of this commitment was the state's biological control programs, which were dismantled in favor of agrichemical and biotechnology research and extension (but see also Altieri et al. 1997; Altieri 2018). The dominance of industrial agriculture practices that largely accept as given high environmental externalities is exacerbated by consistently low investment in research on ecologically-based farming techniques (Miles et al. 2017). The cumulative result is low scientific legitimacy of environmentally sustainable farming practices and limited access to scale-appropriate technologies to facilitate those practices (Montenegro de Wit and Iles 2016).

Food safety

Rules and Standards

Food safety represents the desire to keep people from getting sick due to produce that is contaminated with human pathogens. From a regulatory standpoint, food safety comprises a robust set of policy mechanisms that combine government "meta"-regulation with industry self-regulation (Coglianese and Lazer 2003), forming a hybrid public–private system (Garcia Martinez et al. 2007; Garcia Martinez et al. 2013). Food safety reform in the produce sector began with industry-led voluntary standards for good agricultural practices; examples include the state-level California Leafy Greens Marketing Agreement (LGMA) and the privately maintained GlobalGAP and PrimusGFS metrics. The necessity of complying with voluntary standards is now reinforced by mandatory, nationwide "science-based minimum standards" for growing and handling fresh produce authorized by the US Food Safety Modernization Act of 2011 (FSMA). Compliance with this program is enforced by government inspectors employed by the California Department of Food and Agriculture.

Legal liability

Market pressures are reinforced by the specter of bankruptcy-inducing lawsuits or even criminal charges, as food producers are subject to strict liability in the case of food poisoning (Broad Leib and Pollans 2019). This legal risk forms a potent pressure on operators across the supply chain to lower their food safety liability through any means possible (Baur et al. 2017). Specialized law firms litigate on behalf of foodborne illness victims; the most prominent, Marler Clark, advertises its "food poisoning lawyers" following each outbreak. Moreover, farmers such as the Jensen brothers, whose cantaloupe operation caused the deadly 2011 *Listeria* outbreak, have been jailed following outbreaks (Elliot 2013).

Market and supply chain forces

Powerful retail and foodservice firms, with strong interest in protecting their brand reputations (Havinga 2006; Bain et al. 2013), also push food safety through supply chain management mechanisms (Busch 2007), such as production contracts or third-party audits. In a food economy dominated by a small number of corporate buyers (Howard 2016; IPES-Food 2017), vegetable and fruit farmers—who often have to sell their crops quickly before they rot—have little leverage to resist buyer demands (Baur et al. 2017). Moreover, growers generally default to the strictest food safety requirements to minimize management complexity and satisfy the widest range of customers (Olimpi et al. 2019), suggesting that further long-term erosion of farmer agency may be underway, as previously noted by Stuart (2009).

Social networks and norms

Together, regulations, standards, and liability risks threaten farmers with lost sales, failed audits, regulatory citations, damaged reputations, lawsuits, expensive product recalls, or criminal charges. The combined pressure to act is high, and the industry has worked diligently to spread "food safety culture" to every farm and farm worker (Baur et al. 2017). Several normative institutional mechanisms are in place to facilitate the spread of food safety culture through educational campaigns and networks of food safety professionals. FSMA rules require that every farm covered by the law designate one employee to take an annual agency-approved training on food safety (80 FR 74353), and private standards require that each farm maintain a detailed food safety plan. Organizations such as the Produce Safety Alliance, based at Cornell University, have been granted authority by FDA to conduct training and outreach on food safety to farmers. The FDA's deputy commissioner for food policy and response, Frank Yiannas, has previously published a book on food safety culture (Yiannas 2009), and in 2014, when he served as Wal-Mart's vice president of safety and health, he publicly stated that food safety culture means "getting to the path of food safety as a social norm" (quoted in Baur et al. 2017). Moreover, food safety auditing and consulting have grown into independent industries, strengthening the normative pressure on farmers to put food safety first as a matter of principle.

Scientific knowledge and available technologies

The scientific basis for food safety practices also appreciates strong legitimacy, with widespread consensus among academic, industry, and government scientists that pathogens such as *E. coli*, *Salmonella*, and *Listeria* can and do contaminate crops pre-harvest, posing a serious threat to public health (Painter et al. 2013; Gubernot et al. 2016; Gutierrez-Rodriguez and Adhikari 2018). This legitimacy is underwritten by public health organizations including the US Centers for Disease Control and the World Health Organization, and is lent credibility by mass foodborne disease surveillance data gathered and analyzed by national and global networks of laboratories, medical professionals, and public health agencies (Henao et al. 2015; Wang et al. 2016). Food safety professionals and regulators routinely invoke science as a legitimizing strategy when presenting standards, rules, and best practices. FSMA explicitly calls for "science-based" rules, while LGMA advertises its "rigorous science-based food safety system."

Food safety practices entail monitoring and preventing pathogenic contamination. While farmers may complain of the added paperwork associated with pre-harvest field inspections and monthly water tests (Olimpi et al. 2019), monitoring is a familiar responsibility. Most farmers are used to taking soil samples for laboratory testing; collecting water samples represents more of the same. Likewise, farmers and their irrigation contractors, pest control advisors, ranch, and harvest managers are in and out of their fields almost continuously during the growing season to check on soil conditions, monitor crop health, or watch for pest activity; a field inspection to check for animal tracks, feces, or other signs of potential contamination slots neatly into the list of potential problems to monitor.

Preventing contamination entails isolating fields from animals (wild or domestic) and non-crop vegetation that could harbor those animals (Baur et al. 2017; Olimpi et al. 2019). This task is familiar. Farmers have long battled against weeds and pests: "In the history of progressive agriculture, wild creatures had never counted for much. They failed to conform to the farmer's productive purposes and so were seen as useless when not seen as a threat" (Worster 1994, 268). A decade of intensifying pressure to produce the safest food possible has expanded the customary definition of "pests" to encompass any animal with the potential to contaminate crops. The US Food and Drug Administration, which oversees food safety on produce farms, defines "pest" in its good manufacturing practices as "any objectionable animals or insects including, but not limited to, birds, rodents, flies, and larvae" (21 CFR §110.3.j). By extension, in the context of farming, any non-crop vegetation that may harbor a pest also poses a threat, and in effect becomes a "weed". Although California farmers generally ethically oppose "sterilizing the growing environment" (Stuart 2008, 2009), in practice removing animals and non-crop vegetation fits an accustomed behavioral and cognitive pattern. Food safety professionals reinforce this pattern in grower training workshops: "We're kind of moving into [wild animals'] space," one trainer explained at a workshop on animal intrusion for Oxnard Plain growers in 2017. "And we want them out… But we're doing it for a good reason, to keep people from getting sick." The workshop—part of the FDA-approved Produce Safety Alliance curriculum—advocated removing any water, food, or shelter that will help animals survive near fields, framing preventive action as a simple process of pest and weed removal. In summary, food safety monitoring and prevention require little to no change in dominant farming conventions.

Discussion

Aspirations to farm responsibly, i.e. in ways consistent with socially desirable visions for agriculture, are filtered through institutions that shape, constrain, and drive how farmers act (or do not) in practice, a process I examined by comparing food safety and agricultural sustainability as different systems of constrained choice for California produce growers. Analyzing this space of tension through my institutional framework yields several insights into this case.

Choosing between two "right" ways

Farmers feel multiple, concurrent obligations to farm responsibly. When these obligations conflict, farmers may face a "'right vs. right' dilemma" (Hendrickson and James 2005; Stuart 2009). My analysis suggests that farmers defer to the course of action they perceive as most feasible within the bounds imposed by their institutional environment; in comparing these two cases, food safety aligns more consistently with multiple institutional drivers than does

environmental sustainability. While food safety finds broad support in comprehensive rules, standards, and market mechanisms, sustainability is often implicitly discouraged by market mechanisms and receives only disjointed support from fragmentary rules and standards. As Broad Leib and Pollans (2019, p. 139) observe, "The contrast between prescriptive regulations [for microbial food safety] on the one hand and education or voluntary standards incentives [for reducing environmental externalities of agriculture] on the other, reflects a serious mismatch between the nature and severity of each problem and the solutions brought to bear." Thus, when food safety and sustainability obligations diverge, the former tends to override the latter, especially when farmers incorporate their exposure to legal liability risk (Pollans 2015). Moreover, applied produce safety research frequently assumes tradeoffs between safety and sustainability, with limited research into potential synergies, suggesting the presence of a cultural-cognitive constraint. The cumulative effect sends the message to farmers that food safety is a necessity and environmental sustainability is an option.

Viewed as an institutional logic, food safety also encapsulates a concise, concrete imperative: keep dangerous microbes out of food. Sustainability, in contrast, notoriously defies precise definition; even for my brief analysis, I had to divide sustainability into on-farm (agricultural conservation) and off-farm (environmental protection) obligations to make sense of a wide-ranging constellation of aspirations, practices, incentives, and constraints. The narrow scope of food safety focuses assessment on clearly measurable outcomes, enables explicit designations of responsibility and accountability, and facilitates shared understanding across jurisdictions and supply chains. The nebulous scope of sustainability, however, covers outcomes spread across time and space in a way that resists simple metrics. Who is responsible to whom for what remains ambiguous, leading to contested meanings and fragmentation (see Worrell and Appleby 2000). Food safety glides smoothly through the web of complicating contexts that bog down aspirations to farm sustainably, and is therefore "easier and more politically palatable to regulate" (Broad Leib and Pollans 2019).

Coordinating across structural constraints

Interventions based in "policy silos" not only miss heterogeneity in on-the-ground challenges (Cumming et al. 2006), but can also miss macro-scale conflicts among diverging institutional logics. Advocates for more socially desirable farming—whether more sustainable, safe, secure, or just—should therefore identify and address potential friction among structural constraints faced by land-users and seek coordination across policy spheres. Balancing food safety and agricultural sustainability will require systemic reform across legal, market, and knowledge-production institutions in addition to educating farmers and raising awareness (Broad Leib and Pollans 2019). Without greater protection from food poisoning lawsuits, farmers will likely continue to prioritize food safety over sustainability. Given that the risk of pathogenic contamination is never zero (De Keuckelaere et al. 2015), it may be reasonable to limit farmers' liability for food safety (Olimpi et al. 2019); increased access to liability insurance could also mitigate farmers' legal risk (Armstrong 2014). Greater parity in market incentives for safety and sustainability is also needed. Farmers may feel empowered to equally co-manage safety and sustainability if (1) the market price covers the extra costs of more complex farm management or (2) buyers prioritize sustainable farming practices. These incentives hinge on measurability: robust standards for sustainable farming are likely also needed. Moreover, the apparent success of food safety reform in shifting farm management behavior demonstrates the powerful synergy in pairing 'carrots' with 'sticks', suggesting the need for stronger sustainability enforcement mechanisms— e.g. audits, inspections, and/or mandatory reporting— comparable to those for food safety. Lastly, to facilitate truly holistic co-management of sustainability and safety will require "thicker" scientific legitimacy for agroecological approaches, to shift the cultural-cognitive focus from safety-sustainability tradeoffs toward synergies.

Uneven constraints and scale-dependency

The above analysis tacitly assumes that these institutions shape management decisions independent of farm-level differences. The reality is more complex—different farms may operate in entirely different systems of constrained choice depending on the scale of their operation. Agriculture in the United States seems to be bifurcating into distinct scalar systems: large-scale corporate industrial farms selling through vertically-integrated supply chains and small-scale, diversified farms selling directly to local markets (Lyson et al. 2008, p. xi). Policy interventions can, perversely, exacerbate bifurcation, as observed following establishment of national organic standards (Guthman 2004; Constance et al. 2008; Constance 2009). Farmers in "the middle" increasingly feel pressure to either scale up and "conventionalize" (Buck et al. 1997) or shrink to a level where they can claim artisanal, niche markets as part of the ongoing "quality turn" (Goodman 2003). Recent developments with social network innovations such as food hubs add a further variable to the milieu.

The consequences of bifurcation for aspirations to farm in a more socially responsible manner are mixed. Goldberger (2011), for example, found that conventionalization suppresses environmental stewardship among organic farmers, whereas other scholars find that adopting conservation practices is easier for larger farms (Knowler and Bradshaw 2007). These seemingly contrasting findings result from the differential impact of various institutions on farmer choice. For example, large farms must maximize yield and crop consistency to access the corporate supply chains capable of handling the volume of product they grow, whereas small farms may have greater marketing, and thus managerial, flexibility. However, heavily-capitalized large farms have more resources and may have greater access to the latest conservation research and technologies, making it easier for them to adopt sustainable 'upgrades' such as water-saving micro-irrigation systems.

The effect of farm scale on food safety is hotly contested. Some scholars argue that small-scale producers face food safety challenges "of a distinctly different nature" than those faced by large, conventional producers (DeLind and Howard 2008), thus necessitating scaled interventions (Hassanein 2011) rather than universal food safety standards "developed for productivist and industrial scale agriculture" (McMahon 2013). For this reason, family farm advocates secured a partial exemption for small farms in the FSMA, which drew backlash from consumer advocacy groups concerned that small farms would be allowed to operate unsafely (Strauss 2011). Thus, other scholars advocate that "all food producers should be held to the same food safety standards" (Thomas 2014, p. xiv), based on the argument that "pathogens don't know what size operation they're on," as the lead scientist at one prominent fresh produce trade association phrased it (personal communication, 2014). Consumer response to foodborne illness outbreaks also fails to distinguish by farm scale; in the two largest *E. coli* outbreaks related to produce—from spinach in 2006 and romaine lettuce in 2018—the market price of the affected vegetable plummeted across the board, regardless of farm source (Calvin 2007; Newman and Haddon 2018). Bifurcation is thus insufficient to explain the complex ways in which distinct regulatory, economic, legal, cultural, and technological institutions intersect with farm scale. The effects on farmer 'choice' are both mixed and dynamic, the result of competing pressures to standardize across scales and adapt to scale-specific needs.

Conclusion

Aspirations to farm 'better' fall short in practice due to constraints outside of farmers' control. In both sustainability and food safety, farmers are implored to do the 'right' thing and steward their land and their crop responsibly, for future generations on the one hand, and for downstream consumers on the other. These are powerful societal obligations. Yet farmers face proliferating pressures to adopt practices that align with various societal aspirations for 'better' agriculture that is more sustainable, safe, healthy, just, efficient, resilient, and so on. In this paper, I raise two critical yet understudied questions: What happens when the accumulation of external pressures overwhelm farm management capacity? Or, worse, when different visions of better agriculture pull farmers toward conflicting management paradigms?

As a first step toward addressing these questions, I compared the institutional manifestation of two distinct societal obligations placed on California fruit and vegetable farmers: to practice sustainable agriculture and to ensure food safety. Through this institutional case comparison, my concern has been to assess the practical implications of and limits to interventions that appeal to farmers' sense of responsibility in the context of multiple, sometimes competing institutional logics. Focusing on farm-level choice—including attitudes, values, and beliefs—in isolation, no matter how sophisticated the model, misses the critical influence of institutional constraints and drivers that shape farmer 'choice.' Failure to address divergent institutional logics among multiple societal aspirations for agriculture can lead to impossible situations pitting "right" against "right." Such zero-sum contests not only waste farmers' energy and resources, but also inhibit investment in synergies or creative alternatives that might escape false binaries. I suggest that both researchers and agricultural reform advocates can benefit from systematically considering the positionality of their appeals—both to agricultural communities and to the policy networks that govern them—within broader legal, political, market, and techno-scientific institutional systems.

To do this, I proposed applying tools from institutional theory to the literature on constrained choice. As my framework demonstrates, combining these literatures more precisely reveals the dimensions and mechanisms that bound farmer agency and shape farm management paradigms, in turn facilitating comparison across multiple social objectives for farms. Ample opportunities exist to expand this analysis to cover other types of objectives in other places, as well as to deepen and extend the framework through targeted empirical inquiry. For example, efforts to farm agroecologically may contrast with efforts to promote more fair and safe conditions for agricultural labor (Dumont and Baret 2017), while growing calls to reduce meat consumption and increase plant-based diets may create friction with growing interest in the use of integrated crop-livestock systems to promote sustainable nutrient cycling (Kronberg and Ryschawy 2019). Such tensions also operate across multiple social levels and across regulatory, normative, and cultural-cognitive aspects of institutions.

Improved understanding of the ways in which institutional logics overlap or diverge may lead to novel opportunities for promoting food system transformation in response to

global threats including climate change, biodiversity, and food inequity (Anderson et al. 2019; Willett et al. 2019). With growing scholarly and policy interest in promoting multifunctional agriculture (van Huylenbroeck et al. 2007; Huttunen 2019), social-ecological agrifood outcomes (Oteros-Rozas et al. 2019), and diversified farming systems (Kremen et al. 2012; IPES-Food 2016) as routes toward greater sustainability, health, and equity, there is also a growing need to consider the intersecting institutional dimensions of these envisioned transitions. Moving forward, more attention is needed on evaluating options to mitigate the cumulative detrimental effects of multiplying societal obligations placed on farmers. In particular, interventions should focus on how to distribute responsibility for juggling multiple functional goals more equitably throughout the food system; currently, too much of the burden is placed on farmers, without the requisite resources or freedoms (per Hendrickson and James 2016) to fulfill those responsibilities.

Considering constrained choice as the result of a combination of institutional drivers reveals a conceptual pitfall—a focus only on structures that constrain farmer agency misses the corollary role that structures may play in also enabling farmer choice. Viewing both structural roles through the lens of institutional logics shows how a given institution may both enable one form of farmer behavior while constraining others. In essence, this insight builds intuitively upon Hendrickson and James' (2005) concern that exposure to structural constraints over extended periods of time might erode farmer agency, especially in terms of tacit knowledge and the capacity to navigate multi-faceted ethical decisions. If institutional design has the power to erode farmer agency, then logically it should also be possible to reconfigure institutions to work in the reverse—to *empower* farmer agency. In other words, interventions aimed at reforming constraining institutions should recognize that the task is not to simply remove barriers, but to also provide handholds.

Farmer empowerment is a critical challenge for the future of agriculture and food systems. Transformation hinges on the capacity of people who farm to actually make the necessary farm-level changes. As Carlisle et al. (2019a, b) argue, the farms of a more sustainable and just future will be knowledge-intensive. The corollary to this prediction is that such knowledge-intensive farms will be supported by an institutional environment that facilitates the flexible application of practitioner knowledge, including for the purpose of navigating ethical questions related to social responsibility and societal obligation. Much literature has been devoted to envisioning better institutional configurations—e.g. multi-level, polycentric, multifunctional, participatory, democratic, continually-learning—to empower communities, social movements, and peer networks to shift toward more sustainable, healthful, and equitable food systems (Kloppenburg et al. 2000; Sayer et al. 2013; Carlisle 2014; IPES-Food 2016; Anderson 2019). However, the gap between vision and reality is wide.

My framework can serve as a stepping stone to better understand this gap between institutions that might promote active practitioner agency and the institutions that currently exist. By providing an example roadmap for systematic examination of the various and at times competing logics that farmers must navigate, the framework can be used to incorporate a more precisely delineated set of "macro factors" (Liu et al. 2018) into models that connect structures (e.g. markets, policies) to farmer decisions and to outcomes. It also helps pose further empirical questions, such as, under what conditions might overlapping institutional logics allow farmers to break away from constraints and pioneer new farming modalities (following Battilana 2006), as opposed to withdrawing from active participation? If competing institutional drivers do indeed undermine farmer confidence in expert advice and incentive programs (Higgins et al. 2016; Mercado et al. 2018), through what specific institutional reforms can that confidence be re-gained? The overarching contribution, I argue, is that systematically attending to competing institutional logics and the complex and fluid configurations through which they confront farmers helps re-center analysis on the active agency that farmers exhibit on a daily basis as they continuously assess how and when to engage with, acquiesce to, or embrace institutions, as well as how and when to disengage from, avoid, or resist them. The institutions which ostensibly govern agriculture have much to learn from farmers' experiences multi-tasking, translating, and negotiating among diverse aspirations for agriculture.

Though the precise form of analysis may vary, when considering what it takes to transition toward 'better' agriculture, I contend that it is always worth explicitly posing three overarching questions in tandem: To whom does society expect farmers to bear responsibility? In what tangible forms do these responsibilities manifest on the farm? And to what extent do farmers actually have the capacity and power to fulfill all of those responsibilities in practice? Appeals to farmers' sense of social obligation can help align farm management with societal aspirations for sustainable, healthy, and productive landscapes, but such appeals must recognize and work within (or on) the existing context of institutional drivers and constraints. It may be that the 'choices' which most need to change are located beyond the farm.

Acknowledgements I am grateful to Jennie Durant, Lisa Kelley, Amber Sciligo, Daniel Suarez, Michael Polson, Alastair Iles, Kathryn De Master, and my colleagues in Nancy Peluso's Land Lab for invaluable feedback, suggestions, and advice during the many stages and versions of this manuscript.

References

Ahnström, Johan, Jenny Höckert, Hanna L. Bergeå, Charles A. Francis, Peter Skelton, and Lars Hallgren. 2009. Farmers and nature conservation: What is known about attitudes, context factors and actions affecting conservation? *Renewable Agriculture and Food Systems* 24: 38–47. https://doi.org/10.1017/S1742170508002391.

Altieri, Miguel A. 2009. Agroecology, small farms, and food sovereignty. *Monthly Review* 61 (3): 102–113.

Altieri, Miguel A. 2018. Berkeley: The betrayal of an agricultural legacy. *Food First*. https://foodfirst.org/betrayalofanagriculturall egacy/. Accessed 18 Oct 2019.

Altieri, Miguel A., Peter M. Rosset, and Clara I. Nicholls. 1997. Biological control and agricultural modernization: Towards resolution of some contradictions. *Agriculture and Human Values* 14: 303–310. https://doi.org/10.1023/A:1007499401616.

Anderson, Colin Ray, Janneke Bruil, Michael Jahi Chappell, Csilla Kiss, and Michel Patrick Pimbert. 2019. From transition to domains of transformation: Getting to sustainable and just food systems through agroecology. *Sustainability* 11 (19): 5272. https://doi.org/10.3390/su11195272.

Anderson, Molly. 2019. The importance of vision in food system transformation. *Journal of Agriculture, Food Systems, and Community Development* 9: 55–60. https://doi.org/10.5304/jafsc d.2019.09A.001.

Armstrong, Rachel. 2014. *Farmers' guide to reducing the legal risks of a food safety incident*. Farm Commons. https://farmcommon s.org/resources/farmers-guide-reducing-legal-risks-food-safet y-incident. Accessed 16 June 2020.

Ayars, J.E., A. Fulton, and B. Taylor. 2015. Subsurface drip irrigation in California—Here to stay? *Agricultural Water Management* 157: 39–47. https://doi.org/10.1016/j.agwat.2015.01.001.

Bacon, Christopher M., Christy Getz, Sibella Kraus, Maywa Montenegro, and Kaelin Holland. 2012. The social dimensions of sustainability and change in diversified farming systems. *Ecology and Society*. https://doi.org/10.5751/es-05226-170441.

Bain, Carmen, Elizabeth Ransom, and Vaughan Higgins. 2013. Private agri-food standards: Contestation, hybridity and the politics of standards. *International Journal of Sociology of Agriculture and Food* 20: 1–10.

Barnes, A.P., L. Toma, J. Willock, and C. Hall. 2013. Comparing a "budge" to a "nudge": Farmer responses to voluntary and compulsory compliance in a water quality management regime. *Journal of Rural Studies* 32: 448–459. https://doi.org/10.1016/j.jrurstud.2012.09.006.

Bartel, Robyn, and Elaine Barclay. 2011. Motivational postures and compliance with environmental law in Australian agriculture. *Journal of Rural Studies* 27: 153–170. https://doi.org/10.1016/j.jrurstud.2010.12.004.

Batie, Sandra S. 2009. Green payments and the US Farm Bill: information and policy challenges. *Frontiers in Ecology and the Environment* 7: 380–388. https://doi.org/10.1890/080004.

Battershill, Martin R.J., and Andrew W. Gilg. 1997. Socio-economic constraints and environmentally friendly farming in the Southwest of England. *Journal of Rural Studies* 13: 213–228. https://doi.org/10.1016/S0743-0167(96)00002-2.

Battilana, Julie. 2006. Agency and institutions: the enabling role of individuals' social position. *Organization* 13 (5): 653–676. https://doi.org/10.1177/1350508406067008.

Battilana, Julie, and Silvia Dorado. 2010. Building sustainable hybrid organizations: The case of commercial microfinance organizations. *Academy of Management Journal* 53 (6): 1419–1440. https://doi.org/10.5465/amj.2010.57318391.

Baumgart-Getz, Adam, Linda Stalker Prokopy, and Kristin Floress. 2012. Why farmers adopt best management practice in the United States: A meta-analysis of the adoption literature. *Journal of Environmental Management* 96: 17–25. https://doi.org/10.1016/j.jenvman.2011.10.006.

Baur, Patrick, Laura Driscoll, Sasha Gennet, and Daniel S. Karp. 2016. Inconsistent food safety pressures complicate environmental conservation for California produce growers. *California Agriculture* 70: 142–151.

Baur, Patrick, Christy Getz, and Jennifer Sowerwine. 2017. Contradictions, consequences and the human toll of food safety culture. *Agriculture and Human Values* 34: 713–728. https://doi.org/10.1007/s10460-017-9772-1.

Beretti, Melanie, and Diana Stuart. 2008. Food safety and environmental quality impose conflicting demands on Central Coast growers. *California Agriculture* 62: 68–73.

Bergstrom, Jogn C., and Terence J. Centner. 1989. Agricultural nuisances and right to farm laws: Implications of changing liability rules. *The Review of Regional Studies* 19: 23.

Bigelow, Daniel, Allison Borchers, and Todd Hubbs. 2016. *U.S. farmland ownership, tenure, and transfer.* EIB-161. US Department of Agriculture, Economic Research Service.

Brennan, Eric B., and Richard F. Smith. 2005. Winter cover crop growth and weed suppression on the central coast of California. *Weed Technology* 19: 1017–1024. https://doi.org/10.1614/WT-04-246R1.1.

Brennan, Eric B., and Richard F. Smith. 2017. Cover crop frequency and compost effects on a legume–rye cover crop during eight years of organic vegetables. *Agronomy Journal* 109: 2199–2213. https://doi.org/10.2134/agronj2016.06.0354.

Broad Leib, Emily M., and Margot J. Pollans. 2019. The new food safety. *California Law Review* 107: 1173.

Brodt, Sonja, Karen Klonsky, Louise Jackson, and Stephen B. Brush, and Sean Smukler. 2009. Factors affecting adoption of hedgerows and other biodiversity-enhancing features on farms in California, USA. *Agroforestry Systems* 76: 195–206. https://doi.org/10.1007/s10457-008-9168-8.

Brodt, Sonja, Karen Klonsky, and Laura Tourte. 2006. Farmer goals and management styles: Implications for advancing biologically based agriculture. *Agricultural Systems* 89: 90–105. https://doi.org/10.1016/j.agsy.2005.08.005.

Brown, Sandy, and Christy Getz. 2008. Towards domestic fair trade? Farm labor, food localism, and the 'family scale' farm. *GeoJournal* 73: 11–22. https://doi.org/10.1007/s10708-008-9192-2.

Browne, William Paul. 1988. *Private interests, public policy, and American agriculture.* Lawrence, KS: University Press of Kansas.

Buck, Daniel, Christina Getz, and Julie Guthman. 1997. From farm to table: The organic vegetable commodity chain of northern California. *Sociologia Ruralis* 37: 3–20. https://doi.org/10.1111/1467-9523.00033.

Busch, Lawrence. 2000. The moral economy of grades and standards. *Journal of Rural Studies* 16: 273–283. https://doi.org/10.1016/S0743-0167(99)00061-3.

Busch, Lawrence. 2007. Performing the economy, performing science: from neoclassical to supply chain models in the agrifood sector. *Economy and Society* 36: 437–466. https://doi.org/10.1080/03085140701428399.

Busch, Lawrence. 2020. Contested terrain: The ongoing struggles over food labels, standards and standards for labels. In *Labelling the economy: Qualities and values in contemporary markets,* ed. Brice Laurent and Alexandre Mallard, 33–58. Singapore: Springer. 10.1007/978–981–15–1498–2_2.

Buzby, Jean C., and Paul D. Frenzen. 1999. Food safety and product liability. *Food Policy* 24: 637–651. https://doi.org/10.1016/S0306-9192(99)00070-6.

Buzby, Jean C., Paul D. Frenzen, and Barbara Rasco. 2001. Product liability and food safety: The resolution of food poisoning lawsuits. In *Interdisciplinary food safety research,* ed. Neal H. Hooker and Elsa A. Murano, 232. Boca Raton, FL: CRC Press. https://doi.org/10.1201/9781420039092-11.

Calo, Adam. 2018. How knowledge deficit interventions fail to resolve beginning farmer challenges. *Agriculture and Human Values* 35: 367–381. https://doi.org/10.1007/s10460-017-9832-6.

Calo, Adam, and Kathryn Teigen De Master. 2016. After the incubator: Factors impeding land access along the path from farmworker to proprietor. *Journal of Agriculture, Food Systems, and Community Development* 6: 111–127. https://doi.org/10.5304/jafscd.2016.062.018.

Calvin, Linda. 2007. Outbreak linked to spinach forces reassessment of food safety practices. *Amber Waves* 5: 24.

Carlisle, Liz. 2014. Diversity, flexibility, and the resilience effect: Lessons from a social-ecological case study of diversified farming in the northern Great Plains, USA. *Ecology and Society* 19 (3): 45. https://doi.org/10.5751/ES-06736-190345.

Carlisle, Liz, Maywa Montenegro de Wit, Marcia S. DeLonge, Adam Calo, Christy Getz, Joanna Ory, Katherine Munden-Dixon, et al. 2019a. Securing the future of US agriculture: The case for investing in new entry sustainable farmers. *Elementa: Science of the Anthropocene* 7: 17. https://doi.org/10.1525/elementa.356.

Carlisle, Liz, Maywa Montenegro de Wit, Marcia S. DeLonge, Alastair Iles, Adam Calo, Christy Getz, Joanna Ory, et al. 2019b. Transitioning to sustainable agriculture requires growing and sustaining an ecologically skilled workforce. *Frontiers in Sustainable Food Systems.* https://doi.org/10.3389/fsufs.2019.00096.

Castellari, Elena, Claudio Soregaroli, Thomas J. Venus, and Justus Wesseler. 2018. Food processor and retailer non-GMO standards in the US and EU and the driving role of regulations. *Food Policy.* https://doi.org/10.1016/j.foodpol.2018.02.010.

[CDC] US Centers for Disease Control. 2006. *Multistate outbreak of E. coli O157:H7 infections linked to fresh spinach (FINAL UPDATE).* https://www.cdc.gov/ecoli/2006/spinach-10-2006.html. 16 June 2020.

[CDC] US Centers for Disease Control. 2012. *Multistate Outbreak of Listeriosis Linked to Whole Cantaloupes from Jensen Farms, Colorado (FINAL UPDATE).* https://www.cdc.gov/listeria/outbreaks/cantaloupes-jensen-farms/. Accessed 16 June 2020.

[CDFA] California Department of Food and Agriculture. 2020. Healthy soils program. *The Office of Environmental Farming and Innovation.* https://www.cdfa.ca.gov/oefi/healthysoils/. Accessed 16 June 2020.

Claassen, Roger, Andrea Cattaneo, and Robert Johansson. 2008. Cost-effective design of agri-environmental payment programs: U.S. experience in theory and practice. *Ecological Economics* 65: 737–752. https://doi.org/10.1016/j.ecolecon.2007.07.032.

Coglianese, Cary, and David Lazer. 2003. Management-based regulation: Prescribing private management to achieve public goals. *Law & Society Review* 37: 691–730. https://doi.org/10.1046/j.0023-9216.2003.03703001.x.

Constance, Douglas H. 2009. 2008 AFHVS presidential address. *Agriculture and Human Values* 26: 3–14. https://doi.org/10.1007/s10460-008-9187-0.

Constance, Douglas H., Jin Young Choi, and Holly Lyke-Ho-Gland. 2008. Conventionalization, bifurcation, and quality of life: certified and non-certified organic farmers in Texas. *Southern Rural Sociology* 23: 208–234.

Cox, Craig. 2007. US Agriculture conservation policy and programs: History, trends, and implications. In *US Agricultural Policy and the 2007 Farm Bill*, ed. Kaush Arha, Barton H. Thompson, Tim Josling, and Daniel A. Sumner, 113–146. Stanford, CA: Woods Institute for the Environment, Stanford University.

Crohn, David M., and Mary L. Bianchi. 2008. Research priorities for coordinating management of food safety and water quality. *Journal of Environmental Quality* 37: 1411–1418. https://doi.org/10.2134/jeq2007.0627.

Cumming, Graeme S., David H.M. Cumming, and Charles L. Redman. 2006. Scale mismatches in social-ecological systems: Causes, consequences, and solutions. *Ecology and Society* 11 (1): 14.

Czarnezki, Jason J., Margot J. Pollans, and Sarah Main. 2018. *Eco-Labeling*. SSRN Scholarly Paper ID 3230440. Rochester, NY: Social Science Research Network. 10.2139/ssrn.3230440.

Davies, Ben B., and Ian D. Hodge. 2006. Farmers' preferences for new environmental policy instruments: Determining the acceptability of cross compliance for biodiversity benefits. *Journal of Agricultural Economics* 57: 393–414. https://doi.org/10.1111/j.1477-9552.2006.00057.x.

De Keuckelaere, Ann, Liesbeth Jacxsens, Philip Amoah, Gertjan Medema, Peter McClure, Lee-Ann Jaykus, and Mieke Uyttendaele. 2015. Zero risk does not exist: Lessons learned from microbial risk assessment related to use of water and safety of fresh produce. *Comprehensive Reviews in Food Science and Food Safety*. https://doi.org/10.1111/1541-4337.12140.

DeLind, Laura B., and Philip H. Howard. 2008. Safe at any scale? Food scares, food regulation, and scaled alternatives. *Agriculture and Human Values* 25: 301–317. https://doi.org/10.1007/s10460-007-9112-y.

Doremus, Holly. 2003. A policy portfolio approach to biodiversity protection on private lands. *Environmental Science & Policy* 6: 217–232. https://doi.org/10.1016/S1462-9011(03)00036-4.

Doremus, Holly. 2010. The Endangered Species Act: Static law meets dynamic world new directions in environmental law. *Washington University Journal of Law and Policy* 32: 175–236.

Dowd, Brian M., Daniel Press, and Marc Los Huertos. 2008. Agricultural nonpoint source water pollution policy: The case of California's Central Coast. *Agriculture, Ecosystems & Environment* 128: 151–161. https://doi.org/10.1016/j.agee.2008.05.014.

Drevno, Ann. 2016. Governing water quality in California's Central Coast: The case of the conditional agricultural waiver. *Journal of Science Policy & Governance* 8 (1).

Drevno, Ann. 2018. Central Coast growers' trust in water quality regulatory process needs rebuilding. *California Agriculture* 72: 127–134.

Duff, Alison J., Paul H. Zedler, Jeb A. Barzen, and Deana L. Knuteson. 2017. The Capacity-Building Stewardship Model: assessment of an agricultural network as a mechanism for improving regional agroecosystem sustainability. *Ecology and Society*. https://doi.org/10.5751/ES-09146-220145.

Dumont, Antoinette M., and Philippe V. Baret. 2017. Why working conditions are a key issue of sustainability in agriculture? A comparison between agroecological, organic and conventional vegetable systems. *Journal of Rural Studies* 56: 53–64. https://doi.org/10.1016/j.jrurstud.2017.07.007.

Elliot, Dan. 2013. FDA: Criminal case shows food safety is paramount. *USA Today*, September 27.

[FDA] US Food and Drug Administration. 2011. *Fresh Strawberries From Washington County Farm Implicated In E. coli O157 Outbreak In NW Oregon*. Recall -- State / Local Press Release. https://www.fda.gov/Safety/Recalls/ucm267667.htm. Accessed on 19 January 2017.

Feola, G., A.M. Lerner, M. Jain, M.J.F. Montefrio, and K.A. Nicholas. 2015. Researching farmer behaviour in climate change adaptation and sustainable agriculture: Lessons learned from five case studies. *Journal of Rural Studies* 39: 74–84. https://doi.org/10.1016/j.jrurstud.2015.03.009.

Garbach, Kelly, and Rachael Freeman Long. 2017. Determinants of field edge habitat restoration on farms in California's Sacramento Valley. *Journal of Environmental Management* 189: 134–141. https://doi.org/10.1016/j.jenvman.2016.12.036.

Garcia Martinez, Marian, Andrew Fearne, Julie A. Caswell, and Spencer Henson. 2007. Co-regulation as a possible model for food safety governance: Opportunities for public–private partnerships. *Food Policy* 32: 299–314. https://doi.org/10.1016/j.foodpol.2006.07.005.

Garcia Martinez, Marian, Paul Verbruggen, and Andrew Fearne. 2013. Risk-based approaches to food safety regulation: What role for co-regulation? *Journal of Risk Research* 16: 1101–1121. https://doi.org/10.1080/13669877.2012.743157.

Gennet, Sasha, Jeanette Howard, Jeff Langholz, Kathryn Andrews, Mark D. Reynolds, and Scott A. Morrison. 2013. Farm practices for food safety: An emerging threat to floodplain and riparian ecosystems. *Frontiers in Ecology and the Environment* 11: 236–242. https://doi.org/10.1890/120243.

Goldberger, J.R. 2011. Conventionalization, civic engagement, and the sustainability of organic agriculture. *Journal of Rural Studies* 27: 288–296. https://doi.org/10.1016/j.jrurstud.2011.03.002.

Goodman, D. 2003. The quality "turn" and alternative food practices: reflections and agenda. *Journal of Rural Studies* 19: 1–7. https://doi.org/10.1016/S0743-0167(02)00043-8.

Gubernot, Diane, Marianne Fatica, Cerise Robinson, Sheila Merriweather, Tami Cloyd, and Gary Weber. 2016. A summary of foodborne illness outbreaks investigated by FDA's coordinated outbreak response and evaluation network, August 2011 to December 2015. In *International Association for Food Protection 2016 Annual Meeting*. St. Louis, MI. https://iafp.confex.com/iafp/2016/webprogram/Paper11553.html. Accessed on 21 July 2016.

Guthman, Julie. 2004. The trouble with 'organic lite' in California: a rejoinder to the 'conventionalisation' debate. *Sociologia Ruralis* 44: 301–316. https://doi.org/10.1111/j.1467-9523.2004.00277.x.

Guthman, Julie. 2016. Going both ways: More chemicals, more organics, and the significance of land in post-methyl bromide fumigation decisions for California's strawberry industry. *Journal of Rural Studies* 47: 76–84. https://doi.org/10.1016/j.jrurstud.2016.07.020.

Guthman, Julie. 2017. Life itself under contract: rent-seeking and biopolitical devolution through partnerships in California's strawberry industry. *The Journal of Peasant Studies* 44: 100–117. https://doi.org/10.1080/03066150.2016.1217843.

Gutierrez-Rodriguez, E., and A. Adhikari. 2018. Preharvest farming practices impacting fresh produce safety. *Microbiology Spectrum*. https://doi.org/10.1128/microbiolspec.PFS-0022-2018.

Harrison, Jill Lindsey, and Christy Getz. 2015. Farm size and job quality: Mixed-methods studies of hired farm work in California and Wisconsin. *Agriculture and Human Values* 32: 617–634. https://doi.org/10.1007/s10460-014-9575-6.

Hart, John. 2018. New idea in soil health: Pay farmers for their actions. *Western FarmPress*, March 20. https://www.westernfarmpress.com/land-management/new-idea-soil-health-pay-farmers-their-actions. Accessed 26 Mar 2018.

Hassanein, Neva. 2011. Matters of scale and the politics of the Food Safety Modernization Act. *Agriculture and Human Values* 28: 577–581. https://doi.org/10.1007/s10460-011-9338-6.

Hatanaka, Maki, and Lawrence Busch. 2008. Third-party certification in the global agrifood system: An objective or socially mediated governance mechanism? *Sociologia Ruralis* 48: 73–91. https://doi.org/10.1111/j.1467-9523.2008.00453.x.

Hatanaka, Maki, Jason Konefal, and Douglas H. Constance. 2012. A tripartite standards regime analysis of the contested development of a sustainable agriculture standard. *Agriculture and Human Values* 29: 65–78. https://doi.org/10.1007/s1046 0-011-9329-7.

Havinga, Tetty. 2006. Private Regulation of Food Safety by Supermarkets. *Law & Policy* 28: 515. https://doi.org/10.111 1/j.1467-9930.2006.00237.x.

Heimlich, Ralph E., and Roger Claassen. 1998. Agricultural conservation policy at a crossroads. *Agricultural and Resource Economics Review* 27: 95–107. https://doi.org/10.1017/S1068 280500001738.

Henao, Olga L., Timothy F. Jones, Duc J. Vugia, and Patricia M. Griffin. 2015. Foodborne diseases active surveillance network—2 decades of achievements, 1996–2015. *Emerging Infectious Diseases* 21: 1529–1536. https://doi.org/10.3201/eid2109.15058 1.

Hendrickson, Mary K., and Harvey S. James. 2005. The ethics of constrained choice: How the industrialization of agriculture impacts farming and farmer behavior. *Journal of Agricultural and Environmental Ethics* 18: 269–291. https://doi.org/10.1007/ s10806-005-0631-5.

Hendrickson, Mary K., and Harvey S. James. 2016. Power, fairness and constrained choice in agricultural markets: A synthesizing framework. *Journal of Agricultural and Environmental Ethics* 29: 945–967. https://doi.org/10.1007/s10806-016-9641-8.

Henke, Christopher. 2008. *Cultivating science, harvesting power: Science and industrial agriculture in California*. Cambridge, MA: MIT Press.

Higgins, Vaughan, Jacqui Dibden, and Chris Cocklin. 2008a. Building alternative agri-food networks: Certification, embeddedness and agri-environmental governance. *Journal of Rural Studies* 24: 15–27. https://doi.org/10.1016/j.jrurstud.2007.06.002.

Higgins, Vaughan, Jacqui Dibden, and Chris Cocklin. 2008b. Neoliberalism and natural resource management: Agri-environmental standards and the governing of farming practices. *Geoforum* 39: 1776–1785. https://doi.org/10.1016/j.geofo rum.2008.05.004.

Higgins, Vaughan, Melanie Bryant, Marta Hernández-Jover, Connar McShane, and Luzia Rast. 2016. Harmonising devolved responsibility for biosecurity governance: The challenge of competing institutional logics. *Environment and Planning A: Economy and Space* 48 (6): 1133–1151. https://doi. org/10.1177/0308518X16633471.

Howard, Philip H. 2016. *Concentration and power in the food system: Who controls what we eat?* Contemporary food studies: Economy, culture and politics. London: Bloomsbury Academic.

Hughes, Thomas. 1987. The evolution of large technological systems. In *The social construction of technological systems: New directions in the sociology and history of Technology*, ed. Wiebe Bijker, Thomas Hughes, and Trevor Pinch. Cambridge, MA: MIT Press.

Huttunen, Suvi. 2019. Revisiting agricultural modernisation: Interconnected farming practices driving rural development at the farm level. *Journal of Rural Studies* 71: 36–45. https://doi. org/10.1016/j.jrurstud.2019.09.004.

[IPES-Food] International Panel of Experts on Sustainable Food Systems. 2016. From uniformity to diversity: A paradigm shift from industrial agriculture to diversified agroecological systems. International Panel of Experts on Sustainable Food Systems.

[IPES-Food] International Panel of Experts on Sustainable Food Systems. 2017. Too big to feed: Exploring the impacts of mega-mergers, concentration and concentration of power in the agri-food sector. International Panel of Experts on Sustainable Food Systems.

James, Harvey S. 2003. On Finding Solutions to Ethical Problems in Agriculture. *Journal of Agricultural and Environmental Ethics* 16: 439–457. https://doi.org/10.1023/A:1026371324639.

James, Harvey S., and Mary K. Hendrickson. 2008. Perceived economic pressures and farmer ethics. *Agricultural Economics* 38: 349–361. https://doi.org/10.1111/j.1574-0862.2008.00305.x.

Kaine, Geoff, Justine Young, Ruth Lourey, and Suzie Greenhalgh. 2017. Policy choice framework: Guiding policy makers in changing farmer behavior. *Ecology and Society*. https://doi. org/10.5751/ES-09135-220202.

Karp, Daniel S., Patrick Baur, Edward R. Atwill, Kathryn De Master, Sasha Gennet, Alastair Iles, Joanna L. Nelson, Amber R. Sciligo, and Claire Kremen. 2015. The unintended ecological and social impacts of food safety regulations in California's Central Coast Region. *BioScience* 65: 1173–1183. https://doi.org/10.1093/biosc i/biv152.

Karp, Daniel S., Sasha Gennet, Christopher Kilonzo, Melissa Partyka, Nicolas Chaumont, Edward R. Atwill, and Claire Kremen. 2015. Comanaging fresh produce for nature conservation and food safety. *Proceedings of the National Academy of Sciences of the United States of America* 112: 11126–11131. https://doi. org/10.1073/pnas.1508435112.

Karp, Daniel S., Rebekah Moses, Sasha Gennet, Matthew S. Jones, Shimat Joseph, Leithen K. M'Gonigle, Lauren C. Ponisio, William E. Snyder, and Claire Kremen. 2016. Agricultural practices for food safety threaten pest control services for fresh produce. *Journal of Applied Ecology* 100000: 10000. https://doi. org/10.1111/1365-2664.12707.

Kelley, Christopher R. 1994. Agricultural production contracts: Drafting considerations. *Hamline Law Review* 18: 397.

Klonsky, Karen, and Catherine Greene. 2005. Widespread adoption of organic agriculture in the US: Are market-driven policies enough? In *American Agricultural Economics Association Annual Meeting, July 24–27*. Providence, RI. Selected Paper 132570. https://doi.org/10.22004/ag.econ.19382.

Kloppenburg, Jr., Jack, Sharon Lezberg, Kathryn De Master, George Stevenson, and John Hendrickson. 2000. Tasting food, tasting sustainability: Defining the attributes of an alternative food system with competent, Ordinary people. *Human Organization* 59: 177–186. https://doi.org/10.17730/humo.59.2.8681677127 123543.

Knowler, Duncan, and Ben Bradshaw. 2007. Farmers' adoption of conservation agriculture: A review and synthesis of recent research. *Food Policy* 32: 25–48. https://doi.org/10.1016/j.foodp ol.2006.01.003.

Knowler, Duncan, Benjamin Bradshaw, and Elizabeth Holmes. 2014. Conservation agriculture: Farmer adoption and policy issues. In *Encyclopedia of Food and Agricultural Ethics*, ed. Paul B. Thompson and David M. Kaplan, 385–393. Dordrecht: Springer Netherlands. https://doi.org/10.1007/978-94-007-0929-4_74.

Kremen, Claire, Alastair Iles, and Christopher Bacon. 2012. Diversified farming systems: An agroecological, systems-based alternative to modern industrial agriculture. *Ecology and Society*. https:// doi.org/10.5751/es-05103-170444.

Kronberg, Scott L., and Julie Ryschawy. 2019. Integration of crop and livestock production in temperate regions to improve agroecosystem functioning, ecosystem services, and human nutrition and health1. In *Agroecosystem diversity*, ed. Gilles Lemaire, Paulo César De Faccio Carvalho, Scott Kronberg, and Sylvie Recous, 247–256. New York: Academic Press. https://doi. org/10.1016/B978-0-12-811050-8.00015-7.

Kross, Sara M., Katherine P. Ingram, Rachael F. Long, and Meredith T. Niles. 2018. Farmer perceptions and behaviors related to wildlife and on-farm conservation actions. *Conservation Letters* 11: 1–9. https://doi.org/10.1111/conl.12364.

Lambert, Dayton M., Patrick Sullivan, Roger Claassen, and Linda Foreman. 2007. Profiles of US farm households adopting conservation-compatible practices. *Land Use Policy* 24: 72–88. https://doi.org/10.1016/j.landusepol.2005.12.002.

Langholz, Jeff A., and Michele Jay-Russell. 2013. Potential role of wildlife in pathogenic contamination of fresh produce. *Human-Wildlife Interactions* 7: 140–157. https://doi.org/10.26077/e5gg-r037.

Liu, Tingting, Randall J.F. Bruins, and Matthew T. Heberling. 2018. Factors influencing farmers' adoption of best management practices: A review and synthesis. *Sustainability* 10: 432. https://doi.org/10.3390/su10020432.

Long, R., K. Garbach, and L. Morandin. 2017. Hedgerow benefits align with food production and sustainability goals. *California Agriculture* 71: 117–119. https://doi.org/10.3733/ca.2017a0020.

Lowell, K., J. Langholz, and D. Stuart. 2010. *Safe and sustainable: Co-managing for food safety and ecological health in California's Central Coast Region*. San Francisco, CA: The Nature Conservancy of California and the Georgetown University Produce Safety Project.

Lubell, Mark, and Allan Fulton. 2007. Local diffusion networks act as pathways to sustainable agriculture in the Sacramento River Valley. *California Agriculture* 61: 131–137. https://doi.org/10.3733/ca.v061n03p131.

Lubell, Mark, and Allan Fulton. 2008. Local policy networks and agricultural watershed management. *Journal of Public Administration Research and Theory* 18: 673–696. https://doi.org/10.1093/jopart/mum031.

Lyson, Thomas A., G. W. Stevenson, and Rick Welsh, ed. 2008. *Food and the mid-level farm: renewing an agriculture of the middle*. Food, Health, and the Environment. Cambridge, MA: MIT Press.

Macaulay, Luke, and Van Butsic. 2017. Ownership characteristics and crop selection in California cropland. *California Agriculture* 71: 221–230. https://doi.org/10.3733/ca.2017a0041.

McElwain, Sean. 2015. The misnomer of right to farm: How right-to-farm statutes disadvantage organic farming. *Washburn Law Journal* 55: 223–268.

McMahon, Martha. 2013. What food is to be kept safe and for whom? Food-safety governance in an unsafe food system. *Laws* 2: 401–427. https://doi.org/10.3390/laws2040401.

Mench, Joy A. 2008. Farm animal welfare in the U.S.A.: Farming practices, research, education, regulation, and assurance programs. *Applied Animal Behaviour Science* 113: 298–312. https://doi.org/10.1016/j.applanim.2008.01.009.

Mercado, Geovana, Carsten Nico Hjortsø, and Benson Honig. 2018. Decoupling from International Food Safety Standards: How small-scale indigenous farmers cope with conflicting institutions to ensure market participation. *Agriculture and Human Values* 35 (3): 651–669. https://doi.org/10.1007/s10460-018-9860-x.

Miles, Albie, Marcia S. DeLonge, and Liz Carlisle. 2017. Triggering a positive research and policy feedback cycle to support a transition to agroecology and sustainable food systems. *Agroecology and Sustainable Food Systems* 41: 855–879. https://doi.org/10.1080/21683565.2017.1331179.

Mills, Jane, Peter Gaskell, Julie Ingram, Janet Dwyer, Matt Reed, and Christopher Short. 2017. Engaging farmers in environmental management through a better understanding of behaviour. *Agriculture and Human Values* 34: 283–299. https://doi.org/10.1007/s10460-016-9705-4.

Montenegro de Wit, Maywa, and Alastair Iles. 2016. Toward thick legitimacy: Creating a web of legitimacy for agroecology. *Elementa: Science of the Anthropocene*. https://doi.org/10.12952/journal.elementa.000115.

Moon, K., and C. Cocklin. 2011. Participation in biodiversity conservation: Motivations and barriers of Australian landholders. *Journal of Rural Studies* 27: 331–342. https://doi.org/10.1016/j.jrurstud.2011.04.001.

Newman, Jesse, and Heather Haddon. 2018. Effects of *E. coli* Outbreak in Lettuce Ripple Through U.S. Food-Supply Chain. *Wall Street Journal*, May 30, sec. Business.

Niles, Meredith T., Mark Lubell, and Margaret Brown. 2015. How limiting factors drive agricultural adaptation to climate change. *Agriculture, Ecosystems & Environment* 200: 178–185. https://doi.org/10.1016/j.agee.2014.11.010.

Niles, Meredith T., Mark Lubell, and Van R. Haden. 2013. Perceptions and responses to climate policy risks among California farmers. *Global Environmental Change* 23: 1752–1760. https://doi.org/10.1016/j.gloenvcha.2013.08.005.

North, Douglass C. 1990. *Institutions, institutional change, and economic performance*. Cambridge: Cambridge University Press.

Olimpi, Elissa M., Patrick Baur, Alejandra Echeverri, David Gonthier, Daniel S. Karp, Claire Kremen, Amber Sciligo, and Kathryn T. De Master. 2019. Evolving food safety pressures in California's Central Coast Region. *Frontiers in Sustainable Food Systems* 3: 100000. https://doi.org/10.3389/fsufs.2019.00102.

Opie, John. 1987. *The law of the land: two hundred years of American farmland policy*. Lincoln: University of Nebraska Press.

Oteros-Rozas, Elisa, Adriana Ruiz-Almeida, Mateo Aguado, José A. González, and Marta G. Rivera-Ferre. 2019. A social–ecological analysis of the global agrifood system. *Proceedings of the National Academy of Sciences of the United States of America* 10000: 10000. https://doi.org/10.1073/pnas.1912710116.

Painter, J.A., R.M. Hoekstra, T. Ayers, R.V. Tauxe, C.R. Braden, and F.J. Angulo. 2013. Attribution of foodborne illnesses, hospitalizations, and deaths to food commodities by using outbreak data, United States, 1998–2008. *Emerging Infectious Disease* 19: 407–415. https://doi.org/10.3201/eid1903.111866.

Parker, Jason, Julia DeNiro, Melanie Lewis Ivey, and Doug Doohan. 2016. Are small and medium scale produce farms inherent food safety risks? *Journal of Rural Studies* 44: 250–260. https://doi.org/10.1016/j.jrurstud.2016.02.005.

Perfecto, Ivette, John Vandermeer, and Angus Wright. 2019. *Nature's matrix: Linking agriculture, biodiversity conservation and food sovereignty*, 2nd ed. Milton Park: Routledge.

Pollans, Margot J. 2015. Regulating farming: Balancing food safety and environmental protection in a cooperative governance regime. *Wake Forest Law Review* 50 (2): 399–460.

Pollans, Margot J. 2016. Drinking water protection and agricultural exceptionalism. *Ohio State Law Journal* 77 (6): 1195–1260.

Price, J.C., and Z. Leviston. 2014. Predicting pro-environmental agricultural practices: The social, psychological and contextual influences on land management. *Journal of Rural Studies* 34: 65–78. https://doi.org/10.1016/j.jrurstud.2013.10.001.

Prokopy, L.S., K. Floress, D. Klotthor-Weinkauf, and A. Baumgart-Getz. 2008. Determinants of agricultural best management practice adoption: Evidence from the literature. *Journal of Soil and Water Conservation* 63: 300–311. https://doi.org/10.2489/jswc.63.5.300.

Raymond, Christopher M., Claudia Bieling, Nora Fagerholm, Berta Martin-Lopez, and Tobias Plieninger. 2016. The farmer as a landscape steward: Comparing local understandings of landscape stewardship, landscape values, and land management actions. *Ambio* 45: 173–184. https://doi.org/10.1007/s13280-015-0694-0.

Rehber, Erkan. 2018. *Contract Farming In Practice: An overview*. Research Report 7. https://doi.org/10.22004/ag.econ.290069.

Reimer, Adam P., Aaron W. Thompson, and Linda S. Prokopy. 2012a. The multi-dimensional nature of environmental attitudes among farmers in Indiana: Implications for conservation adoption. *Agriculture and Human Values* 29: 29–40. https://doi.org/10.1007/s10460-011-9308-z.

Reimer, Adam P., Riva C.H. Denny, and Diana Stuart. 2018. The impact of federal and state conservation programs on farmer nitrogen management. *Environmental Management* 62: 694–708. https://doi.org/10.1007/s00267-018-1083-9.

Reimer, Adam P., Aaron W. Thompson, Linda S. Prokopy, J. Gordon Arbuckle, Ken Genskow, Douglas Jackson-Smith, Gary Lynne,

Laura McCann, Lois Wright Morton, and Pete Nowak. 2014. People, place, behavior, and context: A research agenda for expanding our understanding of what motivates farmers' conservation behaviors. *Journal of Soil and Water Conservation* 69: 57A–61A. https://doi.org/10.2489/jswc.69.2.57A.

Reimer, Adam P., D.K. Weinkauf, and Linda S. Prokopy. 2012b. The influence of perceptions of practice characteristics: An examination of agricultural best management practice adoption in two indiana watersheds. *Journal of Rural Studies* 28: 118–128. https://doi.org/10.1016/j.jrurstud.2011.09.005.

Robertson, G.Philip, Katherine L. Gross, Stephen K. Hamilton, Douglas A. Landis, Thomas M. Schmidt, Sieglinde S. Snapp, and Scott M. Swinton. 2014. Farming for ecosystem services: An ecological approach to production agriculture. *BioScience* 64: 404–415. https://doi.org/10.1093/biosci/biu037.

Rosset, Peter M., and Miguel A. Altieri. 2017. *Agroecology: Science and politics*. Black Point, NS: Fernwood Books Ltd.

Ruhl, James B. 2000. Farms, their environmental harms, and environmental law. *Ecology Law Quarterly* 27: 263–349.

Sayer, Jeffrey, Terry Sunderland, Jaboury Ghazoul, Jean-Laurent Pfund, Douglas Sheil, Erik Meijaard, Michelle Venter, et al. 2013. Ten principles for a landscape approach to reconciling agriculture, conservation, and other competing land uses. *Proceedings of the National Academy of Sciences of the United States of America* 110: 8349–8356. https://doi.org/10.1073/pnas.1210595110.

Schewe, Rebecca L., and Diana Stuart. 2017. Why don't they just change? Contract farming, informational influence, and barriers to agricultural climate change mitigation. *Rural Sociology* 82: 226–262. https://doi.org/10.1111/ruso.12122.

Scott, W.Richard. 2013. *Institutions and organizations: Ideas, interests, and identities*, 4th ed. Los Angeles: SAGE Publications Inc.

Selinske, Matthew J., Benjamin Cooke, Nooshin Torabi, Mathew J. Hardy, Andrew T. Knight, and Sarah A. Bekessy. 2017. Locating financial incentives among diverse motivations for long-term private land conservation. *Ecology and Society* 22: 7. https://doi.org/10.5751/ES-09148-220207.

Sellers, Laurel A., Rachael F. Long, Michele T. Jay-Russell, Xunde Li, Edward R. Atwill, Richard M. Engeman, and Roger A. Baldwin. 2018. Impact of field-edge habitat on mammalian wildlife abundance, distribution, and vectored foodborne pathogens in adjacent crops. *Crop Protection* 108: 1–11. https://doi.org/10.1016/j.cropro.2018.02.005.

Strauss, Debra A. 2011. An analysis of the FDA Food Safety Modernization Act: Protection for consumers and boon for business. *Food and Drug Law Journal* 66: 353–376.

Stuart, Diana. 2008. The illusion of control: industrialized agriculture, nature, and food safety. *Agriculture and Human Values* 25: 177–181. https://doi.org/10.1007/s10460-008-9130-4.

Stuart, Diana. 2009. Constrained choice and ethical dilemmas in land management: Environmental quality and food safety in California agriculture. *Journal of Agricultural and Environmental Ethics* 22: 53–71. https://doi.org/10.1007/s10806-008-9129-2.

Stuart, Diana. 2010. Coastal ecosystems and agricultural land use: New challenges on California's central coast. *Coastal Management* 38: 42–64. https://doi.org/10.1080/08920750903363190.

Stuart, Diana, and Sean Gillon. 2013. Scaling up to address new challenges to conservation on US farmland. *Land Use Policy* 31: 223–236. https://doi.org/10.1016/j.landusepol.2012.07.003.

Stuart, Diana, and Rebecca L. Schewe. 2016. Constrained choice and climate change mitigation in US agriculture: Structural barriers to a climate change ethic. *Journal of Agricultural and Environmental Ethics* 29: 369–385. https://doi.org/10.1007/s10806-016-9605-z.

Taylor, Rebecca, and David Zilberman. 2017. Diffusion of drip irrigation: The case of California. *Applied Economic Perspectives and Policy* 39: 16–40. https://doi.org/10.1093/aepp/ppw026.

Thomas, Courtney Irene Powell. 2014. *In food we trust: the politics of purity in American food regulation*. Lincoln, NB: University of Nebraska Press.

Thornton, Patricia H., William Ocasio, and Michael Lounsbury. 2012. *The institutional logics perspective: A new approach to culture, structure and process*. Oxford: Oxford University Press.

Tidgren, Kristine A. 2017. *Why a Federal Court Dismissed the DMWW Lawsuit*. Center for Agricultural Law and Taxation. https://www.calt.iastate.edu/blogpost/why-federal-court-dismissed-dmww-lawsuit. Accessed 16 June 2020.

Uematsu, Hiroki, and Ashok K. Mishra. 2012. Organic farmers or conventional farmers: Where's the money? *Ecological Economics* 78: 55–62. https://doi.org/10.1016/j.ecolecon.2012.03.013.

van Huylenbroeck, G., V. Vandermeulen, E. Mettepenningen, and A. Verspecht. 2007. Multifunctionality of agriculture: A review of definitions, evidence and instruments. *Living Reviews in Landscape Research* 1: 1–38.

Veissier, Isabelle, Andrew Butterworth, Bettina Bock, and Emma Roe. 2008. European approaches to ensure good animal welfare. *Applied Animal Behaviour Science* 113: 279–297. https://doi.org/10.1016/j.applanim.2008.01.008.

Verbruggen, Paul, and Tetty Havinga. 2017. *Hybridization of food governance: An analytical framework*. Cheltenham: Edward Elgar Publishing.

Vivant, Anne-Laure, Dominique Garmyn, Pierre-Alain Maron, Virginie Nowak, and Pascal Piveteau. 2013. Microbial diversity and structure are drivers of the biological barrier effect against *Listeria monocytogenes* in soil. *PLoS ONE* 8: e76991. https://doi.org/10.1371/journal.pone.0076991.

Wacker, Matthew, Alvin D. Sokolow, and Rachel Elkins. 2001. *County right-to-farm ordinances in California: An assessment of impact and effectiveness*. AIC Issues Brief 15. Davis: Agricultural Issues Center, University of California.

Walker, Richard. 2004. *The conquest of bread: 150 years of agribusiness in California*. New York: New Press. Distributed by Norton.

Wang, Siyun, Daniel Weller, Justin Falardeau, Laura K. Strawn, Fernando O. Mardones, Aiko D. Adell, and Andrea I. Moreno Switt. 2016. Food safety trends: From globalization of whole genome sequencing to application of new tools to prevent foodborne diseases. *Trends in Food Science & Technology* 57(Part A): 188–198. https://doi.org/10.1016/j.tifs.2016.09.016.

Warner, Keith Douglass. 2008. Agroecology as participatory science emerging alternatives to technology transfer extension practice. *Science, Technology & Human Values* 33: 754–777. https://doi.org/10.1177/0162243907309851.

Warren, C.R., R. Burton, O. Buchanan, and R.V. Birnie. 2016. Limited adoption of short rotation coppice: The role of farmers' socio-cultural identity in influencing practice. *Journal of Rural Studies* 45: 175–183. https://doi.org/10.1016/j.jrurstud.2016.03.017.

Wild Farm Alliance. 2016. *Co-managing farm stewardship with food safety GAPs and conservation practices: A grower's and conservationist's* handbook. *Watsonville*: Wild Farm Alliance.

Willett, Walter, Johan Rockström, Brent Loken, Marco Springmann, Tim Lang, Sonja Vermeulen, Tara Garnett, et al. 2019. Food in the Anthropocene: the EAT–Lancet Commission on healthy diets from sustainable food systems. *The Lancet* 393: 447–492. https://doi.org/10.1016/S0140-6736(18)31788-4.

Wollaeger, Heidi M., Kristin L. Getter, and Bridget K. Behe. 2015. Consumer preferences for traditional, neonicotinoid-free, bee-friendly, or biological control pest management practices on floriculture crops. *HortScience* 50: 721–732. https://doi.org/10.21273/HORTSCI.50.5.721.

Woods, Timothy A., and Debra Tropp. 2015. CSAs and the battle for the local food dollar. *Journal of Food Distribution Research* 46: 17. https://doi.org/10.22004/ag.econ.209984.

Worrell, Richard, and Michael C. Appleby. 2000. Stewardship of natural resources: Definition, ethical and practical aspects. *Journal of Agricultural and Environmental Ethics* 12: 263–277. https://doi.org/10.1023/A:1009534214698.

Worster, Donald. 1994. *Nature's economy: A history of ecological ideas*. Cambridge: Cambridge University Press.

Yiannas, Frank. 2009. *Food safety culture: Creating a behavior-based food safety management system*. New York: Springer.

Publisher's Note Springer Nature remains neutral with regard to jurisdictional claims in published maps and institutional affiliations.

Patrick Baur is an assistant professor in the Department of Fisheries, Animal and Veterinary Sciences at the University of Rhode Island. His research is motivated by the search for a better balance among food system livelihoods, ecological resilience, social justice, and human health—put simply, how can people better work together to provide good food more sustainably and equitably? He previously was a postdoctoral scholar in the Department of Environmental Science, Policy and Management at University of California, Berkeley, where he also received his PhD with a designated emphasis in Science and Technology Studies.

Agriculture and Human Values (2020) 37:1195–1206
https://doi.org/10.1007/s10460-020-10126-5

Political economy challenges for climate smart agriculture in Africa

Helena Shilomboleni[1]

Accepted: 4 July 2020 / Published online: 10 July 2020
© Springer Nature B.V. 2020

Abstract

Climate Smart Agriculture (CSA) has gained prominence in global agriculture and climate agendas for its perceived "triple win" contributions to food productivity, adaptation, and mitigation to climate change. This paper highlights three important challenges for CSA activities in Africa which provide insights into contested debates surrounding CSA's ability to respond holistically to the complex realities facing resource-constrained farmers in the global South. These are (1) prevailing neoliberal market policies that emphasize private-sector driven agricultural development in the face of rising input costs and falling commodity prices; (2) an expansion in diversified livelihood strategies amongst smallholder households as a response to the highly unpredictable biophysical environment and economic climate under which they live; and (3) a growing competition for land and other productive resources. A deeper dive into political economy processes surrounding these three issues aims to bring critical attention to factors relevant to African agricultural development that highly impact farm-level practices and carry important implications for rural livelihood outcomes.

Keywords Climate smart agriculture · Political economy · Smallholder farmers · Livelihoods · Productivity · Food security

Abbreviations

AUDA-NEPAD	African Union Development Agency
CSA	Climate Smart Agriculture
ERA	Evidence for Resilient Agriculture
GACSA	Global Alliance for Climate Smart Agriculture

Introduction

Climate Smart Agriculture (CSA) has gained prominence in global agriculture and climate agendas for its perceived "triple win" contributions to food productivity, adaptation, and mitigation to climate change. The World Bank first introduced CSA in its 2010 *World Development Report: Development and Climate Change*. Around the same time, the Food and Agriculture Organization (FAO) formally launched CSA at the first global policy conference dedicated to the topic in the Hague, Netherlands (31 October–5 November 2010) (Taylor 2018, p. 93). In 2014, the Global Alliance for Climate Smart Agriculture (GACSA) was launched at the UN Climate Summit. GACSA has since brought together a coalition of over 450 members comprising multilateral agencies, governments, research institutes and corporations to support the scaling up of CSA activities around the world (GACSA 2020). CSA is set to become a major vehicle for agricultural development finance, particularly in low-and-middle income countries. Overall, international donor funding for climate adaptation is expected to reach $100 billion per annum in 2020 (Amadu et al. 2020). The World Bank alone has pledged at least US$ 8 billion of its annual spending on agriculture to support CSA activities since 2018 (Dinesh et al. 2017, p. 4). The African Union Development Agency (AUDA-NEPAD) is similarly leading continental efforts to have at least 25 million farming families in Africa practicing CSA by 2025 (Vision 25×25).

Over the last few years, CSA scholars and practitioners have developed various tools and frameworks to identify and prioritize context-specific climate smart options which have a potential to improve agricultural productivity, support increases in income and food security, and enhance farmers' adaptive capacity to climate change (Thornton et al. 2018; CIAT 2014; Khatri-Chhetri et al. 2017; Andrieu et al. 2017). Such tools and frameworks are intended to generate evidence-based data on promising CSA technologies and practices and to support agricultural

✉ Helena Shilomboleni
    h.shilomboleni@cgiar.org

[1] CGIAR Research Program on Climate Change, Agriculture and Food Security (CCAFS) East Africa, International Livestock Research Institute, Box 30709, Nairobi, Kenya

decision making at different scales, i.e., farm, project and program, and policy levels (see Rosenstock et al. 2016; ICRAF 2020). This growing body of work provides conceptual clarity to CSA, including empirical guidance for its interventions in diverse farming systems in developing countries. Despite efforts to delineate criteria and boundaries around CSA, there remains much controversy and debate over how agricultural problems and solutions are framed in its discourse, including scarce attention to issues of power, authority and equity (Karlsson et al. 2018; Newell and Taylor 2018; Clapp et al. 2018; Anderson 2014).

Critical scholars take issue with CSA's bias towards farm-level technologies and management practices, and a re-branding of almost any agricultural technology or practice that contributes to food productivity, low-inputs, or resource conservation as climate smart (Newell and Taylor 2018; Chandra et al. 2018). Such a narrow focus on farm-level responses and outcomes, according to critics, run the risk of characterizing the nature (and causes) of vulnerabilities to climate change and food insecurity as technical and managerial, solvable through "appropriate technologies" and "best practices" (Clapp et al. 2018; Eriksen et al. 2015). In addition, little engagement with the socio-political and economic processes that shape the historical specificities of agrarian change in particular contexts (see Schnurr 2019), raise questions about the potential of CSA to alleviate the burden of hunger and poverty for millions of smallholder farmers in low-and-middle income countries.

This paper examines the debates surrounding CSA in Africa's smallholder agricultural systems. Critical scholars view most new agricultural development initiatives in Africa, including CSA, as a continuation of broader trends focused on addressing low agricultural productivity or "the yield gap" between what is realized on farmers' fields and those possible under ideal conditions, e.g., on demonstration plots (Schnurr 2019; Moseley et al. 2015; Schurman 2017; Ollenburger et al. 2019). As such, CSA arguably represents the latest of productionist approaches narrowly focused on yield enhancement but pay little attention to larger socio-political and economic realities that profoundly impact farmers' livelihood outcomes. In contrast, CSA supporters see it as a transformational approach to sustainable agriculture, not another reincarnation of the Green Revolution, that aims to meet the three challenges of productivity, adaptation and mitigation in a manner that pays attention to the outcomes and implications of interventions beyond the farm (CCAFS n.d.; see also FAO 2019). Thornton et al. (2017) explains that among the reasons that CSA is not business-as-usual is its emphasis on implementing context-driven sustainable agricultural solutions that support equitable increases in income, food security and development among other things (p. 149).

These claims and counter claims surrounding CSA warrant serious systematic analysis to shed light on crucial considerations that will play a central role in the potential success or failure of CSA interventions, and to move beyond prominent reporting of "success stories" seen in numerous CSA pilot programs and projects (Cavanagh et al. 2017, p. 115; Giller et al. 2017). A useful starting point is a look at the methodological orientations and ontological underpinnings of both the CSA literature and the political economy scholarship surrounding agricultural development trends in sub-Saharan Africa.

On one hand, CSA is largely rooted in empiricism, focused on addressing current problems and investing in evidence-based solutions. Specifically, CSA supports a global food security agenda that calls for food production to (sustainably) double by 2050 in order to feed a population expected to reach 9 billion (FAO 2009). In sub-Saharan Africa, CSA's response to this objective has been an overwhelming focus on increasing agricultural productivity, utilizing a wide range of technologies and agronomic practices. On the other hand, critically minded scholars call into question this empiricist approach, arguing that it perpetuates a "technocratic and deceptively apolitical agenda for agricultural development and hunger alleviation in Africa" (Moseley et al. 2015, p. 3). Such scholars draw on a broad range of theoretical perspectives under political ecology to chart a political economy analysis of agricultural development trends in Africa (Schnurr 2019; Schurman 2017; Ignatova 2017; Patel 2013; Holt-Giménez and Altieri 2013; Sumberg et al. 2012). This scholarship provides not only a historical grounding of agricultural change in specific contexts but bring to the fore important power relations and political interests behind agricultural interventions involving donors, (inter)national and research organizations, NGOs, private-sector actors, local partners and farmers.

The political economy analysis of this literature highlight at least three critical concerns relevant in the CSA agenda in Africa that will be explored in this paper, namely: (1) the prevailing neoliberal market policies that emphasize private-sector driven agricultural development in the face of rising input costs and falling commodity prices; (2) an expansion in diversified livelihood strategies amongst smallholder households as a response to the highly unpredictable biophysical environment and economic climate under which they live; and (3) a growing competition for land and other productive resources. Together, these three topics pay attention to issues of power, authority and equity, which critical scholars rightly argue are essential for evaluating the potential of agricultural development interventions to respond holistically to the realities and challenges facing Africa's resource-constrained farmers (see Schnurr 2019). Given CSA's purported transformative agenda to bring about fundamental change in agricultural systems, engaging these issues is especially pertinent, and could help galvanize targeted action to address some of the structural constrains that smallholders face.

This paper's review of CSA activities in Africa was prompted by an ongoing big-data project analyzing over 1,400 scientific agricultural studies conducted in Africa over the last 40-plus years to explore the performance of agricultural technologies against a set of indicators on productivity, resilience (adaptation) and mitigation to climate change (ICRAF 2020; see also Rosenstock et al. 2016). Work on this Evidence for Resilient Agriculture (ERA) database, formerly known as the CSA Compendium, is being conducted under the leadership of the CGIAR World Agroforestry Centre (ICRAF). My engagement with aspects of the ERA and the broader CSA literature is not an attempt to assess the effects of CSA outcomes.[1] Rather, it is meant to draw attention to broader political economy processes of agrarian change that farm-level practices are deeply embedded in. The next section outlines the paper's review criteria and makes brief remarks about the ontological and epistemological underpinnings of CSA in relation to the political economy of agricultural development trends in Africa. The subsequent sections address the three critical concerns (neoliberal market policies, diversification of rural livelihoods and competition for productive resources) in a successive order. There follows a short conclusion.

CSA activities in Africa and links to theory

The ERA database comprises an impressive record of data detailing expected performance of over 250 agricultural technologies and management practices on the three pillars of CSA (ICRAF 2020). This data is intended to inform agricultural decision-making among different users, including development practitioners, extension officers and policymakers in specific locations. Thus far, data points have been compiled from 1, 453 studies undertaken in Africa dating as far back as 1971, following inclusion criteria of primary, quantitative data of both "a conventional technology (a control) and an 'improved' agricultural technology," as well as information on at least one of CSA three pillars (ICRAF 2020). The database has filters to show study outcomes on (1) productivity sub-indicators on yields (1, 239 studies)

and economics (204 studies); (2) resilience/adaptation sub-indicators on efficiency (344 studies), physical (546 studies) and social (33 studies); and (3) mitigation sub-categories on carbon stocks (13 studies); cookstoves (20 studies) and emissions (29 studies).

Similarly, I undertook a review of scientific literature covering theory and evidence on CSA, published from 2009 to 2019. The qualitative analysis methods to CSA in this paper was informed by similar reviews on the topic conducted by Chandra et al. (2018) and Karlsson et al. (2018). Academic articles on CSA in Africa were retrieved from several major academic publishers' databases including ScienceDirect (Elsevier), Springer Link, Wiley Online Library and Emerald Insights, based on inclusion criteria of CSA activities that target Africa's smallholder agriculture, a focus on at least one of CSA's three pillars (productivity, adaptation and mitigation) and engagement with one of three political economy considerations. The database searches on 'climate smart agriculture', AND 'Africa', AND 'smallholder', AND productivity gave the largest total number of studies (1049 articles), additional keywords were added to these: 'mitigation' (670 articles), 'resilience' (504 articles), 'neoliberal' (44 articles), 'livelihood diversification' (514 articles), and 'land consolidation' (180 articles).

Both of these results show an overwhelming response to increasing agricultural yields, utilizing a wide range of technologies and agronomic practices, including (bio)physical enhancements around the farm (e.g., planting living fences) that can help buffer systems against climate shocks and stresses (see also Rosenstock et al. 2016). CSA's focus on intensifying agricultural productivity reflects broader agricultural development trends, which critical scholars argue favor particular agricultural technologies and impact pathways over others (Schurman 2017; Andersson and Sumberg 2017; Westengen et al. 2018).

In sub-Saharan Africa, multiple initiatives and funding arrangements are focused on addressing low agricultural productivity and the threats posed by climate change, often emphasizing reliance on the best of scientific discoveries (see Schurman 2017, p. 441). Among the most prominent endeavors on the continent is the New Green Revolution, being advanced by powerful actors under various initiatives. These include the Alliance for a Green Revolution in Africa, established by the Bill & Melinda Gates and Rockefeller Foundations in 2006; the Grow Africa Partnership of the World Economic Forum and AUDA-NEPAD which was launched in 2011; and the G7's New Alliance for Food Security and Nutrition formed in 2012. Most of these initiatives were created following the 2007/2008 global food crisis. The food crisis gave rise to a powerful 'feed the world' narrative that calls for global food production to (sustainably) double to meet the world's food security needs for a population expected to reach 9 billion by 2050 (see FAO 2009). CSA

---

[1] Empirical evidence on smallholder farmers' sustained adoption of CSA practices remain subject to ongoing debate and the data on adoption success is at times complicated by the provision of free or subsidized inputs (e.g., improved seeds, fertilizers and herbicides) and/or access to finance by agricultural development projects or national governments, as is the case in studies assessing farmers' uptake of conservation agriculture in southern Africa (see Andersson and D'Souza 2014; Giller et al. 2015).

emerged in this context and its activities largely aim to support this agenda (see Lipper et al. 2014). Critics however raise concerns about the economic and material interests that are at play in these various agricultural development initiatives, and the implications they carry for rural livelihoods.

In the New Green Revolution for Africa discourse, for instance, Ignatova (2017) reveals how the framing of Genetically Modified (GM) crops as "pro-poor" and their proprietary genetic seed materials as "gifts" donated to agricultural development in Africa facilitate their entry into new markets (p. 2259). Evidently, controversial agro-corporations such as Syngenta (merged with ChemChina), and Monsanto (acquired by Bayer) that are developing "pro-poor" GM technologies have embedded themselves into various partnerships to earn legitimacy and reputational benefits while expanding their market reach (Ignatova 2017, p. 2267). Westengen et al. (2018) similarly illustrate how narratives to address low productivity in smallholder agriculture and adapt to climate change have led to a push for high-input CSA interventions, comprising high usage of herbicides, pesticides and mineral fertilizer, and an emphasis on private-sector led modernization in Zambia. There, agricultural development projects promoting conservation agriculture and CSA are closely tied to several multinational agro-input suppliers, including Syngenta, BASF and Pioneer DowDupont (now Corterva), which recruit former lead farmers to work as their "community sales agents" (Westengen et al. 2018, p. 261). CSA supporters largely welcome private-sector involvement in smallholder agricultural development. Evidently, business actors provide not only much needed financial support to facilitate the development of high performing innovations (e.g., genomics and phenomics technologies) but ensure sustainable scaling through markets (see World Bank 2016; Das et al. 2019; Kadzamira and Ajayi 2019).

The above sentiments surrounding strong ties between CSA and the private sector, although appear logical or necessary solution to Africa's agricultural challenges, carry important ideological dimensions rooted in empiricist ontological and epistemological traditions. Currie-Alder (2016) explains that empiricism took hold in the development field starting in the late twentieth century, when neoliberal policies saw state budget cuts across key sectors (agriculture, healthcare, education, etc.) and growing public pressure to demonstrate the value created by foreign aid spending in terms of real-life outcomes, e.g., on income generated and lives saved (p. 11). This period also coincided with the onset of the Gene Revolution (starting in the late 1990s and early 2000s) which saw a significant rise in private-sector investments in agricultural research and development (Parayil 2003). The combination of new research actors and funders together with increased pressure to use development funds more effectively has resulted in a shift towards empiricist

development interventions that put emphasis on problem-solving. As a result, projects' net returns are now increasingly assessed based on to their ability to deliver quick wins for large-scale impact at scale as a way to demonstrate "value for money" (Andersson and Sumberg 2017; Sumberg et al. 2012).

The problem with empiricist approaches, as Kapoor (2002) argues, is their inability to sufficiently theorize and politicize development solutions, including the broader political economy forces in which they are embedded in. Doing so can help to not only "illuminate the diversity of interests, postures and underlying asymmetries of power" (Grégoire et al. 2017, p. 171) in agricultural development interventions, but take into account historical processes that highly shape present and future realities of agrarian change (see Schnurr 2019, p. 20). While CSA promote many commendable practices that can help poor rural households increase their agricultural production, e.g., low-cost sustainable agroecological methods such as intercropping, mulching, soil cover and farmer-managed agroforestry, these alone are unlikely going to lift them out of poverty (see Andersson and Giller 2019).

Prevailing neoliberal market policies

As mentioned, CSA activities in Africa are part of a larger agricultural development initiatives that aim to address low agricultural productivity, predominately funded by powerful networks of multilateral institutions, philanthropies, and private-sector actors. Critics argue that such network actors largely favor neoliberal market policies in smallholder agriculture, emphasizing private-sector pathways to deliver farm inputs and extension and services as opposed to public systems (Dawson et al. 2016; Westengen et al. 2018; Ollenburger et al. 2019).

Multinational corporations have enthusiastically embraced the CSA agenda and some have re-branded their technologies, e.g., biochar, transgenics and chemical fertilizers, as not only climate-smart, but imperative to address climate change and food insecurity (Karlsson et al. 2018, p. 163). Various corporations have also joined CSA partnerships and platforms, including the GACSA. This trend is welcomed by global governance and development actors who seek to leverage and scale up private-sector finance in agricultural development (Newell and Taylor 2018, p. 114). Poor countries looking to attract CSA development finance have also put in place CSA adaptation agendas that align with donor priorities. However, such priorities largely cast aside serious political concerns in agriculture such as anti-dumping provisions, subsidy regimes in international trade and market concentration (Clapp et al. 2018).

Various scholars raise questions about how smallholder farmers will benefit from market-driven agricultural solutions considering the steady rise of consolidation and concentration across the agri-business industry which enable corporations to set private standards in the sector (Clapp 2016; IPES-Food 2017; Chandra et al. 2017). Today, a handful of corporations wield enormous power in the global food system due to merger and acquisition strategies that have increased their market concentration rates well above what economists deem competitive: less than 40% market share for top four firms in a sector (Clapp 2016). Already, as much as 70% of the seed and agrochemical industry is controlled by three recently merged companies (DowDuPont, Bayer-Monsanto and ChemChina-Syngenta) (IPES-Food 2017, p. 13). This level of concentration puts farm input buyers (notably farmers) at a disadvantage when bargaining prices with such powerful suppliers. Indeed, farmers across the world have seen a steady rise in the cost of inputs but a real term reduction in farm-gate prices for most staple food commodities over the last two decades (Whitfield 2017). There are concerns that market-driven CSA solutions in Africa could drive smallholder farmers into further livelihood marginalization through mounting commercial input costs and associated effects of reduced autonomy over what food to grow and how to do so (see McMichael 2013; Chandra et al. 2017).

While the list of solutions proposed under CSA extend beyond market-led yield-increasing technologies, critics still take issue with the conceptualization and application of the term—the types of actions and activities that are rendered climate smart and what triple wins mean for whom are shaped by who interprets them (Karlsson et al. 2018 p. 151). Indeed, the praxis of labelling any agricultural techniques and practices that advance one or more of the triple win objectives as climate smart has partly contributed to incongruencies around what CSA entails on the ground. Often, CSA is framed differently to appeal to various agendas and audiences or to build consensus and alliances across multiple governance levels and stakeholder groups (Faling and Biesbroek 2019).

For example, multilateral actors such as the World Bank at times frame CSA as carbon offsets to promote sustainable agricultural land management entailing agro-forestry and conservation agriculture, and to participate in the selling and buying of carbon credits for CO_2 emission reductions (see Cavanagh et al. 2017). Agri-corporations have embraced CSA as sustainable intensification of agriculture to expand their agro-chemical inputs and technologies to new markets. The government of Kenya, according to Faling (2020), has shifted its policy frames around CSA-related issues over time in order to attract funding as well as to comply with various bi-lateral and multilateral agreements [e.g. UN Framework Convention on Climate Change (UNFCCC) and the Comprehensive Africa Agriculture Development Program (CAADP)]. In the mid-to-late 2000s, the government emphasized agricultural industrialization, private sector investments and trade, which appealed to international lending institutions' preferences towards liberalization and privatization. After 2009, Kenya shifted its policies to give greater attention to climate change adaption and to transform smallholder farming to modern and commercial agriculture following multiple UN agreements and programs (e.g., UNFCCC) that support adaptation activities in smallholder agriculture (Faling 2020).

The (re)framing of CSA to align with various interests and agendas raises questions about whether CSA can respond effectively to smallholder farmers' own livelihood priorities and constrains. Donors' objectives to mitigate and adapt to climate change may be quite different from some of the complex realities that farmers face. In a Kenyan CSA carbon project funded by the World Bank and Swedish International Development Agency (2009–2030), low-income smallholder farmers were unable adopt agro-forestry practices at the scale initially planned (45, 000 ha) due to small land holdings and the perceived inefficiencies for water and space of some tree species (Cavanagh et al. 2017, p. 121).

In recent years, CSA scholars and practitioners have recognized a need to identify and prioritize 'suitable' climate smart options with a potential to improve agricultural productivity, resilience and adaptability to climate change for farmers in specific contexts (CIAT 2014; Khatri-Chhetri et al. 2017; Andrieu et al. 2017). The CGIAR's Research Program on Climate Change, Agriculture and Food Security (CCAFS) together with the International Center for Tropical Agriculture (CIAT) have introduced a participatory planning tool, the CSA Prioritization Framework, that seeks to narrow down long lists of CSA technologies and practices and focus attention to end-user priorities (Campbell et al. 2016). This tool and similar ones (see Thornton et al. 2018) provide much needed conceptual and empirical guidance for CSA interventions across different geographic scales. A few empirical studies that pilot-tested the CSA Prioritization Framework[2] found that the tool was helpful to (1) quantify the contribution of CSA to food productivity, adaption and resilience using limited time and resources, (2) prioritize 'best-bet' CSA options for intervention contexts and (3) enhance key stakeholder engagement in the selection of locally-relevant options (Andrieu et al. 2017; Khatri-Chhetri et al. 2019).

While these efforts are a step in the right direction, they still largely follow an 'impact-at-scale philosophy' (Whitfield 2017, p. 260) focused on technical solutions to increase

---

[2] The CSA Prioritization Framework has been pilot-tested in several countries in Africa, Latin America, and Asia since 2014 (CIAT 2014).

agricultural productivity. Development agencies and donors that fund the bulk of CSA activities in Africa tend to prioritize impact-oriented solutions that can demonstrate efficiency results, e.g., in yield increase or numbers of beneficiaries reached (see Giller et al. 2017). The World Bank, as mentioned is committed to a CSA agenda focused on scaling up carbon capture practices, as well as high- efficiency/low-energy use irrigation programs; energy solutions for agribusiness; livestock productivity and mainstreaming of risk management (in Taylor 2018, p. 95). Norway, which co-chaired the establishment of GACSA and supports AUDA-NEPAD's Program on Agriculture and Climate Change, puts greater emphasis on private-sector led CSA initiatives, including on the need for smallholder agriculture to become "more consumer and market oriented" (MFA 2016 in Westengen et al. 2018, p. 265). As such, the selection of 'best-bet' options generally involve undertaking cost–benefit analysis to quantify the value preposition of CSA innovations, for example based on their ability to be scaled up or to attract additional investments from private partners (Campbell et al. 2016). Meanwhile, farmers themselves who are the intended beneficiaries are often left with little to no say in influencing the design, objectives, targets, implementation, and evaluation of CSA solutions.

Thus, whereas CSA prioritization tools are designed to promote participatory selection of CSA options that answer to local conditions (see Andrieu et al. 2019), there is scarce critical reflection on the challenging nature of unequal power relations between different stakeholders. International organizations and donors that set development priorities increasingly want researchers to "specify the outcomes and impacts of their work long before it is undertaken" (Giller et al. 2017, p. 157). Such impacts are generally expected to align with pre-determined pathways and themes around closing yield gaps through sustainable intensification and the use of specific technologies in smallholder agricultural systems (Ollenburger et al. 2019, pp. 290–291). Yet, the CSA discourse offers little guidance on what the responsibilities of donors, implementing agencies, and other stakeholders might involve, and how the agency of beneficiaries (and local stakeholders) could be harnessed to articulate their needs and priorities more effectively. Questions around how smallholders' own livelihood priorities might differ from the objectives of donors are also largely left unaddressed (Ollenburger et al. 2019).

Livelihood diversification among rural households

An important growing trend in Africa's smallholder agriculture is increased diversification in livelihood strategies. The multiple risks that characterize smallholder agricultural

environments (e.g., market price volatility, economic and labor constrains, farm size, crop loss, tenure security, etc.) has partly catalyzed livelihood diversification as a coping strategy amongst rural households. Bryceson (2002) explains that livelihood diversification has grown steadily in Africa since the 1980s as a response to structural adjustment policies which triggered widespread erosion of rural economies, including exposure to unfavorable terms of trade (p. 737). In contemporary times, farmer decisions to diversify their livelihoods is no longer only driven by risk factors, but by new market opportunities in agriculture, such as increased food demand from Africa's fast urbanizing centers and improvements in roads and rural infrastructure (Snyder et al. 2019). Additional pull factors driving diversification practices include opportunities in small and informal business enterprises, mining, and livestock production—these activities are often more lucrative than many traditional agricultural commodities (Ollenburger et al. 2019, p. 305; Nagler and Naudé 2017).

The CSA literature clearly recognizes that livelihood diversification is an important adaptation strategy, especially in rain-fed agricultural systems where climate change shocks are expected to have negative effects on production and farm incomes (Arslan et al. 2018; Campbell et al. 2014). Indeed, CSA activities promoted in CCAFS' climate smart villages, or climate 'vulnerability hotspots' in East and West Africa comprise broad-based agricultural diversification strategies, including crop and livestock production, agroforestry, and market access (Radeny et al. 2018; Partey et al. 2018; Ouédraogo et al. 2018). CSA's efforts to offer a wide range of options indicate an appreciation of, and build upon, smallholder farmers' own diversification practices. The focus on investing in crops and practices that matter to such farmers, as Schnurr (2019) explains, also illustrate a compelling redress to single-event products and technologies that are still promoted in some agricultural development interventions.

With growing evidence on the potential of diversified CSA agricultural practices to positively impact yields, there is also significant emphasis on scaling up their uptake amongst a broader rural base (Amadu et al. 2020; Makate et al. 2019a, b; Mutenje et al. 2019). This discourse calls for various policy actions and institutional efforts to aid the process, comprising simultaneous access to credit and extension services (Makate et al. 2019b); market access and value chain integration (Makate 2019; Westermann et al. 2018); and documenting the cost-effectiveness of different CSA practices (e.g., in relation to land-use, labor and capital requirements) at the farm-level (Amadu et al. 2020; Mutenje et al. 2019). As with broader agricultural development interventions in Africa, CSA has adopted sustainable intensification is an integral cornerstone of its activities (Lipper et al. 2014, p. 1069), often citing imperatives to address low crop productivity, achieve food security and reduce poverty (Van-lauwe et al. 2014; Jayne et al. 2019).

An important challenge here is that the emphasis on scaling CSA practices to intensify productivity, albeit with provisions for on-farm diversification, may still not fully appreciate the limits to how much resource-constrained smallholder farmers can effectively sustain and profit from such activities (see Harris 2019; Nagler and Naudé 2017). Like broader agricultural development interventions in Africa, CSA tends to assume, without question, the attractiveness of market incentives and the potential livelihood gains in intensifying agricultural productivity for disparate smallholder farmers. This logic, informed by ideological dimensions of modernity and rationality, also tends to portray farmers as monolithic actors who will choose to maximize yields and increased profits for their livelihoods every time (Schnurr 2019, p. 199). However, various studies have shown that sustainable intensification and commercialization is generally not a priority for smallholders particularly when less effective at improving the welfare status of their households (Harris 2019, p. 275; Ollenburger et al. 2019; Nagler and Naudé 2017; Poole et al. 2013). This literature finds that in most contexts, the primary goal for agriculture amongst smallholder farmers is food self-sufficiency—crop diversification serves as an important risk reduction strategy, albeit at an expense for increased profits.

In southern Mali, for instance, Ollenburger et al. (2019) found that sustainable intensification was relatively ineffectiveness in changing household poverty levels and food self-sufficiency status (p. 291). The authors assessed the potential benefits of agricultural intensification from staple crops on food self-sufficiency and gross profit margins of three intensification levels, represented by typical yields, best farmer yields, and attainable yields (p. 293). Their results show that intensifying productivity for the various crops[3] grown by smallholder farmers provided relatively less benefits under existing price regimes, unless yields (from cropland expansion) or prices increased dramatically. At the same time, off-farm livelihood strategies such as gold mining, family businesses and sale of firewood and charcoal outperformed income earnings from sustainable intensification (Ollenburger et al. 2019, p. 297).

In Zambia, similarly, Poole et al. (2013) illustrate with empirical research from a cassava marketing project, the All ACP Agricultural Commodities Program (AACP) of the European Union, that increased commercialization was not an important objective for smallholder farmers. Cassava markets generally offered small financial benefits and some farmers saw better livelihood changes from exploiting new sources of income, including being integrated into urban economies (Poole et al. 2013). The authors call for a better understanding of farmers' perceptions and willingness to adopt new agricultural innovations based on both the external macro-influences that condition the policy environment for smallholder agriculture and the micro constrains which determine farmers response to intervention initiatives (p. 157). This is particularly important in contexts where diversified livelihood options are available to smallholder farmers—a focus on sustainable intensification can potentially divert their investments (e.g., capital, labor) from more attractive economic options (Harris 2019, p. 275).

Farmers' complex livelihood diversification objectives and how they may differ from donors' development goals for yields and profits poses important challenges for CSA's rhetoric of participatory selection of best-bet practices but promote sustainable intensification in practice. As such, CSA sustainable intensification activities might undermine its own efforts to build climate adaption and resilience in smallholder agricultural systems. In Zambia and Malawi, for instance, Arslan et al. (2018) found that the provision of some support services and market access for certain crops and technologies have affected incentives for diversification in opposing ways, potentially at the expense of long-term livelihood resilience (p. 557). Both countries have long-standing input support programs for maize and fertilizer subsidies, and out-grower schemes (cotton and tobacco) that offer important agronomic services (extension, credit, and market access) but are less conducive to building agricultural diversification as an effective adaptation strategy to risks. Overall, the new rural realities of expansion in livelihood options and the unpredictable ways in which households participate in them suggest that CSA interventions must be targeted to respond more closely to these circumstances rather than to pre-defined impact pathways largely set on intensifying agricultural productivity (Ollenburger et al. 2019).

A growing competition for land and productive resources

Over the last two decades, there has been a proliferation in market transfers of land across Africa. The rise in land commodification is driven by a wide diversity of factors, including the "global land rush" of the 2000s that saw large-scale land concessions for food-crop and biofuel production; tree planting schemes for carbon trading; and local agricultural enterprises growing food for domestic urban centers (Peluso and Lund 2011; Borras and Franco 2012; Cotula 2013; Yaro et al. 2017). These activities have seen a resurgence in three broad production models: plantation and large-scale estates; contract farming and out-grower schemes; and medium-scale commercial farming (Hall et al. 2017, p. 517). Although the

---

[3] These were cotton, groundnuts, maize, millet, rice and sorghum.

precise scale and nature of land deals remain opaque due to limited transparency and poor data availability, the Land Matrix initiative (https://landmatrix.org/) has some of the most comprehensive data on large-scale land acquisitions in middle-and-low-income countries. The initiative tracks data on intended, concluded and failed land deals since the year 2000 that cover an area of 200 hectares or more and involve a transfer of land use rights from smallholder farmers and communities to commercial actors (Nolte et al. 2016). The data show that Africa is the most targeted region for large-scale land acquisitions in the global south, with over 580 concluded land deals covering over 15 million hectares and an additional 103 intended deals covering nearly 10 million hectares (Land Matrix 2020).

Multiple African governments and business elites welcome these investment opportunities as part of new development cooperation in Africa that goes ' "beyond aid", offering not only technology transfer and capacity training, but joint commercial alliances (e.g. state-business) that finance and manage such enterprises (Scoones et al. 2016). There has been particularly a rapid rise in agricultural investments from emerging powers, especially China and Brazil. Amanor and Chichava (2016) explain that both countries focus on a gradual expansion of agricultural markets for their technologies (e.g., seeds, farm machinery and other inputs) and technical services to be used on African commercial farms rather than engage in direct large-scale land acquisitions. Despite the frequent rhetoric of South-South cooperation based on mutual learning and joint benefits, these activities tend to operate within the dominant frameworks of agribusiness penetration in developing countries' agriculture sectors and the integration of smallholder farmers into market value chains (Amanor and Chichava 2016, p. 21).

Overall, growing commercial investments in Africa's agriculture has put pressure on the competition for productive resources as state institutions, traditional authorities and farmers sell or lease land-use rights to the highest bidder (Yaro et al. 2017; Paul and Steinbreacher 2013; Kelly and Peluso 2015; Leonardi and Browne 2018). The ensuing processes of land-and-resource accumulation for some (including increased income opportunities from diversified production) and dispossession for others (entailing landlessness or resettlement in high-risk agro-ecological zones) usher in social differentiation, particularly along gender lines (Yaro et al. 2017, p. 548; James and Woodhouse 2017). As women's land rights are often subordinate (or tied) to those of their husbands and male relatives, they tend to be more vulnerable to losing their land-use rights, and benefit less from land market regimes, as customary systems of land tenure shift towards private property (Colin and Woodhouse, 2010, p. 8).

The CSA literature recognizes that growing pressures on land-use from new commercial investments can pose challenges for the sustainability of smallholder farmers' livelihoods as well as on equity considerations (see Vanlauwe et al. 2014; World Bank 2015; Jayne et al. 2019). For example, farmers are less likely to invest scarce economic resources (and labor) towards improving lands for which their tenure rights are insecure (Snyder et al. 2019). Further, socially marginalized groups may be forced to occupy high-risk agroecological zones where they experience more pronounced effects of climate change (e.g., droughts) (see Scoville-Simmonds et al. 2020). As such, the CSA discourse advocates for stronger land tenure rights and other pro-poor institutional and policy approaches that promote equitable access and management of productive resources (World Bank 2015; Barnard et al. 2015). In their *Gender in Climate-Smart Agriculture Sourcebook*, the World Bank, FAO and IFAD (2015) espouse gender-responsive and gender transformative approaches that not only address the different needs and realities of women and men, but challenge and foster behavioral and institutional change around underlying causes of social inequalities. Among these are socio-cultural norms that restrict women's land and tenure rights, and ability to fully participate in decision-making in accessing and managing CSA technologies and practices on and off the farm (World Bank 2015, pp. 24–26).

Efforts to tackle socio-cultural barriers surrounding people's equitable access to and control over resources are undoubtedly an important step towards generating more pro-poor forms of agricultural development. However, this strict focus on local structural constrains leaves several critical factors unexamined (see Collins 2018). Foremost, state policies, institutions and government officials often play key roles in facilitating the conditions for land grabbing and commodification through formal laws and/or coercion (see Borras and Franco 2013, p. 1728; Cotula 2013; Kelly and Peluso 2015; Dell'Angelo et al. 2017). This suggests that land tenure and governance processes are generally characterized by unequal power relations and at times conflict, which pose difficulties for meaningful participation and equitable outcomes surrounding marginalized people's access to and control over productive resources. Failure to confront some of these power relations has important implication for CSA interventions. As Karlsson et al. (2018) argues, limiting the assessment of equity to procedural questions of *who* and *what* are vulnerable is unlikely going to address the structural constraints of *why* farmers are vulnerable. Further, limited attention to power relations might also undermine the legitimacy of participatory decision-making processes that are frequently championed as essential to bringing about best-bet, locally relevant CSA solutions.

Moreover, in a broader context where CSA increasingly rely on intensifying agricultural production with market-led innovations (e.g., fertilizers, herbicides and mechanization), land titling and formalization processes might further serve to exacerbate structural inequalities. In northern Ghana, Kansanga et al. (2018) found that agricultural commercialization, entailing increased uptake of high-input technologies and a high demand for land from local elite farmers and multinational investors, is driving intra-familial land grabbing among smallholders, a term they refer to as 'intimate dispossession' (p. 215). Previously, weaker/poorer landless family members relied on sharecropping arrangements to access land for farming. However, rising land values have resulted in marginalized family members being pushed off fertile lands by more powerful farmer relatives who bring it under continuous cultivation using mechanized technologies or lease it out to paying tenants (Kansanga et al. 2018; Yaro et al. 2017).

Conclusion

Over the last decade, CSA has garnered remarkable prominence in agricultural development policy, business, and research arenas. However, CSA's salience and legitimacy has also come under scrutiny from critically minded scholars and actors who question the different interests that shape its agenda and how scientific knowledge is used to build consensus around seemingly value-neutral technologies, best practices, and appropriate investments (Clapp et al. 2018; Newell and Taylor 2018). The productive moments of these debates are especially relevant for sub-Saharan Africa, where CSA activities are being implemented in a context involving diverse stakeholders with different priorities as well as unequal power relations between them and intended beneficiaries. The broader lineage of African agricultural development is predominately focused on intensifying agricultural productivity using particular technologies and predetermined impact pathways as evident in the New Green Revolution for Africa (Ollenburger et al. 2019; Moseley et al. 2015). CSA's empiricist approach largely fits within this agenda despite a rhetoric of bringing about transformative change to low-income agricultural systems. What is novel is its relatively greater appreciation and support for farmers' own agricultural diversification strategies compared to mainstay interventions.

Without denying CSA's important contributions to raising agricultural productivity, its narrow focus on farm-level technologies and management practices can overlook some crucial socio-political and economic realities that profoundly impact farmers' livelihood outcomes. This paper outlined three political economy concerns that need to be grappled with if CSA's efforts are to respond holistically to the complex realities and challenges that Africa's resource-constrained smallholder face. One is that the network of donors and development partners that support CSA emphasize greater private sector-led agricultural initiatives that can help scale up efficiency results, i.e., in yields, at a time when farmers are seeing rising costs in agricultural inputs and low and volatile farm-gate prices. As such, their preference for sustainable intensification and reliance on a select set of productivity enhancing technologies, does not always align well with CSA common themes of offering a wide diversity of best-bet local options selected through participatory processes.

The second challenge relates to the growing importance of livelihood diversification (on and off farm activities) in meeting farmers' household needs in a context of highly unpredictable biophysical environments and economic climate under which they live. While CSA is increasingly focused on scaling up its practices to stimulate large-scale uptake among smallholder farmers, rural household have their own complex livelihood objectives that often differ from donors' development goals for yields and profits. Further, CSA sustainable intensification activities might undermine efforts to build climate adaption and resilience where support services favor particular technologies and impact pathways. The third political economy concern is a growing competition over productive resources—brought on by a proliferation in land market transfers as well as increased uptake of high-input technologies, which is exacerbating the marginalization of weaker/poorer groups (Kansanga et al. 2018). CSA engagement with equity concerns over land rights largely addresses procedural constrains around local tenure security, but neatly sidesteps the structural drivers of land dispossession, and the roles that sustainable intensification can play in this process.

Whereas CSA interventions in Africa makes attempts to bring about transformative and fundamental change in climate and agricultural systems, questions remain whether CSA actors on the ground can push for a more critical agenda that tackles some of the concerns raised here, and openly debate how authorities, knowledges and subjectivities coalesce to frame problems and responses in particular ways both at the local and global levels (Eriksen et al. 2015). Such an approach can help to make CSA interventions more equitable and potentially improve adoption rates of proven agricultural technologies and practices in smallholder agricultural contexts where their long-term uptake and effective scaling remain a formidable challenge. As Collins (2018) argues, however, the prospects for transformative CSA agendas (e.g., in gender relations, access and control over resources, justice and power relations, etc.), are likely to be lost in the shuffle as more powerful corporate actors and multilateral donors with deep pockets scale up market-led climate-smart agriculture in the near future (p. 189).

Acknowledgements I would like to thank three anonymous reviewers and the editor for very helpful feedback on an earlier version of this paper.

References

Amadu, F.O., P.E. McNamara, and D.C. Miller. 2020. Understanding the adoption of climate-smart agriculture: A farm-level typology with empirical evidence from southern Malawi. *World Development* 125: 104692.

Amanor, K.S., and S. Chichava. 2016. South–south cooperation, agribusiness, and African agricultural development: Brazil and China in Ghana and Mozambique. *World Development* 81: 13–23.

Anderson, T. 2014. Clever name, losing game? How climate smart agriculture is sowing confusion in the food movement. Action Aid International, September, 1–8.

Andersson, J.A., and S. D'Souza. 2014. From adoption claims to understanding farmers and contexts: A literature review of conservation agriculture (CA) adoption among smallholder farmers in southern Africa. *Agriculture, Ecosystems and Environment* 187: 116–132.

Andersson, J.A., and K.E. Giller. 2019. Doing development-oriented agronomy: Rethinking methods, concepts and direction. *Experimental Agriculture* 55 (2): 157–162.

Andersson, J.A., and J. Sumberg. 2017. Knowledge politics in development-oriented agronomy. In *Agronomy for development: The politics of knowledge in agricultural research*, ed. J. Sumberg, 1–13. Abingdon: Routledge.

Andrieu, N., F. Howland, I. Acosta-Alba, J.-F. Le Coq, A.M. Osorio-Garcia, D. Martinez-Baro, C. Gamba-Trimiño, A.M. Loboguerrero, and E. Chia. 2019. Co-designing climate-smart farming systems with local stakeholders: A methodological framework for achieving large-scale change. *Frontiers in Sustainable Food Systems* 3: 37.

Andrieu, N., B. Sogoba, R. Zougmore, F. Howland, O. Samake, O. Bonilla-Findji, M. Lizarazoa, A. Nowak, C. Dembele, and C. Corner-Dolloff. 2017. Prioritizing investments for climate-smart agriculture: Lessons learned from Mali. *Agricultural Systems* 154: 13–24.

Arslan, A., S. Asfaw, R. Cavatassi, L. Lipper, N. McCarthy, M. Kokwe, and G. Phiri. 2018. Diversification as part of a CSA strategy: the cases of Zambia and Malawi. In *Climate smart agriculture: Building resilience to climate change*, ed. L. Lipper, N. McCarthy, D. Zilberman, S. Asfaw, and G. Branca, 527–562. New York: Springer.

Barnard, J., H. Manyire, E. Tambi, and S. Bangali. 2015. *Barriers to scaling up/out climate smart agriculture and strategies to enhance adoption in Africa*. Accra: Forum for agricultural research in Africa.

Borras, S.M., and J.C. Franco. 2012. Global land grabbing and trajectories of agrarian change: A preliminary analysis. *Journal of Agrarian Change* 12 (1): 34–59.

Borras, S.M., and J.C. Franco. 2013. Global land grabbing and political reactions 'from below'. *Third World Quarterly* 34 (9): 1723–1747.

Bryceson, D.F. 2002. The scramble in Africa: Reorienting rural livelihoods. *World Development* 30 (5): 725–739.

Campbell, B.M., P.K.R. ThorntonZougmore, P. van Asten, and L. Lipper. 2014. Sustainable intensification: What is its role in climate smart agriculture? *Current Opinion in Environmental Sustainability* 8: 39–43.

Campbell, B.M., S.J. Vermeulen, P.K. Aggarwal, C. Corner-Dolloff, E. Girvetz, A.M. Loboguerrero, J. Ramirez-Villegas, T. Rosenstock, L. Sebastian, P.K. Thornton, and E. Wollenberg. 2016. Reducing risks to food security from climate change. *Global Food Security* 11: 34–43.

Cavanagh, C.J., A.K. Chemarum, P.O. Vedeld, and J.O. Petursson. 2017. Old wine, new bottles? Investigating the differential adoption of 'climate-smart' agricultural practices in western Kenya. *Journal of Rural Studies* 56: 114–123.

CCAFS. n.d. Climate smart agriculture. https://csa.guide/. Accessed 15 June 2019.

Chandra, A., K.E. McNamara, and P. Dargusch. 2017. The relevance of political ecology perspectives for smallholder climate-smart agriculture: A review. *Journal of Political Ecology* 24: 821–842.

Chandra, A., K.E. McNamara, and P. Dargusch. 2018. Climate smart agriculture: Perspectives and framings. *Climate Policy* 18 (4): 526–541.

CIAT. 2014. *Climate-smart agriculture investment prioritization framework*. Cali: CIAT.

Clapp, J. 2016. *Food*, 2nd ed. Cambridge: Policy Press.

Clapp, J., P. Newell, and Z.W. Brent. 2018. The global political economy of climate change, agriculture and food systems. *Journal of Peasant Studies* 45 (1): 80–88.

Colin, J.P., and P. Woodhouse. 2010. Introduction: Interpreting land markets in Africa. *Africa: Journal of the International African Institute* 80 (1): 1–13.

Collins, A. 2018. Saying all the right things? Gendered discourse in climate-smart agriculture. *Journal of Peasant Studies* 45 (1): 175–191.

Cotula, L. 2013. *The great African land grab? Agricultural investments and the global food system*. London: Zed Books.

Currie-Alder, B. 2016. The state of development studies: Origins, evolution and prospects. *Canadian Journal of Development Studies* 37 (1): 5–26.

Das, B., F. Van Deventer, A. Wessels, G. Mudenda, G. Key, and D. Ristanovic. 2019. Role and challenges of the private seed sector in developing and disseminating climate-smart crop varieties in eastern and southern Africa. In *The climate-smart agriculture papers: Investigating the business of a productive, resilient and low emission future*, ed. T.S. Rosenstock, A. Nowak, and E. Girvetz, 67–78. New York: Springer.

Dawson, N., A. Martin, and T. Sikor. 2016. Green revolution in sub-Saharan Africa: Implications of imposed innovation for the wellbeing of rural smallholders. *World Development* 78: 204–218.

Dell'Angelo, J., P. D'Odorico, M.C. Rulli, and P. Marchand. 2017. The tragedy of the grabbed commons: Coercion and dispossession in the global land rush. *World Development* 92: 1–12.

Dinesh, D., P. Aggarwal, A. Khatri-Chhetri, A.M. Loboguerrero-Rodríguez, C. Mungai, M. Radeny, L. Sebastian, and R. Zougmoré. 2017. The rise in climate-smart agriculture strategies, policies, partnerships and investments across the globe. *Agriculture for Development* 30: 4–9.

Eriksen, S.H., A.J. Nightingale, and H. Eakin. 2015. Reframing adaptation: The political nature of climate change adaptation. *Global Environmental Change* 35: 523–533.

Food and Agriculture Organization of the United Nations (FAO). 2009. How to feed the world in 2050. https://www.fao.org/filea dmin/templates/wsfs/docs/expert_paper/How_to_Feed_the_ World_in_2050.pdf. Accessed 10 June 2019.

Food and Agriculture Organization of the United Nations (FAO). 2019. Agriculture and climate change—challenges and opportunities at the global and local level—collaboration on climate-smart agriculture. https://www.fao.org/3/CA3204EN/ca3204en. pdf. Accessed 18 Oct 2019.

Faling, M. 2020. Framing agriculture and climate in Kenyan policies: A longitudinal perspective. *Environmental Science and Policy* 106: 228–239.

Faling, M., and R. Biesbroek. 2019. Cross-boundary policy entrepreneurship for climate smart agriculture in Kenya. *Policy Science* 52: 525–547.

GACSA. 2020. Global alliance for climate smart agriculture. https://www.fao.org/gacsa/members/members-list/en/ Accessed 10 March 2020.

Giller, K.E., J.A. Andersson, M. Corbeels, J. Kirkegaard, D. Mortensen, O. Erenstein, and B. Vanlauwe. 2015. Beyond conservation agriculture. *Frontiers in Plant Science* 6: 1–14.

Giller, K.E., J.A. Andersson, J. Sumberg, and J. Thompson. 2017. A golden age for agronomy? In *Agronomy for development: The politics of knowledge in agricultural research*, ed. J. Sumberg, 150–160. Abingdon: Routledge.

Grégoire, R.E., B. Campbell, and M.C. Doran. 2017. The complex terrain of rights-based approaches: From the renewal of development practices to depoliticisation. *Canadian Journal of Development Studies* 38 (2): 169–183.

Hall, R., I. Scoones, and D. Tsikata. 2017. Plantations, outgrowers and commercial farming in Africa: Agricultural commercialisation and implications for agrarian change. *Journal of Peasant Studies* 44 (3): 515–537.

Harris, D. 2019. Intensification benefit index: How much can rural households benefit from agricultural intensification? *Experimental Agriculture* 52 (2): 273–287.

Holt-Giménez, E., and M.A. Altieri. 2013. Agroecology, food sovereignty, and the new green revolution. *Agroecology and Sustainable Food Systems* 37 (1): 90–102.

ICRAF. 2020. Evidence for resilient agriculture. https://era.ccafs.cgiar.org/. Accessed 09 Jan 2020.

Ignatova, J. 2017. The 'philanthropic' gene: Biocapital and the new green revolution in Africa. *Third World Quarterly* 38 (10): 2258–2275.

IPES-Food. 2017. Too big to feed: Exploring the impacts of megamergers, consolidation and concentration of power in the agrifood Sector. International Panel of Experts on Sustainable Food systems. https://www.ipes-food.org/_img/upload/files/Concentration_FullReport.pdf. Accessed 19 May 2019.

James, P., and P. Woodhouse. 2017. Crisis and differentiation among small-scale sugar cane growers in Nkomazi, South Africa. *Journal of Southern African Studies* 43 (3): 535–549.

Jayne, T.S., S. Snapp, F. Place, and N. Sitko. 2019. Sustainable agricultural intensification in an era of rural transformation in Africa. *Global Food Security* 20: 105–113.

Kadzamira, M.A., and O.C. Ajayi. 2019. Innovative partnerships to scale up climate-smart agriculture for smallholder farmers in southern Africa. In *the climate-smart agriculture papers: Investigating the business of a productive, resilient and low emission future*, ed. T.S. Rosenstock, A. Nowak, and E. Girvetz, 289–299. New York: Springer.

Kansanga, M., P. Andersen, K. Atuoyea, and S. Mason-Renton. 2018. Contested commons: Agricultural modernization, tenure ambiguities and intra-familial land grabbing in Ghana. *Land Use Policy* 75: 215–224.

Kapoor, I. 2002. The devil's in the theory: A critical assessment of Robert Chambers' work on participatory development. *Third World Quarterly* 23 (1): 101–117.

Karlsson, L., L.O. Naess, A. Nightingale, and J. Thompson. 2018. 'Triple wins' or 'triple faults'? analyzing the equity implications of policy discourses on climate-smart agriculture (CSA). *Journal of Peasant Studies* 45 (1): 150–174.

Khatri-Chhetri, A., P.K. Aggarwal, P.K. Joshi, and S. Vyas. 2017. Farmers' prioritization of climate-smart agriculture (CSA) technologies. *Agricultural Systems* 151: 184–191.

Khatri-Chhetri, A., A. Pant, P.K. Aggarwala, V.V. Vasireddy, and A. Yadav. 2019. Stakeholders prioritization of climate-smart agriculture interventions: Evaluation of a framework. *Agricultural Systems* 174: 23–31.

Kelly, A.B., and N.L. Peluso. 2015. Frontiers of commodification: State lands and their formalization. *Society & Natural Resources* 28 (5): 473–495.

Land Matrix. 2020. Africa regional focal point. https://landmatrix.org/region/africa/. Accessed 12 Jan 2020.

Leonardi, C., and A.J. Browne. 2018. Introduction: valuing land in eastern Africa. *Critical African Studies* 10 (1): 1–13.

Lipper, L., P.K. Thornton, B.M. Campbell, T. Baedeker, A. Braimoh, et al. 2014. Climate-smart agriculture for food security. *Nature Climate Change* 4: 1068–1072.

Makate, C. 2019. Effective scaling of climate smart agriculture innovations in African smallholder agriculture: A review of approaches, policy and institutional strategy needs. *Environmental Science and Policy* 96: 37–51.

Makate, C., M.J. Makate, N. Mango, and S. Siziba. 2019a. Increasing resilience of smallholder farmers to climate change through multiple adoption of proven climate-smart agriculture innovations. Lessons from southern Africa. *Journal of Environmental Management* 231: 858–868.

Makate, C., M. Makate, M.J. Mutenje, N. Mango, and S. Siziba. 2019b. Synergistic impacts of agricultural credit and extension on adoption of climate-smart agricultural technologies in southern Africa. *Environmental Development* 32: 100458.

McMichael, P. 2013. Value-chain agriculture and debt relations: Contradictory outcomes. *Third World Quarterly* 34 (4): 671–690.

MFA. 2016. *Endring og utvikling: En fremtidsrettet jordbruksproduksjon*. In Meld. St. 11 (2016–2017) Report to the Storting (White Paper). Oslo: Norwegian Ministry of Foreign Affairs.

Moseley, W.G., M.A. Schnurr, and R. Bezner Kerr. 2015. Interrogating the technocratic (neoliberal) agenda for agricultural development and hunger alleviation in Africa. *African Geographical Review* 34 (1): 1–7.

Mutenje, M.J., C.R. Farnworth, C. Stirling, C. Thierfelder, W. Mupangwa, and I. Nyagumbo. 2019. A cost-benefit analysis of climate-smart agriculture options in southern Africa: Balancing gender and technology. *Ecological Economics* 163: 126–137.

Nagler, P., and W. Naudé. 2017. Non-farm entrepreneurship in rural sub-Saharan Africa: New empirical evidence. *Food Policy* 67: 175–191.

Newell, P., and O. Taylor. 2018. Contested landscapes: The global political economy of climate-smart agriculture. *Journal of Peasant Studies* 45 (1): 108–129.

Nolte, K., W. Chamberlain, and M. Giger. 2016. International land deals for agriculture fresh insights from the land matrix: analytical report II. https://boris.unibe.ch/85304/. Accessed 07 July 2019.

Ollenburger, M., T. Crane, K. Descheemaeker, and K.E. Giller. 2019. Are farmers searching for an African green revolution? exploring the solution space for agricultural intensification in southern Mali. *Experimental Agriculture* 52 (2): 288–310.

Ouédraogo, M., S.T. Partey, R.B. Zougmoré, A.B. Nyuor, S. Zakari, and K.B. Traoré. 2018. Uptake of climate-smart agriculture in west Africa: what can we learn from climate-smart villages of Ghana, Mali and Niger?" Infonote: CGIAR Research Program on Climate Change, Agriculture and Food Security (CCAFS). https://ccafs.cgiar.org/fr/node/56012#.XvrWuSgzZPY. Accessed 20 July 2019.

Parayil, G. 2003. Mapping technological trajectories of the green revolution and the gene revolution from modernization to globalization. *Research Policy* 32: 971–990.

Partey, S.T., R.B. Zougmore, M. Ouédraogo, and B.M. Campbell. 2018. Developing climate-smart agriculture to face climate variability in west Africa: Challenges and lessons learnt. *Journal of Cleaner Production* 187: 285–295.

Patel, R. 2013. The long green revolution. *The Journal of Peasant Studies* 40 (1): 1–63.

Paul, H., and R. Steinbrecher. 2013. African agricultural growth corridors and the new alliance for food security and nutrition, who benefits, who loses? *EcoNexus*. https://www.econexus.info/publication/african-agricultural-growth-corridors-and-new-alliance-food-security-and-nutrition-who-b. Accessed 16 June 2019.

Peluso, N.L., and C. Lund. 2011. New frontiers of land control: Introduction. *Journal of Peasant Studies* 38 (4): 667–681.

Poole, N.D., M. Chitundu, and R. Msoni. 2013. Commercialisation: A meta-approach for agricultural development among smallholder farmers in Africa? *Food Policy* 41: 155–165.

Radeny, M., M.J. Ogada, J. Recha, P. Kimeli, E.J.O. Rao, and D. Solomon. 2018. Uptake and impact of climate-smart agriculture technologies and innovations in east Africa. CCAFS Working Paper no. 251. Wageningen, Netherlands: CGIAR Research Program on Climate Change, Agriculture and Food Security.

Rosenstock, T.S., C. Lamanna, S. Chesterman, P. Bell, A. Arslan et al. 2016. The scientific basis of climate-smart agriculture: A systematic review protocol." CCAFS Working Paper no. 136. CGIAR Research Program on Climate Change, Agriculture and Food Security (CCAFS). Copenhagen, Denmark.

Taylor, M. 2018. Climate-smart agriculture: What is it good for? *Journal of Peasant Studies* 45 (1): 89–107.

Thornton, P.K., P. Aggarwal, and D. Parsons. 2017. Prioritising climate-smart agricultural interventions at different scales. *Agricultural Systems* 151: 149–152.

Thornton, P.K., A. Whitbread, T. Baedeker, J. Cairns, and L. Claessens. 2018. A framework for priority-setting in climate smart agriculture research. *Agricultural Systems* 167: 161–175.

Schnurr, M.A. 2019. *Africa's gene revolution: Genetically modified crops and the future of African agriculture*. Montreal: McGill-Queen's Press-MQUP.

Scoones, I., K. Amanor, A. Favareto, and G. Qi. 2016. A new politics of development cooperation? Chinese and Brazilian engagements in African agriculture. *World Development* 81: 1–1.

Scoville-Simonds, M., H. Jamali, and M. Hufty. 2020. The hazards of mainstreaming: Climate change adaptation politics in three dimensions. *World Development* 125: 1–10.

Schurman, R. 2017. Building an alliance for biotechnology in Africa. *Journal of Agrarian Change* 17 (3): 441–458.

Snyder, K.A., E. Sulle, D.A. Massay, A. Petro, P. Qamara, and D. Brockington. 2019. "Modern" farming and the transformation of livelihoods in rural Tanzania. *Agriculture and Human Values* 37: 33–46.

Sumberg, J., J. Thompson, and P. Woodhouse. 2012. Why agronomy in the developing world has become contentious. *Agriculture and Human Values* 30: 71–83.

Vanlauwe, B., D. Coyne, J. Gockowski, S. Hauser, J. Huising, C. Masso, G. Nziguheba, M. Schut, and P. Van Asten. 2014. Sustainable intensification and the African smallholder farmer. *Current Opinion in Environmental Sustainability* 8: 15–22.

Westengen, O.T., P. Nyanga, D. Chibamba, M. Guillen-Royo, and D. Banik. 2018. Climate for commerce: The political agronomy of conservation agriculture in Zambia. *Agriculture and Human Values* 35: 255–268.

Westermann, O., F. Wiebke, P. Thornton, J. Körner, L. Cramer, and B. Campbell. 2018. Scaling up agricultural interventions: Case studies of climate-smart agriculture. *Agricultural Systems* 165: 283–293.

Whitfield, S. 2017. More vital to our future than we realize?' Learning from Netting's thesis on smallholder farming, 25 years on. *Outlook on agriculture* 46 (4): 258–264.

World Bank. 2016. World bank group climate change action plan. https://openknowledge.worldbank.org/handle/10986/24451. Accessed 20 May 2019.

World Bank, Food and Agriculture Organization of the United Nations (FAO), and International Fund for Agricultural Development (IFAD). 2015. *Gender in climate-smart agriculture: module 18 for gender and agriculture sourcebook*. Washington, DC: World Bank Group, the Food and Agriculture Organization of the United Nations, and the International Fund for Agricultural Development.

Yaro, J.A., J.K. Teye, and G.T. Torvikey. 2017. Agricultural commercialization models, agrarian dynamics and local development in Ghana. *Journal of Peasant Studies* 44 (3): 538–554.

Publisher's Note Springer Nature remains neutral with regard to jurisdictional claims in published maps and institutional affiliations.

Helena Shilomboleni is a Postdoctoral Fellow and Scaling Specialist of the CGIAR Research Program on Climate Change, Agriculture and Food Security (CCAFS) East Africa.

 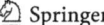

Agriculture and Human Values (2020) 37:1207–1215
https://doi.org/10.1007/s10460-020-10129-2

Sustainability transitions in the context of pandemic: an introduction to the focused issue on social innovation and systemic impact

Geoffrey Desa[1] · Xiangping Jia[2]

Accepted: 13 July 2020 / Published online: 21 July 2020
© Springer Nature B.V. 2020

Abstract

For society to achieve the Sustainable Development Goals, the agri-food industry needs a substantial *sustainability transition* toward food systems capable of delivering greater volumes of nutritious food, while simultaneously lowering the environmental footprint. This issue of AHV focuses on the big picture—on mechanisms of sustainability transition, from social innovation, to models of finance and institutional systems, and calls for business and agricultural researchers to transform the sector together. Contributors to this issue embrace a transdisciplinary outlook, including scientific, technical, social and political dimensions of agroecology. This issue is a call to action: to encourage the community of social entrepreneurs, ecosystem players and researchers to contribute analytical methods, experiences and scientific insights on emerging social innovations related to food, agriculture and rural–urban transformation.

Keywords Sustainability transition · Social innovation · Agroecology · Food · Agriculture

Introduction

The global agriculture and food industry has an enormous economic and environmental impact. Valued at $8 trillion USD (van Nieukoop 2019) and employing over 1 billion people,[1] the industry also accounts for approximately half of all land use, 70% of water use and one-quarter of global greenhouse gas emissions (IPCC 2019). For society to achieve the Sustainable Development Goals, the agri-food industry needs a substantial *sustainability transition* toward food systems capable of delivering greater volumes of more nutritious food, while simultaneously lowering the environmental footprint (FAO 2018; Hawken 2017). Despite this global imperative, the economic incentives are skewed against research and action. While large in real terms, the industry contributed less than 4% to global GDP (FAO 2019a). Public and private R&D resources for food and land use systems together account for 0.1% of global GDP (FOLU 2019, p. 171), with limited funds for open-source research (Heisey and Fugley 2018; FAO 2017). From the perspective of business researchers, the industry is a niche sector with limited potential for academic research.[2] However, this view of a small, siloed research sector with limited investment importance, misses the big picture and a big opportunity. The industry is going through one of the greatest changes since the post-war period, with changing consumer preferences, technology enabled productivity improvements, and turmoil in domestic and international markets (Djanian and Ferreira 2020). To meet these challenges, researchers across multiple disciplines in agriculture and business need to identify food systems that work—systems in which food and agriculture are not mere niche sectors with marginal economic returns, but are key sources of social innovation, of economic and environmental survival. This is not merely an agricultural imperative but a global business imperative,

✉ Xiangping Jia
   jia.xiangping@outlook.com

   Geoffrey Desa
   gdesa@sfsu.edu

1 San Francisco State University, San Francisco, CA, USA

2 Agricultural Information Institute, Chinese Academy of Agricultural Sciences, Beijing, China

---

[1] The food and agriculture industry is responsible for 26.89% of global employment World Bank (2020a), and employs over 1 billion people (Cassidy and Snyder 2019).

[2] A Web of Science literature review yields 608 articles which list "food", "agriculture" or"agri-food" in the title/abstract of the Financial Times Top 50 Business Research Journals (from 2000 to 2020). This, of approximately 59,669 articles published for the time period, indicating a sectoral focus of 1%. For comparison, the search term "innovation", listed 6307 articles, a 10% hit-rate.

and the challenge of our day. This issue of AHV focuses on the big picture—on mechanisms of sustainability transition, from social innovation, to models of finance and institutional systems, and calls for business and agricultural researchers to transform the sector together.

The need for transdisciplinary research and action is felt with ever more urgency in light of the Covid-19 pandemic. Food security and nutrition (FSN), access to a living wage, access to a clean environment, availability of healthcare—all chronic social issues, have been further stressed for the poor and the vulnerable. From a global food systems perspective, the lock-down and social distancing phases have prompted many conversations (WFP 2020; FAO 2020a; GNR 2020; Houngbo 2020) on what sustainable food systems (SFS) will look like in a pandemic constrained, and post-pandemic world. By exacerbating social challenges, and adding complexity to the global food system the pandemic has renewed the focus on agroecological transitions—the "broad and varied processes of experimentation and innovation that often start in niches and have the potential of transforming the dominant agri-food system into a more sustainable one" (Buurma et al. 2017). For global food systems to transition toward sustainability, we need systems that encourage experimentation, innovation and transfer. However, research and policy on sustainability transitions in the food system suffer from three limitations: (a) niche perspectives on innovation, (b) functionalism, and (c) resource-dependencies, all of which limit the viability and longevity of such endeavors.

When food & agriculture is viewed as a distinct, specialized sector, sustainability transition is interpreted as a sectoral and technical configuration, focused on interactions between inputs, outputs and the environments within that sector. Innovations focused on sustainability transition are treated as internal artefacts within existing regimes on food research and policy (HLPE 2020). Rather than tweaking the practices of unsustainable agricultural systems, what does it take to address the root causes of problems in an integrated way and provide holistic and long-term solutions? The FAO's primer on the transition to sustainable food and agricultural systems (FAO 2019b), calls for an explicit focus on the social and economic dimensions of food systems, with a strong focus on the rights of women, youth and indigenous peoples. Cross-disciplinary conversations are essential to integrating specialist insights on food and agriculture sectoral practices, with general insights from the fields of social innovation, and social entrepreneurship.

Even when innovation systems are studied in more detail, sustainability transitions are studied as *functionalist* approaches that dictate prescriptive (e.g. Loorbach 2010) or ontological lenses (e.g. Geels 2010), but tend to neglect normative and ethical dimensions (Schaile et al. 2017). For example, when viewed from a scientific and technical perspective, agroecology has been defined as the "application of ecological concepts and principles to farming systems, focusing on the interactions between plants, animals, humans and the environment, to foster a sustainable agricultural development that ensures FSN" (HLPE 2016). Normative dimensions, however, deeply influence the directionality, legitimacy, responsibility, and their interrelations in innovation systems. Stakeholders have conflicting visions, interests, norms, and expectations, and sustainability transition research that adopts a narrow functionalist approach ignores these norms at its own peril. In line with this caution, the FAO acknowledged that "today's more transformative visions of agroecology integrate transdisciplinary knowledge, farmers' practices and social movements while recognizing their mutual dependence" (HLPE 2019, p. 31), and called for research into broader conceptions of the term. *Resource-dependencies* further weaken the resilience of food systems (Schipanski et al. 2016; Puma et al. 2015). In practice, social finance is fragmented (Haveman and Negra, this issue). The acquisition and continued provision of financial resources (e.g. impact capital, development finance, government financing) is often siloed, with different regional and sectoral priorities, appetites for risk, temporal horizons, and approaches to impact. Without models of capital that work together for all—including women, youth and indigenous peoples, systemic transformations are weakened, and words like 'transition', 'sustainability' and 'innovation' form a coercive rhetoric that perpetuates the political and institutional interests of a powerful few (Voss and Kemp 2006).

The consequences of these niche, functionalist, and resource-dependent perspectives on sustainability transitions are severe. At the policy arena (global and local), transformation of food and agricultural systems is often advocated by scientific communities and implemented through research-led, top-down architectures, with little input from consumers and farmers (Cerf et al. 2012; Prost et al. 2017). Where innovations and sustainability have been recognized in food policy agendas, formulating effective policies remain a daunting challenge (Fukasaku 2005). Emerging agricultural entrepreneurship and alternative agri-food initiatives are dismissed as local niche events if successful, and blamed for not being 'real transition events' if the initiative fails to scale sufficiently to influence the existing regime (Beckie et al. 2012). In contrast, business research on innovation and entrepreneurship emphasizes the role of messy, human-centered action within organizations and nations (Davidsson et al. 2006; Wong et al. 2005; Mason and Brown 2014). Policies that encourage entrepreneurial ecosystems, tend to urge similarly cross-functional and cross-sectoral interactions (see Cavallo et al. 2019 for a review), as recent policies for sustainability transitions (Loorbach et al. 2017; Kivimaa and Kern 2016).

This special issue responds to the call for reflexive research on 'sustainability transitions', and offers contributions from a wide set of disciplines. By integrating normative perspectives into functional mechanisms of sustainability transition, this issue takes up Schaile et al.'s (2017) call to study innovation systems beyond the technological dimension. In line with agroecological approaches that have broadened in recent years to focus on whole agri-food systems, and not only farming systems (Thompson and Scoones 2009), this issue goes beyond separating scientific and technical dimensions of agroecology from the social and political dimensions, and embraces a transdisciplinary outlook. The main goal of the special issue is to engage a community of social entrepreneurs, ecosystem players and researchers to contribute analytical methods, experiences and scientific insights on emerging social innovations related to food, agriculture and rural–urban transformation.

The Covid-19 pandemic and implications for sustainability transition

The COVID-19 pandemic has stressed many parts of the agri-food industry—from supply chains, production and retail (WFP 2020; FAO 2020a). The pandemic has also stressed local food systems that may not have the capacity or labor to respond to growing needs (Woodhill 2020; FAO 2020b; ARC 2020). The pandemic, in other words, will have long-lasting macroeconomic effects. In the context of these systemic societal challenges, neither global agricultural systems nor local food networks may be enough to create environmentally sustainable or economically resilient food-systems. We consider the specific challenges brought on by the pandemic, and the implications for sustainability transition.

Effects on supply chains, production and retail

The main challenges in food and nutrition security are around food delivery (FAO 2020a, b, c), and availability (WFP 2020) as shelves are not restocked due to supply-chain distribution slowdowns (FOLU 2020). Subsequent challenges in food production continue as a proportion of producers (farmers and food processing workers) fall ill, and farms and processing plants are shut down. Migrant farm workers and laborers are stopped at regional borders as countries enforce border shutdowns with the hope of preventing the further spread of the pandemic (IOM 2020). These, and other travel restrictions when coupled with the closing of small and medium enterprises (SMEs), result in severe labor shortages that affect the production of food, and deplete incomes for the service sector, for day-laborers, and for broad swaths of the population at large (HLPE 2020, p. 3).

While the agri-food industry may see an initial surge in business for pre-packaged foods, shelf-stable foods, and other fast-moving consumer goods (FMCGs), distribution challenges create uncertainties in the supply-chain (HLPE 2020, p. 4). From a food and health perspective, the move toward shelf-stable, processed, packaged foods is often less nutritious than fresh-food alternatives (HLPE 2020, p. 5). In contrast, in less industrialized countries, individuals in these situations are likely to spend more time on food harvesting and preparation, likely leading to less economic productivity (Laborde et al. 2020).

Effect on macroeconomic indicators

Countries that are net food importers may not gain access to sufficient supplies, and subsequently food prices may increase (WFP 2020). For countries that are commodity/mineral/oil exporters, foreign currency reserves may be rapidly depleted due to manufacturing and productivity slowdowns, that in turn, lead to an unfavorable balance of trade (WFP 2020, p. 6). Economic challenges may extend to wealthier countries as well, as they extend social safety programs through cash transfers and subsidies (WFP 2020, p. 11). Countries that can, will extend social safety programs through cash transfers and subsidies, which can further increase debt ratios (OECD 2020a). Global financial organizations (e.g. World Bank, OECD), project a general decrease in global economic output with East Asia and the Pacific in need of an immediate stimulus (World Bank 2020b), and with global GDP growth hovering around 0% for 2020 (OECD 2020b).

Implications for the sustainable transition of food and agriculture systems

Transforming 'modern' agriculture into a system that offers FSN and a healthy natural environment is challenging. Much more so during a pandemic. Prior attempts at change have tended to reinforce systemic 'lock-ins' (FAO 2019c; Makard et al. 2012), and indicate the importance of research into 'sustainability transition'. For a long time, agricultural innovations were viewed as extension processes of scientific knowledge through a knowledge architecture of hierarchy (i.e. research-education-extension), and such an intellectual deeply shaped the policy agenda towards science, technology and R&D (Collinson 2000; OECD 2012). The success of the model was context specific to a homogenous production environment, large commercial farm units and stable economic conditions. The growing dependency on external actors in modern agriculture are leaving the subjects of public concern, given the emerging interaction with social and environmental context (Darnhofer et al. 2012). The reconfiguration of food and agriculture system distinguishes itself from controlling processes, planning, standardization, constancy and predictability, characterizing as a dynamic, adaptive, uncertain and complex system. Transition in such a system features socio-technical interactions and goes beyond sectoral approaches (Geels 2004).

The role of alternative food networks (AFNs)

The uncertainties and stressors on industrial supply-chains, cast light on alternative agri-food movements that have been an emerging part of public consciousness, and flourish locally in many parts of the world. These movements, which take on a variety of forms, such as "fair trade", "organic", "agroecology" and "food sovereignty" advocate for social justice and rights to healthy and ecologically appropriate food production and sustainability through innovative methods (Agarwal 2014; Edelman 2014; Misra 2018). These concepts often move beyond a focus on food security—access to sufficient food—to advocate access to knowledge, capacity and resources (Alkon and Mares 2012). Some of these movements explicitly oppose neoliberalism and industrial food systems (*radicalism*) while some propose deliberative reconfiguration within the system (*reformism*) (Roman-Alcalá 2017).

Limitations of AFNs

Critiques to alternative agri-food movements question the movements' immediate relevance, their capacities to scale, and internal philosophical coherence. For example, very few movements address immediate problems (such as hunger, malnutrition and degraded natural resources) while working towards structural changes needed for sustainable and democratic food systems (Martiniello 2015). In addition, the changes brought about through local practice are often small, and hard to scale toward lasting, transformative institutions or social structures (Alonso-Fradejas et al. 2015; Fairbairn 2012; Hinrichs 2003). Alternative agrifood movements (farmers markets, organic grocery stores, CSAs) are also critiqued for being inconsistent with social justice, as the practices inadvertently reproduce exclusions and run the risk of "defensive localism" (Winter 2003).

Agriculture and human values and agri-food transformation

Agriculture and Human Values has critically examined the conflicts and contradictions within contemporary agricultural food systems and the impact of policies, institutions and practices on innovative production, process and empowerment relating with the agrifood transformation. In 2012, the journal convened a special issue to reflect the

global debates on alternative food regime and the critiqued concept of 'food security' (Anderson and Bellows 2012). In 2015, the journal organized another special issue about Community Support Agriculture (CSA), an alternative food network that endorses local supply of agricultural produces and implicates contracts and trust between consumers and producers (Lagane 2015). The journal convened a special issue in 2016 to understand the relationship between discourses on food security and those on food sovereignty and food justice, with a symposium titled *"From Food Security to the Enactment of Change"* (Piatti and Dwiartama 2016). The journal has also been quick to recognize agriculture as a "financial asset class", and has documented the contemporary process of financialization in global food and agricultural markets over the past decade and the blurring of sectoral boundaries (Clapp et al. 2017).

These continuing endeavors deepen our knowledge of what it takes to transform agrifood systems. Perhaps even more starkly, these initiatives also highlight knowledge gaps that need to be bridged across stakeholders in the food system: from entrepreneurs to companies, investors and policy makers. Discourses on agriculture as an asset class (Clapp et al. 2017) remain disconnected from a unified approach to innovative social finance and the emerging asset classes of impact investment, pay-for-success and other result-oriented investment instruments (O'Donohue et al. 2010). Food sovereignty and alternative food networks are cross-cutting issues that need multidisciplinary conversations. Without these dialogues, the discourse of alternative food systems run the risk of being seen as 'anti-business' and gets confounded with political praxis, limiting the potential for transformation (Anderson and Dees 2006).

Emerging and flourishing practices of alternative food networks appeal to a wide range of thematic issues in different research fields including social entrepreneurship, impact investment, development and social finance, and socially responsible investment. In particular, the social innovations and financing mechanisms mentioned in the food system literature, are extensively explored in fields that study organizational theory, social entrepreneurship, finance, and management (Fayolle and Matlay 2010). To bridge the gap between the research fields of social entrepreneurship and development economics, there is a pressing need to promote collaboration between academic colleagues of different fields and practitioners and encourage conversations that cast a broad net. In response to this need, the Swiss Agency for Development and Cooperation (SDC) organized a symposium on "Social Entrepreneurship and Innovative Finance for Rural–Urban Transformation" in October 2018 in Beijing.

The special issue: social innovation and sustainability transition

The symposium highlighted the need to understand social innovation as a set of multi-level mechanisms that facilitate *sustainability transitions:* across sectors, across social finance, across institutional structures, and through social ventures across geographic regions. This special issue is an outgrowth of that initial conversation between the fields of food systems research, policy, and entrepreneurship and social innovation. Scholars who answered the call for papers adopted a variety of perspectives on mechanisms of sustainability transition.

Social innovation across rural–urban landscapes

Jia and Desa (this issue), draw from an extensive literature review on social entrepreneurship, to place the symposium discussion within the context of research. The symposium tapped into a broad range of contemporary issues in the field—from pluralistic definitions and ecosystems of social entrepreneurship, mechanisms of social change (in the context of rural–urban transformation) to measuring and metrics of social impacts, and the landscape of impact investment. Symposium participants viewed social entrepreneurship as a process-based approach to enduring, social and environmental change, in which the balance between social mission and market approaches depended upon the social sector, and geo-political region under consideration. The following articles elaborate on social innovation and transitions in a variety of sectors and regions.

Transitions across AFN's: from local to regional

Tezcan et al. (this issue), in a detailed comparative study of CSA's in Wales, offer us a hopeful vision toward 'workable utopias' built on social inclusion and empowerment. By exploring sources of social innovation: social economy, collective action, social movements and direct public policy, the authors study how Welsh CSA's address three main dimensions of social innovation: (a) the product—in responding to and satisfying alienated needs, (b) community empowerment—by increasing socio-political capabilities and access to resources, and (c) the process dimensions that change governance relations. In so doing, the authors explore the steps needed to transfer these local social innovations into scalable utopias, and identify the limitations that prevent AFNs from replicating, participating in policies, and decision-making at macro level.

Transitions across values embedded in social innovations

Chowdhury (this issue) studies the process of transferring social innovations from one region to another, across differences in country, and from urban to rural environments. In a longitudinal study of the transfer of eyecare from a social venture in India, to two different social ventures in Paraguay, the author reflects upon two often-mentioned dimensions that motivate social innovation: the social mission, and the economic proposition, and points to yet another mechanism, the role of the spiritual. The study emphasizes the role of values as super-ordinate to economic and social mission in facilitating an effective innovation transfer and poses a key reminder for researchers not to discount the roles of the spiritual and values-based logics when studying mechanisms of social innovation.

Transitions across social finance mechanisms

In the case of agriculture as a financial asset class, knowledge gaps surrounding financing mechanisms and subsequent transfers of social innovation limit the investments and entrepreneurial potential of the private sector. This fragmented financing landscape is especially stark, when we look at the resource gap—a "billions-to-trillions"[3] challenge, that will require financing mechanisms that facilitate resource transfer from the private sector, increase effectiveness and accountability. As a symptom and consequence of this fragmentation, alternative mechanisms of finance (e.g. impact investment and blended finance) are appearing across the landscape of development finance. In 2017, $228 billion was under private management with the intention to generate social and environmental impact alongside a financial return (GIIN 2018). More than half (about 57%) of the impact investment deals were conducted through a strategic use of development finance and philanthropic funds to mobilize private capital flows to emerging and frontier markets, termed as "blended finance" (GIIN 2018; WEF & OECD 2015). Havemann and Negra (this issue) address this core issue of blended finance and illustrate that all financing mechanisms are not equivalent, with different institutional priorities, constraints, risk tolerances, and sectoral preferences.

---

[3] While conventional development finance (from developed countries to developing countries) totaled less than USD 150 billion in 2016 (OECD, 2017a, b), studies suggest that eradicating global poverty in line with the Sustainable Development Goals (SDGs) will require additional investments of much as $2.5 trillion per year (Sachs and Schmidt-Traub 2014).

Transitions across institutional structures

Jia (this issue) studies nascent landscapes of social entrepreneurship in China, and identifies a variety of fragmented institutional and resource-based limitations that hinder the growth of effective enterprises. Drawing upon a multi-level perspective of the social innovation landscape, that includes multiple niches with small yet innovative ventures, and larger established regimes—institutional structures with access to resources but limited social innovation 'deal-flow', the author offers guidance for social finance that is appropriate for ventures at various stages of growth and development, and also offers suggestions for institutional structures that facilitate certification and further legitimation of social innovations. In so doing, the study adopts a wide lens of social innovation, and encourages business and innovation ecosystems that emphasize shared value.

Social innovation and systemic impact: a call to transdisciplinary action

55% of the world's population live in urban areas, and that proportion may increase to 68% by 2050 (United Nations 2018). The sustainable transitions of food and agriculture systems is a pressing issue, and from a macro-economic perspective, the Covid-19 pandemic increases the potential of social crisis in economically fragile countries (WFP 2020). Asia, for example, is home to 54% of the world's urban population. Declines in economic output, local incomes and/or food scarcity will prompt further global migrations that stress health care systems and environmental ecosystems (Pearce 2020; Hepburn et al. 2020).

Sustainability transition in food and agriculture need a broader lens than the mere application of innovation and entrepreneurship into food and agriculture subsystems. The process needs a) social and cultural changes that embrace multiple trajectories and pathways (FAO 2019c, and b) systemized interactions of scientific interpretations and symbolic meanings of technologies, and relating institutions (Darnhofer et al. 2012). This calls for a reflexive understandings of shifts in knowledge regime and design practices (Barbier and Elzen 2012; Voss et al. 2006).

The challenges ahead require new regulations, new behaviors, cultural change, and institutional 'hybridity' (Allaire and Wolf 2004). Transdisciplinary research is challenging—requiring the corralling of diffuse research interests and research questions, translations of discipline-specific empirical methods, field-specific research methodologies, and a variety of literatures (Brandt et al. 2013; Bunders et al. 2015; Lang et al. 2012). Transdisciplinary research may also fall into the forgotten-middle (MacCleave 2006; Baker 2006) between specialized disciplines. *Innovation* and *transition*

may be rhetorically hollow to researchers in agricultural science. Business researchers, in contrast, may view the focus on *food and agriculture* as an esoteric contextual domain with little application to theory or practice. However, such transdisciplinary conversations can also foreshadow nascent collaborations across pre-paradigmatic fields, new research streams, and re-prioritize the normative, as glimpsed over the twenty-year evolution of social entrepreneurship research (Kraus et al 2014).

Sustainability transitions that accompany migrations between rural and urban areas will not occur naturally and equally. While urbanization creates employment and entrepreneurial opportunities (Cook et al. 2001; Pingali 2007; Reardon and Barrett 2000), social returns to migration may be less than private returns with distributional consequences being less than optimum (Mazumdar 1987). Questions abound as food systems struggle to meet the needs of appetites that are urbanized and globalized. As the pandemic exacerbates patterns of rural–urban transformation, we need new opportunities for social service delivery in agriculture and rural economy. Researchers, policy makers and entrepreneurs trying to understand these enduring challenges are invited to partake in transdisciplinary conversations—on social innovation, finance, and impact within this fundamental context of agriculture, ecology and human values.

Acknowledgement The authors are thankful to research funding from National Natural Science Foundation of China (71573209; 71661147001) and China National Social Science Foundation (16ZDA021).

References

Agarwal, Bina. 2014. Food sovereignty, food security and democratic choice: Critical contradictions, difficult conciliations. *Journal of Peasant Studies* 41: 1247–1268.

Alkon, Alison Hope, and Teresa Marie Mares. 2012. Food sovereignty in US food movements: Radical visions and neoliberal constraints. *Agriculture and Human Values* 29: 347–359.

Allaire, G., and S. Wolf. 2004. Cognitive representations and institutional hybridity in agrofood systems of innovation. *Science, Technology and Human Values* 29: 431–458.

Alonso-Fradejas, Alberto, Saturnino M. Borras, Todd Holmes, Eric Holt-Giménez, and Martha Jane Robbins. 2015. Food sovereignty: Convergence and contradictions, conditions and challenges. *Third World Quarterly* 36: 431–448.

Anderson, Molly D., and Anne C. Bellows. 2012. Introduction to symposium on food sovereignty: Expanding the analysis and application. *Agriculture and Human Values* 29: 177–184.

Anderson, B.B., and J.G. Dees. 2006. Rhetoric, reality, and research: Building a solid foundation for the practice of social entrepreneurship. In *Social entrepreneurship, new models of sustainable social change*, ed. A. Nicholls. New York: Oxford University Press.

ARC. 2020. Coping with Covid19—The open food network and the new digital order(s). https://www.arc2020.eu/coping-with-covid19-open-food-network-and-new-digital-orders/. Accessed 29 May 2020.

Baker, L.A. 2006. Perils and pleasures of multidisciplinary research. *Urban Ecosystems* 9 (1): 45–47.

Barbier, M., and B. Elzen. 2012. *System innovations, knowledge regimes, and design practices towards transitions for sustainable agriculture*. Paris: Inra.

Beckie, Mary A., Emily Huddart Kennedy, and Hannah Wittman. 2012. Scaling up alternative food networks: Farmers' markets and the role of clustering in western Canada. *Agriculture and Human Values* 29: 333–345.

Brandt, P., A. Ernst, F. Gralla, C. Luederitz, D.J. Lang, J. Newig, F. Reinert, D.J. Abson, and H. Von Wehrden. 2013. A review of transdisciplinary research in sustainability science. *Ecological Economics* 92: 1–15.

Bunders, J.F., A.E. Bunders, and M.B. Zweekhorst. 2015. Challenges for transdisciplinary research. *Global sustainability*, 17–50. New York: Springer.

Buurma, Jan, Anne-Charlotte Hoes, Karel de Greef, and Volkert Beekman. 2017. Role of NGOs in system innovation towards animal friendly pork production in the Netherlands. In *AgroEcological transitions: Changes and breakthroughs in the making*, ed. Boelie Elzen, Anna Augustyn, Marc Barbier, and Barbara Van Mierlo. Gelderland: Wageningen University & Research, Applied Arable and Vegetable Research.

Cassidy, E., and A. Snyder. 2019. Map of the month: How many people work in agriculture? http://blog.resourcewatch.org/2019/05/30/map-of-the-month-how-many-people-work-in-agriculture/. Accessed 29 May 2020

Cavallo, A., A. Ghezzi, and R. Balocco. 2019. Entrepreneurial ecosystem research: Present debates and future directions. *International Entrepreneurship and Management Journal* 15 (4): 1291–1321.

Cerf, M., M.H. Jeuffroy, L. Prost, and J.M. Meynard. 2012. Participatory design of agricultural decision support tools: Taking account of the use situations. *Agriculture for Sustainable Development* 32: 899–910.

Clapp, Jennifer, S. Ryan Isakson, and Oane Visser. 2017. The complex dynamics of agriculture as a financial asset: Introduction to symposium. *Agriculture and Human Values* 34: 179–183.

Collinson, M. 2000. *A history of farming systems research*. Oxon: CABI.

Cook, M.L., T. Reardon, C. Barrett, and J. Cacho. 2001. Agroindustrialization in emerging markets: Overview and strategic context. *International Food and Agribusiness Management Review* 2: 277–288.

Darnhofer, Ika, David Gibbon, and Benoit Dedieu. 2012. Farming systems research: an approach to inquiry. In *Farming systems research into the 21st century: The new dynamic*, ed. Ika Darnhofer, David Gibbon, and Benoit Dedieu, 3–31. Dordrecht: Springer.

Davidsson, P., F. Delmar, and J. Wiklund. 2006. *Entrepreneurship and the growth of firms*. Cheltenham: Edward Elgar Publishing.

Djanian, M., and N. Ferreira. 2020. Agriculture trends disrupting the food value chain | McKinsey. https://www.mckinsey.com/industries/agriculture/our-insights/agriculture-sector-preparing-for-disruption-in-the-food-value-chain. Accessed 29 May 2020

Edelman, Marc. 2014. Food sovereignty: Forgotten genealogies and future regulatory challenges. *Journal of Peasant Studies* 41: 959–978.

Fairbairn, Madeleine. 2012. Framing transformation: The counter-hegemonic potential of food sovereignty in the US context. *Agriculture and Human Values* 29: 217–230.

FAO. 2018. Transforming food and agriculture to achieve the SDGs: 20 interconnected actions to guide decision-makers. https://www.fao.org/3/I9900EN/i9900en.pdf. Accessed 29 May 2020

FAO. 2019b. Sustainable food systems: Concept and framework. https://www.fao.org/3/ca2079en/CA2079EN.pdf. Accessed 29 May 2020

FAO. 2019a. Global trends in GDP, agriculture value added, and food-processing value added (1970–2017). https://www.fao.org/economic/ess/ess-economic/gdpagriculture/en/. Accessed 29 May 2020

FAO. 2019c. The 10 elements of agroecology. guiding the transition to sustainable food and agricultural systems. https://www.fao.org/agroecology/knowledge/10-elements/en/. Accessed 29 May 2020

FAO. 2020a. Keeping food and agricultural systems alive—Analyses and solutions in a period of crises—COVID-19 Pandemic. https://www.fao.org/2019-ncov/analysis/en/. Accessed 29 May 2020

FAO. 2020b. COVID- 19 and smallholder producers' access to markets. https://www.fao.org/3/ca8657en/CA8657EN.pdf. Accessed 29 May 2020

Fayolle, Alain, and Harry Matlay. 2010. Social entrepreneurship: A multicultural and multidimensional perspective. In *Handbook of research on social entrepreneurship*, ed. Alain Fayolle and Harry Matlay, 1–14. Cheltenham, UK: Edward Elgar.

FOLU. 2019. Growing better: Ten critical transitions to transform food and land use. Food and land use coalition. https://www.foodandlandusecoalition.org/global-report/. Accessed 29 May 2020

FOLU. 2020. A call to action for world leaders. food and land use coalition. https://www.foodandlandusecoalition.org/a-call-to-action-for-world-leaders/. Accessed 29 May 2020

Fukasaku, Yukiko. 2005. The need for environmental innovation indicators and data from a policy perspective. In *Towards environmental innovation systems*, ed. Matthias Weber and Jens Hemmelskamp. Heidelberg: Springer.

Geels, Frank W. 2004. From sectoral systems of innovation to socio-technical systems Insights about dynamics and change from sociology and institutional theory. *Research Policy* 33: 897–920.

Geels, Frank W. 2010. Ontologies, socio-technical transitions (to sustainability), and the multi-level perspective. *Research Policy* 39: 495–510.

GIIN. 2018. Annual Impact Investor Survey (8th edn). Global Impact Investing Network.

GNR. 2020. The 2020 Global Nutrition Report in the context of Covid-19. https://globalnutritionreport.org/reports/2020-global-nutrition-report/2020-global-nutrition-report-context-covid-19/. Accessed 29 May 2020

Hawken, P. 2017. *Drawdown: The most comprehensive plan ever proposed to reverse global warming*. New York: Penguin.

Heisey, P., and K. Fuglie. 2018. Agricultural research investment and policy reform in high-income countries. ERR-249. U.S. Department of Agriculture, Economic Research Service; World Bank data, Global GDP.

Hepburn, C., B. O'Callaghan, N. Stern, J. Stiglitz, and D. Zenghelis. 2020. Will COVID-19 fiscal recovery packages accelerate or retard progress on climate change? Oxford Review of Economic Policy, 36.

Hinrichs, C.Clare. 2003. The practice and politics of food system localization. *Journal of Rural Studies* 19: 33–45.

HLPE. 2016. Sustainable agricultural development for food security and nutrition: what roles for livestock? A report by the high level panel of experts on food security and nutrition of the committee on world food security. Rome. https://www.fao.org/3/a-i5795e.pdf. Accessed 29 May 2020

HLPE. 2019. Agroecological and other innovative approaches for sustainable agriculture and food systems that enhance food security and nutrition. https://www.fao.org/3/ca5602en/ca5602en.pdf. Accessed 29 May 2020

HLPE. 2020. Impact of COVID-19 on food security and nutrition (FSN). https://www.fao.org/fileadmin/templates/cfs/Docs1920/Chair/HLPE_English.pdf. Accessed 29 May 2020

Houngbo. 2020. What's needed to protect food security in Africa during COVID-19. https://www.ifad.org/en/web/latest/blog/asset/41945191. Accessed 29 May 2020

IOM. 2020. Covid-19: Policies and impact on seasonal agricultural workers, UN International Office of Migration, Issue Brief. https://www.iom.int/sites/default/files/documents/seasonal_agricultural_workers_27052020_0.pdf. Accessed 29 May 2020

IPCC. 2019. Climate change and land. An IPCC Special Report on climate change, desertification, land degradation, sustainable land management, food security, and greenhouse gas fluxes in terrestrial ecosystems. https://www.ipcc.ch/site/assets/uploads/2019/08/4.-SPM_Approved_Microsite_FINAL.pdf. Accessed 29 May 2020

Kivimaa, P., and F. Kern. 2016. Creative destruction or mere niche support? Innovation policy mixes for sustainability transitions. *Research Policy* 45 (1): 205–217.

Kraus, S., M. Filser, M. O'Dwyer, and E. Shaw. 2014. Social entrepreneurship: An exploratory citation analysis. *Review of Managerial Science* 8 (2): 275–292.

Laborde, D., W. Martin, and R. Vos. 2020. Poverty and food insecurity could grow dramatically as COVID-19 spreads | IFPRI : International Food Policy Research Institute. https://www.ifpri.org/blog/poverty-and-food-insecurity-could-grow-dramatically-covid-19-spreads. Accessed 29 May 2020

Lagane, J. 2015. Introduction to the symposium: Towards cross-cultural views on Community Supported Agriculture. *Agriculture and Human Values* 32: 119–120.

Lang, D.J., A. Wiek, M. Bergmann, M. Stauffacher, P. Martens, P. Moll, M. Swilling, and C.J. Thomas. 2012. Transdisciplinary research in sustainability science: Practice, principles, and challenges. *Sustainability Science* 7 (1): 25–43.

Loorbach, Derk. 2010. Transition management for sustainable development: A prescriptive, complexity-based governance framework. Governance: An international journal of policy. *Administration, and Institutions* 23: 161–183.

Loorbach, D., N. Frantzeskaki, and F. Avelino. 2017. Sustainability transitions research: Transforming science and practice for societal change. *Annual Review of Environment and Resources* 42: 599–626.

MacCleave, A. 2006. Incommensurability in cross-disciplinary research: A call for cultural negotiation. *International Journal of Qualitative Methods* 5 (2): 40–54.

Markard, J., R. Raven, and B. Truffer. 2012. Sustainability transitions: An emerging field of research and its prospects. *Research policy* 41 (6): 955–967.

Martiniello, Giuliano. 2015. Food sovereignty as praxis: Rethinking the food question in Uganda. *Third World Quarterly* 36: 508–525.

Mason, C., and R. Brown. 2014. Entrepreneurial ecosystems and growth-oriented entrepreneurship. *Final Report to OECD, Paris* 30 (1): 77–102.

Mazumdar, Dipak. 1987. Rural-urban migration in developing countries. In *Handbook of regional and urban economics*, ed. E.S. Mills. New York: Elsevier.

Misra, Manoj. 2018. Moving away from technocratic framing: Agroecology and food sovereignty as possible alternatives to alleviate rural malnutrition in Bangladesh. *Agriculture and Human Values* 35: 473–487.

O'Donohue, N., C. Leijonhufvud, Y. Saltuk, A. Bugg-Levine, and M. Brandenburg. 2010. Impact investments: An emerging asset class. https://assets.rockefellerfoundation.org/app/uploads/20101129131310/Impact-Investments-An-Emerging-Asset-Class.pdf.

OECD. 2012. *Improving agricultural knowledge and innovation systems: OECD conference proceedings*. Paris: OECD Publishing.

OECD. 2017a. Development aid rises again in 2016 but flows to poorest countries dip. https://www.oecd.org/dac/development-aid-rises-again-in-2016-but-flows-to-poorest-countries-dip.htm. Accessed 11 Oct 2018

OECD. 2017b. Global private philanthropy for development. https://www.oecd.org/dac/financing-sustainable-development/development-finance-standards/Philanthropy-Development-Survey.pdf. Accessed 13 Nov 2018

OECD. 2020. OECD economic outlook, interim report march 2020. *OECD Publishing, Paris,*. https://doi.org/10.1787/7969896b-en.

Pearce, F. 2020. After the Coronavirus, two sharply divergent paths on climate. Yale E360. https://e360.yale.edu/features/after-the-coronavirus-two-sharply-divergent-paths-on-climate. Accessed 29 May 2020

Piatti, Cinzia, and Angga Dwiartama. 2016. From food security to the enactment of change: Introduction to the symposium. *Agriculture and Human Values* 33: 135–139.

Pingali, Prabhu. 2007. Westernization of Asian diets and the transformation of food systems: Implications for research and policy. *Food Policy* 32 (33): 281–298.

Prost, Lorène, Elsa T.A. Berthet, Marianne Cerf, Marie-Hélène Jeuffroy, Julie Labatut, and Jean-Marc Meynard. 2017. Innovative design for agriculture in the move towards sustainability: Scientific challenges. *Research in Engineering Design* 28: 119–129.

Puma, M.J., S. Bose, S.Y. Chon, and B.I. Cook. 2015. Assessing the evolving fragility of the global food system. *Environmental Research Letters* 10 (2): 024007.

Reardon, Thomas, and Christopher B. Barrett. 2000. Agroindustrialization, globalization, and international development: An overview of issues, patterns, and determinants. *Agricultural Economics* 23: 195–205.

Roman-Alcalá, Antonio. 2017. Looking to food sovereignty movements for postgrowth theory. *Theory & Politics in Organization* 17: 119–145.

Sachs, J and G. Schmidt-Traub. 2014. Financing sustainable development: Implementing the SDGs through effective investment strategies and partnerships: UNCTAD: World Investment Report 2014. Investing in the SDGs: An Action Plan

Schipanski, M.E., G.K. MacDonald, S. Rosenzweig, M.J. Chappell, E.M. Bennett, R.B. Kerr, et al. 2016. Realizing resilient food systems. *BioScience* 66 (7): 600–610.

Schlaile, Michael P., Sophie Urmetzer, Vincent Blok, Allan Dahl Andersen, Job Timmermans, Matthias Mueller, Jan Fagerberg, and Andreas Pyka. 2017. Innovation systems for transformations towards sustainability? Taking the normative dimension seriously. *Sustainability* 9: 1–20.

Thompson, J., Scoones, I. (2009). Addressing the dynamics of agri-food systems: an emerging agenda for social science research. *EnvironmentalScience and Policy*, 12(4): 386–397.

United Nations. 2018. 68% of the world population projected to live in urban areas by 2050, says UN. https://www.un.org/development/desa/en/news/population/2018-revision-of-world-urbanization-prospects.html. Accessed 18 July 2018

van Nieukoop, M. 2019. Do the costs of the global food system outweigh its monetary value? https://blogs.worldbank.org/voices/do-costs-global-food-system-outweigh-its-monetary-value. Accessed 29 May 2020

Voss, Jan-Peter, Dierk Bauknecht, and René Kemp. 2006. *Reflexive governance for sustainable development*. Cheltenham: Edward Elgar.

Voss, Jan-Peter, and René Kemp. 2006. Sustainability and reflexive governance: Introduction. In *Reflexive governance for sustainable development*, ed. Jan-Peter Voß, Dierk Bauknecht, and René Kemp. Cheltenham: Edward Elgar.

WEF & OECD. 2015. Blended finance vol. 1: A primer for development finance and philanthropic funders. Paris: World Economic Forum & OECD

WFP. 2020. COVID-19—Potential impact on the world's poorest people. https://www.wfp.org/publications/covid-19-potential-impact-worlds-poorest-people. Accessed 29 May 2020

Winter, Michael. 2003. Embeddedness, the new food economy and defensive localism. *Journal of Rural Studies* 19: 23–32.

Wong, P.K., Y.P. Ho, and E. Autio. 2005. Entrepreneurship, innovation and economic growth: Evidence from GEM data. *Small business economics* 24 (3): 335–350.

Woodhill. 2020. Responding to the impact of COVID-19 on rural people and food systems. https://www.foresight4food.net/wp-content/uploads/2020/05/Impact-of-COVID-19-on-Rural-Poverty-and-Food-Systems-V2.pdf. Accessed 29 May 2020

World Bank. 2020a. Employment in agriculture (% of total employment) (modeled ILO estimate). International Labour Organization, ILOSTAT database. https://data.worldbank.org/indicator/SL.AGR.EMPL.ZS. Accessed 29 May 2020

World Bank. 2020b. World Bank East Asia and Pacific Economic Update, April 2020: East Asia and Pacific in the Time of COVID-19. Washington, DC: World Bank. https://openknowledge.worldbank.org/handle/10986/33477. Accessed May 29 2020.

Publisher's Note Springer Nature remains neutral with regard to jurisdictional claims in published maps and institutional affiliations.

Geoffrey Desa is a professor at San Francisco State University. He received his Ph.D. from the University of Washington, an M.S from Stanford University, and a B.S from Georgia Tech. He follows two questions: How do social innovations scale and have social impact? Under what conditions does academic learning drive social change? His research adds empirical evidence to the literature on social innovation at the international and local level, and to the pedagogical literature that emphasizes critical thought and learning. He teaches classes at the intersection of business, innovation for sustainability and climate adaptation.

Xiangping Jia is a professor at the Agricultural Information Institute, Chinese Academy of Agricultural Sciences. He received his PhD from University of Hohenheim, Germany. Working on research about agricultural policies and rural development, he has broad interest in the application of new institutional economics to the issues of agri-food market coordination, development of inclusive financial markets, the organization of smallholder farms, knowledge transfer and innovation, and agricultural extension system. At present, he researches innovative institutions that align agricultural universities with stakeholders such as public extension systems and private sectors for a resilient farm system.

Agriculture and Human Values (2020) 37:1217–1239
https://doi.org/10.1007/s10460-020-10133-6

SYMPOSIUM/SPECIAL ISSUE

Social entrepreneurship and impact investment in rural–urban transformation: An orientation to systemic social innovation and symposium findings

Xiangping Jia[1] · Geoffrey Desa[2]

Accepted: 13 July 2020 / Published online: 27 July 2020
© Springer Nature B.V. 2020

Abstract

Migrations from rural to urban areas do not occur equitably. Food, economic, and health systems are strained by this global rural–urban transformation. Climate change exacerbates agricultural shifts and biodiversity loss. The fields of social entrepreneurship and social innovation address these systemic inequities by re-envisioning challenges as opportunities for positive change. Innovative finance models (e.g. blended-financing, public–private partnerships) emerge in support of such initiatives. Despite this transformative potential, social innovators face significant challenges when mobilizing resources, and when moving beyond niche endeavors to scale impacts that facilitate systemic change. This article engages in a sensemaking exercise: we review literature on social entrepreneurship and innovative finance, and report outcomes from a participatory symposium in China with a variety of ecosystem stakeholders. The results presented in this paper help clarify the space and offer next steps for theorists, social entrepreneurs, non-governmental organizations, development agencies, policymakers and investors.

Keywords Social innovation · Agriculture · Ecosystems · Social finance · Equitable transformation

Introduction

Global migration from rural to urban areas is accompanied by many social issues. 55% of the world's population live in urban areas, and the proportion is expected to increase to 68% by 2050 (United Nations 2018). Despite the relatively lower level of urbanization on average and great heterogeneity within, Asia is home to 54% of the world's urban population. However, the transformation process from an agricultural and rural based existence to agglomerated one in urban areas does not occur naturally and equally. Social returns to migration may be less than private returns and the distributional consequences of urbanization is less than optimum (Mazumdar 1987). Besides exacerbating scarcity of resources and pollution (such as fertile soil and water in

agriculture), the process of urbanization is far from inclusive when a sizable population of migrants 'drift' between rural and urban systems in countries such as China (World Bank 2014).

The above challenges appear well positioned for the field of social entrepreneurship, an emerging phenomena focused on dealing with enduring social challenges through innovation. About 65% of the world population live with less than $2.5 a day and the market potentials for the bottom of the pyramid (BOP) is largely untapped (Prahalad 2005; Prahalad and Hammond 2002). Rural–urban transformation creates new opportunities for social service delivery in agriculture and rural economy. For example, a more direct and immediate effect of urbanization is on the food system; as the appetite is urbanized and globalized, the transformation of agri-food chain is accelerated to meet diversified consumers and the value-adding creates additional employment and entrepreneurial opportunities (Cook et al. 2001; Pingali 2007; Reardon and Barrett 2000). In addition, the agglomeration in urban settlement causes great learning externalities, and this fosters great market efficiency as individuals have a wide range of options to acquire skills and to engage in employment or entrepreneurial businesses (Duranton and Puga 2004).

✉ Geoffrey Desa
gdesa@sfsu.edu

Xiangping Jia
jiaxiangping@caas.cn

[1] Agricultural Information Institute, Chinese Academy of Agricultural Sciences, Beijing, China

[2] San Francisco State University, San Francisco, USA

To unleash social entrepreneurship and facilitate scaling impact and social change, innovative finance is rising up and the relative importance of conventional development finance is declining. To eradicate global poverty and reach Sustainable Development Goals (SDGs), resources as much as $2.5 trillion per annum are additionally required (Sachs and Schmidt-Traub 2014). Nevertheless, official development assistance (ODA) and private philanthropy, widely considered an important measure of external development finance from developed countries, totaled less than USD 150 billion in 2016; the majority of financing for development in developing countries is mobilized domestically through government expenditure (OECD 2017a, 2017b).[1] To meet this "billions-to-trillions" challenge, innovative investments are developed and facilitated to leverage resources in sustainable development and to increase the effectiveness and accountability. For example, it was estimated that a total of USD 228 billion assets was under management with the intention to generate social and environmental impact alongside a financial return by 2017 (GIIN 2018). More than half (about 57%) of the impact investment deals were conducted through a strategic use of development finance and philanthropic funds to mobilize private capital flows to emerging and frontier markets, termed as "blended finance" (GIIN 2018; WEF and OECD 2015). Innovative finance vehicles (such as impact investment and blended finance) are transforming the landscape of development finance towards result-based and effective partnerships.

Yet, while social entrepreneurship and its accompanying financial tools have the potential to address social and environmental problems within rural–urban transformation, many obstacles remain. First, the field of social entrepreneurship has antecedents in different disciplines—from management, non-profits, development, sociology and policy among others (Short et al. 2009). These fragmented origins have led to academic pluralism amidst critical debates and are reflected in limited consensus on the scope of research, action, and knowledge development (Dacin et al. 2011; Nicholls 2010; Weerawardena and Mort 2006). Second, the ability of the field to drive social change can be limited by deep challenges of resource constraint. Mobilizing financing for ventures that drive social change also has longer horizons and lower rates for return on investment. Third, the differences between top-down and bottom-up measures of impact have led to investment bottlenecks in many sectors that need capital. Great progress has been made over the past ten years

in standardizing a set of impact investment metrics, yet few investors use these yardsticks, and fewer entrepreneurs still have the resources, awareness, and motivation to measure them well. Finally, while entrepreneurship is seen to be most effective in thriving ecosystems, social entrepreneurship in rural–urban transformation needs to thrive in environments that are specifically less conducive to ecosystem development.

Recognizing the imperatives of filling this knowledge gap, the Swiss Agency for Development and Cooperation (SDC) organized a symposium titled as "Social Entrepreneurship and Innovative Finance for Rural–Urban Transformation" on October 17th and 18th in 2018 at Beijing. The symposium gathered around 100 participants who were pre-identified and structured into an ecosystem with four identities: social entrepreneurs that engage in innovative approaches to local social problems occurred in rural–urban transformation, investors who have interest in investing social ventures and scaling up social impacts, intermediaries that network and facilitate the process, and blue-printers that orient and advocate for the ecosystem.

The SDC symposium aimed to facilitate the creation and communication of knowledge relating with social entrepreneurship and the emerging innovations in social finance. During the symposium, participatory development communication (PDC) was conducted to set the context, elicit information, determine consensus and plan action regarding a wide range of issues about social entrepreneurship (such as definition, sectoral context, enabling technologies, theories of social changes, scaling strategies, the roles of ecosystem, etc.) and innovative finance (including monitoring and impact assessment, enabling policies, etc.).

The article proceeds as follows. We introduce the methodology of PDC and its operation at the SDC symposium. Next, we elaborate upon four aspects of social entrepreneurship in the context of rural–urban transformation: field definition, resource mobilization, impact investment, and ecosystem support. For each aspect, we discuss the results of polling questions and participant discussion at the SDC symposium. Based on these discussions, we offer a set of research questions for an ecosystem centered on social innovation for rural–urban transformation.

Participatory development communication: Organization of the conference and participants

Participatory communication and integrated participation are cornerstones of the development process (Nohlen and Nuscheler 1974). This approach has been used by a variety of international development organizations for governance and agenda setting (Eberlei 2001; Schneider 1999). Early

---

[1] In addition to ODA and private philanthropy, personal remittances and private capital flows, including FDI, are also important external finance for developing countries. For example, the remittance flows to developing countries in 2016 amounted to three times the total ODA that year, or USD 429 billion; foreign direct investment (FDI) was roughly the same size (World Bank Group/KNOMAD 2017).

Fig. 1 Workflow of participatory development communication

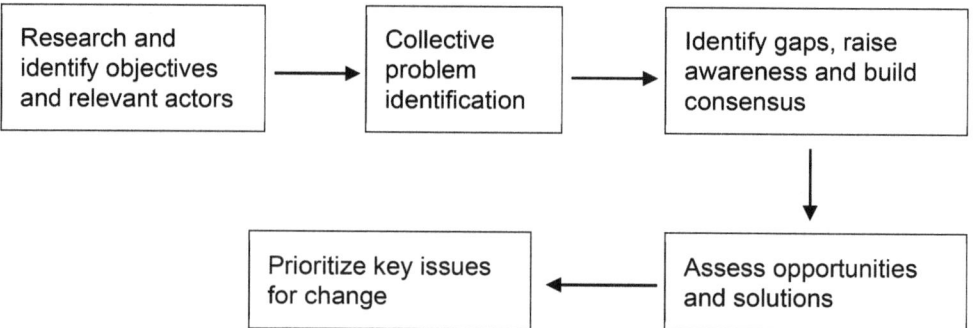

models of communication theory viewed communication as a transfer or dissemination of information and originated from the advertising industry. Changes in individual behaviors through communication are also associated with imposing pre-established business routines or through social marketing (Rogers 1962). In this linear model, the root of the problem is assumed to be a lack of knowledge and information and external agents drive the processes through linear monologue and top-down communication. In contrast to the linear, top-down pedagogy, participatory development communication (PDC) advocates liberating approaches and emphasizes a collective identification of the local problem, decision-making and behavior change (Servaes 1996, 2002). In the participatory model, the root of the problem is considered to lie in the inadequacy of mechanisms for integration, coordination and adaption to context and the complexity, rather than in the lack of knowledge. The role of the development agency is to facilitate and empower the communication process through bilateral horizontal dialog. PDC constructs a shared vision and commitment to action in a community towards transformational leadership in social change (Huesca and Dervin 1994; Jacobson 2003; Thapalia 1996).

Given the fuzzy definition and multiplicity of stakeholders in social entrepreneurship at the global and regional level, PDC was adapted and applied at the SDC symposium. The knowledge workshop followed a flow of action-reflection-action, as illustrated in Fig. 1. As a crucial ingredient of participatory communication, the empowerment process was based on not only reflections of problems and the articulation of awareness but also on integration of action—the attempt to act collectively on the problem identified. The emphasis on 'the ecosystem' speaks to the collective nature of the process and reinforced commitment to change.

To make the communication cycle genuinely participatory, two-way interactions were adopted and applied throughout the implementation of communication activities. Prior to the symposium, a research team interviewed a group of experts in the relating fields to study the problems, needed changes and opportunities. The sample of experts consulted, represented an ecosystem approach and included social entrepreneurs, investors and networkers. Additional selection criteria included work experience in the field for at least 5 years with both local and global perspectives.

On the basis of the survey and feedback, a total of 110 participants relating with social innovation and entrepreneurship, social finance, agricultural and rural development were identified and invited; 91 invitees showed up and participated the knowledge symposium.

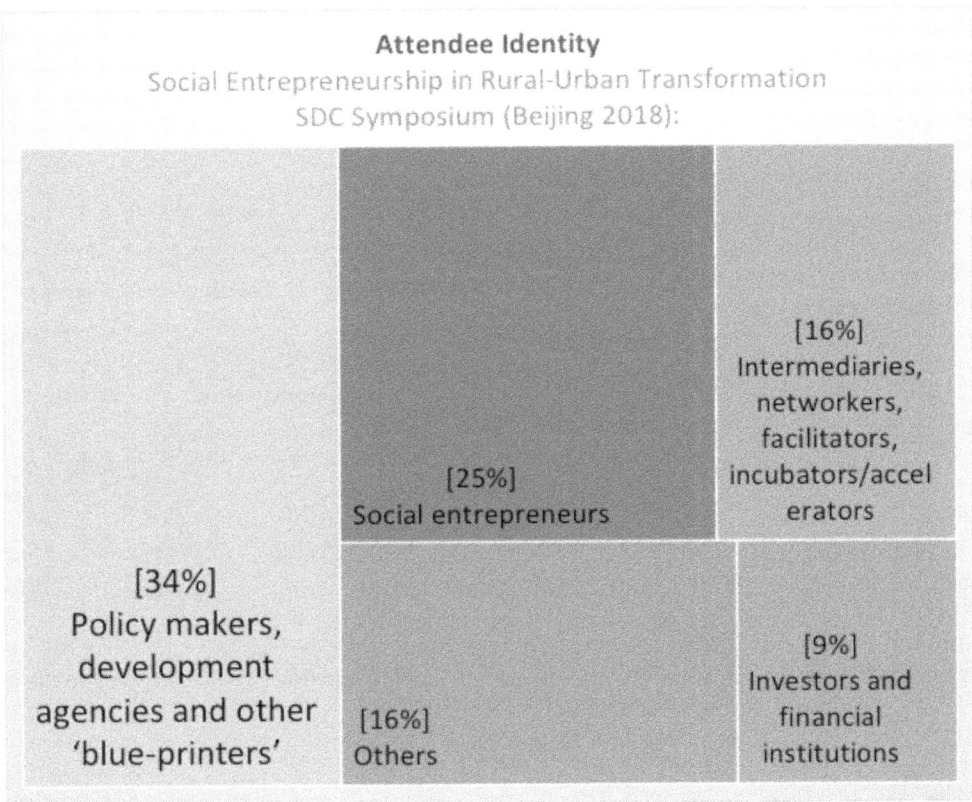

Attendee Identity
Social Entrepreneurship in Rural-Urban Transformation
SDC Symposium (Beijing 2018):

P1/P2/P3: Sample description

Participants were from the entrepreneurial ecosystem and specific development/industry sectors. One quarter of the participants were social entrepreneurs and 9% of the participants represented investors and financial institutions. Networkers and intermediaries accounted for 16%. The participants were diversified in sectoral background, such as environment and biodiversity (17%), food and agriculture (12%), climate change (13%), finance and business services (13%), education and research (12%), health and healthcare (4%) and energy (4%). Participants also represented other industries and sectors such as communication, public administration, development and social work. While conventional donation and grants remained the primary financial tools, mainstream entrepreneurial finance (10%), impact investment (12%) and blended finance (10%) as sources of financing, were also listed.

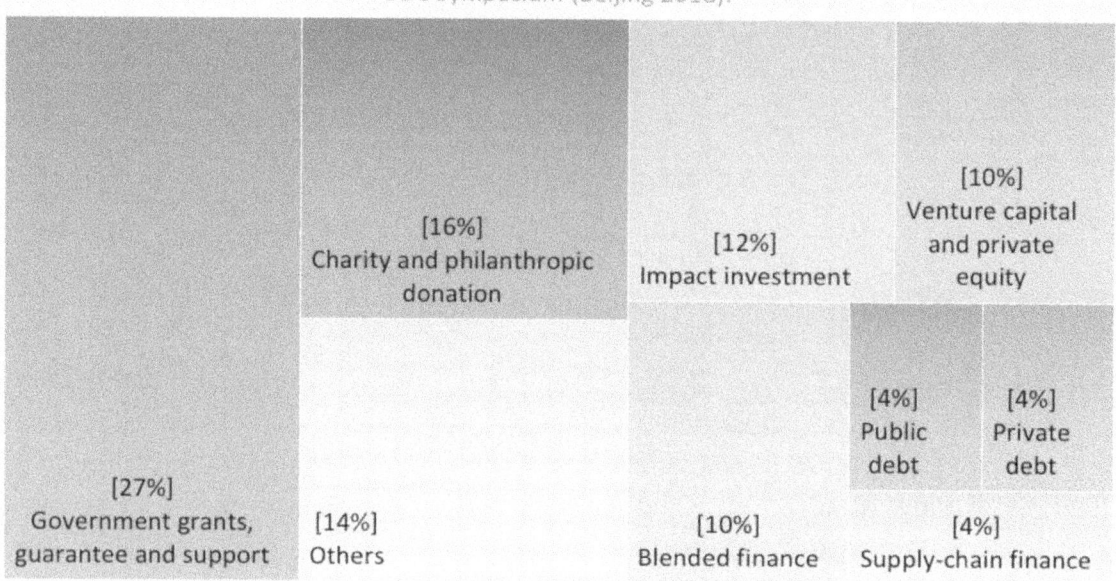

The opening session established a shared conversational space, through monologic and dialogic communication. It started with an icebreaker and live interviews with four pre-identified stakeholders who represented different identities within the ecosystem; each of them shared their stories and perceptions. Following this, a keynote and framing presentation was delivered to facilitate the unpacking of basic concepts related to social entrepreneurship, finance and the ecosystem. The goal in creating the common space was not to build consensus but to allow the audience to engage in a mutual understanding and be involved in more in-depth discussions. The combined approach acknowledged the value of monologic and dialogic communication and derived its strength from a selective and purposeful use of symposium objectives.

To facilitate the participatory communication process, polling clickers were used for the remainder of the symposium. The primary goal of the symposium was not to persuade the audiences to adopt a predefined concept but rather to engage stakeholders to explore the complexity and define the needed change by themselves. To ground the communication on real life problems and facilitate a better discussion, a number of questions and statements were posted and participants were asked to select from multiple options—some opposing each other. The polling results were immediately shown to the audience. A group of panel discussants then elaborated their thoughts on the basis of their experience and also the polling results.

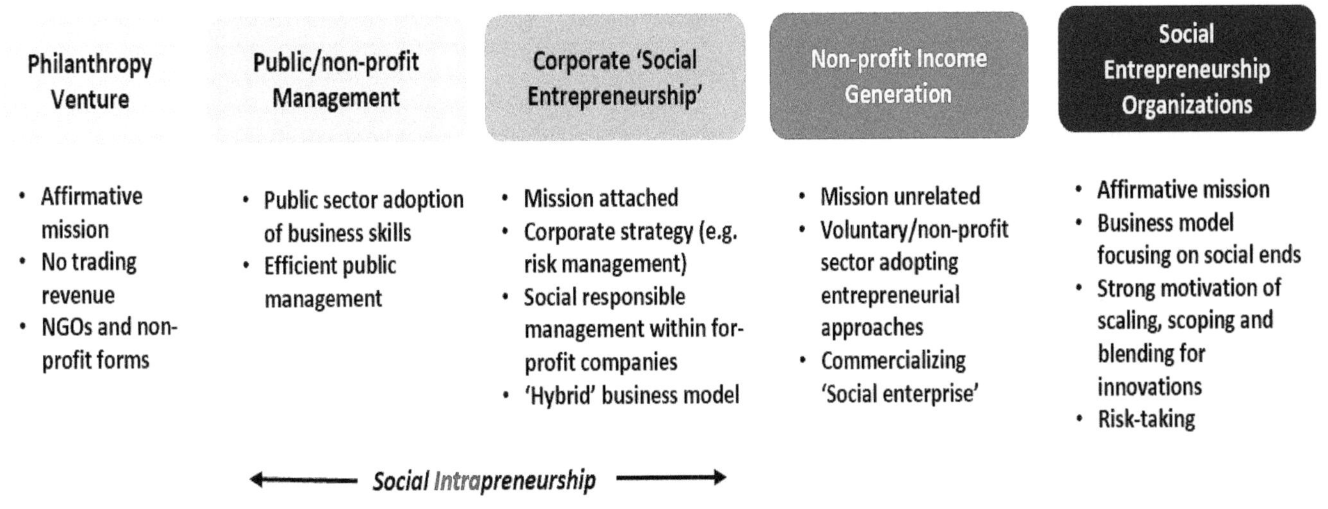

Philanthropy Venture	Public/non-profit Management	Corporate 'Social Entrepreneurship'	Non-profit Income Generation	Social Entrepreneurship Organizations
• Affirmative mission • No trading revenue • NGOs and non-profit forms	• Public sector adoption of business skills • Efficient public management	• Mission attached • Corporate strategy (e.g. risk management) • Social responsible management within for-profit companies • 'Hybrid' business model	• Mission unrelated • Voluntary/non-profit sector adopting entrepreneurial approaches • Commercializing 'Social enterprise'	• Affirmative mission • Business model focusing on social ends • Strong motivation of scaling, scoping and blending for innovations • Risk-taking

◄──────── *Social Intrapreneurship* ────────►

Fig. 2 The pluralistic concept of social entrepreneurship (Sommerrock 2010)

Table 1 Identity and sectoral background of the participants: Polling results (%)

	%
P1. What is your identity in the social innovation ecosystem?	%
A) Social entrepreneurs	25
B) Investors and financial institutions	9
C) Intermediaries, networkers, facilitators, incubators/accelerators	16
D) Policy makers, development agencies and other 'blue-printers'	34
E) Others	16
P2. In which sector do you work (or have a special sectoral interest)?[a]	%
A) Food and agriculture	12
B) Natural resource, environment and biodiversity	17
C) Energy	4
D) Health and healthcare	4
E) Education and research	12
F) Climate change	13
G) Finance and business service	13
H) Others	25
P3. What are the primary means of finance and investment related to your work?[a]	%
A) Charity and philanthropic donation	16
B) Government grants, guarantee and support	27
C) Public debt	4
D) Private debt	4
E) Venture capital and private equity	10
F) Impact investment	12
G) Blended finance	10
H) Supply-chain finance	4
I) Others	14

[a]Two answers are allowed

Table 2 Pluralistic concepts of social entrepreneurship: Participant perceptions (%)

	%
P4. Among the various definitions of SE, with which one do you identify?	%
A) A sub-group of the third sector	0
B) Being an individual or institution, SE is a change maker and innovative leader who alerts public perceptions about (specific) social problems	28
C) Every firm dealing with a "social issue" is a social enterprise	9
D) A process and social action that deals with complex and enduring social problems through innovative, effective and sustainable approaches	44
E) Others and non-response	19
P5. With which of the following do you agree?	%
A) All enterprises are dealing with social problem(s) and are potentially SE	28
B) Only those who first define and address social or environmental problems with rigor	53
C) Others and non-response	19
P6. The differences in the following expressions about SE are seemingly nuanced but (in fact) exemplify distinct mindsets about finance and social value creation. Which one you stand with?	%
A) Do well (first) and do good	25
B) Do good (first) and do well	25
C) Only focus on doing social good	13
D) Only focus on doing well financially	6
E) Others and non-response	31

Table 3 Participant perceptions of innovation as related to social entrepreneurship (%)

	%
P7. SE is said to deal with complex and enduring social problems through innovations. With which of the following phrases do you associate "innovation"?	%
A) New technologies and tools	6
B) New market and consumption	13
C) New resources (including management capacity, finance, etc.)	9
D) New business model	38
E) Others and non-response	34
P8. In your opinion, which among the following technologies, most promote/affect SE in rural–urban transformation? (Two answers allowed)	%
A) Biotechnology	10
B) Chemistry, energy and resources	0
C) Health, nutrition and medical	10
D) Fintech	8
E) Analysis/use of big data	28
F) Automated data collection & analysis, such as IOT	12
G) Machine learning/artificial intelligence	4
H) Augmented reality (AR)/virtual reality (VR)	0
I) Blockchain	12
J) Others	16
P9. In your opinion, which of the following SE fields/sectors have great potential in the context of rural–urban transformation? (Two answers allowed)	%
A) Food system (Farming, Agriculture, Food security and Hunger Alleviation)	19
B) Means of living (Health, Education, Housing, Water, etc.)	22
C) Economic Opportunities and the Leverage Tools (Finance, Small Business Service, Fair-trade products, ICT, etc.)	22
D) Sustainability (Climate Change, Environment, Energy, Natural Resources and Biodiversity Protection)	19
E) Inclusion and Humanity (Human Rights, Migration, Poverty Alleviation, Advocacy)	8
F) None of the above	11

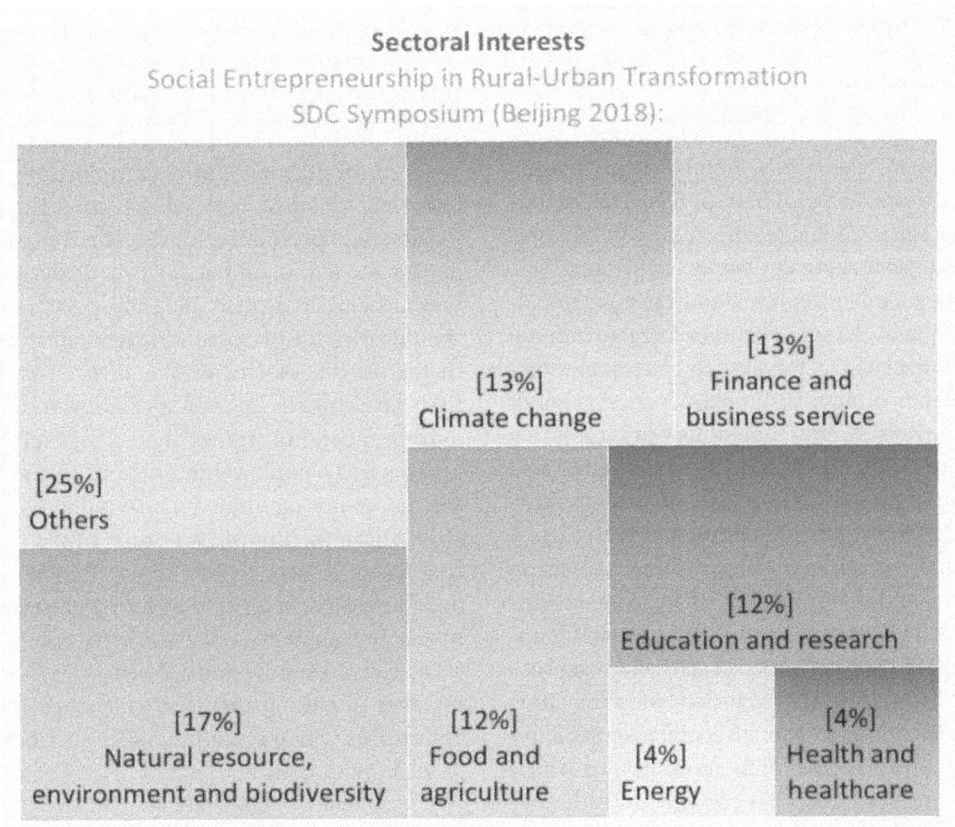

In the sections that follow, the survey results are integrated into a broader discussion of social entrepreneurship in the context of rural–urban transformation. This allows for an emergent set of research topics that are participant-centered, topically relevant and timely.

Social entrepreneurship in rural–urban transformation: From academic pluralism to pragmatic convergence

Social entrepreneurship, with its focus on wicked problems at the intersection of poverty, social change, and environmental degradation, has attracted the attention of a wide range of professionals (Martin and Osberg 2007a, b; Phills et al. 2008; Zhao 2012), network organizations (e.g. SEKN. org, ANDE.org, Ashoka.org) and academics (Mair and Marti 2009; Lepoutre et al. 2013; Dacin et al. 2010; Yu 2011). In the U.S. alone, a Social Enterprise Alliance study found that social enterprises employ over 10 million people and have annual revenues of about $500 billion (which constitutes approximately 3.5 percent of total US GDP (Thornley 2013). These organizations, which may be for-profit or non-profit,

act within a variety of sectors from health and education, to environmental action, social justice, and individual equality.

Products and services are intricately linked to contemporary social movements, empowerment of others, and are often at the forefront of social change to alleviate economic disparity or environmental impact (Nicholls 2008; Mair et al. 2012). Social entrepreneurship is a multi-disciplinary field of practice, and has attracted scholars from a variety of academic domains including economics, social sciences, management, technology and behavioral science (Short et al. 2009).

However, the emerging popularity of social entrepreneurship as a mechanism to address social problems belies the definitional challenges of the field, the lack of phenomenological consensus, and the differences in interpretation and application around the world. Some scholars argue in favor of a broad definition so as not to exclude any form of social engagement from potential resources and support (Cho 2006; Waddock and Post 1991). In this "third-sector" view, individuals and organizations, including philanthropy, non-government, non-profit and for-profit, engage in social mission and deliver social value. Other scholars posit a restrictive definition to enable investors, researchers and other interest groups to identify the phenomenon (Austin et al. 2006; Certo and Miller 2008; Mair and Marti 2006; Martin and Osberg 2007a, b). With nuanced differences among institutions and researchers, social entrepreneurship in this school of thought is defined as "the process of creating and implementing an entrepreneurial solution to a social problem and fulfilling unmet social needs, thereby creating social value and impact" (Sommerrock 2010, p. 68). This definition emphasizes an organizational perspective and highlights evidence-driven change-making and social value creation.

If the practice of social entrepreneurship is conceptualized as a phenomenon at the intersection of organizational effectiveness and social change, it may also be seen in established entities whether private incumbent corporations or in public administrations. A growing number of large-scale corporate incumbents adopt 'hybrid business strategies' and advocate a blending of social commitment and business purpose (Kayser and Budinich 2015; London and Hart 2011). Such organizational *intrapreneurship* views innovation "from the inside out" and existing public and private incumbents play a key role in accelerating the scaling social innovations, and leveraging their networks, brands and resources (Simanis and Hart 2009). Social entrepreneurship may thus be an outgrowth of traditional corporate social responsibility (CSR) efforts or traditional development efforts, in which organizations dedicate resources and commit to furthering social values. However, in traditional CSR, the social mission and values are not necessarily viewed as fundamental to the economic imperatives of the corporation and may be pursued independently within the organization or as a partner foundation. In contrast, social intrapreneurship focuses on mission-driven activities that are integrated into the company's processes of creating economic value. The integration of economic activity and social value creation is similarly reflected in the term "social enterprise", used by NGO's and non-profits that wish to grow/adapt the organizations operations beyond traditional grant-seeking cycles to better serve the social need (Brouard and Larivet 2010) (Fig. 2; Table 1).

Social entrepreneurship manifests differently in different parts of the world, and the differences in context offer deep challenges for practice (Bacq and Janssen 2011; Defourney and Nyssens 2010; Rivera-Santos et al. 2015). There are contextual differences in the legitimacy of pursuing social goals against the backdrop of financial considerations (Thornton et al. 2012; Battilana and Lee 2014), in the specifications of legal regimes (Lasprogata and Cotton 2003), in the nuances of culture (Chaudhary et al. 2017), and in the availability of supportive institutions (Mair and Marti 2009; Ventresca and Fu 2017). To explore how the pluralism of the social entrepreneurship concept was viewed by actors within a nascent ecosystem, we polled conference participants with a sequence of questions (see Table 2: P4-P6; Table 3: P7-P9).

P4/P5: Describing social entrepreneurship

More than one quarter of the participants (28%) selected an individualized description of social entrepreneurship, as related to the activities of a heroic social entrepreneur. Almost half the participants (44%) selected a process-based description that accounted for approaches that encouraged enduring, sustainable social change. A small fraction (9%) of participants picked the most inclusive description of social entrepreneurship, in that any firm dealing with a social issue was a social enterprise. Notably, participants did not select the description of social entrepreneurship as a sub-group of the third-sector of civil society. The polling indicated that participants favored an inclusive approach to social entrepreneurship, rather than a restrictive sectoral-based approach. To explore the limits of this inclusive approach, the moderator facilitated a discussion with a case on Smart Bike Sharing. Almost two thirds of the audience disagreed that the case was an exemplar of social entrepreneurship and suggested instead, that social entrepreneurship should first define specific social and environmental problems and take rigorous innovations. Polling (*P5*) also indicated that a majority of participants (53%) favored a description of SE as enterprises which placed social or environmental challenges at the core of their activity.

P6: Do well or do good?

Given the inherent balance between social value creation and economic value capture, participants were asked to prioritize their preferences for doing social good and doing well financially. The poll revealed that participants were equally divided in their priorities. For example, 25% of participants prioritized social good, and 25% of participants prioritized financial gain. While most chose to accept the premise that social entrepreneurship required both value creation and value capture, a small fraction of participants emphasized the sole eminence of social good (13%), and financial gain (6%). It is interesting that nearly one third of the participants declined to respond to this polling. In a following panel discussion, the audience argued that the priority of social and financial goals is dynamic and contingent on sectors and regions.

P7: Sources of innovation

SE deals with complex and enduring social problems through innovations. What is the source of these innovations? A majority of participants (38%) indicated that the key social innovations needed to be around business models. Participants also suggested that social innovations could address new markets and consumption patterns (13%), while smaller fractions focused on innovations in resources and financing (9%) and technologies and tools (6%). A significant share of people (34%) considered that innovations should be a combination of the listed elements.

technology as 'very important' for impact investing. At the conference 28% of participants declared perceived potentials of the analysis and use of big data, while smaller fractions of participants indicated interests in automated data collection and IoT (12%), blockchain (12%), biotech (10%),

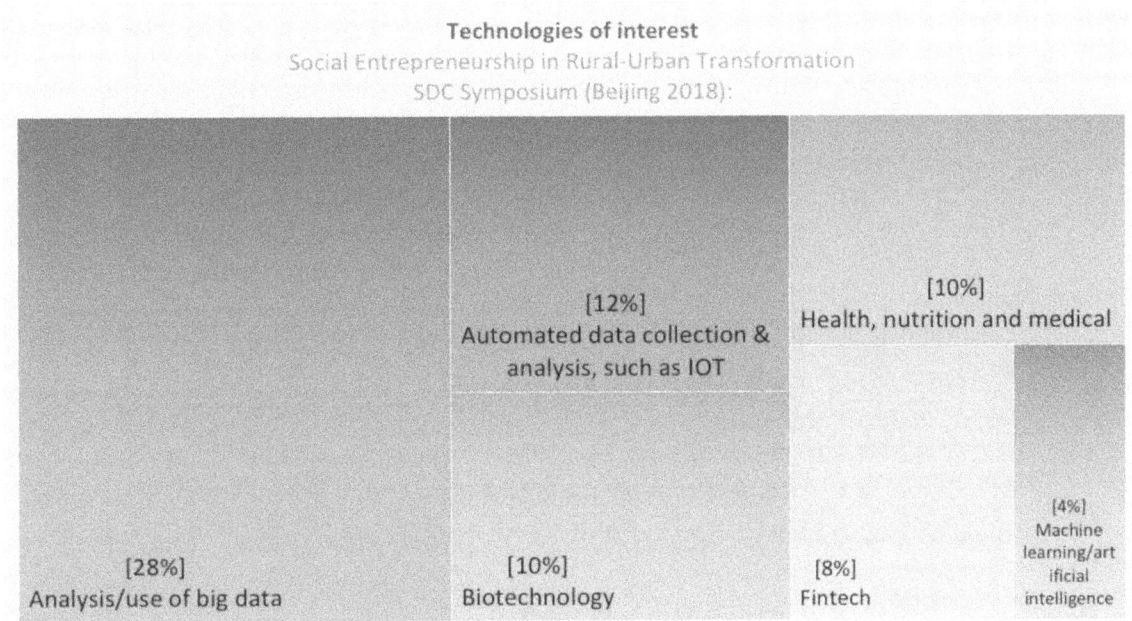

P8-P9: Technologies and sectors for social change

High-tech seems to be a primary source of social innovations during rural–urban transformation. In the GIIN (2018, p. 17) survey of 229 impact investing organizations around the world, respondents when asked about technology, prioritized the analysis and use of big data (56% rating this 'very important') and automated data collection and analysis (45%). Twenty-nine percent also cited blockchain

health-tech (10%) and fintech (8%). When being asked about the sectoral potential of social entrepreneurship, respondents indicated equivalent support for Means of Living [health, education, housing, water—22%], and Economic Opportunities [finance, small-business services, fair-trade, ICT—22%]. Respondents also saw potentials for SE in Food systems [farming, agriculture, food security—19%], and Environmental Sustainability [climate change, natural resource and biodiversity protection—19%]. A smaller fraction of participants (8%) saw potential for social entrepreneurship in the fields of inclusion and humanity (human rights, migration, poverty alleviation, advocacy).

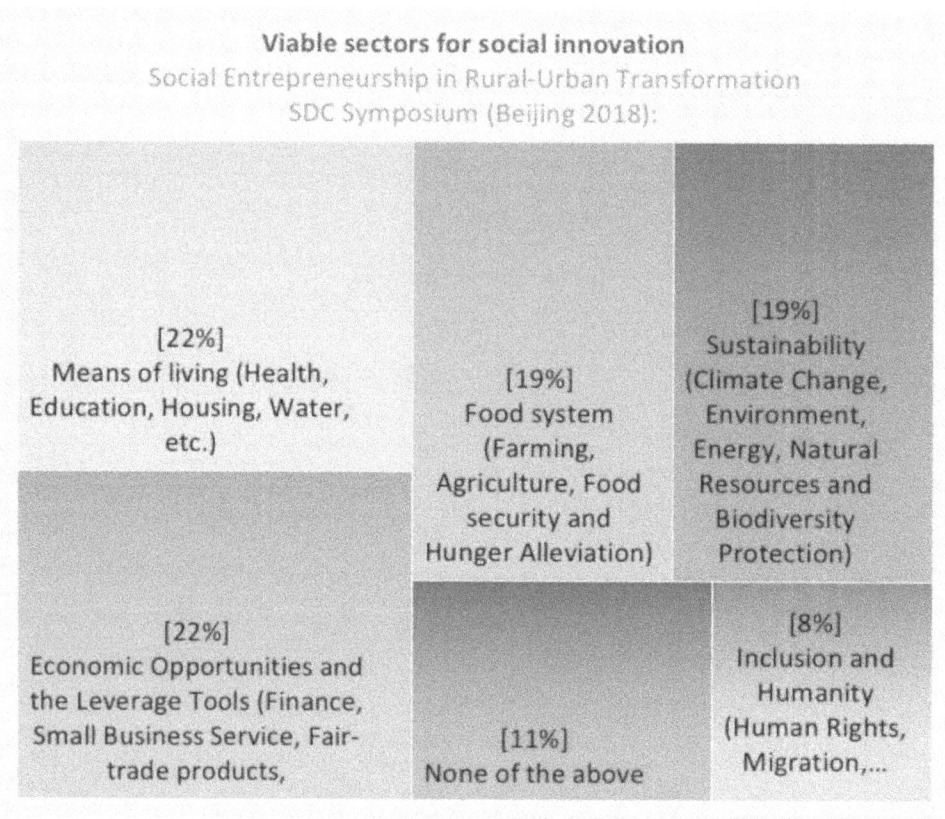

In summary, participants selected an inclusive definition of social entrepreneurship that favored process-based approaches to enduring, sustainable social change. This definition was agnostic to the type of organization or individual, but emphasized placing the social and environmental mission at the core of the activity. Participants also accepted the inherent dynamic between financial and social motives in SE, but were divided on whether to start with economic value capture or create social value first. The moderated follow-up discussion revealed that these choices were deeply contextualized by the social sector which participants had in mind. In certain sectors (economic development for example), financial gain was seen as a pre-requisite, while in other sectors (environmental sustainability), the social issue was seen as pre-eminent. Participants also emphasized the role of business-model innovations in SE as facilitating new mechanisms and incentives for exchange to encourage addressing social issues. The participatory communication process indicates a convergence toward a pragmatic pluralism, a formulation that accepts various approaches to social entrepreneurship across a variety of sectors as long as the emphasis on social value is clear. SE is viewed as a way to address market imperfections (Cohen and Winn 2007) while emphasizing social value creation (Santos 2012).

Mechanisms of change: Rural–urban social transformation through entrepreneurship

Narratives of social entrepreneurship offer dynamic accounts of social change. By taking business approaches to addressing poverty and inequality at global and regional levels, social entrepreneurs apply relatively small interventions and manage to achieve disproportionately significant results (Bornstein 2007; Nicholls 2006; Sachs 2005). While the initial stimuli are often barely visible, over time they gain strength and in the long run achieve immense impact and systemic change within the old system. On closer scrutiny, the underlying processes of change are often communicative and bottom-up, empowering communities and societies before changes are enacted and impacts made visible (Ervaes et al. 1996; Mair et al. 2012; Praszkier and Nowak 2012).

How then in practice, does social change take place? Given that social ventures are embedded in broader cultural and institutional contexts, Dacin et al (2010) strongly

Table 4 Mechanisms of social change in rural–urban transformation through social entrepreneurship (%)

P10. Of the following two schools of thoughts about Social Change, with which do you agree?	%
A) When dealing with a complex and enduring social problem, SE has to be "disruptive" and introduce a dramatic change to the existing system	26
B) Given the complex system that SE normally deals with, the role of SE is not to aim to directly make changes (and deliver impacts) but to catalyze changes in the old system	52
C) Other	22
P11. Entrepreneurship often requires resources and investment capital to get started. This can be particularly hard in social entrepreneurship and in rural–urban agricultural contexts where the possibility of high return on investment is harder to achieve	%
A) SE should start without waiting for external funding. Make do with whatever materials, labor and skills are at hand	70
B) External investment is important before getting a project started. Only then can the right people be hired and tools be used to address the social problem or opportunity	13
C) I disagree with both answers	17

recommend a consideration of the enabling and constraining factors that influence the social change potential of the venture. The sociology and institutional literature offers a variety of ways to look at social change (Praszkier and Nowak 2012; Tina Dacin et al. 2002), including evolutionary conflict-based, and structural–functional theories of change. Evolutionary models of social change suggest that a system continuously progresses toward greater levels of complexity—adapting to the environment or overwhelming existing ecologies as it attempts to decrease internal entropy and dissonance (Praszkier and Nowak 2012). Conflict-based theories of change examine phenomena of deinstitutionalization—the processes by which institutions weaken, disappear and are subsequently replaced by alternative institutions (Scott 2001, p. 182). Structural–functional theories view social changes as "boundary destruction" and social entrepreneurship functions as an exogenous force to break the boundary (Parsons 1970). These structural–functional theories are embodied in the disruptive innovation models that advocate social change through entrepreneurially-driven innovation (Christensen et al. 2006; Saul 2010). These models of social change parallel the opportunity-driven models of change in commercial entrepreneurship (Shane and Venkatraman 2000).

An alternative approach to social change is to examine the underlying goals embedded within social entrepreneurship. Is the goal to create cultural change? Cultural entrepreneurs, for example, seek to change attitudes, beliefs, and values—the normative milieu of a given context (Dacin et al. 2010). These cultural entrepreneurs approach an organization or social group as "new members with backgrounds and experiences that differ from existing members [and] bring different interpretive frameworks and social definitions of behavior to the organization that act to diminish consensus and unquestioning adherence to taken for granted practices" (Dacin et al. 2002, p. 47). Is the goal instead to create institutional change? Battilana et al. (2009, p. 87), suggest that a combination of field characteristics, social position and

network alliances offer the greatest opportunities for entrepreneurs seeking to create divergent institutional change.

Embedded within these theories of change is a recognition of the complexity of social issues. Social problems are highly embedded with the local context, and any individual efforts, notwithstanding their significance in the short term, do little to move an old system to a new structure. Social change then becomes a process characterized by an accumulation of changes on the micro level (Coleman 2000), augmented through social capital and networks at the community and regional level (Praszkier et al. 2009). Social entrepreneurs play a catalytic role in social changes, assembling an ecosystem of component drivers that gradually move a system toward an intentional or unconscious tipping point until an enduring change can take place.

P10: Disruptive or incremental change?

Given the alternative approaches to social change as disruptive or catalytic, conference participants were asked about how they saw the change processes in social entrepreneurship.

A majority (52%) favored approaches that were catalytic, responding to the statement that given the complex system that SE normally deals with, the role of SE is not to make direct changes, but to catalyze changes in the old system. Another group of respondents (26%) took a more direct approach, and suggested that when dealing with a complex and enduring social problem, SE has to be disruptive and introduce a dramatic change to the existing system.

Social change is intrinsically linked to resource-mobilization. Given that social ventures can operate under conditions of resource-constraint, in sectors with low economic opportunity or even market failures, the role and source of resources is often called into question. Battilana et al. (2009, p. 63) suggest that financial resources and resources related to social position, such as formal authority and social capital, play a key role in helping institutional entrepreneurs

Table 5 Poll responses on measuring social impact (%)

P12. Social Value: Does social value exist and is it measurable?	$
A) Can and should be objectively measured	13
B) Subjective existing, but can and should be measured through a set of proxy indicators by a third-party	30
C) Subjective existing; can and should be measured through a set of proxy indicators by people in the social problems	30
D) Subjective existing, and no way for measuring	9
E) Others	17
P13. Impact assessment: is impact assessment important, and if so, by whom should assessments be made?	%
A) Important and should be made by SE	13
B) Important and (should be) institutionalized by external third-party entities	39
C) Alternative results (such as outcomes) instead of outputs	26
D) Makes no sense as it is time consuming, misleading and methodologically flawed	17
E) Others	4
P14. Which of the following legal frameworks and policy tools are crucial for SE and social innovations? (Two answers allowed)	%
A) Legal identity and certification	26
B) Credit enhancement (such as guarantees or first-loss capital)	0
C) Subsidies for investors (such as tax credits)	18
D) Streamlined and clearly defined regulations for investment offerings	15
E) Funding for technical assistance on investees	21
F) Co-investment by government	3
G) Macro advocacy on SDGs	10
H) Others	8

convince other actors to endorse and support the implementation of a vision for divergent change. Resources enable individual social entrepreneurs (Meyskens et al. 2010) and social ventures across institutional contexts (Desa and Basu 2013) and sectors (Zahra et al. 2009). However, given the paradox of mobilizing resources while struggling through the crosswinds of existing institutional regimes and resource constraint, scholars have also looked at how social entrepreneurship gets started in the absence of investment capital (DiDomenico et al. 2010; Desa 2012). Bricolage, the process of mobilizing resources at hand—be they material, labor and/or skills, allows social entrepreneurs to "make-do" and overcome the constraints that may act as a barrier in relation to project inputs (Fisher 2012). A fundamental aspect of bricolage involves the combination and, potentially, recombination of resources to solve problems, while at times opportunistically incorporating additional resources to serve an emergent need. This "trove" of resources (Baker and Nelson 2005) allows the social entrepreneur to enact creative solutions to social problems, and acts as the medium of social change.

Given the alternative approaches to resource mobilization, participants were asked about how they view the process in social entrepreneurship. Specifically, considering the fact that resources and investment capital to get started could be particularly hard in social entrepreneurship and in rural–urban agricultural contexts where the possibility of high return on investment is harder to achieve.

P11: Resource optimization or bricolage?

A majority of participants (70%) suggested that SE adopt the bricolage approach and start without waiting for external funding. Make do with whatever materials, labor and skills are at hand. Only 13% of participants suggested that external investment was important before getting a project started, in order to hire the right people and access the tools needed to address the social problem or opportunity. A follow-up discussion revealed that the 17% of the non-respondents who disagreed with the dichotomous choice placed an emphasis on the sectoral context—saying that while mobilizing resources through bricolage might be feasible in some sectors, it would be completely infeasible in other social sectors.

Measuring impact: The impact investing landscape

How does one decide if a given social intervention has had an impact? The decision about whether a social venture has impact can have direct consequences on the welfare of the beneficiary population, the venture's access to resources, its ability to scale, and its ability to inspire similar initiatives in other regions around the country and the world. Setting an impact target helps motivate, test, iterate, and validate theories of social change. Respondents to the GIIN 2018 survey who set impact targets identified the need to drive social/

environmental impact management (81%) as the primary motivator. As one respondent explained, "Only by setting impact targets—and subsequently tracking performance—can we test our impact hypotheses and improve our understanding of impact creation for future investments." (GIIN 2018, p. 36).

Consider organizational effectiveness—the degree to which an organizations' outputs benefit the society to which they belong" (Kroeger and Weber 2014, p. 515). Measurements of organizational effectiveness and interpretations of impact outcomes are weighted with socio-political legitimacy, and are often contested and debated by concerned stakeholders. The social value created as part of an organization's activities is a core topic discussed within the field, yet manifests differently in different domains. The entrepreneurship domain, for example, focuses on opportunity recognition and venture growth. The public and nonprofit management domain measures the extent to which an organization's outputs achieve intended social outcomes. The corporate domains social responsibility focus on effective and socially equitable allocation of scarce and valuable resources, while minimizing adverse impacts to commercial activity.

Measures of organizational effectiveness suffer from two distinct challenges: First, of attributing the effects and comparing social value created across different interventions/sectors. The second challenge is to ensure that social value is uniformly interpreted and enacted within the entire value chain of a given intervention/sector.

Assessing impact across sectors

How does one compare social interventions in different sectors (improving health vs. improving environmental outcomes), across different socio-economic and institutional contexts (a service for the disabled in the U.S vs. a microfinance program in Ghana), and with different interpretations of social value (direct social-welfare or indirect welfare through economic outcomes)? While standards bodies set vocabularies of social interventions, social innovations, impacts and outcomes, these assessments are most effective when co-ordinated. We describe three initiatives: SPM, ESG and IRIS.

SPM

As banks sought to evaluate and measure their portfolios' social and environmental performance, the U.S. National Community Investment Fund (NCIF) developed a Social Performance Metrics (SPM) methodology to identify and evaluate the financial and non-financial services of community financial institutions to low-income and vulnerable communities (Narain and Schmidt 2009). NCIF is the largest investor in the mission-oriented banking industry, with

$205.6 million of assets under management (NCIF 2019). As an impact investor, NCIF pursues a triple bottom line strategy that maximizes social, environmental, and financial returns.

ESG

The Environmental, Social and Governance (ESG) standards were led by an informal group of financial leaders, lawyers and environmental stewardship NGOs to finetune the relationship between environmental and social standards and financial performance of investment and invested companies. ESG was further developed by the United Nations Environment Programme Finance Initiative and the UN Global Compact to offer companies a framework for responsible investment practice. The financial returns of top publicly listed companies was found to be positively associated with their commitment to ESG over 1984–2009 (Edmans 2011).

IRIS

The Impact Reporting and Investment Standards (IRIS) framework was initially created by the Rockefeller foundation, Acumen and B Lab, to create common metrics to report the performance of impact capital (GIIN 2018). Supported by a large coalition of investors in the Global Impact Investors Network (GIIN), IRIS incorporated and built upon a variety of established sector-specific efforts to enable comparison and communication across the breadth of organizations that prioritize social or environmental impact. The metrics in the IRIS catalog are built on over 40 sector-specific standards and reporting frameworks (available at: https://iris.thegiin.org/about/history). The NCIF SPM metrics for example, are aligned to the IRIS framework (https://iris.thegiin.org/metrics/sets).

Un-coordinated impact metrics pose challenges to social innovation by creating fragmented ecosystem and limiting organizational effectiveness. In a broader response, the Sustainable Development Goals (SDGs), adopted by the member states of the United Nations in 2015, established an ambitious set of goals for progress against a wide range of social and environmental factors. Central to the achievement of these goals was collaboration between private, public, and philanthropic sectors, and there have been encouraging results. In a survey of 229 global impact investors, 90% of respondents indicated that the SDGs were a useful way to communicate their impact externally since it was a widely recognized framework. 73% indicated that it helped the impact investor integrate into the global development paradigm. 53% indicated that incorporating the SDG's helped attract investors. 43% indicated that the SDGs helped them refine their theory of change and set appropriate impact objectives and impact targets (GIIN 2018, p. 38). Just two

Table 6 Poll responses on scaling social impact through social entrepreneurship ecosystems (%)

	%
P15. SE in agricultural and rural–urban context is often "local" in nature, making scaling difficult. Among the following answers, with which do you agree?	%
A) Have to be local and hardly be trans-regional and scaling	0
B) Can be scaling and capital investment vehicles this	17
C) Capital can hardly vehicle scaling in SE of this context. But can be scaling through a loose arrangement, such as franchising, alliance, networking or open resource	65
D) Others	17
P16. How should SE in rural–urban context deal with the regional and (multi-) sectoral focus?	%
A) SE should stay local and focusing on specific sector	13
B) SE should stay local, but deal with a wide range of multi-sector issues (through maybe alliance and partnerships)	13
C) Focus on specific sector, and then scale that up to as many people/regions as possible	61
D) Others	13
P17. What are the major constraints for scaling SE?	%
A) Lack capital	17
B) Insufficient know-how and management skills	74
C) Founder of the SE programs or social enterprises	9
D) Others	0
P18. What roles do you anticipate intermediaries (such as SDC) to play in the future? (Two answers allowed)	%
A) Promoting knowledge generation and exchange	21
B) Policy advisory and advocacy	18
C) Networking financial resources	21
D) Strategical mapping and business model relating with SE	12
E) Technical assistance (such as impact measurement and management, risk mitigation, etc.)	9
F) SE education and training	9
G) Others	12

years since the adoption of the SDGs, 55% of impact investors track their investment performance to them and another 21% plan to do so in the future (GIIN 2018, p. 38). 71% of the ESG metrics have also been mapped to the SDGs, and investors are paying closer attention to mission alignment (Calvert 2018).

Nonetheless, the challenges with standardizing impact metrics across different social activities are also well documented. Nearly 70% of impact investors responding to the GIIN survey used proprietary metrics or frameworks that were not aligned to external methodologies, and another 66% used qualitative information. These numbers illustrate how impact standards, while well-meaning, are often framed by resource-granting bodies and administered by third-party certification, which can be less conducive to social entrepreneurs on the ground.

Assessing impact consistently along the value chain

The second challenge to organizational effectiveness is to ensure that social value is uniformly interpreted and enacted within the entire value chain of a given intervention/sector. How does one ensure that high-level SDG impact metrics (for example), translate into appropriate implementation and measurement within grassroots social ventures?

The discrepancy among the various regimes that define impact metrics and grassroots classifications of impact can be counterproductive to organizational effectiveness—with investments that do not achieve intended goals. A recent report on the "missing middle"—small and growing businesses in emerging and frontier markets, suggests that these businesses are often passed over by impact investors for not having clearly defined impact measures, and by commercial investors for not having clearly defined exit options (Hornberger and Chau 2018). The missed opportunities for social impact are vast. Entrepreneurs from this vast group of enterprises at the heart of the global economy require over $930 billion in investment but classify themselves into groups with very different impact targets: high growth, niche, dynamic, and livelihood sustaining (Hornberger and Chau 2018, p. 6).

The dangers of not communicating across these different interpretations of impact can be seen for example in the agri-food industry. Standardized certifications required by investors, and purchasers in the apple industry, are not necessarily understood by the small farmers actually growing the produce (Ding et al. 2019). Impact metrics can have the adverse consequence of being lost in translation—of being adopted at the investor end of the value chain, but not implemented at the producer end of the chain.

As an alternative approach, Kroeger and Weber (2014) suggest returning to subjective measures of social well-being that are primarily concerned with a respondent's own (perceptional) judgment of meaning and well-being. Evaluations of social well-being take people's values and preferences into account, and include emotional affect and cognitive measures of life satisfaction (Kroll and Delhey 2013). Social well-being depends specifically on meaningfulness and experienced domain satisfaction: the satisfaction felt in a perceived situation as related to a particular need.

Given the alternative approaches to impact and impact measurement, participants were asked about how they viewed impact in social entrepreneurship. Specifically, considering the fact as to whether social value was objectively measurable, and if not, whether the subjective value could be captured by a set of proxy indicators. Participants were also asked about who should certify social impact.

P12: Is social value measurable?

As shown in Tables 4 and 5, 13% of participants indicated that social value should be objectively measured. 30% of participants considered that social value was subjective, but could be measured by proxy indicators by an independent third-party. 30% concurred that social value could be measured through proxy indicators, but felt that the social entrepreneurs closest to the action should set those measurement standards. Nine percent of the participants caveated the dangers and limitations of measurement, suggesting instead, that given the subjective nature of social value, there was no way of measuring real transformation.

P13: Impact assessment

The findings were paralleled in a follow-up question about impact assessment. While 13% of participants indicated that impact should be self-measured by the social entrepreneur, 39% indicated a preference for impact measurement by a third-party. 26% of participants pushed for thinking more about outcomes than impact, and 17% of participants averred that impact measures were time-consuming, misleading and methodologically flawed.

P14: Legal and policy frameworks

Given the emphasis on well-crafted proxy indicators for measuring impact, the follow-up question on legal and policy frameworks was particularly revealing, with an approximately equal split between the need for legal clarity, and capitalization. With regard to legal clarity, 26% of participants prioritized the development of legal identities and certifications for SE, and 15% indicated a need for regulation for investment offerings. With regard to capitalization,

21% requested funding for technical assistance, and 18% requested subsidies (such as tax credits) for investors. While 10% indicated the need for macro-advocacy that connected grassroots impact to SDGs, a notably low percentage of respondents requested government co-investment (3%) or financial guarantees (0%).

Ecosystem views of social entrepreneurship

While the individual social entrepreneur is the undeniably important driver of social impact, case studies of social entrepreneurship consistently and gently remind us that social impact is less about the heroic efforts of a single actor and more about networks of actors and contexts that help drive the process of social change (Hamschmidt and Pirson 2018; Bornstein 2007). Scholarly attention in the field has similarly shifted the level of analysis from the single individual (Mair and Noboa 2006), to the social venture (Miller and Wesley 2010), to communities (Haugh 2007) and more recently to the regional ecosystems (Thompson et al. 2018) that engender social change. These social entrepreneurial ecosystems may be defined as systems of co-located elements where a variety of actors, functions, and institutions interact to support the creation and growth of new ventures focused on social impact (Isenberg 2010; Thomas 2013).

The potential social and economic impact of ecosystem formation and replication has attracted scholars from different fields, including management science, sociology, international development and policy. Research on entrepreneurial ecosystems has traditionally adopted a top-down focus on how ecosystems are structured. Whether through the formation of industry clusters (Baum and Haveman 1997; Delgado et al. 2010), network linkages (Bresnahan et al. 2001), or authoritative action by governments or powerful other actors (Autio et al. 2014; Etzkowitz 2008), researchers emphasize ecosystem designs for efficiency, value-creation potential, and scaling impact. In these streams of research, the analytical focus is on the structural elements comprising an ecosystem, the critical linkages among them, and how best to design and implement ecosystems.

Social entrepreneurship ecosystems, however, can be harder to design as top-down structures since the incentives for creating social impact are different from the high-returns expected in a market-based capitalistic system. Recent research suggests instead, that rather than being created through the actions of governments and other powerful actors, social entrepreneurial ecosystems form through the everyday interactions of individuals striving to create shared meaning, resources, and infrastructure needed to support their new ventures (Thompson et al. 2018, p. 96). Unlike the more elaborated commercial sector, the organizational and contextual conditions that social entrepreneurs face compound the difficulties of scaling impact and accessing resources (Austin et al. 2006; Bradach 2003; Dees et al. 2004).

Drawing upon sociological theories of field emergence, recent research has identified how ecosystems of social impact can be created in cities (Thompson et al 2018). As individuals repeatedly interact over time, these interactions create interdependencies and shared meanings that give rise to a set of collective values and beliefs that support a broader set of ecosystem participants. The initial activities of disparate individuals and groups can suddenly coalesce into more coordinated, integrated, and durable patterns of social interaction, creating the methods, resources, and legitimacy needed for an entrepreneurial ecosystem for social impact businesses to coexist (Thompson et al. 2018, p. 112).

To accomplish the social mission of targeting underserved or disadvantaged population, scaling is at the heart of social entrepreneurship, and ecosystem development is a key enabler of scale. Ecosystem development while particularly important, is particularly challenging for rural–urban social transformation. Investments in agriculture are riskier than in other industries (manufacturing, for instance) because the agricultural cycle is long and product is usually harvested on an annual basis. Organic product harvesting can take even longer (Kleeman et al. 2014). Cash-crop sectors like coffee can have a four-year lag from seed to harvest—and that in the face of clement weather (NCAUSA 2018). When we take climate change into account, the harvest variability crops increases further (Wang et al. 2008; Schlenker and Roberts 2009; Piao et al 2010). Beyond the risks of climate change are the potentials for land conflict due to migration and population growth, political changes, and other exogenous factors that make rural agriculture a risky sector for investment. Risk when added to long recovery horizons, does not allow ventures to fail fast, iterate and improve (Khanna et al. 2016; Pelling 2010), making the sector particularly challenging for entrepreneurs and unattractive to younger generations.

Furthermore, the geographic, cultural, and contextual dissimilarities due to site specific and tacit knowledge make scaling highly costly and risky in agricultural and rural domains (Ghemawat 2001). The last mile gap for agriculture is particularly strenuous (Stone 2011), since geographic distance translates into market distance and communications distance, which in turn increases food storage, brokerage and other transaction costs. Costly and slow iterative cycles lower the potential for learning and development. While geographic isolation can be improved with internet connectivity, and the digital revolution is connecting people in unprecedented ways, cultural distance puts farmers and rural producers on the far side of the digital divide. Most farmers have cell-phones, but don't really communicate and share information with other parts of the region or the world, as they lack the cultural and social norms to do so (Wyche and Steinfeld 2016; Idlemudia and Raisinghani 2014). Rural–urban transformation social transformation also has to contend with aid and subsidy-based norms, since developing economic regions that have larger percentages of their population engaged in agriculture, have for a very long time, been under the purview of developmental aid organizations and government (Bates 2014; Long et al 2016; Parish 2016; Franko 2018).

The above factors compound the challenges for adopting an entrepreneurial mindset when engaging in agriculture. Whether through top-down structural approaches, or through the repeated interactions of interested individuals, the spontaneous emergence of a social-entrepreneurial ecosystem can be harder to identify and sustain. And when additional knowledge and specific resources are needed for replicating social impact at a different site, the transferability of social entrepreneurship is rather low. All of which can be particularly challenging for viewing social entrepreneurship as a dynamic business that leads to social impact.

P15/P16/P17: Scaling social impact

At the symposium, conversations on scaling social entrepreneurship were facilitated and framed in the rural–urban context. As shown in Table 6, only a small amount of participants (17%) agreed that scaling can be enabled by capital investment; the majority of participants (65%) recognized the dissimilarities and contextual complexity in rural–urban context and believed that scaling can be achieved through business model, such as innovative partnership or innovative alliance. However, 61% of the audience voted for horizontal scaling instead of vertical partnership across multiple sectors (Polling 16). Insufficient know-how and management capabilities were shown as the primary constraints for scaling. As shown from the results of Polling 17, 74% of the participates believed insufficient know-how and management skills were the primary constraints of SE scaling. It is interesting that about 9% of the participants considered that the founders were the major obstacle to scale. A similar argument was made by Kayser and Budinich (2015) who argued that the primary obstacle for scaling was the founder syndrome—the founders who do not accept letting go.

The review and survey results suggest the role of multi-theoretical perspectives to explain formal and relational governance issues in interfirm networks (Combs and Ketchen 1999; Windsperger et al. 2015). A wide range of institutional options and variants, such as franchising, cooperatives and strategic alliances, need to be adopted for scaling and scoping the food and agricultural sector. Compared to formal mechanisms and centrality in authoritative and hierarchical governance, interfirm networks—referring to relational governance, such as trust, norms and solidarity in contractual relations—become organizational alternatives as a safeguard mechanism against opportunism risk under high uncertainty and contextual complexity (Dyer 1997; Nooteboom et al. 1997). While a small tranche of social entrepreneurship literature underscores "partial replicability" and standardization as an appropriate means to facilitate scaling (Bradach 2003; Ratliff and Moy 2004), a major body of literature caveats replication given the dissimilarities between the context in which social entrepreneurs are active before scaling and the context they aim to scale into (Dees et al. 2004; Jenkins and Ishikawa 2010; Desa and Koch 2014).

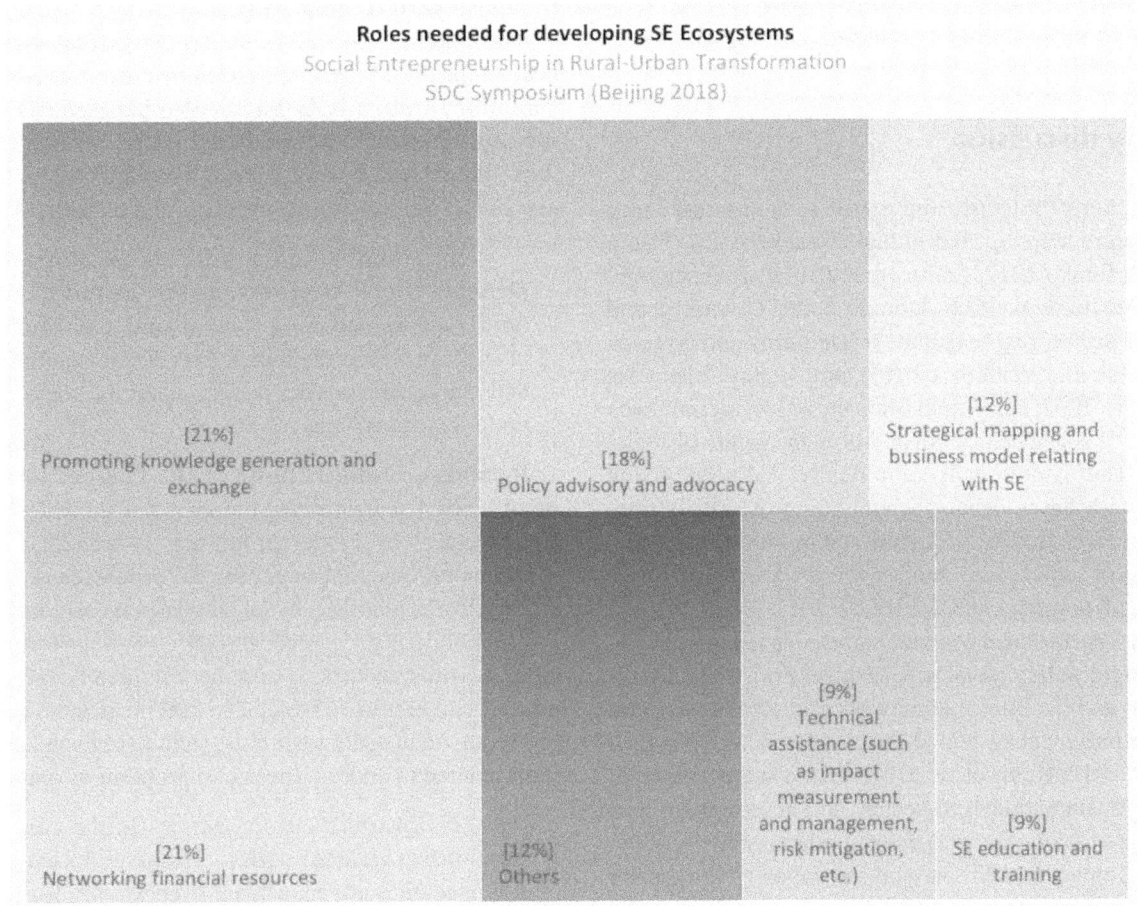

Roles needed for developing SE Ecosystems
Social Entrepreneurship in Rural-Urban Transformation
SDC Symposium (Beijing 2018)

[21%]
Promoting knowledge generation and exchange

[18%]
Policy advisory and advocacy

[12%]
Strategical mapping and business model relating with SE

[21%]
Networking financial resources

[12%]
Others

[9%]
Technical assistance (such as impact measurement and management, risk mitigation, etc.)

[9%]
SE education and training

P18: The role of intermediary organizations

About 21% of the conference participants indicated that knowledge generation and exchange was needed to build out the SE ecosystem for rural–urban transformation, while another 21% indicated the importance of networked financial resources. Of the remaining responses, 12% indicated that the importance of strategic mapping and business models, 18% indicated the importance of policy advocacy, 9% asked for technical assistance, and SE education and training (9%).

Summary discussion

A vast number of intrinsic and extrinsic factors influence social entrepreneurship. Individual characteristics (Dacin et al. 2010; Renko 2012; Smith et al. 2012), organizational identity (Austin et al. 2006; Dorado 2006; Townsend and Hart 2008), sector, regional context (Defourny and Nyssens 2010; Dorado and Ventresca 2013), and venture lifecycles (Lichtenstein 2007; Ruvio and Shoham 2011), all influence the process of social change. In addition, the nature of social change in itself is politically, empirically, and normatively weighted with fierce debates surrounding the attribution, assessment (Hjorth 2013; Sawhill and Williamson 2001), and process of social value creation (Christensen et al. 2006; Desa 2012; Korsgaard and Anderson 2011; Waddock and Post 1991). Practitioners demand knowledge about business model design, policy advocacy, sectoral priorities, investment tools, and facilities that network and mediate a social entrepreneurship ecosystem (Mair and Schoen 2007; Battilana et al. 2012; Kania et al. 2014). With the emergence of innovative finance models, investors and development professionals are looking for well-defined relationships between grassroots impact and SDGs, and the capacity to monitor and assess that impact (OECD 2018).

The SDC symposium, modeled upon a participant-centered discussion model, offers some directions and priorities for social entrepreneurship in the context of rural–urban transformation. We summarize four aspects that emerged as particularly relevant: field definition, resource mobilization, impact investment, and ecosystem development.

Field Definition: Participants selected an inclusive definition of social entrepreneurship that favored process-based approaches to enduring, sustainable social change. This definition was agnostic to the type of organization or individual, but emphasized placing the social and environmental mission at the core of the activity. The participatory communication process indicated a convergence toward a pragmatic pluralism, a formulation that accepts various approaches to social entrepreneurship across a variety of sectors as long as the emphasis on social value is clear. SE is viewed as a way to address market imperfections (Cohen and Winn 2007)

while emphasizing social value creation (Santos 2012). The inherent dynamic between financial and social motives was deeply contextualized by the social sector which participants had in mind.

Participants discussed the challenges of objectively measuring social value. A majority concurred that proxy-indicators can help measure social value, as long as they were set by social entrepreneurs closest to the action, and measured by an independent third-party.

The emphasis on well-crafted proxy indicators for measuring impact was also emphasized by the discussion on legal and policy frameworks that created legal clarity, allowed for capitalization, and regulated social investment offerings. Impact investment requests were focused on technical assistance, on investment subsidies, and on metrics that connected grassroots impact to SDGs.

Defining social entrepreneurship: A process based approach to enduring, social and environmental change, in which the balance between social mission and market approaches depend upon the social sector under consideration.

Resource mobilization: In the context of resource mobilization, participants emphasized the role of action and execution over external funding. A bricolage approach was deemed essential to getting SE processes started. This approach was qualified by an emphasis on sectoral context. While mobilizing resources through bricolage might be feasible in some sectors, it may be completely infeasible in other social sectors in which external investment might be a pre-requisite in order to hire the right people and access the tools needed to address the social problem or opportunity.

Process of resource mobilization: In the context of rural–urban transformation, bricolage-based action and execution are prioritized over waiting for impact investment, with the caveat that some sectors require large-scale investment to develop.

Impact investment: Participants discussed the challenges of objectively measuring social value. A majority concurred that proxy-indicators can help measure social value, as long as they were set by social entrepreneurs closest to the action, and measured by an independent third-party.

The emphasis on well-crafted proxy indicators for measuring impact was also emphasized by the discussion on legal and policy frameworks that created legal clarity, allowed for capitalization, and regulated social investment offerings. Impact investment requests were focused on technical assistance, on investment subsidies, and on metrics that connected grassroots impact to SDGs.

Measuring impact: Impact assessment by an independent third-party is preferable, as long as there are clear certifications, legal identities and well-crafted indicators that draw from social entrepreneur input and are correlated to social outcomes.

Ecosystem development: Participants actively discussed the nature and structure of ecosystems that facilitated social innovation. The emerging discussion touched upon on a wide range of institutional options and variants, such as franchising, cooperatives and strategic alliances that needed to be adopted for scaling and scoping the food and agricultural sectors. Participants attributed equivalent levels of importance to knowledge generation and exchange (through training and technical assistance), and networked financial resources (including strategic mapping and business models), when it came to building out the SE ecosystem. The roles of trust, norms and solidarity—of learning from peers were underlying themes in the discussion.

Ecosystems for social innovation: Social innovation is at a nascent stage, and ecosystems should encourage experimentation across a variety of institutional options—from franchising, co-operatives and alliances. Knowledge generation, exchange, and networked financial resources are key elements for a viable ecosystem, and forums that nourish conversation should be encouraged.

Social entrepreneurship in rural–urban transformation is a long-term process of social change. We encourage scholars to build on a variety of methodologies to deepen the understanding of social change. Especially in regions with nascent conceptions of social entrepreneurship, the SDC symposium emphasized the need for legal identities at the regional and local levels. These legal identities in turn, could encourage business and innovation ecosystems that emphasized shared value, offered financial models that suited social ventures at various stages of growth and development, and forums of mentorship in which social entrepreneurs could communicate their own sector and stage specific needs. We encourage a variety of perspectives on social entrepreneurship in the rural–urban context.

Funding The first author acknowledges funding from the National Natural Science Foundation of China (Nos. 71573209, 71661147001); National Key Research and Development Program of China (No. 2016YFD0201303) and National Social Science Foundation (16ZDA021). The authors are thankful to the Swiss Agency for Development and Cooperation (SDC) and the China Office for facilitating the workshop.

References

Austin, J.E., H. Stevenson, and J. Wei-Skillern. 2006. Social entrepreneurship and commercial entrepreneurship: Same, different, or both. *Entrepreneurship Theory and Practice* 30 (1): 1–22.

Autio, E., M. Kenney, P. Mustar, D. Siegel, and M. Wright. 2014. Entrepreneurial innovation: The importance of context. *Research Policy* 43 (7): 1097–1108.

Bacq, S., and F. Janssen. 2011. The multiple faces of social entrepreneurship: A review of definitional issues based on geographical and thematic criteria. *Entrepreneurship & Regional Development* 23 (5–6): 373–403.

Baker, T., and R.E. Nelson. 2005. Creating something from nothing: Resource construction through entrepreneurial bricolage. *Administrative Science Quarterly* 50 (3): 329–366.

Bates, R.H. 2014. *Markets and states in tropical Africa: The political basis of agricultural policies.* California: Univ of California Press.

Battilana, J., and M. Lee. 2014. Advancing research on hybrid organizing–Insights from the study of social enterprises. *The Academy of Management Annals* 8 (1): 397–441.

Battilana, J., B. Leca, and E. Boxenbaum. 2009. 2 how actors change institutions: Towards a theory of institutional entrepreneurship. *Academy of Management Annals* 3 (1): 65–107.

Battilana, J., M. Lee, J. Walker, and C. Dorsey. 2012. In search of the hybrid ideal. *Stanford Social Innovation Review* 10 (3): 50–55.

Baum, J.A., and H.A. Haveman. 1997. Love thy neighbor? Differentiation and agglomeration in the Manhattan hotel industry, 1898–1990. *Administrative Science Quarterly* 42: 304–338.

Bornstein, D. 2007. *How to change the world: Social entrepreneurs and the power of new ideas.* Oxford: Oxford University Press.

Bradach, J.L. 2003. Going to scale: The challenge of replicating social programs. *Stanford Social Innovation Review* 1 (1): 19–25.

Bresnahan, T., A. Gambardella, and A. Saxenian. 2001. 'Old economy' inputs for 'new economy' outcomes: Cluster formation in the new Silicon Valleys. *Industrial and Corporate Change* 10 (4): 835–860.

Brouard, F., and S. Larivet. 2010. Essay of clarifications and definitions of the related concepts of social enterprise, social entrepreneur and social entrepreneurship. In *Handbook of research on social entrepreneurship*, ed. A. Fayolle and H. Matlay. Cheltenham, UK: Edward Elgar.

Calvert. 2018. Incorporating SDGs into ESG investment research. https://www.calvert.com/includes/loadDocument.php?fn=28860.pdf&dt=fundpdfs%27. Accessed 26 January 2019.

Certo, S.T., and T. Miller. 2008. Social entrepreneurship: Key issues and concepts. *Business Horizons* 51 (4): 267–271.

Chaudhury, A.S., T.F. Thornton, A. Helfgott, M.J. Ventresca, and C. Sova. 2017. Ties that bind: Local networks, communities and adaptive capacity in rural Ghana. *Journal of Rural Studies* 53: 214–228.

Cho, A.H. 2006. Politics, values and social entrepreneurship: A critical appraisal. In *Social entrepreneurship*, ed. J. Mair, J. Robinson, and K. Hockerts, 34–56. London: Palgrave Macmillan.

Christensen, C.M., H. Baumann, R. Ruggles, and T.M. Sadtler. 2006. Disruptive innovation for social change. *Harvard Business Review* 82 (12): 94–101.

Cohen, B., and M.I. Winn. 2007. Market imperfections, opportunity and sustainable entrepreneurship. *Journal of Business Venturing* 22 (1): 29–49.

Coleman, J.S. 2000. *Foundations of social theory.* Cambridge, MA: Belknap Press.

Combs, J.G., and D.J. Ketchen. 1999. Explaining interfirm cooperation and performance: Toward a reconciliation of predictions from the resource-based view and organizational economics. *Strategic Management Journal* 20 (9): 867–888.

Cook, M.L., T. Reardon, C. Barrett, and J. Cacho. 2001. Agroindustrialization in emerging markets: Overview and strategic context. *International Food and Agribusiness Management Review* 2 (3/4): 277–288.

Dacin, M.T., P.A. Dacin, and P. Tracey. 2011. Social entrepreneurship: A critique and future directions. *Organization Science* 22 (5): 1203–1213.

Dacin, P.A., M.T. Dacin, and M. Matear. 2010. Social entrepreneurship: Why we don't need a new theory and how we move forward from here. *Academy of Management Perspectives* 24 (3): 37–57.

Dees, J.G., B.B. Anderson, and J. Wei-Skillern. 2004. Scaling social impact: Strategies for spreading social innovations. *Stanford Social Innovation Review* 1 (4): 14–32.

Defourny, J., and M. Nyssens. 2010. Conceptions of social enterprise and social entrepreneurship in Europe and the United States: Convergences and divergences. *Journal of Social Entrepreneurship* 1 (1): 32–53.

Delgado, M., M.E. Porter, and S. Stern. 2010. Clusters and entrepreneurship. *Journal of Economic Geography* 10 (4): 495–518.

Desa, G. 2012. Resource mobilization in international social entrepreneurship: Bricolage as a mechanism of institutional transformation. *Entrepreneurship Theory and Practice* 36 (4): 727–751.

Desa, G., and S. Basu. 2013. Optimization or bricolage? Overcoming resource constraints in global social entrepreneurship. *Strategic Entrepreneurship Journal* 7 (1): 26–49.

Desa, G., and J.L. Koch. 2014. Scaling social impact: Building sustainable social ventures at the base-of-the-pyramid. *Journal of Social Entrepreneurship* 5 (2): 146–174.

Di Domenico, M., H. Haugh, and P. Tracey. 2010. Social bricolage: Theorizing social value creation in social enterprises. *Entrepreneurship Theory and Practice* 34 (4): 681–703.

Ding, J., P. Moustier, X. Ma, X. Huo, and X. Jia. 2019. Doing but not knowing: How apple farmers comply with standards in China. *Agriculture and Human Values* 36: 61–75.

Dorado, S. 2006. Social entrepreneurial ventures: Different values so different process of creation, no? *Journal of Developmental Entrepreneurship* 11 (4): 319–343.

Dorado, S., and M.J. Ventresca. 2013. Crescive entrepreneurship in complex social problems: Institutional conditions for entrepreneurial engagement. *Journal of Business Venturing* 28 (1): 69–82.

Duranton, G., and D. Puga. 2004. Microfoundations of urban agglomeration economics. In *Handbook of region and urban economics (Vol 4: Cities and Geography)*, ed. V. Henderson and J.-F. Thisse, 2063–2111. Amsterdam: Elsevier.

Dyer, J.H. 1997. Effective interim collaboration: How firms minimize transaction costs and maximise transaction value. *Strategic Management Journal* 18 (7): 535–556.

Eberlei, W. 2001. *Institutionalised participation in processes beyond the PRSP*: Deutsche Gesellschaft für Technische Zusammenarbeit (GTZ) GmbH.

Edmans, Alex. 2011. Does the stock market fully value intangibles? Employee satisfaction and equity prices. *Journal of Financial Economics* 101: 621–640.

Ervaes, J., T.L. Jacobson, and S.A. White (eds.). 1996. *Participatory communication for social change*. New Delhi: Sage.

Etzkowitz, H. 2008. *The triple helix: University-industry-government innovation in action*. London: Routledge.

Fisher, G. 2012. Effectuation, causation, and bricolage: A behavioral comparison of emerging theories in entrepreneurship research. *Entrepreneurship Theory and Practice* 36 (5): 1019–1051.

Franko, P. 2018. *The puzzle of Latin American economic development*. Lanham: Rowman & Littlefield.

Ghemawat, P. 2001. Distance still matters: The hard reality of global expansion. *Harvard Business Review* 79 (8): 137–147.

GIIN. 2018. *Annual impact investor survey*, 8th ed. New York: Global Impact Investing Network.

Hamschmidt, J., and M. Pirson (eds.). 2018. *Case studies in social entrepreneurship and sustainability: The oikos collection (Vol. 2)*. London: Routledge.

Haugh, H. 2007. Community-led social venture creation. *Entrepreneurship Theory and Practice* 31 (2): 161–182.

Hjorth, D. 2013. Public entrepreneurship: Desiring social change, creating sociality. *Entrepreneurship and Regional Development* 25 (1–2): 34–51.

Hornberger, K and Chau, V. (2018). Segmenting enterprises to better understand their financial needs, the missing middles, Omidyar Foundation, Palo Alto CA. https://www.omidyar.com/sites/default/files/Enterprise_Segmentation_Summary_Report.pdf. Accessed 16 January 2019.

Huesca, R., and B. Dervin. 1994. Theory and practice in Latin American alternative communication research. *Journal of Communication* 44 (4): 53–73.

Idemudia, E.C., and M.S. Raisinghani. 2014. The influence of cognitive trust and familiarity on adoption and continued use of smartphones: An empirical analysis. *Journal of International Technology and Information Management* 23 (2): 6.

Isenberg, D.J. 2010. How to start an entrepreneurial revolution. *Harvard Business Review* 88 (6): 40–50.

Jacobson, T.L. 2003. Participatory communication for social change: The relevance of the theory of communicative action. *Annals of the International Communication Association* 27 (1): 87–123.

Jenkins, B., and E. Ishikawa. 2010. *Scaling up inclusive business: Advancing the knowledge and action agenda*. Cambridge: Harvard Kennedy School.

Kania, J., M. Kramer, and P. Russell. 2014. Strategic philanthropy for a complex world. *Stanford Social Innovation Review* 12 (3): 26–33.

Kayser, O., and V. Budinich. 2015. *Scaling of business solutions to social problems*. London: Palgrave Macmillan.

Khanna, R., I. Guler, and A. Nerkar. 2016. Fail often, fail big, and fail fast? Learning from small failures and R&D performance in the pharmaceutical industry. *Academy of Management Journal* 59 (2): 436–459.

Kleemann, L., A. Abdulai, and M. Buss. 2014. Certification and access to export markets: Adoption and return on investment of organic-certified pineapple farming in Ghana. *World Development* 64: 79–92.

Korsgaard, S., and A.R. Anderson. 2011. Enacting entrepreneurship as social value creation. *International Small Business Journal* 29 (2): 135–151.

Kroeger, A., and C. Weber. 2014. Developing a conceptual framework for comparing social value creation. *Academy of Management Review* 39 (4): 513–540.

Kroll, C., and J. Delhey. 2013. A happy nation? Opportunities and challenges of using subjective indicators in policymaking. *Social Indicators Research* 114 (1): 13–28.

Lasprogata, G.A., and M.N. Cotten. 2003. Contemplating"enterprise": The business and legal challenges of social entrepreneurship. *American Business Law Journal* 41 (1): 67–114.

Lepoutre, J., R. Justo, S. Terjesen, and N. Bosma. 2013. Designing a global standardized methodology for measuring social entrepreneurship activity: The Global Entrepreneurship Monitor social entrepreneurship study. *Small Business Economics* 40 (3): 693–714.

Lichtenstein, B.B., N.M. Carter, K.J. Dooley, and W.B. Gartner. 2007. Complexity dynamics of nascent entrepreneurship. *Journal of Business Venturing* 22 (2): 236–261.

London, T., and S. Hart (eds.). 2011. *Next-generation business strategies for the base of the pyramid: New approaches for building mutual value*. Upper Saddle River, NJ: Financial Times Press.

Long, H., S. Tu, D. Ge, T. Li, and Y. Liu. 2016. The allocation and management of critical resources in rural China under restructuring: Problems and prospects. *Journal of Rural Studies* 47: 392–412.

Mair, J., and Marti, I. 2006. Social entrepreneurship research: A source of explanation, prediction, and delight. *Journal of World Business* 41 (1): 36–44.

Mair, J., and Marti, I. 2009. Entrepreneurship in and around institutional voids: A case study from Bangladesh. *Journal of Business Venturing* 24 (5), 419–435.

Mair, J., and E. Noboa. 2006. Social entrepreneurship: How intentions to create a social venture are formed. In *Social entrepreneurship*, ed. J. Mair, 121–135. London: Palgrave Macmillan.

Mair, J., and O. Schoen. 2007. Successful social entrepreneurial business models in the context of developing economies: An explorative study. *International Journal of Emerging Markets* 2 (1): 54–68.

Mair, J., I. Marti, and M.J. Ventresca. 2012. Building inclusive markets in rural Bangladesh: How intermediaries work institutional voids. *Academy of Management Journal* 55 (4): 819–850.

Martin, R.L., and S. Osberg. 2007a. *Social entrepreneurship: The case for definition (Vol 5, No. 2)*, 28–39. Stanford, CA: Stanford Social Innovation Review.

Martin, R.L., and S. Osberg. 2007b. Social entrepreneurship: The case for definition. *Stanford Social Innovation Review* 1: 29–39.

Mazumdar, D. 1987. Rural-urban migration in developing countries. In *Handbook of regional and urban economics (Vol II)*, ed. E.S. Mills. Amsterdam: Elsevier.

Meyskens, M., C. Robb-Post, J.A. Stamp, A.L. Carsrud, and P.D. Reynolds. 2010. Social ventures from a resource-based perspective: An exploratory study assessing global Ashoka Fellows. *Entrepreneurship Theory and Practice* 34 (4): 661–680.

Miller, T.L., and C.L. Wesley. 2010. Assessing mission and resources for social change: An organizational identity perspective on social venture capitalists 'decision criteria. *Entrepreneurship Theory and Practice* 34 (4): 705–733.

Narain, Saurabh, and Joseph Schmidt. 2009. NCIF social performance metrics: Increasing the flow of investments in distressed neighborhoods through community development banking institutions. *Community Development Investment Review* 2: 65–75.

NCAUSA. 2018. National Coffee Association. https://www.ncausa.org/Industry-Resources/Webinars/NCDT-2018.

NCIF. 2019. National Community Investment Foundation. https://ncif.org/connect/about-ncif#.XEO8UhMzbdQ.

Nicholls, A. 2006. *Social entrepreneurship – New models of sustainable social change*. Oxford: Oxford University Press.

Nicholls, A. (Ed.). 2008. *Social entrepreneurship: New models of sustainable social change*. OUP Oxford.

Nicholls, A. 2010. The legitimacy of social entrepreneurship: Reflexive isomorphism in a preparadigmatic field. *Entrepreneurship Theory and Practice* 34: 611–633.

Nohlen, D., and Nuscheler, F. 1974. *Handbuch der Dritten Welt*: Band 1, Hamburg.

Nooteboom, B., H. Berger, and N.G. Noorderhaven. 1997. Effects of trust and governance on relational risk. *Academy of Management Journal* 40 (2): 308–338.

OECD. 2017a. Development aid rises again in 2016 but flows to poorest countries dip. https://www.oecd.org/dac/development-aid-rises-again-in-2016-but-flows-to-poorest-countries-dip.htm. Accessed 11 October 2018.

OECD. 2017b. Global private philanthropy for development. https://www.oecd.org/dac/financing-sustainable-development/development-finance-standards/Philanthropy-Development-Survey.pdf. Accessed 13 November 2018.

OECD. 2018. *Making blended finance work for the sustainable development goals*. Paris: The Organisation for Economic Co-operation and Development OECD Publishing.

Parish, W.L. 2016. *Chinese rural development: The great transformation: The great transformation*. London: Routledge.

Parsons, T. 1970. *The social system*. London: Routledge & Kegan Paul Ltd.

Pelling, M. 2010. *Adaptation to climate change: From resilience to transformation*. London: Routledge.

Phills, J.A., K. Deiglmeier, and D.T. Miller. 2008. Rediscovering social innovation. *Stanford Social Innovation Review* 6 (4): 34–43.

Piao, S., P. Ciais, Y. Huang, Z. Shen, S. Peng, J. Li, et al. 2010. The impacts of climate change on water resources and agriculture in China. *Nature* 467 (7311): 43.

Pingali, P. 2007. Westernization of Asian diets and the transformation of food systems: Implications for research and policy. *Food Policy* 32 (33): 281–298.

Prahalad, C.K. 2005. *The fortune at the bottom of the pyramid*. Upper Saddle River: Wharton School Publishing.

Prahalad, C.K., and A. Hammond. 2002. Serving the world's poor, profitably. *Harvard Business Review* 80 (9): 48–57.

Praszkier, R., and A. Nowak. 2012. *Social entrepreneurship: Theory and practice*. New York: Cambridge University Press.

Praszkier, R., A. Nowak, and A. Zabocka-Bursa. 2009. Social capital built by social entrepreneurs and the specific personality traits that facilitate the process. *Psychologia Spoleczna* 1–2 (10): 42–54.

Ratliff, G.A., and K.S. Moy. 2004. New pathways to scale for community development finance. *Profitwise News and Views* 12 (4): 1–23.

Reardon, T., and C.B. Barrett. 2000. Agroindustrialization, globalization, and international development: An overview of issues, patterns, and determinants. *Agricultural Economics* 23 (3): 195–205.

Renko, M. 2012. Early challenges of nascent social entrepreneurs. *Entrepreneurship, Theory and Practice* 37 (5): 1045–1069.

Rivera-Santos, M., D. Holt, D. Littlewood, and A. Kolk. 2015. Social entrepreneurship in sub-Saharan Africa. *Academy of Management Perspectives* 29 (1): 72–91.

Rogers, E. 1962. *The diffusion of innovations*. Glencoe, IL: Free Press.

Ruvio, A., and A. Shoham. 2011. A multilevel study of nascent social ventures. *International Small Business Journal* 29 (5): 562–579.

Sachs, J. (2005). The end of poverty: How we can make it happen in our lifetime. Penguin UK.

Sachs, J., and Schmidt-Traub, G. 2014. *Financing sustainable development: Implementing the SDGs through effective investment strategies and partnerships*: UNCTAD: 'World Investment Report 2014. Investing in the SDGs: An Action Plan'.

Santos, F.M. 2012. A positive theory of social entrepreneurship. *Journal of Business Ethics* 111 (3): 335–351.

Saul, J. 2010. *Social innovation Inc: Five strategies to drive business value through social change*. Hoboken: Wiley.

Sawhill, J.C., and D. Williamson. 2001. Mission impossible? Measuring success in nonprofit organizations. *Nonprofit Management and Leadership* 11 (3): 371–387.

Schlenker, W., and M.J. Roberts. 2009. Nonlinear temperature effects indicate severe damages to US crop yields under climate change. *Proceedings of the National Academy of sciences* 106 (37): 15594–15598.

Schneider, H. 1999. *Participatory governance: the missing link for poverty reduction*: OECD Development Centre Policy Brief No. 17.

Scott, W.R. 2001. *Institutions and organizations*. Thosands Oaks: Sage.

Servaes, J. 1996. Linking theoretical perspectives to policy. In *Participatory communication for social change*, ed. J. Servaes, T.L. Jacobson, and S.A. White, 29–43. New Delhi: Sage.

Servaes, J. (ed.). 2002. *Approaches to development communication*. Paris: The United Nations Educational, Scientific and Cultural Organization (UNESCO).

Shane, S., and S. Venkataraman. 2000. The promise of entrepreneurship as a field of research. *Academy of Management Review* 25 (1): 217–226.

Short, J.C., T. Moss, and G. Lumpkin. 2009. Research in social entrepreneurship: Past contributions and future opportunities. *Strategic Entrepreneurship Journal* 3: 161–194.

Simanis, E., and S. Hart. 2009. Innovation from the inside out. *Sloan Management Review (Summer)* 9: 77–86.

Smith, W., M. Besharov, A. Wessels, and M. Chertok. 2012. A paradoxical leadership model for social entrepreneurs: Challenges, leadership skills, and pedagogical tools for managing social and commercial demands. *Academy of Management Learning and Education* 11 (3): 463–478.

Sommerrock, K. 2010. *Social entrepreneurship business models: Incentive strategy to catalyze public goods provision*. London: Palgrave Macmillan.

Stone, G.D. 2011. Contradictions in the last mile: Suicide, culture, and E-Agriculture in rural India. *Science, Technology, & Human Values* 36 (6): 759–790.

Thapalia, C.F. 1996. Animation and leadership. In *Participatory communication for social change*, ed. J. Ervaes, T.L. Jacobson, and S.A. White, 150–161. New Delhi: Sage.

Thomas, L. D. W. 2013. Ecosystem emergence: An investigation of the emergence process in six digital services ecosystems. (Unpublished Ph.D. dissertation). London, U.K.: Imperial College. https://spiral.imperial.ac.uk/bitstream/.10044/1/18315/1/Thomas-LDW-2013-PhD-Thesis.pdf.

Thompson, T.A., J.M. Purdy, and M.J. Ventresca. 2018. How entrepreneurial ecosystems take form: Evidence from social impact initiatives in Seattle. *Strategic Entrepreneurship Journal* 12 (1): 96–116.

Thornley, B. 2013. The Facts on U.S. Social Enterprise. https://www.huffingtonpost.com/ben-thornley/social-enterprise_b_2090144.html. Accessed June 5 2018.

Thornton, P.H., W. Ocasio, and M. Lounsbury. 2012. *The institutional logics perspective: A new approach to culture, structure, and process*. Oxford: Oxford University Press on Demand.

Tina Dacin, M., J. Goodstein, and W. Richard Scott. 2002. Institutional theory and institutional change: Introduction to the special research forum. *Academy of Management Journal* 45 (1): 45–56.

Townsend, D.M., and T.A. Hart. 2008. Perceived institutional ambiguity and the choice of organizational form and social entrepreneurial ventures. *Entrepreneurship: Theory & Practice* 32: 685–700.

United Nations. 2018. 68% of the world population projected to live in urban areas by 2050, says UN: https://www.un.org/development/desa/en/news/population/2018-revision-of-world-urbanization-prospects.html. Accessed July 18 2018.

Ventresca, M., and Fu, X. 2017. Meeting in the middle: A multi-level analysis of Chinese HIV civil organisations (Doctoral dissertation, University of Oxford).

Waddock, S.A., and J.E. Post. 1991. Social entrepreneurs and catalytic change. *Public Administration Review* 51 (5): 393–401.

Wang, H.L., Y.T. Gan, R.Y. Wang, J.Y. Niu, H. Zhao, Q.G. Yang, and G.C. Li. 2008. Phenological trends in winter wheat and spring cotton in response to climate changes in northwest China. *Agricultural and Forest Meteorology* 148 (8–9): 1242–1251.

Weerawardena, J., and G.S. Mort. 2006. Investigating social entrepreneurship: A multidimensional model. *Journal of World Business* 41 (1): 21–35.

WEF & OECD. 2015. *Blended finance Vol. 1: A primer for development finance and philanthropic funders*: World Economic Forum & OECD.

Windsperger, J., G. Cliquet, T. Ehrmann, and G. Hendrikse (eds.). 2015. *Interfirm networks - Franchising, cooperatives and strategic alliances*. Heidelberg: Springer.

World Bank Group/KNOMAD. 2017. Migration and remittances: Recent developments and outlook: Migration and Development Brief, No. 27, World Bank Group/ Global Knowledge Partnership on Migration and Development (KNOMAD), Washington, DC. https://pubdocs.worldbank.org/en/992371492706371662/MigrationandDevelopmentBrief27.pdf.

World Bank. 2014. *Urban China: Toward efficient, inclusive, and sustainable urbanization*: World Bank and the Development Research Center of the State Council, P. R. China. Washington, DC: World Bank.

Wyche, S., and C. Steinfield. 2016. Why don't farmers use cell phones to access market prices? Technology affordances and barriers to market information services adoption in rural Kenya. *Information Technology for Development* 22 (2): 320–333.

Yu, X. 2011. Social enterprise in China: Driving forces, development patterns and legal framework. *Social Enterprise Journal* 7 (1): 9–32.

Zahra, S.A., E. Gedajlovic, D.O. Neubaum, and J.M. Shulman. 2009. A typology of social entrepreneurs: Motives, search processes and ethical challenges. *Journal of Business Venturing* 24 (5): 519–532.

Zhao, M. 2012. The social enterprise emerges in China. *Stanford Social Innovation Review* 10 (2): 30–35.

Publisher's Note Springer Nature remains neutral with regard to jurisdictional claims in published maps and institutional affiliations.

Xiangping Jia is a professor at the Agricultural Information Institute, Chinese Academy of Agricultural Sciences. He received his Ph.D. from University of Hohenheim, Germany. Working on research about agricultural policies and rural development, he has broad interest in the application of new institutional economics to the issues of agri-food market coordination, development of inclusive financial markets, the organization of smallholder farms, knowledge transfer and innovation, and agricultural extension systems. At present, he researches innovative institutions that align agricultural universities with stakeholders such as public extension systems and private sectors for a resilient farm system.

Geoffrey Desa is a professor at San Francisco State University. He received his Ph.D. from theUniversity of Washington, an M.S. from Stanford University, and a B.S. from Georgia Tech. Hefollows two questions: How do social innovations scale and have social impact? Under whatconditions does academic learning drive social change? His research adds empirical evidence tothe literature on social innovation at the international and local level, and to the pedagogicalliterature that emphasizes critical thought and learning. He teaches classes at the intersectionof business, innovation for sustainability and climate adaptation.

Agriculture and Human Values (2020) 37:1241–1260
https://doi.org/10.1007/s10460-020-10141-6

'Workable utopias' for social change through inclusion and empowerment? Community supported agriculture (CSA) in Wales as social innovation

Tezcan Mert-Cakal[1] · Mara Miele[2]

Published online: 18 August 2020
© The Author(s) 2020

Abstract

The focus of this article is community supported agriculture (CSA) as an alternative food movement and a bottom-up response to the problems of the dominant food systems. By utilizing social innovation approach that explores the relationship between causes for human needs and emergence of socially innovative food initiatives, the article examines how the CSA projects emerge and why, what is their innovative role as part of the social economy and what is their transformative potential. Based on qualitative data from four different models of CSA case studies in different regions of Wales, UK, and by using concepts from an alternative model for social innovation (ALMOLIN) as analytical tool, the article demonstrates that the Welsh CSA cases play distinctive roles as part of the social economy. They satisfy the needs for ecologically sound and ethically produced food, grown within communities of like-minded people and they empower individuals and communities at micro level, while at the same time experiment with how to be economically sustainable and resilient on a small scale. The paper argues that in order to become 'workable utopias', the CSA initiatives need to overcome the barriers that prevent them from replicating, participating in policies and decision-making at macro level, and scaling up.

Keywords Community supported agriculture · CSA · Social innovation · Alternative food · Grassroots initiatives · Food sustainability

Introduction

The growth in alternative food networks (AFNs) in late 1980s and early 2000s is indicative of the bottom-up responses to the unsustainable food systems that are increasingly unable to address the needs and demands of food producers and consumers alike (Sage 2014). Farmers' markets, box schemes, community supported agriculture (CSA), producer and consumer co-operatives, and community gardening initiatives are all examples of such AFNs (Jarosz 2008).

✉ Mara Miele
  mielem@cardiff.ac.uk

  Tezcan Mert-Cakal
  tezcanmert@yahoo.com

[1] Cardiff University Alumna, 48 St James Court, Altrincham WA15 8FG, Greater Manchester, UK

[2] School of Geography and Planning, Cardiff University, Glamorgan Building, King Edward VII Avenue, Cardiff CF10 3WA, Wales, UK

Bos and Owen (2016) argue that these types of food provisioning systems are significantly different to conventional counterparts as they can redefine relations between producers and consumers through transparent short(er) food supply chains. These are founded upon quality and provenance and point towards more sustainable modes of production (Marsden et al. 2000; Renting et al. 2003; Sage 2003; Goodman 2004; Ilbery and Maye 2005; Morris and Kirwan 2011).

The recent COVID-19 pandemic exposed not only the vulnerabilities and risks of the current food systems, specifically of the longer supply chains, but also its deep inequalities and injustices (Anderson 2020). On the one hand, many places around the world have faced empty supermarket shelves while on the other, crops were left to rot on the field due to restriction of movement, causing shortage of seasonal workers (Hendrickson 2020; Gustin 2020). These were further exacerbated by disruption in logistics and shortage of animal feed and fertilizers (Roy Chaudury 2020; Fertilizers_Europe n.d.). On international level, since most of the grains are traded across borders as a result of the trade liberalization of 1980s, there have been fears of shortage

of staples as some of the exporting countries restricted the supply, which reminded of 2007–2008 food crisis (IPES-Food 2020). However, as much as the COVID-19 pandemic exposed how fragile and unsustainable the global food systems are in the event of shock, it also demonstrated the resilience of the local food initiatives and short food supply chains. CSA schemes in many countries saw increased customer numbers while the interest to local box schemes grew dramatically (Schmidt et al. 2020; URGENCI 2020). For example, a survey of the vegetable box schemes in the UK revealed an overall increase in sales by 111% during the pandemic, with increase in small boxes by 134%; 10% of these schemes created systems to help those who were economically vulnerable (Wheeler 2020). Furthermore, local initiatives created platforms to collaborate with each other, exchange produce and help people in need. For instance, The CSA UK Network published online resources to help the initiatives cope with the COVID-19 conditions and increased demand (CSA 2020).

In this paper, we examine the role and transformative potential of the CSA in Wales (UK) by utilizing an alternative social innovation approach that explores the relationship between causes for human needs and emergence of socially innovative food initiatives (González et al. 2010). The CSA is an innovative idea bringing consumers and producers together, where consumers share the risks and benefits of production (Hinrichs 2000; Hayden and Buck 2012) and both sides mutually resolve some uncertainties (Lamine 2005). It started in 1970s in Japan and Switzerland and spread later to other parts of the world. The International CSA Network URGENCI was launched in 2008, and the CSA Network UK followed in 2013. The existing literature does not offer much evidence about how these initiatives emerge, what resources are mobilized and what needs they aim to address. Secondly, little research has focused on how the CSA empowers individuals and communities. And finally, the existing studies examining the CSA from social innovation perspective tend to favour socio-technical transition frameworks based on niche-regime interaction (Brunori et al. 2010; Marsden 2013; Rossi 2017), where niches are spaces for experimenting with innovative ideas and regime is the complex of settled institutions, policies, regulations, actors and relations (Kemp et al. 1998; Smith 2007; Schot and Geels 2008). Therefore, our threefold aim is to address these gaps in the literature: first, by examining how the CSA initiatives in Wales emerge and why; second, by exploring their innovative role as part of the social economy; and third, by scrutinizing their transformative potential through an alternative social innovation framework. We use data from four CSA schemes in Wales as qualitative case studies based on participant observations and semi-structured interviews. The paper proceeds in the following way: first we review the literature about social innovation, CSA and AFNs; secondly,

we explain our methodological approach, then we present the four cases in Wales and the results that correspond to the research aims; and finally, we discuss their theoretical and practical implications before concluding with areas for further research.

Social innovation, CSA and AFNs

Social innovation is defined by Gilles Deleuze as "opportunity spaces at micro scales [that] may make creative strategies possible at macro scales" and as a way of "building 'workable utopias'"; it is about countering the conservative forces eager to preserve social exclusion situations, thus being "an ethical position of social justice" (Moulaert et al. 2013, p. 17). A source for social innovation is the social economy (Howaldt et al. 2014), which is an alternative vision for the economy, different from its neoliberal representation as 'a monolithic entity' detached from social life (Jessop et al. 2013). MacCallum et al. (2009, pp. 1–2) explain that social innovation emerges through collective action, social movements or public policy to address social problems like exclusion, deprivation and lack of wellbeing, and improve human conditions through satisfaction of needs, empowerment and improvement of social relations, also defined as the three main dimensions of social innovation (Moulaert et al. 2005), namely (1) product dimension, which focuses on the satisfaction of 'alienated' needs that have not yet been satisfied or are no longer considered as important by mainstream actors, (2) empowerment dimension, about increasing the socio-political capabilities and access to resources required to satisfy those needs, and (3) process dimension, about the change in social and governance relations.

Regarding the product dimension, firstly, identifying the needs is essential to understand what triggers the emergence of social innovations. According to Parra (2013), needs in social innovation can be material and existential and may be collectively defined by communities. Therefore, we ask why people participate in the CSA schemes. The closest answer that the existing literature offers is about motives for participation rather than needs. For the consumers, food safety concerns and knowing the source of their food (Cooley and Lass 1998; Goland 2002), acquiring quality and nutritious produce (Sharp et al. 2002; Farmer et al. 2014), addressing environmental concerns, and supporting local farmers (Goland 2002; MacMillan Uribe et al. 2012; Farmer et al. 2014) are primary motives while for the producers, these range from providing organic and seasonal produce for local people (Cox et al. 2008) to accessing larger markets, increasing awareness of the food systems, and building stronger community (Sharp et al. 2002). But although motives demonstrate why people participate to the schemes, there is lack of evidence in the literature about the needs that create deprivation and exclusionary circumstances in the case of the CSA. Furthermore, identifying the needs will enable us to establish if the product dimension of social innovation has been addressed. Again, despite an abundance of research on the various benefits of the CSA, such as health benefits (Ostrom 2007; Cohen et al. 2012; Minaker et al. 2014; Wilkins et al. 2015; Wharton et al. 2015; Allen IV et al. 2017), lifestyle changes (Ostrom 2007), impact on the participants' environmental ethics (Hayden and Buck 2012), higher benefits for lower-income members (Galt et al. 2016), meeting psychological needs (Zepeda et al. 2013), and being powerful approach to food justice (Gottlieb and Joshi 2010, p. 149), the existing studies do not provide much evidence about the role of the CSA initiatives in satisfying the unmet needs causing their emergence.

The empowerment dimension of social innovation is about increasing the socio-political capabilities of individuals and communities (Moulaert et al. 2005) by including people in decision-making and service provision and creating common visions for change (González et al. 2010). Empowering people means increasing the recognition, access and voice rights of marginalised groups (Martinelli 2010). Renting et al. (2012) suggest that access to healthy food in a socially inclusive way and engagement in food growing is a way of empowerment. In addition, building strong community is a way of increasing their socio-political capabilities of the CSA initiatives, also considered "a major selling point" in attracting more members (Schnell 2007, p. 559). There is a positive correlation between community capital and the retention of members (Flora and Bregendahl 2012). Many studies report difficulties about building strong community in the CSA, mainly related to attracting and retaining members, attributed to disappointment of the type and amount of produce (Hinrichs 2000; Ostrom 2007; Janssen 2010; Hayden and Buck 2012), or little interest among the members in participating to community events. Consequently, maintaining the community side is left to "already overworked CSA farmers" (Hinrichs 2000, p. 300), and finding and retaining members in many cases happens at the expense of the producers' self-exploitation (Galt 2013). Studies suggest that targeting people who are committed to environmental values (Goland 2002) and better communication between producers and consumers are ways of increasing consumers' commitment in longer term (Cox et al. 2008). However, as much as the existing research examines the issues related to social capital, it does not offer much evidence about other ways of empowerment in the CSA, such as learning or participating in decision-making, which can enhance the capabilities and the voice rights of people.

The process dimension of social innovation is not only about changing relations between individuals, but at macro-level it is about changing the governance relations between market economy and social economy and reorganising the power dynamics between the state, civil society and the market (González et al. 2010). It is related to the transformative potential of the CSA and AFNs in general. On the positive side, these networks are seen as a response to the growing problems of the conventional food system (Mount et al. 2013), as innovative means of the social economy juxtaposed to the market economy, named 'seeds of change' (Seyfang 2009, p. 74), as 'diverse economies' (Gibson-Graham 2008, p. 2), or as 'the new moral economy' based on ethical values in contrast to the neo-liberal economy (Morgan et al. 2006, pp. 166–167). Moreover, AFNs are believed to possess the ability to "reconvene trust between food producers and consumers" and "articulate new forms of political association and market governance" (Whatmore et al. 2003, p. 389). One of the criticisms that AFNs face is their inability to tackle social injustices by serving predominantly middle-class consumers (Renting et al. 2012) and white people (Guthman 2008a) and perpetrating social inequalities instead of including disadvantaged populations (Matacena 2016), defined as "narrow 'class diet' of privileged income groups" (Goodman 2004, p. 13). It is suggested that the obscure inequalities and injustices created by the AFNs are caused by their focusing exclusively on local values, named 'defensive localism' (Hinrichs 2003; Winter 2003) or 'un-reflexive' localism (DuPuis and Goodman 2005). Particularly the CSA schemes are criticized for serving mainly those who have the necessary education, income and time to commit (Cone and Myhre 2000) or affluent consumers (Selfa and Qazi 2005) by being predominantly located in areas populated by middle and upper middle class (Schnell 2007), and for creating "marginalization and powerlessness" by excluding certain groups (Farmer et al. 2014, p. 323, emphasis original). A counter-argument states that both 'social inclusion' and 'social exclusion' are contested terms, as not participating can also mean a choice (Shortall 2008). And although some studies suggest that the CSA schemes can become more inclusive by being sensitive to ethnicity and class positions (Caraher and Dowler 2014; Galt et al. 2016), the existing literature does not offer concrete examples of how these initiatives address the social injustice and exclusion.

Another criticism regarding the transformative potential of the AFNs is about their failure to counter the corporate food regime and transform the food systems. Food movements are accused of trying to solve social problems by placing responsibilities on individuals, which results only in changes at market level rather than state level, or changes in local policies rather than national policies (Guthman 2008b; Alkon and Mares 2012; Fairbairn 2012). Therefore, regarding the position of alternative food initiatives against the dominant food system, Watts et al. (2005) distinguish between 'weak' alternatives, that put emphasis on 'local' and can be subordinated by the conventional food supply chains, and 'strong' alternatives that can provide spatial, social and economic alternatives to conventional networks. Follett (2009) suggests that 'weak' alternatives adopt customs of both alternative and conventional networks, while 'strong' alternatives are guided by the customs of moral economy, such as human and animal welfare, community-building, supporting small scale farmers, ecological sustainability, trust and transparency. When it comes to the ways in which the alternative food movement can scale up, Wiskerke (2009) offers a holistic approach by bringing the concept of alternative food geographies, which combine public procurement, urban food strategies and AFNs. In a similar way, Matacena (2016) states that urban food policies can provide outlets and growing spaces for the AFNs through infrastructure, spatial planning and public procurement. And finally, Blay-Palmer et al. (2016, p. 39) suggest building a "System of Sustainable Food Systems (SoSFS) as a counter-point to the corporate food regime" and based on community food networks connected via sharing good practices and knowledge. However, the literature does not provide much insight about how the actors of the CSA initiatives view their position and values against the main food system. Yet another gap is the lack of comparative evaluation of the barriers and opportunities for scaling up different CSA models.

In sum, the review of the literature on AFNs and CSA revealed several gaps related to each dimension of social innovation. Regarding the product dimension, there is a lack of studies about the needs that trigger the emergence of the CSA initiatives and the way these needs are satisfied. Regarding the empowerment dimension, apart from building strong community, there is not much evidence about learning and decision-making as other processes of empowerment. And the process dimension is under-researched in terms of the actors' perspective on the position and values of the CSA against the main food economy, the ways of addressing the social injustice and exclusion, and comparative evaluation of the barriers and opportunities for transformation of different CSA types. We address these gaps by searching an answer to the main question of this study: how the CSA initiatives in Wales can become 'workable utopias' for food systems' change through social inclusion and empowerment. More specifically, we ask (1) what are the needs that cause exclusionary circumstances and lead to the emergence of the CSA initiatives in Wales and how are these needs satisfied; (2) how do learning and participating in decision-making at various levels empower people; (3) what are the actors' perceptions about the position and values of the CSA against the main food system; (4) how do CSA farms and gardens tackle social injustice; and (5) what are the barriers and

possibilities for transformation of different CSA types in Wales that create 'path-dependency'.[1]

Methodology

To carry out this research, we used qualitative case studies based on semi-structured interviews and participant observation in four CSA initiatives in Wales. Case studies are particularly suitable as a method in understanding 'how' and 'why' the CSA initiatives operate (Yin 2003) and the 'real-life' phenomena unfold in practice (Flyvbjerg 2006, p. 235). Moreover, they are well suited to understand the systems, everyday practices, and the relations between the actors of the CSA initiative, that requires a "qualitative, context-sensitive [and] interactive" type of approach (Hamdouch 2013, p. 260). And lastly, case studies allow some general conclusions to be drawn through 'theoretical reasoning' (May and Perry 2011, p. 223). The four cases were selected based on two criteria: different locations and different ownership models. We used two sources to identify the potential cases. The first was the Federation of City Farms and Community Gardens (the Federation), which is the umbrella organisation of all community food growing projects in the UK (now renamed Social Farms and Gardens), and the second was the Soil Association,[2] which at the time listed on its website the Welsh CSA projects, later transferred to the website of the newly-launched CSA UK Network.[3]

The data was collected between July 2014 and February 2015 with 3–5 days spent in each CSA by being actively involved as volunteer in the daily works of the initiatives and doing semi-structured interviews and observations. Interview was chosen as method due to its usefulness in explaining the complexity of processes and systems in detail and to investigate personal approaches and perceptions (Ritchie 2003; Cloke et al. 2004, pp. 150–151). Participant observation of daily activities at the CSA sites allowed us to gain additional insights by having direct access to the phenomenon (Laurier 2010) and provided a more objective perspective since interviews

---

[1] 'Path dependency' or 'path dependence' is the tendency of institutions or technologies to become committed to develop in certain ways as a result of their structural properties or their beliefs and values (Encyclopaedia Britannica). It can create either 'lock-in' situation that blocks the change, or 'path-paving' when it supports the social innovation, or 'path-breaking' when it leads to a sudden transformation (González et al. 2010).

[2] The UK's leading organisation for promoting sustainable food and farming and the largest organic certification body in the country.

[3] Finding the cases was facilitated by the participation of the leading author to the CSA Network's launch conference in December 2013 in Stroud where she made connections with some of the Welsh initiatives.

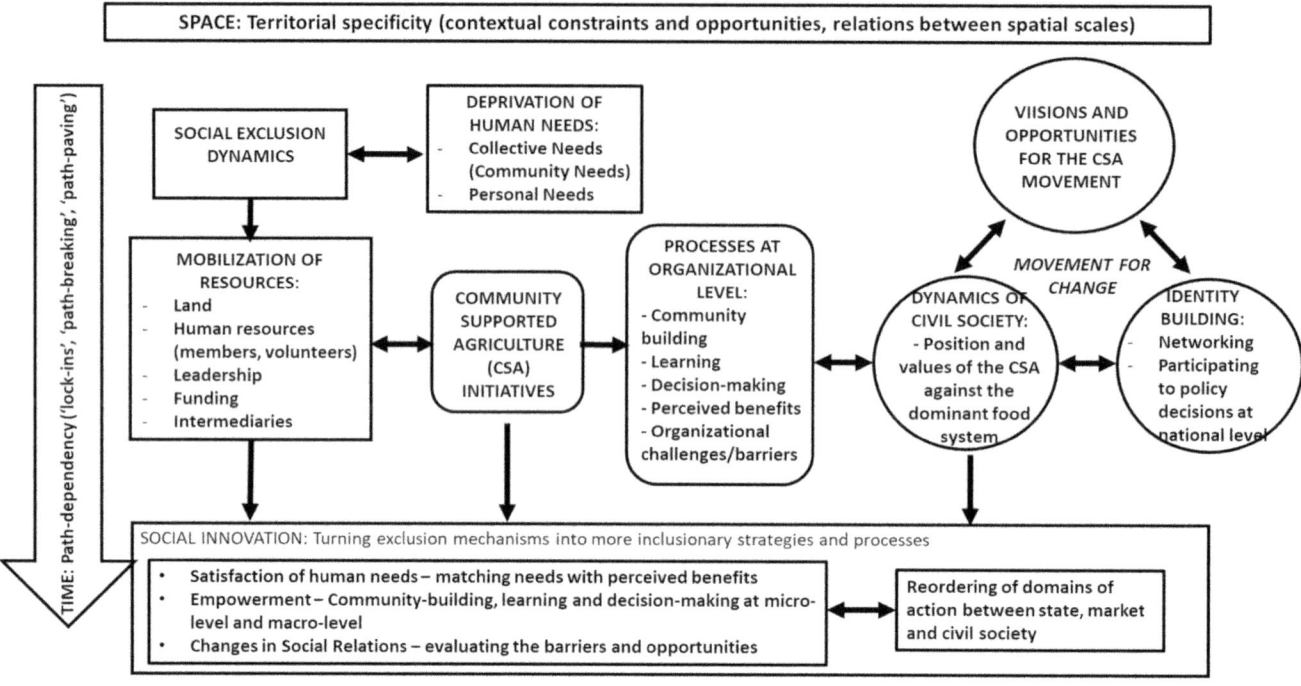

Fig. 1 ALMOLIN—themes of analysis for the CSA initiatives. Adapted from González et al. (2010, p. 52)

can sometimes be biased (Hancock and Algozzine 2006, p. 46) and cannot be replicated (Valentine 2005). We conducted 19 interviews in total with various actors involved in these initiatives, e.g. growers, members and volunteers and combined these with the observation notes during the analysis. We also revised related policy documents, reports, and media articles to establish a general picture of the CSA in Wales.

We based our thematic analysis on ALMOLIN—alternative model for local innovations—an innovative tool that enables mapping the relationship between causes of deprivation of human needs and the way resources are mobilised to create social economy initiatives and a bigger movement for change (Moulaert et al. 2005; González et al. 2010). We utilized the main concepts from the model and added or specified some processes and inputs that were key for evaluating the CSA initiatives in terms of social innovation dynamics, as shown on Fig. 1: (1) 'Needs'—we specified these as collective and personal; (2) In 'Mobilization of resources' we specified the resources used in founding the CSAs; (3) we added 'Processes at organizational level'; (4) we specified the 'Dynamics of the Civil Society'; (5) In 'Identity Building' we specified two key processes. The names of both the CSA projects and the interviewees were changed for confidentiality and anonymity purposes, in accordance with the research ethics procedures.

Results

How the CSA initiatives emerge

In this section we introduce the cases and provide a detailed picture of how resources such as leadership, human capital, land, revenue/funding, and technical support have been mobilized, thereby demonstrating how social innovations emerged and how different CSA models followed different strategies to acquire resources. Summary of the main characteristics of the cases is presented in Table 1 at the end of the section.

Bont Market Garden (Bont MG)

Situated on a 5-acre' land in Southeast Wales about 10 miles away from a big urban centre, Bont MG is a social enterprise registered as industrial provident society for community benefit and has about 100 shareholders—members of the enterprise. The actual growing on the site started in 2010. The garden operates as a box scheme with weekly deliveries to subscribers. Member-shareholders are not necessarily subscribers and for those who are, there is a small discount. Bont MG also sells its produce at a farmers' market in the nearby urban centre and delivers to

Table 1 Main characteristics of the CSA cases

Cases	BONT	TYDDEWI	CLWYD	OFFA
Model	Social enterprise with share-holders	Producer-community partner-ship	Community-led enterprise	Producer-led, owned by a couple of growers
Location	Southeast Wales (10 miles away from a big city)	Southwest Wales (rural farm 5 miles away from a town)	North Wales (3 different sites, close to 2 towns)	Mid-Wales (rural farm between two towns)
Starting Year	2010	2010	2011	2008
Sales	Box scheme (delivery) and farmers' market	Weekly boxes for the members (picked from the farm)	No sales; food shared weekly among members	Farmer's markets and shop
Land	5 acres rented	2 acres (of total 70) allocated for the CSA; land owned by the producer	3 acres + 600 m2 from private landlords + 4 polytunnels and an orchard from the local university (free use)	6 acres (4 acres rented + 2 acres owned)
Leaders	Board of 4 directors	Farm owner + core group from community	Chairperson, board of directors, site managers	Couple of growers
Paid staff	Main grower and assistant grower (both part-time)	Grower (part-time)	None	Worker (part-time)
Members	About 100 shareholders	40 members	About 20 members	About 20 members
Volunteers	Regular volunteering days	Regular volunteering days for members + hosting international volunteers	All members involved	Regular volunteering group from the local community
Revenue	Veg box subscription, sales at farmers' market + some restaurants	Annual membership fee + sales of salad to cafees	Symbolic amount of annual membership fee	Sales at two farmers' markets and a shop
Financial support	Grants for knowledge transfer, rent (land), growers' wages, and basic infrastructure	Grants for start-up, grower's wage, polytunnels, seeds, and a caravan	Grants for start-up and mentoring	Grant for organic conversion

local restaurants. The land is rented from a neighbouring community running a forest garden project and hosting events. Part of the initial financial capital of the garden comes from selling one-off shares of £50 each, which can be withdrawn if the shareholder does not wish to continue the membership. There is no monthly or annual subscription fee. Reportedly, more than 100 investors contributed with almost £10.000 in total. This enabled the purchase of a second-hand tractor and building two polytunnels. The enterprise received funding from the Welsh Government via the Rural Development Plan (RDP) under the EU Common Agriculture Policy (CAP), which was used for 'knowledge transfer' and another grant from an independent foundation to rent the land, employ a grower-horticulturalist and purchase some basic infrastructure. The project's revenue comes from sales at the farmers' market and the weekly box deliveries/restaurant deliveries. However, in order to keep producing, it needs more investment for better machinery, two more polytunnels, and a cold storage. Bont MG is managed by a board of four directors, who brought a range of expertise and skills to the scheme. One of the directors has experience in establishing sustainable businesses and social enterprises. Another provides networking and marketing support, but also has horticultural skills. A third director is a popular figure in the city and has access to a wide range of community groups. And the fourth person is the founding director who is a sustainable food mentor and is directly involved in the garden's daily issues. Additionally, he is actively involved in the local food council and with other community projects. The directors used their networks and connections to find shareholders and customers, and secure grants from various organisations. The tasks of growing food and organising the weekly boxes is predominantly done by a paid grower and assistant grower, both working on a part-time basis but usually putting more hours than what they are paid for. There are regular volunteering days but very few people come. Reportedly, the engagement level is very low at the annual general meeting (AGM), too. According to the management, those who became members-shareholders were driven by ethical motives rather than desire to invest.

> People … have not done the investment for a return; it's more eco-social investment. In other words, they

believe in the values, the ethics and the goal that we have. (Terry, founding director)

Tyddewi Community Organic Farm (TYCOF)

Located in Pembrokeshire, Southwest Wales, Tyddewi (TYCOF) was initially a private dairy and potato farm since 1938, owned by the father of the current owner and converted in 1996–1997 into organic farm. Inspired by one of the leading CSA farms in the UK, the owner established the CSA scheme in 2010 by allocating land of 2-acres out of total 70. Despite running on a privately-owned farm, the initiative has a distinctive model of producer-community collaboration in terms of management. The CSA produces vegetables and fruits for 40 members and delivers salad bags to some local cafes. This is the only CSA among the four cases that has revenue from annual membership fee, paid on a monthly basis either for a whole share or half share, depending on the membership. Although TYCOF had the advantage of being an established farm with its own land and some equipment, it still needed tools and additional equipment, polytunnels and seeds because it was not a horticultural farm in the past. In addition, it needed to promote the community scheme to attract more members. Establishing a CSA project only became available with a dedicated leadership, particularly from the farm owner who is passionate about food sustainability, and the networking and support of many people and organisations. For example, the local Eco City Group provided a start-up fund and grants for a grower's salary, a caravan for the international volunteers, and technical support. Other grants came from the National Parks UK and a local CAP group for establishing polytunnels. A bank sponsored seeds as a start off for the first 2 years and the Federation provided technical support. The human capital consists of a paid grower, members, and volunteers. The son of the farm owner is recruited as a grower and paid for 20 h/week, but he works more than 40 h/week. At the start-up stage, the founding members paid for the initial fund, first without receiving any vegetables, and put a lot of time and effort in the project.

> We signed up to start paying £30 a month towards [the CSA] as if we were getting vegetables. … We had weeding sessions every week, we had weekly directors' meetings at that point. It was quite intense as time and energy we put into and lot of enthusiasm as well. (Paul, founding member)

Members can be involved at various levels; there is a voluntary core group, formed of elected members and at least one grower (either the farm owner or his son), responsible for the day-to-day running of the farm. Members can also take part in other groups, e.g. growing, distribution, membership, events or governance and finance. Although volunteering is not essential for the membership, there are regular volunteering days for all members. But most of the help comes from the international volunteers, who come through two organisations: WWOOF UK[4] and UNA Exchange.[5] They stay in fully equipped caravans and yurt and are paid weekly for food. At the time of the fieldwork, there were six WWOOF volunteers and about ten UNA volunteers. The growers acknowledge their work as essential for the farm's success.

> If we did not have WWOOF-ers to help us grow and maintain the vegetables, and have the quality we need for our members, we could not succeed. (Roger, farm owner)

Clwyd enterprise

Clwyd (pronounced Clue-ed) is a community-run social enterprise in North Wales, producing fruits and vegetables for its 20 members. Differently from the other CSA cases, it has three growing places spread over the county with a total area of about 4 acres. Started first as a community garden, Clwyd established the CSA project in 2011 and ran both schemes in a parallel way. However, in 2014 the CSA scheme experienced difficulties and the Clwyd community decided to put it on hold. The CSA model is based on a symbolic annual membership fee and a separate fee for opening a veg account, which then can be credited either with money or time, by doing any kind of work for the CSA. Therefore, some people who spend time growing veg may not have to add money; and the system is based on trust. One of the sites is part of a wooded hillside in proximity to a village. It is a private property, but it was given by the landlord to the use of the Clwyd community for free with a five-year' agreement. The area was covered in bracken and members had to clear it and prepare it for growing. The second site belongs to the local university and is used for training purposes but four polytunnels and an orchard were allocated with an agreement to the Clwyd enterprise. Situated between the other two sites, it is also used as a hub, and food from all three sites is weighed and distributed here to the members.

---

[4] The WWOOF—World Wide Opportunities on Organic Farm—is an organisation that was established in 1971 in the UK but quickly grew and became an international organisation. It brings together organic farms and gardens throughout the world with volunteers who want to do practical work They are not expected to have farming or gardening skills prior to coming to work because they learn it on the site (WWOOF_UK n.d.).

[5] UNA Exchange is a Wales-based international volunteering organisation founded in its present form in 1973, but with a history of almost 100 years as a volunteering movement. It works with partners in 70 different countries and provides volunteering opportunities in projects in different areas of work, designed with local organisations (UNA n.d.).

The third is part of a farm estate that belongs to a local landlord but was given to Clwyd for growing purposes. The enterprise did not use much financial support from other organisations.

> We've had bits and pieces, but we never managed to get a big chunk. They have tried a few times but not successfully. (Debbie, Chairperson and member)

They used a grant of initial £7.000 from the Welsh Government via the county's Rural Development Agency for starting up the CSA with a further extension of £3.000 for mentoring. The only income of the enterprise is the symbolic annual membership fee, for the model is not based on an upfront payment from the members. Leadership played a key role in setting up the CSA; the idea and mentoring came from its initial founder who is a professional horticulturalist. Additionally, the CSA had various support from intermediary organisations: The Federation provided 'moral' help and has facilitated their networking; Organic Centre Wales, the local Città Slow, Keep Wales Tidy, and the town council provided technical support. Human capital of Clwyd CSA is formed only of its members. From overall 30 community garden members, 20 were in the CSA. Therefore, although the two models were running in parallel, there was overlapping between the members, a group of highly skilled people who provided all necessary support to the CSA, including the financial, marketing and legal advice. At the time of the fieldwork, the paid grower of the CSA had resigned shortly before to start her own enterprise, and many of the members moved with her due to more convenient location. This left the CSA initiative with very few members. The people who remained were trying to re-organise themselves, continuing with the community garden.

Offa Market Garden (Offa MG)

Producer-led organic garden on a 6 acre' land, Offa MG is situated between two towns in Mid-Wales. It is run by a couple of growers who first established the garden in 2008 on 4 acres of rented meadow with a shed, where they lived in a caravan with their children. Three years later, they bought 2 more acres of land to expand the initiative to 6 acres with three polytunnels, a greenhouse, a purpose-built packing shed, and an eco-house for their family. Half of the investment was financed personally by the family, and the rest was paid with small grants and interest-free loans from the landlord and a trust that lends to small-scale organic growers. Certified organic since 2010, Offa MG received a grant from the Welsh Government through the Organic Centre Wales and technical support as part of a package for conversion to organic growing. Apart from a small grant from the local council, the initiative did not use other financial support. The garden sells its fresh produce at the farmers' markets in the nearby towns and to local pubs and restaurants. In 2015, Offa MG also started selling its produce at a local shop opened jointly with another retailer. The CSA scheme is based on a voucher system rather than membership fee. Community members receive free introductory box of fruit and veg when joining the CSA and then buy voucher books of £200 to purchase their food from the farmers' markets or the shop. The overall number of the CSA members is about 20. Part of the members only support the grower financially by buying vouchers and do not help on the field. The other part is a group of 6 to 12 people who call themselves the 'Weeding Group' coming on a regular weekly basis to help with any growing tasks. One of the volunteers was offered later a part-time job for 1 day a week or 2–3 days in the high season. Separately from the 'Weeding Group', there are volunteers who help with organising the volunteering sessions or promoting the initiative. Volunteers are predominantly retired people.

> We are coming from the older end of the spectrum and quite many of us are grannies …, because we have the time. Maybe financially we are a bit more solvent." (Lynn, volunteer)

But there are also younger people in the group who bring their children, and that is why the 'Weeding Group's sessions are usually organised later in the evening, after the workday or school. The grower explained that it was very easy for him to find volunteers to help, which demonstrates the existence of a supportive community in the area, willing to give their time and work for the cause.

How the CSA initiatives satisfy unmet needs

To address the first gap related to the product dimension about identifying the needs, we explored the *collective needs* that started the initiative, and the *personal needs* for people's involvement. Then we identified the *perceived benefits* that people get from these initiatives and compared all three for each case to address the second gap and find out if the needs have been met. We summarized the results in Table 2. According to the results, the most important need that triggered the emergence of the initiatives and people's involvement was the need for good quality local and organic vegetables, which is similar to the findings in studies about the CSA motives (Cox et al. 2008; Sharp et al. 2002; Farmer et al. 2014). The exclusionary circumstances that this need created was articulated as deficit in terms of vegetable production in Wales, or lack of local and organic produce in the area. However, behind the need for local and organic food, there were more complex drives, such as wider concern for sustainability and the environment.

Table 2 Matching needs with benefits

	Community needs	Personal needs	Personal benefits
BONT	Supply of organic vegetables		Fresh organic food
	Transferring knowledge and skills about growing food	Learning with the aim to set up own horticultural business	Learning
	Creating employment	Doing something different from their usual job/ gardening as a hobby	Earning modest wage (grower)
	Contributing to the environmental sustainability	Attracted to the ethical aspects of the project	Supporting organic food growing
		Meeting other people	Social contacts
		Having free time due to retirement or unemployment/desire to help	Fresh air and being outdoors/staying fit
TYDDEWI	Need for local and organic produce in the area	Need for local and organic vegetables	Good food, vegetable share
	Having people care about the farm	Growing food together in a community	Sense of community, social environment
		Provide livelihood for the farmer	Contributing towards sustainability
	Reconnecting people to the source of their food without the	Learning about sustainable, organic agriculture	Learning, sense of achievement
		Working outdoors in a farm	Therapeutic benefits, being in a beautiful environment, access to a real farm
			Accommodation (volunteers)
CLWYD	Need for local and organic vegetables in the area	Need for local and organic vegetables	Good quality, fresh vegetables/cheap, properly grown, nice food
	Growing food in a community	Socialising, being in the community	Being in a community/teamwork
		Learning about growing food/growing own food	Learning
			Physical and mental health
OFFA	Need for a good quality local and organic food producer in the area	Good quality, organic food	Good quality, fresh organic food
		Supporting the values of growing local and organic food, helping a good cause	Sense of fulfilment of doing something positive and productive
			Sense of achievement/seeing the results of the labour
		Social side, growing food as part of the community	Meeting others in the community
			Nice place to spend time
			Physical and psychological wellbeing/fresh air and physical activity

We weren't really driven by a desire to make a business. … We were interested in being part of contributing to not only a sustainable food chain for Wales, but also a lower carbon Wales. (Terry, Bont MG).

For the producers, it was more important to reconnect people with the source of their food by eliminating the food miles and packaging, and by respecting food as something valuable rather than commodity:

I liked the whole principle of it: feeding local people direct from the field and having people care about the farm and about the source where food came from. (Roger, TYCOF)

Members had broader ethical concerns behind the need for local and organic food, too. But in addition, they also wanted to be part of a community and to gather, talk, and share experiences with like-minded people, while at the same time provide livelihood for the grower.

It fits my values and I have always been fascinated by food and the idea of coming together with people to grow substantial amount of food; but also, to provide the livelihood of [the grower's] family was a good idea. (Paul, TYCOF)

Other, less articulated need for the founders was to create employment and educate people who wanted to establish their own small-scale organic food growing projects, while for members, learning with the aim to set up their own business and spending time outdoors, were other personal needs.

Comparing the needs with benefits clearly demonstrates that each CSA initiative is a medium for satisfying the needs that caused deprivation and triggered the social innovation. Firstly, local and organic fresh food was both community and personal need. And respectively, food is the main benefit that people get from their involvement, which participants describe as 'fresh', 'organic', 'good quality', 'cheap', 'nice', 'decent', and 'grown properly'. The need for 'growing food together in a community', 'socialising', or 'meeting other people' was also reported as a benefit. For some participants, it was even more important than getting quality food. Ethical concerns expressed as needs for 'supporting local farmer', 'supporting the values of growing local and organic food', and 'supporting a good cause', were reported as benefits of 'supporting organic food growing', 'contributing toward sustainability', and 'sense of fulfilment of doing something positive'. Meanwhile, although mental and physical health was stated neither as a community need nor as a personal motive, it was among the benefits in all cases, which demonstrates that the CSA initiatives provide additional benefits beyond satisfying unmet needs. Participants described nature as 'beautiful', 'relaxing' and 'real environment', and stated that it helps them connect to themselves and contributes to their mental health. Some spoke about the 'medical benefits of having your hands in the soil'. Others emphasised the physical health benefits of working in fresh air and outdoors. In sum, results show that the Welsh CSA initiatives provide the means of satisfying the needs that created social exclusion and triggered their emergence.

How the CSA initiatives empower people

Related to the empowerment dimension, we explained earlier that the issues about social capital are well documented in the literature but there is no evidence about other ways of empowerment. Therefore, in addition to the sense of strong community, we examined learning and decision-making as two processes in the CSA initiatives that can enhance the capabilities and the voice rights of people. Building strong community and social capital in the CSA initiatives is one way of empowering the community. Our results show that the model of the CSA has an impact on the type and quality of the community. The strong sense of community was more tangible in two of the cases where the grower was the leader of the initiative, namely TYCOF and Offa MG, because the growers play bonding role for the members and volunteers. From the growers' point of view, there is a mutual recognition of the significance of each side to the other side. At TYCOF the community organizes events and gatherings separately from the volunteering times; they have feasts together when a member bakes pizza for everybody in a clay oven. Also, the weekly vegetable boxes are not delivered to the members but picked from the farm. However, some

members argued that the community was not as strong as the farm owner would have liked due to many people not being interested. Low level of engagement was a problem in Bont MG, too. The work was mainly done by paid growers who came on part-time basis. The founding director was uncertain about the strong sense of community due to the low level of engagement both on the field and at the AGMs. The community-led case, Clwyd, had different problems related to social capital. One of the challenges was the distance between the three growing sites, which was partly overcome by designating one of the sites as a hub for food sharing. However, the biggest challenge was to introduce the CSA scheme separately from the existing community garden.

> Those people who had come from community garden roots … were very committed to giving what was asked of them when they moved to the CSA. The people who did not come from that background … took advantage and they did not put that hours in. (Kelly, founder of Clwyd CSA)

After losing considerable part of their members, who left with the resigned paid grower to join her enterprise, only a small group of dedicated and loyal members remained in Clwyd, thus showing the real cohesive group.

Learning as another way of empowerment increases the capacity of individuals and communities to produce their own food, acquire essential life skills, connect with like-minded people, and learn about sustainability issues. Therefore, learning is not only a benefit but a vital process in maintaining the initiatives. Very few people in all four cases had formal horticultural training. The members and volunteers usually learn while working on the garden/farm. Practical food growing is the main skill that they learn. However, it can include some advance knowledge, e.g. different pests and diseases, seasonal changes and organic principles, such as crop rotation, companion planting, and soil care. There is no formal process, and people learn by watching the grower or horticulturalist, then doing it on their own. At Clwyd, the horticulturalist gives members agency by making them teach other members how to do certain things, which helps people memorise well all the process. Additionally, transfer of skills can make people economically powerful by acquiring a job or setting up their own food growing enterprises. Meanwhile, teaching gardening skills is not the main aim for some of the initiatives. For example, at Bont MG the leaders were interested in teaching horticulture at entrepreneurial level, and the garden provided courses on the field to apprentices who wanted to become horticulturalists, through a government-funded training and employment project, Horticulture Wales, which was planned to come to an end in August 2015.

Table 3 Learning in the CSA

Learning	BONT	TYDDEWI	CLWYD	OFFA
Basic gardening/horticultural skills	✓	✓	✓	✓
Advanced horticultural skills (organic gardening growing)	✓	✓	✓	✓
Communication skills	✓		✓	✓
Sustainability Issues	✓		✓	
Managerial skills/planning/dealing with customers	✓			✓
Variety of vegetables		✓	✓	
Running small horticultural business	✓			
Multicultural skills		✓		

Table 4 Decision-making in the CSA

	Daily decisions		Higher level/managerial decisions	
	Who takes the decisions	Members' participation	Who takes the decisions	Members' participation
BONT	Main grower	Informal contribution	Board of Directors	Voting participants at AGM
TYDDEWI	Main grower + growing group	Informal but can take part in the growing group	Core group	Active participation
CLWYD	Core group + site meetings	Active participation	Directors (elected members)	Voting participants at AGM
OFFA	Growers	Informal contribution	Growers	n/a

People learn about some of the difficulties both of trying to grow on a small scale like this and growing wide range of crops, but also difficulties in dealing with customers. (Ryan, paid grower at Bont MG)

Learning also involves communication skills, or 'people skills' and multicultural skills due to meeting different people, specifically volunteers from all around the world in case of TYCOF. At the same time, the practical involvement in producing food teaches people to appreciate the hard work that it takes, as often emphasized by members and volunteers. But more importantly, they also learn about sustainability and environmental issues, and the importance of the CSA, often discussed in informal conversations. This is also referred to as 'second-order learning' (Seyfang and Haxeltine 2012; Marsden 2013). Table 3 summarizes the learning across the cases.

Decision-making is another way of empowerment, as it enhances people's 'voice rights' (Martinelli 2010, p. 42). The ability of people to have their say about the processes or management of the CSA initiatives is empowerment at micro level. In two of the cases—TYCOF and Clwyd—members are given the opportunity to take part in decisions at all levels by being involved in core groups or board of directors. Any willing member can take part in these groups, which is a way of including everyone and 'eases the burden' of the grower.

Everybody is fully involved in the decision-making. … People want to have ideas; they are actually inviting me to give them new ideas. (Ruth, Clwyd)

At Bont MG, daily decisions are left to the growers. At a higher level, these are taken by the directors. Members can participate to the AGM, where they can bring important issues and vote for decisions. And at Offa MG, the decisions at all levels are taken by the growers. But whatever the mechanisms, members and volunteers in all cases feel that their suggestions are taken into consideration, formally or informally, which is a form of democratic governance at micro-level, also described by Defourny and Nyssens (2013) as a recent trend of diversification of the actors in social enterprises working on the same project, where even users and suppliers work and manage together. Comparative summary about decision-making is provided in Table 4.

The potential of the CSA in Wales for social change

Regarding the process dimension, to address the first gap, we explored how the people in the Welsh initiatives see the CSA and its values compared to the dominant food system. Results demonstrate that participants think of the CSA as a very small part of the main food system, described as 'relatively minority', 'fringe', 'tiny part', 'marginal' and 'so small that does not have any impact at all'. One reason is the fact that the CSA is still very limited in numbers—nearly 100 in the UK, of which ten are in Wales. And the second reason is that few people know about the CSA as most people are either unaware or do not care about sustainable food; those who care were described as having "in their souls the idea of growing food, community growing and sharing things" (Lynn, Offa). The CSA is also viewed as complementary to the main food economy by providing inclusive environment for different parts of the society, e.g. people with disabilities or health problems, or low-income families. And despite currently being 'tiny' part of the main economy, people believe it will become bigger because the movement 'is building up knowledge and skills that could be expanded' (Dave, Offa) and 'that is how change happens' (Terry, Bont). When asked to compare the values of the CSA to those of the dominant food system, all interviewees expressed unequivocally that the two have different values. In sum, they explained the values of the CSA as organically, locally and sustainably produced food; polyculture on a small scale; balancing profitability with environmental sensitivity; bringing value to the food; and sharing and cooperation. These were juxtaposed to the conventions of the dominant food system, expressed as intensive farming by using harmful chemicals; profit at any cost by creating externalities; cheap food; and competition rather than cooperation.

To address the second gap related to the process dimension, we explored the difficulties and opportunities for scaling up for the different CSA models, and their visions for movement. We summarized the results in Table 5 at the end of the section. In terms of difficulties at organizational level, *insufficient human capital*, or not having enough members or volunteers was the primarily reported challenge. The reasons behind the need for more members varies between the different models. Bont MG needed more shareholders to finance new equipment and more polytunnels to grow more quantities and speed up the work, while TYCOF needed at least 65–70 members in total (currently has 40) to be able to both provide livelihood for the farmer and feed its members. And Clwyd needed more members to-re-launch the CSA scheme that had been put on hold. Second major difficulty was related to *accessibility*. For instance, Bont MG is situated about 10 miles away from a big city and a few miles away from a town, but this creates a 'psychological barrier' of remoteness. In case of Clwyd, reaching the CSA site even from the nearby village is difficult. Also, none of the CSA initiatives are on a main public transport route. *Lack of initial capital* was another big challenge that can be even worse for the communities not possessing land, which is why some participants spoke of the need for support package from the Government at the start-up phase. *Inadequate equipment*, e.g. machinery and cold storage, *marketing difficulties*, and *time-constraint* were other challenges. All these organizational difficulties create lock-in at micro-level in reaching economic viability. As for the macro-level barriers that relate to the CSA movement, firstly, the place of the CSA in the Welsh food and agriculture policies is almost non-existent. For example, policy documents like the Farming Strategy for Wales (WAG 2009) or the Food Strategy (WAG 2010) do not include any arrangements to promote and support the CSA in Wales. Secondly, the mechanisms that enable growers

and members of the CSA initiatives to directly take part in policy-formulations, which is also empowerment at macro-level, are limited. There were very few occasions when such participation happened, e.g. taking part in the consultations of the Welsh government about reshaping the organic agriculture framework within the EU CAP. Third barrier for the movement is the *inadequate formal horticultural education* in the country and lack of programmes in colleges about growing food. And the final reported macro-level barrier is *low financial reward* for horticultural producers that makes the profession unattractive especially for young people.

In addition to the barriers at micro and macro levels, we examined the networking as an opportunity for scaling up. The launch of the CSA Network UK in 2013 and the Tyfu Pobl (Growing People) program by the Federation have facilitated the networking between the initiatives, which can be interpreted as change in governance relations at meso-level. One way of networking is by taking part in the regular national or regional gatherings organised by these organizations. Two of the cases, TYCOF and Clwyd hosted annual CSA gatherings in the past. Networking also happened by exchanging visits with other CSA farms and gardens to share knowledge and skills, for which the Federation provided travel bursaries.

> It was good to see a much bigger CSA and how they structured themselves … it was a very useful visit. (Will, TYCOF)

Meanwhile, some participants suggested that attracting more farmers to the CSA schemes through farmer unions is another way of scaling up the movement that can help the struggling farms and create employment.

> The community can help the farmer to survive because not many small-scale farmers are making a good living, they are struggling … [They] can get together with enthusiastic communities locally to do it. Then that may be the way forward, which was the original concept of the CSA model. (Trevor, Clwyd)

The final gap in the literature related to the transformative potential of the CSA initiatives was the lack of concrete examples of how they address social injustice and exclusion. The Welsh cases have different mechanisms for making the schemes accessible to everybody. In Clwyd people can participate without paying and can take food in exchange for their work for the initiative, i.e. pay with their time. TYCOF keeps the membership fees at minimum level and allows people to pay 'as much as they can afford',[6] which is openly

---

[6] This CSA farm wants to put into practice a payment system that has been successfully implemented in Freiburg (Germany) according to a member who attends the European and international CSA meetings. The system allows each member to offer anonymously the amount they can afford to pay, and all the sums are revised until they add up

stated on their website. Although the other two initiatives do not have direct ways to make their membership affordable for everyone, they always accept volunteers and offer them fresh produce in exchange for their help. Regarding the participants' visions for the CSA movement, our research highlights that CSA is considered 'definitely' part of the solution for transitioning towards sustainable food systems with the potential to contribute "enormously" to the food sustainability by boosting the local economy, creating jobs and reconnecting people to the source of their food.

Discussion

Related to the product dimension, although the Welsh CSA initiatives have different models and characteristics, they all emerged as social innovations in reaction to similar needs. Secondly, according to ALMOLIN model (González et al. 2010), unmet needs cause deprivation and social exclusion, but in our CSA cases there is no food deprivation per se; the primary need is specifically for locally and sustainably grown, ecologically sound and possibly organic food for people who have broader concerns for the food systems and the environment. All other needs, e.g. being part of a community, supporting sustainable enterprises and learning, are clustered around the food growing as practice. And a third point is that many of the needs are currently not urgent but can become urgent as participants often spoke about the probability of a fuel crisis leading to food scarcity, which is interesting because crisis is considered the second drive for social innovation after needs, and the two are interrelated (Baker and Mehmood 2013). Moreover, the fact that participants in the CSA schemes are people with concern for the environment and sustainability supports Seyfang's argument (2009, pp. 72–74) that ideology can be another driver for social innovation. Therefore, the Welsh CSA community can target people who support environmental sustainability and value the sharing and community aspects, also suggested by Goland (2002). Future research may examine if the antecedent needs change over time and whether these differ in places where ecologically grown, local and organic food is widely available.

Regarding the empowerment dimension, learning as an empowering process increases the capacity of people to produce their own food thus making them more resilient. Moreover, learning equips individuals with skills that makes them economically powerful by acquiring jobs or starting

their own enterprise. One example is the main grower of Bont MG, who had joined as a volunteer but later offered the job and currently has his own horticultural organic enterprise. Another example is the assistant grower of the same initiative who later acquired a job for setting up a new CSA. In a similar way, a founding member of Tyddewi started a new CSA initiative. These are all examples of how learning can lead to transfer of skills that also strengthen the movement by replicating the innovation. In terms of participation in decision-making, the results demonstrated that CSA cases are places for empowerment at micro level. However, when it comes to decision-making at macro level, participants do not feel empowered enough due to two reasons. The first is the lack of national policies or strategies for promoting community food growing, and the second reason is the limited possibility for the CSA communities to participate in the food-related policy decisions at national level. Occasions when leaders of the CSA are invited to discuss policies are extremely rare, and response from community members to government consultations is usually low due to lack of information, lack of time or because these are considered as 'closed-doors'. This brings forward two questions. Firstly, what are the factors that limit the participation of the CSA initiatives in consultations and policy decisions at national level? And secondly, how can this situation improve and what mechanisms can be developed to enable it? To sum up, the CSA in Wales empowers individuals and communities at micro level, within the organization, but the movement must get stronger in order to be empowered at macro level as well. As Miquel et al. (2013) explain, if citizens' political capacity is strong, they can influence institutions in their policy decisions; but if they are not mobilized enough, their influence remain within the boundaries of their community.

Related to the process dimension, we discussed earlier that AFNs have been criticised for their failure both to be more socially just (Goodman 2004; Guthman 2008a; Farmer et al. 2014) and to oppose the neoliberal food regime. We already examined the different ways the Welsh CSA schemes developed to be more socially equitable, such as alternative payment possibilities. Another suggested way for the CSA schemes in addressing social injustice is by playing active role in food deserts, i.e. areas with limited access to fruits and vegetables (Mader and Busse 2011). To do that, the Welsh CSA initiatives first need to overcome the lock-in about acquiring land that is close to cities and towns and accessible to more communities. Land as a resource is considered the biggest challenge for the communities (Armstrong 2000; Henderson and Hartsfield 2009). The Welsh communities used great creativity in gaining access to land by collaborating with local landlords in various forms of agreement, also discussed by Franklin and Morgan (2014). However, they think that the Government and councils should make available land for food growing communities

---

Footnote 6 (continued)

to a level that provides livelihood to the grower. Thus, rather than imposing everybody the same fee, people are given the opportunity for fair distribution of the payment.

Table 5 Process dimension of the CSA

	BONT Shareholding social enterprise	TYDDEWI Community–producer partnership	CLWYD Community-led social enterprise	OFFA Producer-led enterprise
Organizational (micro-level) barriers	Location is difficult to reach on public transport Insufficient equipment (second-hand tractor) Insufficient facilities, e.g. cold storage	Insufficient number of members to provide livelihood for the grower Not enough people working on the farm Lack of demand for the produce from the local shops and restaurants	Remote location that is difficult to access Not enough number of dedicated members	Marketing the voucher scheme and the idea of CSA to more customers Time-consuming, hard work Pests
Current status/needs	Depends on funding to pay grower's wage Needs more shareholders to provide sources for two more polytunnels, a new tractor and cold storage	Has 40 members and provides half of the grower's wage Needs at least 65–70 members/families to be self- sustainable	CSA scheme on hold; growing continues as a community gardening at 3 different sites Needs a new place close to the town and new members to re-launch the CSA	Self-sustainable and owns part of the land but growers work long hours With the help of 20 members it provides livelihood to the growers' family
Vision/aims (organizational level)	To be self-sustainable To be a model for economically viable small- scale horticultural enterprise and inspire others to increase food production in Wales	To be self-sustainable with a bigger community (more members) To improve facilities, e.g. temperature-controlled storage	To re-launch the CSA with more members on a more acessible land To do more marketing and promote the way of eating locally and sustainably	To keep the initiative sustainable and provide livelihood for the growers' family To get more efficient by refining the growing methods and to supply the nearest towns with locally grown, organic vegetables
Barriers on macro-level	Lack of governmental policy (strategy or action plan) regulating or aiming at promoting and developing community food growing in Wales	(1) Lack of support payments from the Government suited to small producers; (2) lack of promotion of the CSA by the government; (3) not recognized by farmer unions	(1) Lack of support of the idea of the CSA; (2) insufficient formal horticultural training	(1) Low financial reward for horticultural producers; (2) insufficient formal horticultural training
Networking	Member of the CSA UK Network, FCFCG, Soil Association and Organic Centre Wales	Member of the CSA UK Network, the Welsh CSA group, FCFCG, Soil Association	Member of the CSA UK Network and FCFCG	Member of the CSA UK Network, FCFCG, the Soil Association and Organic Centre Wales
	Taking part in their events	Hosted annual gatherings	Hosted annual gathering and had visits to/from other gardens/farms through events	Involvement with organic growers' alliance
	Had visits to/ from other community gardens/farms	Minimal collaboration with a local Transition Town initiative	Connections with the local Citta Slow movement and other community groups	Involvement with other local community projects by giving talks
Taking part in decision-making at macro-level	Indirectly, via membership at the city food council	Taking active part in the international gatherings of the CSA network	Taking part in a scoping study about potential places for setting up new CSAs in Wales (funded via the FCFCG)	Taking part in meetings about reshaping the organic framework within CAP and in talks of Welsh organic growers with the minister for agriculture

Table 5 (continued)

| | BONT | TYDDEWI | CLWYD | OFFA |
	Shareholding social enterprise	Community–producer partnership	Community-led social enterprise	Producer-led enterprise
Visions for the movement	CSA is one of the many solutions for transitioning towards more sustainable food systems	CSA has a huge potential to become exponential in numbers and provide jobs for young people who want to get into farming and growing	If more farms get involved, the CSA model can help the struggling farmers because the community can share the risks	Rather than purely community-led food growing, a combination of business and community might result in more successful initiatives
	Need for (1) more coordinated governmental policy toward facilitating people's access to land and (2) support from the communities either by volunteering or buying the produce	Need for (1) promoting by the Government the idea of the CSA as a sustainable way of producing food and (2) support package for establishing CSA projects with financial aid for land and grower's wage	Need for (1) cheap land by the Government and councils for communities to establish CSA with financial support, loans, and equipment- sharing (similar to the AMAPs in France) and (2) more formal education programs about growing food	Need for (1) better financial reward for the horticultural producers and better value for the food produce, and (2) more formal horticultural training with programs about growing food

at more accessible locations. And although the Welsh Government awarded funding in 2018 to Community Land Advisory Service (CLAS) to help at least 50 communities every year to access and own land (CLAS 2018), it barely made difference for the CSA. As for the second criticism about the failure of the food movements to oppose the neoliberal regime, establishing the position of the CSA against the main food economy and comparing the values of the two systems helped us determine the CSA in Wales as a 'strong alternative' for two reasons. Firstly, although the CSA is very small in its position against the corporate food economy, it is not subordinated by the latter (Watts et al. 2005); the initiatives can rather be defined as autonomous food spaces, separate from the corporate food system, as suggested by Wilson (2013). And secondly, the CSA initiatives use the conventions of the moral economy as opposed to the market economy (Follett 2009), e.g. human and animal welfare, community-building, ecological sustainability, and trust and transparency in relations. All these values are also in line with the principles of food sovereignty, which is considered as having the best potential to make real transformation of the food systems through its political stance of rejecting the neoliberal food governance (Fairbairn 2012) and clear opposition to the trade liberalization of food (Alkon and Mares 2012). Moreover, Welsh CSA Network is part of the International Network for CSA Urgenci, which openly states its involvement with the European movement for Food Sovereignty since 2011 with its focus on countering the expansion of supermarkets via short supply chains and providing food for everybody regardless of their income (URGENCI n.d.).

The major limitation of this study is that the results cannot be generalized despite the big number of cases. In addition, they are predominantly based on perceptions and observations reflecting a relatively short period of time. Nevertheless, this study contributes to the literature by studying the processes and factors that enable and constrain the transfer of social innovations from micro-scale opportunity spaces into macro-scale "workable utopias". Additional contribution is the use of ALMOLIN as an analytical tool and its adaptation to the CSA cases, which allowed the evaluation of the social innovation initiatives from their emergence to their transition to a bigger movement and impact on social change at various levels. We also identified several questions that might contribute to a future research agenda: One question is whether needs change over time or differ in areas with access to sustainably produced food. Second question is about the factors limiting the participation of the CSA initiatives in consultations and policy decisions at national level. Third question is about the reasons for the slow replication of the Welsh CSAs given the funding and support of intermediary organizations. Another relates to the reasons for non-participation of local communities to the CSA. And the final questions are about the attitudes of Welsh farmers towards the CSA and possible ways for scaling up by establishing collaborations with other types of social innovation.

To have an impact on the food policies at local and national level, negotiate support from the Government and fairer prices for producers and get promoted, the CSA movement in Wales needs to grow through replicating the initiatives and scaling up. Replication of the projects in a horizontal way is metaphorically compared by Deleuze to rhizomes with underground network of roots linked to each other (Scott-Cato and Hillier 2010). It is also defined as 'expanding de-commodified spaces', referring to places that challenge the corporate regime (Calvário and Kallis 2017, p. 598). Replication of the CSA is already happening in Wales, although at slow pace. The number of initiatives was expected to grow since a scoping study commissioned by the Federation identified 20 potential places for setting up new projects (Groves 2015). It is worth exploring why these projects have not emerged for five years since the results of the study, especially considering the support with land provided by CLAS. In addition, further research is needed to survey the reasons for non-participation to the CSA by the local communities. On the matter of scaling up, hybrid strategies might be one possible way for the CSA. Although different types of hybridity are suggested for scaling up the CSA, e.g. through commodity practices such as labour, seasonality, and addressing customer expectations (Nost 2014), our argument is about hybridity by involving mainstream actors, which was reiterated by many participants of the CSA cases and also discussed by Seyfang and Haxeltine (2012) as a way of promoting grassroots community initiatives. Further research is needed to establish what is the attitude of the Welsh farmers towards the CSA. Additional question is how the CSA initiatives can preserve their alternative values in the case of involving farms as mainstream actors. Also, an idea worth researching is the possibility of scaling up the CSA by establishing collaborations with other types of social innovations, e.g. community energy and housing, alternative currencies or transition movement. And finally, some studies suggest that alternative food initiatives can become part of urban food strategies (Wiskerke 2009; Matacena 2016), which can have a huge impact on the promotion and transformative power of the CSA in Wales. The question is, how can the Cardiff example as the only city with urban food policy can be replicated in other areas as regional food strategies.

Conclusion

Reflecting back on our main research question of how the CSA initiatives in Wales can become 'workable utopias' for food systems' change through social inclusion and empowerment, we demonstrated that the Welsh CSA cases analysed here show great variety in their characteristics, processes and possibilities. However, they all play distinctive roles as part of the social economy in satisfying the needs for ecologically sound and ethical food, grown within communities of like-minded people and empowering individuals and communities at micro level. Moreover, the CSA initiatives are places where communities experiment with producing different crops on a small scale and finding ways to become economically sustainable and resilient, thus contributing to gradual transformation by building knowledge and skills and raising awareness, also termed 'quiet sustainability' (Kneafsey et al. 2016). The type of the CSA affects the financial sustainability of the initiatives as the results show that the purely producer-led type of CSA is the most self-sufficient among the cases while on the contrary, the community-led model is the most vulnerable. In order to become 'workable utopias', the CSA initiatives need to overcome the barriers that prevent them from replicating, participating in policies and decision-making at macro level, and scaling up.

The COVID-19 pandemic is increasingly regarded as an opportunity for transforming the unsustainable and unjust global food systems. Food scholars focus on various features that the new food systems must incorporate. Resilience and the ability to 'bounce back' in the event of drastic change is one feature that is repeatedly articulated (Worstell 2020). In addition, the call is for more equitable, healthy and ecologically-sound, decentralized and distributive systems based on democratic governance at all levels, (Moragues-Faus 2020; Blay-Palmer et al. 2020). It is also suggested that the transition must be towards systems found upon the principles of agroecology with solidarity and circular economies and strengthening the local food value chains (Gemmill-Herren 2020). Social capital, cooperation of people and communities, and collective management of resources as well as re-orienting policies to support communities and protect livelihoods is regarded as essential for the way forward (Pretty 2020; Graddy-Lovelace 2020). It seems that community supported agriculture has an important role to play in the future as it embodies all the features considered for more sustainable food systems: it is solidarity-based, equitable, ecologically sound, and healthy. But most importantly, the CSA has demonstrated for now that it is resilient in times of crisis and not only provides food but nurtures communities and cares for the vulnerable people.

Acknowledgements The authors are grateful to the UK Economic and Social Research Council (ESRC) for funding this research. The authors would like to thank Paul Milbourne and Christopher Bear for their contributions to the project, to the CSA initiatives and individuals who took part in the research, and to the anonymous reviewers for their constructive criticism and helpful suggestions.

Funding The research was funded by the UK Economic and Social Research Council (ESRC).

Compliance with ethical standards

Ethical approval The project has been approved by the Research Ethics Committee of the Cardiff University School of Geography and Planning on 2nd June 2014. All procedures have been performed in accordance with the ethical standards of the above committee and with the 1964 Helsinki declaration and its later amendments or comparable ethical standards.

Informed consent Informed consent was obtained in writing from all individual participants included in the study.

Open Access This article is licensed under a Creative Commons Attribution 4.0 International License, which permits use, sharing, adaptation, distribution and reproduction in any medium or format, as long as you give appropriate credit to the original author(s) and the source, provide a link to the Creative Commons licence, and indicate if changes were made. The images or other third party material in this article are included in the article's Creative Commons licence, unless indicated otherwise in a credit line to the material. If material is not included in the article's Creative Commons licence and your intended use is not permitted by statutory regulation or exceeds the permitted use, you will need to obtain permission directly from the copyright holder. To view a copy of this licence, visit http://creativecommons.org/licenses/by/4.0/.

References

Alkon, Alison Hope, and Teresa Marie Mares. 2012. Food sovereignty in US food movements: Radical visions and neoliberal constraints. *Agriculture and Human Values* 29: 347–359.

Allen, I.V., E. James, Jairus Rossi, Timothy A. Woods, and Alison F. Davis. 2017. Do community supported agriculture programmes encourage change to food lifestyle behaviours and health outcomes? New evidence from shareholders. *International Journal of Agricultural Sustainability* 15 (1): 70–82.

Anderson, Molly D. 2020. Pandemic shows deep vulnerabilities. *Agriculture and Human Values*. https://doi.org/10.1007/s10460-020-10108-7.

Armstrong, D. 2000. A survey of community gardens in upstate New York: Implications for health promotion and community development. *Health & Place* 6 (4): 319–327. https://doi.org/10.1016/s1353-8292(00)00013-7.

Baker, Susan, and Abid Mehmood. 2013. Social innovation and the governance of sustainable places. *Local Environment: The International Journal of Justice and Sustainability*. https://doi.org/10.1080/13549839.2013.842964.

Blay-Palmer, Alison, Roberta Sonnino, and Julien Custot. 2016. A food politics of the possible? Growing sustainable food systems through networks of knowledge. *Agriculture and Human Values* 33: 27–43. https://doi.org/10.1007/s10460-015-9592-0.

Blay-Palmer, Alison, Rachel Carey, Elodi Valette, and Matthew R. Sanderson. 2020. Post COVID 19 and food pathways to sustainable transformation. *Agriculture and Human Values*. https://doi.org/10.1007/s10460-020-10051-7.

Bos, Elisabeth, and Luke Owen. 2016. Virtual reconnection: The online spaces of alternative food networks in England. *Journal of Rural Studies* 45: 1–14.

Brunori, Gianluca, Adanella Rossi, and Vanessa Malandrin. 2010. Co-producing transition: Innovation processes in farms adhering to solidarity-based purchase groups (GAS) in Tuscany, Italy. *International Journal of Sociology of Agriculture and Food* 18 (1): 28–53.

Calvário, Rita, and Giorgos Kallis. 2017. Alternative food economies and transformative politics in times of crisis: Insights from the Basque Country and Greece. *Antipode* 49 (3): 597–616.

Caraher, M., and E. Dowler. 2014. Food for poorer people: Conventional and alternative transgressions. In *Food transgressions: Making sense of contemporary food politics*, ed. M. Goodman and C. Sage, 227–246. Farnham: Ashgate.

CLAS. 2018. Welsh Government funding boost for community land use in Wales. https://wl.communitylandadvice.org.uk/en/news/12072018-1754/welsh-government-funding-boost-community-land-use-wales. Accessed 14 Sept 2019.

Cloke, Paul, Ian Cook, Philip Crang, Mark Goodwin, Joe Painter, and Chris Philo. 2004. *Practising human geography*. London: Sage.

Cohen, J.N., S. Gearhart, and E. Garland. 2012. Community supported agriculture: A commitment to a healthier diet. *Journal of Hunger & Environmental Nutrition* 7: 20–37.

Cone, Cynthia Abbott, and Andrea Myhre. 2000. Community-supported agriculture: A sustainable alternative to industrial agriculture? *Human Organization* 59 (2): 187–197.

Cooley, Jack P., and Daniel A. Lass. 1998. Consumer benefits from community supported agriculture membership. *Review of Agricultural Economics* 20 (1): 227–237.

Cox, Rosie, Lewis Holloway, Laura Venn, Liz Dowler, Jane Ricketts Hein, Moya Kneafsey, and Helen Tuomainen. 2008. Common ground? Motivations for participation in a community-supported agriculture scheme. *Local Environment: The International Journal of Justice and Sustainability* 13 (3): 203–218.

CSA. 2020. COVID-19: Resources for CSAs. Accessed 8.06.2020. https://communitysupportedagriculture.org.uk/help-advice/covid-19/.

Defourny, Jacques, and Marthe Nyssens. 2013. Social innovation, social economy and social enterprise: what can the European debate tell us? In *The international handbook on social innovation: Collective action, social learning and transdisciplinary research*, ed. Frank Moulaert, Diana MacCallum, Abid Mehmood, and Abdelillah Hamdouch, 40–52. Cheltenham: Edward Elgar.

DuPuis, E.Melanie, and Davi Goodman. 2005. Should we go "home" to eat?: Toward a reflexive politics of localism. *Journal of Rural Studies* 21: 359–371.

Fairbairn, Madeleine. 2012. Framing transformation: The counter-hegemonic potential of food sovereignty in the US context. *Agriculture and Human Values* 29: 217–230.

Farmer, James R., Charles Chancellor, Jennifer M. Robinson, Stephanie West, and Melissa Weddell. 2014. Agrileisure: Farmers' markets, CSAs, and the privilege in eating local. *Journal of Leisure Research* 46 (3): 313–328.

Fertilizers_Europe. n.d. COVID-19: Implications for the European fertilizer industry. https://www.fertilizerseurope.com/covid-19-implications-for-the-european-fertilizer-industry/. Accessed 8 June 2020.

Flora, Cornelia Butler, and Corene Bregendahl. 2012. Collaborative community-supported agriculture: Balancing community capitals for producers and consumers. *International Journal of Sociology of Agriculture & Food* 19 (3): 329–346.

Flyvbjerg, B. 2006. Five misunderstandings about case-study research. *Qualitative Inquiry* 12 (2): 219–245. https://doi.org/10.1177/1077800405284363.

Follett, Jeffrey R. 2009. Choosing a food future: Differentiating among alternative food options. *Journal of Agricultural Environmental Ethics* 22: 31–51. https://doi.org/10.1007/s10806-008-9125-6.

Franklin, Alex, and Selyf Morgan. 2014. Exploring the new rural–urban interface: Community food practice, land access and farmer entrepreneurialism. In *Sustainable food systems: Building a new paradigm*, ed. Terry Marsden and Adrian Morley, 166–185. London: Routledge.

Galt, Ryan E. 2013. The moral economy is a double-edged sword: Explaining farmer earnings and self-exploitation in community supported agriculture. *Economic Geography* 89 (4): 341–365.

Galt, Ryan E., Katharine Bradley, Libby Christensen, Cindy Fake, Kate Munden-Dixon, Natasha Simpson, Rachel Surls, and Julia Van Soelen Kim. 2016. What difference does income make for community supported agriculture (CSA) members in California? Comparing lower-income and higher-income households. *Agriculture and Human Values* 34 (2): 435–452.

Gemmill-Herren, Barbara. 2020. Closing the circle: An agroecological response to covid-19. *Agriculture and Human Values*. https://doi.org/10.1007/s10460-020-10097-7.

Gibson-Graham, J.K. 2008. Diverse economies: Performative practices for 'other worlds'. *Progress in Human Geography* 32 (5): 613–632.

Goland, Carol. 2002. Community supported agriculture, food consumption patterns, and member commitment. *Culture & Agriculture* 24 (1): 14–25.

González, Sara, Frank Moulaert, and Flavia Martinelli. 2010. ALMOLIN: How to analyse social innovation at the local level? In *Can neighbourhoods save the city? Community development and social innovation*, ed. Frank Moulaert, Flavia Martinelli, Erik Swyngedouw, and Sara González, 49–67. London: Routledge.

Goodman, David. 2004. Rural Europe redux? Reflections on alternative agro-food networks and paradigm change. *Sociologia Ruralis* 44 (1): 3–16.

Gottlieb, Robert, and Anupama Joshi. 2010. *Food Justice*. Cambridge: MIT Press.

Graddy-Lovelace, Garrett. 2020. Re-orienting policy for growing food to nourish communities. *Agriculture and Human Values*. https://doi.org/10.1007/s10460-020-10112-x.

Groves, Jim. 2015. FCFCG Community supported agriculture scoping study moves into phase two. https://www.tyfupobl.org.uk/fcfcg-community-supported-agriculture-scoping-study-moves-phase-two/. Accessed 24 Mar 2017.

Gustin, Georgina. 2020. Empty grocery shelves and rotting, wasted vegetables: Two sides of a supply chain problem. *Inside Climate News*.

Guthman, Julie. 2008a. Bringing good food to others: Investigating the subjects of alternative food practice. *Cultural Geographies* 15 (4): 431–447.

Guthman, Julie. 2008b. Neoliberalism and the making of food politics in California. *Geoforum* 39: 1171–1183.

Hamdouch, Abdelillah. 2013. Introduction: 'Reality' as a guide for SI research methods? In *The international handbook on social innovation: Collective action, social learning and transdisciplinary research*, ed. Frank Moulaert, Diana MacCallum, Abid Mehmood, and Abdelillah Hamdouch, 259–263. Cheltenham: Edward Elgar.

Hancock, Dawson R., and Bob Algozzine. 2006. *Doing case study research: A practical guide for beginning researchers*. New York: Teachers College Press, Columbia University.

Hayden, Jennifer, and Daniel Buck. 2012. Doing community supported agriculture: Tactile space, affect and effects of membership. *Geoforum* 43 (2): 332–341. https://doi.org/10.1016/j.geoforum.2011.08.003.

Henderson, Bethany Rubin, and Kimberly Hartsfield. 2009. Is getting into the community garden business a good way to engage citizens in local government? *National Civic Review*, Winter

Hendrickson, Mary K. 2020. Covid lays bare the brittleness of a concentrated and consolidated food system. *Agriculture and Human Values*. https://doi.org/10.1007/s10460-020-10092-y.

Hinrichs, C. Clare. 2000. Embeddedness and local food systems: Notes on two types of direct agricultural market. *Journal of Rural Studies* 16: 295–303.

Hinrichs, C. Clare. 2003. The practice and politics of food system localization. *Journal of Rural Studies* 19: 33–45.

Howaldt, Jürgen, Anna Butzin, Dmitri Domanski, and Christoph Kaletka. 2014. *Theoretical approaches to social innovation—A critical literature review*. Dortmund: Social Innovation: Driving Force of Social Change (SI-DRIVE) Project.

Ilbery, Brian, and Damian Maye. 2005. Food supply chains and sustainability: Evidence from specialist food producers in the Scottish/English borders. *Land Use Policy* 22: 331–344.

IPES-Food. 2020. COVID-19 and the crisis in food systems: Symptoms, causes, and potential solutions.

Janssen, Brandi. 2010. Local food, local engagement: Community-supported agriculture in eastern Iowa. *Culture & Agriculture* 32 (1): 4–16. https://doi.org/10.1111/j.1556-486X.2010.01031.x.

Jarosz, Lucy. 2008. The city in the country: Growing alternative food networks in Metropolitan areas. *Journal of Rural Studies* 24 (3): 231–244.

Jessop, Bob, Frank Moulaert, Lars Hulgard, and Abdellilah Hamdouch. 2013. Social innovation research: A new stage in innovation analysis? In *The international handbook on social innovation: Collective action, social learning and transdisciplinary research*, ed. Frank Moulaert, Diana MacCallum, Abid Mehmood, and Abdellilah Hamdouch, 110–130. Cheltenham: Edward Elgar.

Kemp, R., J. Schot, and R. Hoogma. 1998. Regime shifts to sustainability through processes of niche formation: The approach of strategic niche management. *Technology Analysis & Strategic Management* 10 (2): 175–195. https://doi.org/10.1080/09537329808524310.

Kneafsey, Moya, Luke Owen, Elizabeth Bos, Kevin Broughton, and Margi Lennartsson. 2016. Capacity building for food justice in England: The contribution of charity-led community food initiatives. *Local Environment* 22 (5): 621–634.

Lamine, Claire. 2005. Settling shared uncertainties: Local partnerships between producers and consumers. *Sociologia Ruralis* 45 (4): 324–345.

Laurier, Eric. 2010. Participant observation. In *Key methods in geography*, ed. Nicholas Clifford, Shaun French, and Gil Valentine. London: Sage.

MacCallum, Diana, Frank Moulaert, Jean Hillier, and Serena Vicari Haddock (eds.). 2009. *Social innovation and territorial development*. Farnham: Ashgate.

MacMillan Uribe, Alexandra L., Donna M. Winham, and Christopher M. Wharton. 2012. Community supported agriculture membership in Arizona. An exploratory study of food and sustainability behaviours. *Appetite* 59: 431–436.

Mader, Erin, and Heidi Busse. 2011. Hungry in the heartland: Using community food systems as a strategy to reduce rural food deserts. *Journal of Hunger & Environmental Nutrition* 6 (1): 45–53.

Marsden, Terry. 2013. From post-productionism to reflexive governance: Contested transitions in securing more sustainable food futures. *Journal of Rural Studies* 29: 123–134. https://doi.org/10.1016/j.jrurstud.2011.10.001.

Marsden, Terry, Jo Banks, and Gillian Bristow. 2000. Food supply chain approaches: Exploring their role in rural development. *Sociologia Ruralis* 40 (4): 424–436.

Martinelli, Flavia. 2010. Historical roots of social change: Philosophies and movements. In *Can neighbourhoods save the city? Community development and social innovation*, ed. Frank Moulaert,

Flavia Martinelli, Erik Swyngedouw, and Sara González, 17–48. London: Routledge.

Matacena, Raffaele. 2016. Linking alternative food networks and urban food policy: A step forward in the transition towards a sustainable and equitable food system? *International Review of Social Research* 6 (1): 49–58.

May, Tim, and Beth Perry. 2011. Case study research. In *Social research: Issues, methods and process*, ed. Tim May, 219–242. Buckingham: Open University Press.

Minaker, Leia M., Kim D. Raine, Pat Fisher, Mary E. Thompson, Josh Van Loon, and Lawrence D. Frank. 2014. Food purchasing from farmers' markets and community-supported agriculture is associated with reduced weight and better diets in a population-based sample. *Journal of Hunger & Environmental Nutrition* 9 (4): 485–497.

Miquel, Marc Pradel, Marisol Garcia Cabeza, and Santiago Eizaguirre Anglada. 2013. Theorizing multi-level governance in social innovation dynamics. In *The International handbook on social innovation: Collective action, social learning and transdisciplinary research*, ed. Frank Moulaert, Diana MacCallum, Abid Mehmood, and Abdelillah Hamdouch, 155–168. Cheltenham: Edward Elgar.

Moragues-Faus, Ana. 2020. Distributive food systems to build just and liveable futures. *Agriculture and Human Values*. https://doi.org/10.1007/s10460-020-10087-9.

Morgan, Kevin, Terry Marsden, and Jonathan Murdoch. 2006. *Worlds of food: Place, power, and provenance in the food chain. Oxford Geographical and Environmental Studies*. Oxford: OUP.

Morris, Carol, and James Kirwan. 2011. Ecological embeddedness: An interrogation and refinement of the concept within the context of alternative food networks in the UK. *Journal of Rural Studies* 27 (3): 322–330.

Moulaert, F., F. Martinelli, E. Swyngedouw, and S. Gonzalez. 2005. Towards alternative model(s) of local innovation. *Urban Studies* 42 (11): 1969–1990. https://doi.org/10.1080/0042098050027989 3.

Moulaert, Frank, Diana MacCallum, and Jean Hillier. 2013. Social innovation: Intuition, percept, concept, theory and practice. In *The international handbook on social innovation: Collective action, social learning and transdisciplinary research*, ed. Frank Moulaert, Diana MacCallum, Abid Mehmood, and Abdelillah Hamdouch, 13–24. Cheltenham: Edward Elgar.

Mount, Phil, Shelley Hazen, Shawna Holmes, Evan Fraser, Anthony Winson, Irena Knezevic, Erin Nelson, Lisa Ohberg, Peter Andree, and Karen Landman. 2013. Barriers to the local food movement: Ontario's community food projects and the capacity for convergence. *Local Environment* 18 (5): 592–605.

Nost, Eric. 2014. Scaling-up local foods: Commodity practice in community supported agriculture (CSA). *Journal of Rural Studies* 34: 152–160.

Ostrom, Marcia Ruth. 2007. Community supported agriculture as an agent of change: Is it working? In *Remaking the North American food system*, ed. C. Clare Hinrichs and Thomas A. Lyson, 99–120. Lincoln: University of Nebraska Press.

Parra, Constanza. 2013. Social sustainability: A competing concept to social innovation? In *The international handbook on social innovation: Collective action, social learning and transdisciplinary research*, ed. Frank Moulaert, Diana MacCallum, Abid Mehmood, and Abdelillah Hamdouch, 142–154. Cheltenham: Edward Elgar.

Pretty, Jules. 2020. New opportunities for the redesign of agricultural and food systems. *Agriculture and Human Values*. https://doi.org/10.1007/s10460-020-10056-2.

Renting, Henk, Terry Marsden, and Jo Banks. 2003. Understanding alternative food networks: Exploring the role of short food supply chains in rural development. *Environment and Planning A* 35: 393–411.

Renting, Henk, Markus Schermer, and Adanella Rossi. 2012. Building food democracy: Exploring civic food networks and newly emerging forms of food citizenship. *International Journal of Sociology of Agriculture and Food* 19 (3): 289–307.

Ritchie, Jane. 2003. The applications of qualitative methods to social research. In *Qualitative research practice: A guide for social science students and researchers*, ed. Jane Ritchie and Jane Lewis. London: Sage.

Rossi, Adanella. 2017. Beyond food provisioning: The transformative potential of grassroots innovation around food. *Agriculture* 7 (1): 6.

Roy Chaudury, Nandini. 2020. Covid-19: The impact on the animal feed industry. *All About Feed*.

Sage, Colin. 2003. Social embeddedness and relations of regard: Alternative 'good food' networks in south-west Ireland. *Journal of Rural Studies* 19: 47–60.

Sage, Colin. 2014. The transition movement and food sovereignty: From local resilience to global engagement in food system transformation. *Journal of Consumer Culture* 14 (2): 254–275.

Schmidt, Claudia, J. Stephan Goetz, J. Sarah Rocker, and Zheng Tian. 2020. Google searches reveal changing consumer food sourcing in the COVID-19 pandemic. *Journal of Agriculture, Food Systems, and Community Development* 9 (3): 9–16.

Schnell, Steven M. 2007. Food with a farmer's face: Community-supported agriculture in the United States. *Geographical Review* 97 (4): 550–564.

Schot, Johan, and Frank W. Geels. 2008. Strategic niche management and sustainable innovation journeys: Theory, findings, research agenda, and policy. *Technology Analysis & Strategic Management* 20 (5): 537–554. https://doi.org/10.1080/0953732080229265 1.

Scott-Cato, Molly, and Jean Hillier. 2010. How could we study climate-related social innovation? Applying Deleuzean philosophy to transition towns. *Environmental Politics* 19 (6): 869–887. https://doi.org/10.1080/09644016.2010.518677.

Selfa, Theresa, and Joan Qazi. 2005. Place, taste, or face-to-face? Understanding producer–consumer networks in "local" food systems in Washington State. *Agriculture and Human Values* 22: 451–464.

Seyfang, Gill. 2009. The new economics of sustainable consumption: Seeds of change. In *Energy, climate and the environment*, ed. David Elliott. Houndmills: Palgrave Macmillan.

Seyfang, Gill, and Alex Haxeltine. 2012. Growing grassroots innovations: Exploring the role of community-based initiatives in governing sustainable energy transitions. *Environment and Planning C-Government and Policy* 30 (3): 381–400. https://doi.org/10.1068/c10222.

Sharp, Jeff, Eric Imerman, and Greg Peters. 2002. Community supported agriculture (CSA): Building community among farmers and non-farmers. *Journal of Extension* 40 (3). http://www.joe.org/joe/2002june/a3.php.

Shortall, Sally. 2008. Are rural development programmes socially inclusive? Social inclusion, civic engagement, participation, and social capital: Exploring the differences. *Journal of Rural Studies* 24: 450–457.

Smith, Adrian. 2007. Translating sustainabilities between green niches and socio-technical regimes. *Technology Analysis & Strategic Management* 19 (4): 427–450. https://doi.org/10.1080/09537320701403334.

UNA. n.d. What we do? https://www.unaexchange.org/whatwedo/. Accessed 14 Sept 2019.

URGENCI. 2020. Community supported agriculture is a safe and resilient alternative to industrial agriculture in the time of Covid-19.

URGENCI. n.d. Food sovereignty. https://urgenci.net/food-sovereignty/. Accessed 14 Sept 2019.

Valentine, Gill. 2005. Tell me about…: Using interviews as a research methodology. In *Methods in human geography: A guide for students doing a research project*, ed. Robin Flowerdew and David Martin. Pearson: Edinburgh.

WAG. 2009. *Farming, food and countryside: Building a secure future—A new strategy for farming*. Cardiff: WAG.

WAG. 2010. *Food for Wales, food from Wales 2010–2020: Food strategy for Wales*. Cardiff: WAG.

Watts, David C.H., Brian Ilbery, and Damian Maye. 2005. Making reconnections in agro-food geography. *Progress in Human Geography* 29 (1): 22–40.

Wharton, Christopher M., Renee Shaw Hughner, Lexi MacMillan, and Claudia Dumitrescu. 2015. Community supported agriculture programs: A novel venue for theory-based health behavior change interventions. *Ecology of Food and Nutrition* 54 (3): 280–301.

Whatmore, Sarah, Pierre Stassart, and Henk Renting. 2003. What's alternative about alternative food networks? *Environment and Planning A* 35: 389–391.

Wheeler, Amber. 2020. COVID-19: UK Veg Box Report. Food Foundation.

Wilkins, Jennifer L., Tracy J. Farrell, and Anusuya Rangarajan. 2015. Linking vegetable preferences, health and local food systems through community-supported agriculture. *Public Health Nutrition* 18 (3): 2392–2401.

Wilson, Amanda DiVito. 2013. Beyond alternative: Exploring the potential for autonomous food spaces. *Antipode* 45 (3): 719–737.

Winter, Michael. 2003. Embeddedness, the new food economy and defensive localism. *Journal of Rural Studies* 19: 23–32.

Wiskerke, Johannes S.C. 2009. On places lost and places regained: Reflections on the alternative food geography and sustainable regional development. *International Planning Studies* 14 (4): 369–387. https://doi.org/10.1080/13563471003642803.

Worstell, Jim. 2020. Ecological resilience of food systems in response to the COVID-19 crisis. *Journal of Agriculture, Food Systems, and Community Development* 9 (3): 23–30. https://doi.org/10.5304/jafscd.2020.093.015.

WWOOF_UK. n.d. WWOOF United Kingdom. https://wwoof.org.uk/. Accessed 14 Sept 2019.

Yin, Robert K. 2003. Case study research: Design and methods. In *Applied social research methods*, 3rd ed, ed. Leonard Bickman and Debra J. Rog. Thousand Oaks: Sage.

Zepeda, Lydia, Anna Reznickova, and Willow Saranna Russell. 2013. CSA membership and psychological needs fulfillment: An application of self-determination theory. *Agriculture and Human Values* 30: 605–614.

Publisher's Note Springer Nature remains neutral with regard to jurisdictional claims in published maps and institutional affiliations.

Tezcan Mert-Cakal is a doctoral graduate from the Cardiff University, School of Geography and Planning. Her research focused on community food growing initiatives, social innovation and sustainable agrifood systems.

Mara Miele is a Professor of Human Geography at Cardiff University, School of Geography and Planning. Her research addresses the geographies of ethical food consumption, and the role of animal welfare science and technology in challenging the role of farmed animals in current agricultural practices and policies. In recent years her work has developed in conversation with cultural geography and STS scholars. Recent publications appeared in Environment and Planning A, Environment and Planning D, Theory, Culture & Society, Food Ethics.

Agriculture and Human Values (2020) 37:1261–1279
https://doi.org/10.1007/s10460-020-10132-7

Bridging the rural–urban divide in social innovation transfer: the role of values

Imran Chowdhury[1] ⓘ

Accepted: 13 July 2020 / Published online: 5 October 2020
© Springer Nature B.V. 2020

Abstract

This study examines the process of knowledge transfer between a pair of social enterprises, organizations that are embedded in competing social and economic logics. Drawing on a longitudinal case study of the interaction between social enterprises operating in emerging economy settings, it uncovers factors which influence the transfer of a social innovation from a dense, population-rich setting to one where beneficiaries are geographically dispersed and the costs of service delivery are correspondingly elevated. Evidence from the case study suggests that institutional bricolage—the crafting of improvised solutions in resource-constrained settings—can serve as potent driving force in driving innovation transfer, and that this process of re-combining available resources may be facilitated by the extent to which the values between partner social enterprises are aligned. With such alignment, social enterprise partners may be able to increase trust, develop a smoother knowledge-transfer process, and find practical solutions which facilitate the transfer of life-enhancing social innovations to neglected rural settings.

Keywords Social entrepreneurship · Institutional complexity · Institutional logics · Knowledge transfer · Partnerships · Values

Introduction

In recent years, the growth of social enterprises, which provide vital services across diverse sectors of the economy such as education, healthcare, and enterprise development, has received significant popular and academic attention. For instance, the growth of microfinance organizations, both for-profit and not-for-profit, has captured the imagination of government and non-governmental organization officials and generations of students in public policy and business schools across the world (Zhao 2014; Battilana and Dorado 2010). The rapid increase in attention to these firms reflects not only the tremendous demand for the services they provide, but also the potential for these services to reach a wider audience. In short, there is a clear and compelling case for social enterprises to "scale" their successful models and practices beyond their local area to maximize their impact on society.

While a number of options for scaling the impact of this knowledge exist, including developing franchises or diffusing best practices via mediating bodies such as multilateral organizations, partnering with other social enterprises remains among the most popular options (Bloom and Chatterji 2009; van Wijk et al 2018). Traditional research on inter-organizational partnerships has considered the value of partnerships in gaining access to new markets and technologies, sharing knowledge, engaging in organizational learning, and developing greater levels of mutual dependence between partners (Horowitz and McGahan 2019). More recently, research on partnerships has examined organizations in a cross-sector setting (Vurro et al. 2010; Ahmadsimab and Chowdhury 2019), or in the context of public–private partnerships (Saz-Carranza and Longo 2012; Bishop and Waring 2016). In both streams of research, a focus on organizations which operate with primarily profit-maximizing or primarily social logics, has shifted in recent years to organizations where profit-making and social motivations co-exist (Greenwood et al 2011).

What these studies have not considered in depth, however, is situations in which different forms of social logics co-exist, where the rationality for generating social impact

✉ Imran Chowdhury
  chowdhury_imran@wheatoncollege.edu

[1] Diana Davis Spencer Chair of Social Entrepreneurship,
  Wheaton College, Norton, MA 02766, USA

varies between organizations in a partnership (Friedland 2013). Social logics vary widely, and interact in distinct ways with the commercial logic. For instance, research by Peifer et al. has examined how actors navigate logics such as religion in the context of science commercialization and in the mutual fund industry (Peifer et al. 2019; Peifer 2014). In the case of religious mutual funds, organizations operating at the intersection of the competing logics of religion and finance engage in specific kinds of boundary work which leads to situations of enduring institutional complexity (Peifer 2014). Other studies have looked at social logics in terms of structural roles, for instance, farmers operating as artists in commercial wine production (Voronov et al. 2013), or social logics manifested as social welfare logics in the case of work-integration enterprises (Pache and Santos 2013) or development logics in the microfinance sector (Battilana and Dorado 2010). Partnerships between social enterprises thus offer an ideal setting for the study of these phenomena, wherein different social logics may interact with each other in addition to their interactions with the commercial logic.

Organizations operating with a multiplicity of institutional demands, or in situations of institutional complexity, have varying responses to these demands (Oliver 1991; Mars and Schau 2017; Piatti and Dwiartama 2016). When these demands conflict, as may be the case with organizations that balance social and economic goals, an organization's response may be a function of the nature of the conflicting demands or the way this conflict is dealt with by organization members (Pache and Santos 2010, 2013; Wijers 2019). This thus paper seeks to better understand partnership between social enterprises (Mair and Marti 2006) where the goal is transferring knowledge (Boxenbaum and Battilana 2005; Lounsbury 2007) for the purpose of scaling up social innovations (Bloom and Chatterji 2009).

In focusing on factors which influence the transfer of a social innovation from a dense, population-rich setting to one where beneficiaries are geographically dispersed and the costs of service delivery are correspondingly elevated, this article helps to develop a better understanding of the ways in which organizations manage relationships with partners that have different logics. Further, by examining the spiritual and social welfare logics as distinct rationalities for generating social impact (Gümüsay 2017), and the ways in which these logics may align or clash in the context of a partnership, this study may also be able to better elucidate factors which facilitate or inhibit knowledge transfer between social enterprises and other hybrid organizations. Finally, by examining innovation transfer from urban to rural areas, the article focuses on a long-neglected domain in studies of social enterprise which may have significant policy impacts (Jia and Desa 2018).

In order to address these questions I draw from a fieldwork-based case study of the transfer of affordable eye-care services from the Aravind Eye Hospitals system in India to an eye hospital in Paraguay. Evidence from this case study suggests that institutional bricolage may serve as potent force in driving innovation transfer, and that the process of re-combining available resources can be facilitated by the extent to which the values between partner social enterprises are aligned. Such alignment aided the social enterprise partners in this study to increase trust and to develop a smoother knowledge-transfer process, wherein potential areas of conflict were pre-emptively managed by focusing on points of commonality between the partner organizations. The paper makes three principal contributions. First, it provides evidence that alignment between the logics of social enterprise partners may influence the strategies firms use to transfer their knowledge while retaining the fidelity of that knowledge (Ansari et al. 2010). Second, it points to the importance of such logics alignment in facilitating the institutional bricolage (Baker and Nelson 2005; Desa 2012; Clough et al. 2019) which enables a search for practical solutions geared towards bringing a life-enhancing social innovation to neglected rural settings. The third contribution brings together the first two and points towards an enhanced understanding of how social logics can co-exist in the context of a partnership, and how they may impact broader processes of organizational efficiency and economic concerns as organizations attempt to scale their social impact beyond their local areas of operation.

The remainder of the article is organized as follows. In the next section, I offer a short overview of the literatures on institutional complexity and bricolage, concentrating in particular on how these literatures converge upon the phenomenon of innovation transfer. Subsequently, in the third section of the article, I outline the study's methods. I present the case study upon which I draw for this article, including the broader study of which it is part. I also describe the data collection and data analysis process that I undertook. In the fourth section of the article, I present the findings from this study, including the diverse social logics that were identified in the case and the ways in which the social enterprise partners managed the innovation transfer process. In the fifth, and concluding, section, I discuss the implications of these findings to the broader fields of institutions and entrepreneurial bricolage, and for the literature on "scaling" social innovations. Some practical applications of the study are also discussed for managers, consultants, and others working with social enterprises.

Knowledge transfer: an institutional complexity and bricolage perspective

Recent research in organizational theory has looked at the embeddedness of organizations in pluralistic institutional environments where multiple logics prevail, one that is fundamental to the world in which social enterprises operate (Mair and Marti 2006; Greenwood et al 2011; Skelcher and Smith 2014). This research builds on prior work on institutional dynamics, which tends to see institutional change as driven by changes in a single, dominant logic (Thornton and Ocasio 2008; Thorton and Ocasio 1999). While a move from a focus on dominant institutional logics to multiple, co-existing logics offers numerous opportunities to re-conceive organizations and their broader environment, a number of challenges also arise. When new institutional logics prevail in the environment, organizations tend to imitate the most successful organizations embedded in and identified with these logics. When there are multiple logics in the environment, however, organizations have greater freedom with respect to the institutional pressures that they choose to comply with. In such instances, organizations may choose to adopt and re-interpret successful practices and standards as exemplified by institutional logics by means of a process of translation (Czarniawska and Sevon 1996) or editing (Sahlin-Andersson 1996).

A plurality of institutional logics impacts upon questions of identity as well. As scholars have noted, identity plurality in organizations, often emerging from environments with different sets of norms, may lead to tensions and conflicts within organizations (Wry and York 2017; Smith et al 2014; Fiol et al. 2009). While greater attention is now being focused on multiple institutional logics and their impact on organizational fields and within organizations themselves, for the most part researchers haven't focused on how multiple logics can influence *inter-organizational* processes within the context of organizational hybrids such as social enterprises. This is important because the inter-organizational setting allows us to see how alignment or non-alignment of logics between organizations may impact organizational processes. This setting also allows researchers to observe how organizational processes unfold and are interpreted by the parties in these interactions.

Additionally, an examination of organizational processes in social enterprises must take into account some of the fundamental mechanisms used by entrepreneurs to manage their environments and to ensure their survival. Prior work in the realm of entrepreneurship and organizational studies has discussed the importance of bricolage in actors' attempts to craft solutions to existing problems under situations of resource constraint (Baker and Nelson 2005; Lévi-Strauss 1966). In essence, by making do with the materials and structures and processes available at hand, entrepreneurs are able to overcome constraints which seemingly limit their freedom of action and ability to achieve their goals (Fisher 2012). More recently, scholars of social entrepreneurship have observed the applicability of bricolage concept in resource-constrained settings where organizations are attempting to move forward social goals and social value

creation using processes and techniques borrowed from the business world (Di Domenico et al 2010; Desa 2012; Desa and Basu 2013; McDermott et al 2018; Chowdhury 2019). This research has started to bridge the institutional complexity and bricolage perspectives in the social enterprise context.

For instance, in a study of the challenges encountered by international technology social entrepreneurs, Desa (2012) finds that the conditions for resource mobilization are simultaneously enabled and constrained by the cognitive, normative, and regulative institutional pillars. Though social enterprises embedded in supportive institutional contexts more easily gain legitimacy and access to standard resources, less-embedded social entrepreneurs are freer to engage in bricolage activities that challenge the norms of existing institutional arrangements to their benefit. Relatedly, a recent study of Australian social enterprises finds that entrepreneurial bricolage proceeds differently in rural versus urban settings. While rural social enterprises made relatively greater use of financial and physical assets accessed through networks within their communities, urban social enterprises were more likely to draw on assets via corporate partnerships and structured philanthropic ventures (Barraket et al 2018).

The present study builds this prior work on institutions and bricolage in order to gain a deeper understanding of the ways in which social enterprises manage relationships with partner organizations to transfer innovations that cross the rural-to-urban divide, a domain that heretofore remains under-studied. In addition, it examines the role played by the alignment of institutional logics in facilitating innovation transfer processes. I highlight some of the principal differences for innovation transfer between the traditional commercial entrepreneurship space and the emerging social entrepreneurship space in Table 1 below.

Methods

Research setting

In order to understand how social enterprises transfer innovations from population-rich settings to rural areas with dispersed populations, I draw on a 2-year-long qualitative field study of partnerships between social enterprises in developing countries that encompasses 83 semi-structured interviews, 9 weeks of on-the-ground observations, and the examination of extensive archival material and documents including emails, meeting minutes, annual reports, project reports and updates, briefs and monographs, books written about the social enterprises being studied, consulting evaluations, and survey data. These organizations were drawn from a sample of sixteen finalist organization pairs

Table 1 Innovation transfer in commercial entrepreneurship vs. social entrepreneurship

	Commercial entrepreneurship	Social entrepreneurship
Economic motivation	Maximize	Satisfice
Social motivation	Satisfice	Maximize
Incentives	Capturing value	Creating value
Who are the principals?	Shareholders, managers	Local community, populations neglected by market and government
What is delivered to principals?	Economic benefits (profits)	Social impact, social recognition
Context of transfer	Regulatory frameworks (e.g., tax and legal), macroeconomy, sociopolitical environment	Normative and cognitive frameworks, sociopolitical environment, macroeconomy (with decreased salience of market-selection mechanisms)

from the proposal submission process to a well-known European Foundation. Organizations submitted proposals jointly (generally in pairs) in the hopes of receiving funding for the transfer of a social innovation between a source and target organization.

Five winning proposals were selected from the sixteen finalist proposals which submitted applications, and I had access to the full proposals of all these organizations. The applications were a rich source of information regarding the organizations, the innovation being transferred, and the expected parameters and scope of their partnerships. I also had access to the full list of applicants and letters of interest for the competition, and administered a survey to all 16 finalist source organizations (with 15 responses) to capture information on their social innovations and their past, extant, and future knowledge transfer partnerships. Subsequently I completed field visits to four of the five winning sets of organizations (the fifth set of visits was not undertaken due to logistical and financial constraints), starting with a pilot study between a pair of Indian social enterprises. The bulk of interviews, observations, and archival material and documents were collected during

these field visits. Further details regarding the data collection process for the full study are included in Table 2.

Sampling procedure

From the full sample of organizational partnerships described above, I selected a set of three social enterprises that had partnered for the purpose of transferring a system of affordable eye-care services. I specially chose this partnership for further study for two reasons. First, the innovation transferred was in the domain of healthcare within developing countries, a domain with broad potential implications in a broad range of geographies facing resource constraints. Prior research in both urban and rural settings has pointed to the importance of health-focused social enterprises in delivering care to a range of disadvantaged communities (Nirmalan et al. 2004; James 2014; Chowdhury 2015; McNamara et al. 2018). Second, I selected this set of organizations as the eye care system transferred by the source organization can be configured for a range of different health care environments across different geographic regions, in particular in resource-constrained emerging (developing) economy settings. Due to the relatively repetitive nature of the refraction

Table 2 Data collection process

Stage	Name	Description
1	Preliminary survey and site selection	Short survey was sent out to the 16 finalist organizational pairs; 15 completed the survey. Data from surveys used to construct a list of the four most promising organizational pairs for follow-up
2	Pilot study	An initial case study was conducted on-site at one pair of source and target organization partners, both located in India, to gain insight into the social innovation transfer process. This pilot study allowed the "field-testing" of interview protocol which was refined for use subsequent data-gathering
3	Field visits and interviews	Researcher visited source and target social enterprises in the 3 remaining partnerships, located in Bangladesh, India, Paraguay, and Sri Lanka
4	Follow-up and data analysis	Following and concurrent to field data collection I gathered updates on the progress of the knowledge transfer processes via email exchanges and telephone calls. I also had access to the periodic updates sent by each pair of organizations to the funding agency, up to and including the final reports. Subsequently, final phone calls with both source and target organizations were conducted

techniques, examination procedures, cataract surgeries, and administrative and support operations associated with the system, it can potentially be deployed across a range of settings with varying endowments of local resources. This contextual modularity may thus have important practical considerations for the eye care setting and, more broadly, the public health domain as well.

As noted above, the organizations I focused on in this study included one source (transferring knowledge) organization and one primary target (receiving knowledge), and operated within the field of eye-care services and ophthalmology. In addition, the given the greatly varied geographical scope (dense population vs. dispersed, rural population) between the source and target organizations, a third (local) partner joined the organization to help with the implementation of the social innovation (see Fig. 1 below). The transferring organization, the Aravind Eye Hospital (Aravind) is located in Madurai, India and has been working since 1976 to eliminate needless blindness by providing comprehensive eye care services to the poor. Its technology has allowed millions of poor people to work, support their families, and lead fuller, more productive lives. One of the most productive eye care facilities in the world, Aravind has reached impressive scale; in 2017–2018 it recorded nearly 4.2 million outpatient visits, and performed over 478,000 surgeries. Of these surgeries, nearly 49% were delivered free of charge to the patient. Aravind also has proven methodology for transferring its model to other eye care providers in developing countries through the Lions Aravind Institute of Community Ophthalmology (LAICO). This system was being transferred via the partnership examined in this article:

> Designing services—both range and the volume, based on the community need is the key for an effective eye care program. Aravind, through LAICO tries to design services based on community need by sharing its model of high volume, high quality and affordable eye care through structured Consultancy & Capacity building processes. This starts with gap analysis to facili-

tate eye care programs, developing a good strategic plan, and providing need-based assistance during plan implementation. This process is based on four core principles in eye care—Demand generation, Resource utilization, Quality of services, and becoming Financially viable. Over a period of time, Fundación Visión will become a similar resource center for hospitals from Latin and South American countries.

The receiving, or target, organization was Fundación Visión. Founded in 1992, this hospital system is the leader in blindness prevention in Paraguay, and operates a 6000 m^2 hospital in in central Asuncion, the capital of Paraguay. Fundación Visión is the only institution in the country that provides regular monthly ophthalmologist care in rural areas of the country and trains "eye health promoters" to seek out persons in need of treatment for blindness and other eye problems. The partnership with Aravind was undertaken so that Fundación Visión could increase the "quality, volume and sustainability of the eye care services" and improve the quality of its ophthalmology training programs.

At the same time, Fundación Visión also wished to increase the options for financing of eye care procedures in Paraguay, and, despite a strong in-country network of clinics, was constrained in its ability to reach the most remote parts of the country. Paraguay is country with relatively low population density of about 15 people per square kilometer (versus 416 per square kilometer in India), and outside of its four largest cities (Asuncion, Ciudad del Este, San Lorenzo, Capiata) and surrounding areas much of the population lives in small towns and villages that are very dispersed. As a result, it partnered with Fundación Paraguaya, one of the leading microfinance providers in Paraguay, to gain access to its deep network of offices and broad contacts with the rural poor in the most remote parts of Paraguay, and also to access the organization's expertise in finance to help patients pay for what could be very expensive surgeries. Founded in 1985, Fundación Paraguaya generates most of its revenue through its microfinance operations, but is also involved in

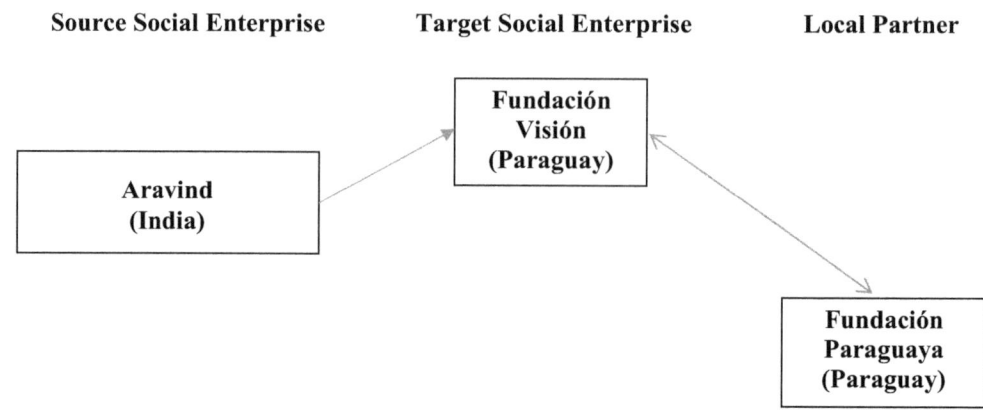

Fig. 1 Source social enterprise, target social enterprise, and local partner

Source Social Enterprise **Target Social Enterprise** **Local Partner**

other operations such as the Junior Achievement program and an agricultural school for youth which teach the values of entrepreneurship throughout Paraguay and beyond.

Data collection

This study uses four data sources: (1) semi-structured interviews; (2) field observations; (3) results from a preliminary survey of the partner organizations; and (4) archival data and documents. The primary source is semi-structured interviews with individual respondents. A total of 32 individual interviews were conducted. These interviews were conducted in-person in India and Paraguay with senior managers and program implementation and monitoring staff at Aravind (11 interviews), Fundación Visión (10 interviews), and Fundación Paraguaya (11 interviews). Interview questions focused on the innovation being transferred, the focal organization's history, operations and objectives, and the dynamics of the partnership being examined (see Appendix for full list of questions). The interviews were recorded and transcribed, and I took extensive field notes during the course of my site visits to each of the three organizations. Interviews and notes were transcribed and summarized within 24 h of the interview's completion. Table 3 provides further information regarding the interviews.

I supplemented my interview findings with field observation, including visits to field offices, screening camps, and project sites. I jotted notes on a paper pad during these visits, including notes from discussions with beneficiaries and local staff going about these work. These notes were later re-written into the field journal I kept on my laptop computer, and served as an important means for me to triangulate data obtained from in-person interviews and from company documents. Finally, as part of my collection of archival material and documents at Aravind, Fundación Visión, and Fundación Paraguaya, I was able to access both publicly-available and internal organizational records. These included emails, meeting minutes, annual reports, project reports and updates, briefs, monographs, and consulting evaluations.

Data analysis

As befits the exploratory nature of this study, no testable hypotheses were formed prior to data collection (Glaser and Strauss 1967). Rather, the goal of the study to use the data collected to develop specific theoretical constructs and related propositions which advance current theory in novel directions (Suddaby 2006). This approach is particularly suitable for studies of social entrepreneurship and bricolage as it allows the development of concepts in close connection to previous theorizing, important given the dearth of prior empirical work at the intersection of these two fields (Miles and Huberman 1994). The data was analyzed iteratively,

Table 3 Description of Informants

Organization	Formal position
Aravind	Faculty associate (training)
	Managing director (Aurolab)
	Executive director (LAICO)
	Administrator
	Administrator—LAICO
	Faculty./senior manager
	Assistant Manager (Field Services)
	Head ophthalmologist
	Medical resident (1st year)
	Assistant administrator (Theni Hospital)
	Chief medical officer (Theni Hospital)
Fundación Visión	Coordinator, mobile visión camps (Asuncíon)
	Medical resident (2nd year)
	Medical resident (3rd year)
	Director of operations
	Manager, IT and purchasing
	Clinic coordinator
	Manager, social work
	Manager's assistant (administration and finance)
	Manager, administration and finance
	Medical director
Fundación Paraguaya	General manager, Fundación Paraguaya
	Director of operations
	Microfinance coordinator
	Director of planning
	Director, Paraguani (branch office)
	Director, Itá (branch office)
	Director, Mariano Roque Alonso (branch office)
	Microcredit Group Manager, Mariano Roque Alonso (branch office)
	Director, agricultural school
	Director, regional microcredit offices (Asuncíon)
	Director, business development

followed the principles of open-ended, inductive theory building, as I kept going back and forth between theory development and empirical data analysis. I worked between interviews, field notes, company documents, archival records, and relevant literature to develop themes and codes in order to categorize findings related to the management of partnerships by social enterprises, including the impact of institutional logics and their alignment (or non-alignment) on organizational processes.

The principal unit of analysis for understanding the impact of institutional complexity (Greenwood et al. 2011) on the innovation transfer partnership between Aravind and its collaborators in Paraguay, Fundación Visión and

Fundación Paraguaya. I attempted to make sense of the data using "temporal bracketing," in line with recommendations from Langley (1999). For the sake of clarity, I present the three steps of data analysis in greater detail below, and in summary form in Table 4.

Step one: identifying key moments in the partnership

In the first step of analysis, I identified key moments in the partnership, including partnership formation and development. By extracting a chronology of events for each case in this manner, I was able to develop narratives documenting how the interactions between Aravind and Fundación Visión and Fundación Visión and Fundación Paraguaya evolved as the tri-partite collaboration developed.

Step two: coding institutional logics and key interactions

During the second stage of data analysis, I coded and compared moments in the partnerships which were associated with the social and commercial logics, the initial theoretical frame drawn from the institutional complexity literature I used to guide my analysis, cycling iteratively between data and emerging concepts related to the tensions in both cases (Suddaby 2006). What I found was that the simple "social" logic was insufficient to explain what was going on. Rather, it became clear that the social logic was actually manifesting as distinct spiritual logic and social welfare logics. Whereas the social welfare logic obtains legitimacy by making products and delivering services to address local social needs, the spiritual logic obtains legitimacy by linking the provision of social services and goods to an association with deities and faith as its fundamental guiding mechanism (Pache and Santos 2013; Gümüsay 2017). These spiritual and social welfare logics, along with the commercial logic, became the organizing frame for my understanding of the interactions between the social enterprises in this study.

Subsequently, I used "pattern matching" (Miles and Huberman 1994) to categorize the interactions between

the based on the dominant institutional logic at play: (1) spiritual logic; (2) social welfare logic; and (3) commercial logic. These categories were derived from my data and existing theory (Gümüsay 2017; Pache and Santos 2013), and I attempted to identify how the three social enterprises dealt with alignment or non-alignment of logics by examining themes in the interviewees' comments. I found that senior executives, mid-level managers, and front-line employees at all three organizations identified tensions in their relationships, but at the same time a subset of these individuals identified specific ways in which these tensions were dealt with in order to advance the innovation transfer process.

Step three: tying together logics, interactions, and emerging constructs

My third step was to tie the emergent spiritual logic, the social welfare logic, and the economic logic to the interactions between the three organizations in the innovation transfer partnership examined in this study. This was not a linear process. Rather, my analysis proceeded recursively (Pratt 2009; Langley 1999) until I had a grasp of the emerging constructs related to the management of the operations of the partnership. Codes consisting of several of the primary constructs of entrepreneurial bricolage (e.g., seeking resources, "making do", avoiding challenges) were used to capture the operations and activities used by the social enterprises in this study to negotiate the alignment or non-alignment of logics and the worldviews which influenced them. Additionally, I identified one important emergent conceptual category, or mechanism, used by Aravind and Fundación Visión to ensure the impact of Aravind's model as it was transferred from India to Paraguay, and also identified sources of tension that emerged in the course of the innovation transfer process. These constructs and interactions are outlined in detail below.

Table 4 Data analysis steps

Step	Name	Outcomes identified
1	Identifying key moments in the partnership	(1) Partnership formation (2) Partnership development (3) Narratives around partnership maintenance
2	Coding institutional logics and key interactions	(1) Identifying three dominant institutional logics: spiritual, social welfare, and economic (2) Describing points of tension in partnership related to logics non-alignment or alignment
3	Tying together logics, interactions, and emerging constructs	(1) identifying entrepreneurial bricolage processes (e.g., seeking resources, "making do", avoiding challenges) (2) Innovation fidelity: developing and defining the construct

Findings

Prior work on multiple or conflicting logics in organizations has highlighted the role of a dominant logic or competing logics in guiding or constraining organizational action (Pache and Santos 2013; Battilana and Dorado 2010). This study points to broader and richer sets of logics which prior work has not systematically considered. For instance, I found that social enterprises are not uniformly impacted by one or two dominant logics of action (e.g., a social logic and an economic logic). Rather, a rich set of logics, multiple in nature, impact their actions. While other scholars have focused on social and economic logics in the past, only recently has research started to examine how spiritual logics can guide the actions of enterprises (Gümüsay et al 2019; Gümüsay 2017; Tracey 2012). My research attempts to characterize how the values linked to a spiritual logic are manifested not only within the focal organization (Friedland 2013), but also how they link to inter-organizational interactions (Ahmadsimab and Chowdhury 2019; Vurro et al 2010) with partner organizations in the innovation transfer context.

Transfers of innovations between social enterprise partners offer a unique setting for studying institutional complexity. I found that logics may channel and guide the process of innovation transfer by impacting micro-processes of transfer as they are enacted at the organizational level by social enterprises, including the role played by institutional bricolage in this process (Baker and Nelson 2005; Desa 2012). Specifically, different social logics and the commercial logic impact the ways in which social enterprises incorporate different practices and innovations from partners, the solutions they craft to "make do" with the resources at their disposal, and the nature of the conflicts they encounter in the course of their partnerships. This, in turn, impacts the potential for "scaling" social innovations beyond the focal organization in which it was developed. These findings are discussed in detail in the sections which follow.

Organizations' commitment to multiple logics

Recent work (Battilana and Lee 2014; Ebrahim et al. 2014) has examined the conditions that encourage the persistence of multiple logics in a field, in particular when the organizations contained within the field are characterized by hybridity. Among the factors proposed are multiple local contexts for practice diffusion (and thereby local adaptation) and the lack of a dominant, overarching regulatory or professional framework that is able to impose field-level standards. Both these factors exist in varying degrees for the organizations examined in this paper. As a result, the actions of organizational actors in this study seem to be influenced by multiple logics, including one which has only recently been characterized in the institutions literature. At the level of social logics, I found the most variation between the organizations. For, Aravind and Fundación Visión, I identified an underlying "spiritual logic" rooted in their own organizations' histories which, in turn, influenced their social logic. For Fundación Paraguaya, the social welfare logic aligned with the economic logic, in line with prior work on microfinance organizations (Zhao 2014; Jia et al 2015). Below, I discuss on findings related to these logics in detail, highlighting both points of convergence and divergence for the three organizations in this study. Table 5 characterizes the spiritual, social welfare, and economic logics in summary fashion.

Spiritual logic

I found that Aravind, and Fundación Visión were guided by a core spiritual logic in their actions. This was in addition to the social logic common to all organizations in this study. Aravind is a pioneer in the provision of comprehensive eye care services to the poor. Its late founder, Dr. G. Venkataswamy, was fifty-eight years old and recently retired from the Indian civil service when he started the organization as an 11-bed hospital founded in a private residence in 1976. Dr. Venkataswamy—or "Dr. V." as he was affectionately known—was inspired by the teachings of Sri Aurobindo, one of the leading Hindu sages of southern India. This spiritual

Table 5 Comparison of spiritual, social welfare, and economic logics

Characteristic	Spiritual logic	Social welfare logic	Economic logic
Goals	Relieve suffering of beneficiaries while adhering to core religious principles	Deliver services and produce products to address local social needs	Maximize surplus revenue through efficiency of operations
Target population	Beneficiaries to be served, particularly the poor and marginalized	Beneficiaries to be served, particularly those who are seen as more or less "deserving" of "support"	Clients to be served with a focus on efficiency and on generating demand for future interventions
Operational principles	Design interventions to serve beneficiaries, to relieve beneficiaries' suffering, and to serve a "higher power"	Design interventions to maximize impact of available funds on social impact and positive social outcomes	Maximize surplus revenue through efficiency of operations and demand generation

This table is based on Pache and Santos (2013) and Gümüsay (2017), and on interviews and other data collected for this study

Bridging the rural–urban divide in social innovation transfer: the role of values

commitment is a core guiding principle of the organization, this philosophy was clearly reflected in Dr. V.'s writings and in the interviews that he gave. Some of these words are quoted each year in Aravind's annual reports. One of the clearest statements of the link between Aravind's spiritual roots and its activities is found on the third page of the 2008 Annual Report, which quotes Dr. V.:

> Our effort is to make Aravind an instrument of the Divine Will. We strive to forget our limitations and work with the direction of the Divine Will, not in a vain superficial way but with a deep commitment and faith that guidance comes from a higher level of consciousness. Then one is able to work with the great confidence that comes only with that faith and realization that we are all part of a spiritual capacity or spiritual power. It is then that all of nature works with you. You don't feel that you are a superior being but you are an instrument in the hands of a higher force and it is in that spirit that we meet our day to day struggles and successes.

More recently, in the 2018 annual report, Dr. V.'s words are again used to reflect the importance of the Divine, especially as reflected in Hindu philosophy, in Aravind's work:

> The Bhagavad Gita became popular and people started reading it to understand Kharma yoga. I remember well reading it in those days. At the same time Swami Vivekananda became very popular with us. His speeches were so powerful and inspiring, they made me look forward to doing something challenging and great.
> I also read the teachings of Sri Ramakrishna Paramahamsa, who had very little schooling, but who had known God in person. All these contacts influenced our thinking in those days. We were not thinking of amassing money as our goal in life. We always aspired to some perfection in our lives, like the realization of God, or reaching of higher level consciousness in Yoga.

This organization-level commitment is manifest in the daily operations of the organization, and influences the actions of organizational members. For instance, the Chief Medical Officer of an Aravind district hospital similarly referenced a "higher power" guiding the organization:

> … I don't know whether you believe it or not, but in this part of country, and on the Indian subcontinent, I think it's believed everywhere, that we are instruments only. Somebody else has decided that what I should do. Whatever we are doing, I don't think it's our effort only. Many people put the same amount of effort or more, but they don't achieve that. Somebody wants us

to do that much, so we are able to do it. Whatever we are doing, that divine force is there, a higher power is pushing us.

Similarly, Fundación Visión displayed a commitment to Christian principles in its work:

> Our mission is "to be a leading institution in the prevention of blindness, in the promotion of ocular health and in the delivery of high-quality services, as well as in the training of new professionals for community health." The work of the organization rests on Christian principles and relies on the Bible as the sufficient rule of faith and conduct. (Fundación Visión web site; accessed: 30 May 2018).

This commitment to a Christian God was confirmed by the physician who served as the Director of Operations at Fundación Visión, who noted how its influence extended to the care and treatment of patients at the organization:

> FV is also a Christian organization. Dr. Reinhold started in a loaned operating theatre in another hospital; he was driven by his Christian values to do this. FV now has 6000 patients per year, and the result of the care that they receive here means that patients can have a better life with God. The way we interact with patients is the way God or Jesus would act with patients. The staff are devoted to the patients.

Parboteeah et al. (2009) have noted that religion has a strong norm-setting influence with respect to work, which is seen as an obligation to society. In this way religion creates important work expectations for individuals, expectations which grow stronger when this work is situated in religious contextual environments. Within these two hospital systems spiritual values were an integral part of the work environment. This was manifested not just in the statements of organizational members and in official documents, but also in the physical symbols and objects which reinforced these values within these organizations. For instance, statues and pictures of Sri Aurobindo are found in the main entrance of all the major buildings within the Aravind system that I visited. The spiritual logic is important in this context as it helps to create a sense of common purpose and unity not just within the organization itself and between organizational members, but also between other organizations which share the same or similar values.

The spiritual logic also informed the way actors at Aravind and Fundación Visión approached their tasks within the health care realm, with their patients, and with respect to their interactions with members of other organizations. Aravind's Executive Director saw this connection in terms of spirituality and purpose—though Aravind is named after and inspired by the teachings of a Hindu sage and Fundación

Springer   335   Reprinted from the journal

Visión is inspired by Christian teachings—as a point of commonality to build upon, one that would help overcome differences in "detail" and level of maturity between the two organizations:

> At some level they [Fundación Visión] and Aravind share the same foundational mission. At one level there is similarity in terms of the purpose and the leadership orientation and those kinds of things, a lot of similarity. For instance, we are both in some way fundamentally committed to some higher values, you know? We are inspired by the teachings of Sri Aurobindo, and they have a lot of connection to the Christian church. Some of the difference is more on the detail, I think we probably have a lot of alignment than they have within the organization. And I think they are -- they are not very young but they probably have a lot more maturing to go through in terms of systems and processes and all of that.

Social welfare logic

At Fundación Paraguaya, the social welfare logic, rather than a spiritual logic, informed decision-making vis-à-vis the social impact of the organization's activities. This set Fundación Paraguaya apart from Fundación Visión, its in-country partner, and led to tensions in the relationship which were primarily manifested as clashes around the economic logic as outlined below. The social welfare logic at Fundación Paraguaya was structured around providing products, services, and support to address local social needs. As detailed in Table 5, in terms of target populations served what separated the social welfare logic from the spiritual logic was the former's emphasis on identifying those who were more "deserving" of support. For instance, one of my main interlocuters at the Fundación Paraguaya head office told me that while the main focus of the organization was "social," it was nevertheless focused on "selecting people for success" using different criteria took into account loan recipients' ability to "help themselves" and to build and grow successful businesses (my field notes).

Similarly, at one of Fundación Paraguaya's branch offices, the local manager told me that his field officers visited potential clients' homes to determine their level of cleanliness, the orderliness of their living conditions, and the general level of progress and order in their lives (my field notes). These criteria were seen as important determinants of individuals' worthiness for different social and credit programs. In this sense, the social welfare logic is more tightly coupled to the economic logic than a spiritual logic, which seeks to help the poor and marginalized without taking into account who might or might not be more deserving based on economic potential (though presumably there might be a bias towards those who show more devotion or faithfulness). For Fundación Paraguaya, the social welfare logic fed directly into the economic resources derived from an organization's various activities, including micro-financing of business opportunities and selling various services and products, and thus provided a framework for the operational principles need for the organization to achieve its goals (Pache and Santos 2013).

Economic logic

Beyond the spiritual and social welfare logics logic, and consistent with theory in social entrepreneurship, the economic logic was embedded in the necessities of the sector in which the organizations operated, and were related to issues of operational efficiency and demand generation for the services that they provide (Battilana and Lee 2014). In addition, Fundación Paraguaya had a "strong" form of this economic logic impacting its actions, whereas Fundación Visión had a "weak" form. Aravind's economic logic strength was somewhere in-between as the organization focused heavily on efficiency, but was at the same time foundationally committed to the idea of serving a "higher power" in serving patients, especially the ones least able to pay.

My point of departure from prior work is in the way in which I found that economic logics and the spiritual and social welfare logics mentioned above interacted in the inter-organizational setting. Where differences appeared in the interactions between the organizations in this study was at the level of the strength of the economic logic, which was more closely aligned with the social welfare logic. Specifically, even when there was no alignment between the strength of the economic logic (e.g., medium–weak in the case of Aravind-Fundación Visión), clashes at the level of operations related to the partnership were muted due to alignment on the spiritual logic aspect. However, a mismatch between the strength of the economic logic between Fundación Visión and Fundación Paraguaya (i.e., weak-strong) exacerbated already existing tensions at the level of social logic, where the lack of alignment between the spiritual logic and the social welfare logic meant that there was no means to diminish tensions between the organizations.

In prior work on social enterprises, the economic logic has been found to counterbalance the social welfare logic and focus organizations on questions such as demand generation and efficiency (Pache and Santos 2013; Battilana and Dorado 2010). In this study, the economic logic differed in strength across companies—for Fundación Paraguay, primarily a microfinance organization, it was a major driving force; for Fundación Paraguay it was far less important or even de-emphasized; for Aravind, social and spiritual commitments were balanced with a focus on delivering services efficiently. This focus on operational efficiency, as much as any spiritual element, was considered by Aravind's Executive Director as the fundamental pre-requisite for transferring knowledge to partners:

> I think most essential [element of the Aravind model to be transferred] is the mindset. Because I think the process... you can say one thing is more important than the other. For a given hospital something can be more important than the other, you know, like certain hospitals they could be already having a tremendous number of patients, you know, but then their conversion rate, they are having very low acceptance or so there we do not focus too much on marketing, you know, that will be institution-specific.
>
> But fundamentally the most important thing is the mindset, how they start thinking.... the market focus, customer focus, all of that... wanting to become efficiency focused or wanting to become self-supporting. You know, if you are always having the mindset that you can always raise money, you are never going to become efficient.

This efficiency focus was mirrored at Fundación Paraguaya, but in the opposite direction: the company's deep social and community connections were seen as its "differentiating" factor against versus other microfinance operators in Paraguay. Here, the emphasis was on the business or economic side, and the company's social programs—such as its agricultural school for disadvantaged but entrepreneurial young Paraguayans—fed into its money-making microcredit initiatives. The company's Director of Planning noted:

> We see our programs as separate in budgetary financial terms, but they are integrated at the operational level. The principles of how to run a sound microfinance program are the same principles that we used to run the agriculture school, we used the Junior Achievement team methodology to teach entrepreneurship at the agriculture school. The agriculture school takes the sons and daughters of microfinance clients as their students, and the microfinance program gives graduates of the agriculture school lines of credit...

On the other hand, for staff at Fundación Visión, the social service of their business was paramount, even for staff at the operational rather than executive level. For instance, the manager of information technology noted to me that patient care and the spiritual side were the key drivers of the organization. The coordinator of one of the clinics at the base hospital said that the focus on caring for patients was what separated Fundación Visión from other hospitals and clinics in Paraguay. Finally, the organization's manager of social work noted that Fundación Visión is committed to its patients both inside and outside of the hospital setting. This stemmed, she felt, from its foundation in "Christian

values": Fundación Visión was deeply involved in charity for patients; it is not a business, like so many other eye clinics and hospitals in Paraguay.

Multiple logics and bricolage

That a spiritual logic was found at Aravind and Fundación Visión is noteworthy, but unsurprising given the preponderance of faith-based organizations working on health and development issues. What is interesting, however, is how organization-level action can be influenced and guided by such spiritual values. My case data suggest that Aravind and Fundación Visión developed emergent strategies to ensure that their business models and innovations got scaled up. These strategies are deeply rooted in not only the process of scaling, but also the way different institutional logics at the source and target enterprise interact, and the extent to which the logics of the source enterprise and the target enterprises are aligned. They are also tied to what prior work has called "making do" with resources available in the local environment, or entrepreneurial bricolage (Desa and Basu 2013; Durand et al. 2013; Desa 2012; Baker and Nelson 2005).

Maintaining innovation fidelity

For instance, my case data indicate that the source social enterprises may be able to actively manage their partners using inter-related strategies related to the underlying balance between multiple institutional logics. These tactics can be broadly organized into one major process: *maintaining innovation fidelity*. In other words, the source enterprise, Aravind in this case, attempts to *maintain* the fidelity of their original model or innovation at the target site to preserve the intended social impact. They achieve this goal through a number of tactics, including bargaining with managers at the target organization to adopt practices and techniques developed by the source entrepreneur to ensure that the success of the original model and innovation in its new locale.

As described below, the process of maintaining innovation fidelity is related to separate aspects of the social enterprises' indirect ability to influence other organizations. This process enables source organizations with limited resources and a constrained direct ability to control of their partners' actions to exercise significant influence despite the absence of an ownership stake or even a principal-agent relationship at the target organization (Ansari et al. 2010). For Aravind, a primary way to ensure that the impact of the knowledge transferred was maintained was to induce adherence to the fidelity of their inter-linked eye care management methodology and practices (see "Methods" section for detailed description of the social innovation being transferred). Thus, their focus was on maintaining basic operating principles which could be shared with Fundación Visión and which would allow their system to transfer over to Paraguay.

Several Aravind managers commented to me about these principles, including the importance of starting the day early and keeping a focus on being organized to do good work. This was true of staff members both at headquarters (e.g., at LAICO, the Lions Aravind Institute of Community Ophthalmology, Aravind's training arm, and at the Madurai base hospital) and in field locations. For instance, a Faculty member at LAICO commented:

> … we are trying to share principles, you know, it may not be exactly that you can replicate all our procedures but the principles can always be replicated anywhere … for example, I will tell you, we start everyday at 7 o'clock, 7 AM in the operating theatre, 7:30 AM in the out-patient department.... We don't say, you also should start at 7, if you start at 7 well and good, but not necessary. You can start at 8 o'clock, 8:30, 9 o'clock, fine… but how you start is what's important, how you are organized to do good work…. in the places where it is possible we suggest that they should start little early or something… Fundación Visión , time is not a problem for them, they start early.

At the Theni field hospital, the Medical Director made a similar comment:

> … we follow certain principles, the basic principle on with the hospital Aravind Eye care to ensure those things. For instance, like discipline. We start 7:30 in the morning… Whatever we have committed has to be honored. It's written that at 7:30 the hospital starts, and a patient coming at 7:30 should be able to see it.

For Aravind, at the root of transferring technical procedures related to information technology, patient management, and community engagement is a focus on developing values which help organizations improve in each of these areas. This was the feeling of Fundación Visión's Chief Operating Officer as well:

> Aravind helps us to improve our procedures and improve the way we are attending the community. They are helping us to improve the attention we give to our patients, and they are telling us to correct some procedures, so we can have a better flow of patients or we can attend more volumes. So they are helping us technically but they are also saying that we have to develop our values, and our discipline too. So they are saying we have to do both.

Logic mis-alignment and conflict

While aligned spiritual logics were a source of partnership strength for Aravind and Fundación Visión, the mis-aligned spiritual and social welfare logics were a source of conflict

between Fundación Visión and Fundación Paraguaya. These difficulties arose as Fundación Visión sought to work with Fundación Paraguaya to fully implement one aspect of Aravind's model: demand generation for eye care services through screening camps. At these camps, patients are screened for cataract and other diseases, and those that are identified as needing follow-up treatment are referred for further follow-up. In southern India, with its great population density even at the village level, Aravind is able to use these camps to generate a constant flow of patients to its base and district hospitals. In Paraguay, with a much more dispersed population, this level of demand generation was not possible. Rather, camps have to be set up in the most rural locations to find potential patients.

This situation provided an opportunity for institutional bricolage (Desa 2012; Clough et al. 2019). While Fundación Visión had its own infrastructure in place for performing this screening function prior to the collaboration with Aravind, they did not have the reach of Fundación Paraguaya, which offers its microfinance and related products in the most remote regions of Paraguay. Facing resource constraints which did not allow it to operate these camps regularly beyond the major urban centers of Paraguay, Fundación Visión thus entered into partnership with Fundación Paraguaya to access its network of contacts across the country. Thus, Fundación Visión "made do" with the resources at their disposal (Desa and Basu 2013; Baker and Nelson 2005) by reaching potential patients needing cataract surgery and other services in rural areas where Fundación Paraguaya was able to use its resources to perform vision screenings for its customers and other community members.

However, this use of Fundación Paraguaya's network by Fundación Visión also led to conflict between the organizations. In particular, Fundación Paraguaya's staff felt that Fundación Visión was not concerned enough about promoting financing of different surgeries for patients screened at these camps. They felt that these surgeries would have been a "win–win" in the sense that Fundcion Visión would earn revenue from performing them, and Fundación Paraguaya would earn revenue by financing the operations. In addition, staff at Fundación Paraguay felt that Fundación Visión should have been more flexible about having more camps, by working in closer coordination the two organizations could build a future cataract surgery and eye care financing business in Paraguay.

The conflict which emerged between Fundación Paraguaya felt that Fundación Visión vis-à-vis the vision camps and associated promotion of financing options can be tied to the mis-alignment between the social welfare logic and "strong" form of the commercial logic at Fundación Paraguaya and the spiritual logic and "weak" commercial logic at Fundación Visión. In this sense, tensions which emerged between the two organizations mirror issues encountered by social enterprise partners where values and identities are mis-aligned (Chowdhury and Santos 2010; Smith et al. 2014). This finding relates to recent work by Arjaliès and Durand (2019), which suggests that values are embedded in the choices made by market actors in choosing investment product categories such as socially responsible investment funds, and that tensions may arise when there is a mis-match between the normative values of producers and consumers. Similarly, Chowdhury and Santos (2010) discuss a case wherein differing approaches to "social impact" between two partnering social enterprises, a kind of mis-alignment, led to difficulties in the transfer process and to a partial failure of the collaboration.

Logics alignment and muted cultural differences

Finally, and quite interestingly, the impact of cultural differences on the innovation transfer process was relatively muted compared to the institutional factors discussed above. Instead, what I found was that organizations tended to refer to geographic differences only with respect to actual or potential pitfalls in the transfer process; otherwise, the issue was for the most part in the background. This finding tended to go counter to expectations, as I assumed that cultural differences would be greatest when the members of the source-target pair were located in different countries, and especially where the "cultural difference" between the organizations' home countries was greatest—i.e., between India and Paraguay in the case of Aravind and Fundación Visión.

However, in the case of Aravind and Fundación Visión, which are located nearly ten thousand miles (16,000 km) apart, geographic separation provided an opportunity rather than a barrier for transfer, according to Aravind's Executive Director:

> I saw an opportunity to create synergy because Latin America is a place where there isn't any place to just do high volume affordable care. I think it is largely a mindset kind of a thing, their models are very heavily driven by the US model. So they practice very expensive medicine, which benefits 2% of the population, and a vast majority cannot afford that price. So, but this guy really genuinely wanted to help the poor people, Reinhold Dirks [Head of Fundación Visión]. That's how that idea came up and Martin [Burt, head of Fundación Paraguaya] was willing, Reinhold was willing… we were trying to create a win-win model.

It might be the case the "foreignness" of Fundación Visión relative to Aravind actually spurred effort on the part of both parties to make the relationship work. That is, due to the potential for miscommunication organizational members actually made a greater effort to communicate effectively with their counterparts from across the world.

Discussion

A number of scholars have called for research which highlights the distinctive nature of organizational processes in a social entrepreneurial context, where multiple institutional logics operate (Battilana and Lee 2014; Pache and Santos 2010; Dees et al. 2004), and in particular for those areas outside of metropolitan centers which are less conducive to developing thriving social enterprise ecosystems (Jia and Desa 2018). The present research addresses these calls by examining how social enterprises manage partnerships with other organizations that have related or distinct guiding logics and, ultimately, how social enterprises use partnerships to transfer innovations across geographic boundaries. In this sense, the article interrogates knowledge transfer processes across inequities of power, resources, market access, and governance (Wijers 2019; Piatti and Dwiartama 2016; Doering 2016; Seelos and Mair 2010; Hodge and Greve 2007), though it explores these issues in the relatively unique context of a "South-to-South" collaboration between organizations in India and in Paraguay. The article makes three contributions.

First, this work helps to illuminate the extent to which organizations which operate in environments with multiple institutional demands and with multi-dimensional goals manage their organizational processes and behaviors in order to achieve increased impact nationally and internationally (Dacin et al 2010). In particular, by illuminating the heretofore under-examined spiritual logic, it provides opportunities for researchers to develop these concepts further. Recent work in institutional theory has pointed to the relatively unexplored domain of religion in the study of modern organizations, and in the domain of international development in particular (Gümüsay 2017; Tracey 2012; Parboteeah et al 2009; Ver Beek 2000). For instance, Gümüsay (2017) argues that a heterogeneous intra-institutional religious logic may help scholars to theorize across different contexts about the impact of religion on organizational practices and values. While this macro-level focus is welcome, the present study goes beyond such broad-level constructs to illuminate the role that a spiritual logic may play in facilitating the transfer of ideas and practices between organizations, even when they are located at different ends of the earth. Both Aravind and Fundación Visión had prior collaboration experience with other spiritually-oriented hospitals. Where there were potential pitfalls due to language barriers, Aravind sent staff to work extensively on-site to ensure that practices were transferred, and this resulted in the successful transfer of their patient care and patient management practices to Fundación Visión.

Beyond this, what is particularly noteworthy about the partnership is the fact that while Fundación Visión is a

Christian organization, Aravind's spiritual dimension is rooted in Hinduism. Nevertheless, despite vastly different religious traditions, with one organization rooted in Western, Christian, traditions even if located in a developing country setting, and the other linked to an ancient "Eastern" religion, the organizations' underlying commitment to a "higher power" served as an important point of commonality which seemed to have facilitated their collaboration. Importantly, such a focus on common values between Aravind and Fundación Visión links to the "old institutionalism" idea that values and norms, versus scripts and other taken-for-granted processes as prescribed by neo-institutional theory, can serve as a primary guide for organizational action (Gehman 2020; Gümüsay 2017; Tracey 2012; Dorado 2006). This values turn in the study of institutions can be linked to the recent work of Friedland, who argues for greater focus on "the internal institutional order" of organizational actions as opposed to the "external conditions of their possibility" (2013). Fundamentally, the notion of "internal institutional order" links to the idea that values are embedded in an organization's practices (Klein Jr. 2015), a notion that this study seemingly substantiates in its exploration of organizational practices focused on care for patients as constituted by spiritual organizations such as Aravind and Fundación Visión. In these two hospital systems, the act of delivering eye care is intimately linked to the idea of a "higher power" and serving humanity as confirmed by a range of informants, organizational documents, and artifacts reflecting the influence of the Divine in structuring organizational action. What this study didn't do, however, is examine how the distinct religious traditions underlying the two principal organizations in this study, Hinduism and Christianity, differ in the ways that they influenced Aravind and Fundación Visión, respectively. Future research might explore these differences (Gümüsay 2017; Peifer 2014; Ver Beek 2000).

A second contribution of this paper is to highlight the tactics used by source firms to manage knowledge transfer partnerships with geographically distant partners that are operating in contexts very different to their own. A number of recent studies suggest that different institutional logics guide organizational behavior by providing specific scripts for action and by establishing core principles for organizing activities and channeling interests (Ahmadsimab and Chowdhury 2019; Ebrahim et al. 2014; Battilana and Dorado 2010). These studies provide little insight into the role played by divergent or convergent social logics on the unfolding of a partnership. Because values determine how partnership goals may be accomplished, this is an area of both theoretical and practical concern. For organizations with social goals, partnership is a means to "create social value" and indeed to spread the organization's key values

to other organizations and locales. The organizational field plays a major role in shaping these values and the nature of the partnership (DiMaggio and Powell 1983; Friedland and Alford 1991; Thornton and Ocasio 1999). Alignment of partner social enterprises' values may thus facilitie inter-organizational collaboration. When alignment isn't there, however, cultural and institutional factors become more important in determining the direction and success of a partnership (Palis 2006), and bricolage mechanisms (Desa 2012; Baker and Nelson 2005) have the potential to gain heightened importance.

For instance, while Fundación Visión entered into the partnership initially to gain access to new technologies and resources related to eye care delivery and process management, they nevertheless encountered difficulties vis-à-vis their cooperation with Fundación Paraguaya, their in-country (local) partner. Specifically, these difficulties stemmed from divergent perceptions relating to efficiency with respect to their vision camps operated distant rural areas using Fundación Paraguaya's networks but which combined both organizations' personnel. In addition, Fundación Paraguaya staff members were "disappointed" with the approach of Fundación Visión towards revenue-generating operations such as the financing of different surgeries. Here, the strong economic logic driving the microfinance organization, Fundación Paraguaya, came into conflict with the weaker economic logic of the eye hospital. Thus, what emerged from the rich case data which constitute this study is that the form of inter-organizational partnership is influenced by the extent to which the source organization manages to balance its multiple logics with those of its partners. While the balance between the two hospital systems was maintained via a strong spiritual logic alignment, differences in the prioritization of the economic logic between Fundación Visión and its in-country partner, Fundación Paraguya, speak to the challenges encountered between organizations which may be operating even in the same local context when their values are not aligned (Bacq and Janssen 2011; Friedland 2013; Klein Jr. 2015; Jia et al 2015).

Building on this second contribution, by situating institutional complexity research in an inter-organizational setting, this study also begins to unpack the processes which emerge when the logics of different organizations have to be simultaneously considered. In doing this, it addresses the following questions: Are these mechanisms indeed different from the traditional knowledge transfer mechanisms studied by organizational scholars? How do partners in a dyadic transfer relationship coordinate their efforts and to what extent to they ensure the fidelity of the innovation is maintained across settings (Ansari et al. 2010)?

By answering these questions, scholars may get closer to an understanding of institutional logics as strategic resources as proposed by Durand et al. (2013); this

340

promising line of inquiry may have much do contribute to the ways in which the logics of different organizations interact in an inter-organizational context. Additionally, the study advances knowledge of institutional bricolage mechanisms (Desa 2012; Clough et al. 2019) by linking these processes to settings where a dyadic transfer relationship has to be taken into account. While institutional bricolage mechanisms may enable a search for practical solutions in the social entrepreneurship setting (Desa and Basu 2013), clashes between institutional logics may also hinder the implementation of such solutions when two or more social enterprise partners are involved.

Third, and finally, this research contributes to the emerging literature on "scaling" entrepreneurial innovations in social settings (Seelos and Mair 2017; Desa and Koch 2014; Chowdhury and Santos 2010; Bloom and Chatterji 2009). It does this by: (1) introducing the concept of "innovation fidelity" as a mechanism for managing the transfer process; and (2) by offering possible a view of "scaling" solutions in the social sector which takes into account the notion of logics alignment between partner organizations. With respect to the former, this study highlights how source social enterprises may use refer to the maintenance of the core features of a focal social innovation to manage partners during the transfer process. Innovation fidelity thus relies on using inter-related tactics related to the underlying balance between multiple institutional logics, and has implications in terms of convergent (matching) and divergent (non-matching) social logics in the context of urban–rural transfer settings. Such tactics include including bargaining with managers at the target organization to adopt practices and techniques developed by the source entrepreneur, and emphasizing the principles underlying the impact of the social innovation being transferred. Ultimately, source organizations attempt to maintain innovation fidelity to ensure that the success of the original model and innovation in its new locale.

Where the notion of scaling solutions in the social sector links with values is in the degree of alignment between logics. For instance, alignment vis-à-vis the spiritual logic between Aravind and Fundación Visión, between the source and the target organization, helped to facilitate the process of maintaining innovation fidelity and, ultimately, the scaling of Aravind's model of delivering eye care. This was noted by managers at both organizations in terms of how organizing principles (such as discipline) were linked to the values that the organizations shared. On the other hand, attempts to maintain innovation fidelity may be less successful when logics are mis-aligned. This situation was apparent in the conflict which emerged between Fundación Paraguaya and Fundación Visión vis-à-vis the vision camps example referenced above, wherein divergent perceptions, rooted in different levels of strength (strong vs. weak) of the economic logic

at the two organizations, relating to efficiency and revenue generation led to tensions between the two organizations.

With respect to the broader debate on scaling in the social sector, much discussion has focused on "scaling organizations" as the primary means to expand the scope and reach of innovations developed by social entrepreneurs. For instance, Desa and Koch (2014) suggest three underlying requirements for scaling a venture across regions: compatibility of the social innovation, a market penetration strategy, and a design for affordability. Whereas the affordable design aspect and compatibility of the eye care services delivered by Fundación Visión and Aravind was a primary driver for the development of the partnership, Fundaction Visión had to "make do" (Baker and Nelson 2005) with the resources available to it in Paraguay in terms of delivering its services to under-served rural markets in Paraguay.

In particular, the on-the-ground partnership between Fundación Visión and Fundación Paraguay allowed the former organization to utilize and country-wide network developed for the microfinance context to access greater numbers of potential patients for its eye care services. While there were clashes with respect to the social motivation of Fundación Paraguaya's approach, which sought to promote micro-enterprise approaches to addressing poverty (Doering 2016), versus the more spiritually-grounded social motivation of Fundación Visión, ultimately this "making do" with available resources allowed the expansion of Aravind's approach to delivering eye care services to Paraguay. In this sense, by highlighting the possibility of instead "scaling solutions" (Dees et al 2004) through inter-organizational partnership as a means to achieve similar impact, the study puts forward novel approaches that organizations may undertake to address the issue of increased impact.

Finally, in addition to the theoretical contributions highlighted, the study also has practical implications for program managers, consultants and other actors working with social enterprises to scale the impact of their innovations. Recent work by the psychologist Jonathan Haidt (Treviño et al. 2017; Haidt and Treviño 2017) highlights the importance of developing and strengthening ethical organizational cultures as a way to deal with conflicting sets of foundational beliefs, including divergent political beliefs. The present study, by how examining varying social motivations (emerging from varied social logics) can manifest as tensions or conflicts in partnerships, builds on Haidt's notions by providing a basis for developing mechanisms for resolving these organizational tensions as they emerge in practice. For instance, it defines some of the tactics used by social entrepreneurs to manage the scaling of social innovations via partnering with other social enterprises (e.g., maintaining innovation fidelity). While these tactics may be less efficacious when values are not aligned, negotiating in good faith with partners may result in a positive outcome that both parties can live with.

As the world deals with the current global pandemic of covid-19, two final practical implications from this study may be particularly relevant. First, managers must understand the importance of modifying practices to fit the reality of contextual conditions where a partner (target) organization is located. While this study took into account the unique resource constraints encountered by organizations seeking to expand their social impact to rural settings, in a covid-19-affected world those constraints could well apply to the urban areas which have been hardest hit by the pandemic. Second, in the context of the pandemic, the issue of value alignments and mis-alignments has played out on a global scale. One short illustration of this phenomenon can be seen in the relative willingness to wear face masks as a preventive measure. In countries such as Taiwan and South Korea, universal acceptance of masks (values alignment) has led, up to this point, relatively rapid declines in infection rates and overall disease burden. In contrast, in countries where mask wearing has been contested (values mis-alignment), infection rates have taken longer to decline (Taleb 2020).

Acknowledgements I thank the guest editors of the special issue, Geoffrey Desa and Xiangping Jia, as well as three anonymous reviewers for their guidance during the review process. I also thank Filipe Santos, Anca Metiu, Ignasi Martí, and Wim van Lent for their comments on earlier versions of this paper. Finally, I am indebted to the social enterprises discussed in this article for the access which allowed me to complete this project.

Appendix

Interview guide

Introduction

1. Background of researcher. Offer thanks for agreeing to interview.
2. *Research Purpose* To investigate how organizations scale-up innovations through transfer to other organizations.
3. *Research Approach* I am tracking the transfer of innovations by social organizations to partners and analyzing the underlying decisions involved in this process and the rationale.
4. I want to get as broad a perspective of [SOURCE ENTREPRENEUR] as possible, and I one way to do this I think is to speak with people at all levels of the organization. I'd like you to be as free and frank as possible with your answers, but do tell me if any particular remarks might be problematic, etc.

5. I usually transcribe all the interviews for my research—it helps me be more attentive to research during our talk. Do you mind if I record our conversation (turn voice recorder on).
6. How much time do we have? (Usually 60 min–90 min)

General background and personal information

I would like to start by betting a broad picture of your personal background.

1. Please tell me about your role at [SOURCE ENTREPRENEUR]? Specifically, can you tell me about what your position entails and which areas are under your responsibility? (For the Founder / President: Has your role evolved over the years? How and why? What is your general management philosophy?)
2. Who do you see as the most important people in [SOURCE ENTREPRENEUR]? Could you please give me at least two or three adjectives to describe them?
3. Please summarize the strategy of [SOURCE ENTREPRENEUR] in impacting society and improving the lives of people. How do you think this strategy came about? How is it different from the strategies used by other organizations in your area? How is it the same?

The innovation

Description:
  Source unit:
  Target unit:

1. The transfer of this innovation from [SOURCE ENTREPRENEUR] to [TARGET ORGANIZATION]? was:

   a. Mandated by top management
   b. Strongly encouraged
   c. Favored
   d. Optional
   e. Entirely spontaneous

2. Who, in your opinion, initiated the transfer of this innovation from [SOURCE ENTREPRENEUR] to [TARGET ORGANIZATION]? (tick one or more)?

   a. Source
   b. Target
   c. Funding agency
   d. Government body
   e. Beneficiaries (villagers)
   f. Other

3. What was your role in this process? How much of your time did you spend working on the [TARGET ORGANIZATION]? staff (ask for percentage)? Did you go to [TARGET ORGANIZATION]?? How many times? When?

4. What are the elements of this innovation? Which is the most important (i.e., the innovation wouldn't work without it)? What has been transferred to [TARGET ORGANIZATION]?—whole or part of the innovation? Why? How might this innovation be applicable to other contexts?

5. Was this innovation changed for [TARGET ORGANIZATION]? in some way vs. how it was done at [SOURCE ENTREPRENEUR] originally? How so?

6. I'd like to establish a chronology of this innovation. When was it developed? Can you tell me how it was developed within your organization? When did you feel you had made the desired impact in your local area? When was the decision to expand beyond your local area made?

7. How did you come to select [TARGET ORGANIZATION]? as a partner for the transfer of this technology? Please describe this process for me.

8. On a scale of 1 to 5, how well do you thing the transfer process has been going? Are the villagers at the pilot [TARGET ORGANIZATION]? site using this technology per your conversations with [TARGET ORGANIZATION]??

9. What do you think will be the most important determinant of whether the transfer of this technology to [TARGET ORGANIZATION]? is successful?

10. What are the greatest challenges you've faced during the scaling-up process? Have you encountered different challenges in your local area versus expanding to areas further away?

11. Reflecting back on the transfer of sanitation technology between [SOURCE ENTREPRENEUR] and [TARGET ORGANIZATION]?, what would have been the single most important action to facilitate the transfer.

Mission and values

1. Does [SOURCE ENTREPRENEUR] have a precise Visión about how to conduct social change? If so, what do you think it is?

2. What values drive [SOURCE ENTREPRENEUR]? Have these changed over time?

3. How would you describe the culture of [SOURCE ENTREPRENEUR]? What is the guiding force behind the organization's actions? Social mission? Science?

Organizational evolution (top executives only)

1. Imagine that you're writing a history of [SOURCE ENTREPRENEUR]. Tell me what you want to write. What are the major changes the organization has faced? Note: Follow story, be clear about time-line, keeping in mind issues of internal and external identity.

2. In 10 years, where do you see [SOURCE ENTREPRENEUR]? What will [SOURCE ENTREPRENEUR] have accomplished during this period?

3. What was the founding mission of [SOURCE ENTREPRENEUR]? Is this still the mission of the organization? If not, what has changed?

Documents to request

- Annual Reports
- Financial Reports
- Strategic planning documents, past and present
- *Organizational chart* Number of employees, number of departments, change over time
- External media coverage of organization—articles, web sites, mentions in reports, etc.

Demographic, health, and other relevant data for geographic areas served by the organization (e.g., at the district, town, village levels)

References

Ahmadsimab, A., and I. Chowdhury. 2019. Managing tensions and divergent institutional logics in firm-NPO partnerships. *Journal of Business Ethics*. https://doi.org/10.1007/s10551-019-04265-x.

Ansari, S., P.C. Fiss, and E.J. Zajac. 2010. Made to fit: How practices vary as they diffuse. *Academy of Management Review* 35: 67–92.

Arjaliès, D.L., and R. Durand. 2019. Product categories as judgment devices: The moral awakening of the investment industry. *Organization Science* 30: 885–911.

Bacq, S., and F. Janssen. 2011. The multiple faces of social entrepreneurship: A review of definitional issues based on geographical and thematic criteria. *Entrepreneurship & Regional Development* 23: 373–403.

Baker, T., and R.E. Nelson. 2005. Creating something from nothing: Resource construction through entrepreneurial bricolage. *Administrative Science Quarterly* 50: 329–366.

Barraket, J., R. Eversole, B. Luke, and S. Barth. 2018. Resourcefulness of locally-oriented social enterprises: Implications for rural community development. *Journal of Rural Studies*. https://doi.org/10.1016/j.jrurstud.2017.12.031.

Battilana, J., and S. Dorado. 2010. Building sustainable hybrid organizations: The case of commercial microfinance organizations. *Academy of Management Journal* 53: 1419–1440.

Battilana, J., and M. Lee. 2014. Advancing research on hybrid organizing: Insights from the study of social enterprises. *Academy of Management Annals* 8: 397–441.

Bishop, S., and J. Waring. 2016. Becoming hybrid: The negotiated order on the front line of public-private partnerships. *Human Relations* 69: 1937–1958.

Bloom, P.N., and A.K. Chatterji. 2009. Scaling social entrepreneurial impact. *California Management Review* 51: 114–132.

Boxenbaum, E., and J. Battilana. 2005. Importation as innovation: Transposing managerial practices across fields. *Strategic Organization* 3: 355–383.

Chowdhury, I. 2015. Resilience and social enterprise: The case of Aravind eye care system. In *Summit on resilience II: The next storm*, ed. J. Ryan, 19–26. New York: Pace University.

Chowdhury, I. 2019. Social entrepreneurship, water supply, and resilience: Lessons from the sanitation sector. *Journal of Environmental Studies and Sciences* 9: 327–339.

Chowdhury, I., and F. Santos. 2010. Scaling social innovations: The case of Gram Vikas. In *Scaling social impact: New thinking*, ed. P. Bloom and E. Skloot, 147–168. New York: Palgrave Macmillan.

Clough, D.R., T. Pan Fang, B. Vissa, and A. Wu. 2019. Turing lead into gold: How do entrepreneurs mobilize resources to exploit opportunities. *Academy of Management Annals* 13: 240–271.

Czarniawska, B., and G. Sevón. 1996. Travels of ideas. In *Translating organizational change*, ed. B. Czarniawska and G. Sevón, 13–48. Berlin: de Gruyter.

Dacin, P.A., M.T. Dacin, and M. Matear. 2010. Social entrepreneurship: Why we don't need a new theory and how we move forward from here. *Academy of Management Perspectives* 24: 37–57.

Dees, J.G., B.B. Anderson, and J. Wei-Skillern. 2004. Scaling social innovation: Strategies for spreading social innovations. *Stanford Social Innovation Review* 1: 34–43.

Desa, G. 2012. Resource mobilization in international social entrepreneurship: Bricolage as a mechanism of institutional transformation. *Entrepreneurship Theory and Practice* 36: 727–751.

Desa, G., and S. Basu. 2013. Optimization or bricolage? Overcoming resource constraints in global social entrepreneurship. *Strategic Entrepreneurship Journal* 7: 26–49.

Desa, G., and J.L. Koch. 2014. Scaling social impact: Building sustainable social ventures at the base-of-thepyramid. *Journal of Social Entrepreneurship* 5: 146–174.

Di Domenico, D., H. Haugh, and P. Tracey. 2010. Social bricolage: Theorizing social value creation in social enterprises. *Entrepreneurship Theory and Practice* 34: 681–703.

DiMaggio, P., and W. Powell. 1983. The iron cage revisited: Institutional isomorphism and collective rationality in organizational fields. *American Sociological Review* 48: 147–160.

Doering, L. 2016. Necessity is the mother of isomorphism: Poverty and market creativity in Panama. *Sociology of Development* 2: 235–264.

Dorado, S. 2006. Social entrepreneurial ventures: Different values so different processes of creation, no? *Journal of Developmental Entrepreneurship* 11: 1–24.

Durand, R., B. Szostak, J. Jourdan, and P.H. Thornton. 2013. Institutional logics as strategic resources. In *Institutional logics in Action, Part A*, ed. M. Lounsbury and E. Boxenbaum, 165–201. Bingley: Emerald Group Publishing Limited.

Ebrahim, A., J. Battilana, and J. Mair. 2014. The governance of social enterprises: Mission drift and accountability challenges in hybrid organizations. *Research in Organizational Behavior* 34: 81–100.

Fiol, C.M., M.G. Pratt, and E.J. O'Connor. 2009. Managing intractable identity conflicts. *Academy of Management Review* 34: 32–55.

Fisher, G. 2012. Effectuation, causation, and bricolage: A behavioral comparison of emerging theories in entrepreneurship research. *Entrepreneurship Theory and Practice* 36: 1019–1051.

Friedland, R. 2013. God, love, and other good reasons for practice: Thinking through institutional logics. In *Institutional logics in action, part A*, ed. M. Lounsbury, 25–50. New York: Emerald Group Publishing Limited.

Friedland, R., and R.R. Alford. 1991. Bringing society back in: Symbols, practices, and institutional contradictions. In *The new institutionalism in organizational analysis*, ed. W. Powell and P.D. Maggio, 232–266. Chicago: University of Chicago Press.

Gehman, J. 2020. Searching for values in practice-driven institutionalism: Practice theory, institutional logics, and values work. *Research in the Sociology of Organizations*.

Glaser, B.G., and A.L. Strauss. 1967. *The discovery of grounded theory: Strategies for qualitative research*. Chicago, IL: Aldine Publishing Company.

Greenwood, R., M. Raynard, F. Kodeih, E. Micelotta, and M. Lounsbury. 2011. Institutional complexity and organizational responses. *Academy of Management Annals* 5: 317–371.

Gümüsay, A.A. 2017. The potential for plurality and the prevalence of the religious institutional logic. *Business & Society*. https://doi.org/10.1177/0007650317745634.

Gümüsay, A.A., M. Smets, and T. Morris. 2019. 'God at work': Engaging central and incompatible institutional logics through elastic hybridity. *Academy of Management Journal*. https://doi.org/10.5465/amj.2016.0481.

Haidt, J., and L. Treviño. 2017. Make business ethics a cumulative science. *Nature Human Behavior* 1: 1–2.

Horowitz, J.R., and A.M. McGahan. 2019. Collaborating to manage performance trade-offs: How fire departments preserve life and save property. *Strategic Management Journal* 40: 408–431.

James, H. 2014. A new institutional economics perspective on the relationship among societal values, governance structure and access to rural health care services. *Southern Business & Economic Journal* 37: 27–55.

Jia, J., & Desa, G. 2018. Social entrepreneurship and impact investment in rural-urban transformation: Summary symposium findings. https://sfsu.app.box.com/s/saw022dcfwk0sqtbknplxl919v7xnywm. Accessed 10 Jan 2019.

Jia, X., H. Luan, J. Huang, and Z. Li. 2015. Comparative analysis of microfinance and formal and informal credit use by farmers in less developed areas of rural China. *Development Policy Review* 33: 245–263.

Klein Jr., V.H. 2015. Bringing values back in: The limitations of institutional logics and the relevance of dialectical phenomenology. *Organization* 22 (3): 326–350.

Langley, A. 1999. Strategies for theorizing from process data. *Academy of Management Review* 24: 691–710.

Lévi-Strauss, C. 1966. *The savage mind*. Chicago: University of Chicago Press.

Lounsbury, M. 2007. A tale of two cities: Competing logics and practice variation in the professionalizing of mutual funds. *Academy of Management Journal* 50: 289–307.

Mair, J., and I. Martí. 2006. Social entrepreneurship research: A source of explanation, prediction, and delight. *Journal of World Business* 41: 36–44.

Mars, M.M., and H.J. Schau. 2017. Institutional entrepreneurship and the negotiation and blending of multiple logics in the Southern Arizona local food system. *Agriculture and Human Values* 34: 407–422.

McDermott, K., E.C. Kurucz, and B.A. Colbert. 2018. Social entrepreneurial opportunity and active stakeholder participation: Resource mobilization in enterprising conveners of cross-sector social partnerships. *Journal of Cleaner Production* 183: 121–131.

McNamara, P., F. Pazzaglia, and K. Sonpar. 2018. Large-scale events as catalysts for creating mutual dependence between social ventures and resource providers. *Journal of Management* 44: 470–500.

Miles, M.B., and A.M. Huberman. 1994. *Qualitative data analysis*, 2nd ed. Thousand Oaks, CA: Sage Publications.

Nirmalan, P.K., J. Katz, A.L. Robin, R. Krishnadas, R. Ramakrishnan, R.D. Thulasiraj, and J. Tielsch. 2004. Utilisation of eye care services in rural south India: The Aravind comprehensive eye survey. *British Journal of Ophthalmology* 88: 1237–1241.

Oliver, C. 1991. Strategic responses to institutional processes. *Academy of Management Review* 16: 145–179.

Pache, A., and F. Santos. 2013. Inside the hybrid organization: Selective coupling as a response to competing institutional logics. *Academy of Management Journal* 56: 972–1001.

Pache, A., and F.M. Santos. 2010. When worlds collide: The internal dynamics of organizational responses to conflicting institutional demands. *Academy of Management Review* 35: 455–476.

Palis, F.G. 2006. The role of culture in farmer learning and technology adoption: A case study of farmer field schools among rice farmers in Luzon, Philippines. *Agriculture and Human Values* 23: 491–500.

Parboteeah, K., M. Hoegl, and J. Cullen. 2009. Religious dimensions and work obligation: A country institutional profile model. *Human Relations* 62: 119–148.

Peifer, J.L. 2014. The institutional complexity of religious mutual funds: Appreciating the uniqueness of societal logics. *Research in the Sociology of Organizations* 41: 339–368.

Peifer, J.L., D.R. Johnson, and E.H. Ecklund. 2019. The moral limits of the market: Science commercialization and religious traditions. *Journal of Business Ethics* 157: 183.

Pratt, M.G. 2009. From the editors: For the lack of a boilerplate: Tips on writing up (and reviewing) qualitative research. *Academy of Management Journal* 52: 856–862.

Piatti, C., and A. Dwiartama. 2016. From food security to the enactment of change: Introduction to the symposium. *Agriculture and Human Values* 33: 135–139.

Sahlin-Andersson, K. 1996. Imitating by editing success. The construction of organizational fields and identities. In *Translating Organizational Change*, ed. B. Czarniawska, and G. Sevón, 69–92. Berlin: de Gruyter.

Saz-Carranza, A., and F. Longo. 2012. Managing competing logics in public-private joint ventures. *Public Management Review* 14: 331–357.

Seelos, C., and J. Mair. 2017. *Innovation and scaling for impact: How successful social enterprises do it.* Redwood City: Stanford Business Books.

Seelos, C., & Mair, J. 2010. Organizational mechanisms of inclusive growth: A critical realist perspective on scaling. *IESE Business School Working Paper.*

Skelcher, C., and S.R. Smith. 2014. Theorizing hybridity: Institutional logics, complex organizations, and actor identities. *Public Administration* 93: 433–448.

Smith, B., M. Meyskens, and F. Wilson. 2014. Should we stay or should we go? 'Organizational' relational identity and identification in social venture strategic alliances. *Journal of Social Entrepreneurship* 5: 295–317.

Suddaby, R. 2006. From the editors: What grounded theory is not. *Academy of Management Journal* 49: 633–642.

Taleb, N. 2020. The masks masquerade. https://medium.com/incerto/the-masks-masquerade-7de897b517b7. Accessed 15 June 2020

Thornton, P., and W. Ocasio. 1999. Institutional logics and the historical contingency of power in organizations: Executive succession in the higher education publishing industry 1958–1990. *American Journal of Sociology* 105: 801–843.

Thornton, P., and W. Ocasio. 2008. Institutional logics. In *The sage handbook of organizational institutionalism*, ed. R. Greenwood, C. Oliver, R. Suddaby, and K. Sahlin-Andersson, 99–129. New York: Oxford University Press.

Tracey, P. 2012. Religion and organization: A critical review of current trends and future directions. *Academy of Management Annals* 6: 87–134.

Treviño, L.K., J. Haidt, and A.E. Filabi. 2017. Regulating for ethical culture. *Behavioral Science & Policy* 3: 56–70.

van Wijk, J., C. Zietsma, S. Dorado, F.G.A. de Bakker, and I. Martí. 2018. Social innovation: Integrating micro, meso, and macro level insights from institutional theory. *Business & Society.* https://doi.org/10.1177/0007650318789104.

Ver Beek, K.A. 2000. Spirituality: A development taboo. *Development in Practice* 10: 31–43.

Voronov, M., D. De Clercq, and C.R. Hinings. 2013. Institutional complexity and logic engagement: An investigation of Ontario fine wine. *Human Relations* 66: 1563–1596.

Vurro, C., M.T. Dacin, and F. Perrini. 2010. Institutional antecedents of partnering for social change: How institutional logics shape cross-sector social partnerships. *Journal of Business Ethics* 94: 39–53.

Wijers, G.D.M. 2019. Inequality regimes in Indonesian dairy cooperatives: Understanding institutional barriers to gender equality. *Agriculture and Human Values* 36: 167.

Wry, T., and J. York. 2017. An identity-based approach to social enterprise. *Academy of Management Review* 43: 437–460.

Zhao, E. 2014. Mission drift in microfinance: An exploratory empirical approach based on ideal types. *Research Methodology in Strategy and Management* 9: 77–109.

Publisher's Note Springer Nature remains neutral with regard to jurisdictional claims in published maps and institutional affiliations.

Imran Chowdhury, Ph.D., is the Diana Davis Spencer Chair of Social Entrepreneurship and Associate Professor of Business & Management at Wheaton College in Norton, Massachusetts (USA). He teaches courses in entrepreneurship, strategic management and international management, and conducts research at the intersection of business and society, encompassing domains such as social entrepreneurship and innovation, corporate social responsibility, philanthropy, and community-focused organizations. He received his Ph.D. from L'École supérieure des sciences économiques et commerciales (ESSEC Business School) in France.

Agriculture and Human Values (2020) 37:1281–1292
https://doi.org/10.1007/s10460-020-10131-8

Blended finance for agriculture: exploring the constraints and possibilities of combining financial instruments for sustainable transitions

Tanja Havemann[1] · Christine Negra[2] · Fred Werneck[1]

Accepted: 13 July 2020 / Published online: 27 July 2020
© Springer Nature B.V. 2020

Abstract

Transitioning to sustainable agricultural systems is imperative to meet the global Sustainable Development Goals (SDGs). Achieving more sustainable agricultural production systems will require significant additional capital, however this cannot be covered by the current financial market setup, which dissociates public and private funders. Blended finance, where concessionary development-oriented funding is used to mobilize additional private capital, is essential. To ensure that the limited pool of concessionary funding is used efficiently and effectively, a shared understanding of the roles and limitations of public and private funders is necessary. In this paper, we describe the high-level funding gap for sustainable agriculture, the general landscape of agricultural finance, and the concept and potential roles of blended finance in this context. This paper introduces the conditions under which different financing mechanisms can contribute to addressing barriers related to sustainable agriculture investments. It highlights that multiple funding modalities must be utilized in order to achieve agricultural investment at a meaningful level and encourages greater exploration of the range of blended financing structures to increase SDG-related agriculture investments.

Keywords Blended finance · Investment · Finance · Agriculture · Agribusiness · Impact investment · SDGs · Sustainable agriculture · Climate smart agriculture

Background

The agricultural sector is central to achieving many of the United Nations 17 Sustainable Development Goals (SDGs). According to the United Nations Food and Agriculture Organization (FAO), for example, nearly 821 million people faced chronic food deprivation in 2017 (FAO et al. 2018). The number of extreme climate-related disasters has also doubled since the early 1990s, negatively affecting agricultural production and food availability. The environmental footprint of human population growth and dietary shifts has contributed to an over-exploitation of resources by the agricultural sector. For example, agriculture accounts for over 70% of global freshwater use, 23% of total anthropogenic greenhouse gas (GHG) emissions, and to rapidly declining biodiversity (IPCC 2019). Agriculture-induced environmental changes undermine agricultural productivity itself triggering profound socio-economic and political repercussions.

As an engine of socio-economic growth in emerging and developing economies, the agriculture sector is central to development. According to the FAO, while agriculture currently contributes circa 3% to global Gross Domestic Product (GDP), in Least Developed Countries (LDCs) the agricultural sector can account for as much as 60% of national GDP (FAO, World Bank Data). In countries with large rural populations, agriculture is critical for domestic food and income security.

To increase sustainability, a wide range of investments is necessary across the diverse set of agricultural systems and stakeholders, addressing a variety of socio-economic and environmental challenges. Across the world, challenges range from poverty inequality, rural depopulation (ESPON 2017), ageing, and obesity to soil degradation and inadequate access to inputs, technology, and infrastructure.

✉ Tanja Havemann
th@clarmondial.com

✉ Christine Negra
christine@versantvision.com

[1] Clarmondial AG, Zürich, Switzerland

[2] Versant Vision LLC, New York, USA

Expected to be home to 9 out of a projected 11 billion people in 2100, South Asia and Sub-Saharan Africa will demand particular support for sustainable rural economic development (FAO 2017). For example, as highlighted by a recent report by the International Fund for Agricultural Development (IFAD 2019), the majority (65%) of the world's rural youth (those aged 15–24) live in Asia and the Pacific region, but the number of rural youths in Africa is expected to rise by almost 20% in the next 3 decades—these youths are approximately three times more likely than adults to be unemployed. Significant additional investment is required to ensure that these individuals are given meaningful socio-economic opportunities, to ensure that their local societies meet basic livelihood needs, their nations prosper and that they contribute to the sustainable management of the world's natural resources.

While significant geographical differences exist, human diets have become more similar across the world by an average of 36% over the last 5 decades (CGIAR 2014). Global production is now centered on a genetically homogenous set of crops: wheat, rice, maize, potato, soybean, sunflower oil, and palm oil (CGIAR 2014). Efforts to reduce yield gaps could go some way to addressing increased demand for these products (FAO and DWFI 2015), however such efforts may become increasingly strained due to climate change (Zhao et al. 2017) and low resilience associated with poor agricultural biodiversity (Bioversity International et al. 2017).

Climate Smart Agriculture (CSA), "an approach to developing the technical, policy, and investment conditions to achieve sustainable agricultural development for food security under climate change" (FAO 2013) should be center stage when considering which investments to prioritize for investment, noting that suitability of a specific technology or practice will be site-specific. According to the FAO (2013), CSA "aims to enhance the capacity of the agricultural systems to support food security, incorporating the need for adaptation and the potential for mitigation into sustainable agriculture development strategies". It is composed of three pillars: (1) sustainably increasing agricultural productivity and incomes; (2) adapting and building resilience to climate change; and (3) reducing and/or removing greenhouse gases emissions, where possible. The pathways to achieving sustainable land use (and agricultural) systems are increasingly known i.e. as recently described in the Report of the FABLE Consortium (2019), but pursuing these pathways requires significant additional investment. The FABLE 2019 Report summarizes ten global targets across seven intervention areas, each of these global targets will require changes to three fundamental pillars; (1) Efficient and resilient agriculture systems, (2) Conservation and restoration of biodiversity, and (3) Food security and healthy diets. In order for this to happen, it must be accompanied by changes to policy and regulations as well as significant additional capital (investment) from both public and private sectors, including in research and development, infrastructure, capacity building and accompanying credit.

Although agriculture investment has been rising in high, middle, and low income countries since the 1990s, it has been growing at different rates and capital intensities (FAO 2017). For example, in the People's Republic of China, agricultural growth increased by 9% annually between 1991 and 2014, compared to 2% in high income countries and around 4% for other low and middle income countries (FAO 2017). The capital intensity of agricultural production diverges across high, medium and low income countries, with generally more capital intensive agriculture in high income countries. This trend is being followed by certain emerging markets, including China (FAO 2017). The FAO, the International Fund for Agricultural Development (IFAD), and the World Food Programme (WFP) estimate that USD 265 billion per year of additional investments are required to end hunger by 2030. However, these agencies acknowledge that the types of investments vary greatly—from basic infrastructure investments in LDCs to development and distribution of more resilient crop varieties (FAO 2017). Thus, prioritizing agricultural investment needs should be done on a context-specific basis.

While there is broad agreement that agriculture should be 'more sustainable,' as noted in the Brundtland Report, it should be done in a manner that meets the needs of the present without compromising the ability of future generations to meet their own needs (United Nations 1987). The seminal Brundtland Report, also known as *Our Common Future,* was published by a high-profile international group and was critical in proposing the concept of sustainable development that is still in use today, namely that such development requires a balance between the environment, social equity and economic growth (United Nations 1987). The Brundtland Report, and the global development community thus recognizes the need for balancing *inter alia* agricultural productivity (for food, fuel, fiber, and feed), health and nutrition, employment opportunities, equality, human rights, environmental goods and services, and economic wealth. In many cases, the information and information systems required to properly assess and calibrate potential trade-offs among these diverse needs is generally lacking (Clarmondial et al. 2019). This paper does not seek to define sustainable agriculture, as many other authors tackle this topic (Velten et al. 2015). Rather, it focuses on how blended finance approaches could be applied to support the development of a sustainable agricultural sector.

Despite ongoing definitional challenges, it is clear that sustainable agriculture will remain a critical area for international cooperation in order to meet globally agreed objectives such as the SDGs. In addition to an improved understanding of the agricultural sector, it will be necessary to mobilize additional funding to implement relevant investment strategies (Global Center on Adaptation and World Resources Institute 2019). Yet, public resources are insufficient: the funding gap to achieve the SDGs is estimated at nearly USD 4 trillion annually, with at least USD 300 billion required to meet the SDGs related to food security (UNCTAD and Convergence). The SDGs will not be reached without significant additional investments by the private sector, including private funders. Small and Medium sized Enterprises (SMEs) in particular are expected to play a critical role in supporting the transition to sustainable agriculture, but face challenges in accessing appropriate financial services (AGRA 2019). Mobilizing additional finance from the private sector—in particular towards relevant agricultural SMEs—will be critical to address the agriculture sector challenges. Blended finance encompasses potentially useful approaches to fill this funding gap.

Agricultural finance: investment approaches and institutions

As with other sectors, a wide variety of financial instruments is potentially available to finance agricultural projects. The type of financial instrument used should be appropriate for *inter alia* the development stage, amount required, cash flow profile, risk-return estimates, regulatory landscape, profile of potential funders, owners, and beneficiaries. Private sector actors, both agri-businesses (including cooperatives) and funders, expect that an investment provides an adequate level of return for the risk taken. Key considerations include the potential liquidity, structure, size, and other values that the proposed investment may contribute to an investment portfolio (e.g. correlation with existing assets).

Agricultural investments may span a wide range of opportunities, including early stage technology investments. Some common investments include agricultural technology ('agtech') venture capital, participation in established companies (i.e. private equity), shares of large publicly traded companies (i.e. listed equities), and publicly and privately issued debt (e.g. bonds, notes). Alternative strategies include financial derivatives[1] and CAT bonds[2] as well as new forms

of more decentralized finance such as security token offerings (STOs)[3] and crowdfunding.[4] These different types of investment modalities suit different types of financing needs.

Investments, or funding more generally, may originate from a range of sources that weigh non-financial (i.e. environmental or social impact) and financial outcomes, and derive from public or private sources. Private sector ('commercial') investors will typically seek risk-adjusted financial returns in line with comparable investments, based on fiduciary responsibility to their beneficiaries (e.g. customers, pensioners, shareholders) to pursue a set of investment strategies and targets. While some initiatives seek to broaden the scope of fiduciary responsibility and better consider environmental, social, and governance (ESG) issues in decision making, these can still be considered nascent. The public sector may provide funding in the form of concessionary and, or commercial capital.

A range of funders sit 'in-between', balancing their expectation of financial return with the intent to achieve measurable positive developmental impacts (i.e. non-financial returns). Thus, 'development finance' is a broader term encompassing Overseas Development Assistance (ODA) and non-concessionary funding by Development Finance Institutions (DFIs), which primarily provide finance for investments that promote development objectives (Swiss Sustainable Finance Glossary). The term 'impact investment' is used in a similar context, referring to "investments made with the intention to generate positive, measurable social and environmental impact alongside a financial return" (GIIN). As a result, the expected financial return may be reduced or the risk appetite increased.

Blended finance

The OECD refers to blended finance as the "strategic use of development finance for the mobilization of additional finance towards sustainable development in developing countries" (OECD 2018a, b) to address the SDG funding

---

Footnote 2 (continued)

if the special event protected by the bond triggers the payout to the insurance company, the obligation to pay interest and repay the principal is either deferred or completely forgiven.".

[3] These are financial securities that have a digital wrapper. They include standard underlying investments (e.g. stocks, bonds, real estate) but represented by a token which investors can buy. These are actual assets, linked to the standard regulatory and compliance architecture, though they are banned in some countries (notably China and South Korea).

[4] According to Investopedia, this refers to the application of small amounts of capital from a large number of individuals to finance a new business venture, typically through using social media platforms.

---

[1] According to Investopedia, "A financial derivative refers to a contract between two or more parties whose value is based on agreed-upon underlying financial asset or set of assets. Common underlying instruments include bonds, commodities, currencies, interest rates, market indexes, and stocks… [the] value is derived on the value of the primary security that they are linked to…futures contracts, options, swaps and warrants are commonly used derivatives.".

[2] According to Investopedia, CAT bonds (catastrophe bonds) are a "high-yield debt instrument designed to raise money for companies in the insurance industry in the event of a devastating natural disaster. A CAT bond allows the issuer to receive funding from the bond only if specific conditions occur such as an earthquake or tornado. However,

gap. ODA providers and DFIs are increasingly considering the role of blended finance in their mandates.

There are important nuances within and between the aforementioned definition and other definitions that can make the term 'blended finance' deceptively complex. For example, there is disagreement if philanthropic capital, e.g. grants from foundations can be classified as 'blended finance'. For a discussion of the definitional aspects, refer to ODI (2019).

Two central themes in blended finance concern determining and monitoring additionality and ensuring data comparability (ODI 2019). According to the Business Dictionary, the term "additionality" refers to the "extent to which a new input (action or item) adds to the existing inputs (instead of replacing any of them) and results in a greater aggregate." In the context of blended finance, additionality is sought both in the form of new funding mobilized and development outcomes. According to the OECD Glossary of Statistical Terms, comparability is "the extent to which differences between statistics from different geographical areas, non-geographical domains, or over time, can be attributed to differences between true values of the statistics. Different stakeholder groups have been collecting and disclosing varying information on development finance, and this can make it difficult to form an accurate picture of trends and thus to define priorities and recommend actions. Practically, blended finance refers to the combination of capital that has commercial risk-return expectations with funding that is concessionary in some form (typically from a public funder), in order to generate additional measurable developmental impact (ODI 2019). Financial and developmental additionality is critical (i.e. more capital flows to development strategies than otherwise would), but often challenging to determine due *inter alia* to a lack of comprehensive data sets and sectoral complexity (Pereira 2015; Carter et al. 2018).

According to ODI (2019) the three key pillars of blended finance are (1) The use of concessionary capital. This form of "below market" capital typically comes from the public sector; (2) That such concessionary capital should mobilize additional finance with non-concessionary objectives, i.e. from the private sector; and, (3) That additional, measurable, development impact is generated as a result. For example, a development agency that provides a partial guarantee to an investment fund that lends to selected agricultural businesses in emerging markets that manage for additional social and environmental impact as well as financial performance, in order to reduce the perceived risk and attract private investment in such a fund.

For the purpose of this paper, we propose the following definition: "the strategic use of concessionary funding mechanisms in order to mobilize additional private finance to achieve additional, measurable, non-financial development (impact) outcomes". This has three aspects that are important to highlight: First, it is inclusive as to the source of concessionary funding, recognizing that such funding can be broader than traditional development funders and instruments. For example, concessionary funding could also come from a private foundation. Second, it clarifies that the purpose is to mobilize additional finance (capital), rather than other goods and/or services. And, third, it highlights that the intent is to have measurable additionality of non-financial (social, environmental) impacts—typically related to achieving the Sustainable Development Goals (SDGs).

The concept of 'blending' is not new. For example, the first government-supported export credit insurance agency in the UK was established in 1919 to support the private sector to undertake more international business (UK National Archives). The term 'blended finance', however, was first adopted by the United Nations in 2015 at the Addis Ababa Action Agenda (AAAA) of the Third International Conference on Financing for Development (IDFC 2019). In 2016, the High Level Meeting of the OECD Development Assistance Committee (DAC) secured commitments by national governments to explore the role of blended finance in delivering the SDGs, notably focusing on the evidence base, best practices, and policy guidance (OECD).

Many historical transactions using a blended finance approach may not have been classified as such. Currently, approximately USD 140 billion worth of transactions have been identified and these types of structures have grown steadily over the past few years (while remaining a small fraction of overall development assistance) (Convergence 2019). Convergence, a global network focused on blended finance, notes that the median transaction size is USD 64 million, and that funds or collective investment vehicles (CIVs) are the most common transaction type. Transactions have mostly focused on Sub-Saharan Africa and have had relatively small deal sizes (ca. USD 55 million), however Asia has demonstrated greater growth. Agriculture has commanded on average 15–21% of blended finance resources. To date, most blended finance transactions have utilized concessional debt or equity, followed by technical assistance funds, then guarantee/risk insurance, and lastly grants.

While evidence is still emerging, early observations from OECD (2018a, b) and others indicates that the blending of commercial and concessional capital does not always lead to superior development results. Development finance experts and practitioners also emphasize the need to adapt the financial structure to the development intervention rather than choosing a particular financial structure because it is *en vogue*. This also requires that the motivations and thus incentives of different stakeholders need to be understood from the start. Where the appropriate financial structure is employed, positive effects on development have been observed. The amount of concessionary capital (subsidy) required to entice private investors has also been observed to be linked to the level of perceived investment risk—and thus the potential for each public sector dollar to result in additional private capital. However, calibrating additionality, and thus the value of pursuing a blended finance approach, is challenging as data are insufficient for evaluating and building the necessary evidence base. More effort is required to strengthen accountability, both on financial and non-financial i.e. development performance (OECD 2018a, b).

Important analytic work on blended finance in the agriculture sector is being done, notably by the Council on Smallholder Agricultural Finance (CSAF), "an alliance of financial institutions serving small- and medium-sized agricultural enterprises (SMEs) in Africa, Asia and Latin America… to share learning and develop industry standards and best practices for agricultural SME finance" (CSAF 2019). In 2019, CSAF facilitated an analysis of 4000 agricultural SME loans totaling USD 2.7 billion made by 20 financial institutions to determine loan-level profitability. As a result, it was possible to identify areas where subsidies may be justified: to support smaller loans, loans to African businesses, informal value chains, loans to new borrowers, and long-term (more than 12 month) loans. This research, and the collaborative loan monitoring and information sharing enabled by CSAF, may create an opportunity for determining additionality and thus contribute to developing and refining blended finance structures.

Blended finance can improve the risk-return characteristics of an investment by mixing capital flows with different financial and non-financial return expectations within an investment structure. Where public budgets are limited, such investment structures may help mobilize more funding from the private sector in support of the SDGs. In the context of agriculture, this may mean using public funding to enable private investors to make investments they may otherwise perceive as too risky, for example with new investment counterparts or in new funding structures. In particular, much attention given to increasing the participation of private sector investors—and in particular 'mainstream' institutional investors such as pension funds in such blended finance structures, yet this appears to be relatively nascent to date. According to the Global Impact Investing Network (GIIN) 2019 annual report, pension funds, insurance companies, and diversified financial institutions only represent 18% of all impact investments. Agriculture represented 10% of invested capital with energy, microfinance, and financial services in the lead. Tepid engagement by more 'mainstream' investors and relatively low private capital flows into sustainable agriculture is largely due to the lack of government support, suitable exit options, appropriate capital across the risk/return spectrum, and deal structures that can accommodate investors' or investees' needs. While the growth of the impact investing market is very encouraging, it remains small in the context of global financial markets, or indeed when compared to more passive' strategies such as negative screening. Despite the relatively small current market size, many stakeholders are excited about the potential role of blended finance in mobilizing more investment in support of the SDGs, including for sustainable agriculture.

There are a variety of reasons for taking a blended finance approach. These reasons do not only include public sector budget constraints, but also other motivations such as building investor confidence in new investment counterparts and approaches. For example, including a risk-absorbing tranche in an investment fund may give commercial investors the confidence to invest larger amounts in a strategy than they may otherwise have done, thus increasing the budget for pro-development outcomes associated with such investment. In the case of a new investment fund manager or financing approach, a partial guarantee supported by the public sector may help mitigate presumed investor risk in the initial deployment phase. Blended finance may also be to reward additional development results achieved, i.e. in the case of Results Based Financing, which can be defined as "any program where the principal sets financial or other incentives for an agent to deliver predefined outputs or outcomes and rewards the achievement of these results upon verification" (Musgrove 2011). This may result in the provision of additional resources alongside a commercial financing structure to help increase or ensure development outcomes, e.g. by adding a technical assistance or monitoring and evaluation component or facility to an investment fund focused on small and medium-sized enterprises.

There are many financial instruments, and combinations of financial instruments, that can be used to 'blend'. These include grant funding (e.g. for technical assistance) as well as concessionary versions of existing financial instruments including concessionary debt (i.e. debt provided at softer terms such as longer grace periods and lower interest rates), risk absorbing equity, and subsidized guarantees and insurance mechanisms. The potential roles and financial instruments are described in Table 1.

However, there are limitations of using blended finance. It is important to note that agreement on the boundaries of blended finance remains elusive. This is a challenge for the agriculture sector given its links to the 'real economy', which is defined by the Cambridge Dictionary as "the part of a country's economy that produces goods and services, rather than the part that consists of financial services such as banks, stock markets, etc." The agricultural sector is heavily influenced by government policies in most countries. Policy approaches to expand agricultural finance include national guarantee funds, subsidized lending, forced lending, and interest rate caps. Government interventions that support additional agricultural investment are at the boundary of what might be considered blending. For example, governments can support local subsidized lending programs to

Table 1 Potential roles of concessionary capital in financial instruments

Role of development finance	Sample instruments	Additionality aspects
Identify and enable new financing structures	Grants, concessional loans	Research to identify opportunities Facilitate the design of new investment structures
Seed new structures—first (anchor) capital	Equity, debt	Test new types of intermediation structures (i.e. proof of concept funding) and help to bring a financial instrument to scale, so additional private capital can engage (Milken Institute and OECD 2018) Conduct professional due diligence that can be shared with potential investors Act as a transaction lead and reference source to other investors
Risk mitigation	Guarantees, first loss tranches, subordinated loans, risk absorbing equity	Change the risk-return perception for private investors
Technical support	Grants	Provide grant funding alongside an investment to help increase chances of success (e.g. technical assistance)
Reward additional development impacts	Grants	Assigns a financial value to an additional non-financial outcome
Market development	Grants	Research and publish reports on the success of different interventions Support the development of investor incentives (e.g. policy changes) Support the development of consistent ways to monitor financial and developmental impact, to appropriately subsidize additional support, and regional approaches to harmonize relevant data

specific sectors or activities by issuing government bonds that attract private capital. Additional private capital may thus be mobilized and may generate additional development benefits (in particular where the use of proceeds of such a government debt issuance are clearly defined and monitored). Local governments may also implement fiscal incentives such as tax holidays and tax rebates to spur private investment in sustainable agriculture interventions. In this case, there may be foregone income to the public purse or public funding may reward certain private investors operating in the agricultural sector after they have done a specific activity that generates a development impact.

Challenges and opportunities for using blended finance for sustainable agriculture

The wide range of financing needs for sustainable agriculture includes early stage technology investments (i.e. venture capital), long term investments in greenfield production, new infrastructure and processing equipment, investments in transport and utilities, working capital, and trade finance. Such funding mechanisms can be designed to better address sustainability concerns, although doing so may incur additional costs (e.g. for impact monitoring) without direct financial returns.

Some of the challenges that investors face in making investments in emerging and developing markets, including in the agricultural sector, are summarized in Table 2. These are based on the practical experience of the authors, and some of these challenges pertain to investments in emerging markets more generally e.g. as described in GIIN (2019). Some of these challenges are heightened in particular types of agricultural investments. For example, greenfield agriculture investments in emerging markets often carry higher risk and are more challenging to finance given long investment terms and thus higher exposure to political, market and weather risks. While the challenges listed in Table 2 below are not necessarily unique to agriculture, many of these challenges are inter-related and also are exacerbated in developing and emerging markets, in the absence of a supportive policy environment, and where customers and shareholders are price sensitive.

Development Finance Institutions (DFIs) in OECD countries maintain an extensive overview of direct private finance mobilization rates. According to their most recent data assessments, mobilization rates in agriculture are highest when using guarantees, syndicated loans, and credit lines respectively (OECD 2019). In terms of banking and financial services, mobilization rates tend to be highest for guarantees and credit lines—however, it is unclear how much of this segment also tackles agriculture. Data are unavailable for

Table 2 Selected challenges related to sustainable investments in emerging and developing markets

Challenge	Description	Implications
Quality of data for decision making	There is a lack of up-to-date, dynamic data in many contexts	Investors and investees may not be able to credibly measure and demonstrate impact
Lack of precedents or comparable investments	There are few example transactions that investors can use to benchmark an opportunity	Novelty and uncertainty make it difficult to assess the proposed transaction
Unsupportive or unpredictable policy environment	International or domestic policies that can significantly change the economics of an investment	Investors attribute additional risk to an investment
Creditworthiness of potential investment counterparties	Potential investees have unknown or poor creditworthiness	Risk might be too large for most investors, or terms unattractive for such counterparties
Inefficient transactions size and high intermediation cost	Relatively small transactions that are resource-intensive for investors to properly assess	Transaction cost makes the proposed return unattractive
Investment term, expected time to profitability	Long time required for repayment, in particular for greenfield or infrastructure investments	Long term commitment results in higher return requirement
Investment liquidity	The difficulty (or inability) of selling or exiting an investment, e.g. transferring a loan	Investors attribute higher risk to more illiquid investments
International currency movements	Investments in a different (local) currency results in additional risks to the (foreign) investor	Investors may avoid certain currencies due to associated exchange risk

mobilization rates in structures that are not clearly labeled as 'blended finance' or where OECD DFIs do not participate.

Opportunities for using blended finance for sustainable agriculture

Equity investments

Ownership of shares in agriculture-related companies can either occur through the public markets ('listed equities') or through private dealings ('private equity'). A company's business model may relate to the ownership and operation of land, technology, or provision of goods and services. Equity investments may be minority or majority (controlling) positions and may have different levels of business-related risk. Investors usually realize financial returns through dividend payments by the company or by exiting the investment (i.e. sale of shares). Shares can be valued in many forms (e.g. according to the price paid by a third-party in a later transaction). An introduction to public versus private equity investments can be found at Investopedia. An overview of equity based blended finance instruments for agriculture can be found in SAFIN (2019).

Most development-oriented equity investments have been in the form of mezzanine/quasi-equity (preferred shares, subordinated debt) (GIIN 2018; ADB et al. 2017; Blue Orchard 2018). Equity in agricultural investments are generally expected to generate higher returns, due to higher risk and lower liquidity. Illiquidity is considered to be a particular challenge to mobilizing more equity investment. While

it would, in theory, be possible to structure various forms of liquidity facilities and equity derivatives (e.g. options, futures) to help attract private capital, this area has not been widely addressed. While some governments have created incentives that encourage public listing of smaller companies, including agricultural companies, barriers such as illiquidity and relatively high costs persists in many emerging and developing markets, hindering capital flows between investors and companies (e.g. Oliver Wyman and World Federation of Exchanges, African Capital Markets News).

Debt investments

Various forms of debt (i.e. credit) may be provided directly to users (e.g. agricultural companies) or through third-party institutions (e.g. banks or non-bank financial institutions—NBFIs). Debt can be relatively short-term credit (e.g. working capital, trade finance, supply chain finance) as well as longer-term finance for greenfield and capital investments. Investors are repaid according to a pre-agreed timeframe and interest rate. In the agricultural sector, it can be difficult to properly assess and administer loans as borrowers may be geographically remote or informally structured, as well as highly variable in their business models. In addition to relatively common financial products (e.g. direct loans and credit lines through local financial institutions), more concessionary types of financial instruments are possible such as soft loans (e.g. interest-free advances), impact bonds, and subordinated loans. An overview of debt based blended finance instruments for agriculture can be found in SAFIN

(2019). Table 3 below provides an overview of debt-based blended finance instruments for agriculture.

Note that other tools exist that sit between debt and equity e.g. quasi-equity structures (EIB 2017) and the 'demand dividend' (Santa Clara University 2013) structures (debt vehicles where payments are tied to cash flows, including a grace period, fixed payoff amount, and term sheet covenants and business plan focused on cash to align incentives).

Guarantees and insurance

Various commercial and concessionary guarantee and insurance products are available. These include instruments that cover political risk, production-related events (including weather), price fluctuation, and performance risk (see Table 4). Different risk mitigation products for agriculture are explored in Dalberg Global Development Advisers (2018) and World Bank Group (2014).

Grants

Grant funding can be used to support product development costs, to deliver technical assistance (TA), and to reward performance. In some cases, this takes the form of challenges or prizes. An overview of grant based blended finance instruments for agriculture can be found in SAFIN (2019), and is summarised in Table 5.

Table 3 Overview of debt-based blended finance instruments for agriculture

Approach	Role of development funder
Bonds, notes and other direct loan, including credit lines e.g. for trade, export,	Provide a direct loan, typically below market terms, to a counterparty. This also includes the provision of a dedicated agriculture credit line through an existing financial institution
Soft loan (interest free advances)	Provide a direct loan that bears no interest. A development funder may advance payment for a good or service, ahead of that good or service being delivered, effectively providing credit at no cost
Impact bonds	Provide upfront investment, or act as the outcome funder to subsidize private investment into an instrument
Subordinated loans	Provide funding in a more junior position in the capital stack compared to other private funders, thus accepting lower returns or higher risk, or both

Table 4 Overview of guarantees and insurance blended instruments for agriculture

Approach	Role of development funder
Credit guarantee	Cover certain potential losses incurred by agricultural lenders
Production insurance	Cover certain potential production-related losses, e.g. due to weather, climate, pests or disease. This may be done directly (i.e. via insurance policy), or through financial instruments
Subsidized market and price insurance	Cover certain potential market-related losses, including on volumes, price fluctuation and currency. This may be done directly (i.e. via insurance policy), or through financial instruments
Payment, performance, surety bonds	Commonly used in real estate and trade finance, these bonds de-risk a transaction between providers and buyers of goods/services. Development funders may subsidize these directly or through a third party (e.g. insurance company) and may participate directly (e.g. provide letters of credit (LCs) and reserve accounts as a form of guarantee)

Table 5 Overview of grant based blended finance instruments for agriculture

Approach	Role of development funder
Technical Assistance (TA)	Pay for TA to farmers, local companies, or intermediaries (e.g. for agronomic or business management expertise) in order to reduce credit risk
Performance-based grants and Results Based Financing	Pay project developers or business owners based on achievement of pre-agreed non-financial (developmental impact) outcomes, typically once these have been verified by an independent third party. This could increase, directly or indirectly, the return of other funders
Design funding	Provide grants to entities that develop and implement new business models or financial instruments to mobilize additional capital to sustainable agriculture
Challenges and prizes	Provide a sum of money to an entity that has won a competition to achieve a specific pre-defined result. This differs from performance-based grants in that it is competitive, and that performance-based grants do not necessarily require third-party verification

Discussion

While private capital providers increasingly seek sustainability impact alongside financial returns, the challenges they face in deploying more funding to sustainable agriculture need to be considered. There are technical challenges (e.g. information ecosystem gaps for evaluating trade-offs) as well as operational challenges on how to mobilize appropriate additional financial resources for developmental impact. Taking a blended finance approach may be useful in this regard. However, this requires greater flexibility, common understanding, and ability to utilize a wider range of funding tools suitable to different stakeholders.

ODA providers and governments, as well as traditional development partners such as Non-Governmental Organizations (NGOs), also need to appreciate the challenges facing private capital providers—in particular institutional investors who have strict fiduciary responsibilities. At the same time, private investors must appreciate the need for demonstrating social and environmental impact when public resources are utilized. Designing effective blended finance mechanisms to support the transition to sustainable agriculture requires an appreciation of the challenges on both the demand for and supply of capital. It is critical that the design of financing structures, including the selection of blended finance instruments and approaches be tailored to the specific situation, including the development challenge and counterparts.

Across the blended finance, impact investing, and development finance communities pursuing sustainable agriculture, key challenges include efficient and effective impact monitoring of social and environmental performance, testing of new investment strategies, and cultivation of appropriate structures and intermediaries. Within blended finance structures, the strategic roles of concessionary capital will likely need to include the following four types of approaches, as well as modifications and mixtures of these approaches, depending on the context: *'Permanent blended finance'*—financial structures that will always need to rely on concessionary finance within the capital mix for example in cases where research has been done to indicate this requirement, e.g. in the case of Aceli Africa based on the aforementioned work by CSAF; *'Transitional blended finance'*—concessionary capital element can taper down as the investment moves past proof of concept, e.g. where a government agency may offer a partial guarantee for an agricultural investment fund to help mobilize private capital, such government-backed guarantees typically only cover a certain quantity of transactions; *'Adjustable blended finance'*—inclusion of concessionary capital varies based on relevant risk or impact creation, e.g. the Social Impact Incentives (SIINC)

mechanism developed by the Swiss Agency for Development and Cooperation (SDC) with Roots of Impact (Roots of Impact 2016), and *'Impact monitoring and verification blended finance'*—concessionary capital covers the cost of monitoring or verifying impact, for example where government donors provided a technical assistance facility to the African Agriculture Fund (Marchand).

Efficient, effective agricultural finance has a number of persistent challenges especially in rural emerging market areas. Sustainable agriculture is a particularly relevant target for blended finance given its significant GDP contribution in many countries, and the need to overcome barriers such as the remote location of counterparties, lack of information, and high opportunity costs. Interesting entry points may be found through novel partnerships, for example with agribusinesses, government agencies, technology companies, private capital providers, and NGOs.

There is widespread agreement that we are at a tipping point in many respects, in all three of the dimensions noted in the Brundtland Report: inequality has risen in the past half-century (Credit Suisse 2019), according to the OECD ODA is stagnant (OECD 2018a, b), and many planetary boundaries are at a critical point (Stockholm Resilience Centre). There is widespread agreement, both in the international development community and also the private sector that more needs to be done to address the SDGs, and that this will require significant additional investment. Blended finance is an important lever in mobilizing this additional investment, and although the engagement of private capital has been rather nascent, the rate of growth in the sustainable and impact investing market has been impressive (GSIA 2018), and there have been increasing impact investing allocations to agriculture (GIIN 2019). If used properly, the public sector can encourage more investment to achieve the SDGs, including in the agricultural sector.

The COVID-19 Pandemic has arguably hastened the need to explore such blended finance approaches. The Pandemic, and responses to it are negatively impacting the agriculture sector, including capital flows notably access to critical working capital for agricultural SMEs in emerging and developing markets (ISF et al. 2020). Public funders should consider immediate steps they can take to help facilitate increased access to working capital for vulnerable groups, as well as planning for how to support the medium to long-term investments that will be required to rebuild agricultural-dependent economies. Public sector budgets are likely to become even further stressed as a result of the Pandemic, and thus the need for blended finance solutions may be heightened in both the near and long-term.

Conclusions

While policy action to address the SDGs is critical, both from a macro-economic and an agriculture sector perspective, significant private investment in sustainable agriculture must be mobilized in order to meet the SDGs. Historical data makes it clear that the public sector and traditional ODA can contribute to only a fraction of the funding gap to transition to sustainable agriculture. Financing approaches that are based on the 'blending' of public and private capital sources will be necessary to amplify public investment and to attract private capital. Blended finance transactions can help address many different challenges, but, given the scarcity of concessionary finance, should be used in a manner that maximizes impact. Building on the existing basic tools and initial examples, growth of the blended finance market will depend on better understanding of successes and failures by quickly and transparently testing a wide range of partnerships, solutions, and approaches.

References

ADB, ICD, IDB, EDFI, AfDB, AIIB, EBRD, IFC, EIB. 2017. DFI Working Group on Blended Concessional Finance for Private Sector Projects. Summary Report. https://www.ifc.org/wps/wcm/connect/a8398ed6-55d0-4cc4-95aa-bcbab e39f79f/DFI+Blended+Concessional+Finance+for+Priva te+Sector+Operations_Summary+R....pdf?MOD=AJPER ES&CVID=lYCLe0B. Accessed 15 June 2020.

African Capital Markets News website https://www.africancapitalm arketsnews.com , e.g. the news story published 19th, 2018: Liquidity and cost of trading on Africa's stock exchanges—Bright Africa 2018. https://www.africancapitalmarketsnews .com/3785/liquidity-and-cost-of-trading-on-africas-stock-excha nges-bright-africa-2018/. Accessed 14 June 2020.

AGRA. 2019. Africa Agriculture Status Report: The Hidden Middle: A Quiet Revolution in the Private Sector Driving Agricultural Transformation (Issue 7). Nairobi, Kenya: Alliance for a Green Revolution in Africa (AGRA). https://agra.org/wp-conte nt/uploads/2019/09/AASR2019-The-Hidden-Middleweb.pdf. Accessed 26 May 2020.

Blue Orchard. 2018. Blended Finance 2.0: Giving voice to the Private Sector. Insights from a BlueOrchard survey on private investors, October 2018. https://www.blueorchard.com/ wp-content/uploads/181016_BlueOrchard_Blended_Finance-2.0.pdf. Accessed 26 May 2020.

Bioversity International, CGIAR, Clarmondial. 2017. Reducing risks and seizing opportunities: integrating biodiversity into food and agriculture investments. Bioversity International, Rome, Italy. https://www.clarmondial.com/wp-content/uploads/2017/07/ ABD_Index_Investor_summary.pdf. Accessed 26 May 2020.

Cambridge Dictionary. Real economy. https://dictionary.cambridge. org/dictionary/english/real-economy. Accessed 26 May 2020.

Carter, P., Van De Sijpe, N. and Calel, R. 2018. The elusive quest for additionality. Working Paper 495. Washington, D.C.: Center for Global Development. www.cgdev.org/sites/default/files/elusi ve-quest-additionality.pdf. Accessed 26 May 2020.

CGIAR. 2014. Increasing homogeneity of world food supplies warns of serious implications for farming and nutrition. ScienceDaily. www.sciencedaily.com/releases/2014/03/140303154102.htm. Accessed 26 May 2020.

Clarmondial, Wageningen University & Research, Versant Vision. 2019. EIRA: Environmental Impact Reporting in Agriculture. Creating a link between agricultural investments and environmental impact. https://static1.squarespace.com/static/56cb8cc14d 088ea8995336de/t/5ca13ceb6e9a7f3e80ba8c5b/1554070764244/ EIRA_Report_April_2019.pdf. Accessed 26 May 2020.

Convergence. 2019. The State of Blended Finance 2019. https://asset s.ctfassets.net/4cgqlwde6qy0/58T9bhxExlNh2RilxWxSNe/ba56f a36c81349640179779ddd68cc99/Convergence_-_The_State_of_ Blended_Finance_2019.pdf. Accessed 26 May 2020.

Council on Smallholder Agricultural Finance (CSAF). 2019. State of the Sector 2019. https://csaf.org/wp-content/uploads/2019/07/ CSAF_State_of_Sector_2019_Full_Final.pdf. Accessed 26 May 2020.

Credit Suisse. October 2019. Research Institute. Global Wealth Report 2019.

Dalberg Global Development Advisers. 2018. Blended finance tools to catalyze investment in agricultural value chains: An initial toolbox.

EIB. 16 January 2017. Quasi-equity: A new financial structure for a new challenge. https://www.eib.org/en/stories/quasi-equity-a-new-financial-structure-for-a-new-challenge. Accessed 26 May 2020.

ESPON. 2017. Policy Brief. Shrinking rural regions in Europe. Towards smart and innovative approaches to regional development challenges in depopulating rural regions. Luxembourg, European Union, ESPON. https://www.espon.eu/sites/default/files/attac hments/ESPON%2520Policy%2520Brief%2520on%2520Shrink ing%2520Rural%2520Regions.pdf. Accessed 26 May 2020.

FABLE. 2019. Pathways to Sustainable Land-Use and Food Systems. 2019 Report of the FABLE Consortium. Laxenburg and Paris: International Institute for Applied Systems Analysis (IIASA) and Sustainable Development Solutions Network (SDSN).

FAO. https://www.fao.org/3/i2490e/i2490e01c.pdf. Accessed 26 May 2020.

FAO. 2013. Climate Smart Agriculture Sourcebook. Rome, FAO. https ://www.fao.org/3/i3325e/i3325e.pdf. Accessed 26 May 2020.

FAO. 2017. The future of food and agriculture. Trends and challenges. Rome, FAO. https://www.fao.org/3/a-i6583e.pdf. Accessed 26 May 2020.

FAO and DWFI. 2015. Yield gap analysis of field crops: Methods and case studies, by Sandras V.O., Cassman, K.G.G., Grassini, P., Hall, A.J., Bastiaanssen, W.G.M., Laborte, A.G., Milne, A.E., Sileshi, G., Steduto, P. FAO Water Reports No 41, Rome, Italy. https://www.fao.org/3/a-i4695e.pdf. Accessed 26 May 2020.

FAO, IFAD, UNICEF, WFP and WHO. 2018. The State of Food Security and Nutrition in the World 2018. Building climate resilience for food security and nutrition. Rome, FAO. https://www.fao. org/3/I9553EN/i9553en.pdf. Accessed 26 May 2020.

GIIN. 2018. Annual Impact Investor Survey 2018. The Eighth Edition. https://thegiin.org/assets/2018_GIIN_Annual_Impact_Inves tor_Survey_webfile.pdf. Accessed 14 June 2020.

GIIN. 2019. Annual Impact Investor Survey 2019. The Ninth Edition. https://thegiin.org/assets/GIIN_2019%2520Annual%2520Impact %2520Investor%2520Survey_webfile.pdf. Accessed 26 May 2020.

Global Center on Adaptation and World Resources Institute. 2019. Adapt Now: A Global Call For Leadership on Climate Resilience. https://cdn.gca.org/assets/2019-09/GlobalCommission_Repor t_FINAL.pdf. Accessed 26 May 2020.

Global Impact Investing Network (GIIN). All You Need to Know about Impact Investing. https://thegiin.org/impact-investing/. Accessed 26 May 2020.

Global Sustainable Investment Alliance (GSIA). 2018. Trends Report 2018. https://www.gsi-alliance.org/trends-report-2018/. Accessed 14 June 2020.

IDFC. 2019. Blended Finance: A Brief Overview. https://www.idfc.org/wp-content/uploads/2019/10/blended-finance-a-brief-overview-october-2019_final.pdf. Accessed 15 June 2020.

IFAD. 2019. Creating opportunities for rural youth. 2019 Rural Development Report. Rome, IFAD. https://www.ifad.org/documents/38714170/41133075/RDR_report.pdf/7282db66-2d67-b514-d004-5ec25d9729a0. Accessed 14 June 2020.

Investopedia. Private equity vs. Public equity: What's the difference? https://www.investopedia.com/articles/investing/030415/difference-between-private-and-public-equity.asp Accessed 15 June 2020

IPCC. 2019. IPCC Special Report on Climate Change, Desertification, Land Degradation, Sustainable Land Management, Food Security, and Greenhouse gas fluxes in Terrestrial Ecosystems. Summary for Policymakers. Approved Draft. https://www.ipcc.ch/site/assets/uploads/2019/08/4.-SPM_Approved_Microsite_FINAL.pdf. Accessed 26 May 2020.

ISF, Mastercard Foundation and Rural & Agricultural Finance Learning Lab. 2020. COVID-19 Emergency Briefing. https://www.raflearning.org/post/covid-19-emergency-briefing-agri-smes-operating-uncertain-financial-operational-and-supply Accessed 14 June 2020.

Marchand, Sarah. Reflections on the effectiveness of TA provided by facilities linked with investment funds. https://www.inclusivebusiness.net/ib-voices/reflections-effectiveness-ta-provided-facilities-linked-investment-funds. Accessed 14 June 2020.

Milken Institute & OECD. 2018. Guaranteeing the goals: Adapting public sector guarantees to unlock blended financing for the UN Sustainable Development Goals. https://assets1c.milkeninstitute.org/assets/Publication/Viewpoint/PDF/Guaranteeing-the-Goals-FINAL-4.pdf. Accessed 26 May 2020.

Musgrove, P. 2011. Financial and Other Rewards for Good Performance or Results: A Guided Tour of Concepts and Terms and a Short Glossary. World Bank Group. Online: https://www.rbfhealth.org/sites/rbf/files/RBFglossarylongrevised_0.pdf. Accessed 14 June 2020.

ODI. 2019. Blended finance in the poorest countries—The need for a better approach. https://www.odi.org/sites/odi.org.uk/files/resource-documents/12666.pdf. Accessed 26 May 2020.

OECD. Blended Finance: Bridging the Sustainable Development Finance Gap. https://www.oecd.org/dac/financing-sustainable-development/development-finance-topics/Blended-Finance-Bridging-SDG-Gap.pdf. Accessed 26 May 2020.

OECD. 2018a. The next step in blended finance: Addressing the evidence gap in development performance and results. As part of the consultation on the OECD DAC Blended Finance Principles for unlocking commercial finance for the SDGs. Copenhagen, Denmark, Monday 22 October 2018. Workshop report. https://www.oecd.org/dac/financing-sustainable-development/development-finance-topics/OECD-Blended%2520Finance-Evidence-Gap-report.pdf. Accessed 26 May 2020.

OECD. 2018b. Making Blended Finance Work for the Sustainable Development Goals, OECD Publishing, Paris. https://read.oecd-ilibrary.org/development/making-blended-finance-work-for-the-sustainable-development-goals_9789264288768-en#page3. Accessed 26 May 2020.

OECD. Development aid drops in 2018, especially to neediest countries: https://www.oecd.org/development/development-aid-drops-in-2018-especially-to-neediest-countries.htm. Accessed 14 June 2020.

OECD. February 2019. Amounts mobilised from the private sector by development finance interventions in 2016–2017. Preliminary insights from the data. https://www.slideshare.net/OECDdev/amountsmobilisedfromtheprivatesectorbydevelopmentfinanceinterventionsin201217. Accessed 26 May 2020.

Oliver Wyman & World Federation of Exchanges. Enhancing liquidity in emerging market exchanges. https://www.world-exchanges.org/storage/app/media/research/Studies_Reports/liquidity-in-emerging-market-exchanges-wfe-amp-ow-report.pdf. Accessed 14 June 2020.

Pereira, J. 2015. *Leveraging aid: A literature review on the additionality of using ODA to leverage private investments*. London: UK Aid Network.

Roots of Impact. SIINC in Practice. https://www.roots-of-impact.org/siinc-programs/ and SDC with Roots of Impact. 2016. SIINC – White Paper. https://www.roots-of-impact.org/wp-content/uploads/2019/01/Social-Impact-Incentives-SIINC-White-Paper-2016.pdf. Accessed 14 June 2020.

SAFIN. March 2019. Landscape Report: Blended Finance for Agriculture. https://5724c05e-8e16-4a51-a320-65710d75ed23.filesusr.com/ugd/7f0ffd_d48e2795163446d88b574c2c5c3ade0a.pdf. Accessed 16 June 2020.

Santa Clara University. 2013. Demand dividend: Creating Reliable Returns in Impact Investing. https://thegiin.org/assets/Santa%2520Clara%2520U_Demand-Dividend-Description.pdf. Accessed 26 May 2020.

Stockholm Resilience Centre. The nine planetary boundaries. https://www.stockholmresilience.org/research/planetary-boundaries/planetary-boundaries/about-the-research/the-nine-planetary-boundaries.html. Accessed 14 June 2020.

Swiss Sustainable Finance Glossary. https://www.sustainablefinance.ch/en/glossary-_content---1--3077.html. Accessed 26 May 2020.

UK National Archives. January 2015. Operational Selection Policy OSP49. Records of UK Export Finance, 1978 onwards. https://www.nationalarchives.gov.uk/documents/information-management/OSP49.pdf. Accessed 15 June 2020.

UNCTAD and Convergence. Blended Finance. https://www.convergence.finance/blended-finance. Accessed 26 May 2020.

United Nations. 1987. Our common future, Brundtland Report. https://www.are.admin.ch/are/en/home/sustainable-development/international-cooperation/2030agenda/un-_-milestones-in-sustainable-development/1987--brundtland-report.html. Accessed 26 May 2020.

Velten, S., J. Leventon, N. Jager, and J. Newig. 2015. What is sustainable agriculture? A systematic review. *Sustainability* 2015 (7): 7833–7865.

World Bank Data: Agriculture, forestry and fishing, value added (% of GDP) website. https://data.worldbank.org/indicator/NV.AGR.TOTL.ZS?most_recent_value_desc=true. Accessed 15 June 2020.

World Bank Group. 2014. *Introduction to agricultural insurance and risk management. Manuel 1*. Washington DC: International Finance Corporation.

Zhao, C., B. Liu, S. Piao, X. Wang, D.B. Lobell, Y. Huang, M. Huang, Y. Yao, S. Bassu, P. Ciais, J.L. Durand, J. Elliott, F. Ewert, I.A. Janssens, T. Li, E. Lin, Q. Liu, P. Martre, C. Müller, S. Peng, J. Peñuelas, A.C. Ruane, D. Wallach, T. Wang, D. Wu, Z. Liu, Y. Zhu, Z. Zhu and S. Asseng. 2017. Temperature increase reduces global yields of major crops in four independent estimates. Proceedings of the National Academy of Sciences of the United States of America (PNAS) in PNAS August 29, 2017 114 (35): 9326–9331. https://www.pnas.org/content/114/35/9326. Accessed 26 May 2020.

Publisher's Note Springer Nature remains neutral with regard to jurisdictional claims in published maps and institutional affiliations.

Tanja Havemann is the co-founder and director of Clarmondial AG, a Swiss advisory firm focused on investments in sustainable natural resource management. She holds academic qualifications in environmental sciences (University of Aberdeen), environmental economics (Imperial College London) and environmental law and policy (University of Kent, Canterbury). Tanja has authored articles and white papers on sustainable investments, primarily on climate smart agriculture and blended finance in emerging markets.

Dr. Christine Negra, Versant Vision LLC is an international consultant specializing in climate change, sustainable agriculture, and integrated landscape management. As a soil chemist with over 25 years of experience as an Extension agent, a researcher, and a program director, she collaborates with development, research, and finance organizations on multi-disciplinary initiatives that link science to policy and markets.

Fred Werneck is the co-founder of Clarmondial AG, a Swiss advisory firm focused on investments in sustainable natural resource management. He holds an MBA from London Business School and a bachelor in economics and finance (IBMEC, Brazil). Fred has extensive experience in private investments in emerging markets, and he has published papers on sustainable investments.

Agriculture and Human Values (2020) 37:1293–1311
https://doi.org/10.1007/s10460-020-10130-9

Priming the pump of impact entrepreneurship and social finance in China

Xiangping Jia[1]

Accepted: 13 July 2020 / Published online: 22 July 2020
© Springer Nature B.V. 2020

Abstract

In recent years, a significant revolution has begun in social entrepreneurship, with public funding organizations and private investors beginning to pursue social purpose. At the heart of the revolution is the ambition of these entities to use innovation and entrepreneurship to sustain and scale impacts in support of social and environmental objectives. This article provides a framework that conceptualizes social innovation and finance as a multi-level and evolving complex system. Two trajectories of transformation, *niche-regime* and *within-regime*, are used to frame a broad range of examples of impact entrepreneurship and social financing. Social finance is viewed as the key to release the 'lock-in' of social innovation, as it is able to gather and coordinate capital, resources, knowledge and ability. Using meta-analysis, the article explores the characteristics, dynamics and ecosystem of social innovation, impact entrepreneurship in China. The study maps the landscape of social finance in China and estimates there are untapped social finance opportunities worth from US$93 to US$208 billion.

Keywords Social innovation · Sustainability transition · Multi-level perspectives · Niche management · Impact investment · Blended finance

Introduction

Emerging social challenges in China and their implications for innovation and impact entrepreneurship

China has experienced rapid economic growth and rural–urban transformation since it initiated reforms to liberalize the economy in 1978. The country's urban population increased from less than 20% of the total population in 1978 to 52% in 2012, an increase of more than 500 million people (World Bank 2014). This is in line with a global migration from rural to urban areas; 54% of the world's population lived in cities in 2014 compared with only 30% in 1950 (United Nations 2015). The World Bank projected that China's urbanization rate will surpass 70% in 2030, and 76% in 2050. The country's agglomerated cities, with abundant cheap labor, robust infrastructure, and strong public

investment, have created an environment that has been highly conducive to job creation and income growth.

Entrepreneurship and innovation now contribute sizably to China's continued economic expansion. Since the beginning of the 2010s, China has shifted away from relying on massive public investment (such as spending on infrastructure and technology research) and international trade to drive economic growth (McKinsey Global Institute 2017). Since then, the Chinese government has introduced incentives for entrepreneurship and innovation to unleash new growth drivers at the national scale (The State Council 2016). A class of audacious, innovative and globally minded entrepreneurs is emerging. The new generation of entrepreneurs is inventive in disruptive technologies (such as fintech, virtual reality and artificial intelligence). The growing number of start-ups has attracted the attention of global venture capital investors. China has become the second largest market for venture-capital-backed entrepreneurial innovation in the world (The Economist 2017).

In addition to creating economic growth, China needs to innovate to address major challenges in environmental sustainability and social inclusion. An OECD report, citing a long-term national survey, revealed that about 20% of China's cultivated land was polluted from April 2005

✉ Xiangping Jia
   jiaxiangping@caas.cn

[1] Agricultural Information Institute, Chinese Academy of Agricultural Sciences, Beijing, China

to February 2013. The report also showed that agriculture had become a major source of pollutants in water and soil and a significant contributor to the nation's greenhouse gas emissions (OECD 2018a; Zhang et al. 2013). As urbanization has continued, the harm and cost of pollution, be it air, soil or water, has been rising. Illness and even death caused by pollution were estimated to be responsible for economic losses ranging from US$100 to US$300 billion a year in 2012 (World Bank 2014). Even though China has avoided some problems associated with urbanization (such as poverty and unemployment), many of the 260 million migrants that moved to cities between 1978 and 2012 from rural areas (and many more thereafter) have not all had equal access public services, including health, education, and insurance (World Bank 2014). By 2050, more than a quarter of China's population will be over 65 years old, and the old-age dependency ratio could be as high as 44% (Glinskaya and Feng 2018). China is ageing more rapidly than most developed countries in recent history.

In spite of the magnitude of challenges arising from urbanization in China, little is known about how innovation, entrepreneurship and the scaling of their impact can address such issues. Social innovation is a complex process and involves interactions among a variety of actors who represent different and possibly conflicting roles at different levels, which can be both disruptive and stabilizing. As a consequence, social innovation around the world has a hard time of scaling. In order for social innovation to successfully address social challenges in China, it needs to be large-scale and result in systemic social innovation, given the size of the economy. However, the scaling strategies of social ventures remain too small to be effective. Three global advocators for global social entrepreneurship, namely two US-based non-governmental organizations (NGOs), *Ashoka* and the *Skoll Foundation*, and a Swiss-based NGO, the *Schwab Foundation of Social Entrepreneurship*, recognized a total of 653 champion social entrepreneurs worldwide between 2007 and 2017, but only three addressed challenges in China, of which one was based in Hong Kong. Considering the size of China's population and escalating social challenges associated with urbanization, the development of social and impact entrepreneurship in China has greatly lagged behind that in other emerging economies (British Council 2016).[1] While social and impact entrepreneurship has become a global phenomenon that tackles enduring social problems through innovation, its presence and influence are still niche in China (Defourny and Kim 2011; Yu 2011).

The development of impact entrepreneurship and social finance in China is greatly constrained by a lack of knowledge and awareness about the subject. In the period from 1990 to 2017, only 78 articles were published in peer-reviewed journals with a thematic highlight on, or direct related to social entrepreneurship in China (Desa and Jia, of this issue).[2] At the SDC Symposium "Social Entrepreneurship and Innovative Finance" in 2018 in Beijing, the national capital, insufficient information and knowledge was considered to be the primary constraint on the development of social entrepreneurship in China; about 22% of the participants cited knowledge formation and communication as key areas for improvement in order to develop social entrepreneurship in China, in addition to opportunities for networking to access financial resources (22%) and policy advocacy (Jia and Desa, of the issue).

Objective, outline and contribution

This paper proceeds as follows. (A) First, we draw from existing sociology literature to offer a multi-level perspective on systemic social innovation. We identify regime and niche constructs, and conceptualize the flows needed in the landscape to scale social innovation. (B) We review the fragmented legal and regulatory frameworks—the regimes that currently govern social innovation—in China. (C) We review social innovation niche in China and offer a descriptive analysis of its characteristics. (D) We identify opportunities for how, and which types of social finance can facilitate social innovation in the landscape, in the context of MLP theory. (E) We discuss ongoing pilot programs to certify social enterprises. (F) Finally, we discuss the broader implications of blended finance models and certifications as facilitators of scaling social innovation in other regions and sectors, from niches characterized by lock-in to stable institutional practices.

To fill the knowledge gap and go beyond the micro–macro dilemma, this study characterizes social and impact entrepreneurship as evolving interactions between niches and regimes. In this study, impact entrepreneurship and social finance are framed through multi-level and systemic perspectives, in which social innovation towards transformative changes results from the interplay of developments at different analytical levels (i.e. landscape, regime and niche). Legal context and regulatory settings are viewed at the landscape

---

[1] For example, there were two million social enterprises operating in India in 2016 and a total of 59 champions were identified by the three global advocators of social entrepreneurship from 2007 to 2017.

[2] The literature review is taken from the Web of Science Core Collection between 1990 and 2019 (retrieved on Feb. 28th 2019). The terms "social entrepreneurship", "social venture", "social enterprise" or "social innovation" were specified in parataxis with "China". Only articles in English were considered. The indexes included SCI-EXPANDED, SSCI, A&HCI, CPCI-S, CPCI-SSH, BKCI-S, BKCI-SSH, ESCI, CCR-EXPANDED, IC.

and regime levels, through a historical overview. The practices of impact entrepreneurship are explored and analyzed as evolving processes and niche-regime interactions, by referring to sample-based survey studies. By compiling statistics from a variety of sectors and institutions, this study values social finance opportunities for impact entrepreneurs to seek in China between US\$93 and US\$208 billion.

Contribution

Aside from "getting the facts right" on pluralistic notions and practices associated with social innovation and social finance in China, this study deploys multi-level perspectives to enable system-level views and approaches to scale the impact of social finance in pursuing social and environmental purpose. Importantly, this study characterizes social and impact entrepreneurship as evolving interactions between niche and regimes, and goes beyond the well-rehearsed micro–macro dilemma that contrasts distinct transformation pathways, top-down and bottom-up. This study considers actors at every level being creative and conserving, and therefore highlights the dynamism of innovation and rigidity in the system. Social finance is a key to resolve the 'lock-in' of social innovation. The value-added of this study to extant literature lies is that the analysis balances 'getting the facts right' with 'getting the right theory'. The study is potentially useful for (but not limited to) researchers, practitioners and agencies working on development, philanthropy, social innovation, development and social finance (e.g. impact investment and blended finance).

Conceptual framework: examining social innovation and social finance through multi-level perspectives

Social innovation does not come about alone. It is characterized as system change as a result of multi-level interactions. The structural construct of transformation as systems was conceptualized by a broad range of perspectives, such as transition management for sustainable development, also called sustainability transition (Caniels and Romijn 2008; Kemp et al. 2007, 2015), strategic niche management (Caniels and Romijn 2008; Kemp et al. 2015), and the theory of social change (Christensen et al. 2006; Hagen 1963). These framework were applied in the context of agrifood and rural–urban integration, calling for systemic approaches to social innovation and entrepreneurship (Bui et al. 2016; Elzen et al. 2012; Pigford et al. 2018; Thornton et al. 2017). Commonly across all of these perspectives, system innovation towards transformative changes results from the interplay of developments at different analytical levels. Frank Geels made a pioneering elaboration in his early literature

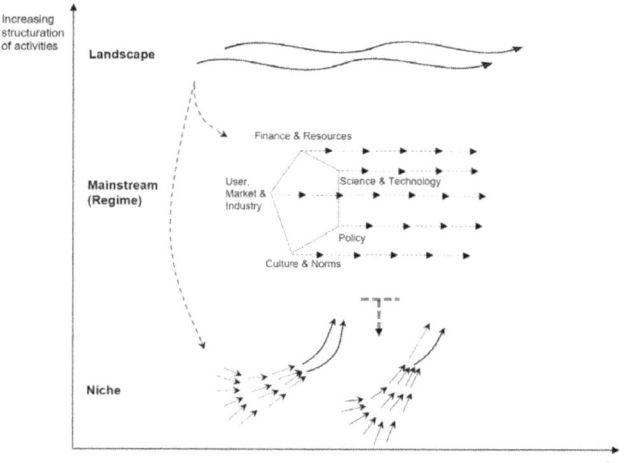

Fig. 1 Multi-level perspective on social innovation and social finance. *Source* (Geels 2002)

(Geels 2004, 2002), using multi-level perspectives (MLP) to frame social innovation and systemic change as interactions among three different analytical levels (Fig. 1): *niches* (the locus for radical innovation), socio-technical *regimes* (the locus of established practices and associated rules that stabilize existing systems), and an exogenous socio-technical *landscape*. The socio-technical landscape forms an external context that actors at niche and regime levels cannot influence in the short run. Regimes are characterized by a set of institutions and rules that have coalesced into stable configurations. Working on radical innovation that deviates from existing regimes, niche actors (such as entrepreneurs, start-ups, spinoffs) articulate expectations and enroll new actors into the system outside the existing regimes.

Social innovation does not come about easily, because existing socio-technical systems are stabilized in many ways. A regime may consist of several subsystems that interpenetrate and co-evolve with each other, such as markets, user preference, science and technology, industrial policy and regulations, social norms and culture (Geels 2002). The notion of a socio-technical regime not only clarifies interactions across different levels and pathways, but also aims to capture the meta-coordination between different sub-regimes, and therefore goes beyond the sectoral system (Geels 2004). When a regime exhibits a high degree of stability, the emergence of innovation within the regime becomes difficult. New socio-technical configurations outside the regime (such as in niches) face resistance because of unalterable and rigid acts in the old system. Social innovation or transformation is defined as shift from one regime to another. System change is often prevented by rigidity and characterized as 'lock-in' in innovation.

The multiple-level perspective of social innovation is used to understand and to frame the structuration and functions of

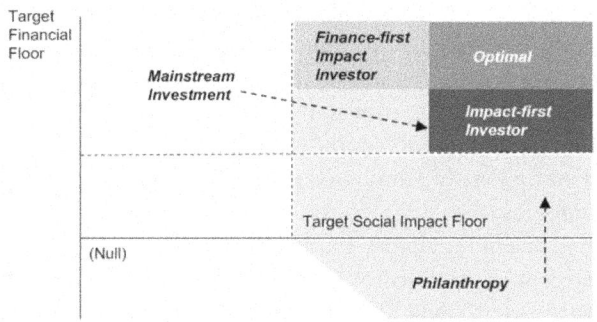

Fig. 2 Segments of social finance and dynamics. *Source* (Freireich and Fulton 2009)

finance in the context of transition. Transition is explained as the co-evolution and interaction between subsystems and the problem-solving capabilities of mobilizing resources (Kemp et al. 2007). As depicted in Fig. 1, niche innovation occurs in small, unstable and precarious networks; entrepreneurial actors put in a lot of work and take risks to uphold the niche. As rules are diffused, there is partial structuration of activities, limited finance and resources, and much uncertainty and fluidity. At the regime level, the institutions and rules become structured into several subsystems, i.e. science, technology, user preference, market, industry, policy, social norms and culture (Fig. 2). Nevertheless, social problems are rooted in complex contexts and solutions to the problems are often locked in the mainstream. Finance and resource mobilization go beyond "fulfilling a function". They are the key to unlock the inertia within regimes and across different levels (Grin et al. 2010).

Over the last few years, a range of new organizations and financial instruments have emerged to balance financial, and social and environmental returns. Mainstream investment seeking high financial returns have started to pay attention to social and environmental benefits and go beyond simply "doing no harm" (Freireich and Fulton 2009). Wealthy individuals and entrepreneurs are using the tools of business to address global problems. Such approaches are transforming conventional philanthropy from "giving" to "impacting" (Bishop and Green 2015; Salamon 2014). Emerging forms of social finance have been developed into a class of financial assets that blend social and financial outcomes, shifting from the 'margins' to the 'mainstream' (Emerson 2003; O'Donohue et al. 2010; World Economic Forum 2013). Together these constitute the spectrum of "impact investing", which pans a large number of diverse types of capital that create social or environmental value with clear intentionality as well as returning invested capital to the investor (Bugg-Levine and Emerson 2011), as depicted in Fig. 2.

The MLP framework synthesizes the duality of pathways related to social innovation and social finance. In researching societal transition, some theories contrast transformation

with "improvement" or "adjustment", and assert the former as a top-down process of *transition management*—the dotted lines connecting landscape to regime and niche (Weber and Hemmelskamp 2005). Such an approach was criticized as being disconnected from autonomy and overstating the role of conspiratorial elites and leadership (Hendriks 2008). Views of transformation as bottom-up participation also received criticism for not clarifying the pathways between the different levels and a bias towards bottom-up change (Berkhout et al. 2004), neglecting broader secular changes that create favorable conditions for revolutionary uprisings and form 'structural opportunities' (Dahle 2007). The bottom-up view of system change also ignores the rigidity of existing regimes that resist and hamper niche innovation, as the double-dotted lines depict in Fig. 1. In the MLP construct of social innovation and social finance, the top-down and bottom-up pathways are synthesized as adaptive cross-level interactions that combine experimental prototypes and learning with continuity.

Legal framework and institutional settings

In the MLP view of social innovation and finance, legal frameworks and regulations are important elements at the landscape and regime levels. Landscape is used as a metaphor that refers to aspects of the wider exogenous environment, while laws define the parts of the exogenous environment that are beyond the direct influence of regime and niche actors. Unlike laws, sectoral policies and regulations are often applied at the regime level, defining the rules and structuration of all actors. While laws and regulations remain in relative stasis in the short term (Geels and Schot 2010), it does not mean humans cannot change the landscape. In this section, legal frameworks and sectoral regulations are reviewed from historical perspectives.

Civil society has been restrained in China throughout the country's modern history. In the period from 2009 to 2017, the total number of registered social organizations tripled (Table 1). The rate of outreach was 1824 people per institution in 2017. The figure is much lower than in most developed countries and in a large number of developing countries, for example, 205 people in the US and 600 in India. In 2006–2007, about 4.6 million Australians volunteered in 600,000 social organizations (ACOSS 2010). The total number of social service organizations in China was about two thirds of that in Australia in 2009. According to a recent study conducted by a group of Chinese research institutions, social sector organizations in China disbursed a total US$98 billion (about Rmb637.3 billion) in 2016, equivalent to just 0.86% of the GDP that year (Ma 2018).

The regulatory framework governing the civil sector in China is fragmented. Dozens of regulations and sectoral

Table 1 The number of registered social sector organizations in China (2009–2018)

	Social service organizations	Foundations	Social groups	Total
2011	204,000	2614	255,000	462,000
2012	225,000	3029	271,000	499,000
2013	255,000	3549	289,000	547,000
2014	292,000	4117	310,000	606,000
2015	329,000	4784	329,000	662,000
2016	361,000	5559	336,000	702,000
2017	400,000	6307	355,000	762,000
2018	428,000	6819	362,000	797,000

Source China Social Sector Organization Public Service Platform (www.chinanpo.gov.cn/)

policies apply to the civil sector (see Table 6 in Appendix). At the core of the legal framework for civil society, the "General Provisions of the Civil Law" defines the civil sector as *social groups*, *foundations* and *social service organizations*" (Article 87 in Section 3).[3] In addition, the Civil Law identifies *rural collective economic organizations* and *self-governing organizations* at the community level as "Special Legal Persons" in the social sector (Article 96 in Section 4). While non-profit organizations are registered with the Ministry of Civil Affairs, rural collective economic organizations (such as farmer cooperatives) are registered with their local Bureau of Industry and Commerce. China does not have a coordinated framework that regulates non-profit organizations, and instead such organisations are regulated by a patchwork of sectoral-specific policies, such as the "Administration of Registering Social Groups" and "Administration of Social Services Organizations", which were first promulgated in 1989 and revised together in 1998 and 2016. Regulations for foundations (the "Administration of Foundations") were drawn up in 2014 as an amendment to "The Charity Law", which came into force on September 1st, 2016.

The various regulatory frameworks that govern the social sector are not coordinated with other sub-regimes in the system. For example, according to "The Charity Law", registered charity trust funds are eligible for preferential taxes. However, such benefits are not synchronized with state and local tax administrations.[4] Under "The Charity Law", public

fundraising for charity can take a variety of forms (such as on-site, informed convening or in the media). However, in the "Administration of Foundations", public fundraising is strictly controlled, especially at the national scale. The existing regulations do not fully take into account the emergence of fundraising through digital and telecommunication technologies.

The current regulatory framework can discourage social investment. Social sector organizations—whatever their identity—are strictly regulated with respect to depositing and distributing returns and profits. For example, according to the "Law of Farmer Specialized Cooperatives", dividends from financial investments are limited to less than 40% of the residuals. In "The Charity Law" and "Administration of Foundations", investments and rewards to managers of foundations/charities are strictly regulated. The benefits of preferential taxes are only valid when a social sector organization limits staffs' salary to no more than twice the average salary of the administrative region they are registered in. Recognizing the importance of preserving and strengthening resources for charity funds and foundations, on Oct. 25th, 2018, the Ministry of Civil Affairs issued "Interim Administration of Value Preserving and Enhancing Investment for Charity Funds" to raise incentives for people to invest social organizations that pursue social and environmental goals. Unfortunately, this industry-specific policy was not aligned with financial regulatory bodies, such as the People's Bank of China (PBOC, the central bank), the China Banking and Insurance Regulatory Commission (CBIRC) or the China Securities Regulatory Commission (CSRC). To ensure transparency in the non-profit industry, information disclosure is mandatory. However, the government's requirement that non-profit institutions disclose all information, while only guaranteeing minimum protection of confidentiality, threatens their competitiveness and can dissuade people from setting up such institutions in the first place.

Characteristics of the social innovation niche in China: a descriptive analysis

Origins

While the emergence of social entrepreneurship in Western Europe and the US is closely associated with the state stepping back from welfare provision and attempts by non-governmental organizations (NGOs) to supplement the loss of welfare that was previously supplied by the state, social

---

[3] Different English translations have been used for the Chinese expressions, for example, social associations (for social groups) and civilian-run non-enterprise units (for social service organizations) in existing studies (Yu 2011).

[4] This gap was filled on Feb. 7th, 2018 with an amendment to Article 26 of "The Enterprise Income Tax Law". Nonetheless, preferential taxes are only available to registered non-profit organizations in which the average salary of staff is no more than twice the salary of

Footnote 4 (continued)

other employees in the region they are registered in. This greatly undermines incentives for people to work for non-profit organizations.

entrepreneurship emerged in China because of the efforts of the international community (Yu 2011). In 2009, the British Council launched its Social Enterprise Program to advocate and inspire the concept of 'social enterprise' and good practices in China. In the period from 2009 to 2016, the British Council program trained over 3200 social entrepreneurs and facilitated US$6.1 million social investment opportunities for 117 social enterprises.[5] Founded in 1998, the *Schwab Foundation for Social Entrepreneurship* began to advocate the concept of 'social entrepreneurship' and related practices at Tianjin, where the Summer Davos Forum has been hosted (alternately with Dalian, Liaoning province) since 2010. The annual meeting of World Economic Forum became another window for the Schwab Foundation to share global experiences of social entrepreneurship with the Chinese community (Han and Zhang 2010). In September 2014, Tianjin hosted the Eighth Annual Meeting of the New Champions at the Summer Davos Forum. Chinese Premier Li Keqiang inaugurated the forum, which was centered on the theme of "Creating Value Through Innovation".[6] These events helped to raise awareness of the concept of social entrepreneurship among high-profile policy makers and mainstream financial institutions.

While there has been broad interest in studying social and impact entrepreneurship practices in China, knowledge gaps exist because of a lack of data. Given the fact that academics and practitioners have not reached a consensus on the definitions of social and impact entrepreneurship, it has become crucial that typology is formulated that embraces its diversity and plurality. To overcome potential biases that could emerge from using anecdotal and normative cases from a small number of sample-based studies, this research has selected two studies that share the same framework and indicators in order to explore the dynamics of social and impact entrepreneurship (CSEIIF 2019; Lane 2013; Yu 2014).[7] One early study was facilitated by Foundation for Youth Social Entrepreneurship (FYSE), a Hong Kong-based organization focusing on youth social entrepreneurship in Asia. The FYSE surveyed 52 social enterprises in China over 2010–2012 (Lane 2013). The other study was carried

out by the China Social Enterprise and Impact Investment Forum (CSEIIF), a Chinese non-profit intermediary promoting social entrepreneurship and investment.[8] The CSEIIF study was conducted in 2017–2018 and surveyed a total of 371 social organizations. By comparing the findings from the two sources, we have conducted a meta-analysis over different timeframes and explored the dynamics of social and impact entrepreneurship in China.

Demographics

Social entrepreneurs in China are mostly young and educated. As shown in Table 2, the FYSE study also found that 42% of the surveyed social entrepreneurs in 2010–2012 were women. The percentage is higher than in other developing countries in Asia. In India it was 24% and in Bangladesh it was 20% (British Council 2016). The majority (63%) of Chinese social entrepreneurs were 31–40 years of age and all had a university degree. This implies that they had acquired post-graduate and professional skills over an average of ten years, before establishing entrepreneurial organizations that pursued social purpose. Interestingly, almost half the surveyed social entrepreneurs had studied and/or acquired occupational experience abroad, indicating that they may have been inspired by social entrepreneurship that they were exposed to in Western countries. About one third of the social enterprises were founded by foreign nationals, reflecting the level of involvement of the international community in promoting social and impact entrepreneurship its nascent years in China. While the CSEIIF study did not include surveys on individual characteristics of the social entrepreneurs, it suggests that they were typically focused on local issues in the region they were based.

Organization profile

Impact entrepreneurship organizations in China are starting to grow, but are far from scaling up. As the FYSE survey shows, 54% of social enterprises had been established within the previous three years. This rate dropped to 14% in the period of 2017–2018, when about 57% of the social enterprises had been established for more than five years, indicating that the social entrepreneurship organizations were becoming progressively more mature. Meanwhile, 54% of social enterprises operated across several provinces and were nationally oriented in 2017–2018, compared to 2010–2012

---

[5] The BC program ceased operating in China in 2016, leaving behind a great legacy in the development of social enterprises in China. More information is available at: https://www.britishcouncil.cn/en/uk-china-social-enterprise-social-investment-partners.

[6] Information retrieved from: https://www.chinatoday.com.cn/english/society/2014-10/13/content_644277.htm; https://english.cri.cn/12394/2014/09/11/53s843716.htm.

[7] The studies conducted by Yu (2014) and Lane (2013) both focused on the period 2009–2012. The first study explored in-depth the governance of social enterprises and while latter looked at social enterprises from a broader perspective. Given the small number of indicators used in Yu (2014), the study by Lane (2013) was selected for further analysis.

[8] CSESIF was founded in 2014 and convenes a community of social enterprises and philanthropic foundations (starting with 17 founding institutions). CSESIF organizes annual meetings. Through its extensive network of members, CSESIF conducts canvas surveys at the national scale to identify good practices by social enterprises.

364 Springer

Table 2 Characteristics of social entrepreneurship organizations in China, %

Source of information	2009–2012	2017–2018
Individual characteristics of social entrepreneurs		
Female	42	–
Age (31–40)	63	–
College education or a university degree	100	–
MBA or doctoral degree	50	–
Educational experience abroad	46	–
Occupational experience abroad	54	–
Being a foreign national in the operating country	33	–
Location		
Beijing	50	33
Shanghai	17	11
Guangdong (including Shenzhen)	–	16
Other	33	40
Territorial focus of social mission		
Learning, education & employment skills	14	33
Healthcare, disadvantaged and poverty alleviation	11	20
Environment and energy	10	10
Miscellaneous	–	37
Age of social entrepreneurship organizations		
Less than 3 years old	54	14
Between 3 and 5 years old	8	29
More than 5 years old	38	57
Annual turnover in RMB		
<100,000	38	14
100,000–one million	37	37
1–10 million	16	42
More than 10 million	8	7
Legal status		
Private company	66	60
Social sector organizations	20	32
Dual-identity	–	5
No identity	14	3
Scale of operation		
Local communities in cities or counties	63	32
Provincial	13	10
Regional (in multiple provinces)	17	54
International	7	4

Source The results of 2010–2012 are adapted from Lane (2013) and the results 2017–2018 are taken from CSEIIF (2019)

when two thirds of social enterprises operated within just one province or county.

Impact entrepreneurship organizations in China tend to be registered as private businesses. Both the FYSE and CSEIF surveys show that a high percentage of organizations had obtained the legal status of private companies (66% and 60% respectively). About 5% of social entrepreneurship organizations had registered a dual status of NGO and private company, indicating a lack of coordination in regulation and possible overlap between policies. The high rate of social entrepreneurship organizations that are registered as private companies is also observed in other developing countries in Africa and Asia and the Pacific. For instance, 58% of social enterprises surveyed in India were registered as private limited companies, while 23% operated as NGOs (British Council 2016). In Ghana, NGO is the least common legal form for social enterprises, and most social enterprises apply for private business status simply because the procedure is simpler and there is less bureaucracy when it comes to registration and administration.

Table 3 Fundraising, financial sustainability and social impact measurement, %

	2010–2012	2017–2018
Fundraising		
Owner's own money	77	86
Grants, sponsorships and awards	12	23
Government	3	12
Mainstream and alternative financing (banks, VC, crowdfunding)	3	22
Social impact investors	4	–
Financial sustainability		
Profitable	42	16
Break-even	33	36
Lossmaking	25	48
Jobs created		
Less than 10 jobs	79	53
More than 10 jobs	21	47
Measurement and valuation		
Social returns	26	26
Financial returns	22	64
No thoughts about evaluation	25	24

Source The results of 2010–2012 are adapted from Lane (2013) and the results 2017–2018 are taken from CSEIIF (2019)

In terms of geographical distribution, a large share of social entrepreneurship organizations in China were registered in densely populated areas, such as Beijing, the national capital, cities eastern coastal provinces (such as Shanghai and Guangdong) and provinces in central and southern China where the business environment is better developed. There were relatively fewer social enterprises in the western and the north-eastern parts of the country.

Sectoral focus

The emergence of impact entrepreneurship in China was triggered in response to a mixture of social challenges and emerging economic opportunities from the bottom of the wealth pyramid. As shown in Tables 2 and 3, an increasing number of impact entrepreneurship organizations focus on improving income-generating capabilities (such as education, training and healthcare) and creating employment; about 47% of social organizations employed more than ten people in the period 2017–2018 versus only 21% in the period 2010–2012. In the context of rural–urban integration, emerging impact entrepreneurship seems to be biased towards the social values of urban areas. The CSEIIF survey indicates that only 14.6% of the surveyed organizations were involved in agriculture and other rural activities. In neighboring countries such as India, food and agriculture was the main focus of 36% of social enterprises, and in the north-east of India, the rate was even higher (64%), according to country studies by

the British Council (2016). MLP suggests that niche-regime interaction for social innovation might be easier in urban China when compared with rural settings, where established regimes are relatively resistant to innovation.

Funding sources

Social and impact entrepreneurship increasingly appeals to both mainstream and alternative investors. As shown in Table 3, social and impact entrepreneurs initially relied mostly on own funds and grants. However, mainstream and alternative investors (such as banks, venture capital and private equity investors) are taking a growing interest in providing funding; about 22% of social entrepreneurship organizations obtained funds from banks, venture capital and equity investors in 2017–2018. Unfortunately, the CSEIIF survey did not break down the sources of fundraising for social impact investors. Anecdotal evidence indicates that global impact investors are present in China.[9]

The majority of emerging impact entrepreneurial organizations in China are far from ready for investment due to lack of financial sustainability. In 2020–2012, 42% of social enterprises reported making a profit, but the figure decreased to 16% in 2017–2018. Almost half the social enterprises were found to be lossmaking in the CSEIF survey. The deteriorating financial performance of the social enterprises reflects insufficient management capacity and skills. This was confirmed in a participatory knowledge workshop involving a variety of stakeholders in impact entrepreneurship in China (Jia and Desa, of this issue).

Performance management

China's growing community of social finance and impact entrepreneurs do not have the necessary capacity for impact measurement and management. As shown in Table 4, in the FYSE study, only 26% of social enterprises had created systemic approaches to measure their social impacts on their targeted beneficiaries. The rate was also low in the CSEIF survey. Impact measurement and management are at the heart of social finance as both internal measures of harnessing social values and important evaluation tools that enhance accountability and transparency (Nicholls et al. 2015). Rather than using conventional evaluation approaches that focus on measuring outcomes, impact investment advocates for a transformation of impact measurement and management (IMM) towards principles of participatory governance

---

[9] For example, in 2018, TPG, a leading global private investment fund, led an investment of US$140 million of series C equity financing in the China Foundation for Poverty Alleviation (CFPA) Microfinance Management, a Chinese social enterprise that provides microcredit in more than 100 impoverished counties in China. Information source: https://press.tpg.com/node/7396/pdf.

Table 4 Simulating opportunities for social entrepreneurship and impact investment (in billions of US$)

Type	Description	Scenario 1	Scenario 2
A	Non-profit social sector	24	24
B	Mainstream finance and alternative investment	10.9	21.7
C	Public–private partnerships	15	15
D	Public intrapreneurship	19	23
E	Private intrapreneurship	9.5	47.7
F	High-net-wealth-individuals (HNWI)	15	77
	Total	93.4	208.5

Source Author

The estimations of different scenarios are held unchanged for Type A and C, given great variations in policy

and informed theory of change (Burd-Sharps et al. 2011). Nonetheless, the reflexive strategy of IMM was found to be missing in China's community of social investors and impact entrepreneurs. Jia and Desa (of this issue) observed that linear views on impact measurement were prevalent among management staffs, during a participatory knowledge workshop that convened a wide range of stakeholders in China's emerging impact entrepreneurship ecosystem.

Opportunities for financing mechanisms to facilitate social innovation in China

To map social finance and impact investment in China, we chose a less-restrictive definition of impact entrepreneurship, given the fragmentation of regulations that govern the sector and the fact that it encompasses a broad range of societal actors. Impact entrepreneurship is defined here as "the process of creating and implementing an entrepreneurial solution to a social problem and fulfilling unmet social needs, thereby creating social value and impact" (Sommerrock 2010, p. 68). This definition emphasizes the organizational perspective and highlights evidence-driven change-making and the creation of social value. In addition, the definition does not exclude impact entrepreneurship by established entities, be they private incumbent corporations or public administrations; both entrepreneurship and intrapreneurship are recognized as sources of social innovation. A poll carried out at a participatory knowledge workshop that convened a group of key stakeholders of China's social entrepreneurship ecosystem indicated the majority of the participants favored an inclusive approach to social entrepreneurship (Jia and Desa, of this issue).

While MLP offer a representation of transition, they do not account for the role of finance. The plurality and dynamics of the concept of social finance and the way it is practiced in China means that it can take two different pathways in the multi-level system (Fig. 3a, b). Niche innovation emerges and subsequently replaces the existing regime, while niche-regime interaction is central to transitions and dynamics.

Social finance of *niche-to-regime* interactions takes a variety of forms such as philanthropy, mainstream banking and alternative investment. Entrepreneurship is developed independently outside established organizations. However, radical

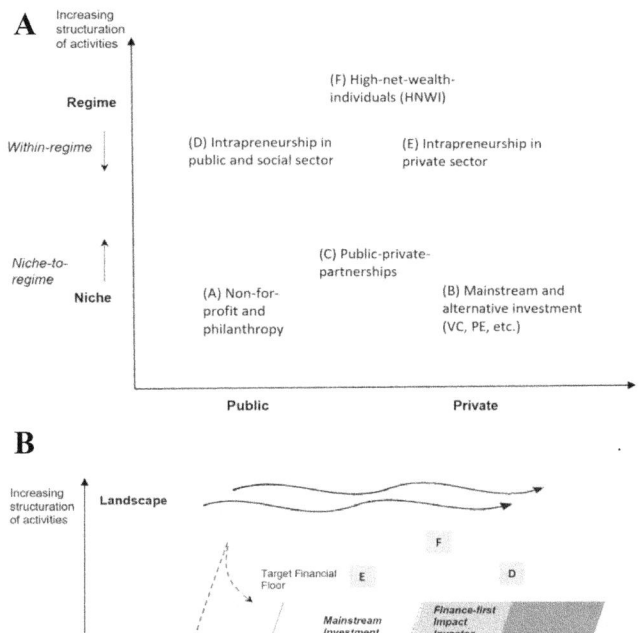

Fig. 3 a Social finance actors: Type (public vs. private) and level (niche vs. regime) of influence on social innovation. **b** A multi-level perspective on blended finance: Mapping the actors and combinatorial dynamics of transformative social innovation

niche innovation has little chance to break through as long as the regime is dynamically stable and rigid. To facilitate radical change, external resources such as finance are required to validate innovation in niche and to develop them in the market. When these niche-innovation reach certain scale and eventually change the rules, creating a new structure and stability, the system shifts to another regime, and this process is called *transition*. This strand of social finance facilitates incremental innovation *within-regime (public or private)*, and the processes proceed in predictable directions with sustained scaling.

'Innovation-from-within' refers to opportunities for scaling impact entrepreneurship and social finance in both the public and private sector in China. In the private sector, a growing number of large incumbents and high-growth technologies start-ups have adopted 'hybrid business strategies', and advocate for blending social commitment and business purpose (Kayser and Budinich 2015; London and Hart 2011). In recent years, impact entrepreneurs who are employed in the public sector are increasingly viewed as valuable sources of social innovation because of their ingrained awareness of social problems and their tacit knowledge of how to mobilize resources and coordinate a variety of social sectors (Krippendorff 2019). Within-regime innovation and intrapreneurship do not view regime as static but as a source of innovation. The discourse of social and impact entrepreneurship in China was originally based on Anglo-American models of entrepreneurship that focused on new ventures. However, the vast majority of innovation that have most impacted society in the world were conceived of by employees, not as independent individuals but in broader communities and inside large organizations, driven not by profit but by a passion to make a difference (Krippendorff 2019). Given that China's culture, history, ideology and social-economic structure that prioritizes institutions over individuals, western notions of social and impact entrepreneurship were adapted by Chinese public and private institutions to fit local practices, namely by being extended to cover intrapreneurship (Curtis 2011).

Social finance of niche-regime in China is slowly moving mainstream entrepreneurship towards sustainable development. As illustrated in Fig. 3, opportunities for social finance in niche-to-regime emerge from civil society (Type A), mainstream finance and alternative investment, e.g. venture capital, private equity and crowdfunding (Type B) and public–private partnerships (Type C). According to the Blue Book of Philanthropy in China (Yang 2008), a series of scandals in 2010–2013 resulted in a collapse in public trust in non-profit charity organizations and philanthropy in China has stagnated since then. Donations totaled US$24 billion (Rmb155.8 billion) in 2017; 54% of the donationa were collected through registered foundations and the remaining were collected by individual philanthropists. China became the second largest country for entrepreneurial investment

in the world in 2016, after the US (The Economist 2017). Had a fraction of the investment been mobilized to finance impact entrepreneurship, it would create opportunities in social finance ranging from US$10.87 to US$21.75 billion (Table 4 and Table 7 in Appendix). To incentivize innovative approaches to providing public services and to increase the accountability of public sector, public–private-partnerships (PPPs) have been used in China (Wang 2013). While the majority of the PPP investments have focused on energy, transport, water and infrastructure, PPP investment in social sectors (such as education, health, aged care and agriculture) amounted to more than US$100 billion in 2014–2018 (see Table 4 and Table 8 in Appendix). While little evidence has been provided about the utilization of private investment in PPPs, if 1% of the aforementioned investment could be directed towards innovative finance, such as 'blended finance', as is advocated for by OECD and World Economic Forum (OECD 2018b), it would yield investment windows for impact entrepreneurship worth US$15 billion. The aggregated opportunities for social finance in niche-regime range between US$49.87 and US$60.75 billion.

Opportunities for *within-public* social finance have been created alongside transformations in the public sector. In 2016, China budgeted US$1.917 trillion to support reforms in a broad range of public sector and institutions (Table 4 and Table 9 in Appendix). To improve accountability, transparency and efficiency in the public sector, within-regime innovation became part of the government's agenda for economic and political reform (Xu 2015).

China is also continuing to undergo rapid urbanization (World Bank 2014), which means there is an ever growing need to provide services to urban residents, especially migrants, through market-based governance. This provides opportunities for social innovation and impact entrepreneurship. Based on a range of scenarios, the untapped opportunities for social finance in China's public sector range between US$19 (i.e. 1% of the government budget for public institutions) and US$23 billion (i.e. an additional US$4 billion of public spending in the civil sector, equivalent to 5% of the budget for the civil sector in 2016).

The opportunities for social finance *within-private* are lie in China's transformative business community. According to traditional notions of corporate social responsibility (CSR), the pursuit of a social mission and values is not necessarily fundamental to the economic imperatives of the corporation and are often managed by public relationship team or through affiliated foundations (Kitzmueller and Shimshack 2012), as part of risk management. In recent years, the pursuit of social values has been viewed by large incumbents as a competitive strategy and has been integrated in the company's processes for creating economic value (Porter and Kramer 2011). For example, the Socially Responsible Investment (SRI) market, which typically screens portfolios to avoid harmful or anti-social stocks by integrating the principles of Environmental, Social and Governance (ESG) in off-balance-sheet lending, accumulated over US$13 trillion in assets under management by end-2013 (Nicholls and Emerson 2015). According to a recent evaluation report from Syntao Green Finance, a consulting company advocating for social responsible investment in China, 70% of publicly listed companies in China were below the B-level indexes of ESG in 2018, indicating substandard social and environmental performance.[10] By end-September 2018, total assets under the management (AUM) of China's listed companies approximated US$36.9 trillion. A conversion of one (or five) in a thousand of the asset value into social and impact investments would yield an opportunity of US$9.5 billion (or US$47.7 billion), as shown in Table 10 in Appendix. It should be noted that small- and medium-sized enterprises (SMEs) contribute 50% of China's GDP (Gao 2019), and so opportunities of within-private social finance are not limited to large listed corporates.

Wealth management has is not geared towards social and impact investment. Between April 2017 and March 2018, a total of US$3.3 billion was donated by the top 100 wealthy individuals in mainland China (AVPN 2019). According to the Global Wealth Report (Credit Suisse 2018), globally, 0.8% of adults owned 44.8% of all wealth in 2017, equivalent to about US$142 trillion. In China, the number of high-net-worth individuals (HNWI) and families will increase from 2.07 million in 2015 to 3.88 million at the end of 2020; the investable assets of HNWI in China are projected to reach US$15.4 trillion (Jiang 2016). China is becoming the largest market for HNWI wealth management in the world. If one (or five) in a thousand of the wealth been invested in social finance in the next year, it would create opportunities of US$15.4 billion (and up toUS$77 billion).

Piloting certification of social enterprises

The blurred regulatory framework governing the social sector in China drives people to look for alternative legal identities to social enterprise. The notion of 'social enterprise' was introduced by the British Council in 2009 and gradually attracted the interest of the government at the local level. On different occasions in 2017 and 2018, the governments

of Beijing, Chengdu and Shenzhen piloted initiatives to institutionalize social enterprises, but institutions and practices varied by location. While Beijing chose a top-down approach to administrating social enterprises, classifying them as upgraded version of organizations in the social sector, Shenzhen adopted an ecosystem approach by promoting niche-regime interactions and outsourcing most of the administration to external agencies and facilities. The government in Chengdu promoted social enterprise certification using a high-profile mandate and coordination at the municipal level; the promotion was systematic throughout niche-regime and landscape. The differences in the approaches of the three cities are summarized in Table 5.

Accreditation of 'social enterprises' is potentially conductive for the development of the social innovation and impact entrepreneurship ecosystem. Accreditation and certification have emerged as a tool to institutionalize performance assessments. Certification, be it mandatory or voluntary, plays an important role in signaling accountability, for example to investors who want to be sure they will be working with reputable providers. In addition, accreditation and certification facilitate the adoption of best practices and build a culture of professionalism in an agency (Teodoro and Hughes 2012). They also affect organizational culture by socializing employees and communicating the agency's priorities to them (Carman and Fredericks 2013; Willems et al. 2017).

Nonetheless, certification also has serious disadvantages. The process requires a tremendous amount of time and considerable resources (Carman and Fredericks 2013). It is challenging for social entrepreneurs to shore up their management capacity and social mission when certification bodies do not have the necessary resources to support them to do so (Carman and Millesen 2005). Most importantly, there is a danger of mission drift when the certification criteria are not aligned with the social enterprise's social values, or when the certification regime is overwhelmed by tech-bureaucracies with little participatory processes to safeguard the interests of beneficiaries of the social entrepreneurial activities. The pilot certification systems in China integrated limited participatory mechanisms to harness accountability, such as goal formulation and constituent presence of beneficiaries in assessment. Without integrating participatory processes, certification of social ventures will face challenges in ensuring the accountability and credibility necessary to facilitate social change and citizenship.

Broader implications: blended finance and certification as facilitators of scaling social innovation

Multi-level perspectives enable system-level orientation and approaches to move social finance towards transformation and sustainability. Transformation, being defined as system

---

[10] Environmental, social and governance (ESG) refers to the three central factors in measuring the sustainability and social impact of an investment in the private sector. It was also developed into a set of standards for a company's operations that socially conscious investors can use to screen potential investments. Environmental criteria examine how a company performs as a steward of nature. Social criteria examine how a company manages relationships with its employees, suppliers, customers and the communities in which it operates. The results of the evaluation were obtained per personal communication with Mr. Pei-yuan Guo (SynTao), and background data are available on the website: https://www.syntaogf.com/Menu_CN.asp?ID=42.

Table 5 Social enterprise certification pilots in China

	Beijing	Shenzhen	Chengdu
Starting date	2018	2016	2018
Government administration	Bureau of Civil Affair (& Social Work Commission)	District Government (Fu-tian)	Municipal Government
Objectives	Strengthening administration of social sector organizations	Developing impact investment	Promoting social innovation and new governance
Facilitating agency	Government attached; political patronage	Independent; knowledge-based; market-based;	Partnership between government and third-party institutions
Legal identity of 'social enterprise'	Mixture of social sector organizations and private companies	Mixture of social sector organizations and private companies	Private entities
Regions	Only Beijing	National (and regional)	Only Chengdu
Number of certified 'social enterprises' in the first year of the pilot project	26	45	39

Source Personal communication with local facilities

shift from one regime to another, implies multi-level interactions among a broad range of actors. Unlike mainstream businesses, social entrepreneurs carry out relatively small interventions with very limited on-hand resources and manage to achieve disproportionately significant results (Bornstein 2007). Importantly, social problems that impact entrepreneurs deal with are often embedded with great local complexity, characterizing transformation as a process of accumulation of changes beyond the micro level (Coleman 2000), and rather at the system level, including social norms and trust between individuals (Dacin et al. 2010; Praszkier et al. 2009). For both within-regime and across-level impact entrepreneurship, social finance plays a harmonizing role and synergizes interactions with other sub-systems, i.e. science and technology; user, market and industry; culture and norms, and policy, as depicted in Fig. 1. Under within-regime finance, be it public or private, regime actors are also viewed as sources of innovation and impact entrepreneurship.

Where landscape pressure and regime rigidity do not immediately welcome changes, social finance becomes the key to release the 'lock-in' of social innovation. Social change often involves conflicts, contestations and power struggles. If niche-innovation is not sufficiently developed and do not become a clear substitute for the status quo, the innovation is locked in and stifled in niche (Geels and Schot 2010). Such a lock-in of niche innovation in stable regimes has been found to be pervasive in sustainability transitions in agrifood systems. For instance, when organic-biologic farming and integrated production were articulated and pioneered in Switzerland in the 1970s, it was considered a niche by the majority of politicians and the scientific community. Organic-biologic farming was carried out mainly by a small group of pioneering and enthusiastic individuals and consumers. The

socio-technological regime of industrialized agriculture was stable and the niche innovation of sustainable alternatives had low visibility until 1990s when a macro-change in cultural values and politics occurred (Belz 2004). In the 1990s, alternative agrifood movements emerged globally and took on variety of forms, such as "fair trade", "community support agriculture", "alternative food network", "agroecology" and "food sovereignty". However, the scaling of sustainable alternatives was confined to the strong technocratic regime of capitalist agriculture (Misra 2018).

When the regime is destabilized from niche innovation, the direction of system change (or transition) is uncertain, given coexistence of multiple sub-regimes, e.g. science and technology, user and market, and culture and norms. When multiple trajectories emerge from the process, social finance plays an anchoring role in harnessing the social value of innovation. To express and reflect the challenges and a desire for a better future, the United Nations sets out 17 Sustainable Development Goals (SDGs) and formulates 169 associated targets and 232 relevant indicators (United Nations 2017). The SDGs also present an opportunity for investors to align finance with this global agenda by allocating capital to high-impact projects that address critical societal challenges. While a number of global initiatives are emerging to coordinate global finance with SDGs, such as the Global Impact Investing Network (GIIN 2018b), OECD (2018b), and Principles for Responsible Investment of the United Nations, the landscape of mainstream finance means that it remains difficult for the many participants to collaborate and coordinate with one another (World Economic Forum 2013). When translating the SDGs into policies and actions, there was unfortunately a prevalence of technocratic and siloed thinking, unjustified simplification and a diversion of attention away from the important overall objectives (Fukuda-Parr et al. 2014).

The global emergence of 'blended finance' is exercising a transformative role in scaling system innovation and promoting interactions in the niche-regime-landscape. Defined as "the strategic use of development finance for the mobilization of additional finance towards sustainable development", blended finance combines capital of different forms at different levels in order to catalyze risk-adjusted and market rate-seeking capital into SDGs (OECD 2018b). Depending on the operational structure, blended finance combines mobilization at the transaction level with catalytic ambition over time. For instance, the presence of development finance provides investors with confidence and incentives to undertake high-impact but also high-risk projects. When combined with the articulation of social values (such as SDGs), blended finance cascades knowledge and resources from the landscape-regime to niche innovation at the project level. As the majority of blended finance is now intermediated through collective and pooled vehicles, these new facilities and funds potentially syndicate a large volume of mobilized capital and resources to lift niche innovation without being locked in, even when facing resistance at the regime level. These intermediaries are important to enable a direct interaction between policy mandates and commercial objectives. Bended finance represents an ecosystem approach to engage a broad range of actors and a variety of legal and financial structures.

While innovative investment vehicles are emerging globally, the notion of impact entrepreneurship and social finance are not universally accepted in China. Impact entrepreneurship is about taking a different approach to deal with social problems. It is more than simply operating start-ups and delivering impacts in a pre-conceived way. It is about redefining problems, testing different solutions, taking risks with unproven ideas, and scaling what works (Baker and Nelson 2005; Dacin et al. 2010). Such an undertaking goes beyond individual capability and calls for strategic partnerships to drive breakthrough change systematically and sustainably. To make that happen, social finance, such as impact investment and blended finance, matters in its coordinating role (Nicholls and Emerson 2015). While the global community has explored and developed a variety of good practices for innovation investment in other parts of the world (GIIN 2018a; Jackson 2012; Thornley et al. 2011; WEF and OECD 2015), such as impact investment, blended finance and pay-for-success (PFS), their presence and application in China are still in niche.

It is necessary that opportunities in social finance in China interact and are syndicated if they are to create a conductive ecosystem and marketplace. Today, there is a dramatic financial shortfall for social finance and impact entrepreneurs, be it in China or globally. Social entrepreneurs' demand for impact investment is often too big for grant programs (such as foundation and philanthropists), and too small and risky for institutional social investors (Le Viet-Clarke 2016). There is a huge need to unlock private capital (including philanthropy) and to redefine the role of public funders to maximize impact and further scale innovation. Impact entrepreneurs go through a life cycle of social innovation, from experiment and founding, to growth, scale and maturity. To match the different development stages of impact delivery, a highly diverse range of investors (as mapped in this study) would benefit each other if they syndicated and coordinated their capital (Nicholls and Emerson 2015). Public funders and philanthropists play an important role in de-risking and promoting investment-ready ventures. As the "deal flow" at this stage, such as due diligence and technical assistance, are often costly, the yet-to-be constructed ecosystem would benefit from aggregating capital and clustering innovation in dense networks of shared information, knowledge, resources and capabilities (Kohler et al. 2011). In addition to aggregation, mainstream businesses (such as secondary markets, equity investors and large incumbent) could provide a sustainable source of liquidity for social innovation and impact entrepreneurship.

Conclusions

This study aims to provide an overview of the landscape and dynamics related to China's evolving impact entrepreneurs and social finance. In spite of the global emergence of impact entrepreneurship as a solution to social issues, be them local or global, little is known about social and impact entrepreneurship in China. The regulatory frameworks that govern social entrepreneurial organizations are fragmented and uncoordinated. The social entrepreneurial sector is growing but in a disorganized way, and often depends upon organizations being able to interpret conflicting regulatory frameworks and policies. Emerging social entrepreneurship is concentrated in urban areas, addressing some issues faced by rural migrants but not on the scale at which problems are present. By framing social innovation as interactive process across different levels, opportunities for social finance are mapped and synergized between external impact entrepreneurship and within-regime intrapreneurship, in the public and private sectors.

Acknowledgement The author acknowledges funding support from National Natural Science Foundation of China (71573209; 71661147001), National Key Research and Development Program of China (2016YFD0201303), National Social Science Foundation (16ZDA021) and Chinese Academy of Agricultural Sciences (CAAS-ASTIP-2016-AII). The author also appreciates support and help from Swiss Agency for Development and Cooperation (SDC) and the China Office. The study has benefited from communications and discussion with a variety of institutions, e.g. UBS, Hystra, SynTao Green Finance, Youcheng Foundation, China Social Enterprise and Impact Investment Forum, China Charity Fair Social Enterprise Certification Center.

Appendix

See Tables 6, 7, 8, 9, 10.

Table 6 Laws, regulations and sectoral policies related to social and impact entrepreneurial organizations and finance

In Chinese	In English
Laws	
中华人民共和国宪法 （第 35 条）	Constitution of the People's Republic of China (Article 35)
中华人民共和国民法通则 （第三章；第四章）	General Principles of The Civil Law of The People's Republic sf China (Chapter 3; Chapter 4)
中华人民共和国资产评估法	Assets appraisal law of the people's Republic of China
中华人民共和国慈善法	Charity Law of the People's Republic of China
中华人民共和国红十字会法	Law of the People's Republic of China on the Red Cross Society
中华人民共和国公益事业捐赠法	Welfare Donations Law of the People's Republic of China
中华人民共和国信托法	Trust Law of the People's Republic of Chin
企业所得税法	The Law of the People's Republic of China on Enterprise Income Tax
中华人民共和国会计法	Accounting Law of the People's Republic of China
中华人民共和国政府采购法实施条例	Government Procurement Law of the People's Republic of China
中华人民共和国招标投标法	Law of the People's Republic of China on Tenders and Bids
农民专业合作社法	Law of the People's Republic of China on Professional Farmers Cooperatives
Regulations and Sectoral Policies	
中华人民共和国境外非政府组织境内活动管理法	The Law of the People's Republic of China on the Administration of Activities of Overseas Nongovernmental Organizations
社会团体登记管理条例	Regulations on the Administration of the Registration of Social Organizations
民办非企业单位登记管理暂行条例	Provisional Regulations on the Administration and Registration of Nonprofit Organizations

Table 6 (continued)

基金会管理条例	Regulation on The Administration of Foundations
民间非营利组织会计制度	Accounting Administration for Nonprofit Organizations
社会组织信用信息管理办法	Administrations on Credit Information of Social Organizations
慈善组织信息公开办法	Administrations on Information Disclosure of Charitable Organizations
关于改革社会组织管理制度促进社会组织健康有序发展的意见	About Reforming Administration on Social Organizations for Development
行业协会商会与行政机关脱钩总体方案	Schemes of Decoupling Professional Consortium and Association with Government
关于公益股权捐赠企业所得税政策问题的通知	Notice about Enterprise Income Tax Issues Relating with Equity Donation of Public Welfare and Philanthropy
关于非营利组织免税资格认定管理有关问题的通知	Notice about Accrediting Tax Exemption for Nonprofit Organization
私募投资基金服务业务管理办法（试行）	Notice on Promulgation of the Administrative Measures on Private Investment Fund (Provisional)
国务院关于促进创业投资持续健康发展的若干意见	State Council Guidelines about Promoting Sustainable Development of Entrepreneurial Investment
（国务院）关于强化实施创新驱动发展战略进一步推进大众创业万众创新深入发展的意见	State Council Advocacy about Strengthening Innovation as Development Strategy and Promoting Mass Entrepreneurship and Innovation

Table 7 Entrepreneurial finance and simulation of social investment

Year	Venture capital (VC)		Private equity investment (PE)		VC&PE for social impact entrepreneurship ($ billion)	
	Imbursement	Investment (US$ billion)	Imbursement	Investment (US$ billion)	Scenario 1 (=5%)	Scenario 2 (=10%)
2010	817	5.5	1180	16.1	1.08	2.15
2011	1505	12.6	2200	39.4	2.60	5.20
2012	1071	7.1	1751	26.2	1.67	3.33
2013	1148	6.2	1808	29.0	1.76	3.52
2014	1917	16.0	3626	67.3	4.17	8.33
2015	3445	19.9	8365	80.8	5.04	10.07
2016	3683	20.2	9124	114.6	6.74	13.48
2017	4822	31.2	10,144	186.3	10.87	21.75

Source VC and PE data from: https://research.pedaily.cn/201802/427464.shtml

Table 8 Investment of public–private partnerships in China, 2014–2018

	Number of cases	Amount (billions of US$)	Amount (%)
Municipal construction	2763	473.65	30.59
Transport	971	449.76	29.04
Urban development	418	223.30	14.42
Environment and ecosystem	598	115.89	7.48
Culture and tourism	426	73.43	4.74
Infrastructure	447	61.90	4.00
Social security and affordable housing	155	42.02	2.71
Education and technology	404	37.97	2.45
Healthcare and elderly care	268	30.13	1.95
Others	114	26.95	1.74
Energy	85	7.13	0.46
Agriculture and forestry	76	6.46	0.42
Total	6725	1548.59	100

Source BRIdata. https://www.bridata.com/project?type=hylx&in_cpppc=1r

Note By Dec.30, 2018

Table 9 Expenditure by public finance institutions and the civil/social sector in China, (billions of US$)

	Total expenditure	Public institutions	Civil sector	% of civil sector
2010	1383	886	42	3.00
2011	1681	1072	50	2.96
2012	1938	1250	57	2.92
2013	2157	1361	66	3.05
2014	2335	1472	68	2.90
2015	2706	1774	76	2.80
2016	2889	1917	84	2.90

Source Ministry of Finance; Ministry of Civil Affairs

Table 10 Assets under the management of companies listed in China before Sept. 20, 2018

	Number of companies	Assets under management	
		(billions US$)	(%)
Manufacturing	2034	3000.0	31.46
Real estate	144	1537.2	16.12
Construction	103	1139.5	11.95
Mining	87	1077.7	11.30
Information technology	588	795.9	8.35
Energy (electricity, gas, water)	116	629.2	6.60
Transportation and warehousing	109	485.7	5.09
Wholesale and retail trade	180	461.8	4.84
Social services	184	268.6	2.82
Communication, media and culture	61	74.0	0.78
Agriculture, forestry and fishery	46	39.5	0.41
Miscellaneous	34	26.2	0.27
Total	3686	9535	100

Source Tai-Kang-An database

Shanghai and Shenzhen Stock Exchange (A & B). Listed companies of financial sector (US$27.3 trillion for 93 cases) are not included

References

ACOSS. 2010. *Contribution of the not-for-profit sector: ACOSS analysis and priorities for future advocacy*. Australian Council of Social Service.

AVPN. 2019. *China: Social investment landscape in Asia*. Asian Venture Philanthropy Network.

Baker, T., and R.E. Nelson. 2005. Creating something from nothing: Resource construction through entrepreneurial bricolage. *Administrative Science Quarterly* 50: 329–366.

Belz, Frank-Martin. 2004. A transition towards sustainability in the Swiss agri-food chain (1970–2000): using and improving the multi-level perspective. In *System innovation and the transition to sustainability: Theory, evidence and policy*, ed. Boelie Elzen, Frank W. Geels, and Ken Green. Northampton: Edward Elgar Publishing Inc.

Berkhout, F., A. Smith, and A. Stirling. 2004. Socio-technological regimes and transition contexts. In *System innovation and the transition to sustainability: Theory, evidence and policy*, ed. B. Elzen, F.W. Geels, and K. Green, 48–75. Cheltenham: Edward Elgar.

Bishop, Matthew, and Michael Green. 2015. Philanthrocapitalism comes of age. In *Social finance*, ed. Alex Nicholls, Rob Paton, and Jed Emerson, 113–129. Oxford: Oxford University Press.

Bornstein, D. 2007. *How to change the world: Social entrepreneurs and the power of new ideas*. Oxford: Oxford University Press.

British Council. 2016. *The state of social enterprise in Bangladesh, Ghanna, India and Pakistan*. London: British Council.

Bugg-Levine, A., and J. Emerson. 2011. *Impact investing: Transforming how we make money*. Hoboken: Jossey-Bass.

Bui, Sibylle, A. Cardona, C. Lamine, and M. Cerf. 2016. Sustainability transitions: Insights on processes of niche-regime interaction and regime reconfiguration in agri-food systems. *Journal of Rural Studies* 48: 92–103.

Burd-Sharps, Sarah, Patrick Guyer and Kristen Lewis. 2011. Metrics matter: A human development approach to measuring social impact. In Community Development Investment Review: Proceedings of the conference on Advancing Social Impact Investments Through Measurement: Center for Community Development Investments, Federal Reserve Bank of San Francisco.

Caniels, Marjolein C.J., and Henny A. Romijn. 2008. Strategic Niche management: Towards a policy tool for sustainable development. *Technology Analysis and Strategic Management* 20: 245–266.

Carman, J.G., and J.L. Millesen. 2005. Nonprofit program evaluation: Organizational challenges and resource needs. *Journal of Volunteer Administration* 23: 36–43.

Carman, Joanne, and Kimberly Fredericks. 2013. Nonprofits and accreditation: Exploring the implications for accountability. *International Review of Public Administration* 18: 51–68.

Christensen, C.M., H. Baumann, R. Ruggles and T.M. Sadtler. 2006. Disruptive innovation for social change. *Harvard Business Review* 82.

Coleman, J.S. 2000. *Foundations of social theory: Cambridge*. Massachusetts: Belknap Press.

Credit Suisse. 2018. Global Wealth Report 2018.

Curtis, Timothy. 2011. 'Newness' in social entrepreneurship discourses: The concept of 'Danwei' in the Chinese experience. *Journal of Social Entrepreneurship* 2: 198–217.

CSEIIF. 2019. China Social Enterprise and Social Investment Landscape Report 2019: Beijing: China Social Enterprise and Impact Investment Forum. https://www.cseiif.cn/index.php.

Dacin, P.A., M.T. Dacin, and M. Matear. 2010. Social entrepreneurship: Why we don't need a new theory and how we move forward from here. *Academy of Management Perspectives* 24: 37–57.

Dahle, K. 2007. When do transformative initiatives really transform? A typology of different paths for transition to a sustainable society. *Futures* 39: 487–504.

Defourny, Jacques, and Shin-Yang Kim. 2011. Emerging models of social enterprise in Eastern Asia: A cross-country analysis. *Social Enterprise Journal* 7: 86–111.

Desa, G., & Jia X. (of this issue). Sustainability Transitions in the Context of Pandemic: An Introduction to the Focused Issue on Social Innovation and Systemic Impact. *Agriculture and Human Values*.

Elzen, Boelie, Marc Barbier, Marianne Cerf, and John Grin. 2012. Stimulating transitions towards sustainable farming systems. In *Farming systems research into the 21st century: The new dynamic*, ed. Ika Darnhofer, David Gibbon, and Benoit Dedieu, 431–455. Dordrecht: Springer.

Emerson, J. 2003. The blended value proposition: Integrating social and financial returns. *California Management Review* 45: 35–51.

Freireich, Jessica, and Katherine Fulton. 2009. *Investing for social and environmental impact: A design for catalyzing an emerging industry*. San Francisco: Monitor Institute.

Fukuda-Parr, Sakiko, Alicia Ely Yamin, and Joshua Greenstein. 2014. The power of numbers: A critical review of millennium development goal targets for human development and human rights. *Journal of Human Development and Capabilities* 15: 105–117.

Gao, Yunlong. 2019. Annual Report on non-state-owned Economy in China: 2917–2018 (中国民营经济发展报告: 2017–2018). In *Annual Report on non-state-owned Economy in China: 2917–2018* (中国民营经济发展报告: 2017–2018): Beijing: All-China Federation of Industry and Commerce Publishing House.

Geels, Frank, and Johan Schot. 2010. The dynamics of transitions: a socio-technical perspective. In *Transitions to sustainable development: New directions in the study of long term transformative change*, ed. John Grin, Jan Rotmans, Johan Schot, Frank Geels, and Derk Loorbach. New York: Routledge.

Geels, Frank W. 2004. From sectoral systems of innovation to socio-technical systems insights about dynamics and change from sociology and institutional theory. *Research Policy* 33: 897–920.

Geels, Frank W. 2002. Technological transitions as evolutionary recon-figuration processes: A multi-level perspective and a case-study. *Research Policy* 31: 1257–1274.

GIIN. 2018a. *Annual impact investor survey*, 8th ed. New York: Global Impact Investing Network.

GIIN. 2018b. *Financing the sustainable development goals: Impact investing in action*. New York: Global Impact Investing Network.

Glinskaya, Elena and Zhanlian Feng. 2018. Options for aged care in china building an efficient and sustainable aged care system. In *Options for aged care in china building an efficient and sustainable aged care system*. Washington: The World Bank.

Grin, John, Jan Rotmans, Johan Schot, Frank Geels, and Derk Loorbach. 2010. *Transitions to sustainable development: New directions in the study of long term transformative change*. New York: Routledge.

Hagen, Everett. 1963. How economic growth begins: A theory of social change. *Journal of Social Issues* 19: 20–34.

Han, Tianyang and Zhao Zhang. 2010. Social innovators' contribution hailed. China Daily Sept. 15th.

Hendriks, C. 2008. On inclusion and network governance: The democratic disconnect of Dutch energy transitions. *Public Administration* 86: 1009–1031.

Jackson, E.T. 2012. *Achievements, challenges and what's next in building the impact invesing industry*. New York: The Rockefeller Foundation.

Jia, X., & Desa, G. (of this issue). Social Entrepreneurship and Impact Investment in Rural-Urban Transformation: An orientation to systemic social innovation and symposium findings. *Agriculture and Human Values*.

Jiang, Xueqing. 2016. China to become largest market of high-net-worth individuals: https://usa.chinadaily.com.cn/business/2016-06/23/content_25814999.htm. Accessed Sept 2018.

Kayser, Olivier, and Valeria Budinich. 2015. *Scaling of business solutions to social problems*. London: Palgrave Macmillan.

Kemp, René, Derk Loorbach, and Jan Rotmans. 2007. Transition management as a model for managing processes of co-evolution towards sustainable development. *International Journal of Sustainable Development & World Ecology* 14: 78–91.

Kemp, Rene, Johan Schot and Remco Hoogma. 2015. Regime shifts to sustainability through processes of Niche formation: the approach of strategic Niche management. In *Innovation, technology and economic change, elgar research collection*. vol. 306, 790–810. Cheltenham, UK and Northampton Elgar: International Library of Critical Writings in Economics.

Kitzmueller, M., and J. Shimshack. 2012. Economic perspectives on corporate social responsibility. *Journal of Economic Literature* 1: 51–84.

Kohler, John, Thane Kreiner and Jessica Sawhney. 2011. *Coordinating impact capital: A new approach to investing in small and growing businesses*. Santa Clara University, Center for Science, Technology, and Society; The Aspen Network of Development Entrepreneurs.

Krippendorff, Kaihan. 2019. *Driving innovation from within: A guide for internal entrepreneurs*. New York: Columbia Business School Publishing.

Lane, Andrea. 2013. 2012 China Social Enterprise Report: Foundation for Youth Social Entrepreneurship (FYSE). https://www.bsr.org/reports/FYSE_China_Social_Enterprise_Report_2012.PDF. Accessed 18 Nov 2018.

Le Viet-Clarke, Caroline. 2016. Unlocking Blended Finance for social entrepreneurs: Ashoka: https://www.ashoka.org/en-us/story/unlocking-blended-finance-social-entrepreneurs.

London, T. and S. Hart. 2011. Next-generation business strategies for the base of the pyramid: new approaches for building mutual value. In *Next-generation business strategies for the base of the pyramid: New approaches for building mutual value*. Upper Saddle River: Financial Times Press.

Ma, Qingyu. 2018. Measuring the size and economic contribution of China's social organization.

McKinsey Global Institute. 2017. China's role in the next phase of globalization. In *China's role in the next phase of globalization*. Discussion Paper.

Misra, Manoj. 2018. Moving away from technocratic framing: Agroecology and food sovereignty as possible alternatives to alleviate rural malnutrition in Bangladesh. *Agriculture and Human Values* 35: 473–487.

Nicholls, Alex, and Jed Emerson. 2015. Social finance: capitalizing social impact. In *Social finance*, ed. Alex Nicholls, Rob Paton, and Jed Emerson. Oxford: Oxford University Press.

Nicholls, Alex, Jeremy Nicholls, and Rob Paton. 2015. Measuring social impact. In *Social Finance*, ed. Alex Nicholls, Rob Paton, and Jed Emerson, 253–281. Oxford: Oxford University Press.

O'Donohue, N., C. Leijonhufvud, Y. Saltuk, A. Bugg-Levine and M. Brandenburg. 2010. Impact investments: An emerging asset class: J. P. Morgan. https://assets.rockefellerfoundation.org/app/uploads/20101129131310/Impact-Investments-An-Emerging-Asset-Class.pdf.

OECD. 2018a. *Innovation, agricultural productivity and sustainability in China*. Paris: The Organisation for Economic Co-operation and Development (OECD).

OECD. 2018b. *Making blended finance work for the sustainable development goals*. Paris: The Organisation for Economic Co-operation and Developmen Publishing (OECD).

Pigford, Ashlee-Ann E., Gordon M. Hickey, and Laurens Klerkx. 2018. Beyond agricultural innovation systems? Exploring an agricultural innovation ecosystems approach for niche design and development in sustainability transitions. *Agricultural Systems* 164: 116–121.

Porter, M. and M. Kramer 2011. Creating shared value. *Harvard Business Review* 1–17.

Praszkier, Ryszard, Andrzej Nowak, and Agata Zabocka-Bursa. 2009. Social capital built by social entrepreneurs and the specific personality traits that facilitate the process. *Psychologia Spoleczna* 1–2: 42–54.

Salamon, Lester M. 2014. *New frontiers of philanthropy: A guide to the new tools and new actors that are reshaping global philanthropy and social investing*. Oxford: Oxford University Press.

Sommerrock, Katharina. 2010. *Social entrepreneurship business models: Incentive strategy to catalyze public goods provision*. London: Palgrave Macmillan.

Teodoro, Manuel P., and Adam G. Hughes. 2012. Socializer or signal? How agency accreditation affects organizational culture. *Public Administration Review* 72: 583–591.

The Economist. 2017. The next wave: China's audacious and inventive new generation of entrepreneurs. Sep. 23rd.

The State Council. 2016. Mass entrepreneurship and innovation as new growth engine. English.GOV.CN https://english.gov.cn/premier/news/2016/03/03/content_281475300571752.htm. Access Feb 2017.

Thornley, Ben, David Wood, Katie Grace, and Sarah Sullivant. 2011. *Impact investing: A framework for policy design and analysis: InSight at Pacific Community Ventures The initiative for Responsible Investment (IRI) at Harvard University*. New York: The Rockefeller Foundation.

Thornton, Phil, Tonya Schuetz, Wiebke Förch, Laura Cramer, David Abreu, Sonja Vermeulen, and Bruce Campbell. 2017. Responding to global change: A theory of change approach to making agricultural research for development outcome-based. *Agricultural Systems* 145–153: 145–153.

United Nations. 2015. The World Population Prospects: 2015 Revision In The World Population Prospects: 2015 Revision Department of Economic and Social Affairs, United Nations. https://www.

376

⚡ Springer

un.org/en/development/desa/publications/world-population-prospects-2015-revision.html Accessed 12 May 2017.

United Nations. 2017. Resolution adopted by the General Assembly on Work of the Statistical Commission pertaining to the 2030 Agenda for Sustainable Development (A/RES/71/313), Annex: Inter-Agency and Expert Group on SDG Indicators (IAEG-SDGs), United Nations.

Wang, Mei. 2013. Public-private partnerships in China: Commentary. *Public Administration Review* 73: 311–312.

Weber, Matthias and Jens Hemmelskamp. 2005. Towards environmental innovation systems. In *Towards environmental innovation systems*. Heidelberg and New York: Springer.

WEF & OECD. 2015. *Blended finance: A primer for development finance and philanthropic funders*, vol. 1. New York: World Economic Forum & OECD.

Willems, Jurgen, Carolin J. Waldner, Yasemin I. Dere, Yuka Matsuo, and Kevin Högy. 2017. The role of formal third-party endorsements and informal self-proclaiming signals. *Nonprofit Reputation Building Nonprofit and Voluntary Sector Quarterly* 46: 1092–1105.

World Bank. 2014. *Urban China: Toward efficient, inclusive, and sustainable urbanization*. World Bank and the Development Research Center of the State Council, P. R. China. Washington: World Bank. https://doi.org/10.1596/978-1-4648-0206-5.

World Economic Forum. 2013. *From the margins to the mainstream: Assessment of the impact investment sector and opportunities to engage mainstream investors*. Cologny/Geneva Switzerland.

Xu, Xiangling. 2015. Social transformation and public governance: Trajectory of China's political reform and policies (shehui zhaunxin yu guojia zhili: zhongguo zhengzhi tizhi gaige xuxiang jiqi zhengce xuanze). *Research in Politics* (Zhengzhixue Yanjiu) 1.

Yang, Tuan. 2008. *Bluebook of Philanthropy* (慈善蓝皮书: 中国慈善发展报告): Beijing: Social Science Academic Press (China).

Yu, X. 2011. Social enterprise in China: Driving forces, development patterns and legal framework. *Social Enterprise Journal* 7: 9–32.

Yu, Xiaomin. 2014. The governance of social enterprises in China. *Social Enterprise Journal* 9: 225–246.

Zhang, Wei-feng, Zheng-xia Dou, Pan He, Ju Xiao-Tang, David Powlson, Dave Chadwick, David Norse, Lu Yue-Lai, Ying Zhang, Wu Liang, Xin-Ping Chen, Kenneth G. Cassmang, and Fu-Suo Zhang. 2013. New technologies reduce greenhouse gas emissions from nitrogenous fertilizer in China. *Proceedings of the National Academy of Sciences* 110: 8375–8380.

Publisher's Note Springer Nature remains neutral with regard to jurisdictional claims in published maps and institutional affiliations.

Xiangping Jia is a professor at the Agricultural Information Institute, Chinese Academy of Agricultural Sciences. He received his Ph.D. from University of Hohenheim, Germany. Working on research about agricultural policies and rural development, he has broad interest in the application of new institutional economics to the issues of agri-food market coordination, development of inclusive financial markets, the organization of smallholder farms, knowledge transfer and innovation, and agricultural extension system. At present, he researches innovative institutions that align agricultural universities with stakeholders such as public extension systems and private sectors for a resilient farm system.

Agriculture and Human Values (2020) 37:1313–1314
https://doi.org/10.1007/s10460-020-10041-9

BOOK REVIEW

Laura-Anne Minkoff-Zern: The new American farmer: immigration, race, and the struggle for sustainability

Massachusetts Institute of Technology Press, Cambridge, Massachusetts, 2019, 195 pp, ISBN 978-0-262-53783-4

Eden Kinkaid[1]

Accepted: 5 May 2020 / Published online: 12 May 2020
© Springer Nature B.V. 2020

In her recently published book, Laura-Anne Minkoff-Zern examines an understudied phenomenon in U.S. agriculture: the rise of Latinx immigrant farm owner-operators. Despite the obstacles of a deeply racialized agricultural system, these farmers have climbed the "agricultural ladder" to own and operate their own farms.

Yet the farmers Minkoff-Zern studies are not merely reproducing the models of agriculture they labored under as farmworkers; rather many of them are pursuing more diverse and "alternative" production methods, drawing on their memories and experiences farming in Mexico and Central America. While these immigrant farmers may represent the "new American farmer," they are often rendered institutionally invisible, leaving them largely unsupported by agricultural extension and support services. Even though their practices reflect a commitment to "alternative" agriculture, they are equally excluded and invisible in alternative food movements and spaces. Minkoff-Zern's study is a much-needed antidote to this invisibility. The book thus makes a unique contribution to agri-food studies while supporting practical efforts to address forms of discrimination and exclusion in agricultural institutions and food cultures.

The new American farmer is based on qualitative fieldwork conducted in five distinct regions of the United States. At each of these sites, Minkoff-Zern interviewed immigrant farmers (70 total), extension agents, USDA staff, and other groups that work with immigrant farmers. She also conducted participant observation at farmers markets and a training course targeting immigrant farmers.

Drawing on insights gleaned from each of these case studies, Minkoff-Zern examines how racial formations have long shaped agrarian class structures, labor regimes, and land access in the U.S. and how these legacies impact immigrant farmers today. While scholars in various fields have considered these themes in depth, Minkoff-Zern's study contributes unique insights in that it shows how some immigrant farmers are able to successfully navigate these structures, suggesting the possibility that these racial and class systems are not entirely fixed. Further, she presents a multi-sited approach that is sensitive to the geographic and historical specificity of these racial formations and agrarian regimes.

In Chapter 2, Minkoff-Zern places the contemporary situation of immigrant farmers within a longer history of the racialized exclusions of US agriculture. She connects the history of Latino labor migration and land access with the history of slavery and Black land ownership to understand the broader relationships between race, labor and land access in the U.S. In doing so, she inserts issues of race into the concept of the "agrarian ladder," a model long used by rural sociologists, agricultural economists, and others to understand how farmworkers become farm owners. By reading race into this "agrarian myth" of social mobility, Minkoff-Zern shows how various racial ideologies and policies have prevented Mexican immigrants from ascending this ladder, retaining them as sources of cheap and available labor for white farmers.

In Chapter 3, Minkoff-Zern expands her analysis to demonstrate how current practices of institutionalized alternative agriculture also operate as forms of racialized exclusion. Even though these immigrant farmers farm in a way that is aligned with many aspects of alternative agriculture, they remain underrepresented in alternative food movements and

✉ Eden Kinkaid
  ekinkaid@email.arizona.edu

[1] Department of Geography, School of Geography
     and Development, University of Arizona, 1064 E. Lowell
     Street, Tucson, AZ 85719, USA

spaces. Minkoff-Zern argues that this is both an institutional and cultural problem. Major institutions like the USDA, which have legacies of racially discriminatory practices, tend to focus on "traditional" (i.e., white) farmers and do not devote resources to cultural and linguistic translation of their processes and services. USDA standardization requirements, like bookkeeping and paperwork, are also barriers to immigrant farmers getting support from the USDA and other bodies. Beyond these institutional barriers, cultural and racial divides exist in alternative food spaces, which are predominately white. If immigrant farmers are to be supported in their endeavors to practice alternative agriculture, institutional and cultural changes are necessary in both "mainstream" and alternative agricultural spaces.

Chapter 4 explores immigrant farmers' personal motivations for running their own farms and traces how these motivations are shaped by their memories and knowledge of farming in their home countries. Here, Minkoff-Zern explores how themes of migration, food, practice, and identity influence farmers' priorities and practices including crop diversity, familial involvement, low inputs, and direct marketing of crops. In Chapter 5, Minkoff-Zern extends this conversation to argue that these principles could be considered an enactment of food sovereignty. While farmers do not identify with social movements around food, their principles do speak to the broader aims of food sovereignty movements. However, there are challenges to enacting such principles; while farmers have ascended the agrarian class and racial hierarchy, they are still part of the same agricultural system which they cannot structurally resist. Yet their determination and commitment to an alternative vision of agriculture evidences some room for maneuver in a highly exploitative and racialized industry.

In her final chapter, Minkoff-Zern reiterates the importance of attending to this new class of American farmers for our understanding of agrarian change: "Only by looking closely at the differences in lived experiences between racialized groups of food producers, and appreciating both their race- and citizenship-based obstacles as well as unique offerings and skills, can we begin to form a new theory of agrarian change" (p. 168). To do so, we not only need to revise our theories of agrarian change and social mobility; additionally, Minkoff-Zern urges that much work needs to be done to change agricultural institutions and cultures to make space for immigrant farmers. In closing, Minkoff-Zern presents a number of recommendations for supporting immigrant farmers in their efforts to pursue alternative agricultural practices. These include institutional issues like technical support, linguistic translation, and more bilingual support staff, as well as cultural issues, like producing more inclusive alternative food spaces.

Minkoff-Zern's study provides crucial insights into the status of immigrant farmers in the U.S., an undercounted and underserved population bringing various forms of knowledge and diverse experiences to sustainable farming. It represents an important starting point for understanding the changing landscape of U.S. agriculture, especially in the current political moment. That being said, more work will need to be done to respond to some of the issues Minkoff-Zern hints at, but does not fully consider, in this book. These issues include the role of immigrant farmers in the industrial agricultural sector and the politics of gendered and family labor on small-scale immigrant farms. Nevertheless, *The new American farmer* will enrich discussions in undergraduate and graduate courses on geographies of food and agriculture, food and migration, and transnational food studies. Minkoff-Zern's study will be an important contribution to ongoing debates on and discussions of agrarian change in geography, rural studies, rural sociology, food studies, and related disciplines.

Publisher's Note Springer Nature remains neutral with regard to jurisdictional claims in published maps and institutional affiliations.

Eden Kinkaid is a doctoral candidate in the School of Geography and Development at the University of Arizona. Eden's research focuses on the governance of organic agriculture in north India and draws on agri-food studies and political ecology.

Agriculture and Human Values (2020) 37:1315–1316
https://doi.org/10.1007/s10460-020-10116-7

Carol Off: Bitter chocolate: anatomy of an industry

The New Press, New York, 2014, 326 pp, ISBN 978-1-59558-980-4

Allison L. Brown[1]

Accepted: 19 May 2020 / Published online: 23 May 2020
© Springer Nature B.V. 2020

"No matter how fierce the fighting, the trucks that transport cocoa beans always seem to get their cargo to the port of San-Pédro [Côte d'Ivoire] and then onward to the candy counters of the Western world" (p. 179). In Carol Off's part-history book, part-exposé, *Bitter Chocolate: Anatomy of an Industry*, she explains the history of cocoa from ancient times to present day, then investigates Côte d'Ivoire's modern role in the cocoa trade by examining the turbulent reality that is cocoa production. Off skillfully puts cocoa in context, as Côte d'Ivoire's major export crop and lifeline, and provides in-country interviews of child slaves tricked into working in cocoa plantations and reporters mixed up in political corruption. The book was originally published in hardcover in 2006, while the paperback version was published in 2014. As a result, most of the research and interviews were completed from 2002 to 2004, effectively encapsulating the Ivoirian cocoa sector during the First Ivoirian Civil War, which took place from 2002 to 2007.

After a sluggish cocoa history introductory section, the original content of the book is presented in a gripping fashion in chapters six through eleven. Off begins this part with a concise history of Côte d'Ivoire, concentrating on the impact of President Felix Houphouet-Boigny's reign from 1960 to 1993. Houphouet-Boigny's policies encouraged in-migration from poorer West African countries, religious tolerance, and above all, cocoa production. His public works projects were responsible for a streamlined cocoa industry that enabled Côte d'Ivoire to become the number one cocoa producer in the world.

This section continues by delving into how the system encourages the transport of child slaves into Côte d'Ivoire. An interview with Abdoulaye Macko, a former Malian

✉ Allison L. Brown
  alb54@psu.edu

1 Department of Food Science, The Pennsylvania State University, University Park, PA 16802, USA

Diplomat, illuminates the child slavery situation and its hopelessness. Macko, who recovered 250 child slaves from Côte d'Ivoire, was later fired by the government of Mali because [according to Macko] trade relations between the two countries were more important than saving the lives of child slaves. Off also interviews Malian child slaves and transcribes their stories, ultimately determining that the middlemen are the only ones profiting from the hopelessness of both the poor Ivoirian farmers and Malian children. This part of the book takes time to grapple with child labor and slavery and its inherent immorality in an interview with Anita Sheth of Save the Children Canada. Off and Sheth conclude that if children want to work and their parents want them to work, then they should be able to work, but only if hazards are minimized and they are paid commensurately.

Off then shifts focus from the growth of the Ivoirian cocoa sector to the rise of Ivoirité, which occurred when President Houphouet-Boigny died and was replaced by President Laurent Gbagbo in 1993. Gbagbo officially supported and adopted Ivoirité, a nationalistic and xenophobic belief, which denounced those who were not born of two Ivoirian parents. Ivoirité divided the nation in multiple ways and ultimately led to the First Ivoirian Civil War, which was fought between Christians and Muslims and North versus South. To understand the implications of the First Civil War and Ivoirité on cocoa production, Off travels to wartime Gagnoa, a central town of a cocoa growing region, where the mayor embodies an Ivoirité attitude although it positions him against his farmers. She interviews the mayor and local farmers, and we hear the xenophobia and terror in their own words. With this insight, Off makes a case for the war being about cocoa, the right to keep one's land on which to grow cocoa, and the money earned from selling cocoa, rather than immigration and religion.

Next, Off describes a different aspect of the corrupt world of Ivoirian cocoa when she introduces Guy-André Kieffer (GAK), a French-Canadian reporter. GAK mysteriously disappeared in April 2004 after unearthing facts about the cocoa trade that tied President Gbagbo to money laundering. Off details GAK's life, investigative journalism, disappearance, and the efforts of the French and Canadian governments to unearth the truth. Off drags the reader to Côte d'Ivoire, where in a midnight meeting with "Kieffer's Network," a group of individuals who are continuing GAK's journalistic work, they discuss GAK and his disappearance. This situation reinforces not only the precariousness of GAK's position, but also the importance of his work.

The book's main strengths are its eyewitness accounts that document child slavery, corruption, and tumult in the Ivoirian cocoa sector. *Bitter Chocolate* transports the reader to Côte d'Ivoire and makes a very strong case for the cocoa sector's systemic support of child slavery.

However, the book begins from too broad an angle, spending the first 100 pages on the history of cocoa and chocolate production by multinational food companies. The history of cocoa has already been well-documented in *The True History of Chocolate* (Coe & Coe, 2013) (cited by Off as her main source), which the reader would benefit from reading in its entirety. In the final chapter, after a deep dive into Ivoirian cocoa, Off somewhat awkwardly shifts to Belize and uses its cocoa industry as a case study to examine how Fair Trade and Organic certifications might be possible methods to overcome cocoa labor issues. While insightful, this space could have been used to more fully digest the situation in Côte d'Ivoire and consider the plausibility of certifications to remedy child slavery in Ivoirian cocoa production.

Another weakness is that the Source Notes section provides only source titles and minimal details, which makes it difficult for the reader to cross-check references.

Because of the positive points, the book is an absolutely worthwhile read for students, academics, or professionals who study or work in the cocoa and chocolate sector. In addition, those who study or work on the topic of slave labor, or the politics and economy of Côte d'Ivoire or West Africa, would also find this book interesting. *Bitter Chocolate* would also interest a broader audience of readers who care to know more about food, food history, and supply chains.

Reference

Coe, S.D., and M.D. Coe. 2013. *The True History of Chocolate*, 3rd ed. London, UK: Thames and Hudson Ltd.

Publisher's Note Springer Nature remains neutral with regard to jurisdictional claims in published maps and institutional affiliations.

Allison L. Brown is a Ph.D. candidate and USDA NIFA predoctoral fellow studying a dual-title degree in Food Science and International Agriculture and Development at The Pennsylvania State University. Allison's research focuses on understanding flavor and quality throughout the cocoa and chocolate supply chain using chemical and sensory analysis, as well as qualitative and quantitative human subjects research.

Agriculture and Human Values (2020) 37:1317–1318
https://doi.org/10.1007/s10460-020-10119-4

Harvey S. James, Jr. (ed.): Ethical tensions from new technology: the case of agricultural biotechnology

CAB International, Boston, Massachusetts, 2018, 184 pp, ISBN 978-1-78639-464-4

Sonja Lindberg[1]

Accepted: 9 June 2020 / Published online: 19 June 2020
© Springer Nature B.V. 2020

Almost 25 years after the first genetically modified (GM) crops entered commodity markets, agricultural biotechnologies and resulting plants and animals remain highly contentious issues within and between societies. Edited by Dr. Harvey James, Jr., *Ethical Tensions from New Technology: The Case of Agricultural Biotechnology* offers insights into why GM debates—which are rooted in ethical tensions—exist and persist. It is a 12-chapter compilation of separate studies and essays on controversies around GM foods and crops. In the introduction, James defines an ethical tension as "when [a] technology creates a conflict of interests, values, or rights" (p. 1). Within the chapters, scholars from social science and humanities disciplines—such as agricultural economics, science communication, philosophy, and policy studies—use this organizing theme to interrogate the strong beliefs held by social actors (e.g. scientists, politicians, nations, organizations, and publics) about the place of GM foods and crops in societies and food systems.

The book does more than offer a glimpse into some of the ethical tensions associated with agricultural biotechnologies in different locations, such as the United States, EU, and Australia. James argues that the broader goal is to teach readers how to identify and predict ethical tensions resulting from new technologies in general, an important step in understanding and mitigating challenges with emergent technologies. Through the scope of "ethical tension" studies, the book successfully achieves that aim. It can help readers to see the social dimensions of other new technologies through fresh lenses.

The first section engages with conflicts linked to public opinions and interests. In Chapter 1, Kolodinksi draws on a science communication ethics framework to contend that a "science alone" approach to communicating GM food risks and benefits is inherently problematic: such approach fails to account for important consumer concerns beyond the scientific safety of the products. In Chapter 2, Jones analyzes the "March Against Monsanto" and "Occupy the Farm" social movements, arguing that public and political resistance against agricultural biotechnologies are progressing and increasingly focused on non-scientific principles. Next, Aerni's chapter critiques narratives opposing GMOs, claiming that "common weal rhetoric" and usage of the term "GMO" are abused to advance private agendas. In Chapter 4, Ankeny et al. draw on three case studies to evaluate GM public-private funding and research partnerships in Australia to determine if the beneficiaries are private corporations or the public.

The next section focuses on tensions around policy and regulation of GM foods and crops. In Chapter 5, Windsor describes a three "value model" framework explaining how public opinions and interests about GM production and consumption reflect differing values across nations. Next, Strauss investigates the GM food labeling debate in the United States, focusing on issues surrounding transparency, consumer rights, and the recently enacted "bioengineered" food labeling law. In Chapter 7, Kolady and Srivastava present a comparison of GM regulatory systems in the USA and India to demonstrate how each nation's social and political contexts contributed to divergent policy outcomes.

Chapters 8 and 9 evaluate tensions caused by "technological fix" approaches to defining problems and solutions. Scott draws from philosophy and social theory to argue for a "technological pragmatism" approach to solving GM debates and food security issues rooted in complex social, economic, and political problems. Next, MacDonald presents a case

✉ Sonja Lindberg
  sonjal@iastate.edu

[1] Department of Sociology, Iowa State University, East Hall #308, Ames, IA 50011, USA

study of the Canadian genetically engineered "Enviropig" to demonstrate why ethical tensions must be considered prior to introducing new technologies which might have implications for entire food and production systems.

The last chapters examine tensions intertwined with the rapidly emerging gene editing technique known as "CRISPR." In Chapter 10, Pirscher et al. discuss the EU's societal debates about CRISPR and GM crops, including divergent concepts of "naturalness" and how such issues present challenges to the development of socially informed and accepted regulatory structures. Next, Ng and James' study analyzes entrepreneurial commercialization of CRISPR and conflicting responsibilities to stakeholders and broader society. In the last chapter, Valdivia et al. tie together the preceding chapters through explaining how translational research engaging stakeholders, such as smallholder farmers, can potentially mediate ethical tensions when developing or introducing new technologies.

The chapters are well-written and effectively organized. However, one notable weakness is the lack of certain voices and critical scholarship. James states outright that none of the book's contributors are opponents of agricultural biotechnologies (p. 5), which is evident in the chapters. Issues of social justice and power are touched upon by some authors, but not sufficiently explored. Readers seeking critical scholarship may find Jones' and Scott's chapters to be the most intriguing.

The book is a useful resource for scholars and graduate students interested in agricultural biotechnologies or technology and society intersections. Because of its broad scope, this book can serve as a useful primer if a reader wants an introduction to an array of some issues and contestations around agricultural biotechnologies. It would also be appropriate for graduate courses in food systems, agriculture and society, or science and technology studies. I caution readers against skipping over the insightful introduction. Within it, James presents a brief history of agricultural biotechnology development. James then goes beyond simply summarizing the arguments and content in each chapter to posing important ethical and social questions for the reader to consider. Asking good questions is a cornerstone for unpacking complex topics and devising solutions to avoid or mitigate ethical tensions from new technologies.

Publisher's Note Springer Nature remains neutral with regard to jurisdictional claims in published maps and institutional affiliations.

Sonja Lindberg is a PhD student in Rural Sociology at Iowa State University. Her research examines social dimensions of agricultural biotechnologies, food systems, and science and technology studies.

Agriculture and Human Values (2020) 37:1319–1320
https://doi.org/10.1007/s10460-020-10120-x

Gina Rae La Cerva: Feasting wild: in search of the last untamed food

Greystone Books, Vancouver, BC, 311 pp., ISBN 978-1-77164 (cloth), 978-1-534-8 (epub)

F. E. Jack Putz[1]

Accepted: 24 June 2020 / Published online: 1 July 2020
© Springer Nature B.V. 2020

Feasting Wild by Gina Rae La Cerva takes readers on a delicious and thought-provoking gastronomic global tour, with emphasis on the social, political, and environmental histories and consequences of what we eat. Given the on-going turmoil about race and privilege, readers will note the author's bare-knuckled approach to how hunting game by white folks is transmogrified into bushmeat poaching by anyone else. Her account is very personal from the get go—she seems like the kind of person many of us would enjoy encountering; self-effacing, knowledgeable, adventuresome, and fun.

The first chapter of *Feasting Wild* skips lightly back and forth across the Atlantic as well as back to a time when all food was from the wild, to the present, when that fare struggles for recognition. Foraging for wild edibles in a cemetery in Copenhagen with a group of famous chefs, she explores why such expeditions fell out of favor for so long. Within the first few pages she reminisces about meeting Bill Murray in a bar, champions wild foods for their nutritional value, points out the dire health consequences of both domestication and urbanization, and quips, "Flavor is a map of our desires. But it is not the territory" (p. 16).

From Copenhagen, the author takes us to Poland's Białowieża Forest, the perfect setting to explore the history of hunting, especially how commoners were excluded by royalty from securing sustenance from formerly open-access lands. Throughout the very readable volume, La Cerva dives deeply into this sort of ethical, conservation-related issue particularly as related to food, with due attention to the gendered roles in foraging, hunting, and the culinary arts. The author's insightful analyses are made in the light

of her struggles with her own roles and responsibilities— she is aware of her privilege and ruthlessly explores its consequences.

A traditional lobster bake on the coast of Maine is the setting for forays into the history of our plundering of wildlife, with an in depth analysis of turtle soup. I was unaware that when Europeans first started to venture outwards, explorers and colonists were warned to avoid foods from warmer climates because their consumption causes "creolean degeneracy" (p. 74) and inspires lust. Fortunately, cooking reportedly reduces the tendency of wild food to provoke barbaric behavior.

In the next chapter, La Cerva takes us on a tour of an abandoned gun factory in New Haven. Between reflections on industrial decay, she describes the fate of the passenger pigeon, bounties on carnivores, and the growth of market hunting. The gradual re-wilding of the decaying factory provides a backdrop for consideration of the concepts of wilderness and wastelands, but at least this reader is not especially compelled by post-wilding.

The second section of *Feasting Wild* includes three chapters about trade in illegal bushmeat from the Congo Basin. Her guide and lover, a sensitive Scandinavian wildlife biologist, helps the author navigate this physically and ethically dangerous territory from a remote outpost in the jungle to the city of Kinshasa; in a later chapter, she visits him in Sweden where they hunt, dress, cook, and eat a moose together. A central point made here and elsewhere is the geopolitical unfairness that brown and black hunters are labeled as bushmeat poachers, while the wild game of white hunters is celebrated with chanterelles and cream sauce. This ethical impasse was missing in *Southern Hunting in Black and White: Nature, History, and Ritual in a Carolina Community* (Marks 1992), my former go-to treatment of this important topic. Another difference is that the author of *Feasting Wild* is an ardent conservationist, concerned about the

✉ F. E. Jack Putz
  fep@ufl.edu

[1] Department of Biology, University of Florida, P.O.
  Box 118526, Gainesville, FL 32641, USA

consequences of defaunation of tropical forests, as driven by market madness for tastes of the wild.

In a book about wild foods, perhaps it is not surprising to have a chapter dedicated to bird spit. For this, the author takes us into the interior of Borneo, a source of the edible nests of cave-nesting swiftlets. While describing the history of trade in birds' nests and the current status of the industry, she reveals that Borneo is a lot less wild than might be imagined; even the birds that make edible nests are being domesticated. The conservation current that runs throughout the volume is strong in this chapter, which explores some of the consequences of the commodification of nature, a form of domestication.

As an increasing number of people around the world seeks alternatives to domesticated foods, *Feasting Wild* explores the consequences of our gastronomy. In an almost overwhelmingly personal way, the author describes the Zen of foraging, ties our tastes to our histories, reveals some of the foibles of a researcher with a social conscience and a sense of history, explores food fetishes, and explains why it makes sense that many of us crave the flavors of the wild. Using her culinary travelogue as a platform, she ruminates on the wild we've irrevocably lost, and celebrates what is redeemable or at least out there to be found by motivated foragers. For former disciples of Euell Gibbons, you'll find much to commend *Feasting Wild*—a savory blend of philosophical queries, ethical quagmires, and ecological insights laced together with a modern love story.

Reference

Marks, S.A. 1992. *Southern hunting in black and white: Nature, history, and ritual in a Carolina community*, 327. Princeton: Princeton University Press.

Publisher's Note Springer Nature remains neutral with regard to jurisdictional claims in published maps and institutional affiliations.

F. E. Jack Putz is a distinguished professor of biology at the University of Florida inGainesville, FL. His research focuses on tropical forest conservation, savanna restoration, andethnobotany.

Agriculture and Human Values (2020) 37:1321–1322
https://doi.org/10.1007/s10460-020-10122-9

Stan cox: the green new deal and beyond: ending the climate emergency while we still can

Jacob A. Miller[1]

Accepted: 24 June 2020 / Published online: 29 June 2020
© Springer Nature B.V. 2020

Stan Cox's *The Green New Deal and Beyond* argues that the realities of our climate crisis require the elimination of fossil fuels from the U.S. economy and a realignment of the unjust system that allows for their exploitation. Cox's thesis is that the Green New Deal legislation (GND) is a good first step, but we in the U.S. must also acknowledge and adhere to the limits of economic growth and material consumption. His evidence-driven analysis builds from the IPCC report's finding that we need to achieve net-zero carbon emissions by 2050 if we are to limit global warming to 1.5 °C.

Cox supports his argument in three major ways. First, he walks through the history of growth and limits from 1933 to the present, with special emphasis on 2016–2020. Second, he explains why limits are inescapable and how to achieve them through a plan called "cap-and-adapt." Finally, Cox argues that any realignment must correct the social inequalities endured by lower- and middle- classes and the Global South.

Initially, to ground the history of limits, Cox discusses the 1933 New Deal, resulting labor movement, and WWII rationing—examples where the Federal Government stepped in to stimulate the economy and impose limits. Cox then reminds us of the 1972 book, *The Limits to Growth*, highlighting its relevance for today. For instance, the books says "when we introduce technological developments that successfully lift some restraint to growth or avoid some collapse, the system simply grows to another limit, temporarily surpasses it, and falls back" (p. 19) and "if you follow those ascending business-as-usual curves to which the world is still adhering out to the year 2030, they show industrial and food production peaking out and then collapsing" (p. 20). What was true fifty years ago is true today: Technology must

adhere to limits, entropy will always prevail, and industrial food production is unsustainable in the long run.

Cox goes on to explain the political pinball that ensued in response to the energy crisis of the 1970s and 1980s, including President Carter's attempt to decrease reliance on foreign oil and President Reagan's National Energy Plan and initiation of federal subsidies for fossil fuel expansion. As Cox ventures into the 1990s and 2000s, he lays out major environmental and political milestones, and concludes each with the U.S. gross domestic product and CO_2 ppm emissions at that point in time (e.g. 1992 Rio Earth Summit: $6.5 trillion (T), 356 ppm; 1997 Kyoto Protocol: $8.6 T, 2008 U.N. Green New Deal: $14.7 T, 385 ppm; 2015 Paris Agreement: $18.2 T, 400 ppm). Cox's parallelism and juxtaposition make the point that no matter the political or social milestone, a rise in GDP, buttressed by unrestrained energy and material consumption (sans unrealistic decoupling), directly correlates with a rise in global emissions.

This historical overview leads Cox into his second main point, that U.S. climate policies must work within limits. He begins by examining popular fixes—including carbon capture and storage, nuclear energy, and claims of 100 percent renewables—and performing a reality check by citing leading research that counters these fixes. In many cases, we do not see their purported benefits because we gloss over key assumptions, such as the embodied energy and extraction required by the steps leading up to the fix. For instance, consider electric car batteries, direct-air capture, and wind machines. Simple calculations reveal the required energy that goes into growing, transporting, and processing cancels out a majority of gained benefits. Cox's demystification supports his book's hard-to-swallow maxim: We must adhere to ecological limits if we are to achieve net zero emissions by 2050.

Achieving this begins with asking the right questions, Cox asserts. Not about what works best in economic models, or what is politically viable, but what ecology *requires of us*. Once we ask that question we can go on an "energy diet" (p.

✉ Jacob A. Miller
  tree11@ksu.edu

1   Department of Sociology, Anthropology, and Social Work, Kansas State University, 204 Waters Hall, 1603 Old Claflin Place, Manhattan, KS 66506, USA

84). Similar to weight loss, there are no quick fixes besides healthy diet and exercise. Cox provides individual actions we should take. These include eating local, eating less meat, and traveling and commuting less. *Less*. As Cox acknowledges, we have long known these will alleviate emissions, but most still avoid them because we "don't want to think or talk about using less energy" (p. 71).

Alongside individual change must come systemic change. After once again debunking unrealistic proposals, *à la* "Eco-modernism" and "Climate Keynesianism," Cox introduces a "Cap-and-Adapt" proposal (p. 97). Called a policy "suggestion," it would place annual, mandatory reductions on fossil fuels themselves, not just carbon or carbon equivalent emitted, thereby capping extraction. To enforce, the government would issue permits, and all imports and exports would be banned. For this suggestion to become policy, a "Coxian" cohort would need to develop a more detailed framework, write the bill, model the legislation's effects, and collect support statements from influential climate change leaders, economists, and politicians.

As Cox has written elsewhere, "Cap-and-Adapt" aims to turn the "Green New Deal" into the "*New* Green Deal" (NGD). The switch is not simply semantic. GND relies on "malignant" "green growth" (Cox 2019) that could promulgate technological dependence (e.g. "slapping solar panels on top of Walmart") and allow the rich to profit from new "green" technologies. In contrast, NGD would ensure that any regulation put green first.

Here is where Cox's third point resides. Any proposal must address the growing economic inequality, domestically and abroad: "If we manage to achieve a fair, effective climate-emergency policy, the 33 percent of American households with highest incomes [> $95,000 annually] will most likely bear the greatest economic burden" (p. 109). The poorest parts of the world in the Global South are responsible for only 15 percent of global greenhouse emissions, yet they are subjected to climate change's worst impacts.

As evidenced by COVID-19 and climate change, we live in a material world. Cox reminds us that no frame, optimism, or flashy proposal will change that fact, and that no technology, market-based policy, or economic growth will save us from the burden of limits. Driving each of Cox's paragraphs is a wide-eyed urgency, best summarized by the book's subtitle. We do not have much time left to take the "off ramp" (p. 83), and while most people ho-hum around, Cox is busy laying out what it will take to ward off climate catastrophe, "while we still can."

Reference

Cox, Stan. 2019. "That green growth at the heart of the Green New Deal? It's malignant. Counterpunch. of 2020https://www.counterpunch.org/2019/01/17/that-green-growth-at-the-heart-of-the-green-new-deal-its-malignant/. Accessed 2 May 2019.

Publisher's Note Springer Nature remains neutral with regard to jurisdictional claims in published maps and institutional affiliations.

Jacob A. Miller is a research assistant for Sociology, Anthropology, and Social Work program at Kansas State University. In the fall of 2020, he enters the program as a Ph.D. student.

Agriculture and Human Values (2020) 37:1323–1324
https://doi.org/10.1007/s10460-020-10175-w

Books received

© Springer Nature B.V. 2020

The Book Review Editor has review copies of the following books. Potential reviewers should contact Carol J. Pierce Colfer to obtain a review copy (cjc59@cornell.edu). **Books not previously listed are in bold-faced type.**

Arnold, Bruce Makoto, Tanfer Emin Tunç, Raymond Douglas Chong (eds.). *Chop Suey and Sushi from Sea to Shining Sea: Chinese and Japanese Restaurants in the United States.* Fayetteville, AR: University of Arkansas Press, 2018. 335pp.

Barnhill, Anne, Mark Budolfson and Tyler Doggett (eds.). *The Oxford Handbook of Food Ethics.* New York: Oxford University Press, 2018. 802pp.

Berardi, Gigi. FoodWISE: A Whole Systems Guide to Sustainable and Delicious Food Choices. Berkeley: North Atlantic Books, 2020. 248pp. (ecopy only now)

Bhullar, Gurbir and Amritbir Riar (eds.). Long-Term Farming Systems Research: Ensuring Food Security in Changing Scenarios. London: Academic Press, 2020. 202pp.

Blanchette, Alex. Porkopolis: American Animality, Standardized Life, and the Factory Farm. Durham, NY: Duke University Press, 2020. 298pp. (ecopy only now)

Campbell, Hugh. Farming Inside Invisible Worlds: Modernist Agriculture and its Consequences. London: Bloomsbury, 2020. 232pp. (ecopy only now)

Clapp, Jennifer. Food. Medford, MA: Polity Press, 2020 (3rd edition). 272pp.

Cohen, Mathilde and Yoriko Otomo (eds.). Making Milk: The Past, Present and Future of Our Primary Food. London: Bloomsbury, 2017. 301pp.

Connerly, Charles E. Green, Fair, and Prosperous: Paths to a Sustainable Iowa. Iowa City: University of Iowa Press, 2020. 201pp.

de Souza, Rebecca. *Feeding the Other: Whiteness Privilege and Neoliberal Stigma in Food Pantries.* Cambridge, Massachusetts: The MIT Press, 2019. 295pp.

Dennis, S. Yael. *Edible Entanglements: On a Political Theology of Food.* Eugene, Oregon: Cascade Books, 2019. 257pp.

Dixon, Beth. *Food Justice and Narrative Ethics: Reading Stories for Ethical Awareness and Activism.* New York: Bloomsbury Publishing, 2018. 177pp.

Dmitriev, Kirill, Julia Hauser, and Bilal Orfali. Insatiable Appetite: Food as Cultural Signifier in the Middle East and Beyond. The Hague: Brill & The Hague Academy of International Law, 2019 (series: Islamic History and Civilization, Volume 163). 362pp. (ecopy only now)

Duncan, Jessica, Michael Carolan, and Johannes S. C. Wiskerke. Routledge Handbook of Sustainable and Regenerative Food Systems. London: Routledge, 2020. 478pp. (ecopy only now)

Earl, Paul D. The Rise and Fall of United Grain Growers: Cooperatives, Market Regulation, and Free Enterprise. Winnipeg, Manitoba, Canada: University of Manitoba Press, 2019. 349pp.

Fairbairn, Madeleine. Fields of Gold: Financing the Global Land Rush. Ithaca, NY: Cornell University Press, 2020. 213pp.

Ferguson, Robert Hunt. *Remaking the Rural South: Interracialism, Christian Socialism, and Cooperative Farming in Jim Crow Mississippi.* Athens, GA: University of Georgia, 2018. 211pp.

Flachs, Andrew. Cultivating Knowledge: Biotechnology, Sustainability, and the Human Cost of Cotton Capitalism in India. Tucson, AZ: University of Arizona Press, 2019. 225pp.

Fowler, Cynthia T. and James R. Welch (eds.). *Fire Otherwise: Ethnobiology of Burning for a Changing World.* Salt Lake City, UT: University of Utah Press, 2018. 236pp.

Fu, Jia-Chen. *The Other Milk: Reinventing Soy in Republican China.* Seattle: University of Washington Press, 2018. 276pp.

Gaddis. Jennifer E. The Labor of Lunch: Why We Need Real Food and Real Jobs in American Public Schools. Oakland, CA: University of California Press, 2019. 291pp.

Gibson, Jane W. and Sara E. Alexander. *In Defense of Farmers: The Future of Agriculture in the Shadow of*

Corporate Power. Lincoln, NE: University of Nebraska

Heide, C. M. van der, W. J. M. Heijman, and J. H. J. Schaminée. *Ecological Economics*. Wageningen, Netherlands: Wageningen Academic Publishers, 2018. 260pp.

James, Harvey S. (ed.). *Ethical Tensions from New Technology: The Case of Agricultural Biotechnology*. Boston, MA and Wallingford, UK: CABI Publishers, 2018. 194pp.

Johnson, Hope. *International Agricultural Law and Policy: A Rights-Based Approach to Food Security*. Cheltenham, UK: Edward Elgar Publishing, 2018. 357pp.

Josephson, Paul R. Chicken: A History from Farmyard to Factory. Medford, MA: Polity Press, 2020. 257pp.

Kaplan, David M. Food Philosophy: An Introduction. New York: Columbia University Press, 2020. 228pp.

King, Samantha, R. Scott Carey, Isabel Macquarrie, Victoria Niva Millious and Elaine M. Power (eds.). *Messy Eating: Conversations on Animals as Food*. New York: Fordham University Press, 2019. 268pp.

Koshy, Kanayathu C. *Sustainability Models for a Better World*. Wageningen, Netherlands: Wageningen University and Research, 2018. 244pp. [e-copy only]

Krasny, Marianne E. (ed.). *Grassroots to Global: Broader Impacts of Civic Ecology*. Ithaca, NY: Cornell University Press, 2018. 264pp.

Lamine, Claire. Sustainable Agri-food Systems: Case Studies in Transitions Towards Sustainability from France and Brazil. New York: Bloomsbury, 2020. 224pp. (ecopy only now)

Lippard, Cameron D. and Bruce E. Stewart (eds.). *Modern Moonshine: The Revival of White Whiskey in the Twenty-First Century*. Morgantown, WV: West Virginia University Press, 2019. 291pp.

Ludington, Charles C. and Matthew Morse Booker (eds.). Food Fights: How History Matters to Contemporary Food Debates. Chapel Hill, NC: University of North Carolina Press, 2019. 293pp.

McCook, Stuart. Coffee is not Forever: A Global History of the Coffee Leaf Rust. Athens, OH: Ohio University Press, 2019. 281pp.

Moreno, Maria Paz. *Madrid: A Culinary History*. Lanham, MD: Rowman & Littlefield, 2018. 203pp.

Muimba-Kankolongo, Ambayeba. *Food Crop Production by Smallholder Farmers in Southern Africa: Challenges and Opportunities for Improvement*. London: Academic Press/ Elsevier, 2018. 368pp.

Murcott, Anne. *Introducing the Sociology of Food and Eating*. London: Bloomsbury Academic, 2019. 223pp.

Nystrom, Justin A. *Creole Italian: Sicilian Immigrants and the Shaping of New Orleans Food Culture*. Athens, GA: University of Georgia Press, 2018. 234pp.

Press, 2019. 422pp.

Ouma, Stefan. Farming as Financial Asset: Global Finance and the Making of Institutional Landscapes. Newcastle upon Tyne: Agenda Publishing, 2020. 208pp.

Ozturk, Munir and Alvina Gul (eds.). Climate Change and Food Security with Emphasis on Wheat. London: Academic Press, 2020.

Pinto-Correia, Teresa, Jørgen Primdahl and Bas Pedroli. *European Landscapes in Transition: Implications for Policy and Practice*. Cambridge, UK: Cambridge University Press, 2018. 286pp.

Plakias, Alexandria. *Thinking Through Food: A Philosophical Introduction*. Peterborough, Ontario, Canada: Broadview Press, 2019. 200pp.

Rankin, Julian. *Catfish Dream: Ed Scott's Fight for his Family Farm and Racial Justice in the Mississippi Delta*. Athens, GA: University of Georgia Press, 2018. 138pp.

Scott, Steffanie, Zhenzhong Si, Theresa Schumilas, and Aijuan Chen. *Organic Food and Farming in China: Top-down and Bottom-up Ecological Initiatives*. New York: Earthscan/Routledge, 2019. 223pp.

Settee, Priscilla and Shailesh Shukla (eds.). Indigenous Food Systems: Concepts, Cases, and Conversations. Toronto: Canadian Scholars, 2020. 284pp.

Sherrard, Cherene. Grimoire. Pittsburgh, PA: Autumn House Press, 2020. 68pp.

Soleri, Daniela, David A. Cleveland, and Steven E. Smith. *Food Gardens for a Changing World*. Oxfordshire, UK: CABI, 2019. 313pp.

Taylor, Keith A. *Governing the Wind Energy Commons: Renewable Energy and Community Development*. Morgantown, WV: West Virginia University Press, 2019. 194pp.

Tippen, Carrie Helms. *Inventing Authenticity: How Cookbook Writers Redefine Southern Identity*. Fayetteville, AR: The University of Arkansas Press, 2018. 215pp.

Tornaghi, Chiara and Chiara Certomà (eds.). *Urban Gardening as Politics*. London: Routledge, 2019. 220pp.

Willes, Margaret. The Domestic Herbal: Plants for the Home in the Seventeenth Century. Oxford, UK: University of Oxford Bodleian Library, 2020. 224pp.

Winders, Bill and Elizabeth Ransom (eds.). Global Meat: Social and Environmental Consequences of the Expanding Meat Industry. Cambridge, MA: MIT Press, 2019. 215pp.

Winne, Mark. *Food Town, USA: Seven Unlikely Cities That are Changing the Way We Eat*. Washington DC: Island Press, Inc., 2019. 210pp.

Publisher's Note Springer Nature remains neutral with regard to jurisdictional claims in published maps and institutional affiliations.

Milton Keynes UK
Ingram Content Group UK Ltd.
UKHW051149141123
432548UK00006B/442

9 783031 185625